# OXFORD WEATHER AND CLIMATE SINCE 1767

# Praise for *Oxford Weather and Climate since 1767*

'*Oxford Weather and Climate since 1767* offers us a detailed analysis of the exceptional work undertaken at the Radcliffe Meteorological Station for more than 200 years...As always, at Oxford, we have the sense that we stand on the shoulders of the generations of scholars and academics who came before us. In this important volume, Stephen Burt and Tim Burt explore their pioneering and painstaking work – work that has led to Oxford having one of the most comprehensive meteorological records in the world.' **Professor Louise Richardson**, *Vice-Chancellor of the University of Oxford*

'Whether your interests are local, regional or global, this book on the Oxford climate record is bound to be of interest to many. The city's remarkable store of long term records has been sensitively analysed and presented by two of the most knowledgeable British authors. This is a delightful place-focused contribution to our understanding of climate change.' **Professor Tim Oke**, *University of British Columbia, Canada*

'Climate is intimately connected to history, and weather to everyday life and new challenges. Tomorrow, the challenge will be global warming; in the past, the challenge was a colder environment. To know the climate of the past is not only pure curiosity, but it is also rediscovering a part of ourselves, the life of our parents and grandparents, and understanding what we should do for our children. This book opens a new, important window to our knowledge.' **Professor Dario Camuffo**, *Emeritus Research Director, National Research Council of Italy*

'This definitive guide to Oxford's climate beautifully documents the variations in the weather over the past two centuries, discussing the extreme highs and lows, and how the long-term view clearly shows the fingerprint of human influence on global temperatures, even at this local scale.' **Professor Ed Hawkins**, *Department of Meteorology, University of Reading*

'Oxford has the longest continuous records of any location in the UK, back to 1811 for air temperature and 1767 for precipitation totals. Stephen Burt and Tim Burt have put together Oxford's history of long-term climate change, while also discussing the daily, monthly and seasonal extremes of the last 200 years. Readers will be able to access all the data online, producing their own results and analyses.' **Professor Phil Jones**, Climatic Research Unit, University of East Anglia

'This is a masterly and thoroughly researched account of the longest remaining continuous record of temperature and rainfall for any one place in England by two experts on UK climate. Although the temperature record is within a city, under future global warming this will become even more valuable, as it will reflect enhanced warming likely to be experienced by many due to both increasing urbanisation and greenhouse gas induced climate change.' **Professor Chris K. Folland**, Universities of East Anglia, UK, Gothenburg, Sweden and Southern Queensland, Australia

'The meteorological records from Oxford's Radcliffe Observatory are among the longest-running and most valuable in the world. Stephen Burt and Tim Burt's new book presents the observations between two covers for the first time, carefully setting them in their historical and scientific context. *Oxford Weather and Climate since 1767* will be essential reading for meteorologists, climatologists, historians of science and all with an interest in the history of the weather.' **Dr Lee Macdonald**, History of Science Museum, Oxford, and author of *Kew Observatory and the Evolution of Victorian Science, 1840–1910*

# Oxford Weather and Climate since 1767

Stephen Burt
Tim Burt

OXFORD
UNIVERSITY PRESS

Great Clarendon Street, Oxford, OX2 6DP,
United Kingdom

Oxford University Press is a department of the University of Oxford.
It furthers the University's objective of excellence in research, scholarship,
and education by publishing worldwide. Oxford is a registered trade mark of
Oxford University Press in the UK and in certain other countries



First Edition published in 2019
Impression: 1



Published in the United States of America by Oxford University Press
198 Madison Avenue, New York, NY 10016, United States of America

British Library Cataloguing in Publication Data
Data available

Library of Congress Control Number: 2018963933

ISBN 978–0–19–883463–2

DOI: 10.1093/oso/9780198834632.001.0001

Printed in Great Britain by
Bell & Bain Ltd., Glasgow

*For Helen and Elizabeth*

# Forewords

## By the Chairman of the Radcliffe Trust

Thomas Hornsby, an 'energetic young Fellow' of Corpus Christi College in Oxford, approached my Radcliffe Trust predecessors in 1768 with his initial proposals for what became Oxford's renowned Radcliffe Observatory. Hornsby clearly was not only an energetic fellow but also a persuasive one, for the Observatory opened in 1773, paid for by the Radcliffe Trust, to a design by James Wyatt and still sometimes described as Oxford's most beautiful building. Hornsby's energy was rewarded with his appointment as the first 'Radcliffe Observer' and, ten years on, he also became the 'Radcliffe Librarian'. In the latter role, he was responsible for the Radcliffe Library (now known as the Radcliffe Camera), but his priorities may be inferred from the lamentable fact that, during his 27-year tenure, the library acquired not a single book while, at the same time, Hornsby was recording tens of thousands of astronomical observations. The Radcliffe Observatory became a world-class astronomical facility which lasted over two centuries, but its entirely unforeseen legacy was the commencement of 250 years (and counting) of meteorological records at the Observatory which forms the subject of this book from two very well-known climatologists.

The astronomical work of the Radcliffe Observatory has always been characterised by a commitment to scientific excellence combined with patient, careful monitoring. The meteorological records are no different. Until as recently as the 1960s, climate was assumed to be steady, unchanging over decades if not centuries: neither Hornsby nor the Radcliffe Trustees could have realised how valuable these records would become as anthropogenic climate change becomes a major challenge to twenty-first-century society. The impeccably documented Radcliffe Observatory records, of vital and continuing importance and in the care of Oxford University since 1935, now represent the longest single-site meteorological time series in the British Isles, and one of the longest and most valuable in the world. It is a pleasure to commend Stephen and Tim's definitive work to you.

**Felix Warnock**

*Chairman, Radcliffe Trust*
September 2018

## By the current Director of the Radcliffe Meteorological Station

Long-term monitoring is widely recognised as crucial to understanding the Earth system. With one of the longest daily weather records anywhere on the planet, the Radcliffe Meteorological Station in Oxford is a fine example of such. In contrast perhaps to their modern-day significance, the history of long-term records tends to be characterised by uncertain beginnings and a sometimes curious mix of serendipity, institutional collaboration and, above all, the determination, dedication and generosity of a few. This book is a fine example of that last ingredient. Stephen Burt and Tim Burt have worked together to provide the most comprehensive analysis yet of this most special of meteorological records. Stephen Burt, an internationally renowned expert on weather stations and author of the best-selling *Weather Observer's Handbook*, has been researching weather records and weather events in the British Isles and Europe for more than four decades. Tim Burt, Director of the Radcliffe Meteorological Station for more than a decade at a time when the observed global warming trend was beginning to acquire the prominence it currently holds in science, has long had a deep research interest in the station and is more familiar with the details of the record than any scientist alive today.

This meticulously researched book begins with the context and changing surrounds of the station in Oxford. Comparisons with other long-term records, such as those from Durham, Uppsala, Padova and Stockholm, help to underscore the significance of the Radcliffe Meteorological Station's length of record in a single, fixed location. It also includes reference to Oxford in relation to broader climate controls such as the Lamb Weather Types and the North Atlantic Oscillation. A great deal of attention is given to the extremes in weather and climate as quantified by the record such as the warmest and driest months on record, the coldest and warmest Decembers, the Great Storm of 1987, the hot summer of 2018 and, of course, the increasing mark that heatwaves are leaving in the data. But it is not only about events marked out by statistical extremes: the work includes quaint references like the weather on Alice in Wonderland Day.

Conceived during the 200th anniversary year of the station, this book underscores the valued contribution and generosity that the Radcliffe Meteorological Station has evoked over the years. The book is a most fitting of tributes to the importance of the record and will help, no doubt, to secure the continued place of the site in long-term monitoring.

**Professor Richard Washington**

*Director, Radcliffe Meteorological Station*
*University of Oxford*
September 2018

# Preface

We first met in May 2015, at the Royal Meteorological Society meeting at Oxford University's School of Geography held to commemorate the 200th anniversary of the Radcliffe Meteorological Station, which one of us organised (SDB) and at which the other was a guest speaker (TPB). We soon agreed to collaborate on extending the formula of Stephen's 2015 book (with Roger Brugge) *One Hundred Years of Reading Weather* to set out a definitive and up-to-date account of the meteorological records collected at the Radcliffe Observatory in Oxford since its construction in the early 1770s. In fact, we have included some of Thomas Hornsby's records from the late 1760s—hence our general claim to span 250 years in this book. Fortuitous and somewhat dusty delving into the original manuscript archives during our research led us to the discovery of, firstly, a previously unknown and near-complete ledger of daily maximum and minimum temperatures recorded at the Observatory from April 1815 and, secondly, to the unearthing of the original meteorological registers showing that the start of the continuous series of daily observations was on 14 November 1813; unknown to us at the time, our 'anniversary conference' was actually a little late. The reconstructed monthly precipitation series dates from 1767 (the first complete year of Hornsby's observations), so 250 years took us to 2017, our original end point when planning started. Fortunately, we were not quite ready at that time and decided to include summer 2018; fortunately because, had we not done so, we should have omitted one of the best summers on record.

We must immediately record our gratitude to those at Oxford who have given us the freedom—and, indeed, positive support—to take this project forward, notably Professor Richard Washington, the current Director of the Radcliffe Meteorological Station, and Professor Heather Viles, Head of the Oxford University School of Geography and the Environment. They and their staff have been most helpful in making the archives available and giving us the opportunity to work through the archival material—and, indeed, to do some tidying up along the way. Ian Curtis generously provided many excellent photographs for which we are most grateful; Senior Observer Amy Creese always responded cheerfully to our numerous requests for data files and other material, whilst Alex Black ensured we had the necessary access to the building's archives whenever it was needed. We must also thank Green Templeton College for allowing us unrestricted access to the Observatory building and gardens, in particular the Bursar, Tim Clayden, and his staff. Green Templeton's long-serving Head Gardener Michael Pirie has generously shared much extremely helpful historical information that only someone with his long and detailed knowledge of the Observatory and its buildings and gardens could provide. Tim Burt would also like to thank Keble College for his appointment as Collaborating Research Scholar during the period when this book was being written.

Renewed association with the College where he was a Fellow for 12 years has been delightful; special thanks to Keble's Senior Tutor Alisdair Rogers.

We also offer special thanks to staff at Oxford University Press, and in particular our Commissioning Editor Ania Wronski. We were delighted when Oxford University Press agreed to publish this book, entirely appropriate given that we have, in effect, written the history of the University's weather station, which is located less than 300 metres from the Press's offices. As we explain in the book, the Radcliffe Observatory was established by the Radcliffe Trust in the early 1770s and funded by the Trust until 1935 when the Observatory migrated to the clearer skies of South Africa. Since then, Oxford University has maintained what is now known as the Radcliffe Meteorological Station, and a University Decree ensures its continuation as long as its records remain useful to science. As we document in this book and its extensive bibliography, the Radcliffe Meteorological Station in Oxford has one of the longest and best-documented weather records anywhere in the world. None of the first Observers back in the reign of George III could ever have envisaged how important the records would become two centuries later and, given ever-increasing concerns about ongoing climate change, we argue for the indefinite continuation of the site and its daily records in its current location. It is good to know that the University recognises the global importance of the site and its record and is fully committed to its preservation and maintenance. Hence, we have two forewords: one from the Radcliffe Trust and one from Oxford University, the two sponsors of the Radcliffe Meteorological Station.

There are many others we need to thank too, for helping us along the way. Stephen's colleagues at the Department of Meteorology at the University of Reading (too many to name individually) provided much useful support and encouragement; Ed Hawkins provided us with his 'climate stripe' graphic depicting Oxford's annual mean temperature since 1814 in shades from blue (cold) to red (warm). The staff of the Bodleian Library in Oxford helped track down numerous obscure volumes during our research, while Helen Drury from the Oxfordshire History Centre in Cowley and Elin Bornemann from the Abingdon County Hall Museum patiently answered our queries regarding the many contemporary photographs we have included. We acknowledge invaluable help and assistance in sourcing data and graphics from Mark McCarthy at the National Climate Information Centre at the Met Office in Exeter, and from Jamie Hannaford, Terry Marsh and Katie Muchan at the Centre for Ecology and Hydrology in Wallingford. From the University of Durham Department of Geography, Michele Allan kindly provided photographs of the Durham Observatory and Gordon Manley's original Central England Temperature graph, David Bridgland supplied information on the Thames terraces, and Chris Orton drew our maps with his usual skill and patience when modifications were requested. Local resident, Jane Buekett (Publishing Editor, *Geography Review*) supplied numerous excellent photographs of Oxford floods, and Simon Collings provided information about the planned Flood Alleviation Scheme. John Butler from Armagh provided photographs and much useful information about Armagh Observatory; similarly, Stefan Gilge from Deutscher Wetterdienst regarding the Hohenpeissenberg Observatory in southern Germany. Judith Curthoys, the Archivist of Christ Church,

Oxford, kindly lent us a copy of E. G. W. Bill's 1965 booklet, *Christ Church Meadow*. Professor Andrew Goudie (former Head of Oxford's School of Geography) provided strong encouragement in the early stages of planning and, more recently, helpful comments on a draft of the manuscript, while Derek Elsom (Emeritus Professor, Oxford Brookes University) helped us out with advice and historical information. Faye McLeod (Archivist and Records Officer at Keble College) provided a photograph of the College chapel. We are most especially grateful to Jonathan Webb (TORRO, Oxford), Julian Mayes (RMetS History Group) and Dennis Wheeler (RMetS NE Centre), who read through early drafts of our manuscript and provided many insightful comments and criticisms which greatly improved the end product you have before you. Thank you all.

By the way, in case you are wondering, we are not related! Nor are we the only UK climatologists with this surname—indeed, we plan a multi-authored 'Burt' paper in due course, once we can find a suitable excuse in the form of a unifying theme . . .

Finally, we have dedicated this book to our long-suffering wives—Helen and Elizabeth, the two Mrs Burts—who have supported our interest in weather and climate over many years.

Stephen Burt
*Stratfield Mortimer, Berkshire*
March 2019

Tim Burt
*Sampford Peverell, Devon*

# Contents

## Part 3 Oxford weather through the seasons



## Part 4 Long-term climate change in Oxford



## Part 5 Chronology of noteworthy weather events in and around Oxford



## Part 6 Oxford weather averages and extremes



## Appendices

*Unless the context requires otherwise, all temperature values are stated in degrees Celsius (°C), and temperature intervals in Celsius degrees (degC). Precipitation records are given in millimetres unless the context requires otherwise, and sunshine durations are quoted in hours.*

*References are shown within the text by square brackets thus: [56]—details of the reference can be found by referring to this number in the References section starting on page 501.*

# Part 1

# Oxford's weather and climate

# 1

# Oxford—Its regional, economic and climatic setting

The city of Oxford, the county town of Oxfordshire, is located in the south Midlands of England. It lies just under 100 km west-north-west of London and 40 km north-west of Reading (Figure 1.1). The rivers Cherwell and Thames (the latter also sometimes known within Oxford as the *Isis*—from the Latinised name *Thamesis*) run through Oxford and meet just south of the city centre.

Located within the zone of temperate westerlies, the south Midlands has a more pronounced continental climate than other parts of England, with warmer summers and colder winters (for its latitude). The south coast of England has the highest annual duration of sunshine, with sunshine amounts diminishing inland with greater average cloud cover. Precipitation—mainly rainfall—is lower than in western and northern districts, although upland areas in the south-east can still receive in excess of 1000 mm per annum. Thunder is more frequent, and snowfall less frequent, than further north and west. Oxford is considerably drier than the upland area of the Cotswold Hills to its west and north-west. These areas of moderate relief provide some shelter from the prevailing south-westerly winds, although cold easterly or north-easterly airstreams crossing the shallow North Sea in winter or spring can easily penetrate across lowland eastern England.

Oxford is situated in the Thames Valley with the Cotswold Hills to the west and north-west, chalk downland to the south. The built-up area of Oxford lies mainly to the east of the Thames floodplain on Quaternary terraces, a few metres above the floodplain itself, which is 56 m above mean sea level (AMSL). Ground level at the Radcliffe Observatory is 63 m AMSL. To the west and east, there are hills underlain by rocks of Jurassic age; to the west, both Old Boars Hill and Wytham Hill rise to over 160 m AMSL. To the east, the Corallian Limestone escarpment just to the east of the River Cherwell means that the suburb of Headington is about 100 m AMSL with a further escarpment of Portland Limestone rising up to 170 m at Shotover Hill.

There is abundant evidence to show that the River Thames is a shadow of its former self, originally having drained large parts of the west Midlands and even North Wales. However, the river underwent significant change during the Pleistocene, partly as a result of glaciation, so that, by the Middle Pleistocene, the river was confined to the south-east of the Cotswold escarpment [1]. Long-term river incision, caused in part by

*Oxford Weather and Climate since 1767*. Stephen Burt and Tim Burt, Oxford University Press (2019).
 DOI: 10.1093/oso/9780198834632.001.0001

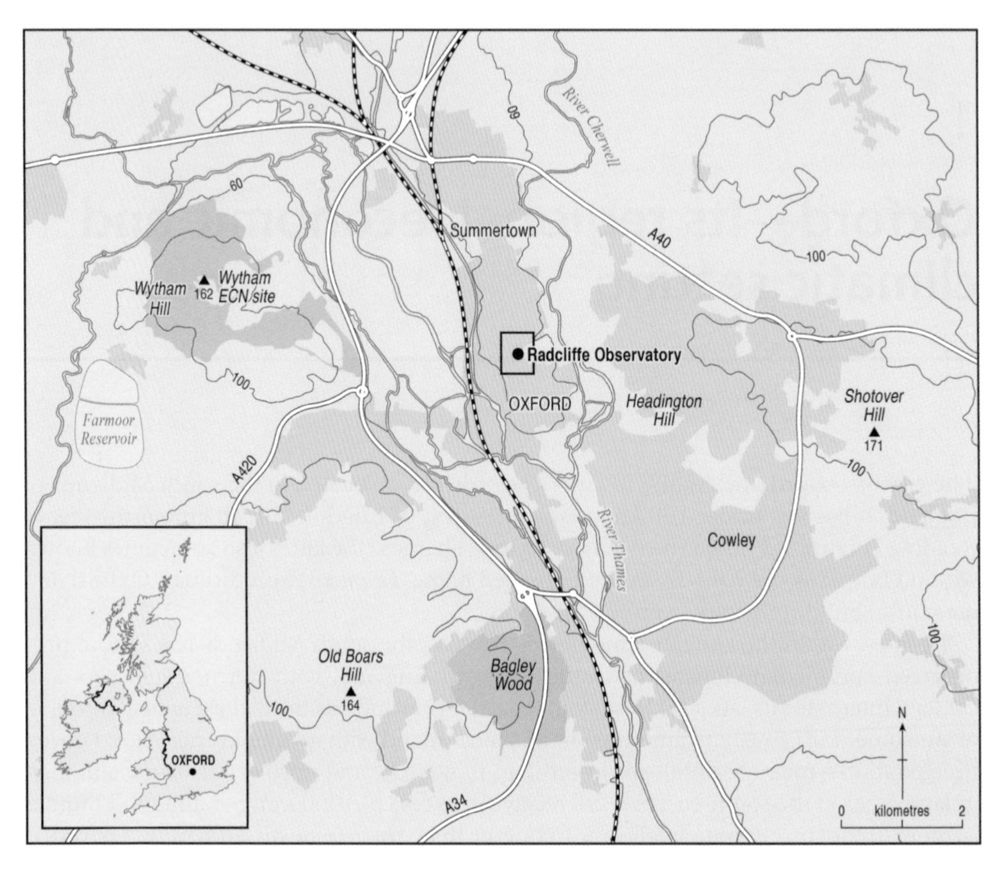


**Figure 1.1** *The Oxford area, showing the city, the River Thames and the 60 m and 100 m contours. The central square shows the Radcliffe Observatory quarter in north Oxford (Chris Orton)*

sea-level change during the Pleistocene, has resulted in the Thames' alluvial deposits being preserved as a staircase of terraces (most of them downstream of Oxford). The Summertown–Radley terrace is a complex series of sediments representing periods of both temperate and cold climate. It is the second youngest terrace in the Oxford area (the younger Northmoor Terrace underlies the modern floodplain), at an elevation of approximately 3–7 m above the present floodplain. The terrace is composed of interbedded gravels, sands and silts and ranges up to 6 m thick. The terrace has been dissected and eroded by younger fluvial incision and mass wasting and it now remains only as a few isolated patches, the largest of which underlies north Oxford, including the site of the Radcliffe Observatory [2]. The Summertown–Radley terrace has a wide range of ages represented amongst the complex sequence beneath its surface, with two interglacials included, with glacial cold-climate deposits preceding and post-dating these at the very base and top (respectively). The age range is 300,000 to 50,000 years ago (David Bridgland, pers. comm.).

The location of the historic core of Oxford on a low terrace next to the floodplain of the River Thames inevitably influences the local climate. The Thames floods regularly and relative humidity is likely to be higher near the river than on adjacent hills. Most importantly, its valley-floor location means that Oxford city centre is susceptible to katabatic cold air drainage on calm nights under clear skies as the air cools. Cold, dense air sinks to the bottom of a valley under relatively calm conditions. Minimum air temperature on some nights will be significantly lower than upslope, therefore, and frosts (ground and air) will be more likely. This means, in turn, that fog will be more likely. Topographic setting affects the detailed record of any climate station, and Oxford is no different; this will all be borne in mind as we compare Oxford to other climatic records in the region and beyond.

The earliest settlement of what is now Oxford dates back to Saxon times; its name in Saxon was *Oxenaforda*, meaning *ford of the oxen*, denoting a passable ford or river crossing. During the tenth century, Oxford became an important military frontier town between the kingdoms of Mercia and Wessex but was largely destroyed in the aftermath of the Norman Conquest, although a Norman castle (Oxford Castle) was quickly built to subdue the region. Monastic communities began to settle in Oxford, and a number of religious houses became established, some providing education to children of the land-owning and ruling classes. The University of Oxford is first mentioned in the twelfth century, the oldest university in the English-speaking world: the earliest colleges were University College (1249), Balliol (1263) and Merton (1264).

Whilst Oxford is known worldwide as the home of the University of Oxford, 'the city of dreaming spires', the city has an ethnically diverse population and a broad and thriving economic base: local industries include motor manufacturing (the BMW plant in Cowley has long been 'the home of the Mini', originally the Morris Motors plant, established in 1910), education, publishing and a large number of information technology and science-based businesses, some being academic spin-outs. Oxford is also home to the Bodleian Library, the largest university library system in the UK and the second-largest library in the United Kingdom, with over 11 million items catalogued. 'Bodley' or 'The Bod' has been in continuous existence since 1602, although its roots date back to the fifteenth century. The population of the City of Oxford in 2015 was 168,270.

# 2

# Weather observations in Oxford

Meteorological observations have been made at what is now known as the Radcliffe Meteorological Station, Oxford, since 1772, and continuously since November 1813. Site details are given below, and more details of the instruments used and the location of the observing sites within the Radcliffe Observatory in the following chapter and in Appendix 1.

Before reviewing the Observatory's history in more detail, it is perhaps worthwhile setting the climatological scene with a brief overview of 'the climate of Oxford'. Over the standard 30-year average period adopted in this book, 1981–2010, the **mean temperature** at the Radcliffe Observatory Oxford site was 10.7 °C (mean daily maximum 14.6 °C, mean daily minimum 6.9 °C). January and February tie as the coldest months of the year on average (mean temperature 4.9 °C), while July is the warmest (mean temperature 17.9 °C). Since continuous records of both daily maximum and minimum temperature commenced in April 1815, the lowest observed air temperature has been −17.8 °C (on 24 December 1860) and the highest 35.1 °C (on 19 August 1932, equalled on 3 August 1990). Within the last 50 years, the lowest observed temperature has been −16.6 °C, on 14 January 1982.

The **annual average precipitation** is 660 mm, falling on 165 days per year, slightly wetter than Reading (635 mm/yr) and London (Kew Gardens, 623 mm/yr). The driest period of the year is late winter to early spring (February average 43 mm) and the wettest period is autumn (October average 70 mm), but with wide variations from year to year. The wettest day on the daily Oxford record (since 1827) was 10 July 1968, when 87.9 mm fell in the 24 hours commencing 0900 GMT. In the composite monthly rainfall record for Oxford extending back to 1767, the wettest month (September 1774) received 224 mm and the wettest year (2012) 979 mm. No calendar month has ever remained completely dry, the driest months since 1767 being April 1817 and April 2011, both of which received just 0.5 mm. The driest years were 1788 (337 mm, just 51 per cent of the current average) and 1921 (381 mm). Snow can be expected to fall on around 16 days in a typical year, with the ground snow-covered on six or seven mornings. The greatest known snow depth has been 61 cm, in February 1888. Thunderstorms occur on around 10 days per annum, most frequently during the summer half-year.

The average **annual sunshine duration** is 1577 hours, about 35 per cent of the possible duration of daylight at this latitude, and slightly sunnier than Reading (1522 hours/year) but less sunny than London (Kew Gardens, 1653 hours/year). Contrary perhaps to perceived wisdom, sunshine is recorded on almost twice as many days per year (300, on average) as measurable rainfall (165).

*Oxford Weather and Climate since 1767*. Stephen Burt and Tim Burt, Oxford University Press (2019).
 DOI: 10.1093/oso/9780198834632.001.0001

**Table 2.1** *Meteorological observations made at the Radcliffe Observatory, Oxford*

| Site name | Period of record | Latitude and Longitude* | Altitude above MSL |
|---|---|---|---|
| Radcliffe Observatory, Oxford | 1772–1811 (incomplete), Nov 1813 to date | 51.7612°N, 1.2640°W NGR SP (42) 509 072 | Rain gauge 63.4 m Barometer 64.6 m |

## Weather observations in Oxford prior to 1767

The majority of this book is based upon the records of the Radcliffe Observatory in Oxford, whose observations began more than 250 years ago. A daily record of air temperature and precipitation which commenced in November 1813 continues to this day—the longest unbroken single-site weather record in the British Isles and one of the longest in the world. It used to be thought that the first complete calendar year of observations was 1815 and this is the year usually identified as the start of the long daily record. However, re-examination of the original ledger shows that the unbroken record runs from 14 November 1813 and so we able here to include the year 1814 in our analyses too. This is important because it allows us to include the very cold winter of 1813/14, one of the most severe on Oxford's long record.

More details of the Radcliffe records are given subsequently, but they were not the earliest weather observations made in Oxford.

### William Merle's weather diary, 1337–1344

William Merle (d. 1347) was a Fellow of Merton College in Oxford and the rector of the parish of Driby in north Lincolnshire. He kept a record of the weather from January 1337 to January 1344, the oldest surviving weather journal in Europe: entries are noted month by month and, in large part, day by day (Figure 2.1). The original manuscript is a small vellum folio volume which has been in the Bodleian Library for over 300 years, catalogued as *Digby MS., Vol. 176* [3–5]. The entries in the diary are in medieval Latin, and translation and facsimile printing of the diary was arranged by George Symons in 1891 [6]. Merle's diary represents the only systematic written account we have of weather in fourteenth-century England and, as such, has been carefully analysed for evidence of how the climate differs from today [7, 8]. Although there are clear references to Oxford in the diary, it is less clear exactly how much of the content relates to Oxford and how much to north Lincolnshire: it has been suggested [5] that Merle was an occasional lecturer at Oxford who travelled to and from Lincolnshire, a distance of some 200 km, several times per year.

* The Met Office has for many years quoted the site details of the Radcliffe Observatory site as 51°46′ N, 1°16′ W (= 51.77°N, 1.27°W), 61 m AMSL: this would place it about 650 m north-north-west of its actual location. The Engineering Science building, where the sunshine recorder has been sited since 1976, is at 51.7602°N, 1.2596°W, NGR (42) $511^{98}$ $070^{94}$, 230 m east-south-east from the instrument enclosure at Green Templeton College gardens.

**Figure 2.1** *A page from William Merle's weather diary, for January 1337 to June 1338 (folio on the right). The translation, from Symons, 1891 [6], is given in the box on the following page. (Stephen Burt/ Bodleian Library Oxford)*

## John Locke: 1666

John Locke FRS (1632–1704) was an English philosopher and physician, widely regarded as one of the most influential of Enlightenment thinkers and whose writings influenced Voltaire and Jean-Jacques Rousseau, many Scottish Enlightenment thinkers and the United States Declaration of Independence. He studied medicine at Christ Church, Oxford, where he met and worked with noted scientists including Robert Boyle, Thomas Willis and Robert Hooke. It was no doubt through the latter's position within the embryonic Royal Society in London that Locke obtained one of the first thermometers in England [9, 10] and made the first instrumental observations of temperature in Oxford, starting in June 1666. Locke's diary, which is preserved in the Bodleian Library (*MS Locke c.25*), contains both medical and meteorological notes, including temperature and barometer observations one to three times daily, with wind direction and force and notes on the sky and weather.

Locke maintained this journal until 1683, although, after he moved to London in 1667 to become personal physician to the 1st Earl of Shaftesbury, his observations mostly relate to London rather than Oxford. He also kept daily barometric pressure

*Merle MS translation—first folio (Bodleian Library Digby MS 176. folio 4)*

TEMPERATURE OF THE AIR AT OXFORD FOR SEVEN YEARS

A.D. 1337.

In **January** there was warmth, with moderate dryness, and in the previous winter there had not been any considerable cold or humidity, but more dryness and warmth.

In **February**, during the first week there was moderate frost, and after an interval of three days there was slight frost for another week.

In **March**, for almost two weeks there was cold, with moderate humidity. The remainder of the month was dry, with moderate warmth.

In **April** there was moderate warmth, with great humidity.

In **May**, four days were humid, with moderate warmth. All the remainder was moderately warm and dry, with moderate showers at intervals, but on the 17th there was sudden heavy rain, with thunder, heavier than any that had fallen at Oxford in so short a time for many years.

In **June** there was great heat with dryness, but on the 17th there was heavy thunder continuously for two hours, and heavier rain in certain parts of Lyndesay [north Lincolnshire] than had fallen in so short a time for a long while. In the last week a moderate wind rose, which lessened the heat and increased the dryness.

In **July** there was moderate heat and wind, with moderate showers occasionally falling, and after the first week, all the remainder of the month was rainy, with moderate wind, except three or four days at the end, on which no rain fell.

In **August** there was very strong wind, and showers fell occasionally. In the middle of the month three or four days were dry and windy, with the wind continually increasing.

In **September** the wind abated and there were moderate showers, but about the 10th and 11th there was very heavy rain. The remainder was moderately rainy, except five or six days at the end.

In **October** there was more warmth than in September, and throughout the month it was moderately humid, but on the 19th there was heavy rain all day.

In **November** there was much rain. On the 6th the air was moderately moist. All the last week was extremely rainy. There was a slight frost beginning at the feast of St Leonard [6 November] and lasting three days.

In **December** the first week was extremely rainy, and the second extremely foggy. In the third there was slight frost, and in the fourth also, but not so much.

A.D. 1338.

In **January** the first week was dark and rather foggy. In the second there was slight frost, and in the third also. The fourth was mild.

In **February** the first week was moderately windy. On the 8th and 10th days there was a slight storm. The second week was rainy and windy, and the third also. The fourth was dry, windy and clear.

In **March** the first week was warm, dry and clear; the second was the same, and the third likewise, with wind. The fourth was warm, dry and clear.

In **April** the first week was windy and dry, with showers occasionally. The second was dry, with moderate wind; the third dry and clear, and the remainder likewise. There was hoar frost throughout the month.

In **May**, in the first week there was heavy rain; the second was moderately dry; the third and remaining weeks were warm and dry.

In **June** the first week was dry and moderately warm, with light showers occasionally. In the second there was more rain than in the first week.

**Figure 2.1** *Continued*

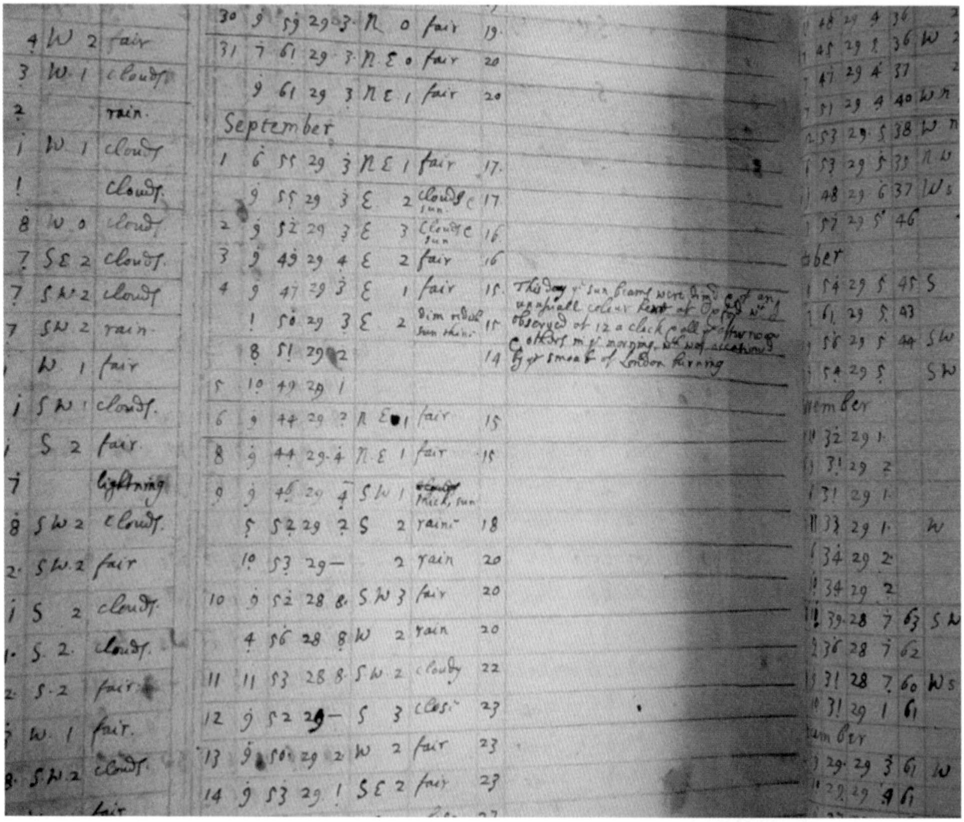

**Figure 2.2** *John Locke's weather record from early September 1666, at the time of the Great Fire of London. The temperature observations (third column, after date and hour) are made according to a Royal Society scale, not in Fahrenheit as they may appear at first glance, for Daniel Gabriel Fahrenheit was not born until 20 years after these entries were written (Stephen Burt/Bodleian Library Oxford)*

observations in Essex in 1694 [11] and these have been used with other pressure records to reconstruct weather patterns across Europe in that year.

Figure 2.2 shows Locke's record from September 1666, at the time of the Great Fire of London; the entry for 4 September at 1 p.m. notes

> *'Dim red or[orange] sun shine'*

and

> *'This day y^e sun beams were dim'd & of an unusual colour Red at Oxford. I observed at 12 o'clock & all afternoon & others in y^e morning, wch was occasioned by ye smoak of London burning'*

There are occasional references to the weather in Oxford in the early volumes of *Philosophical Transactions*, the journal of the Royal Society of London: these include notes of a small earthquake on 19 January 1665 (old style) [12], and of a lightning fatality to two scholars from Wadham College who were boating on the Thames on 10 May 1666 [13].

We also have a surviving account of a 'cloudburst' in Oxford on 31 May 1682 (old style), which is covered in the Chronology, Chapter 25.

## Robert Plot: 1685

Robert Plot FRS (1640–1696) was the first Professor of Chemistry at the University of Oxford, and the first keeper of the Ashmolean Museum. He was also the 'keeper of experiments' at the Oxford Philosophical Society and thus the Oxford equivalent of Robert Hooke at the Royal Society in London*. He was elected a Fellow of the Royal Society

[ 932 ]

*Obſervations of the* Wind, Weather, *and height of the* Mercury *in the* Barometer, *throughout the year* 1684;*taken (after Dr.* Liſters *method,)* *in the* Muſæum Aſhmoleanum, *at the requeſt of the* Philoſophicall Society *of* Oxford; *by* Robert Plot, LLD.

*A Scheme of the weather at* Oxford ; *January* ; 168¾.

| *Day.* | *Weather.* |
|---|---|
| 1 | hard, froſt and fair. |
| 2 | froſty,but yielding a little towards night. |
| 3 | rimy froſt. |
| 4 | hard froſt,and fair. |
| 5 | hard froſt, and fair. |
| 6 | hard froſt, and fair. |
| 7 | hard froſt, but a little yielding at night. |
| 8 | rimy froſt morn.fair all day,windy night. |
| 9 | froſt, but ſnow at night. |
| 10 | cold raw weather toward noon,rain toward night. |
| 11 | moiſt thawing weather. |
| 12 | cloſe thawing weather. |
| 13 | moiſt,cloſe weath. a ſmall froſt at night. |
| 14 | cloſe froſty weather. |
| 15 | cloſe froſty weather. |
| 16 | cloſe froſty weather,at night windy. |
| 17 | froſt, at night ſnow. |
| 18 | ſnow,and wind. |
| 19 | cloſe ſharp weather. |
| 20 | cloſe *ut ſupra* but a little yield. at night. |
| 21 | mild froſt,and fair. |
| 22 | hard froſt, ſnow at night a little. |
| 23 | hard froſt. |
| 24 | hard froſt, and fair. |
| 25 | hard froſt, and ſnow. |
| 26 | froſt, a little ſnow. |
| 27 | froſt, a little thaw about noon. |
| 28 | froſt, and fair. |
| 29 | froſt, a ſmall thaw all the afternoon. |
| 30 | hard froſt, and fair. |
| 31 | froſt, and fair. |

*A Scheme*

**Figure 2.3** *Robert Plot's weather notes made at the Ashmolean Museum in Oxford in January 1684—from Plot (1685). January 1684 is the second-coldest month on the long Central England Temperature series, which extends back to 1659, with a mean temperature of −3.0 °C; only January 1795 was colder, and only by a tenth of a degree (Courtesy of the Royal Society)*

* A fine portrait of Dr Robert Plot FRS by Sylvester Harding exists in the British Museum online collection; unfortunately, owing to copyright restrictions, we are unable to reproduce it here.

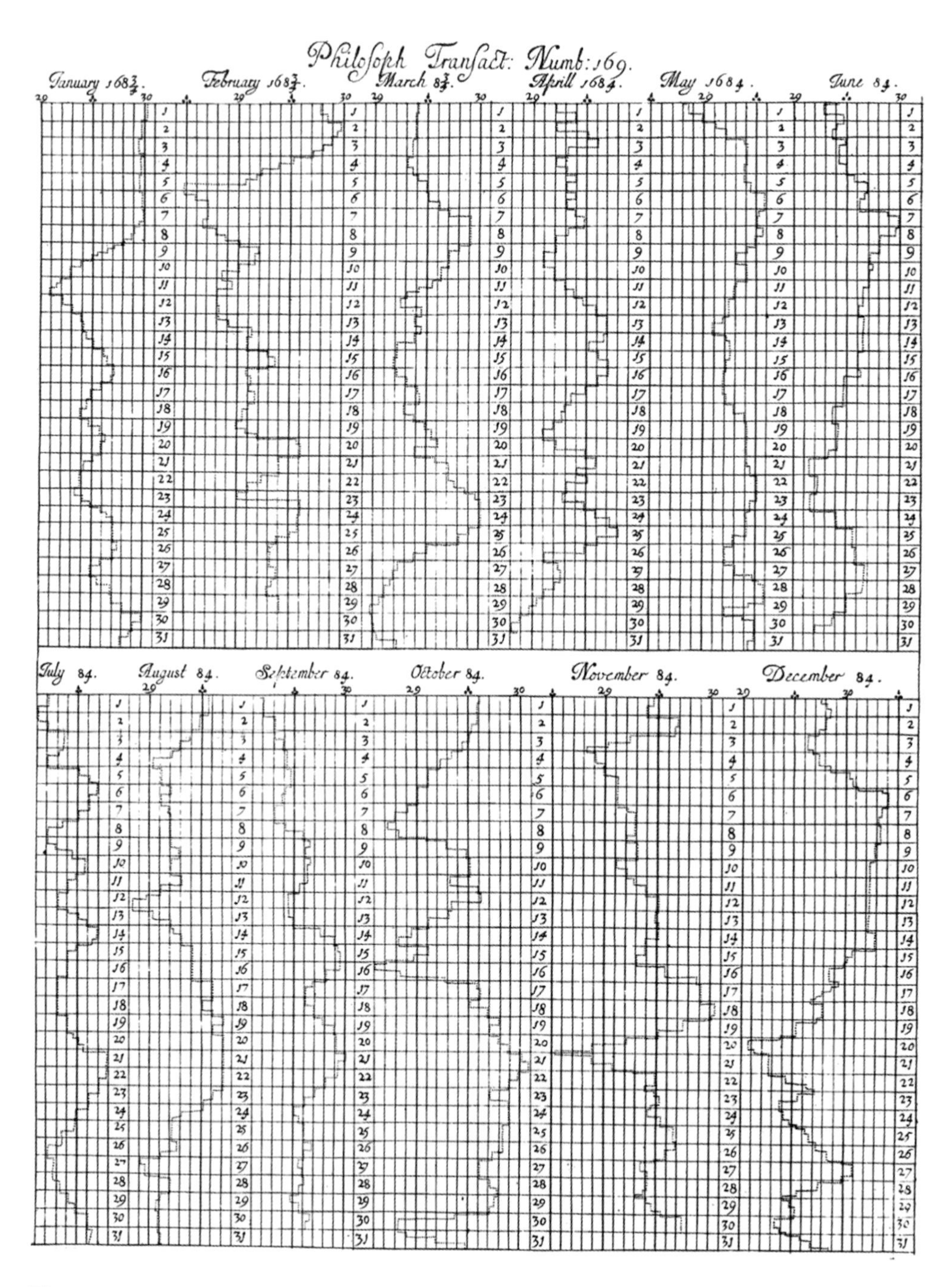


**Figure 2.4** *Plot's daily time series of barometric pressures in Oxford for 1685, as reproduced in* Philosophical Transactions*—from Plot (1685) [15]. The units are inches of mercury (29 inHg = 982 hPa (millibars), 30 inHg = 1016 hPa). The lowest, on 5/6 February, is about 964 hPa and the highest, on 6 December, about 1030 hPa. The Ashmolean Museum is about 66 m above mean sea level, so around 6 hPa would need to be added to reduce these to mean sea level pressures (Courtesy of the Royal Society)*

(FRS) in 1677 and served as the Society's secretary and joint editor of *Philosophical Transactions* between 1682 and 1684. As such, he would have known Boyle, Hooke and others in the Oxford and London scientific circles.

He wrote an account of the great frost in the very severe winter of 1683/84 in *Philosophical Transactions* [14] and commenced daily weather observations at the Ashmolean in January 1684. These observations included daily barometer readings, wind and weather [15] (Figure 2.3).

One of Plot's innovations was to show the daily variation of barometric pressure on a graph (Figure 2.4, from Plot 1685 [15]); indeed, it is sometimes said to be from the form of graphical representation of data employed that the verb 'to plot' (in the sense of 'plotting a graph') originates, although, unfortunately for a good story, the *Oxford English Dictionary* suggests the term actually originates from Late Old English, of unknown origin.

## Thomas Hornsby and the Radcliffe Observatory: 1768

### The construction of the Radcliffe Observatory

The history of observatories in Oxford begins with the foundation of the Savilian Professorship of Astronomy in 1619. However, no purpose-built observatory was provided by the University and over the next century and a half, successive Savilian Professors complained about the inadequate facilities made available to them [16]. It was not until 1768 that Thomas Hornsby petitioned the Radcliffe Trustees for money to build a proper observatory out of the interest on the property left to Oxford University in 1714 by the physician John Radcliffe [17]. The trustees responded promptly and generously [18, 19].

The first step towards the construction of the Observatory was in 1771 when the Lord Chancellor was asked for permission to use the Radcliffe funds to buy 'a piece of ground and to erect a large Observatory Room…and to purchase such mathematical instruments as are proper to be used there' (Radcliffe Trustees' Minute Book; cited in [20]). The leasehold of land immediately to the north of the Radcliffe Infirmary was purchased from the Duke of Marlborough, himself an enthusiastic amateur astronomer. Construction started in 1772 and Hornsby was able to move into the Professor's house the next year, by which time the ground floor of the Observatory building itself was operational. However, 27 years elapsed before the three-storey building was finally finished. The upper levels were designed by architect James Wyatt as a replica of the Tower of the Winds in Athens, a particularly appropriate model for an observatory [20] (Figure 2.5).

Observations of air temperature and atmospheric pressure were taken at astronomical observatories to correct for atmospheric refraction, a phenomenon discovered by Hornsby's immediate predecessor James Bradley. Later, measurements of humidity were added. Given the need to compensate for atmospheric distortions, it was natural that a limited range of meteorological variables were measured alongside the astronomical

**Figure 2.5** *The Observatory from the south-east, about 1814 (about the time the continuous daily series of meteorological records commenced), from R. Ackermann,* A History of the University of Oxford, *1814 (Green Templeton College, Oxford)*

observations. However, Hornsby clearly had a personal interest in the weather, which is why, from the outset, other aspects of the weather were recorded at the Radcliffe Observatory.

## The contribution of Thomas Hornsby

Thomas Hornsby (1733–1810, Figure 2.6), the successor of Edmund Halley and James Bradley as Savilian Professor at Oxford University, was one of a new breed of scholars who introduced experimental science to the University. Appointed Savilian Professor and consequently elected a Fellow of the Royal Society in 1773, Hornsby was a noted pluralist with interests in mathematics and physics (then 'Experimental Philosophy') as well as obviously astronomy [21, 22]. He was appointed Radcliffe Observer in 1772, in effect the Director of the Observatory.

It is clear that Hornsby had more than a passing interest in meteorology. As early as 1758, when he was only 25, he was pondering the best form in which to maintain a meteorological journal and how best to measure rainfall. He was in correspondence with the Revd William Borlase in Cornwall about the design of an 'ombrometer', or rain-gauge. His early observations were made at Corpus Christi College, where he was a Fellow; after 1763, once he became Savilian Professor, these were maintained on the

**Figure 2.6** *Thomas Hornsby. This portrait hangs in the common room within the Observatory building; it is a copy painted by Kenny McKendry in 2013 from an original owned by Hornsby's descendants. The original was probably painted when Hornsby became a Doctor of Divinity in 1785, as shown by the gown he is wearing, so he would be at least 53 in the portrait (By kind permission of Green Templeton College, Oxford)*

roof of the house provided for him in New College Lane. They continued at the Observatory once he had moved to the house there, even before the ground floor rooms of the Observatory itself were completed [23].

Hornsby's earliest surviving weather records, observations of rainfall, began in 1760 and were kept regularly for some months, but had petered out by March 1761. Thereafter, he made numerous if irregular observations of rainfall, temperature, winds and cloud cover from 1767 until 1804, when clearly age and infirmity made writing a burden (these records are now held by the Director of the Radcliffe Meteorological Station at the School of Geography and the Environment). Hornsby's weather journal is only complete for about a third of this period, but the rainfall and temperature observations have been of some value and appear to have been made in a meticulous manner [19]. Craddock & Craddock [24] note that the observations of 1767 show all the signs of a fresh start, and this marks the point from which the long, homogenous monthly rainfall series for Oxford commences. Smith [25] exemplifies Hornsby's interest and enthusiasm for meteorological matters: on 27 January 1776, during a very wintry spell, Hornsby read his thermometer twelve times during the day instead of the normal three daily observations (Figure 2.7). The same day in his journal Hornsby noted ruefully: 'wine beg[an] to freeze in my study' and, shortly afterwards, an outside air temperature of 6 °F (−14 °C). It is not clear whether Hornsby's numerous other duties prevented him from maintaining the journal every day or whether some records have been lost. The existence of annual rainfall totals in Hornsby's own hand for the period 1784–1794 suggest that there may have been daily observations during this period but the original records, if they ever existed, have never been traced [24, 26].

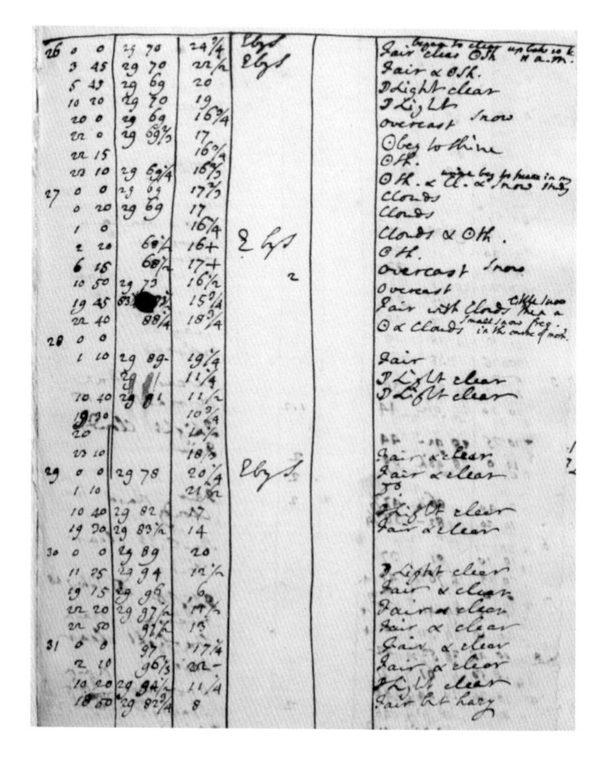

**Figure 2.7** *Part of Thomas Hornsby's weather diary 26–31 January 1776. At 1915h astronomical time on 30 January (7.15 a.m. 31 January 1776,) the outside air temperature is noted as 6 °F (–14 °C); at the bottom of the page, 8 °F (–13 °C) is noted at 1850h astronomical time on 31 January (6.50 a.m. on 1 February 1776), when the weather was 'Fair but hazy', wind E by S, barometer 29.8275 inHg (1010 hPa—approx 1017 hPa at MSL). See also Figure 25.4 (Courtesy of the Director of the Radcliffe Meteorological Station, Oxford)*

## A daily series of weather observations from November 1813

There is a gap in the Oxford meteorological record from 1804, when Hornsby's record peters out, to 1811, when his successors at the Observatory—notwithstanding their lower level of interest in meteorological observations compared to Hornsby—commenced a neat record of pressure, temperature, wind and weather in very similar form to the earlier tabulations. After a false start in 1811 (the daily record only ran as far as May), the daily record is continuous from 14 November 1813, and these records, together with rainfall, are sufficiently well documented and continuous to produce reliable means of monthly rainfall and monthly mean temperatures [19]. Prior to our uncovering new material in the course of researching this book, 1815 was taken as the commencement of the Oxford Radcliffe record but, as noted above, we now know that the continuous record is even older, and 1814 can now be regarded as the first complete calendar year of daily temperature measurements.

Throughout the nineteenth century, as the science of meteorology developed, there were many improvements in the type of instruments in use, and observing methods became standardised so that results from different stations could be reliably compared. Smith [19] notes that all such improvements and changes were quickly adopted at the Radcliffe Observatory: there were frequent exchanges with the observatories at Kew and

**Figure 2.8** *The Radcliffe Observatory building in 1834, from the south. This view was drawn by Frederick Mackenzie, engraved by John Le Keu and published in James Ingram's* Memorials of Oxford *(London & Oxford: Charles Tilt, J. H. Parker; 1837).*

Greenwich and, after 1880, with the Meteorological Office*. A good example of innovation is the erection of the first Stevenson screen at the Observatory in 1878, to house new Meteorological Office standard thermometers [21]—the Stevenson screen was still very much one of many such 'competing' instrument shelters at that time, and was not formally adopted as a standard by the Royal Meteorological Society until 1884 [9]. The Observatory pioneered the use of self-recording or autographic instruments, and some electrical resistance temperature sensors.

For many years, the principal instruments were read at least three times a day, with hourly tabulations transcribed from the main autographic records. Starting in 1873, the Radcliffe Observatory transmitted the details of the morning observation by telegraph to the Meteorological Office in London for inclusion in 'synoptic' weather maps and the Meteorological Office's *Daily Weather Report*. Eventually, the expansion of meteorological observing networks during the Great War, particularly in the early days of military aviation, rendered once-daily reports such as those from the Radcliffe Observatory less operationally useful, and they were discontinued in 1916.

* The term 'Meteorological Office' is used only where required in its historical context; otherwise, the preferred term 'Met Office' is used throughout this book, as this is how the organisation has been known since 1988.

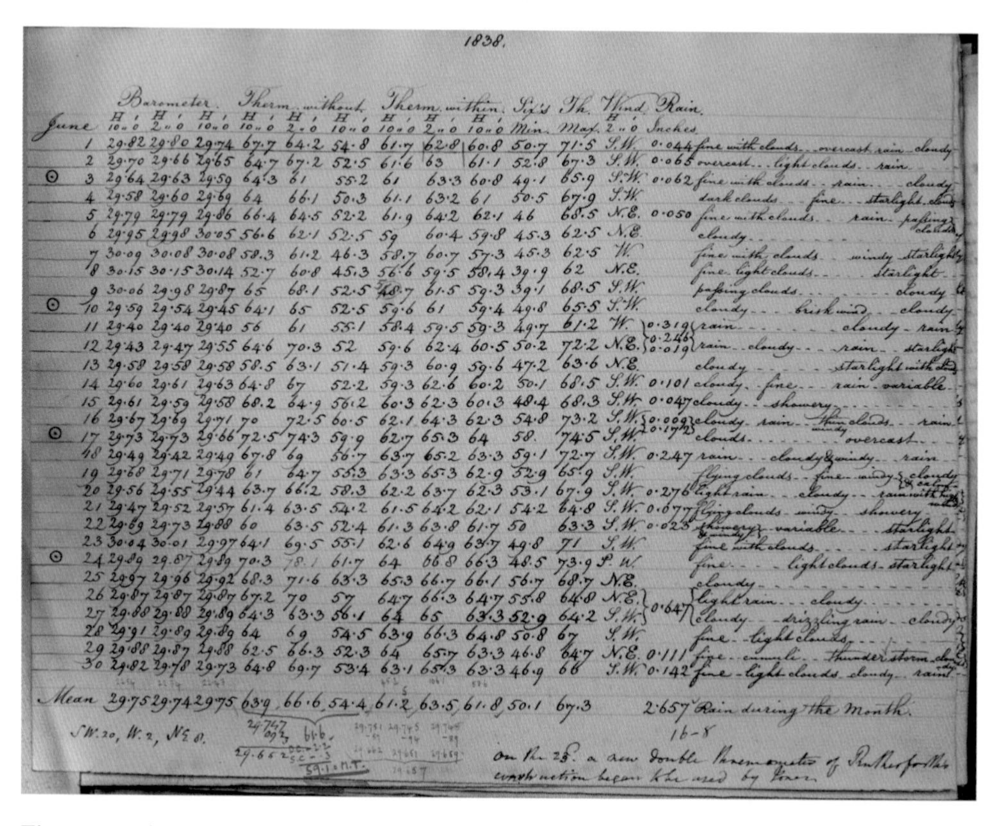

1838.

| | June | Barometer H 10 | Barometer H 2 | Barometer H 10 | Therm. without H 10 | Therm. without H 2 | Therm. without H 10 | Therm. within H 10 | Therm. within H 2 | Therm. within H 10 | Six's Th. Min. | Six's Th. Max. | Wind H 2 | Rain Inches | |
|---|---|---|---|---|---|---|---|---|---|---|---|---|---|---|---|
| | 1 | 29.82 | 29.80 | 29.74 | 67.7 | 64.2 | 54.8 | 61.7 | 62.8 | 60.8 | 50.7 | 71.5 | S.W. | 0.044 | fine with clouds.. overcast rain cloudy |
| | 2 | 29.70 | 29.66 | 29.65 | 64.7 | 67.2 | 52.5 | 61.6 | 63 | 61.1 | 52.8 | 67.3 | S.W. | 0.065 | overcast.. light clouds.. rain |
| ⊙ | 3 | 29.64 | 29.63 | 29.59 | 64.3 | 61 | 55.2 | 61 | 63.3 | 60.8 | 49.1 | 65.9 | S.W. | 0.062 | fine with clouds.. rain.. cloudy |
| | 4 | 29.58 | 29.60 | 29.69 | 64 | 66.1 | 50.3 | 61.1 | 63.2 | 61 | 50.5 | 67.9 | S.W. | | dark clouds.. fine.. starlight |
| | 5 | 29.79 | 29.79 | 29.86 | 66.4 | 64.5 | 52.2 | 61.9 | 64.2 | 62.1 | 46 | 68.5 | N.E. | 0.050 | fine with clouds.. rain.. passing clouds |
| | 6 | 29.95 | 29.98 | 30.05 | 56.6 | 62.1 | 52.5 | 59 | 60.4 | 59.8 | 45.3 | 62.5 | N.E. | | cloudy |
| | 7 | 30.09 | 30.08 | 30.08 | 58.3 | 61.2 | 46.3 | 58.7 | 60.7 | 57.3 | 45.3 | 62.5 | W. | | fine with clouds.. windy starlight |
| | 8 | 30.15 | 30.15 | 30.14 | 52.7 | 60.8 | 45.3 | 56.6 | 59.5 | 58.4 | 39.9 | 62 | N.E. | | fine light clouds.. starlight |
| | 9 | 30.06 | 29.98 | 29.87 | 65 | 68.1 | 52.5 | 48.7 | 61.5 | 59.3 | 39.1 | 68.5 | S.W. | | passing clouds.. cloudy |
| ⊙ | 10 | 29.59 | 29.54 | 29.45 | 64.1 | 65 | 52.5 | 59.6 | 61 | 59.4 | 49.8 | 65.5 | S.W. | | cloudy.. brisk wind.. cloudy |
| | 11 | 29.40 | 29.40 | 29.40 | 56 | 61 | 55.1 | 58.4 | 59.5 | 59.3 | 49.7 | 61.2 | W. | 0.319 | rain.. cloudy.. rain |
| | 12 | 29.43 | 29.47 | 29.55 | 64.6 | 70.3 | 52 | 59.6 | 62.4 | 60.5 | 50.2 | 72.2 | N.E. | 0.246 0.019 | rain.. cloudy.. rain.. starlight |
| | 13 | 29.58 | 29.58 | 29.58 | 58.5 | 63.1 | 51.4 | 59.3 | 60.9 | 59.6 | 47.2 | 63.6 | N.E. | | cloudy.. starlight with clouds |
| | 14 | 29.60 | 29.61 | 29.63 | 64.8 | 67 | 52.2 | 59.3 | 62.6 | 60.2 | 50.1 | 68.5 | S.W. | 0.101 | cloudy.. fine.. rain.. variable |
| | 15 | 29.61 | 29.59 | 29.58 | 68.2 | 64.9 | 56.2 | 60.3 | 62.3 | 60.3 | 48.4 | 68.3 | S.W. | 0.047 | cloudy.. showery |
| | 16 | 29.67 | 29.69 | 29.71 | 70 | 72.5 | 60.5 | 62.1 | 64.3 | 62.3 | 54.8 | 73.2 | S.W. | 0.009 | cloudy.. rain.. thin clouds.. rain |
| ⊙ | 17 | 29.73 | 29.73 | 29.66 | 72.5 | 74.3 | 59.9 | 62.7 | 65.3 | 64 | 58. | 74.5 | S.W. | 0.172 | clouds.. overcast |
| | 18 | 29.49 | 29.42 | 29.49 | 67.8 | 69 | 56.7 | 63.7 | 65.2 | 63.3 | 59.1 | 72.7 | S.W. | 0.247 | rain.. cloudy & windy.. rain |
| | 19 | 29.68 | 29.71 | 29.78 | 61 | 64.7 | 55.3 | 63.3 | 65.3 | 62.9 | 52.9 | 65.9 | S.W. | | flying clouds.. fine.. windy & cloudy |
| | 20 | 29.56 | 29.55 | 29.44 | 63.7 | 66.2 | 58.3 | 62.2 | 63.7 | 62.3 | 53.1 | 67.9 | S.W. | 0.276 | light rain.. cloudy.. rain |
| | 21 | 29.47 | 29.52 | 29.57 | 61.4 | 63.5 | 54.2 | 61.5 | 64.2 | 62.1 | 54.2 | 64.8 | S.W. | 0.077 | flying clouds.. windy.. showery |
| | 22 | 29.69 | 29.73 | 29.88 | 60 | 63.5 | 52.4 | 61.3 | 63.8 | 61.7 | 50 | 63.3 | S.W. | 0.023 | showery.. variable.. starlight |
| | 23 | 30.04 | 30.01 | 29.97 | 64.1 | 69.5 | 55.1 | 62.6 | 64.9 | 63.7 | 49.8 | 71 | S.W. | | fine with clouds.. starlight |
| ⊙ | 24 | 29.89 | 29.87 | 29.89 | 70.3 | 78.1 | 61.7 | 64 | 66.8 | 66.3 | 48.5 | 73.9 | S.W. | | fine.. light clouds.. starlight |
| | 25 | 29.97 | 29.96 | 29.92 | 68.3 | 71.6 | 63.3 | 65.3 | 66.7 | 66.1 | 56.7 | 68.7 | N.E. | | cloudy |
| | 26 | 29.87 | 29.87 | 29.87 | 67.2 | 70 | 57 | 64.7 | 66.3 | 64.7 | 55.8 | 64.8 | N.E. | 0.647 | light rain.. cloudy |
| | 27 | 29.88 | 29.88 | 29.89 | 64.3 | 63.3 | 56.1 | 64 | 65 | 63.3 | 52.9 | 64.2 | S.W. | | cloudy.. drizzling rain.. cloudy |
| | 28 | 29.91 | 29.89 | 29.89 | 64 | 69 | 54.5 | 63.9 | 66.3 | 64.8 | 50.8 | 67 | S.W. | | fine.. light clouds |
| | 29 | 29.88 | 29.87 | 29.88 | 62.5 | 66.3 | 52.3 | 64 | 65.7 | 63.3 | 46.8 | 64.7 | N.E. | 0.111 | fine.. cumuli.. thunder storm |
| | 30 | 29.82 | 29.78 | 29.73 | 64.8 | 69.7 | 53.4 | 63.1 | 65.3 | 63.3 | 46.9 | 66 | S.W. | 0.142 | fine.. light clouds.. cloudy.. rain |
| | Mean | 29.75 | 29.74 | 29.75 | 63.9 | 66.6 | 54.4 | 61.2 | 63.5 | 61.8 | 50.1 | 67.3 | | 2.657 | Rain during the Month. |

SW. 20, W. 2, NE 8.

16-8

On the 28th a new double thermometer of Rutherford's construction began to be used by ...

**Figure 2.9** *A page from the Radcliffe Observatory's meteorological register. This is for June 1838, the month of Queen Victoria's coronation; the entries for Coronation Day (28th) record barometer at 29.91 inches of mercury (1012.9 hPa) at 10 a.m., minimum temperature 50.8 °F (10.5 °C), maximum temperature 67 °F (19.4 °C), weather 'fine, light clouds' (University of Oxford, School of Geography)*

On 1 January 1925, the three daily observations were reduced to a single observation at 0900 GMT, supplemented by chart-based autographic records [19] and, more recently, albeit somewhat intermittently, by an automatic weather station logging system. Of course, the Radcliffe Observatory has remained of paramount importance in this regard, firstly because of the great length and quality of its records, and because they have all been made on the same site throughout with only minor moves of the instruments from time to time, and secondly on account of the careful and meticulous records kept of the instruments, exposure details and observing practices maintained by the Observatory and its successors. This latter point is discussed in more detail below.

In 1930, on the advice of the incumbent Radcliffe Observer Harold Knox-Shaw, the Radcliffe Trustees decided that the (astronomical) observatory should move to Pretoria [17]. Although there was some danger that meteorological observations would have to cease and concern was raised regarding the possible cessation of the record that was already 120 years old [27], an arrangement was made whereby they could continue under the supervision of W. G. Kendrew of the University's School of Geography.

**Figure 2.10** *The front of the Radcliffe Observatory as photographed for* Country Life *on 25 February 1930. This view from the west shows the two screens in front of the building (see also Appendix 1). On the roof can be seen the structure holding the anemometer and the sunshine recorder. This structure was removed after equipment was moved to the Engineering Tower in 1976. The photograph was clearly taken at slow shutter speed (2–3 s) as both anemometers can be seen spinning (Courtesy Country Life Picture Library)*

A University decree has ensured that meteorological observations continue unbroken at the Radcliffe Observatory site to the present day. From 1 July 1935, the station has borne the name 'The Radcliffe Meteorological Station, Oxford'. An experienced member of staff, James Balk, remained behind in Oxford and maintained the meteorological observations until his retirement in 1954 [21]. Since then, observations have been continued

**Figure 2.11** *The Radcliffe Meteorological Station, Oxford—(top) on 11 January 1977 (© Crown Copyright: Information provided by the National Meteorological Library and Archive—Met Office, UK) and (bottom) on 29 June 1995 (Stephen Burt)*

**Figure 2.12** *The Radcliffe Meteorological Station, Oxford, in April 2018 (Stephen Burt)*

by staff of the School of Geography (now the School of Geography and Environment) under the supervision of the Director, assisted by one or more Meteorological Observers. Since 1950, as it happens, all the Directors have been Fellows of Keble College as well as holding appointments in Geography: C. G. (Gordon) Smith (1950–1985 [28], Tim Burt (1986–1996), David Collins (1996–1999) and Richard Washington (1999–present)—further details are given in Appendix 2, *The Radcliffe Observers*. The Radcliffe Meteorological

Station (RMS) has been supported by a generous grant from the Gassiot Fund of the Royal Society since 1980; the grant was transferred to the RMS after the closure of Kew Observatory, with which the original fund was closely tied [29]) but some of the overheads and staff costs have inevitably been met by the University [19].

## Observations and procedures at the Radcliffe Observatory, Oxford

One of the most remarkable aspects of the Radcliffe weather records is the amount of background information ('metadata') available on the site, the instruments and observing practices. Having reviewed the history of rainfall recording at the Observatory in great detail, Craddock & Smith [26], p 270) finished their paper as follows:

> '*In conclusion, we may remark that the fact that this paper is concerned so much with the defects of rain-gauges may give the impression that the Radcliffe observations are slipshod and unreliable. The truth is exactly the opposite, because it is the number of carefully kept and comparable observations which has enabled us to piece the story together, and to bring to light minor discrepancies which in most cases would be incapable of being detected.*'

**Figure 2.13** *Cabinets holding some of the Radcliffe Observatory's archived rainfall charts (Ian Curtis)*

In this regard, the work by Knox-Shaw and Balk [30] to homogenise the Oxford records deserves particular commendation. In order to provide a correction factor between the original roof gauge (records kept 1815–1852) and later standard gauges at ground level, Knox-Shaw and Balk re-installed the roof gauge between 1923 and 1930. Such attention to detail and the extra effort required to obtain overlapping records to adjust earlier data from older instruments and exposures is very rare.

Gordon Wallace's *Meteorological Observations at the Radcliffe Observatory, Oxford: 1815–1995* [21] summarised the principal meteorological instruments in use at the Radcliffe Observatory from 1815 to 1995. Following a brief history, a highly detailed account of instrumentation and record keeping is presented: such information is crucial if we are to understand the calculation and derivation of daily, monthly and annual means for the site. Wallace then gives a full account of the meteorological elements and their measurement, providing a detailed inventory of the instrumentation used and its precise location. For example, before 1800, it was normal to use roof-mounted raingauges; ground-level gauges only became standard in 1874, hence the need to make minor retrospective adjustments between different records to reduce them all to modern-day standards as far as possible. Another significant change was the move to protect thermometers inside a Stevenson screen, instead of simply mounting them on a north wall as was standard until the latter part of the nineteenth century. One aspect of the measurements that has received rather less attention was the move of the Stevenson screen and other instruments from the south lawn to the north lawn in September 1939, following the outbreak of war and the expected expansion of the Radcliffe Infirmary which borders the Observatory site to the south. No measurable effect on readings has been thus far been observed but the potential for inhomogeneity perhaps deserves further attention. The movement of the sunshine recorder and anemometer to the roof of the Engineering Science building in 1976, 230 m east-south-east of the Observatory site, was potentially another move that might have caused a break in the records. No change in the sunshine records has been noted; there has been little analysis of wind records to the authors' knowledge and it is very likely that the wind records are not homogenous as they were made at different heights above ground (Observatory 34 m, Engineering tower 45 m). Visibility records might be similarly affected, but the visibility objects (all located to the north and west of the city) are

**Figure 2.14** *Sunshine recorder on the roof of the Engineering Science building, with observer Ian Ashpole (Ian Curtis)*

within 10 per cent of being equidistant from the two buildings so the change was felt to be inconsequential.

Full details of the air temperature, precipitation and sunshine records, including instruments and their exposure, site metadata and the preparation, quality-control and access details for the full daily, monthly and annual datasets for the Radcliffe Observatory site, are given in Appendix 1.

One other possible source of inhomogeneity relates to the urban development of north Oxford in the near 250 years since the Radcliffe Observatory was established. It is possible that this has compromised the record to some small extent. Both the encroachment of the Radcliffe Infirmary from the south and house-building to the north could both have influenced the temperature record. Gordon Manley and Gordon Smith debated the issue, without reaching agreement [31]. We will return to this topic in Chapter 3.

## Periods of record

Different elements have different periods of record. To avoid confusing the reader which are which, Table 2.2 lists the various periods of (digitised) record that we have used to compile the statistical records in this book; in some cases, earlier records exist in paper format awaiting digitisation. Greater detail on the sources of data, the instruments used and their exposures is given in Appendix 1.

**Table 2.2** *Summary of the various periods of record at the Radcliffe Observatory, by element*

| Element | Frequency | Period of digitised record |
|---|---|---|
| **Temperature** | Daily and monthly mean temperatures | December 1813 to date |
| | Daily and monthly maximum and minimum temperatures, means and extremes | April 1815 to date |
| | Daily grass minimum temperatures | December 1930 to date |
| | Monthly mean soil temperatures at 30 cm | January 1925 to date |
| **Precipitation** | Monthly totals | January 1767 to date |
| | Daily totals | January 1827 to date |
| | Snow depth | December 1959, scattered earlier records |
| **Sunshine** | Monthly totals | February 1880 to date |
| | Daily totals | January 1921 to date |
| Wind speed | Monthly means | January 1881 to December 2014, with some gaps |

## A continuing history

To all intents and purposes, meteorological observations continue at the Radcliffe Meteorological Station today in much the same way as they have done since the latter part of the nineteenth century, when procedures were standardised by the Meteorological Office. An observer attends every day at 0900 GMT (Figure 2.15); this has been the time of observation since 1925 [21]. Whilst an automatic weather station was installed in 1994, little use has yet been made of the data collected, other than providing a check on the daily manual observations.

The School of Geography and Environment at the University of Oxford continues to provide information from the Radcliffe Meteorological Station to local and national media—press, radio and television—during interesting or extreme weather events (Figures 2.16, 2.17, 20.2 and 25.15). The role of long-term weather stations such as the Radcliffe Meteorological Station is more important than ever for managing public

**Figure 2.15** *Amy Creese, University of Oxford DPhil student in Climate Science and Senior Radcliffe Meteorological Observer, doing the 0900 GMT morning observation at the Radcliffe Meteorological Station, Oxford in August 2018. The new (October 2017) Met Office logger system can be seen within the screen (Stephen Burt)*

**Figure 2.16** *Collage of press headlines relating to Oxford's weather over the years, many of which included information from the Radcliffe Meteorological Station at Green Templeton College (Ian Curtis)*

**Figure 2.17** *Met Office regional network manager Phil Johnson (right) presents Professor Richard Washington, the current Director of the Radcliffe Meteorological Station within the School of Geography and the Environment, University of Oxford, with a Met Office award celebrating 200 years of continuous climatological observations—an award that, to date, is unique in Met Office history (Courtesy University of Oxford School of Geography)*

awareness and media impact. In times when 'global warming' is on everyone's lips, media outreach by Oxford University from locations such as the Radcliffe Observatory site, with its long-term perspective, can be particularly effective in providing a balanced context for such reporting.

In May 2015, Professor Richard Washington, the current Director of the Radcliffe Meteorological Station Oxford, was presented with a Met Office award celebrating 200 years of continuous climatological observations—an award that, to date, is unique in Met Office history (Figure 2.17). Given the great length, quality and documentation of the record, and ongoing support from the University and the Royal Society, the future appears positive as we enter a third century of daily observations.

# 3

# Oxford's urban growth and its potential impact on the local climate

One potentially important source of inhomogeneity in the Oxford meteorological record relates to Oxford's urban development since the Radcliffe Observatory was built in 1772, for it is very likely that this has affected the record to some extent. Close to the site itself, the encroachment of the Radcliffe Infirmary from the south, house-building to the north and west and construction of various University buildings to the east could all have influenced the temperature locally. Looking at the broader situation, it would be surprising if the growth of Oxford more generally did not have some effect upon measured air temperatures at the Observatory, as well as other observations such as visibility. Gordon Manley and Gordon Smith debated the urban heat island (UHI) effect in 1975, without reaching agreement [31]. As with almost any settlement of any size, there is undoubtedly some urban influence. In this chapter, we review the issue and chart its likely development over the last two and a half centuries. To do so, it is first necessary to describe developments immediately adjacent to the Observatory as well as further afield.

When the Observatory was built, the choice of site was very deliberately on the northernmost edge of the built-up area, beyond the Radcliffe Infirmary [20, 32]. Of necessity, the Observatory had to be aligned on an east–west axis and, whilst initially the southern side looked out onto open land, the gardens to the north of the building were the private grounds adjoining the Observer's house, into which Hornsby moved with his family in 1773. According to Michael Pirie (pers. comm.), the wall separating the north lawn and gardens from the open land beyond was built at that time, certainly before 1776. Thus, the sky view from the north lawn has always been a little restricted, even before houses were built north of the Observatory. To the south and west, the aspect remained relatively open until the encroachment of the Radcliffe Infirmary from the early twentieth century onwards [17]. Ironically, as we write this book, the land has been cleared immediately to the south-west of the Observatory, with new building developments soon to follow, no doubt. Already, new buildings to the north of the Infirmary have come very close on the south-east side of the Observatory (Figure 3.1). Michael Pirie [32] notes that the encroachment of the hospital during the First World War coincided with the decline of the astronomical observations. Although the raingauges have remained in exactly the same position on the north lawn since 1850 (aside from a short spell on the

*Oxford Weather and Climate since 1767*. Stephen Burt and Tim Burt, Oxford University Press (2019).
 DOI: 10.1093/oso/9780198834632.001.0001

**Figure 3.1** *The view south and south-west from the Radcliffe Observatory tower, April 2018 (Stephen Burt)*

front or south lawn from 1935 to 1939), the screens and their thermometers were located to the south of the main Observatory building from 1878 to 1939 except for 1920 to 1926, when they were located in or close to the current enclosure on the north lawn. They were moved to the north lawn once more in September 1939, where they have remained ever since (Appendix 1 gives more details.) The move of the Observatory to South Africa in 1935 seemed to threaten the continuation of the meteorological observations, but the University wisely insisted on their continuation, whatever use the Observatory itself was put to.

The 1830s saw two important developments close to the Observatory. Oxford University Press was removed from the Clarendon Building to Walton Street in 1830 and its presence there led to the rapid development of small terraced housing in the Jericho district*. On the other side of the north lawn wall, the terraced houses of Observatory Street were built from 1834 (Figure 3.3). A little further to the west, wharves were gradually opened off Walton Street and Hayfield Road, alongside the Oxford Canal which had opened in 1790; terraced housing was gradually built in this area too. North Oxford grew steadily from the 1850s; most of the land was owned by St John's College, which obtained an Act of Parliament in 1855 enabling it to offer 99-year building leases. Large houses were built on farmland either side of Banbury Road and Woodstock Road. Chance *et al* [33] argue that it is a misconception that north Oxford grew up when the dons were released from celibacy. By the time dons were allowed to marry, following the Royal Commission of 1877, the southern part of north

* The authors are pleased to draw attention to the strong 'local' link with our publishers, Oxford University Press, which remains in the same buildings in Great Clarendon Street today, barely 2 minutes' walk from the Radcliffe Observatory.

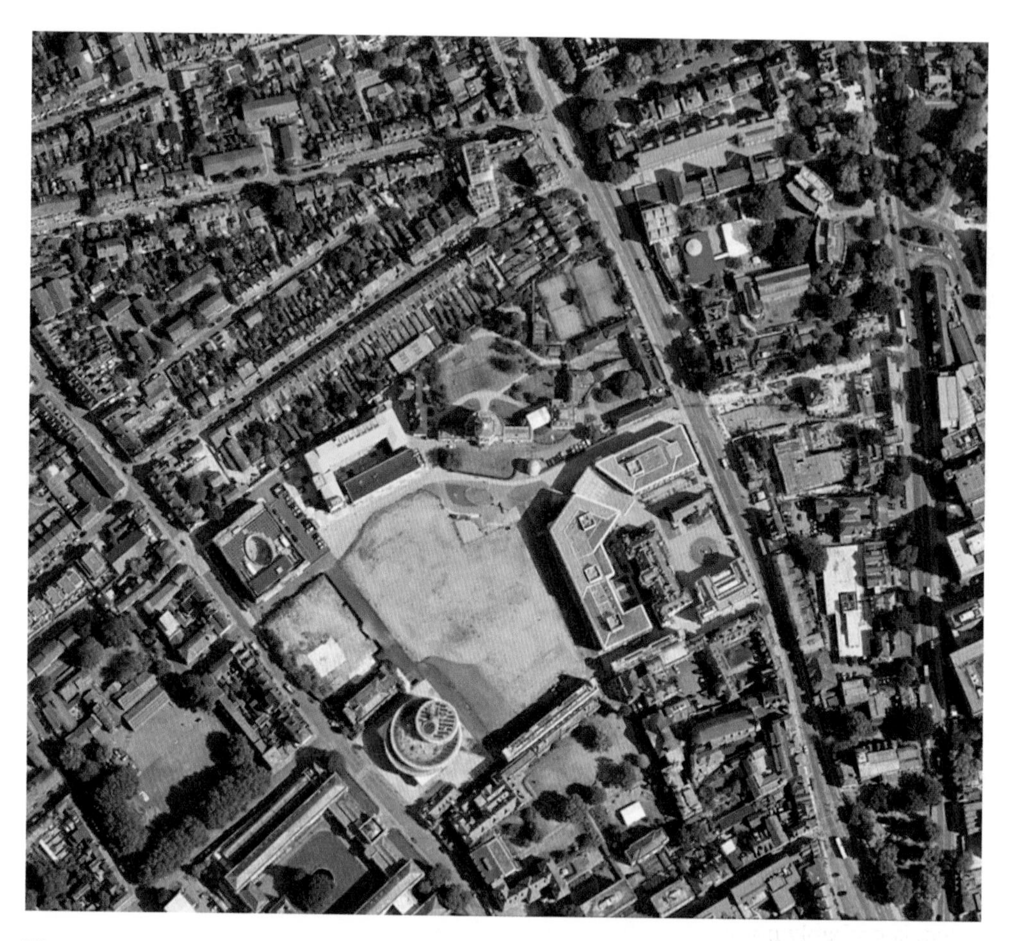

**Figure 3.2** *Google Earth view centred on the Radcliffe Observatory and covering approximately 500 m on each side (0.25 km²). The meteorological enclosure can be seen in the grounds of Green Templeton College immediately north of the Observatory building. The grounds and buildings of the Radcliffe Infirmary to the south and south-west, previously the Observatory grounds (compare Figures 2.5 and 2.8), are in the process of extensive redevelopment (see also Figure 3.1) (Google Earth)*

Oxford was already developed, and the movement of dons out of college was, in any case, a gradual process. Professors and readers had always been allowed to live out, and they accounted for the relatively high concentration of families in Norham Gardens and Park Town, to the east of the Observatory. The new houses to the north of the Observatory were mostly taken by tradesmen, for whom the growth of north Oxford was the first opportunity to move from the city centre into suitable middle-class suburbs. Today, the whole area generally to the north of the Observatory, including Summertown, is completely built up, from Port Meadow to the west to the Cherwell floodplain to the east (Figure 1.1). Whilst this has completely changed the view from the uppermost floor of

**Figure 3.3** *View looking north from the Observatory tower, 25 April 2018. The amount of building that has taken place since the 1770s is obvious, starting with Observatory Street immediately to the north of the garden wall in 1834. The Radcliffe Meteorological Station enclosure is clearly visible on the lawn (Tim Burt)*

the Observatory, the sky view of the north lawn, where most of the instruments are now located, is probably not very much different from what it was in the late 1830s.

Two questions therefore come to mind: what is the current magnitude of Oxford's urban heat island (UHI), and how has the UHI evolved since meteorological observations were begun in the 1760s?

Initially, to gauge the current situation, we had hoped to use 'rural' observations from Wytham Woods, some 9 km west of the Observatory on the other side of the Thames floodplain (located on Figure 1.1). There, in a field close to the woodland, the UK Environmental Change Network (ECN) has run an automatic weather station (AWS) since 1992, at 160 m above sea level (ASL). However, the ECN mean temperatures are calculated from hourly averages whereas the Radcliffe Meteorological Station (RMS) means are based upon maximum and minimum temperatures in the usual way. Until we have sufficiently long automatic weather station records from the Observatory, this comparison will have to wait. In any case, the ECN site is on the top of Wytham Hill (c. 160 m ASL), almost 100 m higher than the RMS in a very different topographic situation, hardly an ideal comparison. An additional AWS somewhere on the Port Meadow floodplain might in the future provide a more useful comparison.

Accordingly, we went further afield to the small town of Wallingford, around 22 km south-east of Oxford, where the Centre for Ecology and Hydrology (CEH) has maintained weather records at a rural site some distance outside the town since 1961. The altitude difference is only 15 m (Wallingford 48 m, Oxford 63 m) so, on lapse rate grounds alone, we would expect Oxford to be 0.1 degC cooler on average. We compared daily records over the three most recent years' record from 2015 to 2017. Figure 3.4 and Table 3.1 show the comparisons for daily maximum and minimum temperatures. As expected, Oxford is somewhat warmer overnight—the mean minimum temperature there averages 0.93 degC above Wallingford, with a slight seasonal variation (summer higher) and a marked positive skew. Oxford is at least 0.2 degC warmer than Wallingford on two nights in three, whereas only one night in five is at least 0.2 degC warmer at Wallingford: 43 per cent of nights are at least 1 degC warmer at Oxford, but only 5 per cent of nights are at least 1 degC warmer at Wallingford. By daytime, the magnitude of Oxford's UHI is reduced—the mean maximum temperature at Oxford is just 0.27 degC higher than Wallingford on average. The daytime UHI also shows a very marked seasonal variation tied to solar angle—close to zero at the midwinter solstice, around +0.7 degC at midsummer. This seasonal variation can be accounted for by a combination of shading of the site by the Observatory buildings in midwinter, and stored heat within the urban fabric in summer. Nevertheless, almost 60 per cent of daily maxima are at least 0.2 degC higher in Oxford (14 per cent are more than 1 degC warmer), compared with just 23 per cent of days which are more than 0.2 degC warmer at Wallingford.

In conclusion, therefore, and based on the Wallingford comparison, to answer our first question, there is clear evidence that the Radcliffe Observatory site is slightly warmer by day than the surrounding countryside, by around 0.4 degC on average (including the expected lapse rate difference due to the altitude difference), and this is more marked during the summer months. By night, the difference is greater, around 1.0 degC (again including the expected lapse rate difference due to the altitude difference), also more marked in the summer months but less so than for the seasonal variation in daytime temperatures. Averaging the two, we can state that the current magnitude of Oxford's urban heat island at the Radcliffe Observatory site averages about 0.7 degC.

The second question is more difficult to answer going back as far as the start of the Radcliffe Observatory record, as there are no single-site records of similar length against

**Table 3.1** *Differences in daily maximum and minimum temperatures between Oxford and Wallingford, based on three years data 2015–17*

| | MINIMUM TEMPERATURE | | MAXIMUM TEMPERATURE | |
|---|---|---|---|---|
| Difference in max or min temperature | Wallingford warmer | Oxford warmer | Wallingford warmer | Oxford warmer |
| ≥ 0.2 degC | 20% | 67% | 23% | 59% |
| ≥ 1.0 degC | 5% | 43% | 5% | 14% |

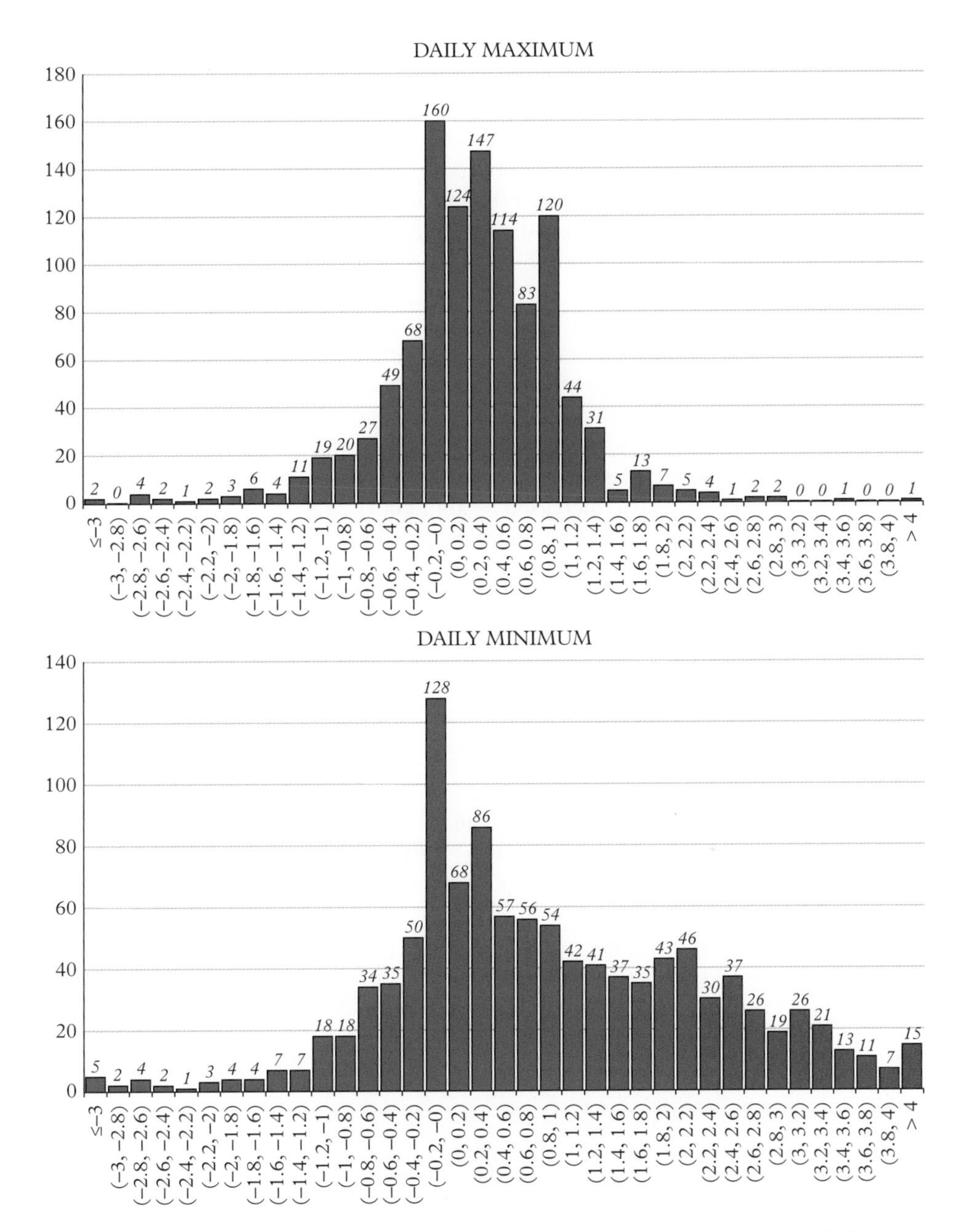


**Figure 3.4** *Frequency of differences in daily maximum and minimum temperatures (degC, 0.2 degC bins, positive indicates Oxford warmer) between Radcliffe Meteorological Station Oxford and CEH Wallingford over three years 2015–17*

which to compare. The obvious comparator would seem to be Gordon Manley's Central England Temperature (CET) series [34, 35], but CET data are themselves very dependent on the Oxford record from 1815 to 1841, so this becomes somewhat of a circular argument and meaningful comparisons are accordingly limited. From the mid-nineteenth century, when other weather station data become available, the CET is less dependent on Oxford, so comparisons over time can then be made. Here, we focus on comparisons with the CET daily mean maximum and mean minimum temperature time series which run from 1878 [36, 37], using monthly data downloaded from the Hadley Centre website.

Table 3.2 compares CET mean daily maximum and minimum temperatures with Oxford's over various 30-year periods since 1901, including the most recent 30 years, 1988–2017 (these are further discussed in Chapter 24). There is no doubt that the urban influence on the long Oxford record continues to increase, although the greatest difference between CET and Oxford is in maximum temperatures*, and this has increased at a greater rate than the difference in minimum temperatures in the last three decades. Mean air temperatures have also increased by about 0.91 degC compared with 1961–90, 0.20 degC greater than the rise in CET over the same period.

Long-term trends in Oxford maximum and minimum temperatures relative to CET are shown by the solid lines in Figures 3.5 and 3.6, which show ten-year running means of the difference between Oxford and CET mean annual maximum (Figure 3.5) and

**Table 3.2** *Comparison of Central England Temperature (CET) and Oxford mean temperatures over various 30-year periods since 1901*

| | | 1901–30 means | 1931–60 means | 1961–90 means | 1988–2017 means | Difference 1988–2017 minus 1961–90 degC |
|---|---|---|---|---|---|---|
| **Mean maximum °C** | CET | 12.79 | 13.24 | 13.07 | 13.95 | +0.88 |
| | Oxford | 13.68 | 14.10 | 13.78 | 14.93 | +1.15 |
| | Difference | +0.89 | +0.86 | +0.71 | +0.98 | **+0.27** |
| **Mean minimum °C** | CET | 5.71 | 5.97 | 5.88 | 6.45 | +0.57 |
| | Oxford | 5.95 | 6.21 | 6.44 | 7.11 | +0.67 |
| | Difference | +0.24 | +0.24 | +0.56 | +0.66 | **+0.10** |
| **Mean air temperature °C** | CET | 9.28 | 9.63 | 9.51 | 10.22 | +0.71 |
| | Oxford | 9.81 | 10.16 | 10.11 | 11.02 | +0.91 |
| | Difference | +0.63 | +0.53 | +0.60 | +0.80 | **+0.20** |

* This is not surprising and is not related to Oxford's UHI—it is simply a matter of geography: Oxford is south of the 'centre of gravity' of Manley's CET region, and is therefore slightly warmer as a result.

mean annual minimum temperature (Figure 3.6), in degrees Celsius. Positive values indicate Oxford warmer than CET.

Gordon Manley used the long record from Rothamsted (Hertfordshire) to assess changes in the urban component in the Oxford record up to 1975 [31]; we decided to repeat the comparison, making use of the additional 40+ years of record since Manley's work [38]. The Rothamsted record commenced in 1878; it is a rural site about 65 km almost due east of Oxford, and at a greater altitude (128 m against Oxford's 63 m); the altitude difference alone would be expected to account for about 0.4 degC difference in mean temperature between the two.

Ten-year running means of the differences between Oxford and Rothamsted's mean annual maximum and mean annual minimum temperatures (dotted lines) and Rothamsted and CET mean annual maximum and mean annual minimum temperatures (dashed lines) are also plotted in Figures 3.5 and 3.6. Given that much work has been done to ensure the CET series remains as homogenous as possible and reflects a proper balance of stations across Central England, and that the Oxford and Rothamsted records

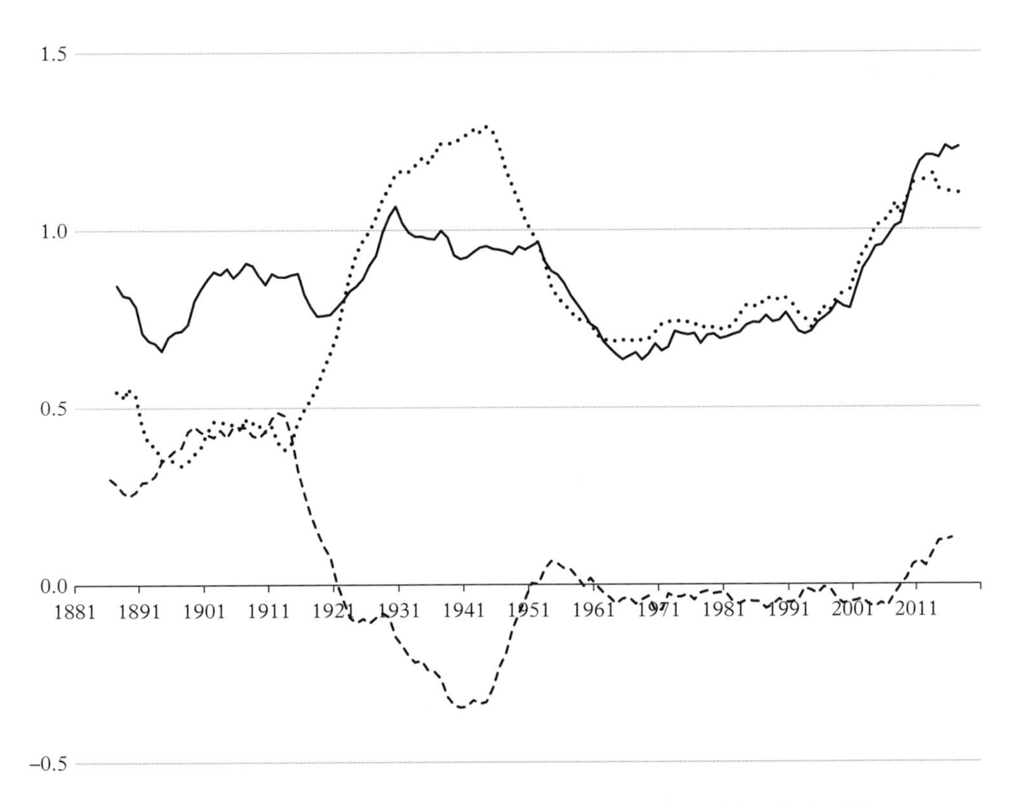


**Figure 3.5** *Ten-year unweighted running means, plotted at year ending, of the relative differences (degC) between Oxford, Rothamsted and Central England Temperature (CET) annual mean maximum temperature over the period 1878 to 2017. Solid line shows Oxford minus CET, dotted line Oxford minus Rothamsted, and dashed line Rothamsted minus CET*

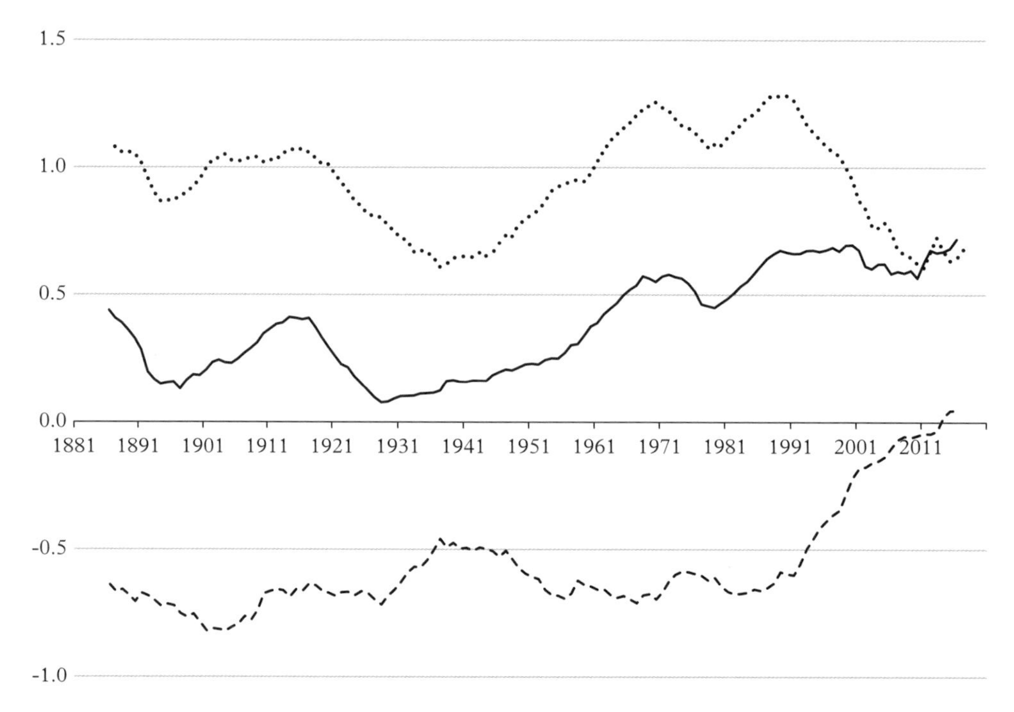


**Figure 3.6** *As for Figure 3.5 but for annual mean minimum temperature. The scales on both plots are identical to facilitate comparison*

reflect much the same regional climate, divergence from the CET and from each other can only arise from changes in observational practice and/or local site factors.

Looking firstly at maximum temperatures (Figure 3.5), there is little overall trend in Oxford's temperatures relative to CET (solid line) until the 1990s, since when the difference has increased quite sharply. The average difference between 1878 and 1990 was 0.76 degC; the last 20 years (to 2017) averaged 1.09 degC and is clearly outside the range of previous inter-decadal variations within the record since 1878. (The significant cooling relative to CET seen in Rothamsted mean maximum temperatures between about 1911 and 1940, dashed line in Figure 3.5, is not reflected in the Oxford record, and may represent an unknown or uncorrected site move or change of observational practice at the former.) There is less evidence at Rothamsted of the relative warming in mean maximum temperatures seen in the Oxford record in recent decades (dashed line), at least until the last decade or so. From this we infer that the increase in Oxford's mean maximum temperatures relative to CET since the mid-1960s is genuine: it most likely relates to an increased urban influence on the Radcliffe Observatory site since then, and particularly since 2000. The observed increase in the duration of bright sunshine in recent decades highlighted in Chapter 24 may also have had some impact on daytime temperatures.

For minimum temperatures (Figure 3.6), the position is simpler—at Oxford there has been a fairly steady warming relative to CET since about 1930 (solid line), averaging about 0.07 degC per decade. Unlike maximum temperatures, there is little evidence of a more rapid warming at Oxford relative to CET over the past two or three decades, although minimum temperatures at Rothamsted (dashed line) appear to have warmed rapidly relative to CET since the 1980s (resulting in a reduction in the mean difference between Oxford and Rothamsted since then, as shown by the dotted line in Figure 3.6). Until about 1990, there is a high degree of similarity between Oxford and Rothamsted mean minimum temperature differences from CET. That this has changed significantly since about 1990 appears to be due to relative warming at Rothamsted rather than Oxford, suggesting that Rothamsted is not itself entirely free of urban effects from the growth of nearby Harpenden.

At a more regional scale, it is possible that synoptic airflow changes may have raised temperatures at both Oxford and Rothamsted more than places further north within the CET area in recent decades. Both sites are in inland south-eastern sites that tend to record larger positive anomalies in generally warm months (Julian Mayes, pers. comm.). Further analysis of the Oxford UHI using rural sites more distant than Rothamsted and further to the north would usefully enhance this picture.

With hindsight, Manley's [31] focus on the difference between Oxford and Rothamsted from the 1950s to the 1960s was prescient, judging by Figure 3.5; he argued for a small urban effect at Oxford of the order of 0.1 degC at that time. To conclude by answering the second question posed at the start of this section, the evidence available 40 years after Smith and Manley's exchange suggests that Oxford's urban effect in comparison to 'background' CET is now 0.2 degC, arising primarily from a more rapid increase in maximum temperatures in recent decades (Table 3.2 and Figure 3.6). The more rapid recent upward trend in Oxford air temperatures is concerning and it may well be that in future some adjustment to Oxford temperature records will need to be made to allow for this more local warming trend: clearly, any inferences regarding climate change at Oxford need to bear this in mind. These increases in urban effects on the long Radcliffe Observatory record are most likely driven partly by changes close to the site itself and partly by wider changes across Oxford's urban area.

# 4

# Oxford's weather in its regional context

Oxford's weather does not happen in isolation. Aside from perhaps the most local of rain showers or thunderstorms, the origins of the atmospheric conditions resulting in any particular day's weather in Oxford may lie much further afield—a winter storm may owe its origin to temperature and moisture gradients over the eastern seaboard of the United States several days earlier, for example, while a thundery outbreak in summer might well result from the collision of hot, dry air moving northwards from north Africa and Spain with moist Atlantic air originating from Bermuda or the Azores. For this reason, any particular day or spell of weather as experienced in Oxford will normally share some characteristics with the weather recorded tens if not hundreds of kilometres distant from Oxford. Of course, beyond the typical scale of individual synoptic-scale weather systems, the main features of the weather elsewhere may be very different, even completely opposite, to those in Oxford. Thus, records of temperature, rainfall and so on recorded in Oxford are correlated to varying degrees with those from more distant locations, the magnitude and sign of the correlation generally varying with the element (sunshine, snowfall and so on), time and distance. For example, a week-long heatwave recorded in Oxford is very likely to feature in weather records from London and Birmingham, and probably Manchester and perhaps Newcastle too, but not necessarily in Edinburgh or Amsterdam, although details such as the intensity and peak of the heatwave will likely differ slightly from place to place as weather systems move and wind directions change accordingly. In contrast, an isolated thundery rain shower one afternoon may affect a single village or suburb for an hour or less, while neighbouring districts remains dry and sunny all day.

The purpose of this chapter is to set Oxford's weather in the broader north Atlantic and European context by examining major influences on the weather of the British Isles and, in doing so, contrasting the wider synoptic circumstances of several multi-week spells of noteworthy weather—a summer heatwave, a wet winter and so on. Although no two sequences or spells are ever exactly the same, wider relationships between particular types of synoptic pattern, Oxford's long weather records and weather and climate records elsewhere within the British Isles can often be extrapolated from multiple such examples. Each occasion is briefly described in terms of Oxford's weather and the wider Atlantic

*Oxford Weather and Climate since 1767*. Stephen Burt and Tim Burt, Oxford University Press (2019).
 DOI: 10.1093/oso/9780198834632.001.0001

or European synoptic context. The examples chosen are mostly selected from the last 40 years or so, most within living memory and thus the experience of the reader.

## Oxford's weather and climate in its regional and global context

The dominant features of the atmospheric circulation across the north Atlantic are the Icelandic Low and the Azores High. These are present in all seasons, although their location and relative intensity vary considerably. The upper air flow pattern remains broadly similar throughout the year, but the strength of the circulation decreases by more than half in summer compared with winter. Another significant pressure system influencing the climate of north-west Europe is the so-called 'Siberian' winter anticyclone, a relatively shallow near-surface feature, the strength and intensity of which is intensified by the marked continentality of Eurasia with its extensive winter snow cover [39]. Oxford's most severe winter weather tends to be associated with an extension of the Siberian anticyclone westwards, producing a 'blocking' high over Scandinavia which, as its name suggests, hinders the easterly progression of maritime air over the British Isles and the European mainland of north-west and western Europe. At such times, the jet stream usually lies well to the south with maritime air moving through the Mediterranean basin. Oxford experiences continental climatic influences in the summer half-year too, being close enough to the south-east of England to benefit from warm continental air masses influencing the weather further east and south.

At a more national scale, by virtue of its location in the south Midlands, Oxford enjoys a mild climate, typical of southern England, but slightly more continental than coastal districts. Air masses moving towards it are somewhat modified by distance from the sea and by upwind topography. Its inland location means that both the diurnal and annual temperature ranges are somewhat larger than at the coast. In summer especially, its inland location increases the likelihood of localised, convective thunderstorms. Oxford's climate is therefore a balance of oceanic and continental influences, the two factors which together determine regional climates throughout the British Isles [40]. Given these influences, it is no surprise that any interpretation of the Oxford climate must rely on weather indices that reflect the origins and track of air masses arriving in the Oxford region. Whilst ocean–atmospheric drivers of hydrology are increasingly investigated using numerical indices such as the North Atlantic Oscillation Index (NAO: [41]—see next section), the Atlantic Meridional Oscillation (AMO) or the Southern Oscillation Index [42], use of weather types to classify patterns of atmospheric circula tion remains a common approach [43]. Reanalyses of historic surface pressure data major operational weather forecasting centres has enabled objective techniques to applied consistently back to the nineteenth century [44–46].

We use Lamb Weather Types (LWT) to analyse changes in Oxford's weathe inset box describes in detail how the LWT scores are derived from the daily classif of weather type. Table 4.1 shows the correlations between seasonal and annual data for Oxford and Lamb Weather Types over 146 years from 1871 to 2016 inclusive. It is immediately clear that westerlies, so important in the uplands and the north-west coastal regions, are much less important at lowland, inland Oxford. Instead, it is the balance of cyclonic (C) and anticyclonic (A) weather that is dominant. For rainfall especially, but also for mean air temperature to an extent, the frequency of A and C weather determines whether Oxford is warm or cool, wet or dry. Thus, in the very wettest periods, such as September 1976 to August 1977, or the last nine months of 2012, cyclonic circulation dominated. In the winter of 2013/14, Oxford's wettest on record, westerlies were important too, but again the influence of cyclonic circulation was paramount. Of the measures of atmospheric flow, flow strength (F) is important for mean air temperature, in autumn and winter especially, but total shear vorticity (Z) controls rainfall and sunshine totals. Burt *et al* [43] concluded that high rainfall over an extended period requires a high frequency of cyclonic weather. Over shorter timescales (i.e. a single season), strong westerlies alone can generate high rainfall in the western uplands whereas a combination of vigorous westerly airflow in tandem with high vorticity is needed to produce exceptional totals in the southern and eastern lowlands. Neither NAO nor AMO are correlated with rainfall totals but both correlate strongly with mean air

## Analysis of synoptic climatology using Lamb Weather Types

Hubert Lamb [47] identified 27 possible Lamb Weather Types (LWTs) which he simplified into seven basic types: anticyclonic (A), cyclonic (C), the four cardinal wind directions—northerly (N), easterly (E), southerly (S) and westerly (W)—together with a distinct north-westerly (NW). Following the objective LWT classifications of Jones *et al* [46], we use Lamb's counting procedure where each pure type (i.e. one of the basic seven) counts one towards the monthly total for the type and then either a half or a third for each of the hybrid types (e.g. north-easterly: NE = 0.5; cyclonic south-westerly: CSW = 0.33). Together with the 'unclassified' type, the totals of the seven basic types add up to the number of days in each month. We calculate two simple indices based on LWTs: the balance westerly minus easterly (W-E) as an indication of zonality, and anticyclonic minus cyclonic (A-C) as an indicator of pressure system dominance. We also include two basic variables derived from mean sea level pressure data used in the objective LWT reanalysis: the resultant mean trength (F) and the total shear vorticity (Z). The flow and vorticity units are geo-each is equivalent to 1.2 knots), expressed as hPa per 10° latitude at 55°N [46]. Our eather types follows Jones *et al* [46] and Burt *et al* [43]. The 'newLWT' dataset erived: from 1871 to 1947 using the '20CR' reanalysis of surface pressure National Center for Environmental Prediction (NCEP) reanalysis from We also include two other indices reflecting climatic conditions in the ) index (www.cru.uea.ac.uk), which reflects the pressure gradient ores (a proxy for the strength of the upper air jet stream) and the n (AMO) index: http://www.esrl.noaa.gov/psd/data/correlation/ s variations in sea surface temperature, an important influ-g the Atlantic Ocean. There is a more detailed analysis section.

**Table 4.1** *Correlations between seasonal and annual Oxford weather data and LWTs (1871 to 2016). The small box following the table shows how significant correlations are indicated in the main body of the table:— in the cell indicates there was no significant correlation.*

| Element | Season | A | C | W | Z | F | NAO | AMO |
|---|---|---|---|---|---|---|---|---|
| **Precipitation** | Winter | **−0.70** | **0.72** | −0.17 | **0.80** | — | — | — |
| | Spring | **−0.51** | **0.55** | — | **0.59** | — | — | — |
| | Summer | **−0.58** | **0.61** | — | **0.63** | — | — | — |
| | Autumn | **−0.48** | **0.63** | — | **0.62** | — | — | — |
| | Annual | **−0.44** | **0.62** | −0.22 | **0.56** | — | — | — |
| **Mean air temperature** | Winter | **−0.30** | — | — | 0.19 | **0.63** | **0.71** | — |
| | Spring | 0.21 | **−0.30** | — | **−0.30** | 0.20 | **0.46** | **0.31** |
| | Summer | **0.47** | **−0.47** | — | **−0.49** | −0.22 | — | **0.46** |
| | Autumn | — | — | — | — | **0.29** | **0.29** | **0.41** |
| | Annual | — | — | — | — | **0.35** | 0.18 | **0.45** |
| **Sunshine hours** | Winter | — | — | — | — | **0.33** | — | — |
| | Spring | **0.44** | **−0.32** | **−0.32** | **−0.48** | — | — | — |
| | Summer | **0.59** | **−0.53** | **−0.53** | **−0.61** | — | — | — |
| | Autumn | — | — | — | −0.18 | — | — | — |
| | Annual | **0.33** | **−0.32** | **−0.32** | **−0.31** | — | — | — |

| Probability $p$ | Font style | Period 1871–2016: 146 years |
|---|---|---|
| $p = 0.05$ | 0.17 | 5% chance of occurring randomly |
| $p = 0.01$ | *0.22* (none in table) | 1% chance of occurring randomly |
| $p = 0.001$ | **0.28** | 0.1% chance of occurring randomly |

temperature, although not in every season: only NAO correlates with winter mean air temperature whereas only AMO correlates with summer mean air temperature. We did investigate whether indices of Pacific Ocean sea surface temperatures (SSTs) might correlate with Oxford; such distant teleconnections can be identified, even as far away as the British Isles but, apart from spring rainfall total (correlation $r = 0.164$ using the NINO34 index), there were no significant correlations.

## Oxford's weather and the NAO

Many aspects of the weather and climate of the British Isles exhibit some correlation with the index of the North Atlantic Oscillation (NAO), particularly in winter. The NAO is one of the major modes of variability of the Northern Hemisphere's atmosphere [48]: a useful index of the strength of the NAO is obtained from the difference between normalised sea level pressures at Ponta Delgada in the Azores or Gibraltar and in south-west Iceland (Reykjavik). By careful examination of historical records from these locations, records of the NAO have been extended back almost 200 years [41].

A strongly positive NAO index (indicated by a greater than normal pressure gradient between Iceland and the Azores) is indicative of an enhanced or progressive zonal flow over the north Atlantic, resulting from a stronger than normal jet stream. In positive NAO conditions, winters in the British Isles are often milder and wetter than normal as a result of frequent Atlantic depressions bringing milder Atlantic air, cloud and rain (and often strong winds) to all districts. Negative or retrograde NAO conditions, in which the westerly flow is weaker or, occasionally, entirely reversed with high pressure over Greenland/Iceland and the main Atlantic low-pressure area much further south than normal, are more likely to see an increased frequency and duration of northerly and easterly air masses from the European continent and/or a higher frequency of anticyclonic blocking. In winter, low or negative NAO conditions are often associated with below normal temperatures, above-normal precipitation in the east (sometimes with significant or prolonged snowfalls) but drier and often sunny conditions in the west; in summer, the weather can be hot and dry. Some of our most extreme spells of weather, winter or summer, arise from prolonged periods of abnormal NAO conditions.

As is the case with most other parts of the British Isles, the influence of the NAO on Oxford's weather is most obvious during the winter half-year. Figure 4.1 shows scatter plots and correlation coefficient $r$ between the NAO index and Oxford's mean temperatures and total precipitation for the months of January and July over the period 1961 to 2018.

The relationship with the NAO is most marked with mean temperatures in January, where the correlation coefficient $r$ is as high as 0.74 (Figure 4.1, top left); the relationship is weaker for January precipitation (Figure 4.1, bottom left, $r = 0.25$). In July, the relationship between the NAO index and Oxford's monthly mean temperature is much weaker than in January (Figure 4.1, top right, $r = 0.28$), while July's precipitation exhibits a weakly negative NAO relationship (Figure 4.1, bottom right, $r = -0.25$); the negative correlation with monthly rainfall probably reflects the increased likelihood of showers and thunderstorms in slack easterly or north-easterly winds with low pressure over France and Spain. The strength of the north Atlantic circulation in the summer months is normally much lower than in winter, and indices of the frequency and strength of cyclonic systems in the vicinity of the British Isles show more significant long-term relationships with temperatures and precipitation than the NAO or AMO index alone during the summer half-year (Table 4.1).

As has already been referred to, Oxford's location in the south Midlands results in a more continental climate, with a greater daily range in temperature than experienced on western or northern coasts. Winds from between north and east can bring bitterly cold conditions in winter, particularly under clear skies and with a snow cover; damp north-easterly or easterly winds blowing off the North Sea can bring prolonged spells of low stratus cloud, particularly in winter and spring. In summer, easterly or south-easterly winds can introduce continental tropical air masses from southern Europe and, under such conditions, away from the ameliorating effects of coastal breezes, temperatures in Oxford can exceed 30 °C. Oxford's inland position also reduces the direct impact of winter rain and gales, although wet spells in the autumn and winter months, particularly

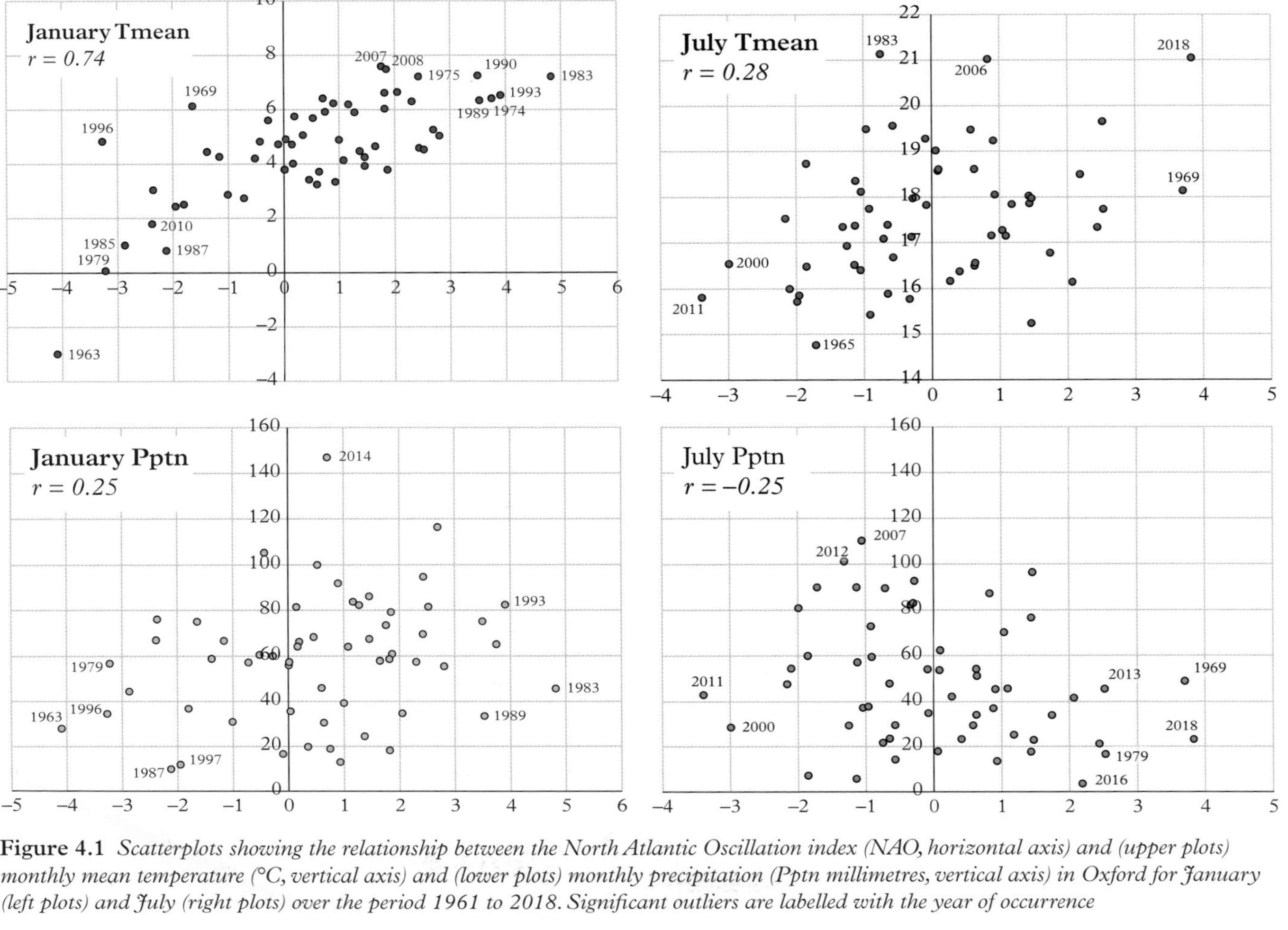


**Figure 4.1** *Scatterplots showing the relationship between the North Atlantic Oscillation index (NAO, horizontal axis) and (upper plots) monthly mean temperature (°C, vertical axis) and (lower plots) monthly precipitation (Pptn millimetres, vertical axis) in Oxford for January (left plots) and July (right plots) over the period 1961 to 2018. Significant outliers are labelled with the year of occurrence*

when the ground is already saturated, often lead to prolonged and extensive river flooding along the rivers Thames and Cherwell (as seen most dramatically in the recent winter of 2013/14, Oxford's wettest on record).

To illustrate the wider synoptic context, several spells of notable weather are examined as follows:

- The prolonged hot, dry and sunny summer of 1976
- The exceptionally dull and wet autumn of 1976
- December 2010—the coldest December since 1890
- April 2011—the warmest and driest April on record
- June 2012—a cool and very wet summer month
- March 2013—very cold and extremely dull
- The very wet winter of 2013/14
- December 2015—the mildest December for at least 350 years
- Summer 2018—the hottest summer on Oxford's records

SOURCES. Unless otherwise stated, all anomalies (difference from normal) are from the current standard 30-year period 1981-2010. The NAO index values are from the Climatic Research Unit, University of East Anglia (monthly values available from https://crudata.uea.ac.uk/cru/data/nao/): mean sea level pressure averages and anomaly charts are from the NOAA/ESRL Physical Sciences Division, Boulder, Colorado (http://www.esrl.noaa.gov/psd/) [44], with overlays by the authors. Plots on all charts are at 2 hPa intervals. Commentaries on UK monthly and seasonal weather have mostly been taken from the Met Office monthly climate summaries (http://www.metoffice.gov.uk/climate/uk/summaries). The British Isles regional rainfall series, including the monthly England and Wales rainfall series, extending back to 1766, can be found on the Met Office Hadley Centre Observation Dataset (https://www.metoffice.gov.uk/hadobs/hadukp/data/download.html). For more information on the various long-series rainfall datasets, see reference [49].

## The prolonged hot, dry and sunny summer of 1976

| Oxford | Mean daily maximum temperature, °C | Precipitation, mm | Sunshine, hours |
|---|---|---|---|
| 19 July to 27 August 1976 | 24.4 °C | 0.1 mm | 330.7 |
| Anomaly | +1.7 degC | 0% | 129% |

The summer of 1976 remains *the* outstanding summer on Oxford's records. Until 2018 this was the equal-hottest summer (June–July–August) in Oxford, and one of the driest and sunniest: summer 1976 still tops the 'summer index' (see Summer, Chapter 22). Following the outstanding and unprecedented 14-day heatwave 25 June to 8 July, each day of which surpassed 30 °C, a prolonged drought set in during July [50–52]. Commencing on 19 July, no measurable rain fell for 40 days, the longest absolute drought on Oxford's records since at least 1827 (Chapter 28). The mean sea level pressure pattern for this 40-day drought period, and the anomaly, are shown in Figure 4.2. A persistent ridge from the Azores anticyclone lay across the British Isles, with a small subsidiary centre of 1024 hPa just off south-west Ireland; the mean pressure was more than 10 hPa above normal to the north-west of Ireland. This pressure pattern resulted in frequent north-easterly or easterly winds over the British Isles. The period was dry, warm and sunny in most of England, although conditions were less extreme in Scotland, which continued to be affected by cloud and occasional rain on the edge of Atlantic weather systems whose main track had shifted well to the north of the British Isles.

Rarely do two such opposite extremes occur almost back to back as occurred with the dramatic breakdown of the prolonged 1975–76 drought in September 1976.

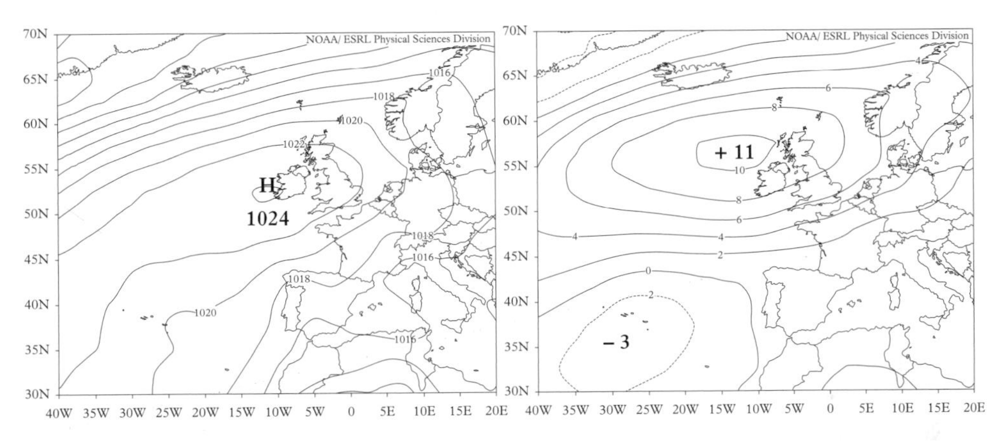


**Figure 4.2** *(Left) mean sea level pressure and (right) anomaly of MSL pressure over the north-east Atlantic, Europe and the British Isles covering the 40-day period 19 July to 27 August 1976, the longest absolute drought on Oxford's daily rainfall record which began in 1827. July's NAO index was weakly negative (−0.57), August weakly positive (+0.62). Isobars are at 2 hPa intervals (NOAA/ESRL Physical Sciences Division)*

## The exceptionally dull and wet autumn of 1976

| Oxford | Mean daily maximum temperature, °C | Precipitation, mm | Sunshine, hours |
|---|---|---|---|
| 22 September to 6 November 1976 | 14.5 °C | 180.6 | 86.0 |
| Anomaly | −0.5 degC | 174% | 52% |

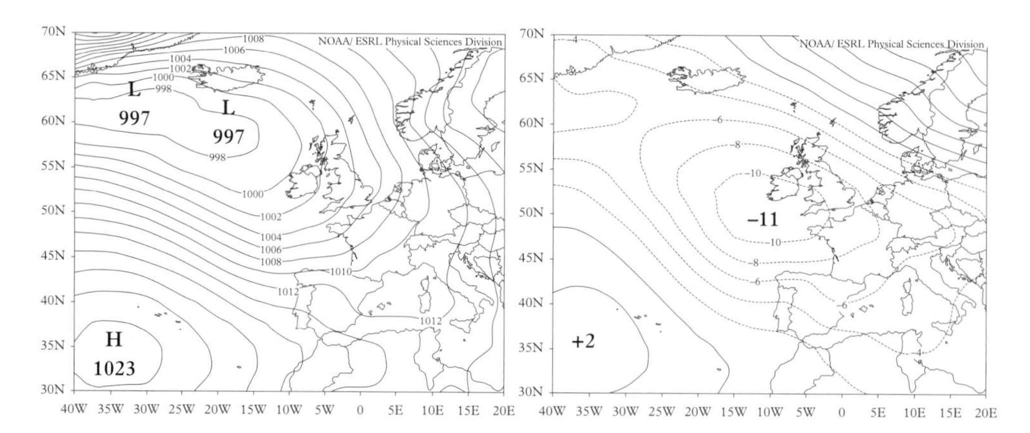


**Figure 4.3** *(Left) mean sea level pressure and (right) anomaly of MSL pressure over the north-east Atlantic, Europe and the British Isles covering the 46-day period 22 September to 6 November 1976, following the dramatic breakdown of the 1975–76 drought. September's NAO index was strongly negative (−3.46), October less so (−0.64) (NOAA/ESRL Physical Sciences Division)*

Although Oxford's absolute drought was convincingly broken by several hours of heavy rainfall on 29 August, amounting to 21 mm (as much as had fallen in the previous 10 weeks combined), and further heavy rain fell on 10 September (17 mm), the really wet weather of autumn 1976 that proved such a vivid contrast to the spring and summer of that year did not reach Oxford until the third week of September. Between 22 September and 6 November (a period of 46 days), rain fell on all but seven days (no more than two dry days were consecutive), and amounted to 181 mm. While the rainfall was almost twice normal over this period, the persistent thick cloud cover resulted in barely half the normal duration of sunshine (October 1976 remains the dullest October on Oxford's record) and a much-reduced range in daily temperatures. For England and Wales as a whole, September–October 1976 with 305 mm of rainfall was at the time the second wettest such period on a long composite record back to 1767, only 1903 (321 mm) being wetter; since 1976, September–October 2000 has equalled 1903's record, 321 mm.

The mean MSL pressure chart for this 46-day period, following on so soon from the above example, shows dramatic differences. The persistent ridge from the Azores anticyclone has been replaced by a trough from the Icelandic low as a string of Atlantic depressions worked their way across north-west Europe: the 11 hPa *positive* anomaly north of Ireland has been replaced by a large area of 11 hPa *negative* anomaly off the south-west of the British Isles.

## December 2010—the coldest December since 1890

| Oxford | Mean temperature, °C | Precipitation, mm | Sunshine, hours |
|---|---|---|---|
| December 2010 | 0.3 °C | 32.7 mm | 20.4 |
| Anomaly | −4.7 degC | 52% | 38% |

December 2010 was the coldest December in Oxford and in most of southern England for 120 years [53, 54]; the mean temperature was almost 5 degC below the 1981–2010 normal, and 20 days remained sunless throughout (see December, Chapter 18). At the Radcliffe Observatory, snow cover was recorded on 12 mornings, while on 19 December the maximum temperature was just −4.4 °C, the coldest December day in over 100 years.

Figure 4.4 (left) shows the monthly mean MSL pressure (left) and the pressure anomaly compared to the 1981–2010 normal (right). High pressure dominated over Iceland and Greenland and low pressure over the Azores, a reversal of the normal pattern, with large positive anomalies (+20 hPa) to the south-west of Iceland contrasting with large negative anomalies (−13 hPa) in the Azores. This strongly NAO-negative month (NAO index −4.61) was bitterly cold throughout the British Isles, owing to frequent incursion of air masses from the north or east, while precipitation—often snow—was above normal in the east and north-east: western districts were much drier than normal. The east, and especially south-east England, was very dull, whereas western districts enjoyed one of their sunniest Decembers on record.

April 2011 was not only the warmest April on Oxford's records but, to that date, the departure of the mean monthly temperature from the normal was the greatest for any

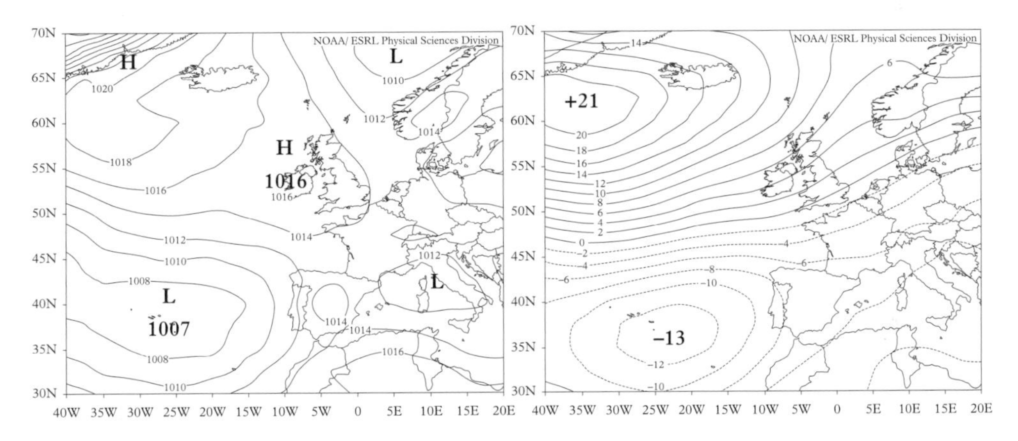


**Figure 4.4** *(Left) mean sea level pressure and (right) anomaly of MSL pressure over the north-east Atlantic, Europe and the British Isles for the calendar month of December 2010 (NOAA/ESRL Physical Sciences Division)*

## April 2011—the warmest and driest April on record

| Oxford | Mean temperature, °C | Precipitation, mm | Sunshine, hours |
|---|---|---|---|
| April 2011 | 13.3 °C | 0.5 mm | 211.4 |
| Anomaly | +4.0 degC | 1% | 131% |

month on record. It was also the equal-driest of any month in almost 250 years records (only April 1817 being as dry), and included the third-longest partial drought in almost 200 years—between 1 March and 5 May (66 days) just under 10 mm of rainfall fell, about one tenth of what would be expected in the same period.

The mean surface pressure for the month and the anomaly compared to the 1981–2010 normal are shown in Figure 4.5. A marked extension of the Azores anticyclone north-eastwards is evident, with a mean separate centre located close to southern England of 1019 hPa, 5 hPa above normal. With dominant anticyclonic conditions, the month was mainly fine and warm in almost all areas. Rainfall was close to or above normal over much of western Scotland, where proximity to a continuing stream of Atlantic depressions gave greater amounts of cloud and more frequent precipitation, but elsewhere it was dry—exceptionally so over much of southern, central and eastern England, where less than 10 per cent of normal rainfall was recorded; a few locations in south-east England recorded no rain throughout the calendar month. For the UK as a whole, this was the warmest April on the Met Office series extending back to 1910; on the Central England Temperature series, this was the warmest April for at least 350 years.

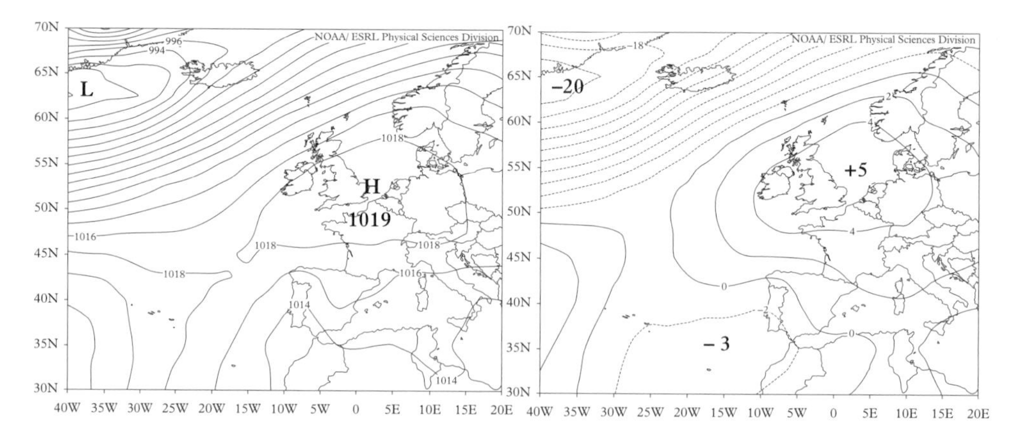


**Figure 4.5** *(Left) mean sea level pressure and (right) anomaly of MSL pressure over the north-east Atlantic, Europe and the British Isles for the calendar month of April 2011 (NOAA/ESRL Physical Sciences Division)*

## June 2012—a cool and very wet summer month

| Oxford | Mean temperature, °C | Precipitation, mm | Sunshine, hours |
|---|---|---|---|
| June 2012 | 15.5 °C | 151.7 mm | 134.5 h* |
| Anomaly | −0.1 degC | 313% | 70% |

* *Sunshine data for June 2012 are missing from the Oxford records; daily data from the CEH site at Wallingford (22 km south-east) have been substituted.*

Very different conditions from April 2011 prevailed in June 2012; the distribution of mean surface pressure for the month (Figure 4.6) reveals that the mean centre of cyclonic activity in the north Atlantic was located close to the British Isles, with a 10 hPa negative anomaly located south-west of Ireland. South-westerly winds were dominant over England and Wales, with frequent rainfall—in Oxford, only seven days in the month remained completely dry, only two of them consecutive, while the month's rainfall total exceeded 150 mm, more than three times normal and the wettest June since 1852. Persistent cloud cover resulted in mean maximum (daytime) temperatures more than a degree below normal, largely compensated by cloudy nights keeping mean minimum temperatures above normal.

Over the UK as a whole, this was the coolest June since 1991, with mean maximum temperatures well below normal, particularly in central and eastern areas, and there were few warm days. Rainfall was well above normal across much of England and Wales, southern and eastern Scotland and Northern Ireland—the equal-wettest June in England and Wales (with June 1860) since the series began in 1766. Sunshine was well below normal in most areas, and this was the equal-dullest June in the Met Office monthly sunshine series which began in 1929: only the far north-west of Scotland was drier and a little sunnier than normal.

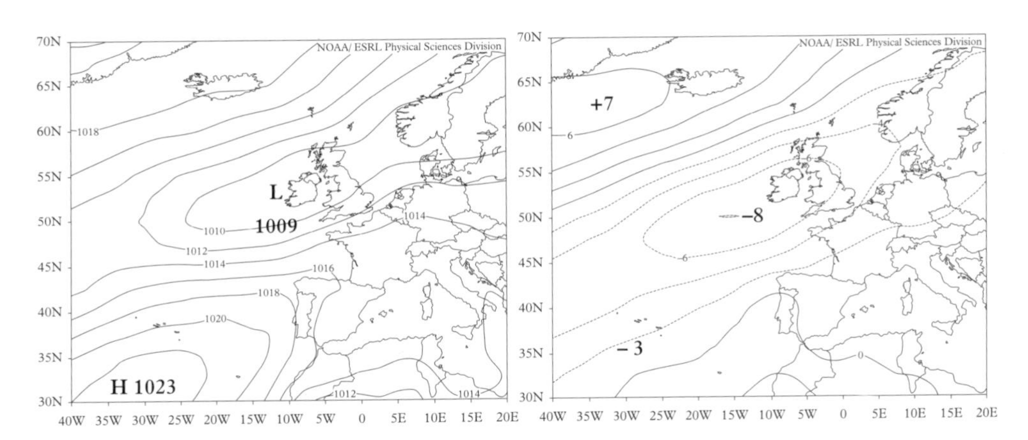


**Figure 4.6** *(Left) mean sea level pressure and (right) anomaly of MSL pressure over the north-east Atlantic, Europe and the British Isles for the calendar month of June 2012, the wettest June in Oxford since 1852 (NOAA/ESRL Physical Sciences Division)*

## March 2013—very cold and extremely dull

| Oxford | Mean temperature, °C | Precipitation, mm | Sunshine, hours |
|---|---|---|---|
| March 2013 | 3.3 °C | 76.6 mm | 60.7 |
| Anomaly | −4.0 degC | 161% | 55% |

March 2013 was the coldest March in Oxford for over 50 years, and the third-dullest on record. The main cause of the poor spring conditions was unremitting north-easterly or easterly winds. Figure 4.7 shows the distribution of mean sea level pressure for the month, and the differences from normal. The main cyclonic centre in north Atlantic was displaced well south of normal, to lie south-west of the British Isles close to where the Azores anticyclone would normally be located; mean barometric pressure was well below normal—around −17 hPa—in the Azores, and well above normal over Iceland and Greenland (anomaly +22 hPa in south Greenland). The NAO index was −3.75, almost identical to March 1962 (−3.78) and close to the most extreme on the near 200 year NAO record (−3.84 in March 2016—another cold and wet March in Oxford).

March 2013 equalled March 1962's mean temperature as the coldest March within the last 100 years across the UK. The month was especially cold during the second half of the month, including substantial late-season snowfalls in places. The persistence of easterly winds led to large east–west variations in conditions across the British Isles: the south and east were wet and very dull, with more than twice normal precipitation and half normal sunshine in places, while parts of western Scotland enjoyed a very dry and extremely sunny March.

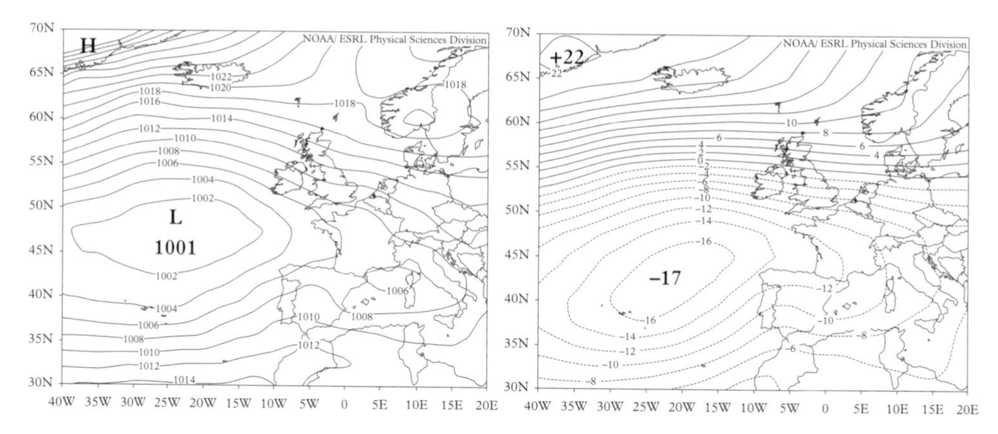


**Figure 4.7** *(Left) mean sea level pressure and (right) anomaly of MSL pressure over the north-east Atlantic, Europe and the British Isles for the calendar month of March 2013 (NOAA/ESRL Physical Sciences Division)*

## The very wet winter of 2013/14

| Oxford | Mean temperature, °C | Precipitation, mm | Sunshine, hours |
|---|---|---|---|
| Winter 2013–14 *(1 December 2013 to 28 February 2014)* | 6.5 °C | 335 mm | 239 h |
| Anomaly | +1.7 degC | 205% | 123% |

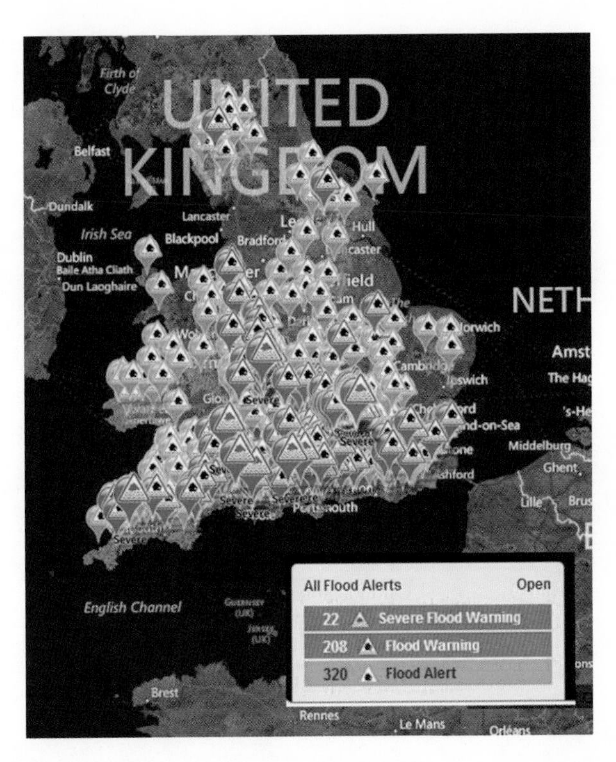


**Figure 4.8** *Environment Agency flood warnings and flood alerts on 14 February 2014—over 500 were in force simultaneously. Note that the Environment Agency remit covers England and Wales only—there were also flood alerts in Scotland and Northern Ireland, but these are not shown on this map (Environment Agency)*

Winter 2013/14 was, by some margin, the wettest winter on Oxford's records and, indeed, across many parts of southern and south-eastern England and the south Midlands, and prolonged and serious flooding became widespread in England and Wales. During the height of the crisis in mid-February, over 500 Environment Agency flood warnings or flood alerts were in place (Figure 4.8), including for the Thames and Cherwell around Oxford (Figure 4.9).

The high rainfall totals were due to a repeated string of Atlantic depressions which deepened rapidly and moved quickly towards and across the British Isles under the influence of a strong and persistent jet stream. In Oxford, wet weather set in in mid-December; between 17 December and 14 February (the date of the flood alert map in Figure 4.8), a period of just 60 days, 291 mm of rainfall was recorded. Only five of those 60 days remained completely dry. Normally, around 113 mm of rainfall would be expected in those 60 days, so this two-month period received more than two and a half times normal rainfall—or, put another way, about five months normal rainfall fell in two months.

Figure 4.10 (left) shows the averaged distribution of mean sea level pressure over this 60-day period in 2013/14, together with the anomaly compared with the 30-year mean

**Figure 4.9** *Flooding on Binsey Lane, Oxford on 9 January 2014—Oxford's wettest January, and its wettest winter, in 250 years' records (Jane Buekett)*

over the same dates in Figure 4.10 (right). The NAO index was strongly positive for the winter, exceptionally so in December (+3.54, the greatest December value since 1924) and in February (+2.32). The intensification of cyclonic activity near and over the British Isles resulted in a displacement of the mean north Atlantic depression centre to be both considerably deeper and closer to the British Isles than normal, with an exceptional negative anomaly of some 25 hPa between Ireland and Iceland. Pressure gradients were particularly steep, a result of the 12 major winter storms which affected the British Isles within the period covered by Figure 4.10: the increased frequency of severe gales resulted in much greater than normal coastal erosion, particularly along England's south and south-west coasts [55–57].

For England and Wales as a whole, this was the wettest winter in a record dating back to 1766. The unsettled westerly conditions resulted in a much milder than normal winter, and mean temperatures were well above normal everywhere. Despite the disturbed conditions, much of central, eastern and southern England was sunnier than average during the winter, although western areas of Wales and Scotland were notably dull.

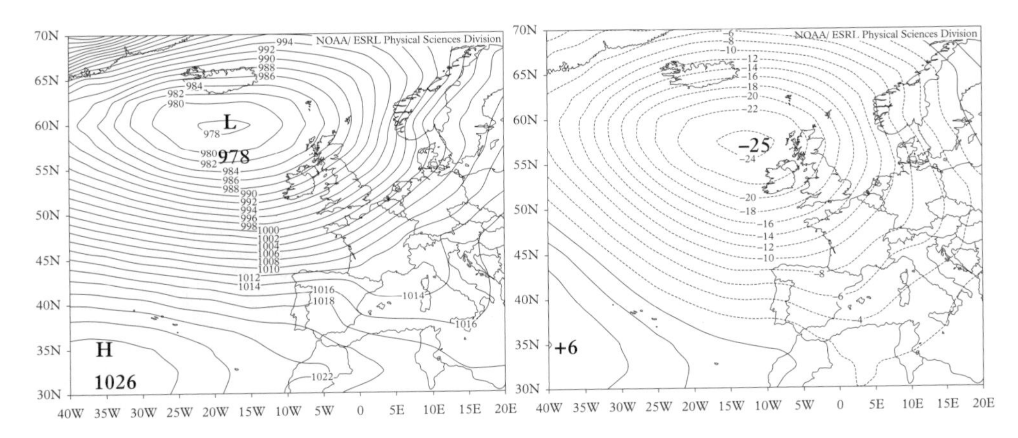


**Figure 4.10** *(Left) mean sea level pressure and (right) anomaly of MSL pressure over the north-east Atlantic, Europe and the British Isles for the 60-day period 17 December 2013 to 14 February 2014, at the height of Oxford's wettest winter on record (NOAA/ESRL Physical Sciences Division)*

## December 2015—the mildest December for at least 350 years

| Oxford | Mean temperature, °C | Precipitation, mm | Sunshine, hours |
|---|---|---|---|
| December 2015 | 10.9 °C | 62.9 mm | 38.3 |
| Anomaly | +5.9 degC | 100% | 73% |

Just five years after the bitter cold of December 2010, December 2015 was to prove as exceptional a month, but in entirely the opposite direction. By almost 2 degC, this was the mildest December on Oxford's records back to 1814—milder, in fact, by a large margin than any previous December on the composite Central England Temperature record, which extends back to 1659 [58].Temperatures were often comparable with those that might be expected in October, April or even May: in Oxford, the air temperature did not fall below 3.5 °C at any time during the month, while, on both 18 and 19 December, daytime temperatures rose to just below 16 °C—the warmest December days on Oxford's record (Chapter 18).

Figures 4.11 (left) and 4.11 (right) show the distribution of monthly mean sea level pressure and pressure anomalies, respectively, for December 2015. Conditions over the north Atlantic were almost the exact reverse of those of December 2010 (compare Figure 4.11 with Figure 4.4). The NAO index was strongly positive (+4.22, the greatest positive value for any December on the CRU record extending back to 1823), and the resultant stronger than normal south-westerly flow introduced unseasonably mild and moist air from much lower latitudes—although Scotland and Northern Ireland were colder at times, particularly in the second week.

Mean temperatures were well above normal everywhere in the British Isles, less so in Scotland and Northern Ireland than in England and Wales. Dominant south-westerly winds resulted in enormous quantities of rain in western and northern areas, including

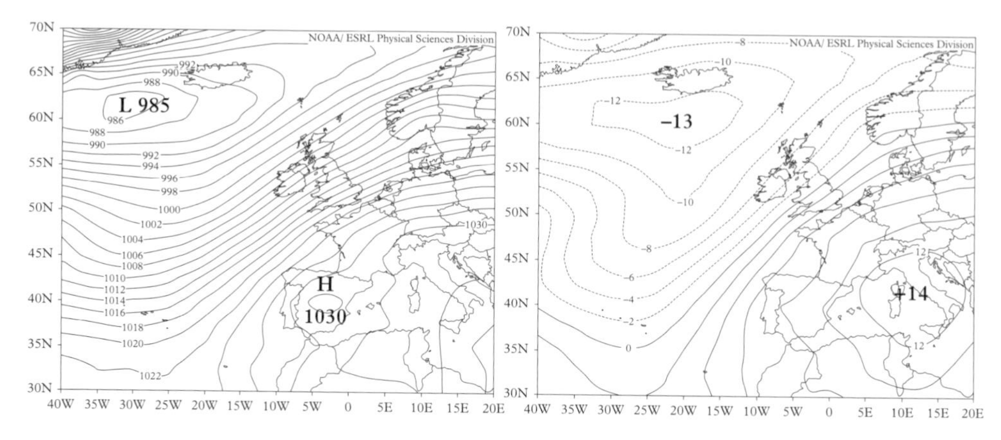


**Figure 4.11** *(Left) mean sea level pressure and (right) anomaly of MSL pressure over the north-east Atlantic, Europe and the British Isles for December 2015, the mildest December for at least 350 years, and the wettest month on record across the UK (NOAA/ESRL Physical Sciences Division)*

new monthly record rainfall totals in places, accompanied by widespread severe flooding [59]. For the UK as a whole, this was the wettest calendar month of any on the Met Office's monthly series, which began in 1910, although extreme eastern areas were a little drier than normal; Oxford was close to the dividing line, and here the month's rainfall was exactly normal. Sunshine was below normal almost everywhere, particularly so in the west.

## Summer 2018—the hottest summer on Oxford's records

| Oxford | Mean daily maximum temperature, °C | Precipitation, mm | Sunshine, hours |
|---|---|---|---|
| 2 June to 26 July 2018 | 25.3 °C | 8.9 mm | 504.9 |
| Anomaly | +3.9 degC | 11% | 140% |

In Oxford, as in many other parts of southern and south-eastern England, the sparkling summer of 2018 was the hottest on record, and established new records for low rainfall (the driest 7 weeks on record) and plentiful sunshine (the sunniest 21-, 30- and 60-day periods on record). On the Davis summer index (see Chapter 22, Summer) for the three summer months, it outranked every other summer except 1976; on the extended summer index (May to September), it was the fourth-best summer on Oxford's records since 1880, eclipsed only by 1911, 1959 and 1989.

The warmest, driest and sunniest weather occurred in June and July, although hot weather persisted into the first week of August; May was also very sunny and dry until the last week, but a succession of heavy showers gave a fall of 36.4 mm on the last day of

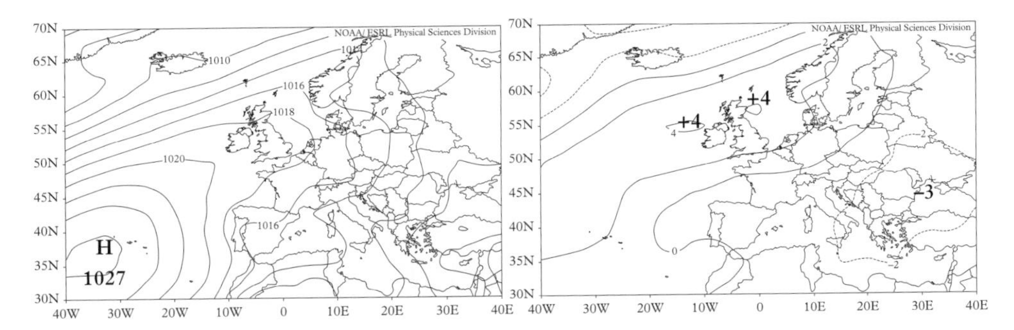


**Figure 4.12** *(Left) mean sea level pressure and (right) anomaly of MSL pressure over the north-east Atlantic, Europe and the British Isles for 2 June to 26 July 2018, the peak of the 2018 summer—the hottest summer on Oxford's long records (NOAA/ESRL Physical Sciences Division)*

the month. Thereafter, little rain fell until the closing days of July. No rain at all fell between 18 June and 19 July, a period of 32 consecutive days, making this the fifth-longest absolute drought since Oxford's daily rainfall records commenced in 1827 (see also Table 28.1 in Chapter 28). Between 2 June and 26 July, the 55-day period in the analysis charts in Figure 4.12, only 8.9 mm of rain fell (7.4 mm of which fell on 20 July) and sunshine amounted to 505 hours, a daily average of just under 9.2 hours daily, while the mean daily maximum temperature was 25.3 °C, almost 4 degC above normal. Several new records for Oxford were established, including the highest monthly mean maximum temperature for any month (July, 27.4 °C). Although there was a long run of hot and sunny conditions, there were relatively few 'heatwave' days (30 °C or more) during 2018—only five days surpassed this threshold, four of them in July and none consecutive: the hottest day of the summer was 26 July, when the temperature reached 32.4 °C.

Figures 4.12 (left) and 4.12 (right) show the distribution of monthly mean sea level pressure and pressure anomalies, respectively, for the peak summer period in 2018, 2 June to 26 July. Although the mean pressure distribution is broadly similar to that for the summer of 1976 (Figure 4.2), with a strong ridge from the Azores anticyclone towards the British Isles, the anomaly maps are quite different—in 1976 there was a strong (+11 hPa) anomaly north-west of Ireland, while pressure was below normal in the Azores; during summer 2018, the main anomaly (+4 hPa) was in a similar position to 1976 but weaker, while pressure was below normal over Europe. The consequences of this slightly different pressure distribution were that summer was relatively warmer, sunnier and drier in the south and south-east of England than further north and west in England, Scotland and Northern Ireland, where Atlantic fronts and weather systems continued to give cooler, cloudier conditions with more frequent rainfall. In south-east England, the mean maximum temperature, sunshine and rainfall anomalies (1981–2010 averages) for July 2018 were +4.2 degC, 140 per cent and 60 per cent, respectively; in Northern Ireland, the corresponding anomalies were +1.8 degC, 123 per cent and 99 per cent. For England as a whole, July was the second warmest (in a series from 1910) and the second sunniest (in a series from 1929).

# 5

# Long-period weather observations elsewhere in the British Isles and Europe

This book is about the Radcliffe Meteorological Station (RMS) in Oxford, and the purpose of this chapter is to place the RMS in historical context, both in the British Isles and further afield across Europe. The detailed history of the RMS itself is presented in Chapter 2.

The earliest instrumental weather records in Europe date back to the middle of the seventeenth century, within a decade or two of the invention of the earliest versions of 'traditional' meteorological instruments (the mercury barometer, early sealed thermometers and various instruments to measure rainfall). At first, such records were experimental and sporadic, often kept by individuals and rarely lasting for more than a few years in any one place, with widely differing standards of exposure, accuracy and (not least) units of measurement. The earliest surviving instrumental weather records in the world are from the Medici Network, based in Florence, Italy, covering the period 1654 to 1670. Thanks not only to painstaking historical detective work but to the quality and craftsmanship of these early instruments, many of the Medici series have been recovered and analysed [60]. Amongst the earliest are daily observations of a mercury barometer in Pisa, Italy, in 1657–58 [11], while for 1694 there are sufficient surviving barometric pressure records across Europe for outline daily synoptic weather maps to be prepared. An almost complete daily pressure record has recently been assembled for locations in Paris back to 1670, and in London since 1692 [61–64].

By the early eighteenth century, regular and systematic weather records began to be kept by institutions at various places in Europe; as in Oxford, many were first noted as part of the observational routine at astronomical observatories. Some of these observatories are still in existence, but the Radcliffe Observatory in Oxford is one of a very few locations in the world where continuous weather observations have been made in much the same location for over 200 years. Despite inevitable changes in instruments, exposure and observational methods over time, and likely urban warming influences on individual temperature records owing to the growth of towns or cities, such observations remain the most detailed and comprehensive records of changes in climate over the last 350 years or so.

*Oxford Weather and Climate since 1767*. Stephen Burt and Tim Burt, Oxford University Press (2019).
 DOI: 10.1093/oso/9780198834632.001.0001

## The longest temperature record in the world: 1659 to date

In 1953, the British climatologist Gordon Manley (1902–80) published his first paper on what became known as the Central England Temperature (CET) series in the Royal Meteorological Society's *Quarterly Journal* [34]. Manley's extensive and painstaking research assembled scattered early instrumental temperature records and descriptive weather diaries to produce a chronology of mean monthly temperatures representative of a roughly triangular area of England enclosed by Lancashire, London and Bristol covering the period 1698 to 1952: the Radcliffe Observatory records became a key component of the series from 1815. A second, longer, paper in 1974 [35] extended the series back to 1659—about the time the earliest thermometers appeared in England—and brought it up to date. Other records that had come to light in the intervening 20 years also allowed for corrections or improved estimates to the existing series. Since Manley's death, the series has been kept up to date by the Hadley Centre, part of the UK Met Office, and today the series forms the longest instrumental record of temperature in the world (Figures 5.1, 5.2). A similar monthly rainfall record, the 'England and Wales

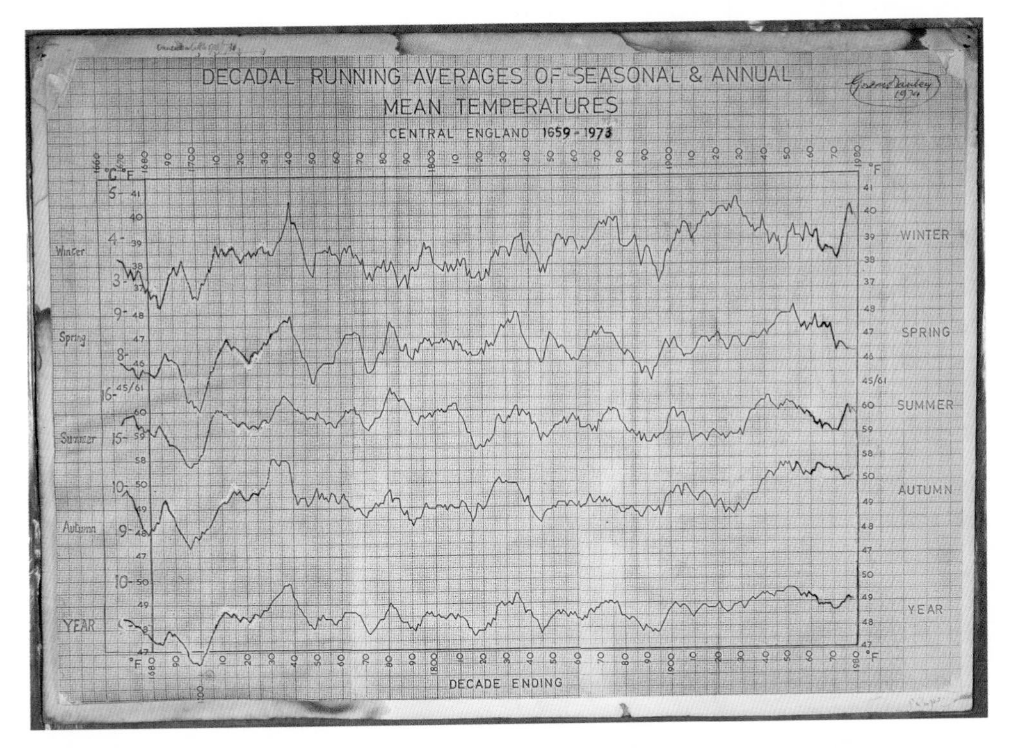


**Figure 5.1** *Gordon Manley's original graph showing 'Central England' seasonal and annual mean temperatures 1659 to 1980; from the framed original held in the Department of Geography at Durham University. Gordon Manley founded the Department of Geography at Durham University in 1928 (Michele Allan, Durham University)*

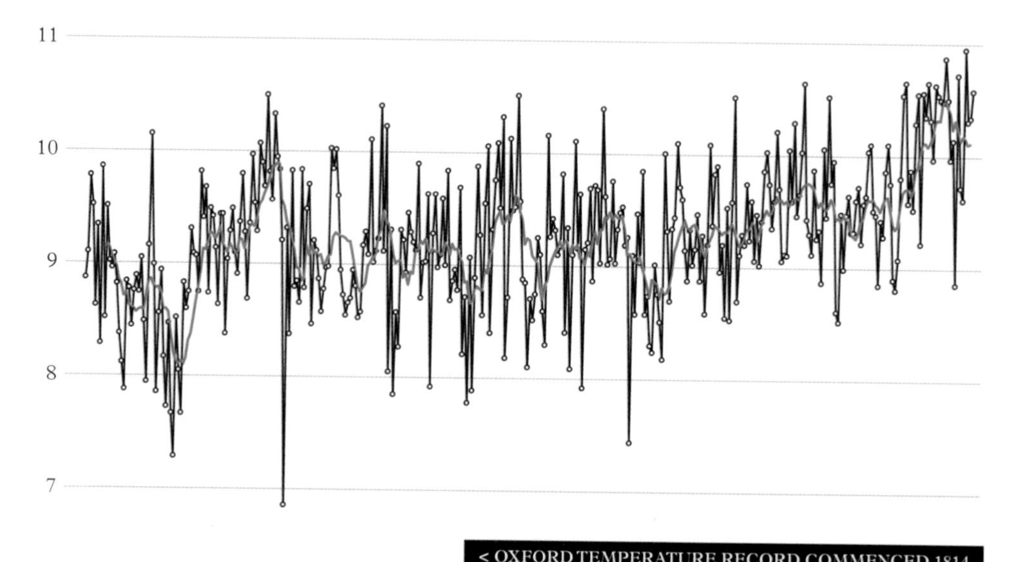


**Figure 5.2** *Central England Temperature series annual means (°C) 1659 to 2017, from work by Gordon Manley updated by the Met Office, with ten-year unweighted running mean ending at the date shown in red*

Precipitation' series, extends back to 1766* and, again, the Oxford records form a key component of the early years within the series.

Manley's work pieced together many disparate records to produce a figure representative of a region rather than a single site. There are other long composite series of temperature, rainfall and/or pressure records representative of other cities or regions in Europe extending back to the eighteenth century. At a few places, instrumental weather records are still made today in the *same* location, or very nearly so, where continuous observations commenced 150–200 years ago or more.

## Uppsala, Sweden—1722 to date

### *59.847°N, 17.635°E, 25 m above sea level*

The oldest mostly continuous records in Europe are those from Uppsala in Sweden, about 65 km north of Stockholm, where records commenced in 1722 [65]. The earliest

* The Central England Temperature (CET) database and the England and Wales Precipitation (EWP) series—and other datasets—are available from the Hadley Centre website: http://www.metoffice.gov.uk/hadobs/index.html.

organised meteorological observations in Sweden were initiated around 1720 by the Society of Science in Uppsala, at the astronomical observatory then in the centre of Uppsala, at that time only a small town. The oldest surviving observatory journal dates from 1722, although there are some gaps in the Uppsala record until 1773. The young Anders Celsius was responsible for the observations from 1729, and it was here that his own thermometer scale, the first Celsius scale, was developed.

The Uppsala record has seen changes of site, of a few hundred metres, in 1853, 1865 and 1952; since 1959 the observational record has been maintained at the University of Uppsala.

## Padova (Padua), Italy—1725 to date

*45.402°N, 11.869°E, 20 m above sea level*

Meteorological observations commenced as part of the astronomical observational routine in Padova (Padua) in northern Italy in 1725 [66]. Until 1767, observations were made by the individual observers in their own dwellings within the town, but from 1768 to 1962 the records were kept at the *Specola* complex in the centre of Padua. Today, the records are maintained at the nearby Botanical Gardens. The temperature record has been carefully reconstructed, making allowance for differing instruments, calibrations, observers, observing sites and practices, making this the oldest record in southern Europe.

## Stockholm, Sweden—1756 to date

*59.342°N, 18.055°E, 38 m above sea level*

Weather observations commenced at the old astronomical observatory in Stockholm in 1754. Complete daily mean series of air temperature and barometric pressure have been reconstructed from the original observational data for the period 1756 to date [67]. The earliest air temperature measurements were made with thermometers exposed immediately outside a second-floor window; the current observation site, dating from 1960, is only about 10 metres away. In 2006 the observatory completed 250 years records, the longest unbroken same-site observation series in the world.

## Milan, Italy—1763 to date

*45.471°N, 9.189°E, 121 m above sea level*

The Astronomical Observatory of Brera in Milan was founded in 1762, and daily meteorological observations have been made here since 1763. It is the oldest scientific institution in Milan and remains one of the top astronomical research institutes in the world. Although observations have always been made at the Observatory, many changes of instruments, station location and observation methods over the years render the original observations series far from consistent. Fortunately, detailed records of the instruments and their exposure were kept, and a meticulous research programme conducted at the University of Milan [68, 69] was able to assemble a complete and homogeneous daily series of mean temperature and barometric pressure from 1763 to date.

## Kew Observatory, West London, England—1773 to 1980

*51.469°N, 0.315°W, 6 m above sea level*

Kew Observatory, or to give it its original title, The King's Observatory at Kew, was built by George III in 1769 as an astronomical observatory, specifically to study the transit of Venus in that year. Meteorological observations were made daily from 1773 but were clearly viewed as of little importance to the Observatory, for as subsequently noted by Whipple in 1937 [70]: 'Our greatest regret is that the meteorological observations, which were started in 1773, were so unsystematic that they are useless for statistical purposes. Otherwise we should have had one of the longest series of meteorological records in the world.'* The government eventually gave up the observatory because George III's successors on the throne took less interest in it (and also because it was expensive to run), and in 1842 the British Association for the Advancement of Science (BAAS) took over the site [72]. More systematic meteorological observations commenced in November of that year, and a continuous weather record was kept for almost 140 years thereafter, apart from a few short breaks (mostly between 1848 and 1853), until the closure of the Observatory by the Met Office on 31 December 1980 [29, 73, 74].

Kew Observatory became an important centre for the checking and calibration of scientific instruments, meteorological and otherwise. Although the observatory became part of the newly formed National Physical Laboratory (NPL) in 1900, instrument-testing continued as before at Kew until its 1910 transfer to the Meteorological Office, after which most of the testing work was transferred to the NPL's main site at Teddington. Thereafter, Kew's calibration facilities were dedicated to meteorological instrumentation. Routine meteorological observations continued alongside pioneering work on atmospheric electricity, geomagnetism, fog, upper-air measurements and atmospheric pollution. The eponymous Kew-pattern barometer was also developed at Kew, and early trials of a prototype of what became the Campbell–Stokes sunshine recorder commenced at the Observatory in 1875.

For many decades, Kew Observatory remained the Met Office's main observatory (Figure 5.3), although its importance declined considerably following relocation of Met

* Perhaps not, after all: in a 1969 article [71] Jacobs notes that Whipple's assessment seems to be based on a logbook containing the Kew observations between 1783 and 1803 that had been mislaid. The logbook was eventually found in 1960.

**Figure 5.3** *Kew Observatory in the mid-1960s: from Galvin [74], Figure 5 (© Crown Copyright 2003. Information provided by the National Meteorological Library and Archive – Met Office, UK)*

Office headquarters and experimental facilities to Bracknell in 1961. Starting in 1866 or 1867, Kew's temperatures were taken from thermometry exposed in a louvred screen mounted on the north wall of the building (Figure 5.4). This non-standard exposure unfortunately limited the comparability and usefulness of Kew's long temperature records, although it is often overlooked that for several decades more conventional records had also been kept of air temperatures recorded within standard Stevenson screens in conventional exposures. Of course, observatory-standard records were also maintained of all other meteorological parameters at Kew.

In 1980, with government budgets under severe pressure in a serious economic recession, the decision was made by the Met Office to close Kew Observatory—a hugely unpopular decision at the time. The 200 year-old Grade I listed building was becoming very expensive to maintain, while the need for two costly, round-the-clock staffed observation sites in west London (Heathrow Airport, just 10 km distant, was also manned 24 hours per day at that time) was increasingly difficult to justify. After an unfortunate gap of several months, once-daily climatological records commenced at a new and well-exposed site in the Royal Botanic Gardens, Kew, just over 2 km to the north-east of the Kew Observatory site, where they continue today.

**Figure 5.4** *The North Wall screen at Kew Observatory, 1960s (with the doors open). In the centre are two control thermometers, one dry bulb and one wet bulb. The photo-thermograph thermometer bulbs are just out of shot: the stems of these thermometers are bent twice at right angles, the horizontal portions passing through the wall of the building so that photographic recording can take place inside the building. Similar photographic recording methods were employed at Oxford between 1854 and 1877, although the surrounding louvred screen was located some distance away from the Observatory's north wall, rather than fixed to the wall as at Kew. The Kew arrangements were certainly influenced by Oxford, because Francis Ronalds had set up experimental self-recording equipment at Kew by the late 1840s and went on to set up similar apparatus at Oxford in the 1850s—so, in a sense, the apparatus originally invented at Kew 'came home' in the 1860s—Kew's photographic recordings did not commence until 1866. On the right, outside the screen, is an additional louvered screen to cut off solar radiation from the setting sun in summer (the main entrance of the Observatory cut off radiation at sunrise). (© Crown Copyright: information provided by the National Meteorological Library and Archive – UK Met Office)*

## Prague, Czechia—1775 to date

### *50.086°N, 14.416°E, 191 m above sea level*

Regular meteorological observations began in the vast Baroque complex of the former Jesuit College in Prague's Old Town, the Clementinum, in 1752, although there are breaks in the record until 1775 [75]. Observations continue at the same site today, in much the same surroundings as they were at the end of the eighteenth century, with observations made at the 'Mannheim hours' of 7 a.m., 2 p.m. and 9 p.m. (see below).

Two thermometer screens are in use, similar to the original eighteenth century models rather than today's standards—a louvred screen located on the first floor of the north side of the south annex and another on the flat roof of the east annex. Rainfall amounts and sunshine duration are also measured here.

## Hohenpeissenberg, Germany—1781 to date

*47.801°N, 11.010°E, 977 m above sea level*

Hohenpeissenberg is the oldest mountain observatory in the world, and possesses one of the longest reliable single-site observational records of any location [76]. It is located about 80 km south-west of Munich at an altitude of just under 1000 m. Meteorological observations were first made here in 1758/59, but regular and uninterrupted records started on 1 January 1781 as one of the stations in the Societas Meteorologica Palatina observation network established by the Meteorological Society of the Palatinate with the support and funding of Karl Theodor, Elector of the Palatinate. This was the world's second international climate observation network (Florence's Medici Network in 1654–70 was the first), consisting of 39 stations extending from eastern America to the Ural Mountains, and from Greenland to the Mediterranean. The Societas Meteorologica Palatina established standardised instruments, observing procedures and observation times (the so-called Mannheim hours of 7 a.m., 2 p.m. and 9 p.m.) for the first time. (Observations made at the standard 'Mannheim hours' are still used for today's climatological records at Hohenpeissenberg.) The observations were initially made by Augustinian monks from the nearby Rottenbuch monastery. Although the Societas Meteorologica Palatina came to an abrupt end in 1792 following the Austrian–French war, fortunately the Augustinian Canons continued the meteorological observations. In 1803 the parish priest of Hohenpeissenberg was appointed as the responsible observer by the Bavarian Academy of Sciences. In 1838, the observatory came under the responsibility of the Royal Observatory of Munich, and in 1878 part of the Bavarian State Weather Service. In 1934, the Meteorological Service of the Third Reich assumed responsibility for the station, which was quickly expanded into a main weather observation site, commencing synoptic observations. In 1940, the station was relocated a short distance from the existing monastery buildings into newly built premises on the western side of the mountain. In the closing days of the Second World War, as southern Bavaria came under attack from Allied armies, observations at Hohenpeissenberg were interrupted by artillery fire on 28 April 1945 (and had to cease altogether on 2 May because of the danger to the observers) but were restarted on 14 May. In April 1946 the station became part of the newly founded West German state weather service, Deutscher Wetterdienst (DWD) and in March 1950 the site was formally upgraded to that of a meteorological observatory. Today, Hohenpeissenberg hosts a huge range of meteorological and atmospheric instruments, a key site in both historical climate change investigations and real-time national and international meteorological networks (WMO station number 10962).

**Figure 5.5** *Hohenpeissenberg Observatory in southern Germany, where records began in 1781 (Stefan Gilge, Deutscher Wetterdienst)*

## Armagh Observatory, Northern Ireland—1794 to date

*54.353°N, 6.648°W, 64 m above sea level*

The astronomical observatory at Armagh, built in 1790, is the oldest scientific institution in Northern Ireland. Intermittent observations of the weather have been made on this site since 1784, prior to the building of the observatory: more systematic daily observations of temperature and barometric pressure commenced in December 1794 [77]. Although there are some gaps in the early years, and numerous changes of instrument and site around the observatory, the records are largely complete from 1833 to the present day. They represent the longest series of continuous weather records on the island of Ireland. All of the records, including scanned copies of the original manuscript records, are available on the Observatory website (http://climate.arm.ac.uk/main.html).

The site lies approximately 1 km north-east of the centre of the small town of Armagh, within an estate of natural woodland and parkland of some 7 ha which has changed little since the foundation of the observatory (Figure 5.6).

**Figure 5.6** *The meteorological enclosure at Armagh Observatory in Northern Ireland, July 2018 (John Butler, Armagh Observatory)*

The third director of the observatory, Thomas Romney Robinson, appointed in 1832, made many experiments in other fields of science. One of his most enduring interests was the study of meteorology and in particular the measurement of wind speed. He invented the cup anemometer, an instrument still widely used throughout the world.

## Durham Observatory, England—1841 to date

*54.768°N, 1.586°W, 102 m above sea level*

In 1839, Durham University (founded in 1832) resolved to establish an astronomical observatory. The Dean and Chapter of Durham Cathedral made an elevated site available, about 800 m south-west of the cathedral, on the other side of the incised River Wear, over 60 m below. The Observatory building was complete by 1841 (Figure 5.7). The University's decision to build an observatory owed much to the Reverend Temple Chevallier, a remarkable Victorian polymath: clergyman, astronomer, mathematician and University administrator, who between 1847 and 1849 made important observations regarding sunspots. Today, the Observatory is no longer required, and its telescope is long gone, but the meteorological station remains, on the gently sloping lawn immediately to the south of the Observatory. Daily observations ceased in 1999 since when data have been acquired by an automatic weather station, operated by the Met Office. Monthly reports continue to be issued by the University's Department of Geography; Tim Burt has been the author of these reports, and annual summaries published in the *International Journal of Meteorology*, since 2000. Gordon Manley, who established the

**Figure 5.7** *The meteorological site at Durham Observatory (Michele Allan, Durham University)*

University's Department of Geography in 1928, made the first detailed analysis of the temperature record [78], intent on establishing the Durham record as comparable to that of Oxford. More recently, the rainfall record has been studied, including analysis of the very wet 1870s [79]. A full account of the Durham Observatory weather station can be found in Kenworthy *et al* [80].

## Valentia Observatory, Ireland—1868 to date

### *51.938°N, 10.240°W, 9 m above sea level*

Meteorological observations began at Valentia, in the far south-west of Ireland, in October 1860 following its establishment as the cable terminus for the Atlantic telegraph [81–83]. The importance of Valentia as a location for meteorological observations was recognised at an early stage: its location on the western extremity of Europe provided useful warning of cyclonic storms approaching from the Atlantic, and its location at the terminus for the Atlantic telegraph cable enabled speedy dissemination of observations and warnings to other parts of the British Isles in the early years of Admiral Fitzroy's Meteorological Office. Valentia Observatory was originally established on Valentia Island in 1868, moving in March 1892 to its current site on the mainland just south-west of the small town of Cahirciveen (Figure 5.8). It was equipped as a meteorological observatory from the outset, and today remains an important surface and upper-air observatory within Met Éireann, the meteorological service of the Republic of Ireland (WMO number 03953).

**Figure 5.8** *Valentia Observatory, County Kerry, Ireland, in October 2010 (Stephen Burt)*

## Why do we need to maintain long weather records?

The importance of long-term observation of the natural environment has long been recognised, and yet 'monitoring' is still sometimes dismissed as low-grade science which can contribute little to our understanding, despite the great increase in awareness and importance of long-period weather records brought about by studies into climate change. The value of long-term study was assessed within the context of the UK's Environmental Change Network (ECN), established in 1992 to provide a minimum of 30 years' data from a network of sites within the UK. It was concluded that long-term study provided an invaluable basis for the development of environmental science, for studying processes whose effects can only be identified over long periods of time, and for revealing new questions which could not have been anticipated at the time the monitoring began [84]. Given significant inter-decadal variability, short records may indicate trends which do not reflect the magnitude, or perhaps even the sign, of longer-term changes. This is especially a problem for analysis of records which are only a few decades long. It is therefore crucial to maintain, document and critically examine very long time series such as those at Oxford and the other stations reviewed above. Using very long records allows us to detect subtle, underlying trends within time series which exhibit large inter-year variability.

## Times of change...

The way we measure weather is changing rapidly. Within the UK and Ireland, and in many other countries, the majority of observing sites are now automated—the observations themselves are made by automatic weather stations. New sensors and measurement methods have evolved and are still evolving, some completely novel, all offering improved ease of use, accuracy and cost-efficiency—although perhaps not longevity—when compared with 'traditional' or manual instruments [85]. Will this change, the most significant since the invention of the thermometer and barometer almost 400 years ago, affect the consistency and homogeneity of long-period records? With care, and a careful overlap of methods to check and assess any method bias or instrumental drift over time, the reliability for future generations of long-period records such as those from Oxford and the other sites reviewed in this chapter can be assured. Such records are more valuable now than ever. Sometimes, the fastest way to obtain reliable and consistent long records is to digitise existing written or published observation registers, and, fortunately, more and more archive materials of this type are being scanned and placed online to make them more widely available to the research community and the interested public (see, for example, [86–88]). A modern homogeneous record of 25 years is useful; one of 250 years in length, or reliable instrumental data from 250 years ago, can be many times more so. Only consistent long records with sound observational practices and instrumental metadata, such as those maintained at Oxford and the handful or so other sites in Europe with similarly long records, can help answer questions such as 'How is our climate changing?' and 'Are extremes of climate becoming more frequent?'

# Part 2

# Oxford weather through the year

# 6

# The annual cycle

This brief chapter provides a prelude to the analysis of months and seasons which follows. The plots show (*top*) the average temperature (daily mean maximum, mean and daily mean minimum, °C), (*middle*) daily mean precipitation (mm) and (*bottom*) daily average sunshine duration (hours) for every day of the year (except 29 February) over the standard averaging period, 1981–2010. Whilst there is considerable variation year to year, these serve to illustrate the general progression of the main weather elements throughout the annual cycle. These plots are referred to in more detail in the chapters that follow.

At first sight, the temperature curves appear remarkably smooth, as might be expected for 30-year averages. Nevertheless, there is some evidence of persistent spells of weather—*singularities* as Hubert Lamb would have called them [89]. Allen Perry [90] defined a singularity as the tendency of some weather characteristic to recur about a specific date in the year. There is a notably cold period in mid-February, the coldest time of the year, which coincides with a distinctly dry spell. No doubt, this is a period dominated by anticyclonic weather, dry, cold and sunny. There is another cool spell in April, when the rate of warming slows visibly, coinciding with another dry spell, quite possibly another period of anticyclonic weather with winds from the east, north-east or even north. There is another dry spell at the beginning of May, but the rate of warming is not affected.

Sunshine hours do not increase greatly from May through August; there are wet spells in mid-May, mid-June and mid-July, meaning cloud cover and a limitation on ground heating. There are dry spells in early June, mid-July and late August. The rate of cooling is greater in autumn than the rate of warming in spring. Whilst the temperature curves are relatively smooth in the latter part of the year, there does seem to be a clear downward shift between November and December. Precipitation, mostly as rainfall, is relatively evenly spread through the last quarter of the year, but with a wetter spell in mid-October and to a lesser extent in late November.

Figure 6.2 shows monthly scores of Lamb Weather Types for the standard averaging period, 1981–2010; the methodology follows Jones *et al* [46]—see also Chapter 4. Cyclonic weather is at a minimum in February, coinciding with the cold, dry spell. The wetter autumn is clearly related to the reduction in anticyclonic weather. A minimum of westerlies in April to June helps explain dry spells at this time of year. December sees an

*Oxford Weather and Climate since 1767*. Stephen Burt and Tim Burt, Oxford University Press (2019).
 DOI: 10.1093/oso/9780198834632.001.0001

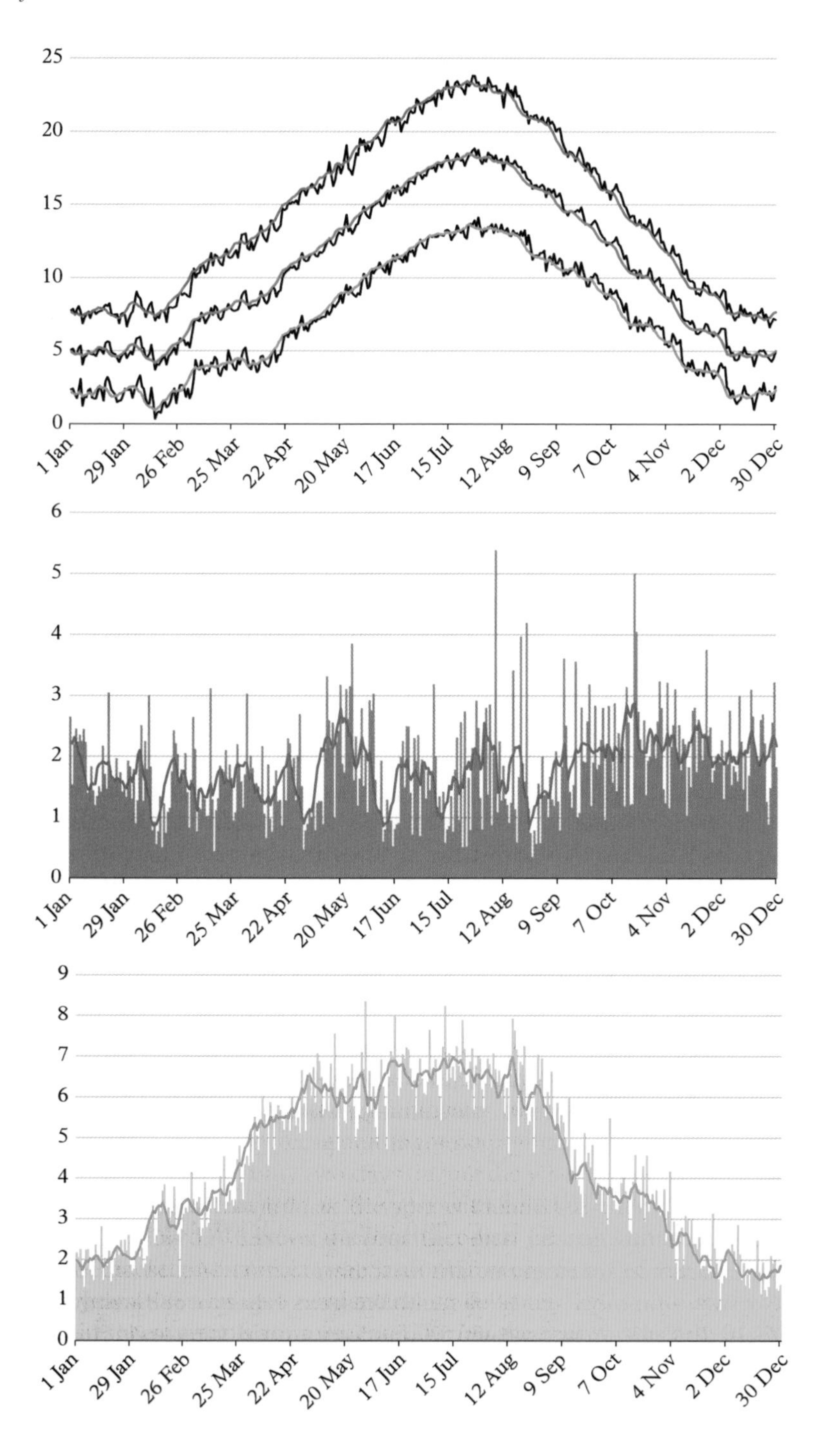


**Figure 6.1** *Daily averages (with weekly averages shown by thicker lines) for the Radcliffe Observatory, Oxford, over the period 1981–2010 of (top) mean maximum (red), mean minimum (blue) and daily mean temperature (green), °C: (middle) precipitation, millimetres: (bottom) sunshine duration, hours. Weekly means are plotted against the starting date of the week. Values for 29 February are excluded.*

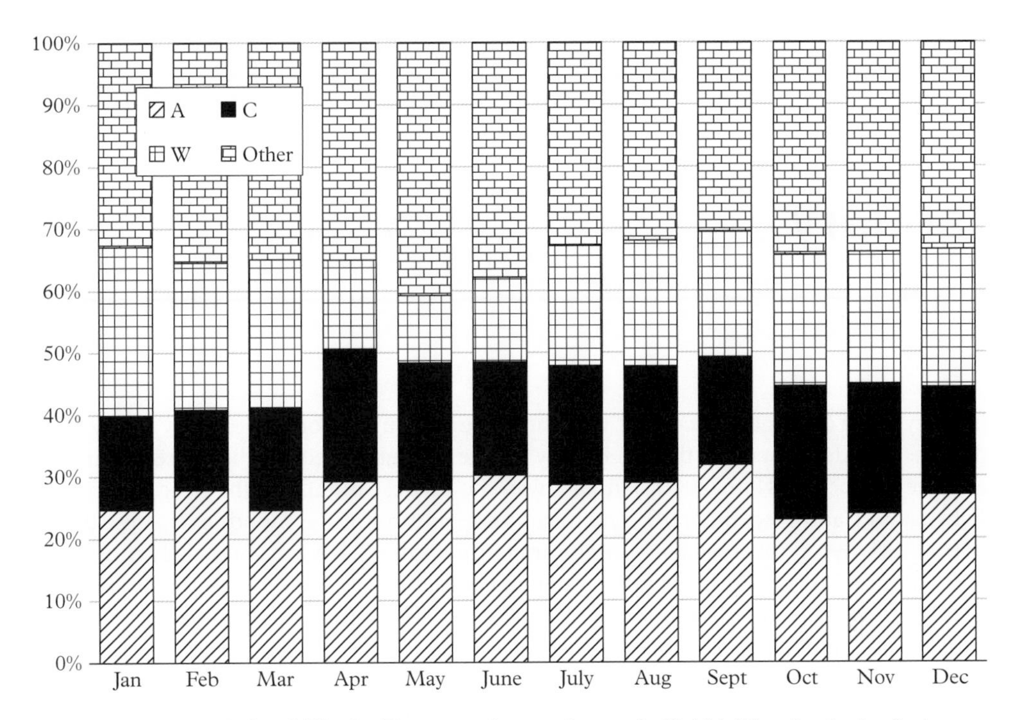


**Figure 6.2** *The main Lamb Weather Type scores by month over the British Isles; A—Anticyclonic, C—Cyclonic, W—Westerly, plus others combined in one category, expressed as a percentage for each month over the standard averaging period 1981–2010*

increase in anticyclonic weather and a reduction in cyclonic weather; together, these provide some rationale for the downward shift in temperature from November to December. Further work would be needed to explore singularities at Oxford in more detail, based on daily data, but it is interesting that the spells identified here no longer closely coincide with those listed by either Lamb [89] or Perry [90]. Not surprisingly, climate change has also seen a shift in when spells tend to occur.

# 7

# January

January is on average the coldest month of the year in Oxford, if only by a few hundredths of a degree colder than February. In the 30 years used for the current standard average period, 1981 to 2010, January was the coldest month of the year nine times, February 11 and December 10. The mean temperature for the month over this 30-year period was 4.9 °C (average daily minimum temperature 2.2 °C, average daily maximum temperature 7.6 °C). There is little systematic variation during the month, although on average the end of the month tends to be slightly colder than the beginning, as can be seen on the 'annual cycle' plots in Chapter 6. The monthly mean precipitation for the month, averaged over the same 30-year period, 1981–2010, is 57 mm. Rain falls on 16 days in an average January (the first week has the highest frequency of rainfall in the year). Although rainfall is frequent and can be persistent, daily totals in excess of 25 mm occur only once every 20 years or so. Some precipitation can be expected to fall as snow in most years, and snow is likely to cover the ground on one or two mornings in an average January. Thunderstorms are infrequent but not unknown at this time of year, occurring once or twice per decade, but rarely last for more than a few minutes. Sunshine duration in January creeps up a little from December's low point, averaging 62 hours during the month—just over 2 hours of bright sunshine daily, although this is slightly misleading as typically 11 or 12 days will remain sunless.

## Temperature

Between 1815 and 2018, air temperatures in January ranged from −16.6 °C on 14 January 1982 to 15.9 °C on 24 January 2016, a range of 32.5 degC (Table 7.1). The coldest day in any month on Oxford's long record was 8 January 1841, when the maximum temperature was only −9.6 °C. This occurred during a short but intensely cold spell; the temperature remained continuously below 0 °C for five consecutive days, falling to −14 °C on both 7 and 8 January (Figure 7.1). At the 10 a.m. observation on 8 January 1841, the sky was clear, the air temperature −13.4 °C, the wind northerly and the weather comment 'brisk'!

Very cold days in January were much more frequent in the first half of the nineteenth century: between 1815 and 1852, 11 January days recorded a maximum temperature

*Oxford Weather and Climate since 1767*. Stephen Burt and Tim Burt, Oxford University Press (2019).
 DOI: 10.1093/oso/9780198834632.001.0001

**Table 7.1** *Highest and lowest maximum and minimum temperatures in Oxford in January, 1815–2018*

| Rank | Mildest days | Coldest days | Mildest nights | Coldest nights |
|---|---|---|---|---|
| 1 | 15.9 °C, 24 Jan 2016 | *−9.6 °C, 8 Jan 1841* | 12.1 °C, 20 Jan 2008 | −16.6 °C, 14 Jan 1982 |
| 2 | *15.1 °C, 31 Jan 1846* | *−7.8 °C, 19 Jan 1823* | 11.8 °C, 6 Jan 1983 | *−16.4 °C, 20 Jan 1838* |
| 3 | 14.7 °C, 19 Jan 1930 | *−7.4 °C, 20 Jan 1838* | 11.7 °C, 21 Jan 1878 and 5 Jan 1957 | *−15.6 °C, 15 Jan 1820* |
| 4 | 14.6 °C, 9 Jan 1998 and 23 Jan 2018 | −7.1 °C, *15 Jan 1838, 19 Jan 1838 and* 13 Jan 1982 | *11.5 °C, 1 Jan 1851* | −15.5 °C, 15 Jan 1982 |
| 5 | 14.5 °C, 6 Jan 1999 | *−6.9 °C, 7 Jan 1841* | 10.8 °C, 31 Jan 1868 | *−15.0 °C, 19 Jan 1823* |
| ***Last 50 years to 2018*** | | | | |
| | 15.9 °C, 24 Jan 2016 | −7.1 °C, 13 Jan 1982 | 12.1 °C, 20 Jan 2008 | −16.6 °C, 14 Jan 1982 |

*Temperatures recorded by unscreened thermometers prior to 1853 are shown in italics; these are likely to be a little too cold by modern (screened thermometer) standards in winter and a little too warm in summer—see Appendix 1 for more details. In January the mean temperature correction for unscreened thermometry is +0.3 degC, although this will vary considerably on a day-to-day basis.*

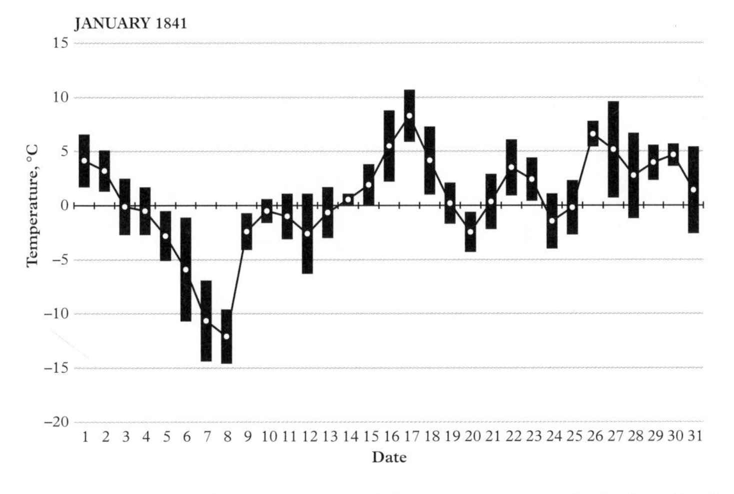


**Figure 7.1** *Daily maximum, minimum and mean daily temperatures at Oxford during January 1841. The coldest day on Oxford's records occurred on 8 January, when the maximum temperature was just −9.6 °C*

of –5 °C or lower, although the thermometer exposures in use at the time made such low readings more likely to occur (see Appendix 1). Since the introduction of screened thermometry at the Radcliffe Observatory in January 1853, such very cold days have been much rarer—only eight January days up to and including 2018 have been as cold, the most recent being 12 January 1987 during an outbreak of bitterly cold air from Europe, when the maximum temperature reached only –6.2 °C. Tim Burt remembers being interviewed beside the weather station that day by a Radio Oxford reporter who was very much underdressed for the occasion! The coldest day of any month in the twentieth century in Oxford was 13 January 1982, with a maximum temperature of −7.1 °C (Figure 7.2).

January is the month most likely to have an 'ice day', one in which the air temperature fails to reach 0 °C: in the 100 years to 2018, 246 days remained at or below freezing throughout, 94 of these in January, an average of a little less than once per year. The longest spell of consecutive 'ice days' since 1815 was 17 days commencing 10 January 1823: the longest spells in January in the twentieth century were those of 17–24 January 1963 and 7–14 January 1982, both lasting 8 days (Table 30.2, Chapter 30). Air frosts (minimum temperature over the 24 hours ending at 0900 GMT of −0.1 °C or below) can be expected on around nine mornings in an average January. January 1963 recorded 28 air frosts, and the Januarys of 1842, 1850 and 1879 recorded 27; in earlier records, January 1830 experienced 30 air frosts in the month. No air frost was recorded in the Januarys of 1884, 1916, 1938 and 2008.

Air temperatures of −10 °C or lower are infrequent in Oxford, but have occurred 63 times in January since 1815, the lowest January temperature being −16.6 °C on 14 January 1982. This occurrence was preceded by an air minimum of −9.7 °C on the 13th and followed by one of −15.5 °C on the 15th—during a spell of 11 consecutive nights with air frost (Figure 7.2). There was a much longer cold spell in January 1963,

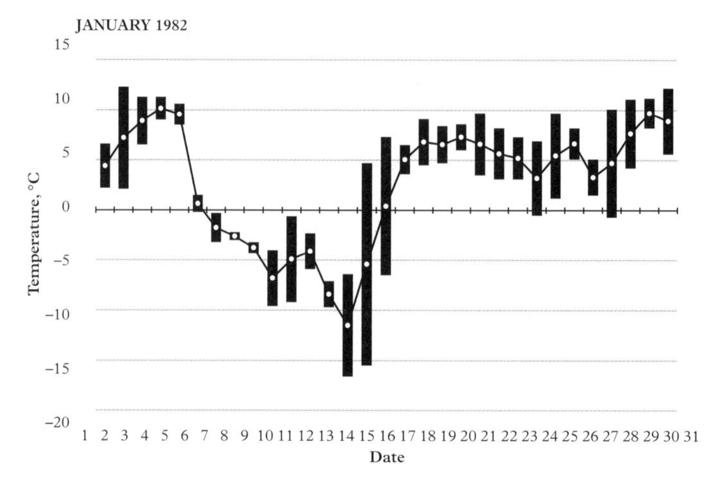


**Figure 7.2** *Daily maximum, minimum and mean daily temperatures at Oxford during January 1982. The coldest January night on Oxford's records occurred on 14th, when the temperature fell to −16.6 °C*

**Figure 7.3** *Two views of Parks Road, Oxford, in snowy January conditions a century apart.* **Top:** *Parks Road in January 1881, when on 18 and 19 January the snow depth at the Radcliffe Observatory was noted as 20 cm (© Oxfordshire History Centre POX 011 2881).* **Bottom:** *Parks Road at 8 a.m. on 14 January 1982—Oxford's coldest morning for over 120 years; at the time of this photograph, the temperature was close to the minimum recorded that morning of −16.6 °C, and the snow depth was 18 cm (Jonathan Webb)*

when air frost was recorded on 36 consecutive nights from 23 December to 27 January, although at its extreme the cold was not as intense—the coldest night in January 1963 was −14.2 °C, on 23rd. At the time of writing, no January morning has been as cold as −10 °C for over 30 years—the most recent occurrence being −10.5 °C on 13 January 1987.

Minimum temperatures above 10 °C in January, over the normal daily 24-hour period 0900–0900 GMT, are equally likely at any time of the month—they depend upon the source of the prevailing air mass and not, generally, on sunny or mild conditions the previous day. On the morning of 20 January 2008, the temperature did not drop below 12.1 °C (Table 7.1).

The largest daily temperature ranges—the difference between the daily minimum and maximum temperatures—occurred on 15 January 1982 (20.2 degC, from −15.5 °C to +4.7 °C) and 26 January 1963 (17.4 degC, from −12.5 °C to +4.9 °C)—both involving a steady thaw following several days of severe cold. Small diurnal ranges—sometimes only 1 degree in 24 hours—are not uncommon in January, not infrequently on cold, foggy days and occasionally during mild, cloudy, windy days.

## Warm and cold months

Despite many mild winters in recent years, January 1916 still holds the title of the mildest January yet recorded in Oxford: January 2007 was almost as mild, with less than a tenth of a degree separating the top two places in Table 7.2. January 1963 remains the coldest of any month yet recorded on Oxford's long records (Table 7.2, Figure 7.4), with snow on the ground every morning throughout the month, although documentary evidence exists of similarly intense cold in January in 1776, 1795 and 1814 (see Winter, Chapter 20). The

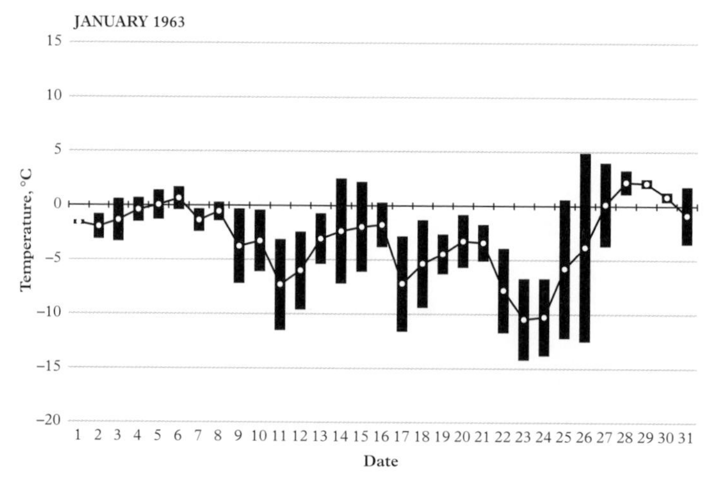


**Figure 7.4** *Daily maximum, minimum and mean daily temperatures at Oxford during January 1963, the coldest month on Oxford's records*

**Table** 7.2 *January mean temperatures at Oxford's Radcliffe Observatory, 1814–2018*

**January mean temperature 4.9 °C** (average 1981–2010)

| Mildest months | | | Coldest months | | |
|---|---|---|---|---|---|
| Mean temperature, °C | Departure from 1981–2010 normal degC | Year | Mean temperature, °C | Departure from 1981–2010 normal degC | Year |
| 7.6 | +2.7 | 1916, 2007 | −3.0 | −7.9 | 1963 |
| 7.5 | +2.6 | 2008 | *−2.6* | *−7.5* | *1814* |
| 7.4 | +2.5 | 1921 | *−2.4* | *−7.3* | *1838* |
| 7.3 | +2.4 | 1990 | *−1.5* | *−6.4* | *1815* |
| 7.2 | +2.3 | 1975, 1983 | −1.4 | −6.3 | 1881 |
| *Last 50 years to 2018* | | | | | |
| 7.6 | +2.7 | 2007 | +0.1 | −4.8 | 1979 |

Thames at Oxford was largely frozen over in January 1940 and January 1963, and many householders remained without water as pipes froze—only then to burst as the thaw arrived.

## Precipitation

Precipitation in this context includes rain, drizzle, snow, sleet, hail and occasionally fog or dew. January is—on average—slightly less wet than October, November or December, with an average monthly precipitation of 57 mm during the 1981–2010 period. Monthly precipitation totals for January over the period since 1767 have varied from 5 mm in 1825 (9 per cent of the current normal) to 147 mm in 2014 (258 per cent of normal) (Table 7.3).

Only 13 Januarys since 1767 have received 100 mm or more during the month, with 2014 being the wettest on record by some margin (Table 7.3). The winter of 2013/14 was, again by some margin, the wettest winter on record; 335 mm of precipitation, just over half a normal year's rainfall, fell between 1 December 2013 and 28 February 2014, of which 296 mm fell in the 61 days commencing 16 December 2013. The heavy and persistent rains led to widespread and prolonged flooding in and around Oxford, its waterways and floodplains (Figure 7.5). The synoptic context of the winter of 2014 is discussed in Chapter 4.

Within the last 100 years, only one January, 1987, has received less than 10 mm of precipitation and, of this, less than 1 mm fell during the final 18 days of the month. Heavy falls of rain are uncommon in January: only nine January days since 1827 have exceeded 25 mm of precipitation (in the 24-hour period commencing at 0900 GMT). The wettest January day on record was 28 January 1958, when 36.1 mm was recorded. The fall of 29.2 mm on 26 January 1940 was noted as entirely rainfall (some of which

**Figure 7.5** *January flooding events in Oxford.* ***Top:*** *Magdalen Bridge in flood, January 1938 (© Oxfordshire History Centre POX 055 0379).* ***Middle and bottom:*** *January 2014 floods—Oxford's wettest January, and its wettest winter, on 250 years' records.* ***Middle:*** *Bankfull at Osney Bridge.* ***Bottom:*** *Flooding and sandbags on Earl Street. Middle and bottom photographs were taken on 9 January 2014 (Jane Buekett)*

**Table 7.3** *January precipitation at the Radcliffe Observatory, Oxford since 1767: extremes of monthly totals 1767–2018, wettest days 1827–2018*

| **January mean precipitation 56.9 mm** (average 1981–2010) | | | | | | | |
|---|---|---|---|---|---|---|---|
| Wettest months | | | Driest months | | | Wettest days | |
| Total fall, mm | Per cent of normal | Year | Total fall, mm | Per cent of normal | Year | Daily fall, mm | Date |
| 146.9 | 258 | 2014 | 5.1 | 9 | 1825 | 36.1 | 28 Jan 1958 |
| 127.3 | 223 | 1948 | 6.3 | 11 | 1855 | 33.5 | 1 Jan 1926 |
| 123.7 | 217 | 1852 | 7.0 | 12 | 1838 | 33.0 | 3 Jan 1887 |
| 116.3 | 204 | 1995 | 7.3 | 13 | 1787 | 32.6 | 13 Jan 1852 |
| 116.1 | 204 | 1877 | 8.7 | 15 | 1802 | 29.2 | 26 Jan 1940 |
| *Last 50 years to 2018* | | | | | | | |
| 146.9 | 258 | 2014 | 9.9 | 17 | 1987 | 21.5 | 1 Jan 2003 |

**Figure 7.6** *The frozen Thames (Isis) at Donnington Bridge in south Oxford during January 1963. The bridge crosses the river on the reach between Iffley Lock and Osney Lock; it had been open only two months when severe cold set in at the end of December 1962, and the river quickly froze. (School of Geography, University of Oxford)*

froze on contact with the cold ground, forming a layer of ice, but no snow, sleet or ice pellets), despite the day's maximum temperature being only 3.3 °C.

## Snowfall and lying snow

Snow or sleet falls on about four days in an average January but, as with any winter month, there are large year-to-year variations. The snowiest Januarys since 1926 were in 1963 (17 days with snowfall—followed by a snowy February), 1945 (16 days) and 1985 (14 days with snowfall). At the opposite extreme, no snow or sleet is observed in about one January in four, and snowless Januarys have been more common in the last three decades.

Snowfall does not always lead to snow cover on the ground. January averages around two days with snow cover, although this figure is hugely skewed by a few very snowy months, as typically less than one January in two will see any lying snow at the morning observation. In January 1963, snow lay throughout the month (as it did also throughout February 1963) in one of the snowiest winters across the UK in the past 120 years (for more on the winter of 1962/63, see Winter, Chapter 20). In January 1940 and again in January 1979, snow lay on 16 mornings. The greatest snow depths observed in Oxford since 1959 have been 27 cm, on the morning of 6 January 2010; 24 cm, on 10 January 1982 (Figure 7.7); and 23 cm, noted on 3–4 and 20–26 January 1963. Since 1959, only

**Figure 7.7** *Snow on Hill View Road, Oxford, on 10 January 1982; the snow depth at the Radcliffe Observatory that morning was noted as 24 cm, the second-greatest snow depth within the last 60 years (© Oxfordshire History Centre POX 003 7301)*

six Januarys have recorded at least one morning with 10 cm or more of snow on the ground at the morning observation (these were 1960, 1962, 1963, 1968, 1982 and 2010).

### Thunderstorms

Thunder is uncommon but not unknown in January, occurring on average less than once in five years, although thunder was heard on two days in Oxford in January 1974.

## Sunshine

January is normally slightly sunnier than December, the monthly average being 62 hours of bright sunshine, although on average 11 or 12 days will remain sunless. The extremes of monthly sunshine duration since 1881, and the sunniest January days since 1921, are shown in Table 7.4.

Winter months have seen a big improvement in sunshine in major towns and cities since Victorian times, owing to the reduction in emissions of smoke and soot from domestic and industrial premises—particularly so since the introduction of the Clean Air Act in 1956. All but one of the sunniest Januarys since 1881 have been recorded within the last 40 years, whereas only one of the dullest has been recorded within the last century. The dullest January within the last 100 years was January 1996 with just 29 hours of sunshine

**Table 7.4** *January sunshine duration at the Radcliffe Observatory, Oxford: monthly extremes 1881–2018, sunniest days 1921–2018*

**January mean sunshine duration 62.3 hours, 2.01 hours per day** (average 1981–2010)
*Possible daylength: 261 hours. Mean sunshine duration as percentage of possible: 23.9*

| Sunniest months | | | Dullest months | | | Sunniest days | |
|---|---|---|---|---|---|---|---|
| Duration, hours | Per cent of possible | Year | Duration, hours | Per cent of possible | Year | Duration, hours | Date |
| 94.1 | 36.1 | 1952 | 14.9 | 5.7 | 1885 | 8.5 | 25 Jan 1986 |
| 87.7 | 33.6 | 2003 | 24.2 | 9.3 | 1917 | 8.4 | 31 Jan 1987 |
| 87.0 | 33.4 | 1994 | 26.8 | 10.3 | 1898 | 8.0 | 26 Jan 1997 |
| 86.1 | 33.0 | 1984 | 27.2 | 10.4 | 1884 | 7.9 | 23 Jan 1986, 28 Jan 1994, 25 Jan 1997, 29 Jan 2004 and 29 Jan 2006 |
| 85.8 | 32.9 | 1959 | 29.3 | 11.2 | 1996 | 7.8 | 11 Jan 2014, 20 Jan 2017 |
| *Last 50 years to 2018* | | | | | | | |
| 87.7 | 33.6 | 2003 | 29.3 | 11.2 | 1996 | 8.5 | 25 Jan 1986 |

(a daily average of just 57 minutes, barely 11 per cent of that possible): 20 days in the month failed to register any sunshine whatsoever. However, the twentieth century's dullest was undercut three times in the last 20 years of the nineteenth century, with January 1885 recording barely half as much as January 1996—an average of just 29 minutes per day, while 17 days remained sunless, including nine consecutive days 13th—21st.

Long sunless spells are not uncommon in January. Since 1921, the two longest runs of sunless days at any time of year have both commenced in January—15 consecutive days without sunshine 26 January to 9 February 1940, and 14 consecutive days 14–27 January 1987 (with a sharply contrasting 26.1 hours of sunshine during the last four days of the latter month).

The sunniest January was in 1952, with 94 hours—a daily average of just over 3 hours of sunshine per day, 36 per cent of the possible duration. Remarkably, January 1952's sunshine total of 94 hours is also not far short of the totals in the dullest summer months—for example, May 1981 (98 hours), June 2016 (100 hours) and July 1944 (97 hours)—despite the much greater length of daylight hours during the summer half-year. At this time of year, 7 hours of sunshine counts as a 'sunny day'.

Owing to the slow increase in daylight hours during January, only at the very end of the month is 8 hours sunshine in a day possible. This has occurred only three times since 1921, the sunniest of all being 25 January 1986 when 8.5 hours of sunshine was recorded.

## Wind

Oxford's mean monthly wind speed (measured at 45 m above ground level) for January is 4.6 metres per second (m/s), or 9.0 knots, the highest of any month in the year, although gales (mean wind speed exceeding 17 m/s, 34 knots, for at least 10 minutes) are uncommon in Oxford. The windiest January on a record back to 1880 occurred in 1983, when the monthly mean wind speed was 6.7 m/s (13.0 knots); the calmest Januarys were in 1964 and 1997, when the monthly mean wind speed was just 2.9 m/s (5.7 knots).

The gale which affected Oxford and many parts of the English Midlands on the evening of 2 January 1976 was almost certainly the worst in the Oxford area since 1881, in which year two gales of similar severity (in January and October) caused widespread damage both locally and nationally. The circumstances of these gales were compared by Gordon Smith [91]. The period of strongest winds at Oxford on 2 January 1976 occurred between 2015 and 2035 GMT during which the mean wind speed reached about 25 m/s (48 knots, or storm force 10). Unfortunately, the wind recorder in use at the time was the ancient Robinson–Beckley anemograph—installed in January 1880—which did not permit the measurement of gusts (with this type of long-obsolete instrument, it is only possible to read off a mean wind speed over a minimum period of about ten minutes; the instrument was finally retired later in 1976, after 96 years' service). Gust speeds recorded elsewhere in Oxfordshire during the January 1976 storm included 36.5 m/s (71 knots) at 2010 GMT at Brize Norton, 22 km to the west of Oxford; 38.5 m/s (75 knots), also at 2010 GMT, at Abingdon, 8 km south-south-west of Oxford; and 34 m/s (66 knots) at 2025 GMT at Benson, 20 km south-east of Oxford. Considerable tree and structural damage resulted from this gale in and around Oxford.

## Oxford temperature, precipitation and sunshine in graphs—January

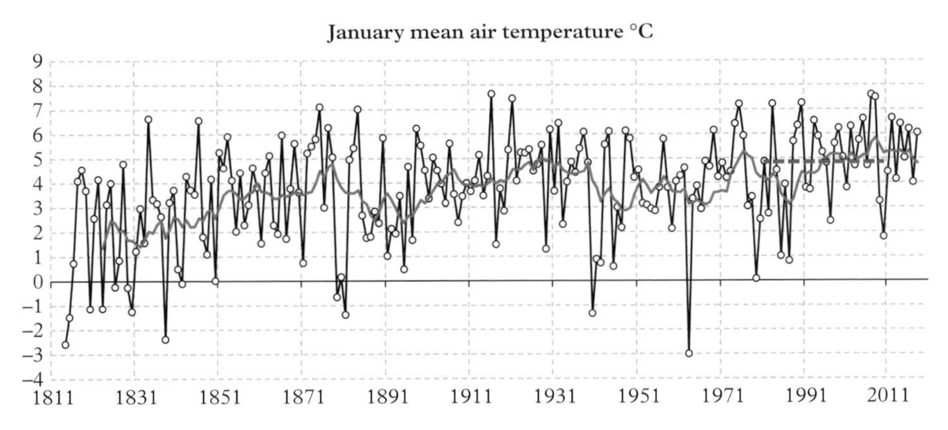


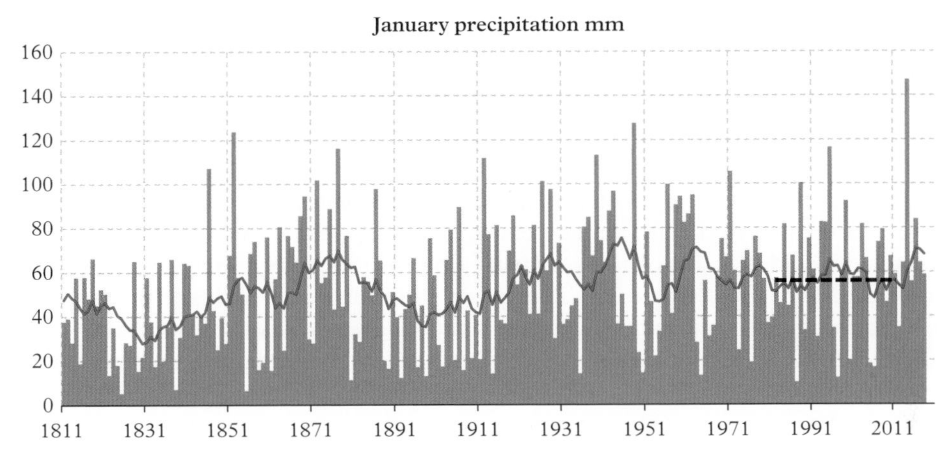


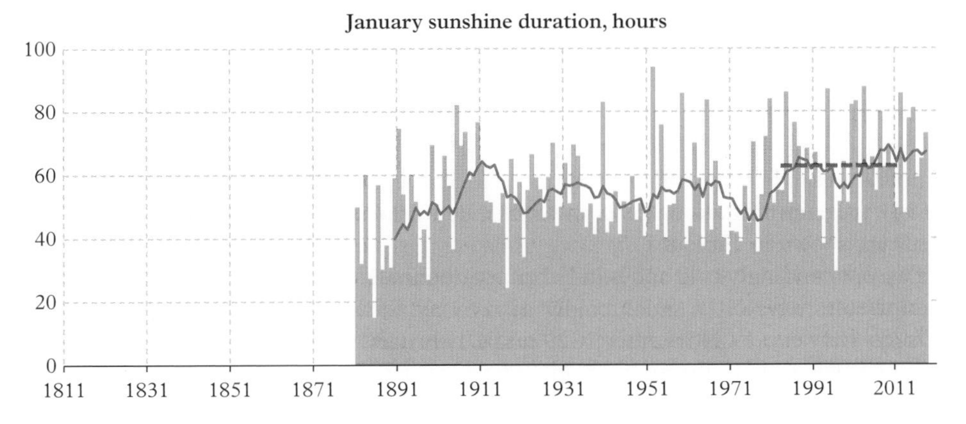


**Figure 7.8** *Monthly values of (from top) mean temperature (°C, since 1814), total precipitation (mm, since 1811) and sunshine duration (hours, since 1881) for January at Oxford. The 1981–2010 averages are indicated by the thick blue dashed line, while the ten-year unweighted running mean ending at the year shown is indicated by the red line*

# 8

# February

February is, on average, the driest month of the year in Oxford, the mean monthly precipitation over the standard 30-year period 1981–2010 being 43 mm, 25 per cent less than January. Part of the reason for February being drier is, of course, that the month has fewer days than other months: in fact, the mean precipitation per day is almost identical in February, March and June. Precipitation can be expected on about 13 days in an average February; three or four of these can be expected to be of snow or sleet. Across the year as a whole, taking an average over many years, the frequency of snowfall tends to be highest in the second week of the month. The lowest mean temperatures of the year in Oxford are also reached during early to mid-February. Although there is, of course, considerable year-to-year variation the week commencing 14 February was on average the coldest week of the year in Oxford during the 1981–2010 averaging period, with a mean temperature of 4.2 °C, while 14 February is on average the coldest night of the year with a mean minimum temperature just three-tenths of a degree above freezing. Thereafter, increasing daylength and greater power of the Sun are reflected in a slow rise in temperatures, as shown in the annual plots (Chapter 6). It is in February that increasing daylength and lighter evenings start to become obvious after the gloomy midwinter months: February is, by daily mean duration of sunshine, on average some 45 minutes per day sunnier than January. The monthly mean duration of bright sunshine is 79 hours, with about 8 days remaining sunless on average.

## Temperature

Temperatures in February in Oxford since 1815 have ranged from a bitterly cold −16.1 °C in 1947 to a summer-like 18.5 °C in 1998. The warmest and coldest February days and nights are shown in Table 8.1.

Two outstandingly cold spells in February dominate the February extremes. At a time when winters were often much colder than today, the winter of 1829/30 has few comparisons. Between 19 November 1829 and 6 February 1830, a total of 80 days, 28 days remained below 0 °C throughout, including the seven consecutive days commencing

*Oxford Weather and Climate since 1767*. Stephen Burt and Tim Burt, Oxford University Press (2019).
 DOI: 10.1093/oso/9780198834632.001.0001

**Table 8.1** *Highest and lowest maximum and minimum temperatures at the Radcliffe Observatory, Oxford in February, 1815–2018*

| Rank | Mildest days | Coldest days | Mildest nights | Coldest nights |
|---|---|---|---|---|
| 1 | 18.5 °C, 13 Feb 1998* | *−7.9 °C, 2 Feb 1830* | 12.0 °C, 4 Feb 2004 | −16.1°C, 25 Feb 1947 |
| 2 | 17.8 °C, 28 Feb 1959 | *−6.2 °C, 3 Feb 1830* | 11.9°C, 5 Feb 2004 | −15.7 °C, 24 Feb 1947 |
| 3 | 17.4 °C, 28 Feb 1891 and 14 Feb 1998 | *−6.0 °C, 9 Feb 1816 and 6 Feb 1830* | 11.4 °C, 2 Feb 2002 | *−13.9 °C, 10 Feb 1816* |
| 4 | 16.8 °C, 27 Feb 1891 | *−5.6 °C, 3 Feb 1841* | 11.2 °C, 3 Feb 2004 | *−13.8 °C, 11 Feb 1847* |
| 5 | 16.7 °C, 18 Feb 1945, 29 Feb 1948, 28 Feb 1960 and 4 Feb 2004 | *−5.1 °C, 5 Feb 1830* | 11.1 °C, 15 Feb 1958 | −13.6 °C, 16 Feb 1855 and 8 Feb 1895 |
| ***Last 50 years to 2018*** | | | | |
| | 18.5 °C, 13 Feb 1998 | −3.6 °C, 9 Feb 1991 | 12.0 °C, 4 Feb 2004 | −10.7 °C, 21 Feb 1986 |

*Temperatures recorded by unscreened thermometers prior to 1853 are shown in italics; these are likely to be a little too cold by modern (screened thermometer) standards in winter and a little too warm in summer—see Appendix* 1 *for more details. In February the mean temperature correction for unscreened thermometry is +0.2 degC, although this will vary considerably on a day-to-day basis*

* In February 2019, 17.6 °C was reached on 25th and 18.8 °C on 26th

31 January. The first six days of February 1830 still stand as four of the five coldest February days on record, including the coldest of all, −7.9 °C on 2 February. (In comparison, the coldest February day of the twentieth century in Oxford was 1 February 1956, with the maximum temperature that day just −4.4 °C.) It is the notoriously cold and snowy February 1947 that dominates the coldest nights, however, with two exceptionally cold nights in succession—24 February with −15.7 °C and 25 February with −16.1 °C (the minimum temperature recorded against the following night, −13.3 °C, was the temperature when the thermometer was reset at 0900 GMT on 25th). Temperatures below −10 °C occurred in February in 1816, 1818, 1827, 1830, 1841, 1845, 1847, 1855 and 1895, 1917, 1929 and 1947, but it is more than 30 years since −10 °C was last recorded in February: in February 1986, four nights fell below this figure, the lowest being −10.7 °C on 21 February. February 1986 was the coldest since 1947, and 1895 before that: February 1986 and February 1895 (Figure 8.1) both recorded air frost on 24 out of 28 nights, the Februarys of 1942 and 1963 both had 25 air frosts, while February 1947 topped the tables with 26.

**Figure 8.1** *Oxford in the very cold February of 1895.* ***Top:*** *Curling on Christ Church Meadows.* ***Middle:*** *Skating on the Cherwell.* ***Bottom:*** *A crowd gathers around a stage coach and team of six horses on the frozen River Thames (All images © Oxfordshire History Centre – top POX 011 2837, middle POX 011 5614, bottom POX 011 3298)*

Until 1947, the lowest grass minimum—or, more likely, above a snow surface—recorded in any month since grass minimum temperature records began in 1857 was –19.3 °C on 7 February 1917; this was finally surpassed on 25 February 1947, when –22.2 °C was recorded above 11 cm snow cover. This remained the lowest 'grass minimum' until 14 January 1982, when it was just exceeded by the reading that morning (also above snow) of –22.5 °C, the lowest value on the Radcliffe Observatory's record to date.

In contrast, when the right conditions occur—a cloudy, mild southerly airflow—February can also produce nights that are warmer than the average June or September night. February 2004 dominates the record books, with three of the five warmest February nights yet recorded, including the mildest of them all, 12.0 °C on 4 February 2004; three consecutive nights in early February 2004 remained above 11 °C throughout. (It has to be said, however, that every other month of the year has recorded warmer 'mildest nights' than February.) Not surprisingly, the longer days towards the end of February can often be relied upon to produce the warmest days of the month, and at this time of year southerly winds with unbroken sunshine can produce warm days which would not be out of place in late May or early June. Notable examples include 17.8 °C on 28 February 1959 (the highest of four consecutive days above 15 °C, and also the sunniest February day on record) and 17.4 °C on 28 February 1891—both very dry and anticyclonic months. The latter date also saw the greatest daily range in temperature on any February day—20.0 degC, an impressive rise from the morning's minimum temperature of –2.6 °C. On 26 February 2019, a minimum temperature of –0.1 °C was followed by a new February record maximum of 18.8 °C, giving a daily range of 18.9 degC. All the more surprising, then, that the *middle* of February in 1998 produced two of the three warmest February days on record, including Oxford's highest February temperature, of 18.5 °C on 13 February 1998.

Mild, cloudy conditions in the winter months can produce day after day with only small variations in temperature between day and night. The very unsettled, stormy and wet month of February 2014 recorded the lowest monthly range in temperature of any month on Oxford's record—just 10.4 degC; the lowest temperature was 1.7 °C on 13th, the highest 12.1 °C 10 days later.

## Warm and cold months

February 1990 was by some margin the mildest February in Oxford's long record, being 3.4 degC warmer than average, with no air frosts recorded during the month—as was also the case in the Februarys of 1872, 1883, 1945, 1961 and, more recently, 2014. In contrast, the *mean* temperature has remained below freezing in eight Februarys, of which 1947 was the coldest and 1986 the most recent. In February 1895 the mean grass minimum temperature was –9.3 °C, the lowest on record for any month back to at least 1881 (Figure 8.1). Particularly worthy of note is February 1986's low absolute maximum temperature of just 3.5 °C for the entire month (Figure 8.2): a temperature of at least 4 °C has been reached on at least one day in every other calendar month back to 1814.

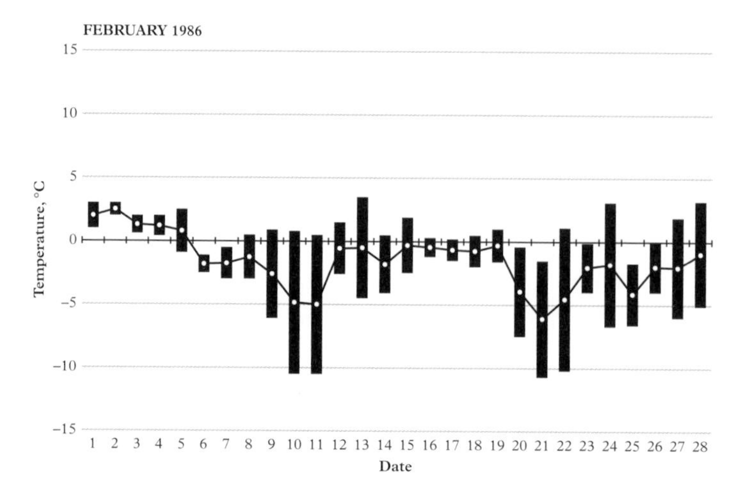


**Figure 8.2** *Daily maximum and minimum temperatures (°C) at the Radcliffe Observatory in February 1986; only the Februarys of 1947 and 1895 have been colder in Oxford since records began in 1814*

**Table 8.2** *February mean temperatures at Oxford's Radcliffe Observatory, 1814–2018*

**February mean temperature 4.9 °C** (average 1981–2010)

| Mildest months | | | Coldest months | | |
|---|---|---|---|---|---|
| Mean temperature, °C | Departure from 1981–2010 normal degC | Year | Mean temperature, °C | Departure from 1981–2010 normal degC | Year |
| 8.3 | +3.4 | 1990 | −2.2 | −7.2 | 1947 |
| 7.8 | +2.9 | 1869, 1945, 1998 | −1.9 | −6.8 | 1895 |
| 7.7 | +2.8 | 1961, 2002 | −1.5 | −6.4 | 1986 |
| 7.4 | +2.5 | 1872, 1903, 1926, 1995, 2011 | −1.4 | −6.3 | 1855 |
| 7.3 | +2.4 | 1997 | −0.7 | −5.6 | 1963 |
| *Last 50 years to 2018* | | | | | |
| 8.3 | +3.4 | 1990 | −1.5 | −6.4 | 1986 |

## Precipitation

February is, on average, the driest calendar month of the year in Oxford: the mean monthly precipitation over the period 1981–2010 is 43 mm, 25 per cent less than in January. The greatest likelihood in the year of snow or sleet falling is in early to mid-February.

Very wet months are uncommon at this time of year, and only five Februarys have accumulated 100 mm of precipitation since 1767 (39 Octobers reached this threshold

**Table 8.3** *February precipitation at the Radcliffe Observatory, Oxford since 1767. Extremes of monthly totals 1767–2018, wettest days 1827–2018*

**February mean precipitation 42.8 mm** (average 1981–2010)

| Wettest months | | | Driest months | | | Wettest days | |
|---|---|---|---|---|---|---|---|
| Total fall, mm | Per cent of normal | Year | Total fall, mm | Per cent of normal | Year | Daily fall, mm | Date |
| 120.3 | 281 | 1937 | 2.2 | 5 | 1959 | 40.8 | 13 Feb 1888 |
| 114.4 | 267 | 1977 | 2.5 | 6 | 1798, 1891 | 30.7 | 25 Feb 1933 |
| 107.0 | 250 | 1900 | 3.5 | 8 | 1932 | 25.9 | 9 Feb 2009 |
| 106.3 | 248 | 1833 | 4.4 | 10 | 1895 | 25.1 | 6 Feb 2014 |
| 103.9 | 243 | 1950 | 4.6 | 11 | 1790 | 24.1 | 1 Feb 1927, 12 Feb 2001 |
| *Last 50 years to 2018* | | | | | | | |
| 114.4 | 267 | 1977 | 7.1 | 17 | 1993 | 25.9 | 9 Feb 2009 |

during the same period), as shown in Table 8.3. February 1937 was the wettest of all, with 120 mm, and flooding occurred frequently along local rivers during the month. Falls in excess of 25 mm of precipitation in a day are almost unknown in February, having occurred on only four occasions since 1827, although two of these have been in recent years, namely 25.9 mm on 9 February 2009 and 25.1 mm on 6 February 2014. The 'wettest' February day of all was on 13 February 1888, when 40.8 mm was recorded—all of it falling as snow during a prolonged snowstorm (see below). The second-wettest day was 25 February 1933, when 30.7 mm fell; this was only part of the story, however, because 23.4 mm had also fallen on the previous day, while 11.4 mm fell on the following day, giving a 72-hour total of 65 mm. (Three such consecutive very wet days are very rare at any time of year, particularly in the driest month of the year.) All three days saw a prolonged fall of cold rain, wet sleet and snow.

At the other end of the scale, February 1959 remains the driest on record, with 2.2 mm.

## Snowfall and lying snow

Snowfall frequency in February is, on average, very similar to January (January averages 4.1 days in a month of 31 days, February averages 3.7 days in 28 or 29 days), but of course, there are very large year-to-year variations in both months. A snow cover can be expected, on average, on two February mornings. About one February in five remains free of snow or sleet, while the Februarys of 1947, 1955, 1956 and 1963 all recorded 15 days with snow or sleet observed to fall. Snow lay throughout the month in February 1963, and on 27 days in the notoriously snowy February 1947; in February 1956, snow cover was noted on 18 mornings. Not surprisingly, these months feature in the 'Top ten coldest months' (Chapter 29, Table 29.4).

**Figure 8.3** *February snowfalls in Oxford.* **Top:** *Rowing, anyone? Hardly the weather for it! This photograph was taken on 18 February 1888, just a few days after Oxford's heaviest snowfall on record. This brave crew was taking part in the Torpids, Oxford University's winter eights event. The Torpids began in the 1820s and are so called because they were originally races for the slower college second-eight boats qualifying for the summer competition (© Oxfordshire History Centre POX 011 3265).* **Bottom:** *Snow-covered bicycles in Oxford city centre, 8 February 2007 (© Oxfordshire History Centre D4243 22a)*

The greatest snow depth known from the Oxford record occurred on 13 and 14 February 1888, when the depth reached 61 cm following a 20-hour snowstorm. Astonishingly, there exists an archive photograph taken shortly afterwards (Figure 8.3). Although notorious as a snowy winter, the greatest depth of snow in Oxford on any morning in February 1947 was 'only' 17 cm—the deepest snow of that winter occurred in early March (see Chapter 9). Since 1959 the greatest snow depth in February has been 15 cm (from 1–6 February 1963); depths of 10 cm or greater were noted in five Februarys between 1959 and 2017 (1960, 1963, 1969, 2007 and 2009).

### Thunderstorms

Thunder is very rare in February, occurring on average about once in ten years, and often just for a few minutes.

## Sunshine

By the middle of February, the strength and duration of sunshine has perceptibly increased from the midwinter minimum. On average, eight of February's 28 or 29 days will remain sunless, but the other days between them amount to around 79 hours of bright sunshine.

**Table 8.4** *February sunshine duration at the Radcliffe Observatory, Oxford: monthly extremes 1880–2018, sunniest days 1921–2018*

**February mean sunshine duration 78.9 hours, 2.79 hours per day** (average 1981–2010)
Possible daylength: 280 hours*. Mean sunshine duration as percentage of possible: 27.9

| **Sunniest months** | | | **Dullest months** | | | **Sunniest days** | |
|---|---|---|---|---|---|---|---|
| **Duration, hours** | **Per cent of possible** | **Year** | **Duration, hours** | **Per cent of possible** | **Year** | **Duration, hours** | **Date** |
| 124.1 | 43.9 | 2008** | 22.8 | 8.1 | 1940 | 9.7 | 27 Feb 1928,***<br>28 Feb 1959,<br>28 Feb 1996 |
| 117.6 | 42.0 | 2018 | 26.3 | 9.3 | 1947 | 9.6 | 28 Feb 1989 |
| 115.9 | 41.0 | 1970 | 27.3 | 9.7 | 1972 | 9.4 | 27 Feb 1986 |
| 115.1 | 40.8 | 1949 | 29.9 | 10.6 | 1897 | 9.35 | 26 Feb 1921 |
| 113.0 | 40.0 | 1998 | 31.1 | 11.0 | 1965 | 9.3 | 26 Feb 1935,<br>25 Feb 2001 |
| ***Last 50 years to 2018*** | | | | | | | |
| 124.1 | 43.9 | 2008 | 27.3 | 9.7 | 1972 | 9.7 | 28 Feb 1996 |

* This is for a 28-day February; for leap years, the figure is 291 hours
** February 2019's sunshine duration amounted to 127.8 hours, 45.6% of possible daylength, a new February record
*** 9.7 hours sunshine was recorded on 25 February 2019 and 10.3 hours on 27 February 2019.

Despite the increasing daylength, sunshine totals do vary enormously in February. February 2008 recorded just over 124 hours of bright sunshine (with only five sunless days), considerably more than some summer months—the dismal 97 hours recorded in July 1944 and 100 hours in June 2016 being prime examples. At the other extreme, long spells of anticyclonic conditions can lead to persistent dry but cloudy conditions at this time of year. The bitterly cold February of 1947 was also exceptionally dull, with just 26 hours sunshine—the three weeks ending 22 February seeing a mere 1.1 hours sunshine, an average of just 3 minutes per day. February 1940 was even duller, with just 23 hours of sunshine during the entire month, only 8 per cent of the possible daylight hours.

As the days lengthen, so the possible duration of sunshine increases, and towards the end of the month close to 10 hours is possible on clear days. Four February days since 1921 have managed more than 9.5 hours—a creditable total for a June day.

## Wind

Oxford's mean monthly wind speed for February is 4.4 metres per second (m/s), or 8.5 knots. The windiest February on a record back to 1880 occurred in 1990, when the monthly mean wind speed was 7.3 m/s (14.3 knots); this month included several gales and near-gales and recorded the second-highest monthly mean wind speed for any month on Oxford's records, just behind October 1967. The calmest February was in 1891, when the monthly mean wind speed was just 2.1 m/s (4.1 knots). A violent gale was recorded on 20–21 February 1861; the anemometer at the top of the Radcliffe Observatory tower (33 m above ground) recorded a mean wind speed of 24.1 metres per second (46.9 knots) over the two hours 2000–2200h on 21 February. The 24-hour mean wind speed for the 21st (at 33 m) was 13.5 metres per second (26.3 knots). Much damage was caused by this gale in Oxford and across southern England.

## Oxford temperature, precipitation and sunshine in graphs—February

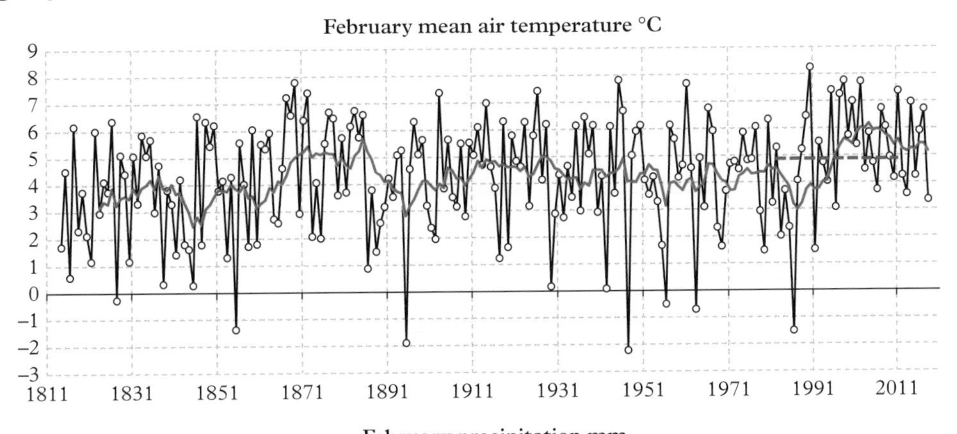


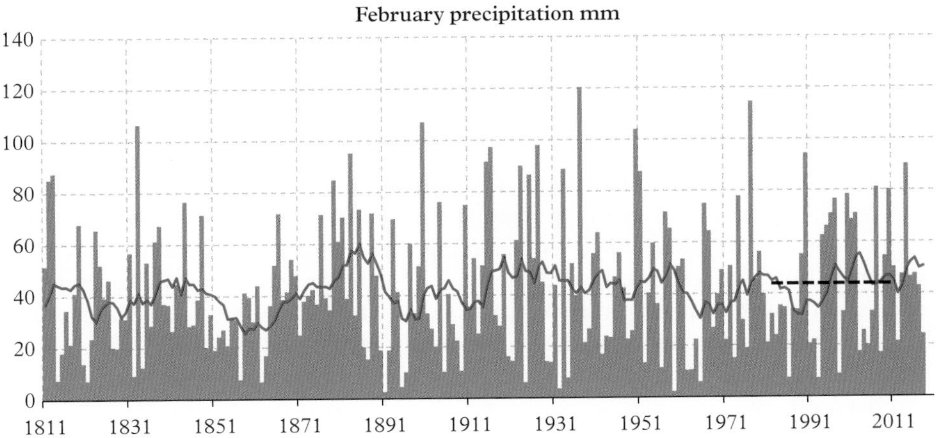


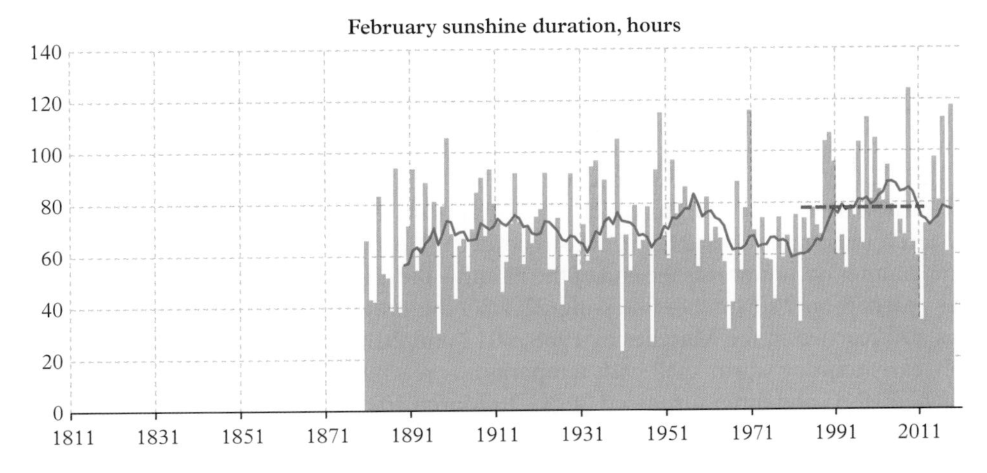


**Figure 8.4** *Monthly values of (from top) mean temperature (°C, since 1814), total precipitation (mm, since 1811) and sunshine duration (hours, since 1880) for February at Oxford. The 1981–2010 averages are indicated by the thick blue dashed line, while the ten-year running mean ending at the year shown is indicated by the red line*

# 9

# March

March marks the transition from winter to spring in most years, and the weather can be very changeable—the old saying 'March comes in like a lion and goes out like a lamb' has more than an element of truth. March weather can feature an unrelenting continuation of winter with frost and snow, or sometimes glorious early spring sunshine and early warmth—occasionally both within a week! Mean daily temperatures increase quite quickly from below 5 °C at the beginning of the month to almost 8 °C in the final week: even so, sharp frosts are not uncommon even at the very end of the month. The monthly mean precipitation for March is slightly higher than February, although the difference is almost entirely due to March having an extra three days in a normal year, two in a leap year. Despite being one of the driest months of the year, widespread and devastating floods on the Thames can occur in March, particularly if a heavy snow cover melts quickly. Snow can be expected on three days in an average March, and thunder about two years in five, more often after mid-month. The duration of bright sunshine in March averages 111 hours, or just over 3½ hours per day, although the sunniest Marches rival or even exceed the normal sunshine duration in the summer months. A daily total of 12 hours of sunshine is just possible by the end of the month, although typically six days will remain sunless.

## Temperature

March marks the transition from winter to spring, and the weather can occasionally be very cold and wintry, or very warm and sunny. The average temperature for the month is 7.3 °C, but 20 °C has been reached at least once per decade over the last century: the earliest date in the year on which 20 °C has been reached in Oxford since 1815 was 9 March (when, in 1948, 21.1 °C was recorded). Normally, though, the highest March temperatures occur towards the end of the month—the highest of all in 1965, when 22.1 °C was reached on 29 March—following 21.1 °C the previous day. More recently, 21 °C was also reached in the Marches of 1968, 2012 and 2017. In 2012, there was a remarkably early ten-day warm spell with temperatures reaching 15 °C every day; seven days reached 18 °C and three surpassed 20 °C. In March 2017, 21.2 °C was reached on 30th. Winter is never far away, however, as was shown during the first half of March 1845

*Oxford Weather and Climate since 1767*. Stephen Burt and Tim Burt, Oxford University Press (2019).
 DOI: 10.1093/oso/9780198834632.001.0001

when seven days failed to reach 0 °C—including the exceptionally cold 13 March 1845, when the maximum temperature was about −4.1 °C*. Much more recently, the first day of March in 2018 became the coldest March day in Oxford since that spell in 1845, with a remarkably low maximum temperature of just −1.9 °C and snowfall all day; this followed the coldest March night in over 50 years. For daytime temperatures to fail to reach 0 °C in March is very rare—this was only the fourth such March 'ice day' on record since screened temperature records began in 1853.

Despite increasing warmth by day, the mildest March nights are only slightly warmer than those in January or February, and a night minimum of 10 °C or more can only be expected in about one March in three. Only four March nights in over 200 years have ever remained above 12 °C—the mildest being 30 March 1998, when the minimum temperature was 12.8 °C, a fraction of a degree below the mean minimum for July and August.

March 1965 saw a very wide range of daytime temperatures, from the coldest March day of the twentieth century (−0.9 °C on 4th) to the highest on March records less than 4 weeks later (22.1 °C on 29th). Occasionally, temperatures go the other way, too—at the end of February 1891 the temperature reached 17.4 °C on 28th, but less than a fortnight later the maximum temperature reached only 0.8 °C in a snowy day on 10 March, followed by severe frost on 12 March with a minimum temperature of −9.7 °C. Mid-March 2018 was also very cold, with a maximum temperature of just 0.5 °C on 17th (a remarkable fall from 13.5 °C the previous day), followed by 0.7 °C on 18th. Sharp frosts are not uncommon in March, although −10 °C has been recorded on only three occasions since 1815, the most recent towards the end of the bitterly cold and snowy winter of 1947, when −10.8 °C was recorded on 7 March. However, we have to go all the way back to 1845 again for the coldest March night on record, when −12.0 °C was observed on 13 March of that year. More recent years rarely experience even −5 °C in March, although the second half of March 2013 was the coldest such period since 1883, with low temperatures and persistent if light snowfalls. Air frost occurred on 23 nights in March 1892, 22 in March 1883 and 21 nights in March 1924. Air frosts at the end of March 1975 led to orchard blossom damage locally—and there were to be more air frosts during early April: blossom had arrived early after one of the mildest winters in the Oxford record. In contrast, the Marches of 1819, 1903, 1912, 1959, 1981, 1994, 1997 and 2017 remained free of air frost.

Very large daily ranges in air temperature can result during clear weather in late March, when the ground is still cold and overnight frosts are likely, and yet the strength of the Sun is the same as that in late September. A daily range of 20 degC occurred on 11 March days between 1815 and 2018, the largest being 21.4 degC on 11 March 1929 (minimum temperature −2.2 °C to maximum temperature 19.2 °C), equalled on 29 March 1965

* The 'maximum temperature' readings of the Six's thermometer then in use are missing for this period, so the day's maximum temperature has been based upon the highest of the three daily observations made at 10 a.m., 2 p.m. and 10 p.m. On 13 March 1845, the 2 p.m. temperature was 23.0 °F, or −5.0 °C, and the maximum temperature for the day was estimated as −4.1 °C (see Appendix 1 for methodology).

**Table 9.1** *Highest and lowest maximum and minimum temperatures at the Radcliffe Observatory, Oxford, in March, 1815–2018*

| Rank | Warmest days | Coldest days | Mildest nights | Coldest nights |
|---|---|---|---|---|
| 1 | 22.1 °C, 29 Mar 1965 | *−4.1°C, 13 Mar 1845* | 12.8 °C, 30 Mar 1998 | *−12.0 °C, 13 Mar 1845* |
| 2 | 22.0 °C, 30 Mar 1929 | −1.9 °C, *16 Mar 1845* and 1 Mar 2018 | 12.3 °C, 22 Mar 2005 | −10.8 °C, 7 Mar 1947 |
| 3 | 21.7 °C 26 Mar 1944 and 29 Mar 1946 | −1.1°C, 6 Mar 1942 | 12.2 °C, *10 Mar 1826* and 30 Mar 2017 | −10.6 °C, 10 Mar 1858 |
| 4 | 21.3 °C 29 Mar 1929 and 28 Mar 2012 | −0.9 °C, 4 Mar 1965 | 11.7 °C, 17 Mar 2005 and 31 Mar 2017 | −9.7 °C, 12 Mar 1891 |
| 5 | 21.2 °C 28 Mar 1929, 29 Mar 1968 and 30 March 2017 | *−0.5 °C, 6 Mar 1845* | 11.3 °C, 11 Mar 1981 | *−8.9 °C, 25 Mar 1850* |
| ***Last 50 years to 2018*** | | | | |
| | 21.3 °C, 28 Mar 2012 | −1.9 °C, 1 Mar 2018 | 12.8 °C, 30 Mar 1998 | −6.1 °C, 1 Mar 2018 |

*Temperatures recorded by unscreened thermometers prior to 1853 are shown in italics; these are likely to be a little too cold by modern (screened thermometer) standards in winter and a little too warm in summer—see Appendix 1 for more details. In March the mean temperature correction for unscreened thermometry is +0.2 degC, although this will vary considerably on a day-to-day basis*

**Figure 9.1** *The quadrangle of the Oxford University Press building on 1 March 2018, the coldest March day in Oxford since 1845. Note the falling and lying snow and the frozen fountain; the maximum temperature on this date was only −1.9 °C (courtesy of Katherine Ward)*

(from 0.7 °C to 22.1 °C). One of the largest ranges in temperature in any month of the year occurred in March 1965, 28.9 degC between the minimum temperature of −6.8 °C on 3rd and the maximum temperature of 22.1 °C on 29th.

## Warm and cold months

Cold Marches were much more common in the nineteenth century; the most recent of the four coldest Marches on Oxford's records occurred in 1892 (mean temperature 3.0 °C), although March 1962 and 2013 both appear in the table of the five lowest mean temperatures for the month since 1814, with a mean temperature of just 3.3 °C (Table 9.2). March 2013 was also a very dull month with very persistent east and north-east winds (see Chapter 4 for more details on the synoptic context) and also had the unusual distinction of being the coldest of any month in the winter half-year 2012/13. There is little to choose between the March warmth of 1938, 1957 and 2017, although in 1938 and 1957 April was to be colder than March.

Surprisingly perhaps, it is not that uncommon for March to be colder than February or even January; March was colder than February 39 times in just over 200 years between 1815 and 2018 (roughly one year in five), most recently in 2013, and colder than January in 35 years in the same period, most recently in 2008, 2013, 2016 and 2018. March 1916 was 3.9 degC colder than January that year, in what must have been a dramatic seasonal reverse—particularly as the month was also the wettest March on record up to that date, with more than three times normal precipitation.

**Table 9.2** *March mean temperatures at Oxford's Radcliffe Observatory, 1814–2018*

| March mean temperature 7.3 °C (average 1981–2010) | | | | | |
|---|---|---|---|---|---|
| Mildest months | | | Coldest months | | |
| Mean temperature, °C | Departure from 1981–2010 normal, degC | Year | Mean temperature, °C | Departure from 1981–2010 normal, degC | Year |
| 9.9 | +2.6 | 1938 | 2.0 | −5.3 | 1837 |
| 9.7 | +2.4 | 1957, 2017 | 2.1 | −5.2 | 1845 |
| 9.4 | +2.1 | 1997 | 2.9 | −4.4 | 1865, 1883, 1892 |
| 8.9 | +1.6 | 1990 | 3.2 | −4.1 | 1814, 1867 |
| 8.8 | +1.5 | 2012 | 3.3 | −4.0 | 1917, 1962, 2013 |
| *Last 50 years to 2018* | | | | | |
| 9.7 | +2.4 | 2017 | 3.3 | −4.0 | 2013 |

## Precipitation

March is, on average, one of the driest months of the year—its monthly mean precipitation total is higher than February only because it has three more days. Snow can be expected on three or four days, often as snow showers, and thunder on about two years in five. Heavy falls of rain in a day are very rare, although rainfall can be persistent at times. March 1947 saw probably the most severe flooding of the Thames Valley in the twentieth century, a result of the sudden thaw of melting snow coinciding with still-frozen soil conditions. The highest *daily mean* flows ever recorded for the Thames at Day's Weir at Dorchester-on-Thames, just downstream of Oxford (where records began in 1938), were 349 cubic metres per second ($m^3/s$) on 17 and 19 March 1947, with 346 $m^3/s$ on 18 March (Figure 9.2). The eight highest daily flows ever recorded were all in the period 14–21 March 1947. The impact of flooding was made much worse by coming at the end of a prolonged cold and snowy winter during a time of post-war austerity.

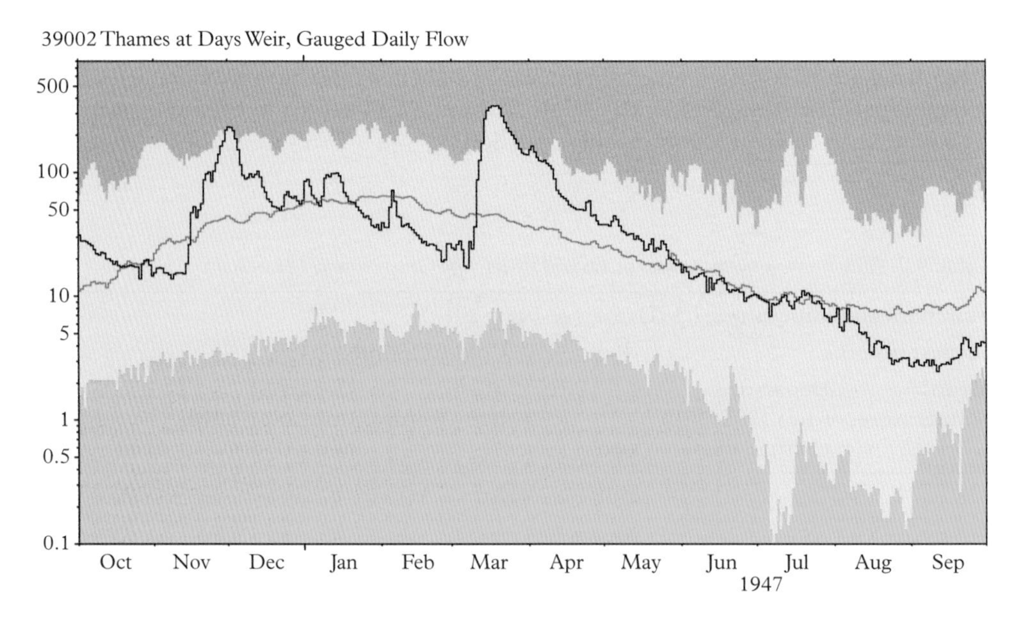


**Figure 9.2** *Thames flood hydrograph for Day's Weir, near Dorchester-on-Thames in Oxford, for water year 1946/47 (October 1946 to September 1947) showing the greatest recorded flood on the Thames in March 1947. River flow hydrographs show the daily mean flows (bold black line, units $m^3/s$), together with the long-term daily average flows in grey and maximum and minimum daily flows prior to 2017 (shown by the blue and pink shaded areas, respectively). Daily flows falling outside the maximum/minimum range are indicated where the bold trace enters or delineates the shaded areas. The pale grey line represents the period-of-record average daily flow (Data from National Rivers Flow Archive – NRFA courtesy Terry Marsh, CEH Wallingford)*

**Figure 9.3** *Flooding in Oxford during March 1947, the greatest recorded flood on the Thames. Top: An ex-wartime Morris army vehicle doing service as public transport in Oxford on 17 March 1947 (© Oxfordshire History Centre POX 012 2327). Bottom: Flooding on Abingdon Road, Oxford (© Oxfordshire History Centre POX 010 1481).*

March has one of the greatest differences between its driest and wettest months of any month in the year—the wettest March, 1947, receiving 95 times as much rainfall as the driest March, in 1929. March 1929 was a remarkably dry month—only three days in the month recorded 0.2 mm or more precipitation, the wettest day (25th) receiving a mere 0.6 mm. The wettest March was in 1947, with 133 mm of precipitation during the

**Table 9.3** *March precipitation at the Radcliffe Observatory, Oxford since 1767. Extremes of monthly totals 1767–2018, wettest days 1827–2018*

| March mean precipitation 47.7 mm (average 1981–2010) | | | | | | | |
|---|---|---|---|---|---|---|---|
| Wettest months | | | Driest months | | | Wettest days | |
| Total fall, mm | Per cent of normal | Year | Total fall, mm | Per cent of normal | Year | Daily fall, mm | Date |
| 132.9 | 279 | 1947 | 1.4 | 3 | 1929 | 42.1 | 15 Mar 2008 |
| 131.8 | 277 | 1916 | 1.9 | 4 | 1781 | 32.6 | 28 Mar 1916 |
| 129.7 | 272 | 1981 | 3.7 | 8 | 1931 | 31.2 | 14 Mar 1964 |
| 127.3 | 267 | 1862 | 4.1 | 9 | 1768 | 29.8 | 6 Mar 1982 |
| 110.3 | 231 | 1979 | 4.5 | 9 | 1961 | 28.5 | 8 Mar 2016 |
| *Last 50 years to 2018* | | | | | | | |
| 129.7 | 272 | 1981 | 9.2 | 19 | 1997 | 42.1 | 15 Mar 2008 |

calendar month: although the first three days of the month were dry, thereafter rain or snow fell on 31 of the following 35 days, accumulating 158 mm by the end of the first week in April. Daily rainfall totals exceeding 25 mm in March are uncommon, with only seven March days since 1827 attaining this threshold, or about once every 25–30 years: the wettest day was 15 March 2008, when 42.1 mm fell.

Many of the longest absolute and partial droughts have commenced in March (see Chapter 28). Notable amongst these is the 31-day absolute drought commencing on 17 March 1893, part of the longest partial drought on Oxford's records—75 days from 2 March to 15 May 1893, during which period of almost 11 weeks only 4.3 mm of rain fell. More recently, 28-day absolute droughts set in during March in 1997 and 2011; in the latter year, a 66-day partial drought commencing on 1 March saw less than 10 mm of rain fall until the first week of May. A prolonged spring drought, especially one that follows a dry winter, can lead to widespread and significant water supply issues later in the summer.

## Snowfall and lying snow

Snowfall can be expected on 3–4 days in an average March, although a morning snow cover is much less frequent, at around once in three years. Many falls of sleet or snow tend to be slight in March, often in short-lived wintry showers, and the awareness of the observers as to the conditions can be crucial in obtaining accurate statistics. In March 1979, snow or sleet fell on 11 days, and ten days with snow or sleet were recorded in March 1937 and March 1962. Snowfall in March has tended to become less prevalent in recent years, although both 2013 and 2018 have been notable exceptions. Lying snow

**Figure 9.4** *Deep snow in Beechcroft Road, Oxford, in early March 1947. The Oxfordshire History Centre photograph caption dates this to 'January/February 1947' but the more likely date is early March 1947—at the Radcliffe Observatory, the greatest depth of snow during winter 1946/47 was 25 cm on the morning of 6 March, and it seems more likely that this photograph was taken on or shortly after that date. See also Oxford's snow depth time series plots for 1947, 1963, 1978/79 and 2009/10 in Winter, Chapter 20. (Image © Oxfordshire History Centre POX 055 0817)*

is usually uncommon in March, although in March 1947 snow lay on the ground for 13 mornings (Figure 9.4), and nine mornings were snow covered in March 1970. In March 2018, snow lay on five mornings. The greatest depth of snow in March since 1959 occurred on 5 March 1970, when it lay 20 cm deep following a heavy fall the previous day; this snow cover persisted for five days. The only other March falls to reach 10 cm since 1959 occurred in 1965 and 1979. More recently, snow lay 7.5 cm deep on 3 March 2018 and 5 cm deep on 18 March 2018.

### Thunderstorms

Thunder is heard about two years in five in March, often associated with a cold front and of the 'one or two rumbles' variety. Thunder was heard on two days in March in 1933, 1977 and 1981.

## Sunshine

The duration of bright sunshine in March averages 111 hours, just over 3½ hours daily, although six days in the month can be expected to remain sunless.

**Table 9.4** *March sunshine duration at the Radcliffe Observatory, Oxford: monthly extremes 1880–2018, sunniest days 1921–2018*

**March mean sunshine duration 111.2 hours, 3.59 hours per day** (average 1981–2010)
*Possible daylength: 369 hours. Mean sunshine duration as percentage of possible: 30.1*

| Sunniest months | | | Dullest months | | | Sunniest days | |
|---|---|---|---|---|---|---|---|
| Duration, hours | Per cent of possible | Year | Duration, hours | Per cent of possible | Year | Duration, hours | Date |
| 198.7 | 53.8 | 1995 | 48.8 | 13.2 | 1984 | 12.4 | 27 Mar 1996 |
| 198.4 | 53.7 | 1893 | 57.8 | 15.6 | 1981 | 11.7 | 27 Mar 1933 and 31 Mar 1997 |
| 190.2 | 51.5 | 1929 | 60.7 | 16.4 | 2013 | 11.5 | 27 Mar 2012 |
| 187.6 | 50.8 | 1933 | 61.9 | 16.8 | 1916 | 11.4 | 26 Mar 1944 and 13 Mar 1995 |
| 185.3 | 50.2 | 1907 | 65.9 | 17.8 | 1964 | 11.3 | Various |
| *Last 50 years to 2018* | | | | | | | |
| 198.7 | 53.8 | 1995 | 48.8 | 13.2 | 1984 | 12.4 | 27 Mar 1996 |

The sunniest Marches can see four times as much sunshine recorded as the cloudiest months. The sunshine total in March 1995, 198.7 hours, not only just exceeded the March record that had stood for over 100 years, namely 198.4 hours in March 1893, but also the average monthly sunshine for both June and August. Remarkably, March in 1907 (185 hours of sunshine) was the sunniest month of that year, the only time this has happened on Oxford's long record. In contrast, March 1984 with only 48.8 hours sunshine was duller than the average December. Increasing daylength during the month means that the sunniest days are much more likely towards the end of the month, when the maximum possible duration of sunshine just exceeds 12 hours per day. A notably sunny spell occurred 24–27 March 2012, when a total of 42.9 hours sunshine was recorded in the four days, an average of 10.73 hours daily, each day with 10 hours or more; a similarly sunny spell during the three days 26–28 March 1933 logged 34.3 hours sunshine, an average of 11.43 hours daily.

## Wind

Oxford's mean monthly wind speed for March is 4.5 metres per second (m/s), or 8.7 knots. The windiest March on a record back to 1880 occurred in 1903, when the monthly mean wind speed was 6.3 m/s (12.3 knots). The calmest March was in 2012, when the monthly mean wind speed was just 1.8 m/s (3.5 knots).

## Oxford temperature, precipitation and sunshine in graphs—March

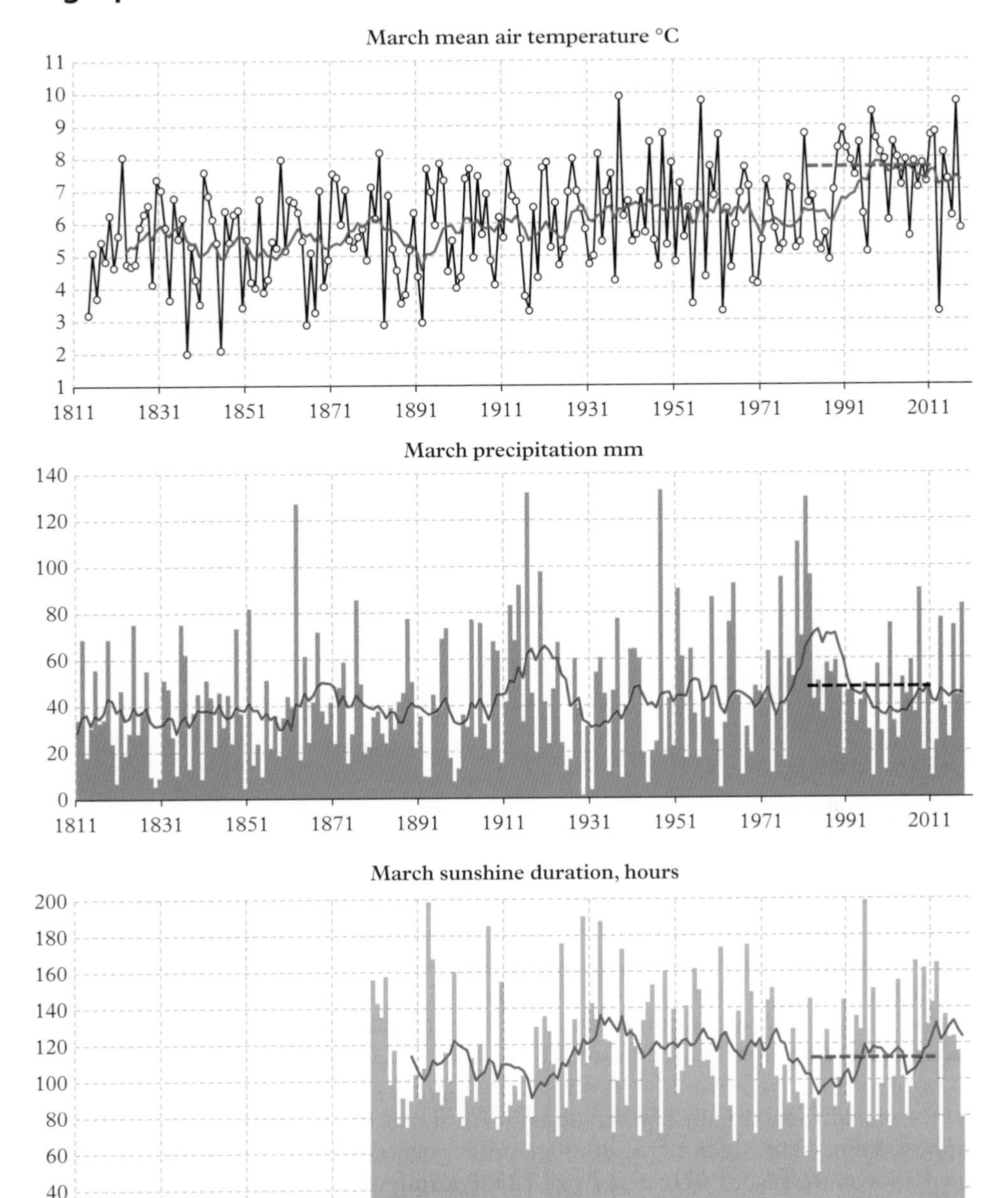


**Figure 9.5** *Monthly values of (from top) mean temperature (°C, since 1814), total precipitation (mm, since 1811) and sunshine duration (hours, since 1880) for March at Oxford. The 1981–2010 averages are indicated by the thick blue dashed line, while the ten-year running mean ending at the year shown is indicated by the red line*

# 10

# April

Despite a recent run of wet Aprils, April remains one of the driest months of the year, and prolonged spells of wet weather are rare. Snowfall and thunderstorms appear with about equal average frequency in April, but typically occur on only one day apiece. The frequency of northerly and north-easterly winds reaches a peak in April; winds off the North Sea at this time of year can bring long spells of cold, overcast weather. On the other hand, April's sunshine sees a noticeable increase on March levels, reaching an average in excess of 5 hours daily; while not quite up to midsummer levels, as a fraction of the hours of daylight at this time of year April is slightly ahead of June. Increasing daylength brings a continuing rise in temperature, daytime values increasing from a little under 12 °C at the beginning of the month to almost 15 °C by the close, but frosts can occasionally still be a hazard for tender outdoor plants even beyond the end of the month, and April can often feel disappointingly cool despite being the middle of spring.

## Temperature

Over the 1981–2010 period, the average date for the last air frost of the winter at the Radcliffe Observatory was 9 April, although with considerable year-to-year variation: in one year in ten, there was no air frost after 11 March, but also in about one year in ten an air frost occurred on or after 28 April. Cold nights are more likely in the first ten days of April, after which the risk of sharp frosts diminishes. April 1837 had the extraordinary number of 15 air frosts, equalled in April 1852, almost twice as many as in a normal January in present times, but in recent years the frostiest April has been 2013 with six. The lowest air temperature yet recorded in Oxford in April was −5.6 °C, on 11 April 1817.

Occasionally, maximum temperatures in April can fail to rise above 3 °C; this has happened nine times since 1815, although not for over 50 years as we write. The coldest April days on the record were 2–4 April 1830; during these three days the highest temperature was just 2.7 °C, while the overnight minimum on the 4th fell to −4.2 °C. Remarkably sharp frosts for the time of year were recorded on 19 and 20 April 1852, the minimum temperatures being −4.3 °C and −4.7 °C, respectively; but to date no April morning has been colder than −3 °C since 1978. Perhaps even more noteworthy is winter cold at the

*Oxford Weather and Climate since 1767*. Stephen Burt and Tim Burt, Oxford University Press (2019).
 DOI: 10.1093/oso/9780198834632.001.0001

**Table 10.1** *Highest and lowest maximum and minimum temperatures at the Radcliffe Observatory in April, 1815–2018*

| Rank | Hottest days | Coldest days | Mildest nights | Coldest nights |
|---|---|---|---|---|
| 1 | 27.6 °C, 23 April 2011 | *0.8 °C, 2 April 1830* | 13.8 °C, 22 April 2018 | *−5.6 °C, 11 April 1817* |
| 2 | 26.9 °C, 19 April 2018 | *1.0 °C, 3 April 1839* | *13.3 °C, 26 April 1821* | *−5.0 °C, 5 April 1830 and 6 April 1851* |
| 3 | 26.8 °C, 20 April 1893 | 1.3 °C, 25 April 1908 and 14 April 1966 | 13.2 °C, 11 April 1981 | −4.9 °C, 15 April 1892 |
| 4 | 26.3 °C, 16 April 2003 | *2.0 °C, 3 April 1830* | 12.9 °C, 26 April 1975 | *−4.7 °C, 20 April 1852* |
| 5 | 26.1°C, 21 April 1893 | 2.1 °C, 12 April 1879 | 12.8 °C, 15 April 1945 and 16 April 1945 | −4.6 °C, 12 April 1868 and 2 April 1922 |
| *Last 50 years* | | | | |
| | 27.6 °C, 23 April 2011 | 4.5 °C, 18 April 1983 and 4 April 2013 | 13.8 °C, 22 April 2018 | −3.2 °C, 11 April 1978 |

*Temperatures recorded by unscreened thermometers prior to 1853 are shown in italics; these are likely to be a little too cold by modern (screened thermometer) standards in winter and a little too warm in summer—see Appendix 1 for more details. In April the mean temperature correction for unscreened thermometry is +0.1 degC, although this will vary considerably on a day-to-day basis*

very end of April: on 25 April 1908 the maximum temperature reached only 1.3 °C during a notable snowstorm (see below, and Chronology, Chapter 25). The latest date in the first half of the calendar year on which the temperature has failed to exceed 5 °C was 27 April 1919, when the maximum temperature reached only 4.6 °C. More recently, the maximum temperature on 29 April 2018 reached just 7.9 °C, almost 20 degrees lower than the near-record 26.9 °C reached just 10 days previously (see below).

The earliest date in the year since daily records commenced in 1814 when 25 °C has been surpassed is 3 April—on this remarkably early date, the temperature reached exactly 25.0 °C in 1946: the only other '25' in the first half of April has been 25.2 °C on 14 April 1869. In April 2003, 26.3 °C was reached on 16 April. The first of two outstanding April heatwaves on the record was in 1893 (the eight days 19–26 April 1893 had a mean daily maximum temperature of 24.0 °C, the lowest in this spell being 21.6 °C). In April 2011, 13 days reached or surpassed 20 °C (the April mean daily maximum is 13.6 °C), with two spells of 20 °C or more of 5 days (6th–10th) and 7 days (19th–25th). During this latter spell, every day reached at least 21.5 °C, the mean daily maximum for the week was 23.9 °C, and the hottest day was 23 April with 27.6 °C—the highest April temperature on record.

In April 2018, a short hot spell just after mid-month saw five consecutive days exceed 21 °C, the warmest day being 19th which reached 26.9 °C.

Clear, sunny days and clear, cool nights in dry weather in April can produce some of the largest daily ranges in temperature. A daily temperature range of 20 degC has been recorded on five days in April, the greatest being 21.8 degC on 24 April 1855 (minimum −2.1 °C, maximum 19.7 °C) and 21.3 degC on 9 April 1909. The largest reduction in maximum temperature from one day to the next for any month on the entire record took place in April. On 4 April 1946, a maximum temperature of 24.4 °C was followed next day by a maximum temperature of just 8.9 °C—a day-to-day reduction of 15.5 degC. As the weather gradually warms during April, at least one night with a minimum temperature above 10 °C can be expected in most Aprils, although 13 °C has been exceeded only three times, most recently in 2018 (Table 10.1).

## Warm and cold months

Table 10.2 shows that April 1837 was by far the coldest April on Oxford's records (following on from the coldest March on record), and April 2011 by far the warmest. The latter month's mean daily maximum temperature was 19.5 °C, less than half a degree below a normal June. It is not particularly uncommon for April to be colder than March—this has happened eight times in the last century or so. In 1938, April was 2.1 degC colder than March, another reminder that April can sometimes feel disappointing even with summer on the horizon.

**Table 10.2** *April mean temperatures at Oxford's Radcliffe Observatory, 1814–2018*

| **April mean temperature 9.3 °C** (average 1981–2010) | | | | | |
|---|---|---|---|---|---|
| **Warmest months** | | | **Coldest months** | | |
| **Mean temperature, °C** | **Departure from 1981–2010 normal degC** | **Year** | **Mean temperature, °C** | **Departure from 1981–2010 normal degC** | **Year** |
| 13.3 | +4.0 | 2011 | 4.6 | −4.7 | 1837 |
| 11.7 | +2.4 | 2007 | 5.9 | −3.4 | 1879 |
| 11.6 | +2.3 | 1943 | 6.0 | −3.3 | 1839 |
| 11.3 | +2.0 | 1865 | 6.1 | −3.2 | 1860, 1917, 1922 |
| 11.0 | +1.7 | 1869 and 1944 | 6.3 | −3.0 | 1817 |
| ***Last 50 years to 2018*** | | | | | |
| 13.3 | +4.0 | 2011 | 6.7 | −2.6 | 1986 |

## Precipitation

Until recent years, April was a reliably dry month; the three driest months since 1767 have all been Aprils, and until 1998, only three Aprils had ever recorded as much as 100 mm in the month. Since and including 1998, three Aprils have exceeded that value, including April 2012, which with 143.0 mm attained the dubious honour of becoming the wettest April in Oxford's records. Not to be outdone, we have also seen the two very dry Aprils in the same period—2007 (1.8 mm, the fifth-driest April, and to that date the equal warmest) and 2011 (just 0.5 mm, the equal-driest month of any on Oxford's 250 years of rainfall records, and also the warmest April on record). Thus, in recent years April's rainfall has become much more variable than it was in most of the twentieth century. There were only two days with measurable rainfall in the Aprils of 1893, 1912 and 2011, but several Aprils have had 23 days with rainfall, most recently 1983, 1986 and 2000.

Only seven April days since 1827 have received more than 25 mm in 24 hours. One of the coldest and wettest spells ever recorded in late spring occurred in April 1908 during Oxford's heaviest snowfall of the twentieth century, when a remarkable 43 cm of snow accumulated within a day (Figure 10.1). Further details of this remarkable late-season snowstorm are given in the Chronology in Chapter 25.

## Snowfall and lying snow

Snow or sleet can be expected to fall on one day in a typical April, but many falls are slight or in showery conditions; snowfall in April is tending to become less prevalent, and

**Table 10.3** *April precipitation at the Radcliffe Observatory, Oxford since 1767. Extremes of monthly totals 1767–2018, wettest days 1827–2018*

| April mean precipitation 48.9 mm (average 1981–2010) | | | | | | | |
|---|---|---|---|---|---|---|---|
| Wettest months | | | Driest months | | | Wettest days | |
| Total fall, mm | Per cent of normal | Year | Total fall, mm | Per cent of normal | Year | Daily fall, mm | Date |
| 143.0 | 292 | 2012 | 0.5 | 1 | 1817, 2011 | 46.7 | 25 April 1908 |
| 137.2 | 280 | 2000 | 0.6 | 1 | 1912 | 38.0 | 11 April 2000 |
| 109.4 | 223 | 1908 | 1.5 | 3 | 1893 | 35.7 | 29 April 1913 |
| 107.7 | 220 | 1998 | 1.6 | 3 | 1984 | 30.3 | 16 April 1918 |
| 107.2 | 219 | 1818 | 1.8 | 4 | 2007 | 26.5 | 29 April 1991 |
| *Last 50 years to 2018* | | | | | | | |
| 143.0 | 292 | 2012 | 0.5 | 1 | 2011 | 38.0 | 11 April 2000 |

**Figure 10.1a** *The aftermath of the extraordinary snowfall of 24–25 April 1908 in Oxfordshire: Oxford, St Giles and High Street (photographer unknown)*

**Figure 10.1b** *The extraordinary snowfall of 24–25 April 1908 in Oxfordshire. Top: The snowstorm in Abingdon, afternoon 25 April 1908. The view is outside the outside the County Hall Museum, looking south-west; the snow depth was reported as 69 cm in Abingdon after the storm. The strength of the wind is apparent from the photograph; at the Radcliffe Observatory in Oxford, the hourly mean wind was north-westerly 13 m/s (25 knots) at 1700. (Photographer Warland Andrew, who had a studio in Abingdon from the 1890s to the 1920s; courtesy of Abingdon County Hall Museum). Bottom: Horses and carts clearing snowfall, dated 26 April 1908 (© Oxfordshire History Centre POX 056 0947)*

many recent Aprils have remained snow-free. In contrast, in April 1936 snow or sleet fell on 7 days, and 6 days in April 1975. Lying snow is rare in April, and only eight April mornings within the past 90 years have been white. The great April snowstorm of 1908 referred to above produced the greatest snow depth measured at Oxford during the whole of the twentieth century, viz. 43 cm at 10.30 p.m. on 25 April (Figure 10.1 and Chapter 25); the deepest snowfall within the last 50 years has been barely half as deep, namely 27 cm on 6 January 2010. The heaviest April snowfall since 1959 occurred on 6 April 2008, 8 cm depth at the morning observation from a succession of heavy early morning snow showers: the latest date since 1959 with a snow cover was 26 April 1981, when 4 cm was recorded at the morning observation.

### Thunderstorms

Thunder can be expected on about one day in most Aprils, often in showery conditions. In April 1948, thunder was heard on as many as five days, and four each in the Aprils of 1934 and 1983.

## Sunshine

The duration of bright sunshine in April averages 161 hours, or just over 5 hours 20 minutes per day: typically, just three days will remain sunless, whereas about one April in eight has sunshine every day. April 1893 still retains the title of the sunniest April on Oxford's records, with 253 hours of sunshine, an average of almost 8½ hours daily, the only April to exceed 60 per cent of the possible sunshine duration—and following an exceptionally sunny March, too. April 1893 remains 9 hours ahead of its nearest contender, 1914, and 15 hours (30 minutes per day) ahead of the sunniest April within the last 50 years, namely 1984 with 237 hours. These sunshine totals are more typical of a sunny summer month, and all surpass 50 per cent of the possible monthly sunshine duration. Occasionally, April turns out to be the sunniest month of the year—this has happened nine times since 1880, most recently in 2007 and 2011. April 1912, still the third-sunniest April on record, was not only the sunniest month of that year, it recorded well over twice the duration of sunshine in the following August—August 1912 registering a dismal 98 hours, the dullest on record and only slightly more than the sunniest January. Most Aprils include at least one day with 12 hours or more sunshine: in April 1984 there were eight such days, and six in April 1990. In April 1893, the six days commencing 21st averaged 11.57 hours sunshine daily, while in April 1912 the six days 20th–25th averaged 12.72 hours sunshine daily (including four consecutive days with more than 13 hours). In April 1984, the ten-day spell commencing 21st averaged 11.37 hours of sunshine: only one day in this period had less than 10 hours sunshine. In April 2011, the 13 days commencing 18 April received a daily average of 9.01 hours sunshine.

**Table 10.4** *April sunshine duration at the Radcliffe Observatory, Oxford: monthly extremes 1880–2018, sunniest days 1921–2018*

**April mean sunshine duration 160.9 hours, 5.36 hours per day** (average 1981–2010)
*Possible daylength: 418 hours. Mean sunshine duration as percentage of possible: 38.5*

| Sunniest months | | | Dullest months | | | Sunniest days | |
|---|---|---|---|---|---|---|---|
| Duration, hours | Per cent of possible | Year | Duration, hours | Per cent of possible | Year | Duration, hours | Date |
| 252.8 | 60.5 | 1893 | 81.5 | 19.5 | 1966 | 13.9 | 30 April 1990 |
| 243.8 | 58.4 | 1914 | 84.2 | 20.2 | 1889 | 13.8 | 30 April 1930 and 27 April 1977 |
| 239.1 | 57.2 | 1912 | 86.9 | 20.8 | 1961 | 13.7 | 27 April 1954, 29 April 1957 and 23 April 2004 |
| 236.8 | 56.7 | 1984 | 87.1 | 20.9 | 1920 | 13.6 | 30 April 1966 |
| 234.4 | 56.1 | 1990 | 91.1 | 21.8 | 1905 | 13.5 | 30 April 1997 and 30 April 1999 |
| *Last 50 years to 2018* | | | | | | | |
| 236.8 | 56.7 | 1984 | 105.8 | 25.3 | 1998 | 13.9 | 30 April 1990 |

## Wind

Oxford's mean monthly wind speed for April is 4.0 metres per second (m/s), or 7.7 knots. The windiest April on a record back to 1880 occurred in 1947, when the monthly mean wind speed was 5.7 m/s (11.0 knots). The calmest April was in 1996, when the monthly mean wind speed was just 2.2 m/s (4.3 knots).

## Oxford temperature, precipitation and sunshine in graphs—April

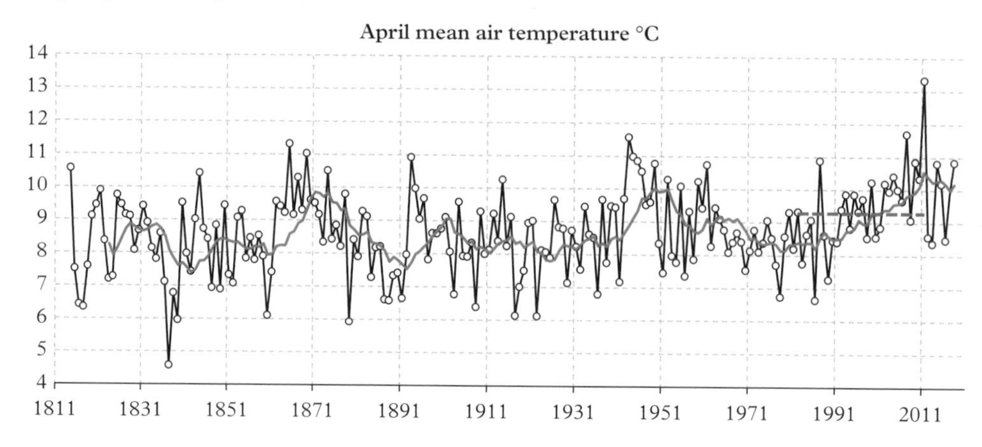


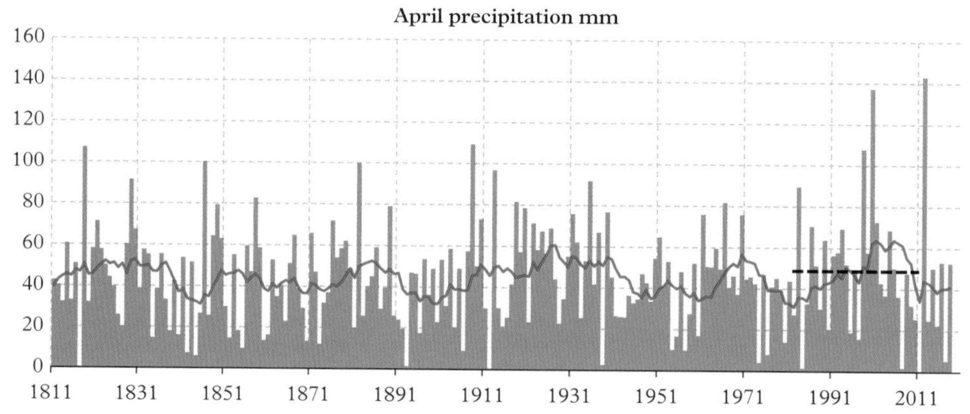


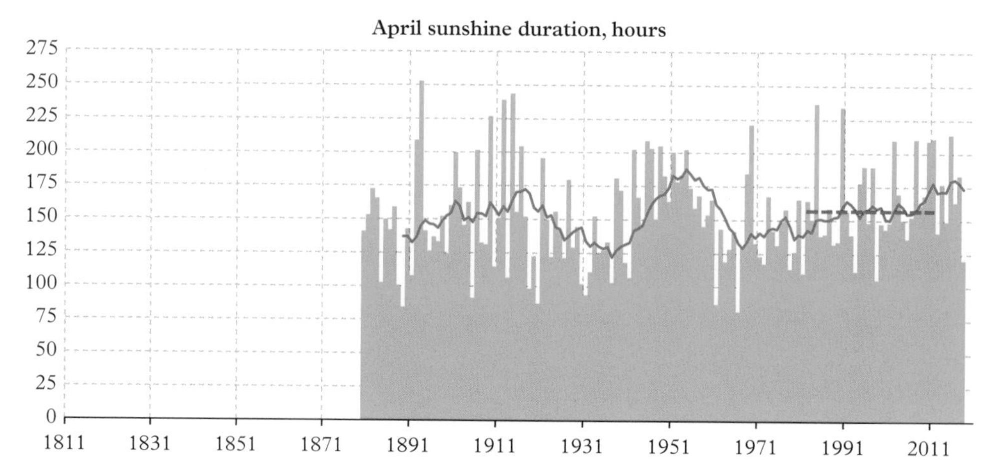


**Figure 10.2** *Monthly values of (from top) mean temperature (°C, since 1814), total precipitation (mm, since 1811) and sunshine duration (hours, since 1880) for April at Oxford. The 1981–2010 averages are indicated by the thick blue dashed line, while the ten-year running mean ending at the year shown is indicated by the red line*

# 11

# May

May is another transition month—it can bring cool spring days and damaging night frosts, early summer heatwaves, and sometimes both. Snowfall this late in the year is very rare, but thunderstorms are not uncommon and can be heavy. Sunshine can be expected on all but two or three days in May, averaging something over 6 hours per day; the last day of May is, on average, the sunniest day of the year in Oxford.

## Temperature

There is a fairly steady rise in temperatures, both by day and by night, in Oxford throughout May: average daytime temperatures increase from around 16 °C at the start of the month to 18 or 19 °C during the closing days, while over the same period average minimum temperatures increase from 6 °C to 9–10 °C. The mean temperature for the month as a whole over the period 1981–2010 was 12.5 °C.

Although air frosts in May were not uncommon in the nineteenth century, today they are distinctly unusual—less than one May in ten will record an air frost, although ground frosts can be expected on a few nights, when they can still cause damage to tender plants. The coldest May night on record was 7 May 1831, when the temperature fell to −2.6 °C, although the lowest in the last 50 years or so has been −0.5 °C, on 9 May 1980. The latest air frost on record occurred on 27 May 1914, when the air minimum was −0.1 °C and the grass minimum −3.1 °C. May 1944 saw the greatest range in temperature of any May, with late frosts early in the month being followed by an intense heatwave during the final week, including Oxford's hottest May day on record—see Figure 11.1.

May nights can be very warm, and occasionally remain above 15 °C throughout; in contrast, the maximum temperature can fail to reach 10 °C on occasion—the coldest May day on record being 1 May 1866, when the temperature reached only 5.4 °C. Even as late as the final week, the maximum temperature has remained as low as 8.3 °C (on 25 May 1891), while in late May 1984 the temperature remained below 10 °C for two consecutive days (26 May 1984 maximum temperature 9.4 °C, 27 May 9.8 °C).

*Oxford Weather and Climate since 1767*. Stephen Burt and Tim Burt, Oxford University Press (2019).
 DOI: 10.1093/oso/9780198834632.001.0001

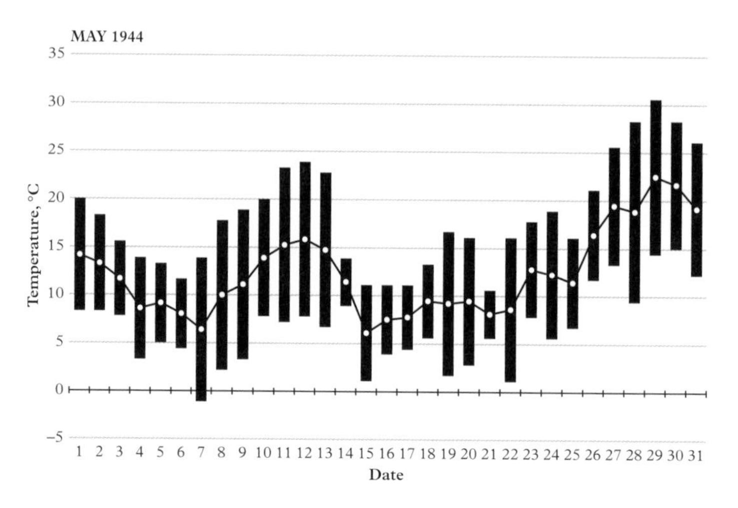


**Figure 11.1** *Daily maximum and minimum temperatures (top and bottom of daily columns) and daily mean temperature (central circle and connecting line) at the Radcliffe Observatory, Oxford, in May 1944, the month with the greatest range in air temperatures in Oxford's records. After a warm start to the month, colder conditions quickly became established, and there was a late air frost on 7 May before temperatures recovered, reaching almost 24 °C on 12 May. Another sharp reversal then took place, with near-frost once more on 15 and 22 May. Thereafter, temperatures climbed quickly, to 30.6 °C on 29 May (the hottest May day yet recorded), followed by one of the warmest May nights on record with a minimum of 15.0 °C before another very hot day on 30 May, at 28.3 °C*

The earliest date in the year on which 30 °C (summer heatwave threshold) has been reached in Oxford is 27 May—this occurred way back in 1841, although only four May days have ever reached this level and, as we write, not for over 70 years, the most recent occasion being in 1947. Since then the hottest May days have been 29.4 °C on 25 May 1953 and, more recently, 28.9 °C on 24 May 2010. The two most notable heatwaves in late May were 27–31 May 1944 (5 days each over 25 °C, reaching 30.6 °C on 29 May, the hottest day of 1944) and 28 May to 3 June 1947 (7 days each over 25 °C, 30.6 °C on 30 May and 32.2 °C on 2 June). In recent years, 25–28 May 2012 saw four consecutive days above 25 °C, reaching 26.5 °C on 27 May: in 2017 27.0 °C was reached on 26 May and in 2018, 27.2 °C as early as 7 May. May has seen the hottest day of the year in ten years since 1815, most recently in 1978. The earliest dates in the year to record the hottest day of the year were 12 May in 1907 (maximum temperature 25.8 °C) and 14 May in 1965 (27.3 °C)—both were very cool summers. Strong sunshine and dry soil conditions can occasionally lead to very large daily ranges in temperature during May, the greatest being 20.4 degC on 3 May 1868 (5.4 to 25.8 °C) and 23 May 1919 (4.7 to 25.1 °C), and 20.1 degC on 11 May 1909 (2.3 to 22.4 °C). The greatest range in daily temperatures on any May day within the last 30 years has been 17.8 degC on 1 May 2007 (4.9 to 22.7 °C).

**Table 11.1** *Highest and lowest maximum and minimum temperatures in Oxford during May, 1815–2018*

| Rank | Hottest days | Coldest days | Warmest nights | Coldest nights |
|---|---|---|---|---|
| 1 | 30.6 °C, 29 May 1944 and 30 May 1947 | 5.4 °C, 1 May 1866 | 16.7 °C, 9 May 1945 | *−2.6°C, 7 May 1831* |
| 2 | 30.1 °C, 31 May 1947 | *6.1 °C, 14 and 15 May 1824* | 15.6 °C, *28 May 1847*, 27 May 1855, 22 May 1922 and 28 May 2001 | −2.4 °C, 5 May 1855 and 4 May 1877 |
| 3 | *30.0 °C, 27 May 1841* | 6.2 °C, 18 May 1891 | 15.4 °C, 18 May 1868 and 25 May 1922 | −2.3 °C, *2 May 1852* and 3 May 1877 |
| 4 | 29.8 °C, 23 May 1922 | 6.7 °C, 1 May 1856 and 3 May 1892 | 15.3 °C, 31 May 1895 | −2.2 °C, *13 May 1838* and 8 May 1861 |
| 5 | 29.6 °C, 29 May 1947 | 7.1 °C, *14 May 1839* and 13 May 1915 | 15.2 °C, 5 May 1862 | *−1.9 °C, 16 May 1839* |
| *Last 50 years* | | | | |
| | 28.9 °C, 24 May 2010 | 8.3 °C, 17 May 1996 | 15.6 °C, 28 May 2001 | −0.5 °C, 9 May 1980 |

*Temperatures recorded by unscreened thermometers prior to 1853 are shown in italics; these are likely to be a little too cold by modern (screened thermometer) standards in winter and a little too warm in summer—see Appendix 1 for more details. In May the mean temperature correction for unscreened thermometry is −0.1 degC, although this will vary considerably on a day-to-day basis*

## Warm and cold months

Monthly mean temperatures in May have varied between 9.0 °C in 1879 and 14.4 °C in 1992 (Table 11.2). There were very warm Mays in 1833, 1841 and 1848, in contrast to the generally much colder conditions in early spring in the early nineteenth century, followed by several very chilly Mays between 1879 and 1902. The three years 1917 to 1919 each saw warm Mays, the very warm May of 1917 following the very cold March and April of that year. The cool May of 1975 was followed by the warm and very dry conditions that characterised many months from the summer of 1975 to the summer of 1976.

Monthly temperature ranges in May have varied between 16.1 degC in 1972 and 16.5 degC in 1983, to 31.7 degC in 1944—the latter the greatest range in air temperature yet observed in any month of the year (Figure 11.1).

**Table 11.2** *May mean temperatures at Oxford's Radcliffe Observatory, 1814–2018*

| May mean temperature 12.5°C (average 1981–2010) | | | | | |
|---|---|---|---|---|---|
| Warmest months | | | Coldest months | | |
| Mean temperature, °C | Departure from 1981–2010 normal degC | Year | Mean temperature, °C | Departure from 1981–2010 normal degC | Year |
| 14.4 | +1.9 | 1848, 1992 | 9.0 | −3.5 | 1817 |
| 14.3 | +1.8 | 1952, 2017, 2018 | 9.1 | −3.4 | 1879 |
| 14.2 | +1.7 | 1833 | 9.5 | −3.0 | 1885 |
| 14.1 | +1.6 | 1841, 1868, 1947 and 1989 | 9.6 | −2.9 | 1941 |
| 14.0 | +1.5 | 1922, 1964 | 9.7 | −2.8 | 1837, 1902 |
| *Last 50 years to 2018* | | | | | |
| 14.4 | +1.9 | 1992 | 9.8 | −2.7 | 1996 |

## Precipitation

May's 1981–2010 monthly mean rainfall in Oxford is 57 mm, the wettest month of all the summer half-year. May's rainfall is less variable than April's, with a lower range in long-term extremes, as can be seen from a comparison of Tables 11.3 and 10.3 for the two months.

Both the wettest and driest Mays on Oxford's records were recorded well over 200 years ago—May 1773, with 155 mm of rainfall, being 10 per cent wetter than any other May since, while May 1795, with just 1.5 mm, received less than half of the rainfall of the next driest May. Within the last 100 years, May 1932 (139 mm) remains the wettest May, closely followed by 2007 with 135 mm. May 1932 was also very cloudy (the dullest May on record to that time), and rain fell on 24 days, the highest for any May aside from the 25 rain days in May 1967. Since 1767, 16 Mays have exceeded 100 mm of precipitation (including three consecutive 2006–2008), while only 10 have received less than 10 mm (including two consecutive years in 1895 and 1896 and in 1990 and 1991). There were only three days with rain in the Mays of 1896 and 1989; earlier records suggest there may have been only two days with rain in the Mays of 1829 and 1833, and three in 1836 and 1844.

Heavy falls of rain at this time of year can often be short-lived but heavy, and sometimes associated with thunderstorms (see Chronology in Chapter 25 for an account of the storm of 31 May 1682 in Oxford). Daily falls of 25 mm or more occur in about one year in ten in May, the wettest May days being 27 May 2007 (40.7 mm—during

**Table 11.3** *May precipitation at the Radcliffe Observatory, Oxford since 1767. Extremes of monthly totals 1767–2018, wettest days 1827–2018*

| **May mean precipitation 57.2 mm** (average 1981–2010) | | | | | | | |
|---|---|---|---|---|---|---|---|
| **Wettest months** | | | **Driest months** | | | **Wettest days** | |
| Total fall, mm | Per cent of normal | Year | Total fall, mm | Per cent of normal | Year | Daily fall, mm | Date |
| 154.9 | 271 | 1773 | 1.5 | 3 | 1795 | 40.7 | 27 May 2007 |
| 139.1 | 243 | 1932 | 3.9 | 7 | 1829 | 39.5 | 14 May 1985 |
| 135.2 | 237 | 2007 | 4.6 | 8 | 1895 | 36.9 | 12 May 1886 |
| 126.7 | 222 | 1878 | 6.1 | 11 | 1896 | 36.5 | 13 May 1915 |
| 126.0 | 220 | 1948 | 6.7 | 12 | 1844 | 36.4 | 31 May 2018 |
| *Last 50 years to 2018* | | | | | | | |
| 135.2 | 237 | 2007 | 8.6 | 15 | 1991 | 40.7 | 27 May 2007 |

a Bank Holiday weekend as chance would have it) and 14 May 1985 (39.5 mm). On 24 May 1989, 27.9 mm fell during a severe thunderstorm at the Radcliffe Observatory; 68.5 mm fell in the same storm at Sandford Sewage Treatment Works, just south of Oxford. More recently, 36.4 mm fell on 31 May 2018, the fifth-wettest May day on record—just days before the onset of the long dry spell that characterised summer 2018.

## Snowfall and lying snow

Snow or sleet is very rare in May, typically less than once in 10 years, although May 1979 saw four days when sleet or snow was recorded, and May 1935 two such days, 14th and 16th. The latest date on which snow has been reliably observed to fall within the last 100 years was 17 May 1955 (when sleet falling during the evening turned to heavy snowfall for an hour from 21h), although there is one earlier mention of snow in July 1888 (see July, Chapter 13). Since at least 1926, no lying snow has been recorded at the morning observation between late April and early November.

## Thunderstorms

Thunder can be expected on about two days in a normal May, although around one May in five remains free of thunder. May 1945 was the most thundery of any month at Oxford since records are available (1926), with nine days recording thunder; May 1969 was close behind with eight, and May 1983 with seven days with thunder heard.

## Sunshine

May's average sunshine duration is 193 hours, a little over 6 hours daily, with typically only two or three days during the month remaining sunless. In about one year in five, May is the sunniest month of the year.

It is remarkable, but no more than coincidence, that two of the five dullest Mays occurred in the three years 1981–1983, while less than a decade later three of the five sunniest Mays also occurred within four years, this time between 1989 and 1992. May 1989 remains one of the sunniest months ever recorded at Oxford, notching up 300.8 hours (a daily average of 9.7 hours sunshine), a remarkable 61.9 per cent of the possible duration—one of only a few months to exceed 60 per cent. In contrast, May 1981 averaged little over 3 hours sunshine per day, just 20 per cent of the possible duration.

Notably sunny spells during May since 1921 have included the 12 days commencing 18 May 1977 (149 hours in total, average 12.41 hours daily), 10 days commencing 10 May 1980 (131 hours sunshine, daily average 13.13 hours, and the only May spell to be included in the all-time sunniest 10-day spells—see Chapter 30), 17 days commencing 15 May 1989 (172 hours sunshine, daily average 10.12 hours), 10 days 28 April to 7 May 1990 (129 hours, 12.89 hours per day) and 15 days commencing 13 May 1992

**Table 11.4** *May sunshine duration at the Radcliffe Observatory, Oxford: monthly extremes 1880–2018, sunniest days 1921–2018*

**May mean sunshine duration 192.9 hours, 6.22 hours per day (average 1981–2010)**
*Possible daylength: 486 hours. Mean sunshine duration as percentage of possible: 39.7*

| Sunniest months | | | Dullest months | | | Sunniest days | |
|---|---|---|---|---|---|---|---|
| Duration, hours | Per cent of possible | Year | Duration, hours | Per cent of possible | Year | Duration, hours | Date |
| 300.8 | 61.9 | 1989 | 98.3 | 20.2 | 1981 | 15.5 | 30 and 31 May 1985 |
| 293.6 | 60.4 | 1909 | 112.4 | 23.1 | 1932 | 15.4 | 30 May 1997 |
| 285.0 | 58.6 | 1990 | 119.8 | 24.7 | 1889 | 15.2 | 30 May 1955, 31 May 1997 |
| 268.8 | 55.3 | 1922 | 126.0 | 25.9 | 1906 | 15.1 | 30 May 1966, 24 May 1997 |
| 262.7 | 54.1 | 1992 | 126.8 | 26.1 | 1983 | 15.0 | 26 May 1957, 29 May 1982 |
| *Last 50 years to 2018* | | | | | | | |
| 300.8 | 61.9 | 1989 | 98.3 | 20.2 | 1981 | 15.5 | 30 and 31 May 1985 |

(171 hours in total, average 11.41 hours daily). In 2018, the early May Bank Holiday weekend saw almost unbroken sunshine, each day receiving 13 hours or more; the three-day total (5–6–7 May) amounted to 40.1 hours.

Over the 30 years 1981–2010, based on daily averages, the sunniest day of the year in Oxford was 31 May, with 8.35 hours sunshine, one of only two days in the year to average more than 8 hours (the other being 11 July, with 8.24 hours).

Sunless days are infrequent in May—less than one day in ten will not receive any bright sunshine, although May 1994 recorded seven such days. Since 1921 there have been only three spells of four consecutive sunless days in May—in 1947, 1960 and 2004—and none lasting as long as five days.

## Wind

Oxford's mean monthly wind speed for May is 3.7 metres per second (m/s), or 7.2 knots. The windiest May on a record back to 1880 occurred in 1977, when the monthly mean wind speed was 5.1 m/s (10.0 knots). The calmest May was in 1978, when the monthly mean wind speed was just 2.7 m/s (5.3 knots).

## Oxford temperature, precipitation and sunshine in graphs—May

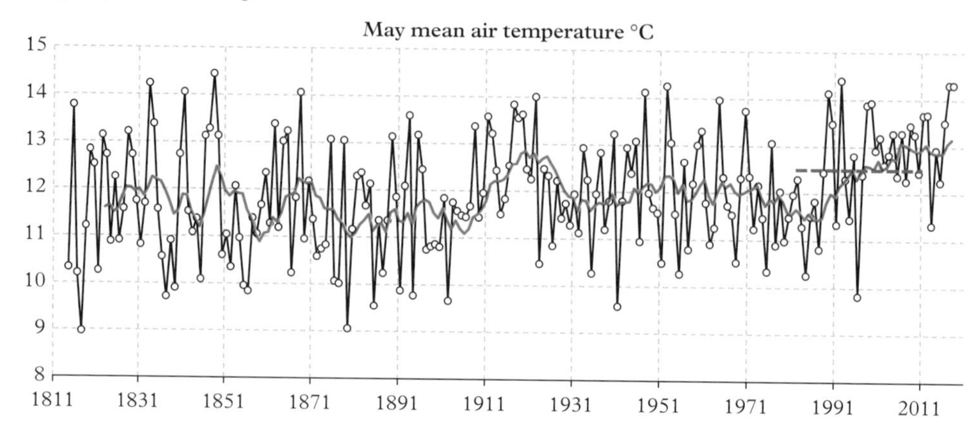


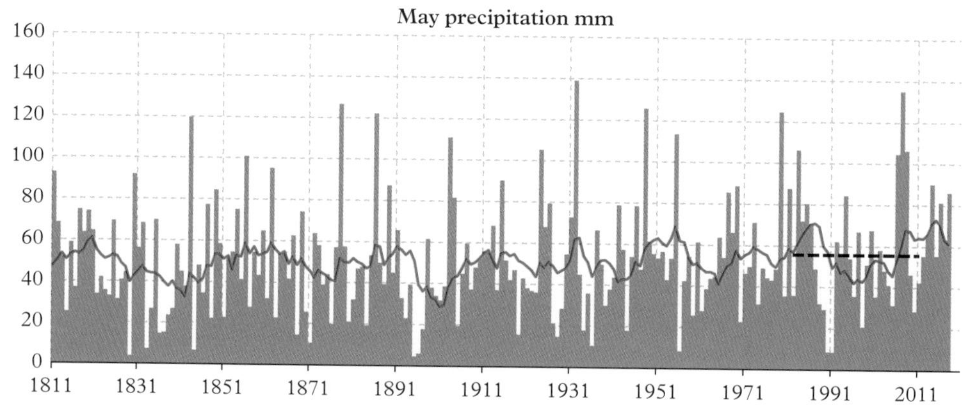


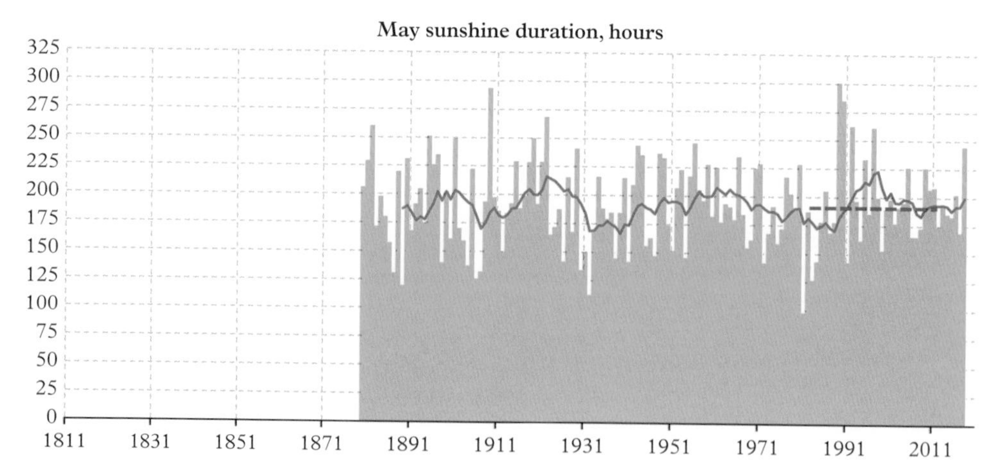


**Figure 11.2** *Monthly values of (from top) mean temperature (°C, since 1814), total precipitation (mm, since 1811) and sunshine duration (hours, since 1880) for May at Oxford. The 1981–2010 averages are indicated by the thick blue dashed line, while the ten-year running mean ending at the year shown is indicated by the red line*

# 12

# June

By normal meteorological convention, June is the first of the three summer months. Temperatures continue to rise quite steadily throughout the month. The first week or two can be rather cool: the short nights can still be on the chilly side, and ground frosts are not unknown, although average daytime temperatures reach 19 °C or so. The second half of the month is usually warmer, the days averaging 22 °C by the end of the month: 30 °C has been reached seven times in June within the last 50 years or so. The average temperature for the month is 15.6 °C. Rain can be expected on about 12 days in the month, but long spells of rainfall are uncommon and mid-June has the lowest frequency of rain throughout the year, although the occasional heavy thunderstorm can deposit as much in a few hours as falls in the average month. Over the period 1981–2010, June's average daily duration of sunshine was 6 hours and 22 minutes. In about one year in three, June is the sunniest month of the year, although only once in the last 50 years has it been the warmest.

## Temperature

The beginning of June can be cool—often cooler than the end of May, and ground frosts are not unknown. No air frost has yet been recorded at the Radcliffe Observatory in June, although some mornings have been close to that level—the coldest June nights on record to date all being below 2 °C. In early June 1991, minimum temperatures below 2 °C were recorded three times in the first week, with three ground frosts in four days (−3.6 °C on 2 June): it is perfectly possible that a slight air frost occurred in rural areas outside the city centre. Cold June days are most often associated with prolonged rainfall on a cyclonic northerly or north-easterly flow. The lowest maximum temperatures yet recorded in June were 9.8 °C on 4 June 1909 and 10.0 °C on 3 June 1953 (following a maximum of just 12.2 °C the previous day—the Coronation Day of Queen Elizabeth II in London). More recently, the highest temperature reached on 1 June 1989—immediately following the sunniest May on record—was just 11.2 °C.

June can also see great heat, and temperatures of 30 °C or more can be expected on at least one day every 5–10 years. The most remarkable heatwave on Oxford's long records, by a long way, was that in late June–early July 1976. The first day to surpass 25 °C

*Oxford Weather and Climate since 1767*. Stephen Burt and Tim Burt, Oxford University Press (2019).
 DOI: 10.1093/oso/9780198834632.001.0001

**Table 12.1** *Highest and lowest maximum and minimum temperatures at the Radcliffe Observatory, Oxford, in June, 1815–2018*

| Rank | Hottest days | Coldest days | Warmest nights | Coldest nights |
|---|---|---|---|---|
| 1 | 34.3 °C, 27 June 1976 | 9.8 °C, 4 June 1909 | 19.6 °C, 27 June 2011 | 1.4 °C, 15 June 1892 |
| 2 | 34.0 °C, 26 June 1976 | 10.0 °C, 3 June 1953 | 18.9 °C, 22 June 1941 | 1.5 °C, 1 June 1893 |
| 3 | 33.1 °C, 28 June 1976 | 10.3 °C, 6 June 1905 | 18.8 °C, 19 June 2000 | 1.6 °C, 1 June 1962 |
| 4 | 32.5 °C, 21 June 2017 | 10.4 °C, 14 June 1903 | 18.7 °C, 28 June 1976 | 1.8 °C, 2 June 1991 |
| 5 | 32.2 °C, *25 June 1820*, 15 June 1858 and 30 June 1995 | 10.6 °C, 2 June 1941 | 18.3 °C, 20 June 2005 and 28 June 2012 | 1.9 °C, 4 and 5 June 1991 |
| ***Last 50 years to 2018*** | | | | |
| | 34.3 °C, 27 June 1976 | 11.2 °C, 1 June 1989 | 19.6 °C, 27 June 2011 | 1.8 °C, 2 June 1991 |

was 22 June, and 30 °C was reached three days later. Thereafter, 14 consecutive days reached or surpassed 30 °C; the 'Top three' hottest June days on record (Table 12.1) were all attained during this exceptional hot spell. Prior to 1976, the longest run of consecutive '30+' temperatures had been just five days, in August 1876 and July 1948 (see 'Prolonged heatwaves' in Chapter 30). Since 1976, a six-day spell has occurred (in July-August 1995), but 1976's record of 14 consecutive days has still not been even remotely challenged in the years since. A hot spell in June 2017 saw three consecutive days reach 30 °C, and a fourth at 29.8 °C, with the hottest day (21 June) attaining 32.5 °C, a value exceeded in June only in 1976.

Minimum temperatures of 15 °C or above occur on average about once each June. The warmest nights in the June records were 19.6 °C on 27 June 2011 and 18.9 °C on 22 June 1941. During the intense June 1976 heatwave, night-time temperatures remained above 16 °C for four consecutive nights, the warmest night being 18.7 °C on 28 June.

A remarkable turnaround from winter to summer took place in a week in June 1975. Following a sharp ground frost (−3.0 °C) and a near air frost (minimum 0.8 °C) on 31 May, there were two further ground frosts in the next four nights, with snow showers as far south as the London area on 2 June. Over the next few days, the unseasonably cold northerly flow was replaced by a warm southerly airstream with plentiful sunshine—on 6 June, the temperature reached 24.8 °C and on 7 June 26.4 °C, both days seeing more than 13 hours of sunshine: a range in temperature of 25.6 degC in 8 days, truly winter to summer in a week! June 1975 went on to become the sunniest June on record (Table 12.4).

Some of the largest daily ranges in temperature occur during clear weather in June, although only two days have surpassed 20 degC difference between night and day—namely 30 June 1995 (21.0 degC range, from 11.2 to 32.2 °C) and 6 June 1855 (20.3 degC, from 7.4 to 27.7 °C).

## Warm and cold months

The Junes of 1846 and 1976 tie for the warmest June on record in Oxford (Table 12.2), although maximum temperatures for June 1846 have had to be estimated based upon 2 p.m. temperatures (see Appendix 1) and there is thus some uncertainty about the exact value for that year. The very hot weather towards the end of June 1976 continued into the first half of July. After a dry start to the year, the continuing hot, dry conditions led to widespread forest and woodland fires in Oxfordshire and Berkshire as elsewhere in southern and south-eastern England. The 69-day period 20 June–27 August was persistently dry (just 14 mm of rain in over 2 months, most of which fell in the two days 13 and 15 July), sunny (631 hours of bright sunshine, a daily average of over 9 hours) and hot (average maximum temperature 26.2 °C). The summer also saw the culmination and exacerbation of a prolonged drought which had started in spring 1975 [52]—see also Chapter 4 for an analysis of the synoptic background to the summer of 1976.

June 1916 was the coolest June on record: the highest temperature reached during the month was a dismal 19.0 °C (it had reached 26.5 °C the preceding May). June 1916 also holds the unenviable record of being the only June on the entire record to have been cooler than the previous May. The exceptional June of 1976 stood out in a decade of miserably cool Junes—those of 1971, 1972 and 1977 all feature in the 'Top 6' coolest Junes on record; June 1977 was the coolest June since 1916. Curiously, the very warm

**Table 12.2** *June mean temperatures at Oxford's Radcliffe Observatory, 1814–2018*

**June mean temperature 15.6°C** (average 1981–2010)

| Warmest months | | | Coldest months | | |
|---|---|---|---|---|---|
| Mean temperature, °C | Departure from 1981–2010 normal degC | Year | Mean temperature, °C | Departure from 1981–2010 normal degC | Year |
| 18.5 | +2.9 | 1846, 1976 | 12.0 | −3.6 | 1916 |
| 18.3 | +2.7 | 1822 | 12.1 | −3.5 | 1821 |
| 17.8 | +2.2 | 1818, 2017 | 12.5 | −3.1 | 1909 |
| 17.7 | +2.1 | 1826, 1858 | 12.6 | −3.0 | 1977 |
| 17.4 | +1.8 | 2006 | 12.7 | −2.9 | 1972 |
| *Last 50 years to 2018* | | | | | |
| 18.5 | +2.9 | 1976 | 12.6 | −3.0 | 1977 |

June of 1822 (still the second-hottest on record) was also 'sandwiched' between much cooler Junes, June 1821 still the second-coolest June on Oxford's records and June 1823 little better, some 4–6 degC colder than the intervening June.

Monthly temperature ranges in June have varied between a mere 14.5 degC in 1954 and 1972 to 28.4 degC in 1893 and 27.2 degC in 2005.

## Precipitation

The monthly mean rainfall for June in Oxford is 49 mm. Since 1767, only two Junes have received more than three times normal rainfall—the wettest being June 1852, with 170.4 mm: only six days remained completely dry, while three received more than 20 mm (Table 12.3). The second-wettest June was in 2012, with 151.7 mm falling on 20 days. The very wet Junes of 1903 and 1971 both featured long cold spells of unbroken rainfall. On 14 June 1903, the Radcliffe Observatory's records show that rain fell for 48 consecutive hours commencing at 4 a.m.: the day's total of 51.1 mm from 0900 GMT was the wettest day on the Observatory's almost 90 years of rainfall records up to that date. The total during the civil day (midnight to midnight GMT) was 62.6 mm (Figure 12.1)—and this remains one of the coldest June days on the record (Table 12.1). This fall also featured heavily in the second-wettest 14-day spell in any month in Oxford's records—in

**Table 12.3** *June precipitation at the Radcliffe Observatory, Oxford since 1767. Extremes of monthly totals 1767–2018, wettest days 1827–2018*

| **June mean precipitation 49.2 mm** (average 1981–2010) | | | | | | | |
|---|---|---|---|---|---|---|---|
| **Wettest months** | | | **Driest months** | | | **Wettest days** | |
| **Total fall, mm** | **Per cent of normal** | **Year** | **Total fall, mm** | **Per cent of normal** | **Year** | **Daily fall, mm** | **Date** |
| 170.4 | 351 | 1852 | 1.7 | 4 | 1925 | 81.3 | 22 June 1960 |
| 151.7 | 313 | 2012 | 2.5 | 5 | 2018 | 67.3 | 27 June 1973 |
| 141.8 | 292 | 1903 | 4.1 | 8 | 1942 | 51.1 | 14 June 1903 |
| 135.8 | 280 | 1971 | 5.3 | 11 | 1818, 2006 | 41.8 | 29 June 1917* |
| 133.9 | 276 | 1795 | 5.5 | 11 | 1962 | 39.4 | 8 June 1955 |
| *Last 50 years to 2018* | | | | | | | |
| 151.7 | 313 | 2012 | 2.5 | 5 | 2018 | 67.3 | 27 June 1973 |

* This is the total for the civil day (midnight to midnight) on 29 June 1917. According to the Observatory notes, heavy rain fell during the early hours of the morning; when reckoned to a standard 0900–0900 GMT rain day, this rainfall would be credited to 28 June rather than 29th.

**Figure 12.1** *Flooding in Oxford, June 1903; the boat in the picture appears to be navigating a flooded roadway between hedges and flooded fields on either side (© Oxfordshire History Centre POX 007 1342)*

all, 141.8 mm fell between 8 and 19 June 1903, more than twice June's normal rainfall in a fortnight (Chapter 28, Table 28.3).

In complete contrast, in June 1925 rain fell on only two days, 19th (0.2 mm) and 26th (1.5 mm) for a monthly rainfall total of just 1.7 mm, the driest of any summer month on record: it was also very sunny, with 267 hours of bright sunshine, one of the sunniest Junes on record. June was something of an exception that year, however, because it was the only dry month from April to October: indeed, July 1925 was very wet, with over twice normal rainfall. Very recently, June 2018 was also very dry with just 2.5 mm recorded, falling on just four days that month.

Falls of 25 mm or more in 24 hours occur in about one year in five in June, and typically about one in three of such falls occur on a day when thunder is heard. The wettest June day on Oxford's record was 22 June 1960 when 81.3 mm fell in a series of overnight thunderstorms, following a maximum temperature of 26 °C the previous afternoon. The second-wettest June day, 27 June 1973 when 67.3 mm fell, was also as a result of heavy thundery rain.

A tremendous rain and hail storm accompanied by a violent thunderstorm affected parts of Oxfordshire on 9 June 1910. Although only 15 mm fell during the day at the Radcliffe Observatory, at Wheatley 110 mm of rain and hail fell during the storm in just 58 minutes, a record which still stands as the highest reliably recorded hourly fall of rain in the British Isles [92]. The total rainfall for the day at Wheatley amounted to 139 mm, although the exact amount remains subject to some doubt because of the unorthodox pattern of raingauge in use. More details are given in the Chronology, Chapter 25.

**Figure 12.2** *Rain didn't stop play in June 2012, the wettest since 1852, although this Oxford cricket match almost had to be called off due to a sodden wicket: it was so wet, the match had to be played on the outfield, with 'net' stumps. The batsman is Peter Carroll (Ian Curtis)*

### Snowfall

Since 1926, snow or sleet has never been recorded in Oxford between mid-May and mid-October.

### Thunderstorms

Most Junes will see one or two days with thunderstorms. In June 1982, thunder was heard on seven days, and six days were noted in June 1933 and June 1936.

## Sunshine

The average duration of sunshine in June is 191 hours, or an average of 6 hours and 22 minutes daily—the monthly total slightly below May's mean owing to the one fewer day in the month, although the percentage of possible sunshine is also slightly lower than in May at 38.3 per cent (May 39.7 per cent). The monthly figure hides a surprising amount of variation in the day-to-day averages, but over the period 1981–2010 15 June was, on daily averages, the third-sunniest day of the year, with an average of 7 hours and 58 minutes of bright sunshine.

June 1975 was the sunniest June on Oxford's records, one of only five months ever to exceed 300 hours, and the only June to surpass 60 per cent of possible sunshine (Table 12.4). In all, 13 days in June 1975 recorded 12 hours or more sunshine duration—although even this impressive total was exceeded in the Junes of 1996 (15) and 1957 (14). The following year, June 1976, was not quite as sunny (261 hours)—but still marked the start of three months of prolonged sunshine in that remarkable summer. More recently, June 2016 received barely 100 hours of sunshine—below even some winter months, and the dullest June on record. Until 2016, the dullest June on record was in 1909, with 109 hours—and this followed a May sunshine total that was not bettered for 80 years: June 1909 saw little more than one-third of the sunshine recorded in the previous month.

**Table 12.4** *June sunshine duration at the Radcliffe Observatory, Oxford: monthly extremes 1880–2018, sunniest days 1921–2018*

**June mean sunshine duration 191.0 hours, 6.37 hours per day** (average 1981–2010)
*Possible daylength: 498 hours. Mean sunshine duration as percentage of possible: 38.3*

| Sunniest months | | | Dullest months | | | Sunniest days | |
|---|---|---|---|---|---|---|---|
| Duration, hours | Per cent of possible | Year | Duration, hours | Per cent of possible | Year | Duration, hours | Date |
| 301.0 | 60.4 | 1975 | 100.4 | 20.2 | 2016 | 15.9 | 7 June 1921 |
| 297.1 | 59.7 | 1957 | 109.4 | 22.0 | 1909 | 15.7 | 17 June 1959 |
| 290.7 | 58.4 | 1996 | 114.0 | 22.9 | 1998 | 15.6 | 29 June 1921, 23 June 1973 |
| 280.3 | 56.3 | 1962 | 121.1 | 24.3 | 1990 | 15.5 | 17 June 1932, |
| 276.9 | 55.6 | 1969 | 122.8 | 24.7 | 1981 | | 5 June 1939, 5 June 1940, 15 June 1957, 28 June 1957, 8 June 1962, 23 June 1995, 30 June 1995, 8 June 2005 |
| ***Last 50 years to 2018*** | | | | | | | |
| 301.0 | 60.4 | 1975 | 100.4 | 20.2 | 2016 | 15.6 | 23 June 1973 |

## Sunny days and spells

Daylength reaches its maximum at the summer solstice on 21 June: in Oxford, midsummer day is 16 hours and 41 minutes long (sunrise at 0346 GMT, sunset 2027 GMT). Allowing for the limited response of the sunshine recorder to very low-angle sunshine, the longest possible duration of sunshine in a day is slightly under 16 hours, and then only on the clearest of midsummer days. The sunniest day on record (since 1921) was 7 June 1921, with 15.9 hours, fittingly the sunniest day in a very sunny summer.

Most years will record at least one day in summer with 15 hours or more of bright sunshine, although only 16 days have recorded 15.5 hours or more (two in May and one in July in addition to the thirteen listed in Table 12.4 for June) since daily records are available (1921). It is therefore not surprising that many of the sunniest spells of between 7 and 60 days listed in Chapter 30 (Table 30.4) are made up of runs of sunny June days. The sunniest week on Oxford's records since 1921 was 12–18 June 1996, when 102.3 hours of sunshine were recorded—a daily average of 14.61 hours. June–July 1976 dominates the 10- and 14-day spells, losing its title to summer 2018 over the 21- to 60-day periods. The 30 days commencing 18 June 2018 saw a remarkable 332.5 hours of sunshine, a daily average of a little over 11 hours.

Despite the long hours of daylight, sunless days are not uncommon in June—typically just one per year, but seven days remained sunless in June 1977, the year following the extraordinarily sunny summer of 1976. Indeed, a remarkable seven days in nine commencing 13 June 1977 remained sunless—only 16 June (9.8 hours) and 19 June (0.1 hours) otherwise relieved the gloom of one of the coolest, wettest and dullest midsummer weeks on record in the Midlands generally [93].

## Wind

Oxford's mean monthly wind speed for June is 3.4 metres per second (m/s), or 6.7 knots. The windiest June on a record back to 1880 occurred in 1907, when the monthly mean wind speed was 5.1 m/s (9.9 knots). The calmest June was in 1996, when the monthly mean wind speed was just 2.0 m/s (3.9 knots).

## Oxford temperature, precipitation and sunshine in graphs—June

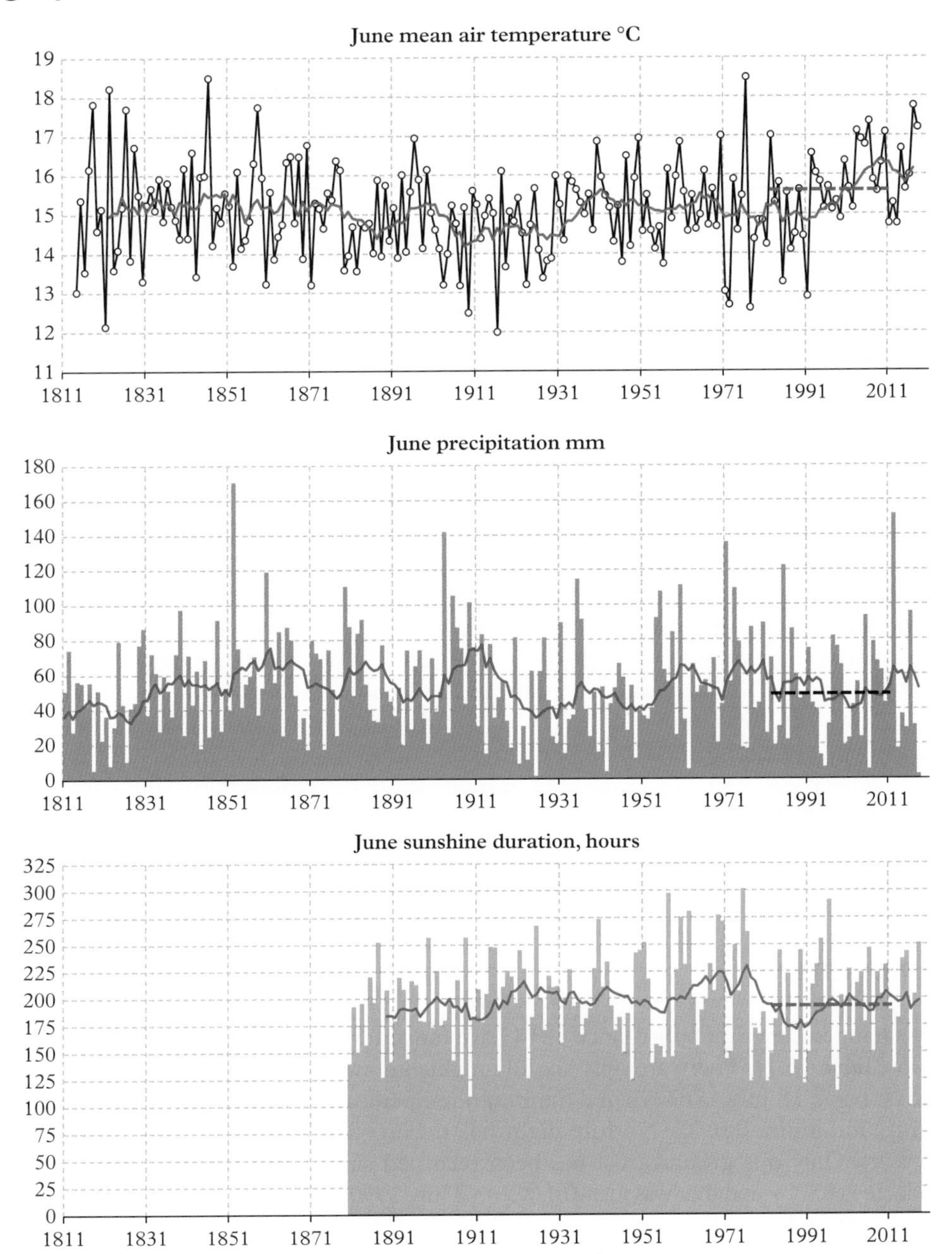


**Figure 12.3** *Monthly values of (from top) mean temperature (°C, since 1814), total precipitation (mm, since 1811) and sunshine duration (hours, since 1880) for June at Oxford. The 1981–2010 averages are indicated by the thick blue dashed line, while the ten-year running mean ending at the year shown is indicated by the red line*

# 13

# July

July is, on average, both the sunniest and warmest month of the year in Oxford, and also the most thundery. The average temperature is 17.9 °C (mean daily minimum 13.0 °C, maximum 22.7 °C, over the period 1981–2010), and mean daily sunshine 6 hours and 41 minutes, about 20 minutes per day on average more than June or August. Over the last century or so, July has been the warmest month of the year in a little more than one year in two, August accounting for the majority of the remaining years. Average temperatures tend to rise as July progresses, with daily mean maximum temperatures around 21–22 °C at the start of the month rising to 23–24 °C towards the end—indeed, the last week or so of July is on average the warmest of the year in Oxford. Minimum temperatures increase more slowly, from about 12 °C in the first week to almost 14 °C at the end of the month (see daily averages plotted in Chapter 6). Although there is, of course, much year-to-year variation, the warmest day of the year in Oxford is, on average, 29 July—mean minimum 13.9 °C, mean maximum 23.8 °C, and a resultant mean temperature of 18.9 °C. No air frosts have ever been recorded during July in Oxford, and (since 1857 at least) only a single ground frost. July's monthly mean rainfall is 49 mm—but the mean *daily* rainfall is very slightly lower than in June. Wet days are infrequent, but occasional downpours—often accompanied by thunder—can deposit more than a month's average rainfall in just a few hours. Only once in over 200 years has there been a record of snowfall in July.

## Temperature

July's temperatures in Oxford since 1814 have ranged from 2.4 °C in 1863 to 34.8 °C in 2006 (Table 13.1). There are only six July nights known to have fallen below 5 °C, the coldest being 18 July 1863 when a minimum temperature of 2.4 °C was recorded, with a grass minimum 0.0 °C. No July night has been colder than 5.6 °C within the last 50 years. Only one ground frost has been recorded since grass minimum records are available (1857)—and that was just -0.1 °C, on 3 July 1990, on the morning when the screen minimum was 6.3 °C. Other low grass minima in July include 1.0 °C on 4 July 1918, 0.6 °C on 7 July 1962 and 1.1 °C on 11 July 1993. The coolest July days are 2–3 degC colder than the mildest January days. On 5 July 1920, the temperature reached only 11.8 °C, the coldest July day on the 'screened' temperature record since 1853 (compare with the

*Oxford Weather and Climate since 1767*. Stephen Burt and Tim Burt, Oxford University Press (2019).
 DOI: 10.1093/oso/9780198834632.001.0001

mildest January day, 15.9 °C), although on earlier records maximum temperatures of 11.1 °C were recorded on 2 July 1821 and 12.1 °C on 1 July 1848. On 11 July 1888, a day when *The Times* reported early morning sleet observed on the fringes of London, the day was only slightly less cold, at 12.4 °C, with a chilly north-westerly wind—while snow fell in Oxford (see below: 'Snow…in July?'). Within the last 50 years, however, no July day has been cooler than 14.3 °C.

Since 1815, temperatures in July have reached or exceeded 33 °C nine times—twice each on consecutive days in 1923, 1976 and 2006, once in July 1825 and July 1943, and again most recently in July 2015. The hottest July day on Oxford's records, at 34.8 °C, was 19 July 2006 (Table 13.1). In July 2015, although the temperature reached 33.5 °C on the first day of the month, the weather turned colder later and a minimum temperature of 5.6 °C (grass minimum 1.5 °C) was recorded on the final day of that month, the coldest July night in 60 years.

By far the longest hot spell in which the maximum temperature reached 30 °C or greater each day occurred in the 14 days commencing 25 June 1976, when the hottest July day was 33.4 °C on 3 July. The next-longest spells by this measure were just six days long in 1995 (29 July to 3 August 1995; hottest day 31.9 °C on 31 July) and five days in 1948 (26–30 July; 32.2 °C on 29 July) and in 1983 (12–16 July; 31.9 °C on 14 July).

**Table 13.1** *Highest and lowest maximum and minimum temperatures in July at the Radcliffe Observatory Oxford, 1815–2018*

| Rank | Hottest days | Coldest days | Warmest nights | Coldest nights |
|---|---|---|---|---|
| 1 | 34.8 °C, 19 July 2006 | *11.1 °C, 2 July 1821* | 21.2 °C, 20 July 2016 | 2.4 °C, 18 July 1863 |
| 2 | 33.9 °C, *19 July 1825* and 12 July 1923 | 11.8 °C, 5 July 1920 | 20.3 °C, 4 July 1976 | *4.3 °C, 17 July 1838* |
| 3 | 33.7 °C, 13 July 1923 | *12.1 °C, 1 July 1848* | 20.0 °C, *30 July 1827* and 20 July 2006 | 4.4 °C, *4 July 1821, 13 July 1826, 2 July 1837* and 4 July 1965 |
| 4 | 33.5 °C, 1 July 2015 | 12.4 °C, 11 July 1888 | 19.9 °C, 1 July 1968 | *5.0 °C, 31 July 1816 and 13 July 1840* |
| 5 | 33.4 °C, 3 July 1976 | *12.8 °C, 3 July 1821* | 19.7 °C, 29 July 2001 | 5.1 °C, 11 July 1888 |
| ***Last 50 years to 2018*** | | | | |
| | 34.8 °C, 19 July 2006 | 14.3 °C, 5 July 1978 | 21.2 °C, 20 July 2016 | 5.6 °C, 31 July 2015 |

*Temperatures recorded by unscreened thermometers prior to 1853 are shown in italics; these are likely to be a little too cold by modern (screened thermometer) standards in winter and a little too warm in summer—see Appendix 1 for more details. In July the mean temperature correction for unscreened thermometry is −0.3 degC, although this will vary considerably on a day-to-day basis*

Only four Julys have seen six or more days in the month attain 30 °C or more—1921 (6 days), 1976 (8 days), 1983 (6 days) and 2006 (7 days). In July 2018, four days exceeded 30 °C, although none were consecutive.

Lowering the threshold to 25 °C, the longest hot spells in July by this measure were 17 consecutive days in 1976 (22 June to 8 July, hottest day of the summer 34.3 °C on 27 June), 15 consecutive days in 2018 (25 June to 9 July, 30.7 °C on 8 July) and 14 consecutive days in 2006 (15–28 July, hottest day 34.8 °C on 19 July—see also Chronology, Chapter 25). In July 2018, 26 days during the month reached or exceeded 25 °C, a record for any month; between 25 June and 27 July (33 days), only two days remained below 25 °C, the lowest maximum temperature being 23.6 °C on 10 July. In July 1983, 22 days during the month reached or exceeded 25 °C, while the Julys of 1868, 1911 and 2013 all attained this threshold on 19 days.

Since 1815, only seven calendar months have reached at least 20 °C on every day of the month in Oxford, all but one of them Julys—namely 1827, 1874, 1983, 2006, 2014 and 2018 (the only other month being August 1826). In July 2006, the lowest daily maximum temperature was 21.5 °C (on 10th), the highest 'lowest daily maximum' on record for any month.

Hot, airless and often humid nights can be a problem in summer heatwaves, with nighttime temperatures remaining above 15 °C at least once in most Julys—although only four July nights have not fallen below 20 °C. The warmest night of any on Oxford's long record occurred in July—on 20 July 2016, when a minimum temperature of 21.2 °C followed a day when the temperature had reached 32.3 °C (Figure 13.1). Spells of minimum temperatures above 15 °C each night for nine consecutive nights have occurred only four

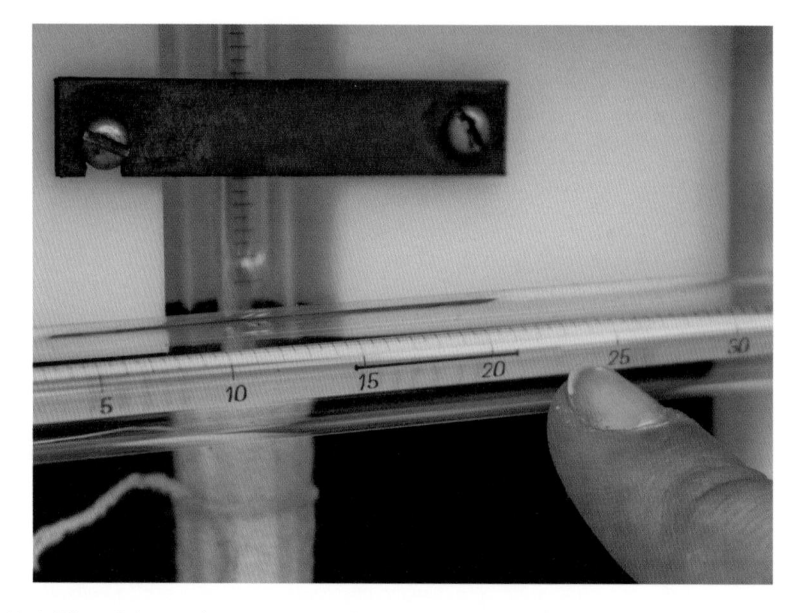


**Figure 13.1** *The minimum thermometer on the morning of 20 July 2016—at 21.2 °C this was the warmest night yet recorded in Oxford (Ian Curtis)*

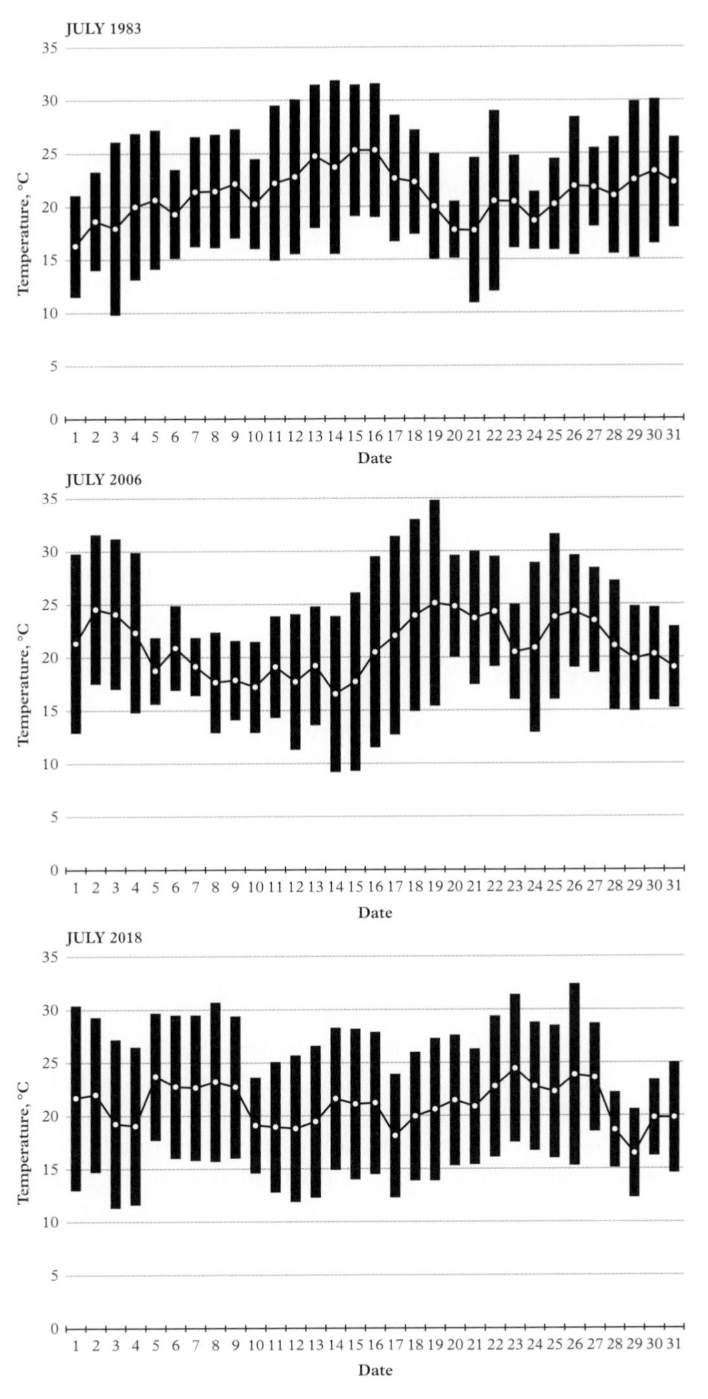


**Figure 13.2** *Daily maximum and minimum temperatures, °C (top and bottom of daily columns) and daily mean temperature (central circle and connecting line) at the Radcliffe Observatory, Oxford, in July 1983 (top), July 2006 (middle) and July 2018 (bottom), the three hottest months on Oxford's records*

times at Oxford since 1815, three times in July—two of those in July 1983 and once in July 2018 (Figure 13.2). The nine days 12–20 July 1983 averaged a night minimum temperature no lower than 16.8 °C (minimum temperature as high as 19.1 °C on 15 July), while only three days later the second nine-day spell from 23–31 July averaged 16.3 °C (minimum temperature 18.1 °C on 27 July). Not far behind was the nine-day spell from 20–28 July 2018, with a mean minimum temperature of 16.2 °C. Not surprisingly, July 1983 still holds the record for the highest mean minimum temperature for any July, at 15.4 °C; only one month (August 1997, mean minimum 15.7 °C) has exceeded it since. July 2018 was also only the second month, and the only July to date, to remain above 11 °C throughout, with a minimum temperature of 11.3 °C (on 3rd); the only previous month to remain above 11 °C throughout was August 1997 (minimum 11.7 °C).

It is evident from Table 13.1 that almost all of the hottest nights in July have occurred within the last 50 years or so—although the minimum temperatures of 20.0 °C on 30 July 1827 and 19.3 °C on 26 July 1830 show that very warm summer nights are not entirely a recent phenomenon.

Diurnal temperature ranges in July occasionally exceed 18 degC, but a 20 degC range has been exceeded only three times—on 19 July 1900 (20.3 degC, from 11.7 to 32.0 °C), on 2 July 1976 (21.1 degC, from 11.9 to 33.0 °C) and the largest yet recorded, 21.2 degC on 15 July 1990 (9.3 to 30.5 °C).

## Warm and cold months

The three Julys of 1983, 2006 and 2018 (Figure 13.2 and Table 13.2) remain the hottest months yet recorded in Oxford, three of only five months in over 200 years to have recorded a mean temperature of 20 °C or greater (the other two months were the Augusts of 1995 and 1997), and the only three to have exceeded 21 °C (Chapter 29, Table 29.4A). Since 1815, only 16 months have recorded a mean monthly maximum temperature of 25.0 °C or greater (twelve Julys, four Augusts), the earliest July 1859 (25.1 °C), the latest July 2018 (27.4 °C)—a complete list is given in Summer, Chapter 22.

At the other extreme, July 1816 and July 1817, the years following the eruption of Tambora in Indonesia in 1815, were very cool indeed—July 1816 managing a mean temperature of just 14.7 °C, 3.2 degC below the current normal for the month, while the succeeding July of 1817 (15.3 °C) was but little better. The Radcliffe observation registers (Figures 13.3 and 13.4) record rain falling on 16 days in July 1816 (although this may well be an underestimate of the actual number of 'rain days'), while midday temperatures exceeded 20 °C on only two days during the whole month. The year 1816 was known for a generation afterwards as 'the year without a summer', and it is not surprising—for, as well as July 1816 being the fourth-coolest July on Oxford's records, August 1816 remains the second-coolest (see also Summer, Chapter 22 and Chronology, Chapter 25). July 1888 was also particularly cool—it remains Oxford's second-coolest July, while the highest temperature reached during the month was just 21.8 °C. It was also the dullest July on record, rain fell on 23 days, and remains the only summer month in more than 200 years to have recorded snowfall (see 'Snow…in July?').

Table 13.2 *July mean temperatures at Oxford's Radcliffe Observatory, 1814–2018*

| July mean temperature 17.9 °C (average 1981–2010) | | | | | |
|---|---|---|---|---|---|
| Warmest months | | | Coldest months | | |
| Mean temperature, °C | Departure from 1981–2010 normal degC | Year | Mean temperature, °C | Departure from 1981–2010 normal degC | Year |
| 21.1 | +3.2 | 1983 | 14.3 | −3.5 | 1919 |
| 21.0 | +3.1 | 2006, 2018 | 14.5 | −3.6 | 1879, 1888, 1922 |
| 19.7 | +2.8 | 1921, 2013 | 14.6 | −3.3 | 1892 |
| 19.6 | +2.7 | 1911, 1976 | 14.7 | −3.2 | 1816, 1910, 1920 |
| 19.5 | +2.6 | 1995 | 14.8 | −3.1 | 1860, 1907, 1965 |
| *Last 50 years to 2018* | | | | | |
| 21.1 | +3.2 | 1983 | 15.2 | −2.7 | 1988 |

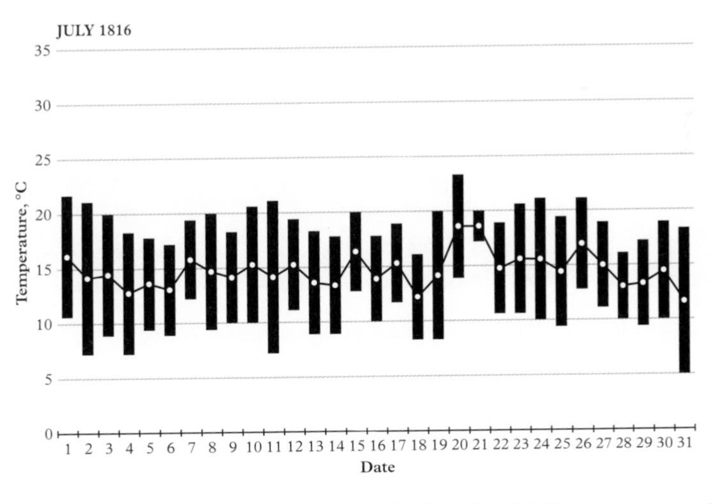


**Figure 13.3** *The dreadful summer of 1816 is revealed in this plot of daily maximum and minimum temperatures, °C (top and bottom of daily columns) and daily mean temperature (central circle and connecting line) at the Radcliffe Observatory, Oxford, in July 1816—see also Figure 13.4*

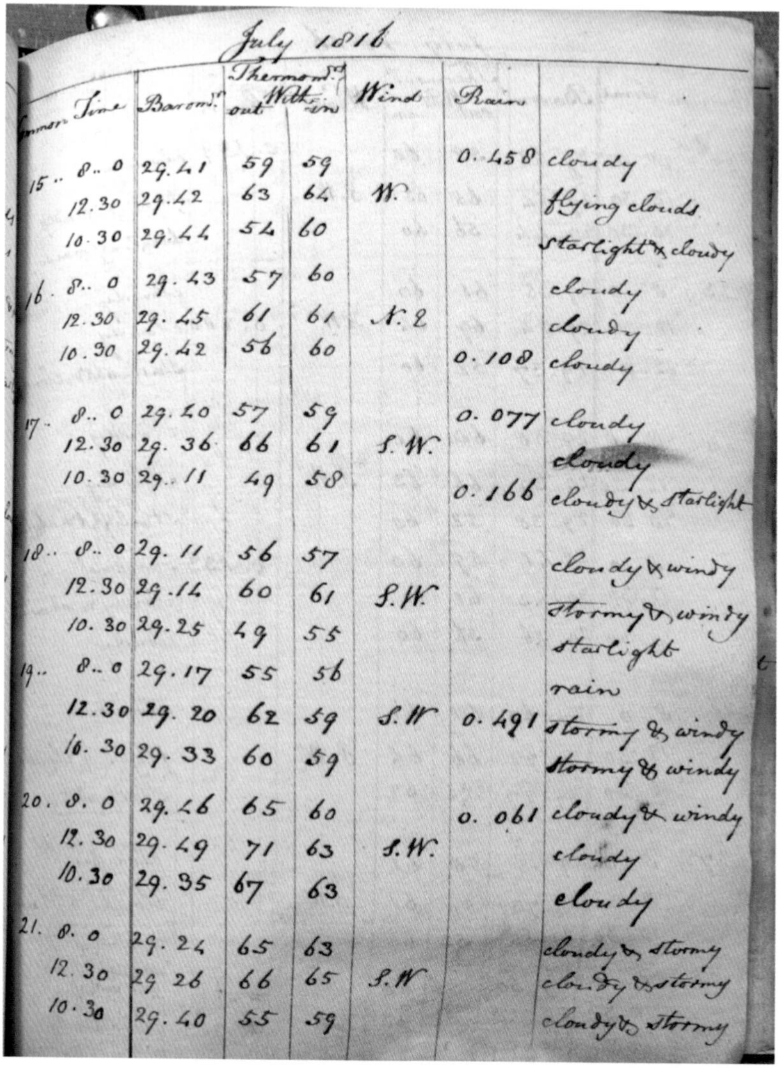

July 1816

| Time | Barom.r | Thermom.r With out | Thermom.r With in | Wind | Rain | |
|---|---|---|---|---|---|---|
| 15.. 8..0 | 29.41 | 59 | 59 | | 0.458 | cloudy |
| 12.30 | 29.42 | 63 | 64 | W. | | flying clouds |
| 10.30 | 29.44 | 54 | 60 | | | starlight & cloudy |
| 16. 8..0 | 29.43 | 57 | 60 | | | cloudy |
| 12.30 | 29.45 | 61 | 61 | N.E | | cloudy |
| 10.30 | 29.42 | 56 | 60 | | 0.108 | cloudy |
| 17.. 8..0 | 29.40 | 57 | 59 | | 0.077 | cloudy |
| 12.30 | 29.36 | 66 | 61 | S.W. | | cloudy |
| 10.30 | 29.11 | 49 | 58 | | 0.166 | cloudy & starlight |
| 18.. 8..0 | 29.11 | 56 | 57 | | | cloudy & windy |
| 12.30 | 29.14 | 60 | 61 | S.W. | | stormy & windy |
| 10.30 | 29.25 | 49 | 55 | | | starlight |
| 19.. 8..0 | 29.17 | 55 | 56 | | | rain |
| 12.30 | 29.20 | 62 | 59 | S.W | 0.491 | stormy & windy |
| 10.30 | 29.33 | 60 | 59 | | | stormy & windy |
| 20. 8.0 | 29.46 | 65 | 60 | | 0.061 | cloudy & windy |
| 12.30 | 29.49 | 71 | 63 | S.W. | | cloudy |
| 10.30 | 29.35 | 67 | 63 | | | cloudy |
| 21. 8.0 | 29.24 | 65 | 63 | | | cloudy & stormy |
| 12.30 | 29.26 | 66 | 65 | S.W | | cloudy & stormy |
| 10.30 | 29.40 | 55 | 59 | | | cloudy & stormy |

**Figure 13.4** *Entries from the Radcliffe Observatory meteorological logbooks for 15–21 July 1816*

In examples of the odd tricks long-period statistics can throw up, the hottest July on Oxford's records to that time (July 1911, mean 19.6 °C) followed the coolest since 1892 (July 1910, 14.7 °C); while several of the coolest Julys on record (1919, 1920 and 1922, all below 15 °C) sandwiched the hottest July on record at that time (1921, 19.7 °C). It is also noteworthy how rare such cool Julys are in modern times—the coolest July in the last 50 years has been 15.2 °C, in 1988 (see also Figure 13.4).

Monthly temperature ranges in July have varied between 13.4 degC in 1960 and 13.8 degC in 1880, to 28.1 degC in 1848. More recently, the temperature range during July 2015 was 27.9 degC.

## Precipitation

The monthly mean rainfall for July in Oxford is 49 mm. Table 13.3 gives the recorded extremes of monthly rainfall since 1767, and daily rainfall since 1827.

July 1825, with just 0.8 mm of rainfall during the month, remains the third-driest month on Oxford's records, and July 1800 the fourth-driest. July 1834, with 175 mm of rainfall, remains the wettest July on Oxford's records by almost 20 mm (Table 13.3). Rainfall was recorded on 14 days, about normal for July, but seven of these days had more than 10 mm—although, as with all of the early Oxford records, it is possible that some of these are multi-day accumulations rather than true daily amounts. We do not have much information on the month's weather other than knowing that there was a short warm spell just after mid-month, the temperature reaching 27 °C on 17 July, and with three consecutive very warm nights (all above 17 °C) 16th–18th; 55 mm fell in the three days 18–20 July, so this may have represented the breakdown of a thundery spell. July 1880 was also very wet, and thundery too, with thunder heard on 10th, 14th (with very heavy hail, 46 mm rainfall), 15th, 16th, 17th and 21st, six days in all. More than half of July 1968's 142 mm of rainfall, the third-wettest July on record, fell in a single day as a result of thunderstorms (see 'Thunderstorms' in this chapter, and the Chronology in Chapter 25). Since 1968, the wettest July has been 2007, with 105 mm, 59 mm of which fell across the two rainfall days 19–20 July during a spell of prolonged heavy thundery

**Table 13.3** *July precipitation at the Radcliffe Observatory, Oxford since 1767. Extremes of monthly totals 1767–2017, wettest days 1827–2018*

**July mean precipitation 48.7 mm** (average 1981–2010)

| Wettest months | | | Driest months | | | Wettest days | |
|---|---|---|---|---|---|---|---|
| Total fall, mm | Per cent of normal | Year | Total fall, mm | Per cent of normal | Year | Daily fall, mm | Date |
| 174.8 | 359 | 1834 | 0.8 | 2 | 1825 | 87.9 | 10 July 1968 |
| 155.1 | 319 | 1880 | 1.2 | 3 | 1800 | 50.0 | 8 July 2004 |
| 141.6 | 291 | 1968 | 3.1 | 6 | 1885 | 47.8 | 25 July 1861 |
| 137.1 | 282 | 1850 | 3.6 | 7 | 2016 | 47.0 | 22 July 1907 |
| 135.5 | 278 | 1950 | 4.0 | 8 | 1905 | 46.3 | 25 July 1886 |
| *Last 50 years to 2018* | | | | | | | |
| 110.2 | 226 | 2007 | 3.6 | 7 | 2016 | 50.0 | 8 July 2004 |

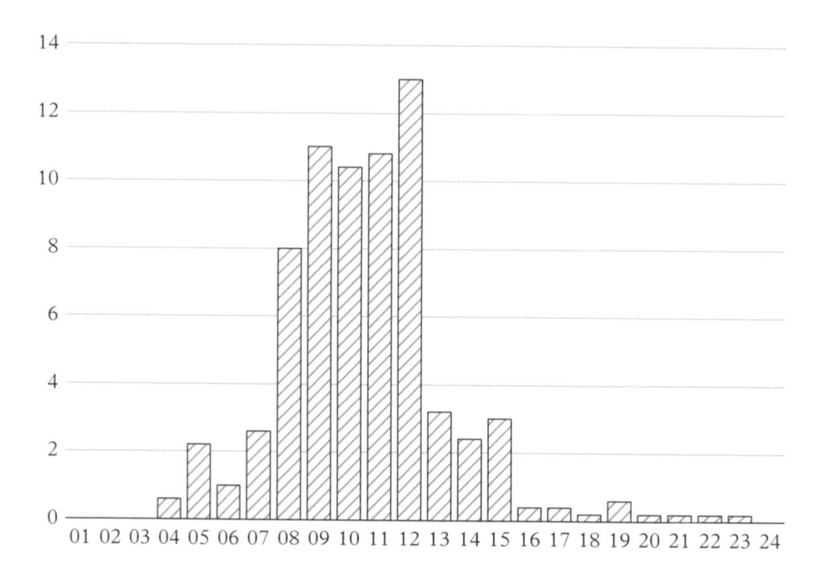

**Figure 13.5** *Rainfall and resulting floods in Oxford on 20 July 2007.* ***Top:*** *hourly rainfall totals (in mm—for the hour ending, GMT) as recorded at Wytham Woods, a little west of Oxford, where the daily total for the civil day (midnight to midnight) was 70.6 mm; 59 mm fell at the Radcliffe Observatory.* ***Bottom:*** *Floods with bystanders outside the Osney Arms pub on 20 July 2007 (© Oxfordshire History Centre POX 003 0082)*

rain: extensive river flooding resulted, an unusual occurrence during the summer half-year (Figure 13.5).

Over the last hundred years or so, daily rainfalls of 25 mm or more occur in about one year in six in July—typically, around half of those days include some contribution from thunderstorms, when rainfall can be very heavy, short-lived and often very localised. On 18 July 1947, 89 mm fell in 1 hour 45 minutes at Sandford St Martin, near Woodstock,

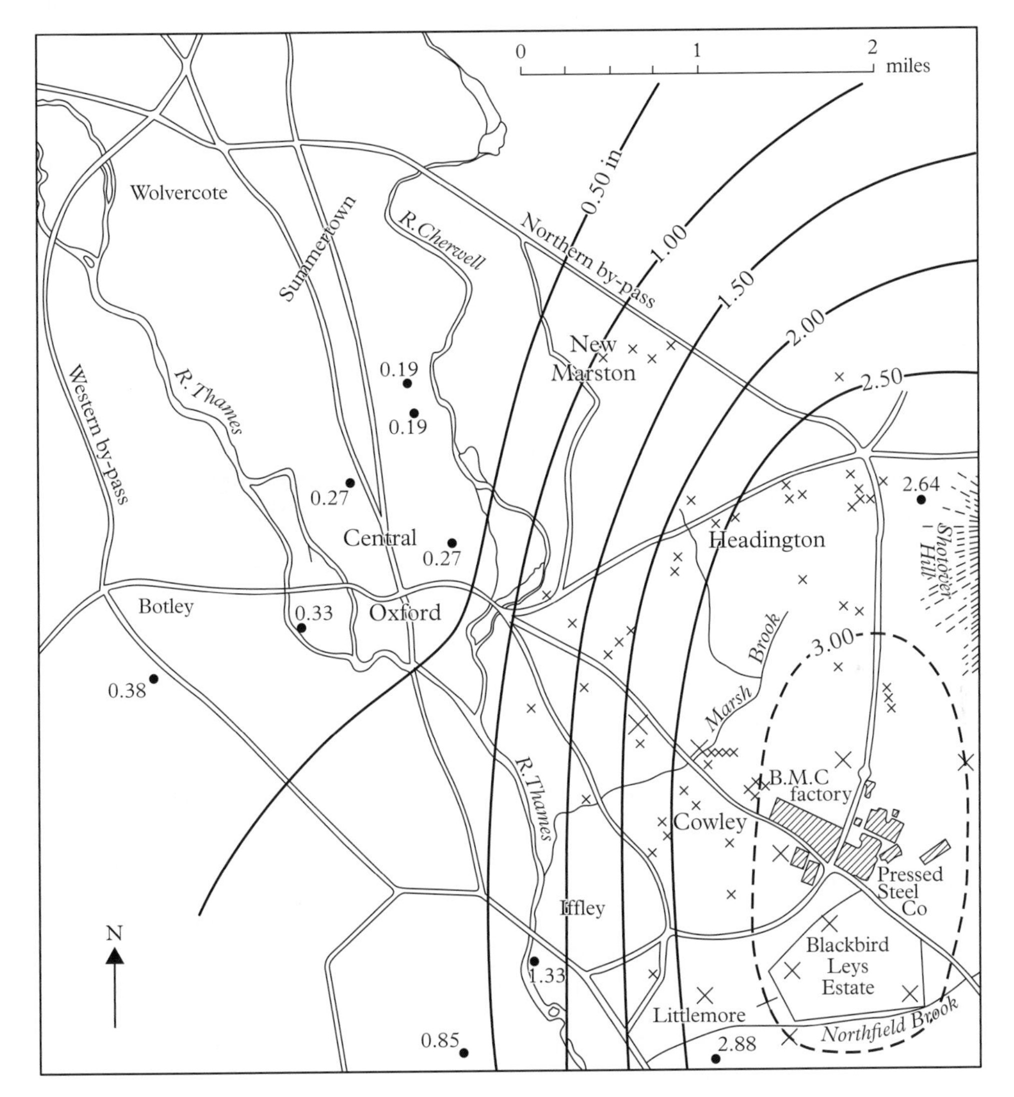


**Figure 13.6** *Distribution of rainfall totals for the very localised thunderstorm on 13 July 1967, when over 75 mm (3 inches) fell in Cowley in 60–70 minutes. Plotted rainfall values are in inches (1 inch = 25.4 mm): isohyets are at 0.5 in (13 mm) intervals; crosses represent flood reports. Adapted from McFarlane and Smith, 1968, Figure 5 [95] (Crown Copyright. Met Office data and information provided under Open Government Licence)*

yet only 3 mm fell at the Radcliffe Observatory in the centre of Oxford [94]; another example occurred on 13 July 1967, when just 7 mm of rain fell at the Radcliffe Observatory, but over 75 mm fell in about an hour not far to the south-east around Cowley, where rapid flooding ensued (Figure 13.6) [95].

Oxford's wettest day in almost 200 years of daily rainfall records was on 10 July 1968. On this date, a small thundery frontal depression moved east-north-east over southern England (see also Chronology, Chapter 25). The Radcliffe Observatory's weather diary noted that the day was 'overcast throughout with heavy showers alternating with longer spells of continuous heavy rain; thunderstorm with heavy rain at 1350; heavy rain, easing off to moderate, continuing into the early morning of 11 July'. The day's rainfall total amounted to 87.9 mm, surpassing the previous wettest day on record in September 1951, when 84.8 mm fell—these are the only two days yet to have recorded 75 mm or more within 24 hours.

At the opposite hydrological extreme, July/August 1976 still holds the record for longest spell of consecutive days without measurable rainfall in Oxford within the last 200 years. The 40 days commencing 19 July 1976 saw only 0.1 mm of rainfall recorded (all on 20 July). Another long dry spell which commenced on 25 June 1979 lasted through most of July, ending on 28 July after 33 days without rainfall; in 2018, a spell of 32 days without rainfall ended on 19 July. A table of the longest droughts in Oxford since 1827 is given in Chapter 28. More recently, July 2016 became the driest July in Oxford for over 130 years, when just 3.6 mm of rain fell during the month. In 2018, the six weeks commencing 5 June became the driest such period on the entire record with just 1.3 mm of rainfall recorded (Chapter 28, Table 28.3), surpassing the previous record established 125 years previously in spring 1893.

## Snow . . . in July?

Remarkably, there is an entry in the weather diary for July 1888 which reads as follows:

> Snow on the 10th, 17¾$^{h}$ to 18¼$^{h}$

The times are given by the astronomical clock then in use at the Observatory, in which reckoning started from noon each day, so this entry 'translates' to 0545–0615 GMT on 11 July. Did snow once really fall in July in Oxford? There are separate entries for hail on other days in 1888, so it is not simply a difference in observing terminology. Following several hours of cold rain overnight, at 8 a.m. the Observatory's weather diary noted 10/10 cloud, a very chilly north-westerly force 4 wind, the temperature just 7.1 °C following the morning's minimum temperature of 5.1 °C*. The 'snow' was probably a mixture of slushy hail and wet snow, as there are other reliable accounts of sleet early that morning; the hills of northern England had turned white with summer snowfall. The synoptic situation at 8 a.m. is shown in Figure 13.7. *Symons's Meteorological Magazine* reported snow observed to fall in the Leicester area and, at Birdlip, between Stroud and

* Using the 'astronomical clock' dating in use at the Radcliffe Observatory at the time, this was entered as the minimum temperature on 10 July, although current attribution would date the minimum temperature to 11 July, as shown here.

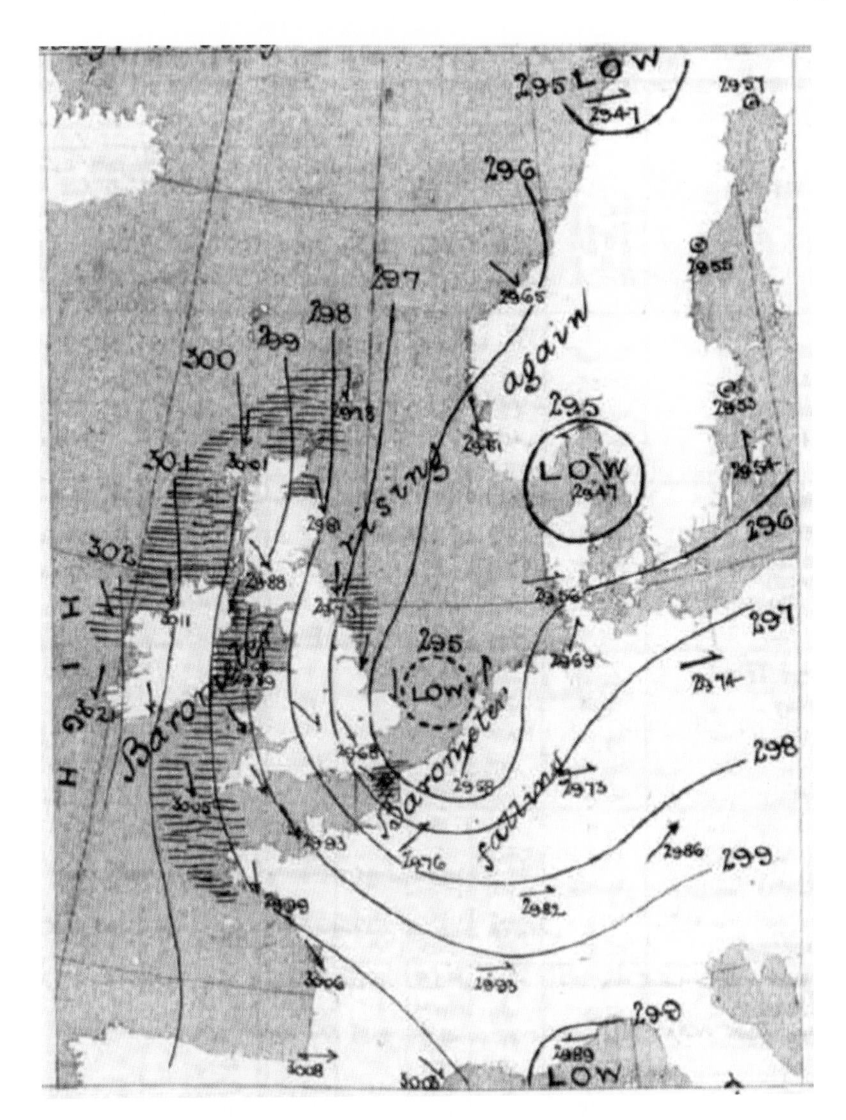


**Figure 13.7** *The Meteorological Office* Daily Weather Report *synoptic map for 8 a.m. on 11 July 1888, when an early morning snowfall was noted at the Radcliffe Observatory, the only occasion of snowfall in Oxford between mid-May and mid-October in over 200 years' records. A vigorous secondary depression formed to the north of Scotland on 10 July and plunged rapidly southwards while deepening, introducing a strong and very cold northerly on its western flanks Isobars are in inches of mercury (inHg): 29.5 inHg = 999 hPa, 30.0 inHg = 1016 hPa. (Met Office Library and Archives)*

Cheltenham. Sleet was observed on the fringes of London, while the Meteorological Office *Monthly Weather Report* commented that snow showers had been observed that day '…as far south as the Isle of Wight'.

### Thunderstorms

Thunderstorms can be expected on two days in a typical July but occurred on eight days in the Julys of 1947 and 1965, and seven days in 1939 and 1960.

## Sunshine

July is, on average, the sunniest month of the year in Oxford with 207 hours of bright sunshine, an average of 6 hours and 41 minutes daily, representing 41 per cent of possible daylight hours—although the latter figure is slightly higher in August, at 43 per cent. Even after more than a century, July 1911 remains the outstandingly sunny month of any name on Oxford's record—with 310.45 hours of sunshine (a daily average just exceeding 10 hours), representing 62 per cent of the possible duration. More recently, July 2006 became the second-sunniest month on record with 304 hours, while July 2013 and July 2018 were not far behind with 297 hours and 282 hours, respectively. In sharp contrast, four Julys have failed to muster 100 hours sunshine—July 1888, with just over 95 hours, remains the dullest of any of the three summer months (June, July and August); the sunshine total for that dismally cold and wet month has been exceeded by several winter months including (remarkably) one December (97 hours, in 2014). In another example of statistical quirkiness, two very dull Julys in 1931 (128 hours) and 1932 (124 hours) were followed by three very sunny Julys (1933, 238 hours, 1934, 269 hours and 1935, 274 hours), then followed by yet another very dull July in 1936 (119 hours).

By early July, daylength is beginning to decrease—ever so slowly at first, but more noticeably by the end of the month. Even so, 15 hours of bright sunshine is possible on the clearest of days past the third week—15.0 hours sunshine was recorded as late as 22 July in 1995. The sunniest July day since 1921 was 8 July 1934, when 15.5 hours bright sunshine was recorded (the same duration was also logged on 12 July 1911). The sunniest fortnight in Oxford since 1921 has been the 14 days 24 June to 7 July 1976, when 183 hours of sunshine were logged, an average of 13.08 hours per day. Of the sunniest periods of a week or more listed in Chapter 30, the most notable entirely within the month of July were 4–10 July 1934 (7 days, 100 hours sunshine, daily average 14.31 hours) and 11–24 July 1990 (14 days, 172 hours, 12.28 hours per day). More recently, the prolonged sunshine of July 2013 resulted in a 14-day total of 179 hours between 6 and 19 July, a daily average of 12.75 hours, surpassed in June–July 2018 with 180.5 hours in 14 days between 21 June and 4 July, a daily average of 12.89 hours. June–July 2018 also established new 21-day and 30-day sunshine records in Oxford, surpassing even the records established during the memorable summer of 1976 (see Table 30.4 in Chapter 30).

The Julys of 1976, 1990, 2006 and 2018 all had 11 days with 12 hours or more sunshine, and July 2013 had 10 such days, but between 1921 and 2018 eleven Julys did not manage a single day recording this amount of sunshine.

**Table 13.4** *July sunshine duration at the Radcliffe Observatory, Oxford: monthly extremes 1880–2018, sunniest days 1921–2018*

**July mean sunshine duration 207.0 hours, 6.68 hours per day** (average 1981–2010)
*Possible daylength: 500 hours. Mean sunshine duration as percentage of possible: 41.4*

| Sunniest months | | | Dullest months | | | Sunniest days | |
|---|---|---|---|---|---|---|---|
| Duration, hours | Per cent of possible | Year | Duration, hours | Per cent of possible | Year | Duration, hours | Date |
| 310.45 | 62.1 | 1911 | 95.3 | 19.1 | 1888 | 15.5 | 8 July 1934 |
| 303.7 | 60.8 | 2006 | 97.5 | 19.5 | 1944 | 15.4 | 6 July 1941, 12 July 1928, 3 July 1968, 11 July 1994 |
| 297.3 | 59.5 | 2013 | 98.6 | 19.7 | 1913 | 15.3 | 7 July 1934, 4 July 1959, 4 July 1989, 20 July 1997, 3 July 2018 |
| 281.9 | 56.4 | 2018 | 105.4 | 21.1 | 1927 | 15.2 | 15 July 1929, 9 July 1967, 3 July 1977 |
| 280.4 | 56.1 | 1989 | 119.2 | 23.8 | 1936 | 15.1 | 15 July 1928, 5 July 1930, 7 July 1959, 5 July 1987, 14 July 1990 |
| *Last 50 years to 2018* | | | | | | | |
| 303.7 | 60.8 | 2006 | 129.0 | 25.8 | 1981 | 15.4 | 11 July 1994 |

A typical July will see only one sunless day, and about one year in four has no sunless days in July—in 1978 there were five sunless days. The longest spell without sunshine in July since 1921 was for three consecutive days 3–5 July 1950.

## Wind

Oxford's mean monthly wind speed for July is 3.4 metres per second (m/s), or 6.7 knots. The windiest July on a record back to 1880 occurred in 2009, when the monthly mean wind speed was 5.1 m/s (9.8 knots). The calmest July was in 2011, when the monthly mean wind speed was just 2.0 m/s (3.8 knots).

## Oxford temperature, precipitation and sunshine in graphs—July

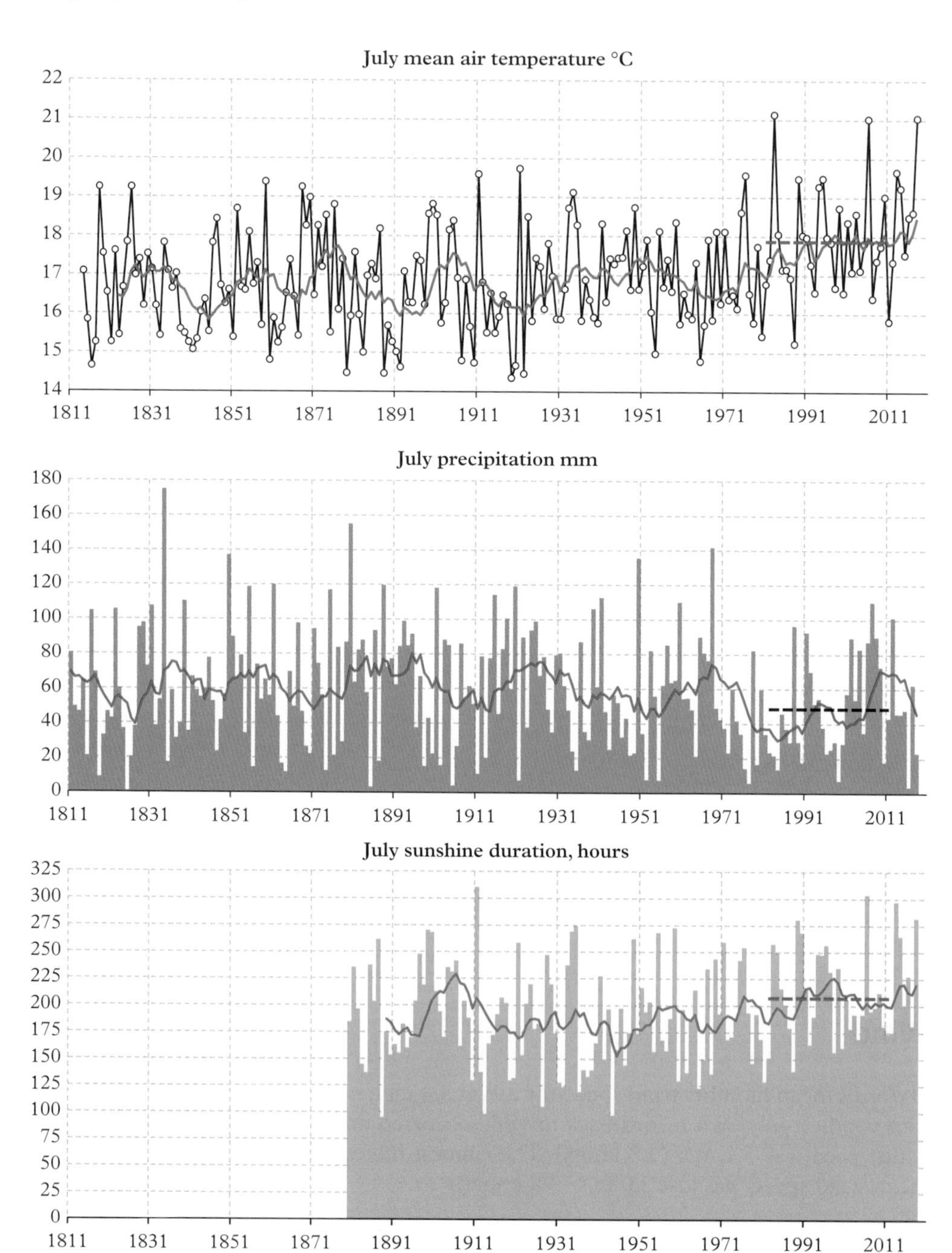


**Figure 13.8** *Monthly values of (from top) mean temperature (°C, since 1814), total precipitation (mm, since 1811) and sunshine duration (hours, since 1880) for July at Oxford. The 1981–2010 averages are indicated by the thick blue dashed line, while the ten-year unweighted running mean ending at the year shown is indicated by the red line*

# 14

# August

August is the typical month of high summer in most people's minds—long school holidays, sunny days at the beach, light and warm evenings. And yet—the decline into autumn is already apparent in Oxford's mean temperatures, which are a few tenths lower than in July, and in sunshine totals—the latter an average of 21 minutes per day less than in July, although mid-August is, on average, the sunniest week of the year. The monthly mean temperature is 17.6 °C, and yet August holds the crown for the hottest days yet recorded in Oxford, with 35 °C recorded twice, and some of the hottest nights too. August also tends to be a wetter month than July, and heavy thundery rains are not uncommon, when once again a month's normal rainfall can fall in the space of just a few hours. Curiously, both the average wettest and driest days of the year over the standard averaging period 1981–2010 occur in August—9 August (5.4 mm) and 28 August (0.4 mm), respectively.

## Temperature

The warmest days of the year tend to be reached around late July on average, after which there is a fairly gentle decline in mean daily maximum temperatures throughout August, from almost 24 °C in the first week to below 21 °C by the end of the month. There is a more irregular fall in mean minimum temperatures, from almost 14 °C in the first week to 11–12 °C towards the end of the month.

Only two days have reached 35 °C in Oxford since daily records began in 1814; the four hottest days in Table 14.1 are separated by just half a degree Celsius. The all-time record in Oxford was set on 19 August 1932 when the screen temperature reached 35.1 °C during a very short hot spell (just the one day above 30 °C) ended by intense thunderstorms*. This was equalled 58 years later, when another short but very hot spell 1–4 August 1990 resulted in four consecutive days reaching 30 °C, including an equal-record 35.1 °C on 3 August of that year (see also the Chronology in Chapter 25 for both events).

* The August 1932 maximum has been quoted for many years as 35.0 °C, the conversion of 95 °F to the nearest degree Fahrenheit. However, in the published Radcliffe Observatory tables for 1931–35 the maximum temperature for the day is given as 95.2 °F = 35.1 °C, and that is the value used here.

*Oxford Weather and Climate since 1767*. Stephen Burt and Tim Burt, Oxford University Press (2019).
 DOI: 10.1093/oso/9780198834632.001.0001

Prolonged hot spells in August, in which the maximum temperature reached at least 25 °C each day for at least 12 consecutive days, occurred in 1947, 1975 and 1997. The longest such spells were 18 days, 12–29 August 1947, an unusually long and late warm spell, and 17 days 6–22 August 1997. In August 1876, 30 °C was reached on five consecutive days 13–17 August (hottest day 33.3 °C on 13th), and in 1995 this value was reached on six consecutive days from 29 July to 3 August (hottest day 33.3 °C on 2 August, Figure 14.1). In 2003, 30 °C was reached on six days in eight 4–11 August—the mean maximum temperature for these eight days was 31.3 °C, and the hottest day 34.6 °C on 9 August. More details on the August 1990 heatwave appear in the Chronology, Chapter 25.

Warm nights can occur at any time during August, although only five August nights have remained above 20 °C—remarkably, two of these five occurred in the same month, August 1997, when the minimum temperature on 11 August was 20.5 °C (the hottest August night on Oxford's record), while a second '20' occurred less than a fortnight later, on 23 August (20.1 °C). The hottest summer nights tend to occur towards the end of a hot spell as the cloud increases, possibly with overnight thunderstorms; each of the

**Table 14.1** *Highest and lowest maximum and minimum temperatures at the Radcliffe Observatory, Oxford in August, 1815–2018*

| Rank | Hottest days | Coldest days | Hottest nights | Coldest nights |
|---|---|---|---|---|
| 1 | 35.1 °C, 19 Aug 1932 and 3 Aug 1990 | 12.1 °C, 17 Aug 1879 | 20.5 °C, 11 Aug 1997 | 0.2 °C, 26 Aug 1864 |
| 2 | 34.8 °C, 9 Aug 1911 | *12.2 °C, 31 Aug 1816 and 31 Aug 1833* | 20.3 °C, 26 Aug 1859 | 1.4 °C, 24 Aug 1864 |
| 3 | 34.6 °C, 9 Aug 2003 | *12.8 °C, 28 Aug 1821 and 19 Aug 1845* | 20.1 °C, 5 Aug 1975 and 23 Aug 1997 | *2.5 °C, 21 Aug 1850* |
| 4 | 33.3 °C, 13 Aug 1876 and 2 Aug 1995 | *13.0 °C, 14 Aug 1844* | 20.0 °C, 18 Aug 1893 | 2.7 °C, 30 Aug 1869 |
| 5 | 33.2 °C, 1 Aug 1995 | 13.1 °C, *15 Aug 1829* and 1 Aug 1917 | *19.7 °C, 13 Aug 1829* | 3.6 °C, *26 Aug 1850* and 11 Aug 1892 |
| ***Last 50 years*** | | | | |
| | 35.1 °C, 3 Aug 1990 | 13.3 °C, 4 Aug 1974 | 20.5 °C, 11 Aug 1997 | 4.5 °C, 30 Aug 2003 |

*Temperatures recorded by unscreened thermometers prior to 1853 are shown in italics; these are likely to be a little too cold by modern (screened thermometer) standards in winter and a little too warm in summer—see Appendix 1 for more details. In August the mean temperature correction for unscreened thermometry is –0.2 degC, although this will vary considerably on a day-to-day basis*

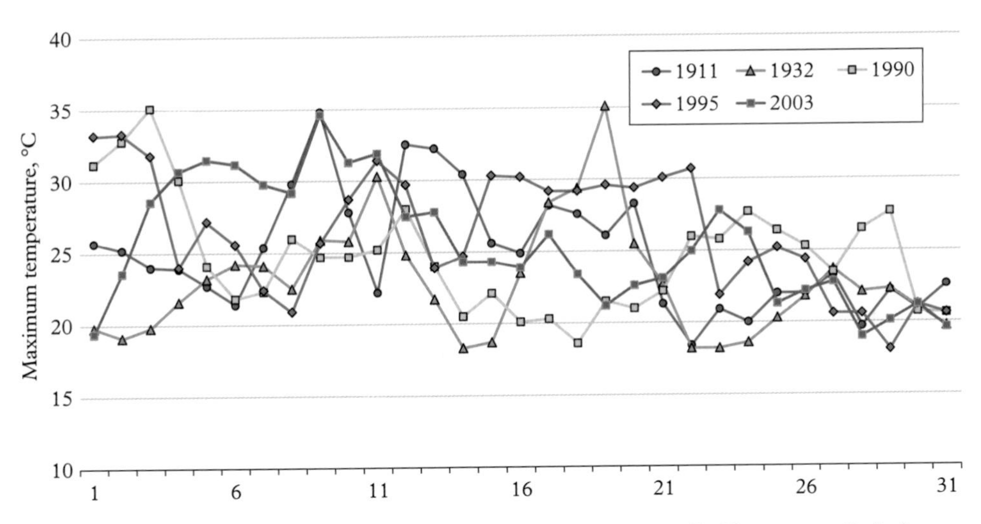


**Figure 14.1** *Daily maximum temperatures (°C) recorded at the Radcliffe Observatory, Oxford during the heatwaves in the Augusts of 1911, 1932, 1990, 1995 and 2003*

four nights that remained above 20 °C was preceded by a day when the temperature reached at least 27 °C, and all but one exceeded 30 °C.

A very long spell of extremely warm nights occurred in August 1997. For the three weeks 5–25 August, the temperature did not fall below 13.8 °C; for seven consecutive nights 19–25 August, the temperature did not fall below 17 °C. August 1997 was also the first month on Oxford's records to remain above 11 °C throughout, with a minimum temperature of 11.7 °C (on 3rd); since then, only July 2018 has come close, with a minimum temperature of 11.3 °C. Not surprisingly, August 1997 still holds the record for the highest mean minimum temperature for any month on Oxford's records, at 15.7 °C. August 1997 also included one spell of nine consecutive nights no lower than 15.0 °C, the equal-longest spell of any month (with July 1983): eight-night spells occurred in the Augusts of 1955, 1982 and 2003.

Cool August days tend also to be wet: three of the five coldest days which also have rainfall records (since 1827—Table 14.1) also recorded more than 20 mm of rainfall. The coolest August day of all was 17 August 1879 which reached only 12.1 °C, a winter-like day in a truly appalling summer, accompanied by 13 mm of rainfall for good measure. The coolest August day within the last 50 years has been 4 August 1974, maximum temperature 13.3 °C, accompanied by 17 mm of rainfall.

Not surprisingly, the coldest nights tend to occur towards the end of August as summer slides into autumn, sometimes all too quickly, particularly in dry weather when the night-time fall in temperature is not ameliorated by abundant soil moisture. All but one of the five coldest August nights have occurred during the last ten days of the month. No air frosts have (yet) been recorded in August, but 26 August 1864 came closest, with a minimum temperature of just +0.2 °C. August 2003 appears twice in Table 14.1—the temperature by day reaching 34.6 °C on 9th, within half a degree of the hottest day on record, and yet on 30th the temperature overnight fell to just 4.5 °C, the lowest in August for 50 years and 30 degC lower than three weeks previously, owing to the very dry ground conditions.

August's mean minimum temperature has been below 10 °C in only ten years, most recently in 1922, but August 1827 holds the dubious distinction of the lowest August mean minimum temperature on record at 8.6 °C, about the current normal for early October. Several remarkably cold summer nights occurred in August 1864: 5.3 °C on 2nd, 4.8 °C on 11th, 3.7 °C on 21st, 1.4 °C on 24th and 4.6 °C on 25th in addition to the record low 0.2 °C on 26 August. Since 1931, 11 nights have recorded a grass minimum of 1 °C or below in August; however, only two ground frosts resulted—and then on two consecutive nights, 20 and 21 August 1964, both merely −0.1 °C. Notable also were 0.0 °C on 12 August 1943 and 0.4 °C on 29 August 1893—the latter towards the end of the blazing summer of that year when once again the ground was dry and heat conducted away rapidly.

Daily temperature ranges in August have reached or surpassed 20 degC on six occasions since 1815, the largest being 21.0 degC on 30 August 1906 (7.2 °C to 28.2 °C). Within the last 50 years, the greatest daily range on any August day was 19.0 degC, on 7 August 1975.

## Warm and cold months

The Augusts of 1911, 1947, 1975, 1990, 1995, 1997 and 2003 were very warm and contained some of the hottest August nights and days, while those of 1912, 1920 and 1922 were very cool with average temperatures similar to those expected in late May or late September (Table 14.2). The contrast between the Augusts of 1911 and 1912 (the

**Table 14.2** *August mean temperatures at Oxford's Radcliffe Observatory, 1814–2018*

| **August mean temperature 17.6°C** (average 1981–2010) | | | | | |
|---|---|---|---|---|---|
| Warmest months | | | Coldest months | | |
| Mean temperature, °C | Departure from 1981–2010 normal degC | Year | Mean temperature, °C | Departure from 1981–2010 normal degC | Year |
| 20.3 | +2.7 | 1997 | 13.5 | −4.1 | 1912 |
| 20.1 | +2.5 | 1995 | 13.8 | −3.8 | 1920 |
| 19.6 | +2.0 | 1975 | 14.1 | −3.5 | 1833, 1844 and 1922 |
| 19.5 | +1.9 | 2003 | 14.3 | −3.3 | 1816 |
| 19.4 | +1.8 | 1911, 1947 and 1990 | 14.4 | −3.2 | 1860, 1885 and 1891 |
| *Last 50 years to 2018* | | | | | |
| 20.3 | +2.7 | 1997 | 14.7 | −2.7 | 1986 |

hottest August on record at that time, followed by the coldest) must have been extraordinary and unwelcome. August 1985 and 1986 both produced a poor end to summer. In August 1933, during a memorably hot summer, the lowest maximum temperature was 20.1 °C on 17th—one of only five months when 20 °C was reached every day. Monthly temperature ranges in August have varied between 15.5 degC in 1952 and 30.1 degC in 2003.

## Precipitation

The monthly mean rainfall for August in Oxford is 55 mm, 13 per cent greater than July's average.

August rainfall totals have ranged from 146 mm in August 2010 (including a fall of 51 mm on 25th) to less than 2 mm in August 1940, one of only six Augusts to have recorded less than 10 mm rainfall. In August 2010, rain fell on 20 days, but in the Augusts of 1912, 1917, 1941, 1956 and 1963, 23 days with rain were recorded; contrast this with the Augusts of 1940 and 1947, when rain fell on only two days in the month (and in 1940, the wettest day received just 1.0 mm of rain). Aside from two small falls on 9–10 August 1940, there was no measurable rain for 43 days commencing 28 July 1940.

Daily rainfall totals of 25 mm or more of rain occur in about one year in five in August, although, since 1827, only four August days have recorded 50 mm or more rainfall—the wettest being 6 August 1922, when 70.8 mm fell, the wettest day on Oxford's records until September 1951 and still the fourth-wettest day in the Radcliffe Observatory records. On 3 August 1971, 31.5 mm of the day's total rainfall of 33.3 mm

**Table 14.3** *August precipitation at the Radcliffe Observatory, Oxford since 1767. Extremes of monthly totals 1767–2018, wettest days 1827–2018*

| August mean precipitation 54.8 mm (average 1981–2010) | | | | | | | |
|---|---|---|---|---|---|---|---|
| **Wettest months** | | | **Driest months** | | | **Wettest days** | |
| **Total fall, mm** | **Per cent of normal** | **Year** | **Total fall, mm** | **Per cent of normal** | **Year** | **Daily fall, mm** | **Date** |
| 146.2 | 267 | 2010 | 1.8 | 3 | 1940 | 70.8 | 6 Aug 1922 |
| 135.0 | 246 | 2004 | 2.4 | 5 | 1778 | 56.1 | 12 Aug 1957 |
| 133.2 | 243 | 1971 | 3.0 | 5 | 2003 | 53.6 | 6 Aug 1962 |
| 130.7 | 239 | 1878 | 3.1 | 6 | 1822 | 51.1 | 25 Aug 2010 |
| 130.3 | 238 | 1922 | 4.4 | 8 | 1995 | 43.4 | 2 Aug 1969 |
| *Last 50 years to 2018* | | | | | | | |
| 146.2 | 267 | 2010 | 3.0 | 5 | 2003 | 51.1 | 25 Aug 2010 |

fell in just under 2 hours. In another of those statistical flukes that beset weather records, both the average wettest and average driest day of the year—at least over the most recent 30 year averaging period—occur in August: the wettest was 9 August, with an average of 5.39 mm owing to several heavy falls on this date in recent years; the driest, 28 August, daily average just 0.36 mm.

The three longest spells of 'absolute drought' on Oxford's records over the past 200 years have all had the majority of their duration within the month of August (Table 28.1, Chapter 28): namely the spells of 37 days without measurable rainfall, which commenced on 5 August 1947, 38 days without rain commencing on 14 August 1959, and the longest rainless period, 40 days, which lasted from 21 July to 27 August 1976 (Figures 14.2 and 14.3). On 28 August 1976, rainfall was just about measurable (0.2 mm) and then the long drought broke on 29th with a total of 20.8 mm, the wettest day for almost a year (13 September 1975: 39.5 mm).

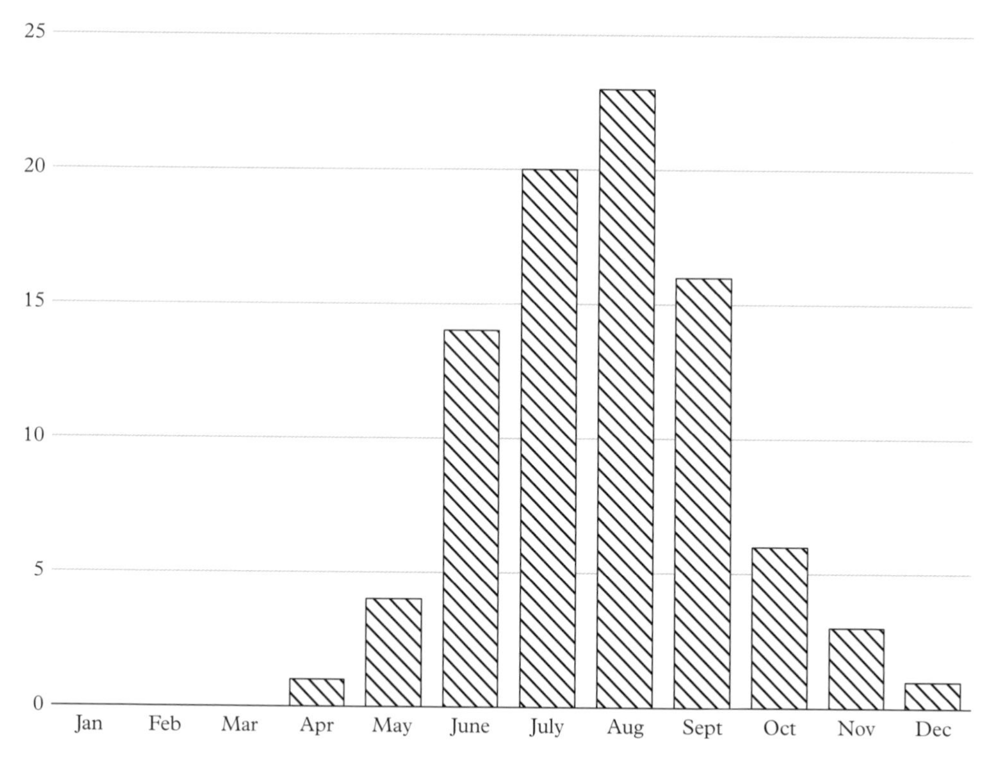


**Figure 14.2** *Frequency of references to 'drought' (number of days in the month including coverage) in the* Oxford Mail *by month during 1976 (Data from Gregory in [52], pp. 77–78). The coverage amounted to 26.2 pages throughout the year, with the longest unbroken run of daily drought references being 19 days, 6–27 August; coverage was greatest on 1 July, amounting to more than 4 pages by total area*

**Figure 14.3** *A stark contrast: brown outfield and well-watered green square in the Parks during the prolonged drought in August 1976 (Andrew Goudie)*

## Thunderstorms

One or two days with thunder can be expected in a typical August, although there were six such days in the Augusts of 1971 and 2004 and five in 1939, 1950 and 1954. About one August in four remains free of thunder.

Although it did not directly affect the city of Oxford, one of the greatest hailstorms recorded in the UK affected parts of north Oxfordshire on 9 August 1843 [96]. The day's weather notes from the Radcliffe Observatory noted 'Cloudy. Rain with thunder and lightning 2.30 to 7.30 p.m.' (lightning was still visible at the 10 p.m. observation), although only 2.4 mm of rainfall fell. A swathe of north Oxfordshire about 10 km across received hailstones of 30–50 mm in diameter from the storm: the *Banbury Guardian* reported that about 50,000 panes of glass were broken in Chipping Norton during the 20-minute onslaught, while slate roofs and crops growing in fields were also much damaged.

## Sunshine

Although August is less sunny than July in average sunshine duration (on average, 197 hours of bright sunshine, 21 minutes per day less than July, making it the second-sunniest month of the year), it is the sunniest month of the year when considered as a percentage of the possible, at 43.5 per cent, compared to July's 41 per cent (Table 14.4). The Augusts of 1995 and 1947 stand out as the two sunniest Augusts on the record, with August 1976 in fourth place (and the third consecutive very sunny month). When considered as a percentage of the possible daylight hours, August 1995 stands out as the sunniest of any month on Oxford's records, achieving a remarkable 63.2 per cent (285 hours out of 451), just ahead of 62.1 per cent in July 1911 and 61.9 per cent in May 1989.

With a single exception, no August within the last 50 years appears in the 'Top 5' dullest on record—that dubious distinction goes to 1977, with only 123 hours of sunshine. Even August 1977, though, managed almost 50 minutes per day more sunshine than the bottom of the table—August 1912, with a miserable 97.7 hours (and that after the sunniest April on record, which logged 239 hours).

The length of daylight in August decreases noticeably, by around 3½ minutes per day, and so it is only to be expected that the sunniest August days occur at the beginning of the month, when more than 14 hours sunshine is still just possible. On 7 August 1988 and 1 August 1989, 14.1 hours of sunshine was recorded, both in excess of 90 per cent of the possible daylight hours for the date, and about the maximum possible with the Campbell–Stokes pattern of sunshine recorder in use.

The longest sunny spells in August—runs of consecutive days each with 9 hours or more sunshine—have included the 12 days commencing 15 August 1976 (total sunshine

**Table 14.4** *August sunshine duration at the Radcliffe Observatory, Oxford: monthly extremes 1880–2018, sunniest days 1921–2018*

**August mean sunshine duration 196.5 hours, 6.34 hours per day** (average 1981–2010)
*Possible daylength: 451 hours. Mean sunshine duration as percentage of possible: 43.5*

| Sunniest months | | | Dullest months | | | Sunniest days | |
|---|---|---|---|---|---|---|---|
| Duration, hours | Per cent of possible | Year | Duration, hours | Per cent of possible | Year | Duration, hours | Date |
| 285.1 | 63.2 | 1995 | 97.7 | 21.6 | 1912 | 14.1 | 7 Aug 1988<br>1 Aug 1989 |
| 274.9 | 60.9 | 1947 | 111.1 | 24.6 | 1922 | 13.9 | 13 Aug 1949<br>3 Aug 1975<br>7 Aug 2005 |
| 269.5 | 59.7 | 1989 | 118.3 | 26.2 | 1894 | 13.7 | *6 occasions* |
| 257.4 | 57.0 | 1976 | 119.6 | 26.5 | 1958 | 13.6 | *5 occasions* |
| 250.3 | 55.5 | 1899 | 123.4 | 27.4 | 1977 | 13.5 | *4 occasions* |
| *Last 50 years to 2018* | | | | | | | |
| 285.1 | 63.2 | 1995 | 123.4 | 27.4 | 1977 | 14.1 | 7 Aug 1988<br>1 Aug 1989 |

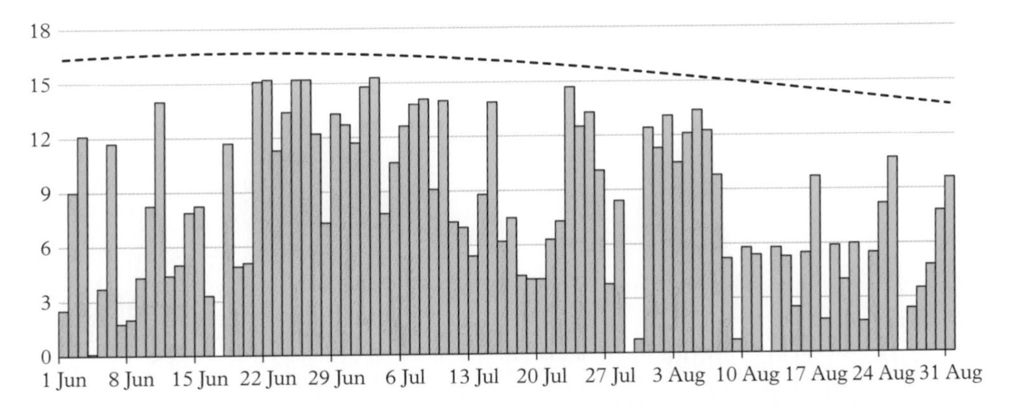


**Figure 14.4** *Daily sunshine duration (in hours) at the Radcliffe Observatory, Oxford, during the summer of 2018. The dashed line shows the length of daylight on each day (sunrise to sunset, hours). The sparkling summer of 2018 established new 140-year records for sunshine duration over 21, 30 and 60 days' duration, surpassing previous records established in the summers of 1976, 1989 and 1990 (Table 30.4)*

total 139 hours, daily average 11.57 hours) and the nine days commencing 29 July 1995 (total sunshine 108 hours, daily average 11.95 hours). The sparkling summer of 2018 came to an end after the first week of August, as both temperatures and sunshine durations declined sharply, but the 60 days ending 8 August 2018 comprised the sunniest two-month period on record in Oxford—with 573 hours of sunshine, a daily average of 9.55 hours, substantially exceeding the previous 60-day record of 561 hours in summer 1989 (Table 30.4, Chapter 30 and Figure 14.4).

A typical August will see sunshine on 30 of its 31 days, although about one year in four will see sunshine every day during the month. August 1977 saw an unenviable eight days without any sunshine and, more recently, August 2015 included five sunless days. The longest spell without any sunshine in August was in 1944, when the four days 19–22 August remained sunless; the Augusts of 1968, 1977 and 1981 all included three consecutive days without sunshine.

## Wind

Oxford's mean monthly wind speed for August is 3.4 metres per second (m/s), or 6.5 knots, equalling September as the lowest of the year. The windiest August on a record back to 1880 occurred in 2008, when the monthly mean wind speed was 5.7 m/s (11.0 knots). August 1998 was the calmest month of any name on record; the monthly mean wind speed was just 1.8 m/s (3.5 knots).

## Oxford temperature, precipitation and sunshine in graphs—August

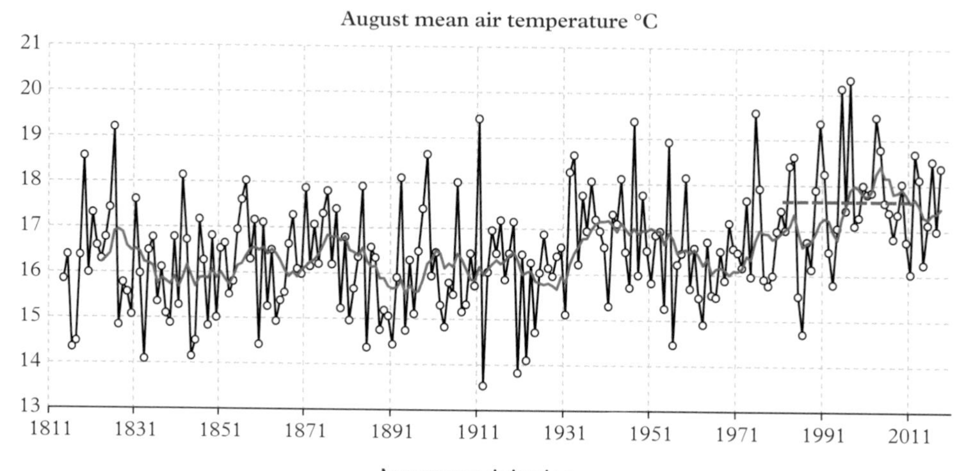


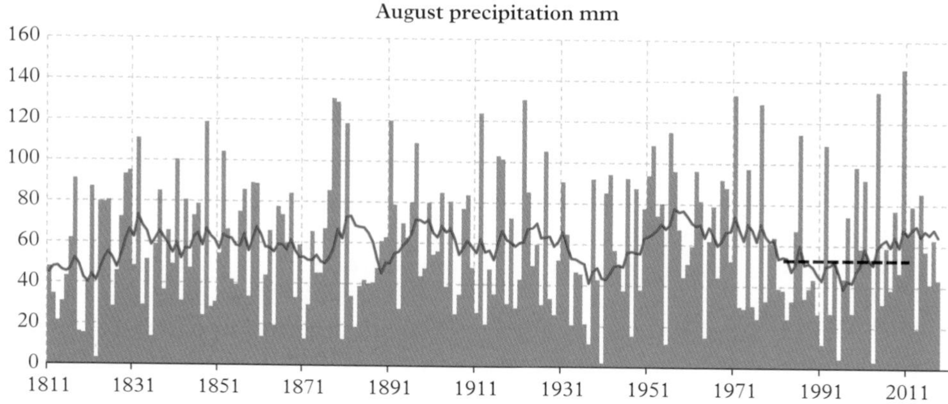


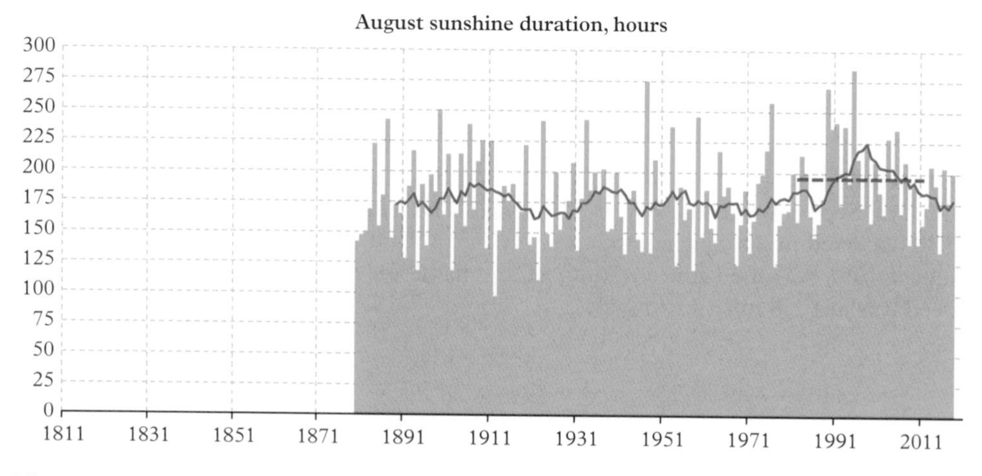


**Figure 14.5** *Monthly values of (from top) mean temperature (°C, since 1814), total precipitation (mm, since 1811) and sunshine duration (hours, since 1880) for August at Oxford. The 1981–2010 averages are indicated by the thick blue dashed line, while the ten-year unweighted running mean ending at the year shown is indicated by the red line*

# 15

# September

September is another 'transition month'. Most Septembers include at least a few warm, sunny days seemingly lingering from high summer—indeed, in some years summer continues well into September—but autumn usually shows its hand in cooler, cloudier and windier conditions by the end of the month. There is a steady and fairly rapid decline in mean temperatures, with average maximum temperatures dropping from 21 °C in early September to near 17 °C by the end; there is a similar decrease in minimum temperatures over the same period, from about 12 °C to about 9.5 °C, although both are subject to considerable year-to-year and day-to-day variability at this time of year. September's average temperature over the period 1981–2010 was 14.9 °C.

September sees a noticeable reduction in the duration and strength of bright sunshine, partly due to declining daylength (almost 4 minutes per day), but mainly due to increased cloud cover as autumn approaches: the percentage of possible sunshine in September declines sharply, from August's 43 per cent to September's 37 per cent. The average monthly sunshine duration for the month is 140 hours, or a little under 4¾ hours per day, down almost 30 per cent from the previous month. In terms of rainfall, September is almost indistinguishable from August—the monthly mean (54 mm) is fractionally lower than August only because September has one fewer day. Although on occasion September has passed by almost rainless, surprisingly one September in the eighteenth century still holds the dubious honour of by far the wettest month yet recorded in Oxford's 250 years of rainfall records.

## Temperature

Not surprisingly, the highest temperatures in September tend to occur at the beginning of the month, although 25 °C is still occasionally reached in early October. The hottest September day in Oxford was 8 September 1911, when 33.4 °C was reached: the previous day also makes it into the hottest September days list, at 31.8 °C (Table 15.1). Just five years previously, 1 and 2 September 1906 also logged two consecutive days above 33 °C, although even 30 °C has not been attained in September since 1929—the closest approaches since being 29.9 °C on 5 September 2004 and 29.2 °C on 13 September 2016. It is now a rare distinction for the hottest day of the year to be recorded in

*Oxford Weather and Climate since 1767*. Stephen Burt and Tim Burt, Oxford University Press (2019).
 DOI: 10.1093/oso/9780198834632.001.0001

**Table 15.1** *Highest and lowest maximum and minimum temperatures at the Radcliffe Observatory, Oxford, in September, 1815–2018*

| Rank | Hottest days | Coldest days | Warmest nights | Coldest nights |
|---|---|---|---|---|
| 1 | 33.4 °C, 8 Sept 1911 | 7.9 °C, 29 Sept 1918 | 20.6 °C, 5 Sept 1949 | *−3.3 °C, 28 Sept 1851* |
| 2 | 33.1 °C, 1 Sept 1906 and 2 Sept 1906 | 9.7 °C, 30 Sept 1918 | 18.4 °C, 8 Sept 1898 | *−1.2 °C, 23 Sept 1845* |
| 3 | 32.2 °C, 8 Sept 1898 | 9.8 °C, 27 Sept 1993 | 18.1 °C, 7 Sept 2016 | −0.4 °C, 29 Sept 1919 |
| 4 | 31.8 °C, 7 Sept 1911 | 9.9 °C, 26 Sept 1885 and 15 Sept 1986 | 18.0 °C, 11 Sept 1999 | 0.1 °C, *29 Sept 1844* and 27 Sept 1885 |
| 5 | 30.0 °C, 4 Sept 1929 | 10.0 °C, *27 Sept 1824* and 23 Sept 1932 | 17.8 °C, 2 Sept 1939, 3 Sept 1939 and 5 Sept 2005 | 0.2 °C, 30 Sept 1969 |
| ***Last 50 years to 2018*** | | | | |
| | 29.9 °C, 5 Sept 2004 | 9.8 °C, 27 Sept 1993 | 18.1 °C, 7 Sept 2016 | 0.2 °C, 30 Sept 1969 |

*Temperatures recorded by unscreened thermometers prior to 1853 are shown in italics; these are likely to be a little too cold by modern (screened thermometer) standards in winter and a little too warm in summer—see Appendix 1 for more details. In September the mean temperature correction for unscreened thermometry is zero, although this will vary considerably on a day-to-day basis*

September; within the last 60 years this has happened only once, in 2004, although this happened six times in just 27 years between 1880 and 1906. (Remarkably, it has also happened—just once—in October.)

The longest spell of hot days at this time of year, those when the maximum temperature reached at least 25 °C, was in September 1959, when the six days 7–12 September each reached this threshold; spells of five consecutive days at or above 25 °C also occurred in September 1868 and September 1999. In 2011, a remarkably late hot spell commenced on 28 September; the five days 29 September to 3 October all surpassed 25 °C (see also October, Chapter 16). High minimum temperatures are not uncommon in early September, particularly following warm summers when the ground temperature is close to its highest, and occasionally the warmest night of the year occurs in this month. The warmest September night occurred on 5 September 1949, when the minimum temperature was 20.6 °C, following a hot day (maximum temperature of 29 °C on the 4th): even the grass minimum temperature was high at 17.8 °C. This remains the only September night to have remained above 20 °C; the next-warmest, almost 2 degC cooler, occurred all the way back in 1898, when the minimum temperature on 8 September was 18.4 °C.

Cold, wet conditions more typical of mid-autumn are a regular feature of September's weather, although fortunately they do not occur every year. They are often a result of rainy cyclonic northerly conditions, which can lead to an abrupt end to lingering summer warmth. Maximum temperatures can struggle to reach 10 °C; the coldest September day on Oxford's record was 29 September 1918, when the temperature reached only a dismal 7.9 °C; the following day was little better, with a maximum temperature of just 9.7 °C. In recent years, the coldest September day has been 27 September 1993, the maximum temperature reaching just 9.8 °C.

Only three air frosts have ever been recorded in September in Oxford, the most recent being almost 100 years ago, although the screen minimum temperature fell to 0.2 °C on 30 September 1969, with a ground frost recorded that morning (the temperature on the grass surface fell to −2.8 °C). The earliest air frost on the Radcliffe Observatory record occurred on 23 September—in 1845, when the minimum temperature fell to −1.2 °C—although a typical year today can expect only the first *ground* frost of autumn to occur towards the end of the month. In September 1851, the air minimum fell to −3.3 °C on 28th—a killing frost.

The largest daily temperature range on a September day was 19.2 degC, recorded on 6 September 1911 and equalled on 17 September 2003. Both were years with hot, dry weather in August, and the dry ground was able to cool quickly by night but warm up rapidly during the day.

## Warm and cold months

The warmest September on record in Oxford was September 2006, surpassing the previous record-holders of 1949 and 1929 by more than half a degree; September 2016 also made the 'Top 5' warmest and, as would be expected, each of these months included

**Table 15.2** *September mean temperatures at Oxford's Radcliffe Observatory, 1814–2018*

| September mean temperature 14.9°C (average 1981–2010) | | | | | |
|---|---|---|---|---|---|
| Warmest months | | | Coldest months | | |
| Mean temperature, °C | Departure from 1981–2010 normal degC | Year | Mean temperature, °C | Departure from 1981–2010 normal degC | Year |
| 17.8 | +2.9 | 2006 | 11.2 | −3.7 | 1840 |
| 17.3 | +2.4 | 1949 | 11.3 | −3.6 | 1860 |
| 17.0 | +2.1 | 2016 | 11.4 | −3.5 | 1952 |
| 16.9 | +2.0 | 1929 | 11.5 | −3.4 | 1912 |
| 16.6 | +1.7 | 1865, 1999 | 11.7 | −3.2 | 1845, 1847, 1863 and 1877 |
| *Last 50 years to 2018* | | | | | |
| 17.8 | +2.9 | 2006 | 12.0 | −2.9 | 1986 |

some unusually warm days and nights (Table 15.2). Only one day in the first 26 in September 2006 failed to reach 19 °C, and only three nights dropped below 10 °C during the entire month; contrast this with September 1833, when no day reached 19 °C.

The coldest Septembers were all in the early Victorian era—the two coldest being 1840 and 1860. Eleven Septembers since 1815 have failed to reach 19 °C on their warmest day, including September 1833, as mentioned above, although this has not happened since 1912. More recently, September 1986 was remarkably cool, with only two days surpassing 20 °C, a minimum air temperature of 2.3 °C on 19th and five ground frosts in all. Notably sharp ground frosts have included −4.5 °C on 27 September 1885 and −4.4 °C on 22 September 1943.

September is, occasionally, warmer than August. This has happened twice in recent years, in 2014 and in 2006, but only three times prior to that since 1814—in 1865, 1929 and in 1956.

Monthly temperature ranges in September occasionally exceed 30 degC; in September 1906, temperatures ranged 31.3 degC, from 33.1 °C on both 1st and 2nd to 1.8 °C on 28th. The Septembers of 1898, 1911 and 1919 also ranged more than 30 degC, although, curiously, no September within the last 50 years has ranged more than 26 degC. In September 1994, the range from the warmest day to the coldest night was only 15.0 degC—a typical September will include one or two *days* which will exceed this *monthly* range.

## Precipitation

The monthly mean rainfall for September in Oxford is 54 mm. Mean daily rainfall increases quite sharply after the first week of the month (see the annual cycle, Chapter 6).

Since 1800, the wettest September on Oxford's records (Table 15.3) has been 1974, when 156 mm fell—on 23 days, a figure equalled in the Septembers of 1866 and 1950 but only exceeded once, in 1918 (25 days with rain). But the closing years of the eighteenth century produced a run of remarkably wet Septembers—three of the 'top four' on Oxford's list all occurring prior to 1800 and the wettest of all, September 1774, whose monthly total was an astonishing 223.9 mm (based upon the reconstruction of Thomas Hornsby's rainfall records—see Appendix 1). More on the truly extraordinary nature of September 1774's rainfall is given in the Chronology, Chapter 25.

In contrast, in September 1929 measurable rain fell on only two days, and only three in the Septembers of 1959, 1986 and 2014—the latter the second-driest on the record, only 1929 being drier.

Daily falls of 25 mm or more of rain occurred 36 times in September in the 192 years between 1827 and 2018, or in about one year in six; only four daily falls have exceeded 40 mm, although one of these remains the second-wettest day on Oxford's long records. For the 24 hours commencing 0900 GMT on 6 September 1951, 84.8 mm was recorded at the Radcliffe Observatory, of which 66.5 mm fell in 1 hour 26 minutes commencing at 0035 GMT (on 7 September) during a severe overnight thunderstorm; this is the heaviest short-period rainfall event yet recorded in Oxford. Extensive flooding of roads and house basements took place in the city, reported as 'the worst in living memory'. This fall was unusual in that the extent of the very heavy rainfall was confined to a narrow belt

**Table 15.3** *September precipitation at the Radcliffe Observatory, Oxford since 1767. Extremes of monthly totals 1767–2018, wettest days 1827–2018*

| September mean precipitation 54.2 mm (average 1981–2010) | | | | | | | |
|---|---|---|---|---|---|---|---|
| Wettest months | | | Driest months | | | Wettest days | |
| Total fall, mm | Per cent of normal | Year | Total fall, mm | Per cent of normal | Year | Daily fall, mm | Date |
| 223.9 | 445 | 1774 | 2.6 | 5 | 1929 | 84.8 | 6 Sept 1951 |
| 170.9 | 339 | 1768 | 4.1 | 8 | 2014 | 46.4 | 26 Sept 1998 |
| 156.1 | 310 | 1974 | 4.6 | 9 | 1865 | 43.7 | 24 Sept 1915 |
| 153.2 | 305 | 1799 | 5.4 | 10 | 1959 | 41.2 | 22 Sept 1992 |
| 152.6 | 303 | 1896 | 5.8 | 12 | 1795 | 39.8 | 20 Sept 1980 |
| *Last 50 years to 2018* | | | | | | | |
| 156.1 | 310 | 1974 | 4.1 | 8 | 2014 | 46.4 | 26 Sept 1998 |

**Figure 15.1** *Wolvercote Bridge in drought conditions at the climax of the 1976 drought on 3 September (© Oxfordshire History Centre POX 021 7306)*

running west-south-west to east-north-east from Cumnor through Botley and North Hinksey across the city centre towards Marston Village, in places barely 2 km wide. The Radcliffe Observatory was close to the centre-line of this belt of intense thundery rain, for less than 500 m north-west of the Radcliffe site the total 24 hour fall was only 47 mm.

In contrast, the beginning of September 1976 marked the end of the prolonged drought of that year; the lowest 12-month total on Oxford's record ended on 18 September 1976, only 297.4 mm had fallen since the same date in 1975 (Figure 15.1). Compare this with September 1774's 223.9 mm in a single month!

### Snowfall

Snowfall has never been recorded in September in Oxford.

### Thunderstorms

Thunder is heard on at least one day in most Septembers. In the Septembers of 1958, 1974 and 1976, thunder was heard on four days. On 17 September 1909, heavy hail during a thunderstorm broke one of the solar radiation thermometers exposed in the open at the Observatory.

## Sunshine

The transition from August to September sees a reduction in mean sunshine by more than 90 minutes per day; and on average, two or three September days can be expected to remain sunless.

The two sunniest Septembers on record were both over 100 years ago—September 1911 lies at the top of Table 15.4 with 223 hours, a creditable figure for any *summer* month. More recently, September 1959 was also exceptionally sunny, recording 214 hours of sunshine. Twelve days in that month recorded 9 hours or more sunshine, including the six consecutive days 9–14 September which averaged 10.25 hours per day: notably even

**Table 15.4** *September sunshine duration at the Radcliffe Observatory, Oxford: monthly extremes 1880–2018, sunniest days 1921–2018*

**September mean sunshine duration 140.4 hours, 4.68 hours per day** (average 1981–2010)

*Possible daylength: 379 hours. Mean sunshine duration as percentage of possible: 37.1*

| Sunniest months | | | Dullest months | | | Sunniest days | |
|---|---|---|---|---|---|---|---|
| Duration, hours | Per cent of possible | Year | Duration, hours | Per cent of possible | Year | Duration, hours | Date |
| 222.6 | 58.8 | 1911 | 64.9 | 17.1 | 1945 | 12.6 | 1 Sept 2002 |
| 214.0 | 56.5 | 1895 | 79.3 | 20.9 | 1936 | 12.5 | 2 Sept 1982 |
| 213.9 | 56.5 | 1959 | 88.3 | 23.3 | 1896 | 12.4 | 2 Sept 2005 |
| 210.4 | 55.5 | 1964 | 92.3 | 24.4 | 1931 | 12.2 | 4 Sept 1928 |
| 205.9 | 54.4 | 1898 | 92.4 | 24.4 | 1983 | 12.1 | 3 Sept 1976 |
| ***Last 50 years to 2018*** | | | | | | | |
| 175.3 | 46.3 | 2003 | 92.4 | 24.4 | 1983 | 12.6 | 1 Sept 2002 |

the final four days of the month each recorded over 10 hours, about the maximum possible at the end of September as the days draw in after the equinox. At the opposite extreme, September 1945 was exceptionally dull—the sunshine duration totalling less than 65 hours (compare this with January's average of 62 hours, with much shorter day lengths), much lower than any other September on record.

Daylength falls below 12 hours after the autumnal equinox, but 12 hours sunshine is still possible in the first week of September; the sunniest September day on record was 1 September 2002, with 12.6 hours. A typical September will see two or three sunless days, but four Septembers since 1921 (1969, 1976, 1977 and 1992) have recorded five days without sunshine. September 1976 opened with a continuation of that summer's brilliant sunshine, with the sunniest September day in almost 50 years on 3rd (12.1 hours); but after the first week the month was exceptionally dull, averaging less than 3 hours sunshine per day for the remainder of the month, a very dramatic come-down from weeks of seemingly unbroken sunshine—coupled with a wet month which brought a decisive end to the major 18-month drought.

## Wind

Oxford's mean monthly wind speed for September is 3.4 metres per second (m/s), or 6.5 knots, the equal-lowest of the year (with August). The windiest September on a record back to 1880 occurred in 1983, when the monthly mean wind speed was 5.2 m/s (10.1 knots). The calmest September was in 1995, when the monthly mean wind speed was just 1.8 m/s (3.6 knots).

## Oxford temperature, precipitation and sunshine in graphs—September

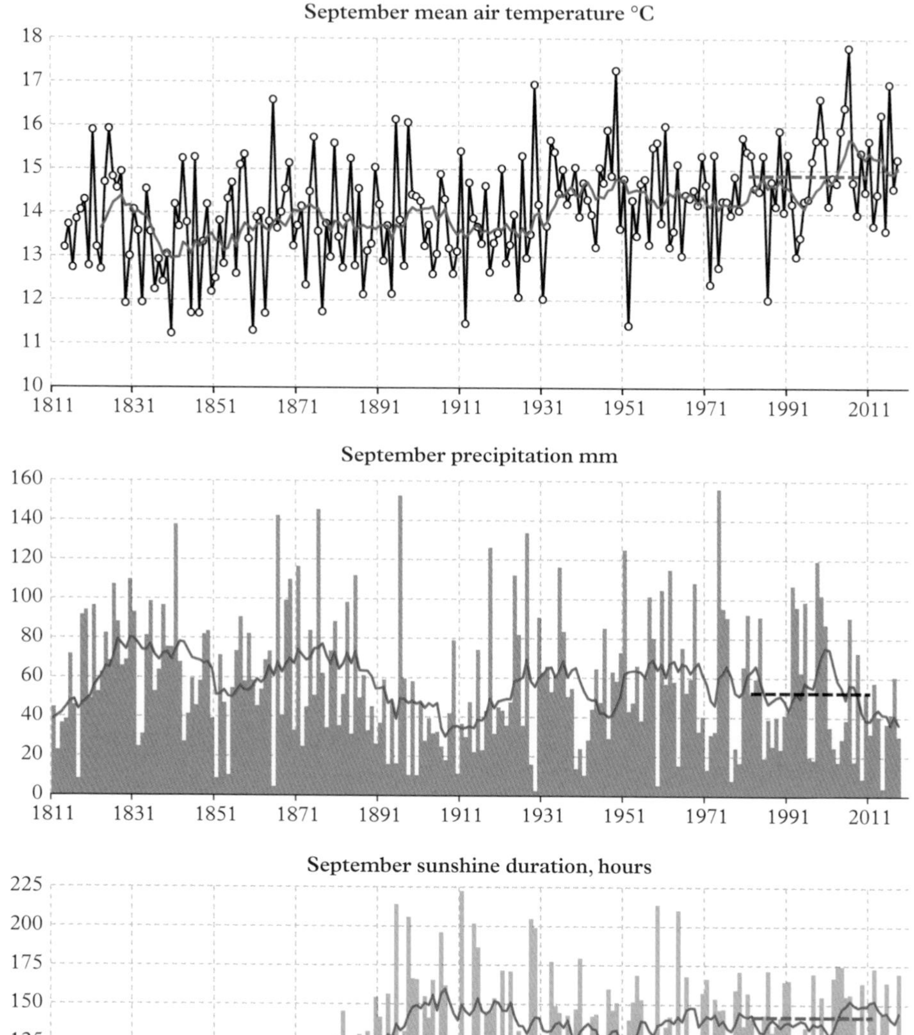


**Figure 15.2** *Monthly values of (from top) mean temperature (°C, since 1814), total precipitation (mm, since 1811) and sunshine duration (hours, since 1880) for September at Oxford. The 1981–2010 averages are indicated by the thick blue dashed line, while the ten-year running mean ending at the year shown is indicated by the red line*

# 16

# October

October is, on average, the wettest month of the year in Oxford, with an increase of about 30 per cent in normal rainfall compared to September (and about three more days with rain during the month). Mid-October sees the wettest week of the year on average, particularly noticeable coming only 6–7 weeks after the average driest week of the year. In some years, the beginning of October can still produce summer-like heat, although the evenings cool quickly. There are further noticeable decreases in both temperature and sunshine duration as the month progresses: the mean daily maximum temperature falls from a pleasant 17 °C at the start of the month to a chillier 13 °C by month-end, while mean daily minimum temperatures fall from a little below 10 ° C to around 6 °C in the same period. The first air frost of autumn can often be expected towards the close of the month, and even the occasional but rare October snowfall. The average temperature for the month as a whole is 11.3 °C. October's mean daily sunshine is more than an hour down on September's, and the percentage of possible sunshine is also further reduced from 37 per cent to 34 per cent, indicating increased cloudiness as well as shorter days. October can also produce occasional very severe gales—most notably the 'Great Storm' of 16 October 1987, although this affected Berkshire, Hampshire and other parts of south-east England much more than Oxfordshire.

## Temperature

Temperatures during October in Oxford have ranged from −5.7 °C back in 1859 to 29.1 °C in 2011. Early October is still capable of producing remarkably late summer heat: temperatures above 25 °C were reached in four Octobers (1921, 1959, 1985 and 2011). The hottest October day on record was 1 October 2011, at a remarkable 29.1 °C (Table 16.1); what was perhaps even more astonishing was that this turned out to be the hottest day of the year, the only time in over 200 years of temperature recording in Oxford that this has occurred in October. The hot spell began in late September 2011, with five consecutive days from 29 September attaining 25 °C, including 27.2 °C as late as 3 October—in fact, the first three days of October 2011 hold the 1-2-3 of the hottest days yet recorded in the month.

*Oxford Weather and Climate since 1767*. Stephen Burt and Tim Burt, Oxford University Press (2019).
 DOI: 10.1093/oso/9780198834632.001.0001

**Table 16.1** *Highest and lowest maximum and minimum temperatures in October at the Radcliffe Observatory, Oxford, 1828–2018*

| Rank | Hottest days | Coldest days | Warmest nights | Coldest nights |
|---|---|---|---|---|
| 1 | 29.1 °C, 1 Oct 2011 | *2.3 °C, 29 Oct 1836* | 17.0 °C, 13 Oct 2018 | −5.7 °C, 24 Oct 1859 |
| 2 | 28.9 °C, 2 Oct 2011 | *2.8 °C, 31 Oct 1823* | 16.8 °C, 3 Oct 1985 | −5.0 °C, 28 Oct 1873, 29 Oct 1873, 28 Oct 1890 |
| 3 | 27.2 °C, 5 Oct 1921, 3 Oct 2011 | 3.3 °C, *31 Oct 1839*, 31 Oct 1934 | 16.7 °C, 12 Oct 2005 | *−4.9 °C, 20 Oct 1842* |
| 4 | 26.7 °C, 6 Oct 1921 | 3.5 °C, *30 Oct 1836*, 27 Oct 1869 | 16.3 °C, 1 Oct 1997 | −4.6 °C, 27 Oct 1873 |
| 5 | 26.4 °C, 1 Oct 1985 | *3.8 °C, 7 Oct 1829, 31 Oct 1836* | 16.2 °C, 9 Oct 1967 | −4.4 °C, *27 Oct 1819* and 23 Oct 1859 |
| *Last 50 years to 2018* | | | | |
| | 29.1 °C, 1 Oct 2011 | 6.0 °C, 30 Oct 1974 | 17.0 °C, 13 Oct 2018 | −3.9 °C, 29 Oct 1997 |

*Temperatures recorded by unscreened thermometers prior to 1853 are shown in italics; these are likely to be a little too cold by modern (screened thermometer) standards in winter and a little too warm in summer—see Appendix 1 for more details. In October the mean temperature correction for unscreened thermometry is +0.2 degC, although this will vary considerably on a day-to-day basis*

With ground temperatures still high in early October, warm nights are not uncommon—eight October nights have remained above 16 °C, the warmest being 13 October 2018 (17.0 °C). Nights as late as 28 October (in 1849) have seen the temperature fall no lower than 16 °C, although, given the fairly rapid decline in temperatures during October, it is no surprise that the lowest temperatures all occur towards the end of the month. The coldest October day was back in 1836, when on 29 October the maximum reached only 2.3 °C, although within the last 50 years no October day has been colder than the 6.0 °C recorded on 30 October 1974 and 6.6 °C on 27 October 2018. As an indication of the temperature contrasts possible in late October, contrast these chilly days with the maximum temperature of 21.6 °C attained on 31 October 2014, the warmest Halloween day on record and an exceptionally high reading so late in the year (had it occurred a single day later, it would have broken the November record by almost 3 degC).

The first air frost of the winter can often be expected towards the end of October; only one year in ten will see an air frost earlier than 19 October, although the average date for the first air frost of the winter at the Radcliffe Observatory over the period 1981–2010 was 8 November. Early autumn frosts were more common in the mid-nineteenth century;

the average date for the first autumn frost during the 30 years 1831–1860 was ten days earlier than currently. There were nine air frosts in the Octobers of 1842 and 1888 and eight in 1895 and 1931, but since 1931 the highest has been six (in October 1997). Ground frosts are more frequent, averaging five mornings in October, although occasionally the month escapes these entirely. Minimum temperatures of −5 °C or below have been recorded four times in October, although the most recent occasion was back in 1890. The last time −4 °C was recorded in October was in 1926, although the temperature fell to −3.9 °C on 29 October 1997. The cold spell in 1997 ran into early November with air frost on seven out of eight mornings from the 25th.

The largest daily range in temperature during October was 19.9 degC, on 13 October 1890, the temperature rising from a morning minimum of 0.6 °C to the day's maximum of 20.5 °C.

## Warm and cold months

The coldest October on record was in 1919, with a mean temperature of 7.1 °C—a value more typical of late November; notably, October 1919 was also the sunniest October on record to that date (Tables 16.2 and 16.4). The Octobers of 1817 and 1887 were almost as cold at 7.2–7.3 °C, but since 1919 the coldest October has been 1974, at 8.0 °C. Octobers have been much warmer in the last three decades, with the top four warmest all occurring since 1995, the warmest on record to date being 2001, with a mean temperature of 14.1 °C, 2.8 degC above the current 30-year average.

The sharp seasonal decline in mean temperatures during the autumn means that no October has ever been warmer than the preceding September, although in both 1847 and 2001 October was only a tenth of a degree Celsius cooler than September. Monthly

**Table 16.2** *October mean temperatures at Oxford's Radcliffe Observatory, 1814–2018*

| October mean temperature 11.3°C (average 1981–2010) | | | | | |
|---|---|---|---|---|---|
| **Warmest months** | | | **Coldest months** | | |
| **Mean temperature, °C** | **Departure from 1981–2010 normal degC** | **Year** | **Mean temperature, °C** | **Departure from 1981–2010 normal degC** | **Year** |
| 14.1 | +2.8 | 2001 | 7.1 | −4.2 | 1919 |
| 14.0 | +2.7 | 2005 | 7.2 | −4.1 | 1817 |
| 13.9 | +2.6 | 2006 | 7.3 | −4.0 | 1887 |
| 13.8 | +2.5 | 1995 | 7.4 | −3.9 | 1840, 1842 |
| 13.4 | +2.1 | 1921, 1968 | 7.5 | −3.8 | 1892, 1905 |
| *Last 50 years to 2018* | | | | | |
| 14.1 | +2.8 | 2001 | 8.0 | −3.3 | 1974 |

temperature ranges in October have varied between 12.7 degC in 2004 and 28.3 degC in 2011.

## Precipitation

October is, on average, the wettest month of the year in Oxford, with a normal monthly fall of 70 mm, 30 per cent more than in September (Table 16.3). The third week of October is, again on average, the wettest week of the year (see annual cycle, Chapter 6), although there is, of course, much variation from year to year.

Since rainfall records commenced in Oxford in 1767, only four Octobers have accumulated over 150 mm of precipitation but, since 1775, no calendar month has been wetter than October 1875. Two days in October 1875 retain the dubious title of the wettest and second-wettest October days on record, just 10 days apart: flooding was extensive in and around Oxford (Figure 16.1). On both days, most of the rain fell in just 9–10 hours. Daily falls of 25 mm or more were recorded on 33 occasions between 1827 and 2018, or about one year in six, but as yet no October day has recorded 50 mm in a day. In October 1875, rain fell on 18 days; the Octobers of 1903 and 1960 both recorded rain falling on 25 days, whereas rain fell on only 5 days in the Octobers of 1834, 1842 and 1884.

The two driest Octobers occurred in 1788 and 1809 but, more recently, 1969 and 1978 were almost as dry. In October 1969 (5.0 mm total) the wettest day of the month saw just 2.0 mm while, in October 1978 (5.4 mm), only one day (17th, 3.1 mm) received as much as 1 mm.

**Table 16.3** *October precipitation at the Radcliffe Observatory, Oxford since 1767. Extremes of monthly totals 1767–2018, wettest days 1827–2018*

**October mean precipitation 70.0 mm** (average 1981–2010)

| Wettest months | | | Driest months | | | Wettest days | |
|---|---|---|---|---|---|---|---|
| Total fall, mm | Per cent of normal | Year | Total fall, mm | Per cent of normal | Year | Daily fall, mm | Date |
| 189.0 | 267 | 1875 | 4.1 | 6 | 1788, 1809 | 48.3 | 9 Oct 1875 |
| 163.3 | 231 | 1903 | 5.0 | 7 | 1969 | 43.0 | 19 Oct 1875 |
| 162.3 | 230 | 1949 | 5.4 | 8 | 1978 | 40.7 | 11 Oct 1910 |
| 150.1 | 212 | 1891 | 6.8 | 10 | 1803 | 40.6 | 17 Oct 1939 and 16 Oct 2007 |
| 147.9 | 209 | 1882 | 7.5 | 11 | 1781 | 38.7 | 9 Oct 1827 |
| ***Last 50 years to 2018*** | | | | | | | |
| 138.8 | 196 | 1987 | 5.0 | 7 | 1969 | 40.6 | 16 Oct 2007 |

**Figure 16.1** *Flooding in and around Oxford in October 1875.* ***Top:*** *Railway flooding (© Oxfordshire History Centre POX 011 3231).* ***Bottom:*** *Abingdon Road (© Oxfordshire History Centre POX 014 9950)*

## Snowfall and lying snow

Snow or sleet is exceedingly rare at this time of year and has been recorded in only four Octobers since 1926 (1933, 1934, 1951 and 2008). The earliest known date when sleet or snow has been observed was 15 October 1934; sleet or snow also fell on 30th and 31st. In 1933, snow fell on 27 October*.

## Thunderstorms

Thunder is infrequent and usually brief in October and occurs on average less than one year in three, although thunder was heard on four days in October 1960 and three days in October 1935.

* Snow certainly fell in the Oxford area during the evening of 28 October 2008, but records of snow or sleet falling are no longer maintained at the Radcliffe Observatory and therefore it is difficult to be conclusive on this point.

## Sunshine

October averages 111 hours of sunshine in Oxford, a decline of more than an hour a day compared to September, and only 34 per cent of the possible. The Octobers of 1921 and 1997 were the sunniest on record, both averaging around 5 hours per day, while October 1976 (following the memorable summer) was exceptionally dull, with less than 90 minutes per day, 20 per cent less than an average December. The reduction in daylength during October means that 10 hours of sunshine in a day is possible only at the beginning of the month: this has occurred several times, the sunniest of all (since 1921) being 1 October 1928 when 10.5 hours of bright sunshine was recorded. The first six days of October 1947 averaged 8.6 hours sunshine daily, while both the first and second days of October 2011 exceeded 10 hours.

A typical October will see around five sunless days, although in 1980 and 2007 there were eleven and in 2015, ten.

**Table 16.4** *October sunshine duration at the Radcliffe Observatory, Oxford, 1880–2018 (daily data 1921–2018)*

**October mean sunshine duration 111.3 hours, 3.59 hours per day** (average 1981–2010)

*Possible daylength: 330 hours. Mean sunshine duration as percentage of possible: 33.8*

| Sunniest months | | | Sunniest days | | | Dullest months | |
|---|---|---|---|---|---|---|---|
| Duration, hours | Per cent of possible | Year | Duration, hours | Per cent of possible | Year | Duration, hours | Date |
| 159.7 | 48.4 | 1921 | 43.6 | 13.2 | 1976 | 10.5 | 1 Oct 1928 |
| 153.8 | 46.6 | 1997 | 49.6 | 15.0 | 1894 | 10.4 | 4 Oct 1994 |
| 147.1 | 44.6 | 1919 | 57.6 | 17.5 | 1915 | 10.3 | 3 Oct 1947<br>4 Oct 1947<br>2 Oct 2011 |
| 144.8 | 43.9 | 1999 | 60.9 | 18.5 | 1960 | 10.25 | 3 Oct 1927 |
| 144.2 | 43.7 | 1959 | 64.0 | 19.4 | 1898 | 10.2 | 2 Oct 1997 |
| *Last 50 years to 2018* | | | | | | | |
| 153.8 | 46.6 | 1997 | 43.6 | 13.2 | 1976 | 10.4 | 4 Oct 1994 |

## Wind

Oxford's mean monthly wind speed for October is 3.8 metres per second (m/s), or 7.4 knots. The windiest October on a record back to 1880 occurred in 1967, when the monthly mean wind speed was 7.4 m/s (14.3 knots)—the windiest month on Oxford's records. The calmest October was in 1951, when the mean wind speed was 2.2 m/s (4.3 knots). During the 'Great Storm' of 16 October 1987, the highest gust in Oxford (recorded at 45 m above ground level) was 62 knots (31.9 metres per second), shortly before 0600 GMT: more details on this event appear in the Chronology, Chapter 25.

## Oxford temperature, precipitation and sunshine in graphs—October

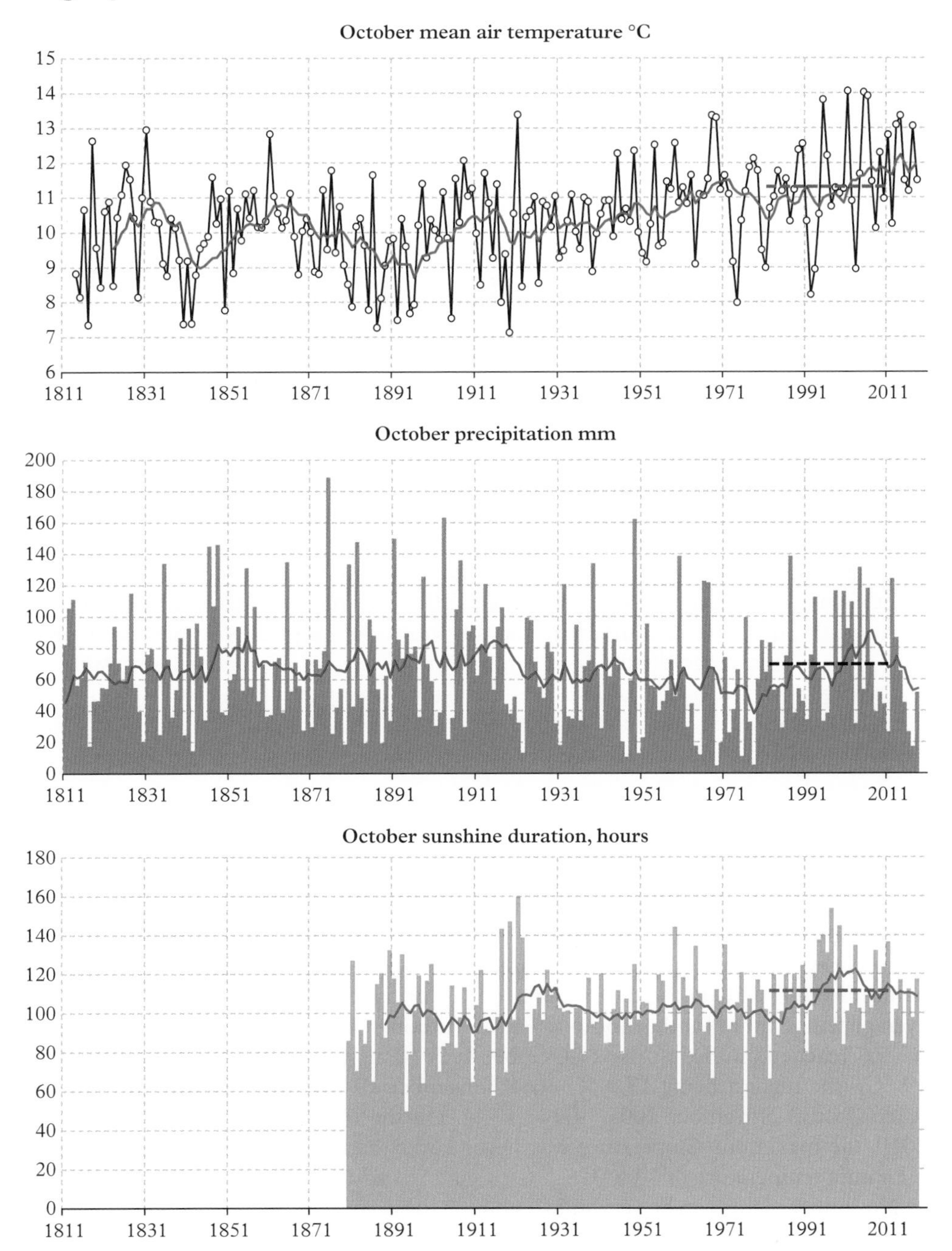


**Figure 16.2** *Monthly values of (from top) mean temperature (°C, since 1814), total precipitation (mm, since 1811) and sunshine duration (hours, since 1880) for October at Oxford. The 1981–2010 averages are indicated by the thick blue dashed line, while the ten-year running mean ending at the year shown is indicated by the red line*

# 17

# November

The rapid decline in temperatures at this time of year, together with abundant moisture following on from October's rainfall and the migration of the time of sunrise towards the morning rush hour, means that foggy mornings are particularly prevalent during November; as noted in Chapter 1, Oxford city centre's topographic location in the Thames Valley makes it susceptible to fog in any case. However, the decrease in air pollution since the introduction of the Clean Air Act in 1956 has dramatically reduced both the frequency and the density of fogs in Oxford in recent decades.

Warm summer-like days are very rare in November, although temperatures of 17–18 °C are occasionally attained. The first sharp frosts of winter also put in an appearance most Novembers and temperatures towards the end of the month can sometimes drop below −5 °C by night and remain below freezing all day. Precipitation—usually rain, but very occasionally snow—occurs on about one day in two in a typical November. Although amounts are on average slightly less than in October, five of the ten wettest months on Oxford's record have been Novembers—the highest of any month in the table. Sunshine duration continues its seasonal decline, to average something over two hours daily; around nine of November's 30 days are likely to remain sunless.

## Temperature

Average daytime maximum temperatures continue their seasonal decline throughout November, falling from a little above 13 °C at the beginning of the month to below 9 °C at the close, while average night minimum temperatures follow a slightly more irregular pattern from 6–7 °C at the beginning to 3 °C during the final week of the month. The average temperature for the month is 7.5 °C. Although a maximum temperature of 21.6 °C was reached on the last day of October in 2014, no November day has yet attained 19 °C—the highest being 18.9 °C on 4 November 1946; more recently, 18.0 °C was reached on 2 November 2005 (Table 17.1). During the first five days of November 2010, the maximum temperature was never lower than 14 °C, while one night had a minimum temperature of 13.6 °C.

November nights can be very mild, although these conditions are much more likely at the beginning of the month than at the end. November nights milder than the average

*Oxford Weather and Climate since 1767*. Stephen Burt and Tim Burt, Oxford University Press (2019).
 DOI: 10.1093/oso/9780198834632.001.0001

**Table 17.1** *Highest and lowest maximum and minimum temperatures in November at the Radcliffe Observatory, Oxford, 1815–2018*

| Rank | Warmest days | Coldest days | Mildest nights | Coldest nights |
|---|---|---|---|---|
| 1 | 18.9 °C, 4 Nov 1946 | −2.7 °C, 28 Nov 1890 | 15.0 °C, 22 Nov 1947 | *−10.1 °C, 17 Nov 1815* |
| 2 | *18.5 °C, 7 Nov 1845* | −1.9 °C, 27 Nov 1915 | 14.7 °C, 3 Nov 1996 | −9.1 °C, 23 Nov 1858 |
| 3 | 18.3 °C, *8 Nov 1852* and 5 Nov 1938 | −1.7 °C, 23 Nov 1858 | *14.4 °C, 1 Nov 1852* | *−8.9 °C, 24 Nov 1816* |
| 4 | 18.0 °C, 2 Nov 2005 | *−1.6 °C, 17 Nov 1815* | 14.0 °C, 20 Nov 1994 | −8.8 °C, 26 Nov 1923 |
| 5 | 17.3 °C, 13 Nov 2011 and 6 Nov 2015 | −1.4 °C, 25 Nov 1923 | 13.9 °C, *2 Nov 1821* and 3 Nov 1969 | −8.4 °C, 24 Nov 1904 and 27 Nov 1923 |
| ***Last 50 years to 2018*** | | | | |
| | 18.0 °C, 2 Nov 2005 | −1.2 °C, 28 Nov 2010 | 14.7 °C, 3 Nov 1996 | −7.0 °C, 23 Nov 1983 |

*Temperatures recorded by unscreened thermometers prior to 1853 are shown in italics; these are likely to be a little too cold by modern (screened thermometer) standards in winter and a little too warm in summer—see Appendix 1 for more details. In November the mean temperature correction for unscreened thermometry is +0.3 degC, although this will vary considerably on a day-to-day basis*

for midsummer have occurred on numerous occasions; the mildest November night on record was remarkably late in the month, when on 22 November 1947 the minimum temperature was 15.0 °C, following three consecutive days which reached 16 °C. Almost as mild, but in the first week of the month, was the minimum temperature of 14.7 °C on 3 November 1996, comparing favourably with the mean daily minimum temperature for July of 13.0 °C. As the days become shorter with the approach of the winter solstice, fog becomes more prevalent and can be slow to clear. Occasionally, maximum temperatures in November fail to rise above 0 °C, often because weak winter sunshine fails to clear overnight freezing fog. The coldest November day on record was 28 November 1890, the start of a prolonged bitterly cold spell, when the maximum temperature reached only −2.7 °C; similar circumstances in 2010 led to a maximum temperature of −1.2 °C on the same date that year.

During 1981–2010 the average date of the first autumn air frost was 8 November—although more than one year in ten did not see an air frost until December. No air frost was recorded in nine Novembers in the 50 years from 1968, or now about one year in five, whereas the first frost-free November on Oxford's records (since 1815) was not until 1877. In November 1851, there were 19 air frosts, and 17 in November 1910; more recently, November 1988 recorded 15 air frosts.

Minimum temperatures of −5 °C or below are now distinctly rare in November, although they were much more frequent in the nineteenth and early twentieth century. At the time of writing, the last November night colder than −5 °C was more than 30 years ago, in 1983 (−7.0 °C on 23rd and −6.7 °C the following night); between 1815 and 1923 a November minimum of −5 °C or lower could be expected about one year in two. The coldest night of the calendar year is occasionally recorded in November, about once every 10–15 years on average; within the last 50 years, this occurred in 1983, 1988, 1989 and 2005.

The largest daily temperature range recorded in November was 19.8 degC on 24 November 1983, the temperature rising from −6.7 °C at 0900 GMT on 23rd to 13.1 °C at 0900 GMT on 25th—the period of 48 hours being a consequence of the terminal hours of the measurement protocols.

## Warm and cold months

The colder spells of the nineteenth and early twentieth century are reflected in the 'coldest Novembers', Table 17.2, where the most recent entry is almost a century old. The coldest of all was November 1815, with a mean temperature of 0.8 °C, almost 7 degC below the current November normal; the coldest November within the last 50 years was 1985, with a mean temperature of 4.6 °C. At the other end of the temperature scale, the mildness of November 1852, coming the year after the very cold November of 1851, is remarkable, and was not surpassed for over 80 years. Recent Novembers dominate the 'warmest' side of Table 17.2, however, with four of the five mildest occurring since 1994.

**Table 17.2** *November mean temperatures at Oxford's Radcliffe Observatory, 1814–2018*

**September mean temperature 14.9°C** (average 1981–2010)

| Warmest months | | | Coldest months | | |
|---|---|---|---|---|---|
| Mean temperature, °C | Departure from 1981–2010 normal degC | Year | Mean temperature, °C | Departure from 1981–2010 normal degC | Year |
| 10.6 | +3.1 | 1994 | 0.8 | −6.7 | 1815 |
| 10.2 | +2.8 | 2015 | 2.6 | −4.9 | 1851 |
| 9.9 | +2.4 | 1938, 2011 | 3.2 | −4.3 | 1816, 1923 |
| 9.7 | +2.2 | 2009 | 3.3 | −4.2 | 1829, 1871 |
| 9.5 | +2.0 | 1852 | 3.4 | −4.1 | 1915, 1919 |
| *Last 50 years to 2018* | | | | | |
| 10.6 | +3.1 | 1994 | 4.6 | −2.9 | 1985 |

November is occasionally slightly milder than the previous October. This has happened six times since 1814, the most extreme being in 1881 when a very mild November followed a very cold October—the mean temperature for November was 1.6 degC greater than October. This reversion of the normal seasonal trend also occurred in 1852, 1895, 1939 and most recently in 1994.

The greatest range in temperature in any November came in 1978, when on 8 November the maximum temperature reached 16.9 °C while, on the last day of the month, the minimum temperature fell to −6.9 °C for a monthly range of 23.8 degC; the last two days of the month remained below freezing all day. In contrast, the Novembers of 1860 and 1878 saw a temperature range during the month of just 12.0 degC.

## Precipitation

The monthly mean rainfall for November in Oxford is 67 mm; the daily mean rainfall is practically identical to that for October (Table 17.3). Rain can be expected to fall on about one day in two.

November is a consistently wet month. Only once since the record commenced in 1767 has less than 10 mm been recorded—in November 1945, just 4.8 mm of precipitation fell, less than half of that of the next-driest November (Table 17.3). In contrast, a monthly total of 150 mm or more has occurred seven times in the same period, the wettest being 192 mm in November 1770. In the Novembers of 1926, 1960, 2002 and 2009, rain fell on 26 of the month's 30 days, although none of these months were as wet as 150 mm.

**Table 17.3** *November precipitation at the Radcliffe Observatory, Oxford since 1767. Extremes of monthly totals 1767–2018, wettest days 1827–2018*

| **November mean precipitation 66.7 mm** (average 1981–2010) | | | | | | | |
|---|---|---|---|---|---|---|---|
| **Wettest months** | | | **Driest months** | | | **Wettest days** | |
| **Total fall, mm** | **Per cent of normal** | **Year** | **Total fall, mm** | **Per cent of normal** | **Year** | **Daily fall, mm** | **Date** |
| 192.4 | 288 | 1770 | 4.8 | 7 | 1945 | 37.9 | 14 Nov 1894 |
| 175.7 | 264 | 1852 | 12.1 | 18 | 1788 | 34.9 | 1 Nov 2008 |
| 175.5 | 264 | 1940 | 13.3 | 20 | 1771 | 33.7 | 13 Nov 1861 |
| 165.9 | 249 | 1929 | 13.4 | 20 | 1901 | 33.3 | 12 Nov 1894 |
| 165.8 | 249 | 1772 | 15.7 | 24 | 1851 | 32.9 | 8 Nov 1906 |
| ***Last 50 years to 2018*** | | | | | | | |
| 150.3 | 226 | 1970 | 20.0 | 30 | 1990 | 34.9 | 1 Nov 2008 |

500 THE ILLUSTRATED LONDON NEWS. [DEC. 4, 1852.

THE INUNDATION OF CHRISTCHURCH MEADOWS, OXFORD.

**Figure 17.1a** *November flooding in Oxford. Top: floods in Christ Church Meadow in November 1852 (from* The Illustrated London News, *4 December 1852 © Oxfordshire History Centre POX 010 2674); flooding in Fisher Row, Oxford, during the great flood of November 1894 (© Oxfordshire History Centre POX 010 2674)*

**Figure 17.1b** *Flooding in Oxford, November 2012 - (Top) Botley Park (Bottom) Oatlands Park (both photographs courtesy of Jane Buekett)*

At this time of year, catchments are normally at or near saturation, and a prolonged spell of wet weather is likely to cause extensive and persistent flooding. This was the case in November 1852, when 79.7 mm of rain fell in the week commencing 11 November, and particularly so in 1894, when 82.8 mm fell in 4 days commencing also on 11 November—this latter occasion producing the largest recorded flood in the Thames Valley (Figure 17.1: see also the Chronology in Chapter 25). The highest recorded mean daily flow for the Thames at Teddington on a record back to 1883 remains 18 November 1894, at 806 cubic metres per second, $m^3/s$ [97]—Figure 25.8 in Chapter 25. This spell includes two of the five wettest November days in Oxford in almost 200 years records, only two days apart (Table 17.3). Other noteworthy November floods occurred in 1940 and 1970. Daily falls of 25 mm in November or more occur slightly less than once per decade on average; only eight November days have reached or surpassed 30 mm in 24 hours since 1827.

### Snowfall and lying snow

Snow or sleet can be expected in November less than one year in two, although in 1952 snow or sleet fell on nine days (and lay on four), and five days in November 1947 and November 1965. Lying snow at 0900 GMT is uncommon, less than once per decade on average, although November 1952 had snow on the ground on four mornings, and November 1947 had three. Since 1952, no November has had more than a single morning with snow cover, the greatest depth being 1 cm on 20 November 1988.

### Thunderstorms

Thunder is rare in November, averaging about once in ten years, although thunder was heard on two days in November 1955.

## Sunshine

November is much less sunny than October, owing to greater cloudiness and sharply reduced daylength (almost 20 per cent less daylight hours). The combination reduces the monthly average duration of bright sunshine to 71 hours, or around 2 hours and 22 minutes daily, 27 percent of possible (down from October's 34 per cent) as shown in Table 17.4. There are, however, large variations from year to year.

With almost 108 hours of bright sunshine, November 1923 remains the sunniest on Oxford's records—and also one of the coldest (Tables 17.2 and 17.4). November 1888 retains the title of the dullest November, with less than 30 hours bright sunshine, although the very dry November 1945 (31 hours sunshine) was almost as dull.

The duration of daylight during November in Oxford decreases from 9.63 hours on the first day of the month to 8.17 hours on the last, an average decrease of just under 3 minutes per day. The sunniest November days at the beginning of the month can thus receive in excess of 8 hours sunshine, and this has been recorded on numerous occasions, the sunniest day of all being 1 November 1990, with 8.9 hours or 92 per cent of the possible sunshine.

**Table 17.4** *November sunshine duration at the Radcliffe Observatory, Oxford, 1880–2018 (daily data 1921–2018)*

**November mean sunshine duration 71.0 hours, 2.37 hours per day** (average 1981–2010)

Possible daylength: 265 hours. Mean sunshine duration as percentage of possible: 26.8

| Sunniest months | | | Dullest months | | | Sunniest days | |
|---|---|---|---|---|---|---|---|
| Duration, hours | Per cent of possible | Year | Duration, hours | Per cent of possible | Year | Duration, hours | Date |
| 107.85 | 40.7 | 1923 | 29.7 | 11.2 | 1888 | 8.9 | 1 Nov 1990 |
| 104.45 | 39.4 | 1925 | 31.0 | 11.7 | 1945 | 8.6 | 3 Nov 1988<br>4 Nov 1995 |
| 104.1 | 39.2 | 1989 | 31.6 | 11.9 | 1912 | 8.5 | 4 Nov 1988<br>7 Nov 2016 |
| 102.4 | 38.6 | 1977 | 32.7 | 12.3 | 1962 | 8.4 | 4 Nov 1961<br>1 Nov 1988<br>4 Nov 2013 |
| 102.0 | 38.4 | 1971 | 33.8 | 12.7 | 2015 | 8.35 | 4 Nov 1930 |
| *Last 50 years to 2018* | | | | | | | |
| 104.1 | 39.2 | 1989 | 33.8 | 12.7 | 2015 | 8.9 | 1 Nov 1990 |

On average, about nine November days can be expected to remain sunless; in November 1962 19 days remained sunless, and more recently 17 remained without any sunshine in November 2015.

## Wind

Oxford's mean monthly wind speed for November is 4.0 metres per second (m/s), or 7.7 knots. The windiest November on a record back to 1880 occurred in 1882, when the monthly mean wind speed was 6.2 m/s (12.0 knots); the calmest November was in 1958, when the mean wind speed was 2.2 m/s (4.3 knots).

## Oxford temperature, precipitation and sunshine in graphs—November

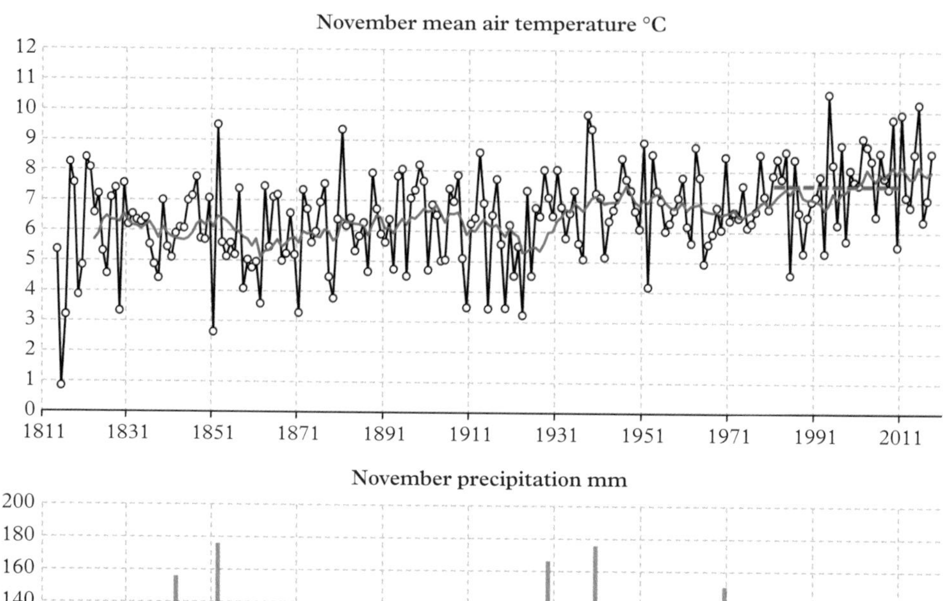


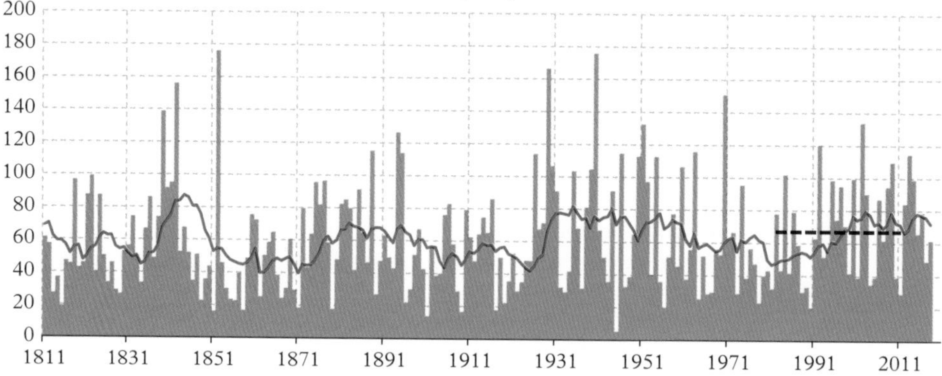


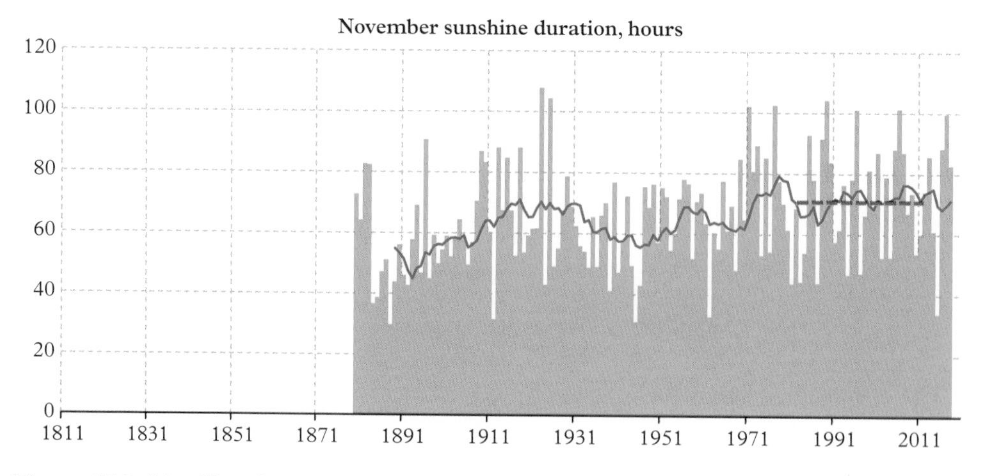


**Figure 17.2** *Monthly values of (from top) mean temperature (°C, since 1814), total precipitation (mm, since 1811) and sunshine duration (hours, since 1880) for November at Oxford. The 1981–2010 averages are indicated by the thick blue dashed line, while the ten-year unweighted running mean ending at the year shown is indicated by the red line*

# 18
# December

December has the unenviable distinction of being, on average, the dullest month of the year, with a daily quota of just over 100 minutes of sunshine. In a typical December, almost half of the month's days will remain sunless, and the week leading up to Christmas is, on average, the least sunny week of the year. Temperatures continue to fall during the month, more slowly after the winter solstice, but the week between Christmas and New Year has the lowest mean maximum temperatures of the year in Oxford. December is, on average, less wet than October or November, yet rain can be expected on one day in two. The first snowfalls of winter normally occur in December.

## Temperature

Average daytime temperatures range from above 9 °C during the first week to below 7 °C towards the end of the month, while average night minimum temperatures are more variable, from 3–4 °C early in the month to 1–3 °C during the rest of the month. The average temperature for the month is 5.0 °C.

The highest December temperatures are normally reached in deep south-westerly airstreams, which can bring very mild conditions, even warm, irrespective of the time of day, although such events are often breezy (reflecting the rapid advection of the mild air responsible). Less often, such mildness can be accompanied by midwinter sunshine and light winds, and these days are perhaps the most pleasant. Daytime temperatures have reached 15 °C and minimum temperatures have exceeded 12 °C several times in December (Table 18.1). The exceptionally mild December of 2015 accounts for the three warmest December days of the last 200 years, and two of the mildest nights. The mildest December day on record was 18 December 2015, when the temperature reached 15.9 °C; the following day was just a tenth of a degree cooler (Figure 18.1, lower). As far back as 1856, however, 15.0 °C was reached on two consecutive days in the first week of December.

Minimum temperatures above 10 °C in December are not particularly uncommon—about one December in two will see at least one 24-hour period remaining at or above this level. The mildest December night followed the highest December temperature on record to that date; the minimum temperature in the 24 hours ending 0900 UTC on

*Oxford Weather and Climate since 1767*. Stephen Burt and Tim Burt, Oxford University Press (2019).
 DOI: 10.1093/oso/9780198834632.001.0001

**Table 18.1** *Highest and lowest maximum and minimum temperatures at the Radcliffe Observatory, Oxford, in December 1815–2018*

| Rank | Mildest days | Coldest days | Mildest nights | Coldest nights |
|---|---|---|---|---|
| 1 | 15.9 °C, 18 Dec 2015 | −6.7 °C, 22 Dec 1890 | 12.5 °C, 3 Dec 1985 | −17.8 °C, 24 Dec 1860 |
| 2 | 15.8 °C, 19 Dec 2015 | −5.7 °C, 30 Dec 1908 | 12.4 °C, 21 Dec 1971 and 5 Dec 1986 | −16.7 °C, 28 Dec 1860 |
| 3 | 15.5 °C, 26 Dec 2015 | −5.4 °C, 17 Dec 1859 | 12.3 °C, 5 Dec 1898 | −16.1 °C, 13 Dec 1981 |
| 4 | 15.2 °C, 2 Dec 1985 | *−5.2 °C, 24 Dec 1830* | 12.2 °C, 7 Dec 2015 | −14.7 °C, 24 Dec 1878 |
| 5 | 15.0 °C, 6 Dec 1856, 7 Dec 1856, 3 Dec 1948 and 4 Dec 1979 | −4.9 °C, 15 Dec 1899 | 12.1 °C, 12 Dec 2000 and 17 Dec 2015 | −13.3 °C, 25 Dec 1860 and 22 Dec 1890 |
| *Last 50 years to 2018* | | | | |
| | 15.9 °C, 18 Dec 2015 | −4.4 °C, 19 Dec 2010 | 12.5 °C, 3 Dec 1985 | −16.1 °C, 13 Dec 1981 |

*Temperatures recorded by unscreened thermometers prior to 1853 are shown in italics; these are likely to be a little too cold by modern (screened thermometer) standards in winter and a little too warm in summer—see Appendix 1 for more details. In December the mean temperature correction for unscreened thermometry is +0.3 degC, although this will vary considerably on a day-to-day basis*

3 December 1985 was 12.5 °C, the second of four consecutive nights that remained above 10 °C. In December 2015, five consecutive nights remained above 10 °C from the 16th.

Within the last 50 years, around one in four Decembers has included at least one day when the maximum temperature failed to reach 0 °C; four such days occurred in 1981 (three consecutive) and eight in 2010 (five consecutive). Four December days have failed to reach -5 °C, the coldest of all being 22 December 1890 when the maximum temperature was -6.7 °C; the most recent of these was more than a century ago in 1908. The coldest December day within the past 50 years was 19 December 2010 at −4.4 °C. Four consecutive days from 16 December 1859 failed to rise even as far as −3 °C.

Minimum temperatures of -10 °C or lower have been recorded on over 100 occasions in Oxford since 1815, but less than a quarter of these occurred in December. Christmas 1860 was exceptionally cold, the temperature falling to −17.8 °C on Christmas Eve and −16.7 °C on 28 December (see also the Chronology in Chapter 25).

The coldest December night since 1860 occurred on 13 December 1981, when the screen minimum fell to -16.1 °C—one of the coldest nights of any month on Oxford's long record and, aside from 6–7 December 1879, the only occasion on which −10 °C has been reached before mid-December. Snow lay 14 cm deep that morning, following heavy snowfalls in previous days. More recently, the temperature fell to −10.9 °C on 20 December 2010 (Figure 18.1, upper).

The largest daily temperature range during December was 20.1 degC on 13 December 1981 (minimum −16.1 °C followed by a maximum of 4.0 °C). Remarkable also were 22–23 December 1855, both days seeing an extreme daily range of 17.2 degC—after a minimum of −9.7 °C on 22nd, the temperature rose to 7.5 °C by day. The following day's minimum temperature was recorded as −8.4 °C, although this may have occurred

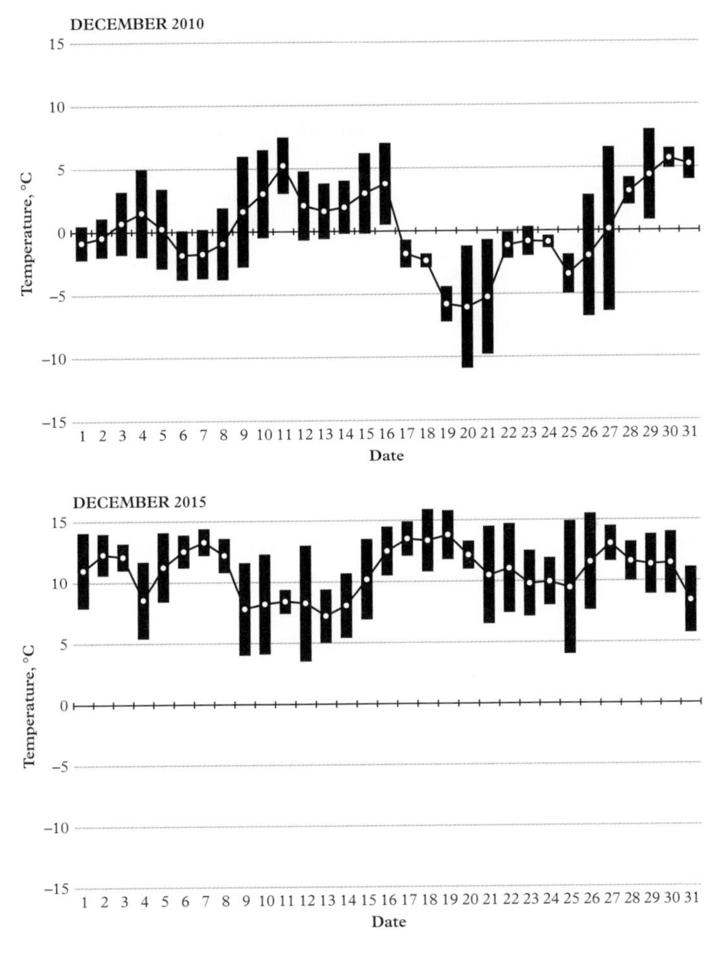


**Figure 18.1** *Daily temperatures (maximum, minimum and mean daily temperature) in two very contrasting Decembers in recent years.* ***Upper****: December 2010, the coldest December since 1890.* ***Lower****: December 2015, the mildest December in at least 300 years. See also Chapter 4 for the synoptic context of both months*

when the thermometer was reset at 9 a.m. on 22nd, and the temperature rose to 8.8 °C later that day.

## Mild and cold months

December 1890 must have been a particularly cheerless month, for not only was it the coldest December on record, it was also the cloudiest month of any name on Oxford's records (see Tables 18.2 and 18.4). Three Decembers have recorded a mean temperature below 0 °C (Table 18.2): December 1879 and December 2010 were both only slightly above 0 °C. December 1829 remains the second coldest December on the record, following a November which also remains the fourth coldest. December 1981 was particularly cold in the second week, remarkably early for severe cold in the winter.

December 2015 stands head and shoulders above any other December as the mildest on record—in fact, the mean temperature of 10.8 °C was milder than any November, December, January, February or March, and it would rank just outside the top five warmest *Aprils*! With no air frost during the month, and just three minor ground frosts, the month saw the greatest positive departure (5.8 degC) from the current 30-year mean temperature for any month—not only on Oxford's record extending back to 1814, but also on the Central England Temperature series since 1659 [58]. December 2015 was a remarkable 10.6 degC warmer in mean air temperature than December 2010, just five years previously (see Chapter 4 for the synoptic context of both months, and Figure 18.1).

**Table 18.2** *December mean temperatures at Oxford's Radcliffe Observatory, 1814–2018*

| **December mean temperature 5.0°C** (average 1981–2010) | | | | | |
|---|---|---|---|---|---|
| **Warmest months** | | | **Coldest months** | | |
| **Mean temperature, °C** | **Departure from 1981–2010 normal degC** | **Year** | **Mean temperature, °C** | **Departure from 1981–2010 normal degC** | **Year** |
| 10.8 | +5.8 | 2015 | −2.0 | −7.0 | 1890 |
| 9.0 | +4.0 | 1852 | −0.2 | −5.2 | 1829, 1840 |
| 8.2 | +3.2 | 1934, 1974 | 0.1 | −4.9 | 1846 |
| 7.8 | +2.8 | 1868 | 0.2 | −4.8 | 1879, 2010 |
| 7.7 | +2.7 | 1988 | 0.4 | −4.6 | 1844, 1870 |
| ***Last 50 years to 2018*** | | | | | |
| 10.8 | +5.8 | 2015 | 0.2 | −4.8 | 2010 |

It is not particularly uncommon for December to be milder than November—within the last 100 years, this has happened 16 times—but December milder than *October* has happened only twice, in 1851 and 1974.

Monthly temperature ranges in December have varied from 12.2 degC in 1934 and 12.4 degC in 2015, to 29.5 degC in 1860 and 26.8 degC in 1878.

## Precipitation

The monthly mean precipitation for December in Oxford is 63 mm, falling on one day in two on average (Table 18.3).

Although only the third-wettest month of the year (after October and November), December is rarely dry—only 10 Decembers have received less than 10 mm of rainfall since records began in 1767, only one of these within the last century (1933): the driest of all was December 1780, with just 1.7 mm or 3 per cent of the current normal for the month. December 1914 remains the wettest of any of the three meteorological winter months (December, January, February), with extensive and persistent flooding along the Thames during the month, a feature also of the Decembers of 1929 and 1989. In December 1934, measurable rain was recorded on 26 of the month's 31 days, but in December 1840 only two days rainfall were recorded.

Daily falls of 25 mm or more in December can be expected about once per decade. The wettest December days were 23 December 1985 (38.3 mm) and Christmas Day in 1927 (34.1 mm); on the first of these, much of the rain fell in about 7 hours overnight, while the day saw spells of sunshine and only a few light showers. Two of the five wettest December days on record occurred only a fortnight apart in December 1979, with great flooding resulting on the Thames after Christmas. In December 1989,

**Table 18.3** *December precipitation at the Radcliffe Observatory, Oxford since 1767. Extremes of monthly totals 1767–2018, wettest days 1827–2018*

| **December mean precipitation 63.1 mm** (average 1981–2010) | | | | | | | |
|---|---|---|---|---|---|---|---|
| **Wettest months** | | | **Driest months** | | | **Wettest days** | |
| **Total fall, mm** | **Per cent of normal** | **Year** | **Total fall, mm** | **Per cent of normal** | **Year** | **Daily fall, mm** | **Date** |
| 147.9 | 235 | 1914 | 1.7 | 3 | 1780 | 38.3 | 23 Dec 1985 |
| 142.2 | 226 | 1934 | 2.4 | 4 | 1775 | 34.1 | 25 Dec 1927 |
| 141.5 | 224 | 1989 | 3.9 | 6 | 1835 | 33.2 | 27 Dec 1979 |
| 131.7 | 209 | 1929 | 4.7 | 7 | 1843 | 33.0 | 26 Dec 1886 |
| 127.7 | 203 | 1979 | 7.5 | 12 | 1788 | 32.8 | 13 Dec 1979 |
| ***Last 50 years to 2018*** | | | | | | | |
| 141.5 | 224 | 1989 | 13.4 | 20 | 1988 | 38.3 | 23 Dec 1985 |

**Figure 18.2** *Christmas flooding, December 2012.* ***Upper:*** *Flooded railway line at Redbridge, 27 December 2012.* ***Lower:*** *Grandpont, Christmas Day 2012 (Both photographs courtesy of Jane Buekett)*

the third-wettest December on record, the entire month's rainfall (141 mm) fell in 15 consecutive days from 11 December—the first ten and the last six days of the month remaining completely dry (see also the Chronology, Chapter 25). More recently, 26.3 mm fell on 23 December 2013, the first of a number of very wet days during the exceptionally wet winter of 2013/14, the wettest winter on Oxford's records (see Winter, Chapter 20).

## Snowfall and lying snow

An average December will see snow or sleet falling on two or three days, although there are wide variations from year-to-year—about one in three Decembers are snowless, and many snowfalls in December are slight. December 1950 saw snow or sleet observed to fall on 12 days, December 1938 on nine, and 1927 and 1981 eight days. The arithmetic mean number of days with lying snow is very skewed towards the occasional snowy month. Although two in three Decembers do not experience any snow cover (i.e. at least 50 per cent lying snow at 0900 GMT), the average is 1.1 mornings with snow lying. How does this come about? In December 1981, there were nine mornings with snow cover and, in December 2010, 12 mornings. These two snowy Decembers alone account for 21 mornings with snow cover of the 33 observed during the standard 30-year averaging period 1981–2010—an average of 1.1 days per year (33/30), although 23 of the 30 years did not record a single morning with snow cover.

December 1981 and January 1982 both had spells of heavy, disruptive snow and low temperatures. Heavy snow fell on 8 December, topped up on 11 December, with the snow depth on the morning of 12 December 1981 standing at 16 cm. Travel was severely disrupted by snowfall and low temperatures during the week, and diesel froze in lorry

**Figure 18.3** *Snow at The Lodge, Wytham, Oxford, 19 December 2010 (Jonathan Webb)*

tanks as the temperature fell to −16.1 °C on 13 December (Table 18.1), the coldest December morning for more than 120 years. Further snow fell just before Christmas and lay for a few days; after a mild spell over New Year, further snowfall in January led to more severe conditions. Since 1959, four Decembers have recorded at least one morning with 10 cm or more of level snow at the morning observation—1961, 1962, 1981 and 2010: the greatest snow depths have been 20 cm on 30 December 1962 (marking the beginning of the bitter winter of 1962/63), 16 cm on 12 December 1981 and 18 cm on 19 and 20 December 2010.

### Thunderstorms

Thunder is uncommon in December, typically about once per decade, and usually of short duration.

## Sunshine

December is, on average, the dullest month of the year, with a daily quota of just an hour and 43 minutes. This limited measure is a result of the combination of reduced daylight hours (the shortest day of the year, the winter solstice on 21 December, sees just 7.78 hours of daylight) and high average cloud amounts. December is, on average, the cloudiest month of the year, resulting in the lowest percentage of possible sunshine for any month—barely one hour in five sees any sunshine (Table 18.4).

**Table 18.4** *December sunshine duration at the Radcliffe Observatory, Oxford, 1880–2018 (daily extremes 1921–2018)*

**December mean sunshine duration 53.2 hours, 1.72 hours per day** (average 1981–2010)
*Possible daylength: 244 hours. Mean sunshine duration as percentage of possible: 21.8*

| Sunniest months | | | Dullest months | | | Sunniest days | |
|---|---|---|---|---|---|---|---|
| Duration, hours | Per cent of possible | Year | Duration, hours | Per cent of possible | Year | Duration, hours | Date |
| 96.9 | 39.7 | 2014 | 5.0 | 2.0 | 1890 | 7.5 | 1 Dec 2016 |
| 79.2 | 32.4 | 2001 | 18.2 | 7.5 | 1956 | 7.3 | 6 Dec 2014 |
| 77.7 | 31.8 | 1952 | 20.4 | 8.4 | 2010 | 7.2 | 2 Dec 1961<br>3 Dec 2014 |
| 76.5 | 31.3 | 1961 | 23.2 | 9.5 | 1884 | 7.1 | 12 Dec 1981<br>6 Dec 1986<br>15 Dec 2000 |
| 75.0 | 30.7 | 1917 | 23.25 | 9.5 | 1930 | 7.0 | *5 occasions* |
| *Last 50 years to 2018* | | | | | | | |
| 96.9 | 39.7 | 2014 | 20.4 | 8.4 | 2010 | 7.5 | 1 Dec 2016 |

December is also the month when year-to-year variations in sunshine are at their most pronounced. Since sunshine records commenced in February 1880, December's sunshine total has varied by almost a factor of 20—from a dismal 5 hours in 1890 (only 2 per cent of possible, by a large margin the dullest of any month on the record) to almost 97 hours in 2014.

It is rare for any day in December to receive more than 7 hours of bright sunshine; this value has been exceeded on only seven occasions since 1921. It is not unusual for one or two sunny days in the winter months to account for half of the month's total sunshine. In a typical December, 13 days will remain sunless—almost one day in two. December 1890 logged 25 sunless days out of 31; December 1953 and December 1956, 21; December 1995, 20; and December 2015, 18 sunless days. Between 1921 and 2018, there were 11 spells of at least seven consecutive sunless days in December, the longest such spells being one of 13 days commencing 3 December 1968, 11 days commencing 1 December 1953, and 10 days commencing 10 December 2015. The first 22 days of December 1953 saw just 0.8 hours of sunshine—the dullest 3 weeks on record since at least 1921. In December 1890, only two days recorded even as much as an hour's sunshine, the second half of the month seeing less than half an hour in all, with 9 minutes sunshine on Christmas Day, and 18 minutes on 27th.

Although December is, on average, the dullest month of the year, in about one year in four over the last century, it has been sunnier than November. The most extreme examples were in 1962, when December (72.6 hours) was more than twice as sunny as November (32.7 hours), and in 2014, when December (96.9 hours) was 60 per cent sunnier than November (60.9 hours). Compare these sunny Decembers with the dullest summer months—for example, July 1888 (total sunshine 95.3 hours) and August 1912 (97.7 hours).

## Wind

Oxford's mean monthly wind speed for December is 4.2 metres per second (m/s), or 8.2 knots. The windiest December on a record back to 1880 occurred in the very mild December of 1974, when the monthly mean wind speed was 6.6 m/s (12.8 knots). The calmest December was in 2005, when the mean wind speed was 2.9 m/s (5.7 knots).

## Oxford temperature, precipitation and sunshine in graphs—December

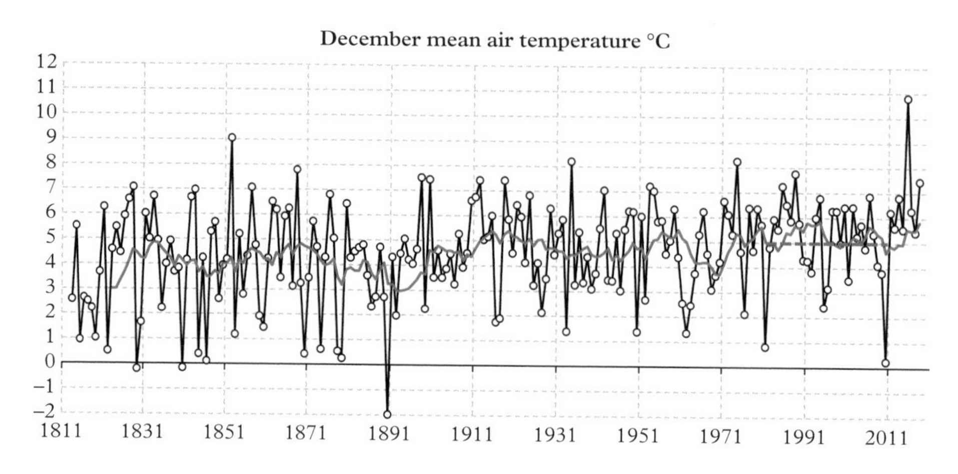


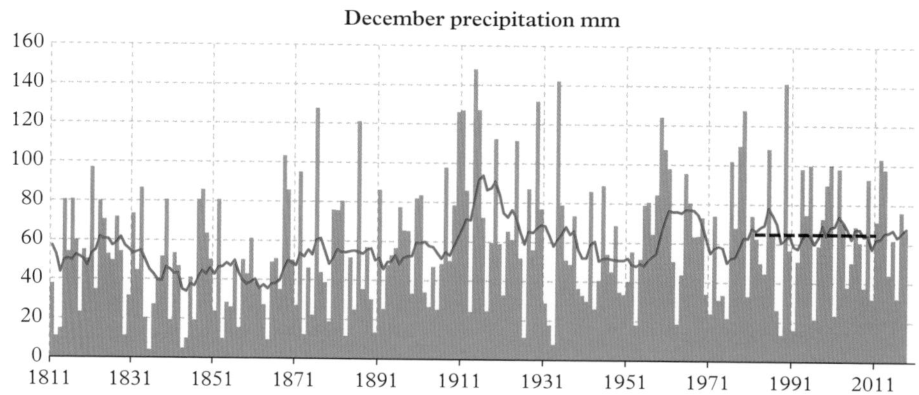


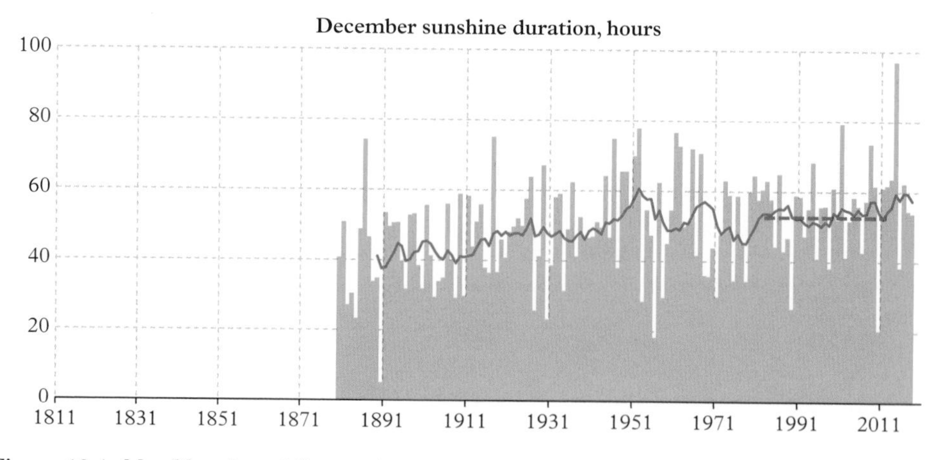


**Figure 18.4** *Monthly values of (from top) mean temperature (°C, since 1813), total precipitation (mm, since 1811) and sunshine duration (hours, since 1880) for December at Oxford. The 1981–2010 averages are indicated by the thick blue dashed line, while the ten-year unweighted running mean ending at the year shown is indicated by the red line*

# 19

# The calendar year

Located as it is in the Thames Valley in southern England, Oxford has a fairly equable climate with few extremes of weather (see Part 1). It is close to the warmest districts of the British Isles, with the nearby continent providing the potential for heatwaves in summer but also occasionally very cold spells in winter, while its position in the south Midlands relatively far from oceanic influence gives its climate a more continental range than other parts of England. Oxford lies far from the wettest areas of the country, and often under something of a rain shadow from the Welsh mountains and Cotswold hills to the west, but its long rainfall record shows that it can still be subjected to prolonged falls of frontal rainfall in autumn or winter, or intense thundery downpours in summer. Snowfall amounts tend to be small, while sunshine amounts in the summer months average out to something over six hours daily.

## Temperature

Since 1815, air temperatures in Oxford have ranged from −17.8 °C (on Christmas Eve 1860) to 35.1 °C (on 19 August 1932, equalled on 3 August 1990). The ten highest and lowest daily maximum, minimum and mean air temperatures are listed in Chapter 29, and more details on these events can be found in the monthly sections.

The largest daily temperature range recorded was 22.5 degC on 23 September 1895 (2.8 to 25.3 °C), equalled on 8 September 1911 (10.9 to 33.4 °C). Within the last 50 years, the greatest daily range has been 21.2 degC on 15 July 1990 (9.3 to 30.5 °C; see Chapter 29, Table 29.6).

The highest temperature of the year is not always recorded during the three summer months (June, July or August): since 1815, it has occurred in May in 10 years and in September in 11 years—an average of about once in twenty years for each month. Only once in over 200 years has it occurred in October. The earliest date on which the hottest day of the year has occurred is 12 May (in 1907, maximum temperature 25.8 °C), and the latest 1 October (in 2011, maximum temperature 29.1 °C). Figure 19.1 shows the highest recorded temperature (°C) in every year since 1815, together with a ten-year unweighted running mean.

*Oxford Weather and Climate since 1767*. Stephen Burt and Tim Burt, Oxford University Press (2019).
 DOI: 10.1093/oso/9780198834632.001.0001

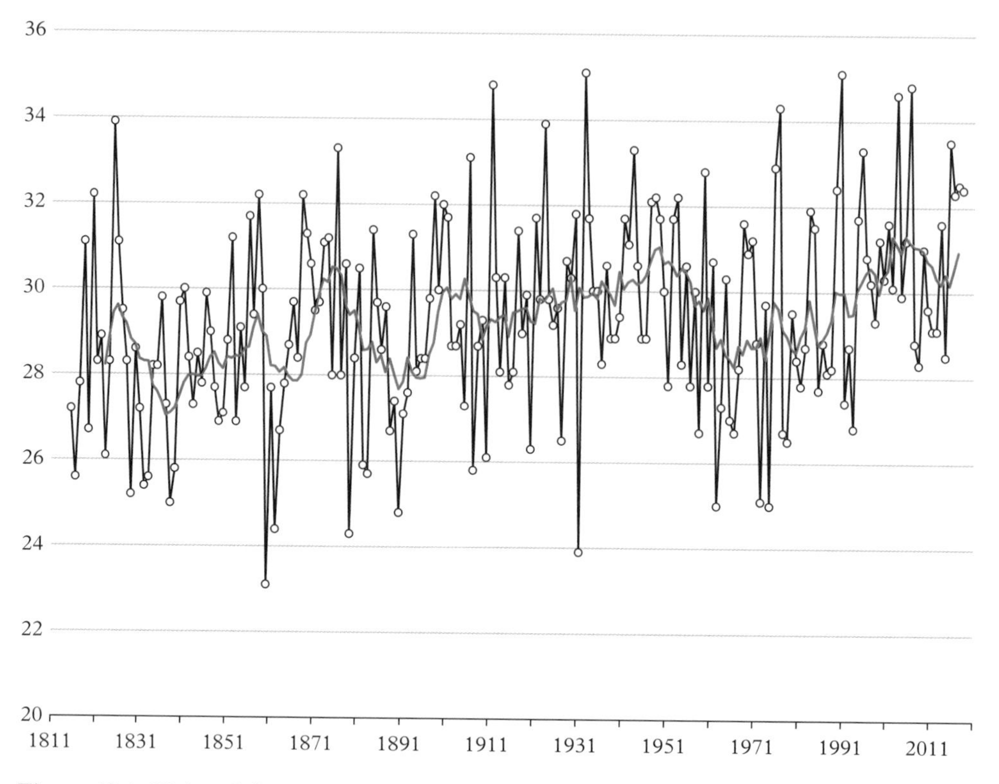


**Figure 19.1** *Highest daily maximum temperature (°C) by calendar year, 1815 to 2018, together with a ten-year unweighted running mean plotted at year ending*

The lowest temperature of the calendar year is recorded outside of the three winter months (December, January and February) about one year in ten, both March and November sharing the honours with 13 occasions each. Only once has the coldest night of the calendar year occurred in April—and that was over 150 years ago, when, on 12 April 1868, the minimum temperature was −4.6 °C. Figure 19.2 shows the lowest recorded temperature (°C) in every year since 1815, together with a ten-year unweighted running mean.

Since 1815, annual mean temperatures in Oxford have varied between 7.7 °C in 1879 and 11.8 °C in 2014, a range of 4.1 degC. The five warmest and five coldest years are shown in Table 19.1. It is noteworthy that all five of the warmest years have occurred since 2006, whereas the most recent year in the 'Top 5 coldest' was in 1879. This is also brought out graphically in Oxford's 'climate stripe', Figure 19.3, where every year's annual mean temperature is represented in colour by a thin vertical stripe, whose colour represents gradation from cold (blue) to warm (red) years. In a simple barcode-like graphic, this shows the progression from cold years (blue) to the series of particularly warm years within the last three decades or so.

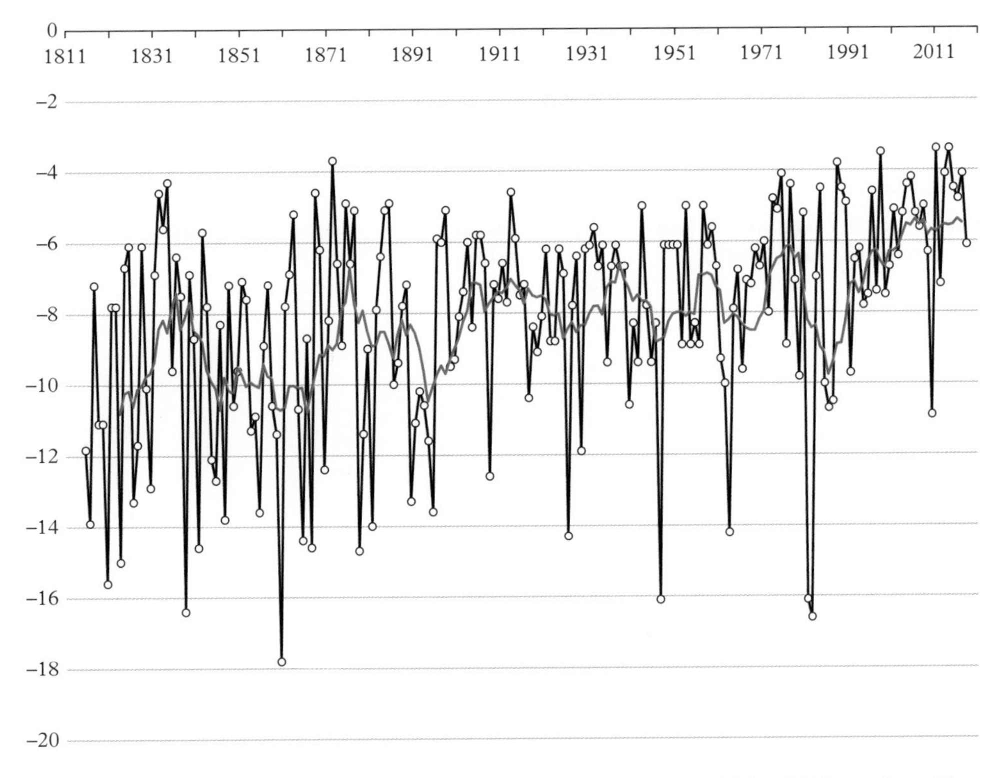


**Figure 19.2** *Lowest daily minimum temperature (°C) by calendar year, 1815 to 2018, together with a ten-year unweighted running mean plotted at year ending*

**Figure 19.3** *Oxford's 'climate stripe', showing the annual mean temperature for every year from 1814 to 2018 by thin coloured vertical stripes, graduated from blue for cold years to red for warm (Ed Hawkins, University of Reading)*

## Warm and cold years

**Table 19.1** *Annual mean temperatures at the Radcliffe Observatory, Oxford, 1815–2018*

| Annual mean temperature 10.7°C (average 1981–2010) | | | | | |
|---|---|---|---|---|---|
| Warmest years | | | Coldest years | | |
| Mean temperature, °C | Departure from 1981–2010 normal °C | Year | Mean temperature, °C | Departure from 1981–2010 normal °C | Year |
| 11.79 | +1.1 | 2014 | 7.69 | −3.0 | 1879 |
| 11.71 | +1.0 | 2006 | 7.81 | −2.9 | 1816 |
| 11.70 | +1.0 | 2018 | 8.15 | −2.6 | 1838 |
| 11.57 | +0.9 | 2011 | 8.32 | −2.4 | 1829, 1860 |
| 11.54 | +0.8 | 2017 | 8.33 | −2.4 | 1815 |
| *Last 50 years to 2018* | | | | | |
| 11.79 | +1.1 | 2014 | 9.44 | −1.3 | 1986 |

## Heatwaves and hot nights

Tables 19.2 and 19.3 and Figure 19.4 show the frequency of hot days (25 °C or more) and heatwave days (30 °C or more) by month and year since 1815. In terms of hot days, July 2018 established a new calendar month record, with 26 days reaching or surpassing 25 °C, well ahead of the previous highest (22 days in July 1983); there were 19 such days in July 1868, July 1911, August 1947 and July 2013. In annual rankings, 2018 comes out top, with 48 days, closely followed by 1976 with 46 days, and then 1911 (45 days) and 1868 (43 days).

The longest unbroken spells of days reaching 25 °C or more were the 18 consecutive days 12–29 August 1947, followed by 17 consecutive days 22 June–8 July 1976 and 6–22 August 1997. In 1976, all but one day of the 21 consecutive days 22 June–12 July reached 25 °C or more (on 9 July the maximum was 'only' 24.0 °C). In 2018, between 25 June and 27 July (33 days), only two days did not attain 25 °C, the lowest maximum temperature being 23.6 °C on 10 July.

Only five years since 1815 failed to reach 25 °C on any day in the year, the most recent occurrence being in 1931; in more recent years, the lowest totals have been a single day in both 1972 and 1974 and four days in 2007. In 1860, the hottest day of the year was on 5 July, with a maximum temperature of just 23.1 °C, the lowest annual maximum temperature in the series.

In terms of heatwave days, those with a maximum temperature of 30 °C or more, July 1976 and August 1995, top the table, with eight days apiece (Table 19.3); 1976 leads the annual table, with 14 'heatwave' days, closely followed by 13 in 1995, although 2018 had 'only' five such days. By far the longest unbroken spell of days reaching 30 °C or more was

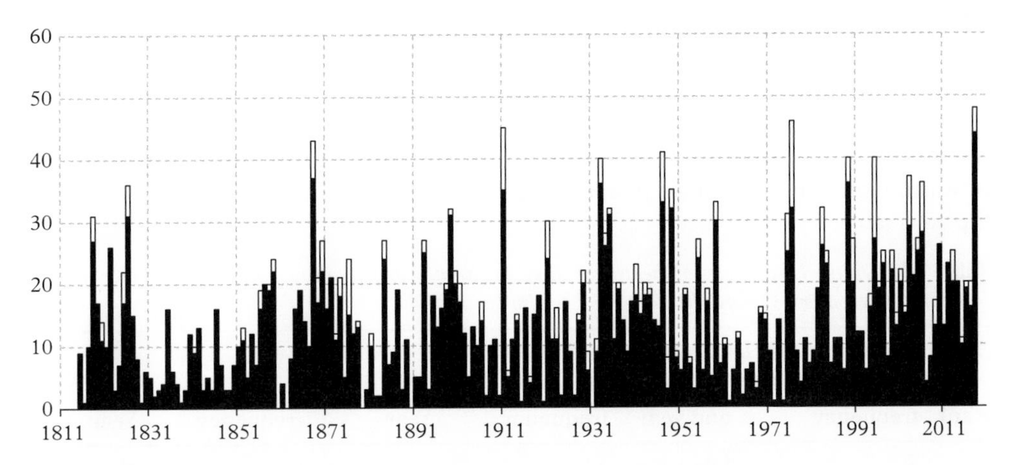


**Figure 19.4** *Combined annual frequency of 'hot days' (maximum temperature 25.0 °C or greater, dark lower column) and 'heatwave days' (maximum temperature 30.0 °C or greater, white upper column) at the Radcliffe Observatory, Oxford, 1815 to 2018, by calendar year*

**Table 19.2** *Annual frequency of 'hot days' (maximum temperature 25.0 °C or greater) at the Radcliffe Observatory, Oxford, 1815–2018 (see also Figure 19.1)*

*Annual average 1981–2010 19 days*

| Rank | Highest monthly frequency | Month and year | Highest annual frequency | Year | Lowest annual frequency | Year |
|---|---|---|---|---|---|---|
| 1 | 26 | July 2018 | 48 | 2018 | 0 | 1860, 1862, 1879, 1890, 1931 |
| 2 | 22 | July 1983 | 46 | 1976 | | |
| 3 | 19 | July 1868<br>July 1911<br>Aug 1947<br>July 2013 | 45 | 1911 | | |
| 4 | 18 | *July 1822*<br>Aug 1995<br>July 2006 | 43 | 1868 | | |
| 5 | 17 | July 1934<br>July 1949<br>Aug 1975<br>Aug 1997 | 41 | 1947 | | |
| 6 | 16 | July 1859<br>July 1921<br>July 2014 | 40 | 1933, 1989, 1995 | | |

the 14 consecutive days 25 June to 8 July 1976; the next-longest such spells were the six days 29 July to 3 August 1995, and five days 13–17 August 1876, 26–30 July 1948 and 12–16 July 1983 (see also Chapter 30, Table 30.1, for more details on hot spells).

Very often, the most uncomfortable aspect of prolonged hot weather is not so much daytime temperatures but a succession of warm, close nights. Table 19.4 shows the incidence of warm nights (minimum air temperature 15 °C or more). The summer of 2018 also tops this table.

**Table 19.3** *Annual frequency of 'heatwave days' (maximum temperature 30.0 °C or greater) at the Radcliffe Observatory, Oxford, 1815–2018 (see also Figure 19.1)*

*Annual average 1981–2010 2.2 days*

| Rank | Highest monthly frequency | Month and year | Highest annual frequency | Year | Lowest annual frequency | Year |
|---|---|---|---|---|---|---|
| 1 | 8 | July 1976<br>Aug 1995 | 14 | 1976 | 0 | Numerous |
| 2 | 7 | July 2006 | 13 | 1995 | | |
| 3 | 6 | July 1921<br>June 1976<br>July 1983<br>Aug 2003 | 10 | 1911 | | |
| 4 | 5 | Aug 1876<br>July 1923<br>July 1948<br>Aug 1975 | 9 | 1876 | | |
| 5 | 4 | Numerous | 8 | 1947, 2003, 2006 | | |

**Table 19.4** *Highest monthly and highest and lowest calendar year totals of 'warm nights' (15.0 °C or more) at the Radcliffe Observatory, Oxford, 1815–2018*

*Annual average 1981–2010 17 nights*

| Rank | Highest monthly frequency | Month and year | Highest annual frequency | Year | Lowest annual frequency | Year |
|---|---|---|---|---|---|---|
| 1 | 23 | July 1983 | 34 | 1983, 2018 | 0 | 1860, 1924 |
| 2 | 16 | Aug 1997, July 2006 | *33* | *1826* | 1 | *1816, 1832,* 1889, 1920, 1963 |
| 3 | 15 | July 2018 | 30 | 2006 | 2 | Numerous |
| 4 | 14 | *July 1828,* Aug 1911 | 27 | 2001 | | |

## Frosts

Over the 1981–2010 period, Oxford averaged 37 air frosts and 96 ground frosts annually. Table 19.5 shows the frostiest months, and Table 19.6 the greatest and least numbers of air and ground frosts in every winter since 1815. The longest spells of consecutive days with air frost on the 200-year record were:

- 37 days between 7 December 1890 and 12 January 1891 inclusive (Figure 19.5)
- 36 days between 23 December 1962 and 27 January 1963
- 34 days—between (i) 1 January and 3 February 1815, (ii) 13 January to 15 February 1816 and (iii) 5 February and 10 March 1947, and
- 33 days between 6 January and 7 February 1830.

Since the winter of 1962/63, the longest spell of consecutive air frosts has been 28 days, 5 February to 4 March 1986.

**Figure 19.5** *Photographs of the frozen River Thames in Abingdon in January 1891, towards the close of the longest spell of consecutive air frosts on record in Oxford.* ***Top:*** *Skaters on the Thames with the weir and lock house in the background.* ***Bottom:*** *The frozen river near East St Helen Street (both images courtesy of Abingdon County Hall Museum)*

**Table 19.5** *Frostiest calendar months at the Radcliffe Observatory, Oxford, 1815–2018: ground frosts 1931–2018*

| Frostiest months | | | |
|---|---|---|---|
| Air frosts | Month and year | Ground frosts | Month and year |
| *30* | *Jan 1830* | 29 | Jan 1940, Jan 1963, Jan 1979, Jan 1985 |
| 28 | Jan 1963 | 27 | Jan 1980, Jan 1984 |
| 27 | *Jan 1842*, *Jan 1850*, Jan 1879 | 26 | Feb 1942, Feb 1947, Feb 1963 |
| 26 | Jan 1940<br>Feb 1947 | 25 | Various |
| 25 | *Jan 1838*, Dec 1890, Jan 1942, Feb 1963, Dec 2010 | | |

**Table 19.6** *Most and least frosty winters at the Radcliffe Observatory, Oxford, 1816–2018 (ground frosts 1931–2018). Total of all air and ground frosts from September to June, dated by January year*

| | Most frosty | | | | Least frosty | | | |
|---|---|---|---|---|---|---|---|---|
| Rank | Air frosts | Year | Ground frosts | Year | Air frosts | Year | Ground frosts | Year |
| 1 | *113* | *1816* | 122 | 1984, 1991 | 7 | 2014 | 62 | 2007 |
| 2 | *101* | *1830* | 119 | 1982 | 11 | 1961 | 70 | 1961, 1975 |
| 3 | *93* | *1820* | 118 | 1956 | 16 | 1975 | 71 | 1935 |
| 4 | 91 | 1888 | 117 | 1980 | 18 | 2016 | 72 | 1967, 2005 |
| 5 | 88 | *1852* | 116 | 1953 | 19 | 1990 | 73 | 1995, 2003 |

The winter of 1946/47 recorded 72 air frosts, and 1962/63, 84. Since 1962/63, the frostiest winter has been 1978/79, with 65 air frosts; 2010/11 recorded 53. Figure 19.6 shows the number of air frosts per calendar year since 1815.

An 'ice day' is one in which the maximum temperature fails to reach 0 °C. During 1981–2010 Oxford averaged 2.2 'ice days' each year, although the distribution is very skewed towards cold winters—more than half of the years within the averaging period

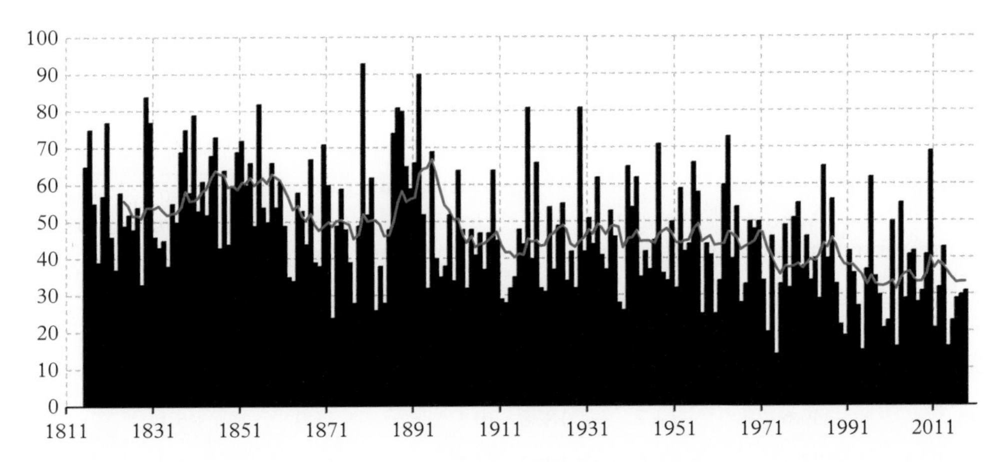


**Figure 19.6** *Number of air frosts (minimum temperature below 0 °C) at the Radcliffe Observatory, Oxford, for every calendar year 1815 to 2018, together with a ten-year unweighted running mean plotted at year ending*

**Table 19.7** *Greatest number of ice days (maximum temperature below 0 °C) for months and winters at the Radcliffe Observatory, Oxford, 1815–2018. Winters dated according to January*

| | Most ice days | | | |
|---|---|---|---|---|
| Rank | In a winter season | Year | In a month | Month |
| 1 | 32 | 1963 | *19* | *Jan 1838* |
| 2 | *28* | *1830* | 18 | Feb 1947 |
| 3 | 25 | *1823*, 1891 | 16 | *Jan 1823*, Dec 1890, Jan 1963 |
| 4 | 23 | *1841*, 1947 | 13 | Jan 1867 |
| 5 | *21* | *1838* | 11 | Jan 1881 |

recorded no days below 0 °C. The greatest numbers of ice days in a month and a year in the period 1815–2018 are shown in Table 19.7. The longest consecutive run of 'ice days' since the winter of 1962/63 has been eight days, in January 1982 and December 2010.

## Monthly and annual ranges in temperature

Only six months since 1815 have recorded a monthly range in temperature of 30 degC or greater, the greatest being 31.7 degC in May 1944 (see May, Chapter 11) and the most recent being 30.1 degC in August 2003. Five *calendar years* have recorded a smaller temperature range than May 1944. The least range of temperature in any month has been just 10.4 degC, in February 2014; only five months, including February 2014, recorded monthly ranges less than 12 degC between 1815 and 2018.

**Table 19.8** *The greatest and least annual temperature ranges (calendar years) at the Radcliffe Observatory, Oxford, 1815–2018*

Average highest maximum 30.4 °C, average lowest minimum −7.1 °C; see also Appendix 3

**Mean annual temperature range 37.5 degC** (average 1981–2010)

| **Greatest annual ranges** | | | | **Least annual ranges** | | | |
|---|---|---|---|---|---|---|---|
| Temperature range, degC | Annual maximum temp °C | Annual minimum temp °C | Year | Temperature range, °C | Annual maximum temp °C | Annual minimum temp °C | Year |
| 48.3 | 32.1 | −16.1 | 1947 | *30.0* | *25.4* | *−4.6* | *1832* |
| *47.8* | *32.2* | *−15.6* | *1820* | 30.0 | 23.9 | −6.1 | 1931 |
| 45.3 | 30.6 | −14.7 | 1878 | 30.1 | 25.0 | −5.1 | 1974 |
| 45.3 | 28.7 | −16.6 | 1982 | 31.1 | 26.7 | −4.4 | 1977 |
| *44.6* | *30.0* | *−14.6* | *1841* | *31.2* | *25.6* | *−5.6* | *1833* |

During the period 1981–2010, the average *annual* temperature range was 37.5 degC. The greatest and least annual ranges since 1815 are shown in Table 19.8.

## Reversal in month-to-month temperature trends

As the days grow longer, the expectation is that, from March to June, the monthly mean temperature should rise each month. Conversely, into autumn we expect September to be cooler than August, October cooler than September, and so on. As the monthly chapters reveal, this is not always the case:

- March has been colder than February roughly one year in five since 1815, most recently in 2013, and colder than January in about one year in six, most recently in 2008, 2013, 2016 and 2018. March 1916 was 3.9 degC colder than January that year, in what must have been a dramatic seasonal reverse.
- It is not particularly uncommon for April to be colder than March, about once every 10–12 years within the last century; as at 2018, the most recent such reversal was in 2012. In 1938, April was 2.1 degC colder than March.
- June 1916 holds the unenviable record of being the only June on the entire record to have been colder than the previous May.
- September is, occasionally, warmer than August, twice in recent years (in 2006 and 2014), but only three times prior to that since 1815.
- No October has ever been warmer than the preceding September, although, in both 1847 and 2001, October was only 0.1 degC cooler than September.

- November is occasionally slightly milder than the previous October; this has happened five times since 1815, the most extreme being in 1881, when a very mild November followed a very cold October, such that the mean temperature for November was 1.6 degC higher than October. This reversal of the normal seasonal trend also occurred in 1852, 1895, 1939 and, most recently, in 1994.
- It is not particularly uncommon for December to be milder than November—within the last century, this has happened 16 times—but December milder than *October* has happened only twice, in 1851 and 1974. December 2015 was a truly exceptional month, with its mean temperature ranking higher than any November, December, January, February or March on the record; indeed, it would rank just outside the top five warmest *Aprils*.

## Precipitation

The average annual precipitation in Oxford is 660 mm, falling on 165 days. The wettest and driest years are set out in Table 19.9. Information on droughts and wet spells over various durations from days to years can be found in Chapter 28, while the wettest and driest months on record are given in Chapter 29, Table 29.7. The wettest days are listed in Table 29.8.

Five months in 2012 received over 100 mm; after a dry start to the year, April (143 mm) was the wettest on Oxford's records, while June (152 mm) became the second-wettest June, and the wettest for 160 years. In 1768, both June (132 mm) and September (171 mm) were exceptionally wet. The year 1960 was persistently wet from early summer—every

**Table 19.9** *Extremes of annual (calendar year) precipitation at the Radcliffe Observatory, Oxford 1767–2018*

| **Average annual precipitation 660 mm** (average 1981–2010) | | | | | |
|---|---|---|---|---|---|
| **Wettest years** | | | **Driest years** | | |
| **Total fall, mm** | **Per cent of normal** | **Year** | **Total fall, mm** | **Per cent of normal** | **Year** |
| 979.5 | 148 | 2012 | 336.7 | 51 | 1788 |
| 964.7 | 146 | 1960 | 380.7 | 58 | 1921 |
| 960.1 | 145 | 1852 | 389.5 | 59 | 1802 |
| 932.7 | 141 | 1768 | 412.2 | 62 | 1854 |
| 913.8 | 138 | 1903 | 418.5 | 63 | 1964 |
| ***Last 50 years to 2018*** | | | | | |
| 979.5 | 148 | 2012 | 470.8 | 71 | 1996 |

month but one from June onwards collected over 100 mm, with extensive and long-lasting flooding during the autumn months. However, the wettest of any 12 month period on record in Oxford since 1767 was that commencing April 2012, when the total amounted to 1088.1 mm, or 165 per cent of the annual average. Chapter 28 gives more details of wet spells by duration.

The year 1788 was exceptionally dry in England; in Oxford, the annual total was just 337 mm, or 51 per cent of the current annual average, although, as with other values, there is inevitably some doubt attached to records made well before both instruments and exposures became standardised during the mid-Victorian era [98, 99]. Only August (82 mm) was significantly wetter than the current normal, while April and May and October to December were all exceptionally dry. In 250 years of rainfall records in Oxford, only 1788 (51 per cent), 1921 (58 per cent) and 1802 (59 per cent) have recorded less than 60 per cent of the annual normal precipitation. However, the driest 12 consecutive month period on record (any 12 consecutive months, not necessarily January to December) in Oxford since 1767 commenced in October 1975, when just 322.6 mm fell, 49 per cent of the 1981–2010 normal (see also Chapter 28). The range between the driest and wettest 12 consecutive months in Oxford since 1767 is thus 322.6 mm to 1088.1 mm, a factor of 3.3.

A 'rain day' is one in which 0.2 mm or more is recorded during the 24 hours commencing at 0900 GMT on the day of measurement. The average number of rain days over 1981–2010 was 165 per annum. The highest and lowest monthly and annual frequency of rain days is given in Table 19.10. (This table relates only to records from 1853 as, before that, it is unclear whether rainfall was measured daily or whether small amounts or days with snow were counted in the record [98]—see Appendix 1.) Although 2012 was the wettest calendar year on record, with 979 mm of rainfall, rain was recorded on 'only' 184 days. The longest unbroken spells of consecutive rain days were the 23 days from 12 March to 3 April 1947, and the 20 days 17 October to 5 November 1960.

**Table 19.10** *Highest and lowest monthly and annual precipitation frequency ('rain days', those with 0.2 mm or more) at the Radcliffe Observatory, Oxford, 1853–2018*

| Greatest monthly | | Fewest monthly | | Greatest annual | | Fewest annual | |
|---|---|---|---|---|---|---|---|
| Rain days | Month and year | Rain days | Month and year | Rain days | Year | Rain days | Year |
| 30 | Jan 2014 | 1 | Feb 1891 | 210 | 1872 | 114 | 1887 |
| 28 | Dec 1934 | | | 209 | 1960 | 124 | 1976 |
| 27 | June 1879, Dec 1911 | 2 | 11 months—most recently April 2011 | 206 | 1954 | 126 | 1855, 1864, 1921 |
| 26 | *Numerous* | | | 202 | 1951 | 130 | 1884 |
| | | | | 201 | 1912 | 131 | 1854 |

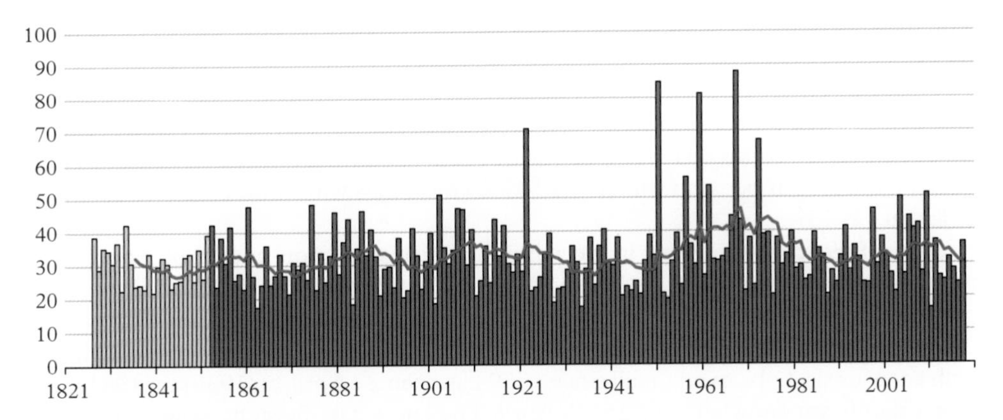


**Figure 19.7** *Highest daily precipitation total (mm) at the Radcliffe Observatory, Oxford, by calendar year, 1827 to 2018, together with a ten-year unweighted running mean plotted at year ending. Daily amounts prior to 1853 may represent multi-day accumulations and are shown in lighter grey*

## Wettest days

Table 29.8 in Chapter 29 lists the highest 24-hour precipitation totals (0900–0900 GMT since 1921) recorded at the Radcliffe Observatory since 1827 (some daily amounts prior to 1853 may represent multi-day accumulations). Figure 19.7 shows a time series of the annual highest 24-hour precipitation totals since 1853, together with a ten-year unweighted running mean. There is a marked cluster in the 1950s and 1960s, with six of the seven highest daily falls occurring in less than 25 years after 1951 but, other than this feature (perhaps no more than statistical chance), and perhaps contrary to popular expectation, there is no significant trend evident on this series.

## Snowfall and lying snow

Snow or sleet can be expected to fall in Oxford on about 16 days in an average year, while lying snow (50 per cent covering or more at the 0900 GMT observation) can be expected on around 7 mornings on average.

The snowiest months during the period of available data (1926–1986) were January 1963 (17 days with snow or sleet observed to fall), 16 days in January 1945 and 15 days in the Februarys of 1947, 1955, 1956 and 1963. The snowiest winters during 1926–1986 in terms of the number of days with snowfall were 1963 (41 days with snow or sleet observed to fall), 1979 (39 days) and 1947 (36 days). Records of the frequency of snowfall have not been kept since 1986 and recent winters are therefore not included.

In terms of snow lying (more than 50 per cent cover at the 0900 GMT observation), the snowiest months were January 1963 (snow cover on all 31 mornings), February 1963 (all 28 mornings with snow cover), and February 1947 (27 mornings with snow cover), followed by February 1956 (18) and January 1940 and January 1979 (both with 16). More recently, December 2010 recorded 15 mornings with snow cover.

The snowiest winters, measured by the duration of snow cover, were 1962/63 (68 mornings with snow cover, 64 of which were consecutive, 27 December 1962 to 28 February 1963) and 1946/47 (52 mornings with snow cover). Other notable winters were 1939/40 and 1969/70, both with 31 mornings with snow cover, and 1978/79, with 30. More recently, calendar year 2010 saw 31 mornings with snow cover, with heavy snowfalls in both January and December, and winter 2012/13 23 mornings with snow cover (13 mornings in January 2013).

Since 1959, the earliest date in the winter half-year with snow or sleet observed to fall was 5 November, in 1980* (further snow fell on the following two days), while the earliest winter half-year date with snow on the ground at the morning observation was 20 November 1988, when 1 cm lay at 0900 GMT. The latest date in the winter half-year with snow or sleet observed to fall since 1959 has been 27 April (in 1985); in 1981, there was a significant snowfall over 25–26 April. The latest date over the same period with snow covering the ground at the morning observation has been 26 April 1981; but see Chapter 10 for details on one of Oxford's greatest-ever snowfalls in late April 1908, when the snow lay 43 cm deep on the morning of 26th. The greatest snow depth known on the Oxford record occurred on 13 and 14 February 1888, when the depth was reported as 61 cm following a 20-hour snowstorm (see Chapter 8). Since 1959, the greatest recorded depth of snow has been 27 cm on 6 January 2010.

### Thunder

Thunder is heard on an average of about 10 days each year in Oxford. A record of days with 'thunder heard' has not been maintained since 1986, but notably thundery months have included 9 days with thunder heard in May 1945, 8 days in July 1947, July 1965 and May 1969 and 7 days in July 1939, July 1960, June 1982 and May 1983. The most thundery years during the period 1931–1985 were 1983 (24 days), 1939 and 1960 (21 days each), and 1932, 1933 and 1982 (20 days each).

## Sunshine

The average annual duration of 'bright sunshine' in Oxford during 1981–2010 was 1577 hours. The sunniest and dullest calendar years since Oxford's sunshine records commenced in February 1880 are shown in Table 19.11. During this period of almost 140 years, monthly sunshine amounts have varied between a mere 5.0 hours in December 1890 to 310.45 hours in July 1911. The months of May to August have similar average sunshine totals and, in about one year in four, the sunniest month of the year occurs outside the three 'summer' months of June, July and August; in 1907, March was, remarkably, the sunniest month of the year. In about one year in eight the dullest month occurs outside the three 'winter' months (December, January and February); in 1894 October was the dullest month of the year.

* Snow fell in the Oxford area on 28 October 2008, but records of snow or sleet falling are no longer maintained at the Radcliffe Observatory and therefore it is difficult to be conclusive on this point.

**Table 19.11** *Calendar year sunshine duration at the Radcliffe Observatory, Oxford, 1881–2018*

**Annual mean sunshine duration 1577 hours, 4.32 hours per day** (average 1981–2010)

*Possible daylength: 4481 hours (leap-year 4492 hours).*
*Mean sunshine duration as percentage of possible: 35.2*

| Sunniest years | | | Dullest years | | | Sunniest days | |
|---|---|---|---|---|---|---|---|
| Duration, hours | Per cent of possible | Year | Duration, hours | Per cent of possible | Year | Duration, hours | Date |
| 1880.3 | 41.9 | 1995 | 1154.5 | 25.7 | 1888 | 15.9 | 28 June 1921 |
| 1875.3 | 41.8 | 1990 | 1204.7 | 26.9 | 1981 | 15.7 | 17 June 1959 |
| 1868.1 | 41.7 | 1989 | 1217.7 | 27.2 | 1889 | 15.6 | 29 June 1921 and 23 June 1973 |
| 1853.4 | 41.3 | 1959 | 1239.4 | 27.6 | 1931 | 15.5 | *12 occasions* |
| 1827.0 | 40.7 | 1949 | 1260.0 | 28.1 | 1884 | | |
| *Last 50 years to 2018* | | | | | | | |
| 1880.3 | 41.9 | 1995 | 1204.7 | 26.9 | 1981 | 15.6 | 23 June 1973 |

**Table 19.12** *Annual frequency of sunless days at the Radcliffe Observatory, Oxford, 1921–2018*

| Most sunless days | Year | Least sunless days | Year |
|---|---|---|---|
| 89 | 1968 | 43 | 2011 |
| 88 | 1963 | 44 | 1935 |
| 85 | 1964, 1979 | 45 | 2014 |
| 84 | 1996 | 46 | 1949, 1990, 2003 |

Figure 19.8 shows that there has been a slight increase in annual sunshine duration over the period of record, and particularly since the early 1980s. Oxford's average annual sunshine over the 30-year period 1921–50 was 1489 hours: the most recent 30-year mean (1981–2010) is 6 per cent greater. The increase is slight, around 14 minutes additional sunshine per day when averaged over the year, but apparent in every month except March and June. The winter months show the greatest increase, January being almost 20 per cent sunnier compared with the earlier period, and this is probably due to reduced pollution and aerosol loading following clean air legislation since the 1950s.

The annual average number of sunless days during the year in the period 1981–2010 was 65 so that, perhaps contrary to popular belief, Oxford averages 300 days of sunshine per year! The number of sunless days in a year has varied between 43 and 89 days since 1921 (Table 19.12). In December 1890, there were 25 sunless days and, more recently, 21 in the Decembers of 1953 and 1956, and 20 in January 1941, December 1995, January 1996 and December 2010.

The annual average number of days with 12 hours or more of sunshine in the period 1981–2010 was 18, most likely in June and July, the number varying in individual years (period 1921–2018) from six in 1931, seven in 1927 and eight in 1946, 1968 and 1982 to 38 in 2018, 36 in 1990, 35 in 1989, and 33 in 1976 and 1996. In July 1911, the sunniest calendar month yet recorded, 12 days recorded at least 12 hours of sunshine; even this figure was exceeded in June 1996 (15 days with 12 hours sunshine or more), June 1975 (13 days) and equalled (12 days) in May 1990. Table 29.11 in Chapter 29 summarises the sunniest and dullest calendar months on record, while Tables 30.4 and 30.5 in Chapter 30 list the sunniest and dullest periods on record from 7 to 60 days duration.

## Barometric pressure

Records of barometric pressure at station level have been kept at the Observatory since 1776, although there are large gaps in the record until 1813 and very little of the data has yet been digitised. Table 19.13 summarises some of the highest and lowest observed pressures on the available dataset since 1820, corrected to mean sea level; these may not represent the full extremes on the record.

**Table 19.13** *Known extremes of barometric pressure on record at the Radcliffe Observatory, Oxford since 1820. See notes for derivations.*

| HIGHEST PRESSURES | | | LOWEST PRESSURES | | |
|---|---|---|---|---|---|
| Date | MSL pressure, hPa | Observation | Date | MSL pressure, hPa | Observation |
| 17 Jan 1882 | 1048.2 | Highest observed, at 2300h | 24 Dec 1821 | 953.3 | Spot reading 2230h |
| 9 Jan 1896 | 1048.6 | At 2100 observation; Daily mean 1047.0 | 13 Jan 1843 | 956.8 | Spot reading 1000h |
| 15 Jan 1902 | 1046.2 | Highest observed, at 0845h | 26 Jan 1884 | 963.7 | Lowest observed, time unstated |
| 29 Jan 1905 | 1050.0 | Highest observed, at 0020h | 8 Dec 1886 | 956.0 | Lowest observed, at 2132 and 2143h 'Violent storm, rain and hail' |
| | | | 22 Feb 1914 | 968.1 | At 0800 observation Daily mean 970.1 |

| HIGHEST PRESSURES | | | LOWEST PRESSURES | | |
|---|---|---|---|---|---|
| **Date** | **MSL pressure, hPa** | **Observation** | **Date** | **MSL pressure, hPa** | **Observation** |
| | | | 19 Nov 1916 | 969.9 | At 0800 observation<br>Daily mean 971.4 |
| | | | 1 Jan 1949 | 958.5 | Minimum at 1430 GMT; lowest since 8 Dec 1886 |
| 23 Jan 1907 | 1046.4 | At 0800 observation<br>Daily mean 1042.7 | 4 Feb 1951 | 956.6 | Minimum at 0000 GMT 4/5 February |
| | | | 9 Dec 1954 | 962.9 | 0900 observation |
| | | | *16 Oct 1987* | *961* | *Approx min, about 04h* |
| | | | 25 Feb 1989 | *955* | *Approx min, about 15h* |
| 23 Dec 1926 | 1041.1 | Daily mean | | | |
| 26 Jan 1932 | **1050.2** | Highest observed at 1030 GMT<br>Daily mean 1049.2 | | | |
| 15 Feb 1934 | 1049.6 | Highest observed at 1100 GMT | | | |
| 15 Jan 1946 | 1045.9 | 0900 observation | | | |
| 16 Jan 1957 | 1041.9 | 0900 observation | | | |
| 7 Feb 1964 | *1047.5* | *Approx peak* | | | |
| 26-27 Jan 1992 | *1047.5* | *Approx peak* | | | |

*NOTES*

*The dates of likely extremes have been extracted from [100, 101] and [102] and the relevant records from the Radcliffe Observatory examined for those dates. As-read mercury barometer readings in inches of mercury (inHg) were converted to hectopascals (hPa = millibars), corrected to 0 °C—where this had not already been done—and reduced to Mean Sea Level (MSL) using the altitude of the barometer cistern (212 ft = 64.6 m) and the external dry bulb temperature (where known—otherwise the mean daily temperature for the date) by standard formulae. In some cases, MSL values derived using modern calculation methods differ slightly (typically within 0.5 hPa) from contemporary observations or published records. Where shown, observations were indicated as the highest or lowest on the relevant dates; other records are daily mean pressures, and as such will slightly underestimate the true extreme—more so in the case of low pressures. Barometric pressures were not noted in the manuscript observation registers after 31 December 1958.*

## Oxford temperature, precipitation and sunshine in graphs—Calendar years

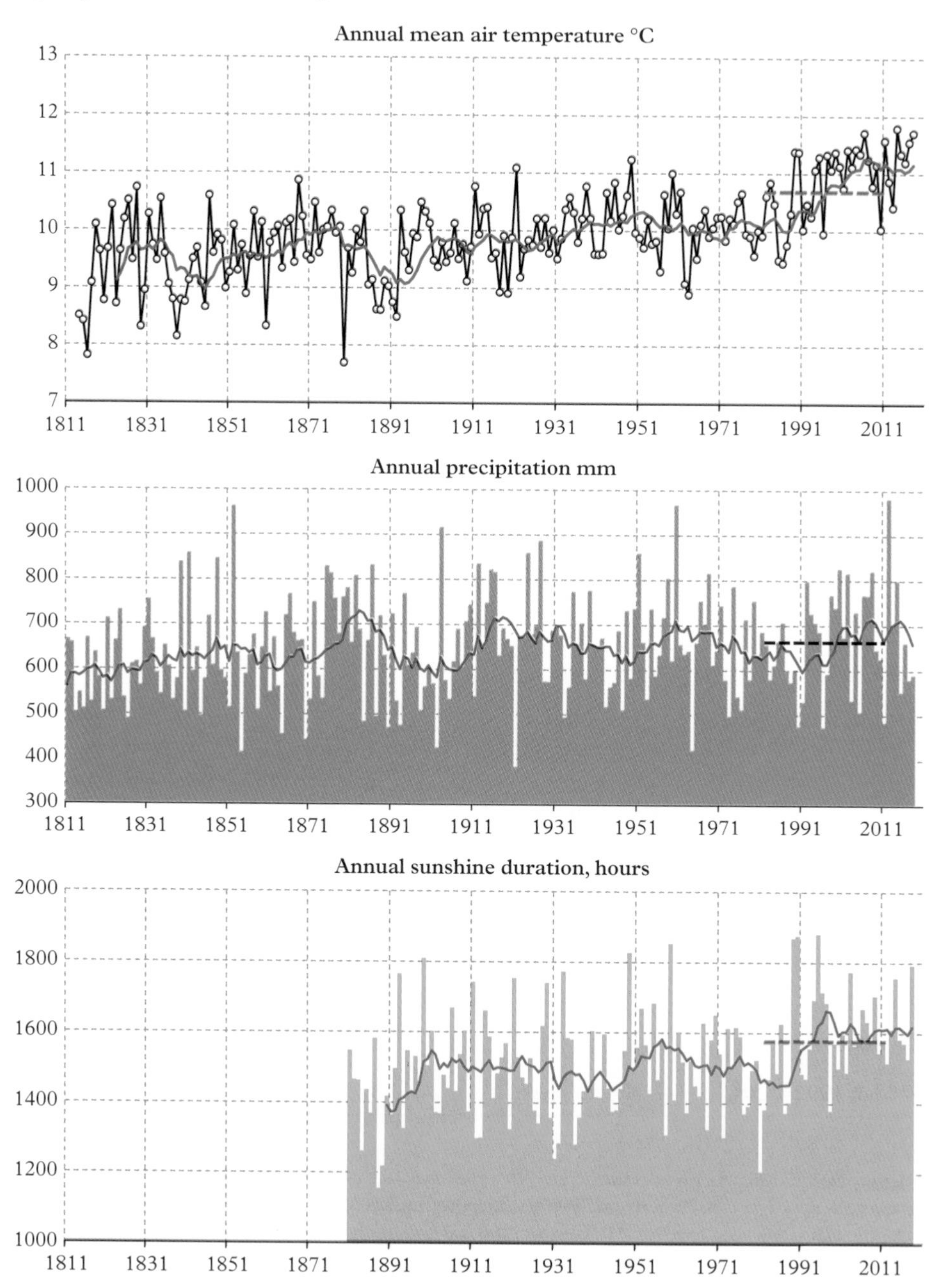


**Figure 19.8** *Calendar year values of (from top) mean temperature (°C, since 1814), total precipitation (mm, since 1811) and sunshine duration (hours, since 1881) at Oxford. The 1981–2010 averages are indicated by the thick blue dashed line, while the ten-year unweighted running mean ending at the year shown is indicated by the red line*

# Part 3

# Oxford weather through the seasons

# 20

# Winter

*December, January and February*

By convention in the UK, the year is subdivided meteorologically into four seasons: winter (the three calendar months December, January and February), spring (March, April and May), summer (June, July and August) and autumn (September, October and November). There are other possible divisions, of course: Hubert Lamb [89] suggested a division of the year into five periods, to be considered as natural seasons, which also encompassed numerous shorter seasonal phases or episodes (singularities). For statistical convenience, however, we will continue to use the conventional three-month seasons, starting with winter.

Winter is the coldest season of the year, and temperatures tend to reach their minimum in early to mid-February, some 7–8 weeks after the winter solstice. Winter air temperatures average 4.9 °C with February having both the mildest days (average daily maximum temperature 8.0 °C) and the coldest nights (average daily minimum temperature 1.8 °C) of the three traditional winter months. In these respects, Oxford is very much representative of a large area of the Midlands and south-east England, a couple of degrees cooler than western coastal districts [103].

Winter averages 163 mm of precipitation, rather less than in autumn and just a little wetter than both spring and summer. December is the wettest winter month (63 mm), a little less than the autumnal months October and November; January (57 mm) and February (43 mm) are both rather drier than December. Some precipitation falls as snow in winter; snow can be expected to cover the ground on an average of about six days between early December and late February.

With daylength being short, sunshine averages just 2 hours and 10 minutes each day (194 hours each winter); about one day in three (an average of 32 of the 90 or 91 days) will remain sunless. December is the least sunny winter month (1 hour 43 minutes bright sunshine daily on average), February the sunniest (average 2 hours 47 minutes daily).

Throughout this chapter, winters are dated by the January year—thus, winter 1962/63 (December 1962 to February 1963) is referred to as 'winter 1963'.

*Oxford Weather and Climate since 1767*. Stephen Burt and Tim Burt, Oxford University Press (2019).
 DOI: 10.1093/oso/9780198834632.001.0001

## Temperature

During the period 1815–2018, winter temperatures in Oxford have ranged from −17.8 °C on 24 December 1860 to 18.5 °C on 13 February 1998. The mildest and coldest winter days and nights on record are shown in Table 20.1.

The mildest winter nights are often in December, when the ground and especially the surrounding seas are still cooling from the warmth of autumn. On the other hand, the mildest winter days tend to occur in the second half of February, as the increasing strength of the Sun begins to be felt. The coldest days tend to occur in January, although the coldest nights can be found anytime between mid-December and the end of February. Mild days or nights in winter usually occur within a mild Atlantic air mass: very low temperatures are usually generated by an easterly airflow off the near-continent around a blocking high over Scandinavia, with winds originating from far eastern Europe or Russia, which is obviously extremely cold at this time of year (see Chapter 4 for the synoptic contexts of the contrasting Decembers of 2010 and 2015). A 'blocked easterly' synoptic situation is more likely in the second half of the season, although occasionally rather earlier, as experienced with devastating effect in the extremely cold winter of 1963.

**Table 20.1** *Highest and lowest maximum and minimum temperatures at the Radcliffe Observatory, Oxford in winter, 1816–2018*

| Rank | Mildest days | Coldest days | Mildest nights | Coldest nights |
|---|---|---|---|---|
| 1 | 18.5 °C, 13 Feb 1998* | *−9.6 °C, 8 Jan 1841* | 12.5 °C, 3 Dec 1985 | −17.8 °C, 24 Dec 1860 |
| 2 | 17.8 °C, 28 Feb 1959 | *−7.9 °C, 2 Feb 1830* | 12.4 °C, 21 Dec 1971 and 5 Dec 1986 | −16.7 °C, 28 Dec 1860 |
| 3 | 17.4 °C, 28 Feb 1891 and 14 Feb 1998 | *−7.8 °C, 19 Jan 1823* | 12.3 °C, 5 Dec 1898 | −16.6 °C, 14 Jan 1982 |
| 4 | 16.8 °C, 27 Feb 1891 | *−7.4 °C, 20 Jan 1838* | 12.2 °C, 7 Dec 2015 | *−16.4 °C, 20 Jan 1838* |
| 5 | 16.7 °C, 18 Feb 1945, 29 Feb 1948, 28 Feb 1960 and 4 Feb 2004 | −7.1 °C, *15 and 19 Jan 1838* and 13 Jan 1982 | 12.1 °C, 12 Dec 2000 and 17 Dec 2015 | −16.1 °C, 25 Feb 1947 and 13 Dec 1981 |
| ***Last 50 years to 2018*** | | | | |
| | 18.5 °C, 13 Feb 1998 | −7.1 °C, 13 Jan 1982 | 12.5 °C, 3 Dec 1985 | −16.6 °C, 14 Jan 1982 |

*Temperatures recorded by unscreened thermometers prior to 1853 are shown in italics; these are likely to be a little too cold by modern (screened thermometer) standards in winter and a little too warm in summer—see Appendix 1 for more details. In winter the mean temperature correction for unscreened thermometry is +0.2 to +0.3 degC, although this will vary considerably on a day-to-day basis*

* In February 2019, 17.6 °C was reached on 25th and 18.8 °C on 26th

## Mildest and coldest winters

The winter of 1963 was by far the coldest winter on record at Oxford, certainly since 1813/14 and very probably since Thomas Hornsby's records began in 1760 its mean air temperature of −0.8 °C was 5.7 degC below the current 1981–2010 average (Table 20.2). The mean air temperature over Central England was the lowest since the winter of 1740. In Oxford, December 1962's mean air temperature was 3.7 degrees below average; January 1963, 7.9 degrees below average; and February 1963, 5.6 degrees below. With a mean temperature of −3.0 °C, January 1963 remains the coldest month of any in Oxford's records (since 1814), followed by January 1838 (−2.4 °C) and February 1947 (−2.2 °C): Table 29.4B in Chapter 29 lists the 'Top 10' coldest months. Whilst researching this book, however, we discovered thermometer readings for January 1814 which suggest that month should be ranked as the second-coldest month recorded at Oxford, for the mean temperature was about −2.6 °C. (The calculation method for 1814 temperatures differs slightly from the rest of the record, however, as it is based upon three temperatures per day rather than the mean of the daily maximum and minimum temperatures, which have been used for all other months since April 1815). To that list we should also add January 1776, for analysis of Thomas Hornsby's records, including the usual corrections [30], also suggests a mean air temperature for that month about −2.6 °C (see the Chronology, Chapter 25) and January 1795, for which the monthly mean temperature was probably close to or slightly below −3.0 °C in Oxford.

Gordon Smith [25, 104] analysed cold winters at Oxford in some detail: 1768, 1776, 1814, 1947 and 1963. Smith did not include 1838, despite its appearance several times in Table 20.1, nor 1830 which includes the second-coldest day on record. Figure 20.1

**Table 20.2** *Winter mean temperatures at Oxford, 1814/15 to 2017/18, dated by January year*

| **Winter mean temperature 4.9 °C (average 1981–2010)** | | | | | |
|---|---|---|---|---|---|
| **Mildest winters** | | | **Coldest winters** | | |
| **Mean temperature, °C** | **Departure from 1981–2010 normal degC** | **Year** | **Mean temperature, °C** | **Departure from 1981–2010 normal degC** | **Year** |
| 7.6 | +2.7 | 2016 | −0.8 | −5.7 | 1963 |
| 7.1 | +2.2 | 1869, 1990 and 2007 | −0.1 | −5.0 | *1830* |
| 6.8 | +1.9 | 1975 and | 0.6 | −4.3 | *1841* |
| | | 1989 | 0.7 | −4.2 | *1814* and *1820* |
| 6.7 | +1.8 | 2014 | 0.8 | −4.1 | *1816, 1823* |
| 6.6 | +1.7 | 1877 | 1.0 | −3.9 | *1838*, 1947 |
| *Last 50 years to 2018* | | | | | |
| 7.6 | +2.7 | 2016 | 2.1 | −2.9 | 1979 |

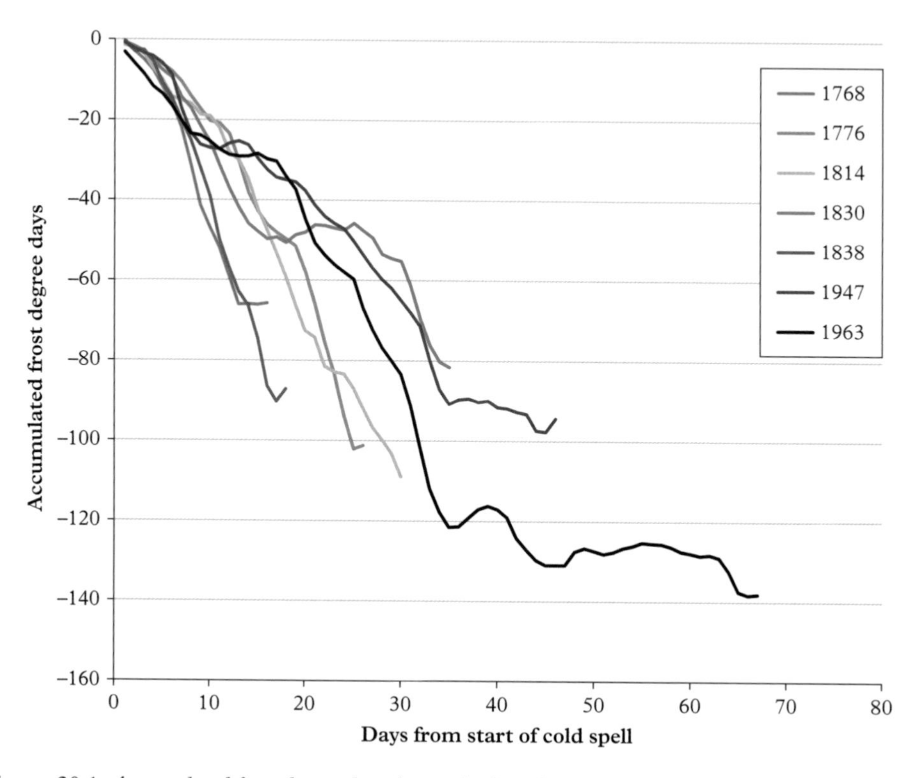


**Figure 20.1** *Accumulated frost degree days for a selection of very severe winters at Oxford since 1767; the winters are dated by the year of the January*

shows accumulated frost degree days for all these winters. Steep gradients indicate a short period of extremely low temperatures; both 1768 and 1838 were relatively short cold spells but very extreme while they lasted. January 1776 included temperatures down to −14 °C and, during the cold snap, Thomas Hornsby's wine began to freeze in his study (see the Chronology, Chapter 25). Winter 1830 had two very cold spells separated by less cold weather, but the overall impact was a relatively protracted cold period of nearly 40 days. The cold winter of 1947 lasted even longer; it has undoubtedly persisted across intergenerational memories because it coincided with post-war austerity including fuel shortages and so was particularly hard to bear.

The most protracted cold winter was easily 1963, which included a middle period of very cold weather throughout January, the coldest month on record at Oxford, as previously noted. Early December 1962 was cold and foggy: the maximum temperature of −2.2 °C on 3 December was the first of three consecutive days with maximum temperatures below zero and was the coldest December day since 1938. After a mild interlude, persistent cold weather set in on 23 December and thereafter north-easterly or easterly winds dominated much of the winter. There was a blizzard across Wales, south-west and most of southern England on 29–30 December. Once the snowstorms at the end of December and start of January were over, it was a mainly dry winter with just 38 mm of precipitation falling in January and February combined—less than half of the normal amount—most of which fell as snow in early January. On 4 March a mild south-westerly flow finally reached the British Isles, and temperatures gradually rose, allowing snow to melt and bringing an end to this harsh winter. Figure 20.1 shows that the cold 1963 winter lasted 65 days, during which time there were very few mild interludes; maximum temperature exceeded 0 °C on half of the days, but minimum temperatures were above zero on only six days in that period. In all, only 13 days recorded mean air temperatures above zero, 8 February reaching a balmy 3.4 °C. Snow cover was protracted. Sunny days towards the end of February and in early March helped clear the snow, preventing the severe flooding along the Thames that can occur after a cold, snowy spell (as happened in March 1947—see March, Chapter 9, and the Chronology, Chapter 25).

Table 20.2 shows that the only winters since 1814 to have recorded a mean air temperature below 0 °C were those of 1830 and 1963. All but two of the coldest winters come from the first half of the nineteenth century; in that respect, the extremely cold twentieth-century winters of 1947 (ranked equal sixth with 1838, with a mean temperature of 1.0 °C) and especially 1963 must now be regarded as very rare indeed, although February 1986 (mean temperature −1.5 °C) and December 2010 (mean temperature 0.3 °C) serve to remind us that some very cold months have been recorded within the last generation or two. Note the inclusion of the winter of 1814, given that we have now discovered data for that period; Gordon Smith included this winter in his analysis of very cold winters at Oxford but nevertheless chose not to start the mean air temperature record until 1815.

In complete contrast, in terms of mildest winters, the only winters to appear from the nineteenth century are 1869 and 1877. All the rest are from 1975 or more recently, with three from the twenty-first century, including the record holder, 2016.

## Frosts

For the period 1981–2010, an average of nine air frosts occur in each of the three winter months, giving a winter total of 27. There were 38 winter air frosts on average during the 1850s but less than 30 from 1890–1920, with similar numbers from the 1970s onwards. So far, the average for the 2010s is only 21: the mild but very wet winter of 2014 recorded only three air frosts, a remarkably low figure for an inland location like Oxford, surpassing the previous record of five in both the winters of 1975 and 1990.

There are, on average, 60 ground frosts each winter, almost equally distributed across the three months. Since records are available (1931), no winter month has ever remained entirely free of ground frosts, the lowest monthly totals being recorded in December 2014 (3), January 1996 (9) and February 1961 (4). Whilst the maximum frequency of ground frosts for December was 28 in 1950, in January (29) and February (26) the highest were both in 1963; together with December 1962's total of 24, the winter of 1963 easily has the highest total of winter ground frosts (82), followed by 1979 (71) and 1991 and 1947, both of which had 69.

## Precipitation

The mean precipitation total for winter in Oxford is 163 mm, falling on an average of 45 days. Some of this usually falls as snow—on 10–11 days in an average winter.

Winter precipitation since 1767 has varied from more than twice the average in 2014 (335 mm) to just less than one-third of the average in 1964 (54 mm, Table 20.3). Winters have tended to become wetter over the past two and half centuries: in the latter part of the eighteenth century, Oxford winter precipitation averaged 126 mm, 137 mm in the nineteenth century, 164 mm in the twentieth century and 173 mm for the winters from 2001 to 2018.

The list of driest and wettest winters in Table 20.3 includes two consecutive winters—1976 and 1977, one very dry and one very wet. After the dry summer of 1975, the following very dry winter provided potential for a major drought if the following seasons were dry too, as indeed they were, the hot dry summer of 1976 especially. Even in the south-east, where groundwater supplies buffer the effects of any short-period deficiency in precipitation, a very dry winter puts water supplies under pressure. Following the 1976 drought, the very wet autumn of that year, and the exceptionally wet winter that followed, ensured an almost unbelievable water supply surplus in the early months of 1977. The winter of 1934 was another dry winter associated with a prolonged drought, although not quite of the scale of 1976. (The 1975/76 drought lasted 16 months and, of all possible 16-month precipitation totals to the end of 2017 (number of possible events $n = 2987$), the period up to and including August 1976 had the lowest total: 458.7 mm.) Other notably dry periods of this length occurred in 1788/89, 1803 and 1855; the winters of 1855 and

**Table 20.3** *Winter precipitation at the Radcliffe Observatory, Oxford since 1767, dated by January year. Extremes of monthly totals 1767–2018, wettest days 1827–2018*

**Winter mean precipitation 162.8 mm** (average 1981–2010)

| Wettest winters | | | Driest winters | | | Wettest days | |
|---|---|---|---|---|---|---|---|
| Total fall, mm | Per cent of normal | Year | Total fall, mm | Per cent of normal | Year | Daily fall, mm | Date |
| 334.7 | 205 | 2014 | 53.8 | 33 | 1964 | 40.8 | 13 Feb 1888 |
| 320.7 | 197 | 1915 | 55.2 | 34 | 1891 | 38.3 | 23 Dec 1985 |
| 310.7 | 191 | 1990 | 59.4 | 37 | 1976 | 36.1 | 28 Jan 1958 |
| 292.6 | 180 | 1912 | 63.3 | 39 | 1934 | 34.1 | 25 Dec 1927 |
| 292.1 | 179 | 1977 | 63.8 | 39 | 1830 | 33.5 | 1 Jan 1926 |
| *Last 50 years to 2018* | | | | | | | |
| 334.7 | 205 | 2013/14 | 59.4 | 37 | 1975/76 | 38.3 | 23 Dec 1985 |

1934 also appear in Table 20.3. At a longer timescale, the protracted rainfall deficit from 1987 to 1992 included the very wet winter of 1990. *The Times* described the 1987–1992 drought as the worst ever, but the Oxford record showed worse 60-month droughts had been experienced in the 1780s, 1800s and early 1900s, albeit at times of much lower water demand [105]. Droughts are discussed in more detail in Chapters 24 and 28, while the synoptic backgrounds to the summer drought of 1976 and the very wet autumn of that year are both examined in Chapter 4.

The wettest winter on record, 2014, was exceptional across wide areas of the country, not just in Oxford (Figure 20.2). Once again, there was serious flooding of parts of west Oxford, so soon after the floods of November 2012. A synoptic description is given in Chapter 4 and additional coverage included in the Chronology, Chapter 25. As a matter of note, winter 2014 was also one of the sunniest on record, just outside the 'top five sunniest' in Table 20.4, suggesting both an absence of fog and 'anticyclonic gloom' together with the broken cloud cover typical of post-frontal polar maritime air masses in winter.

**Figure 20.2** *Ian Ashpole being interviewed for Radio 4 on the morning of 25 February 2014, as the winter's rainfall total surpassed the previous highest accumulation to become the wettest winter on Oxford's 250 years of rainfall records (Ian Curtis)*

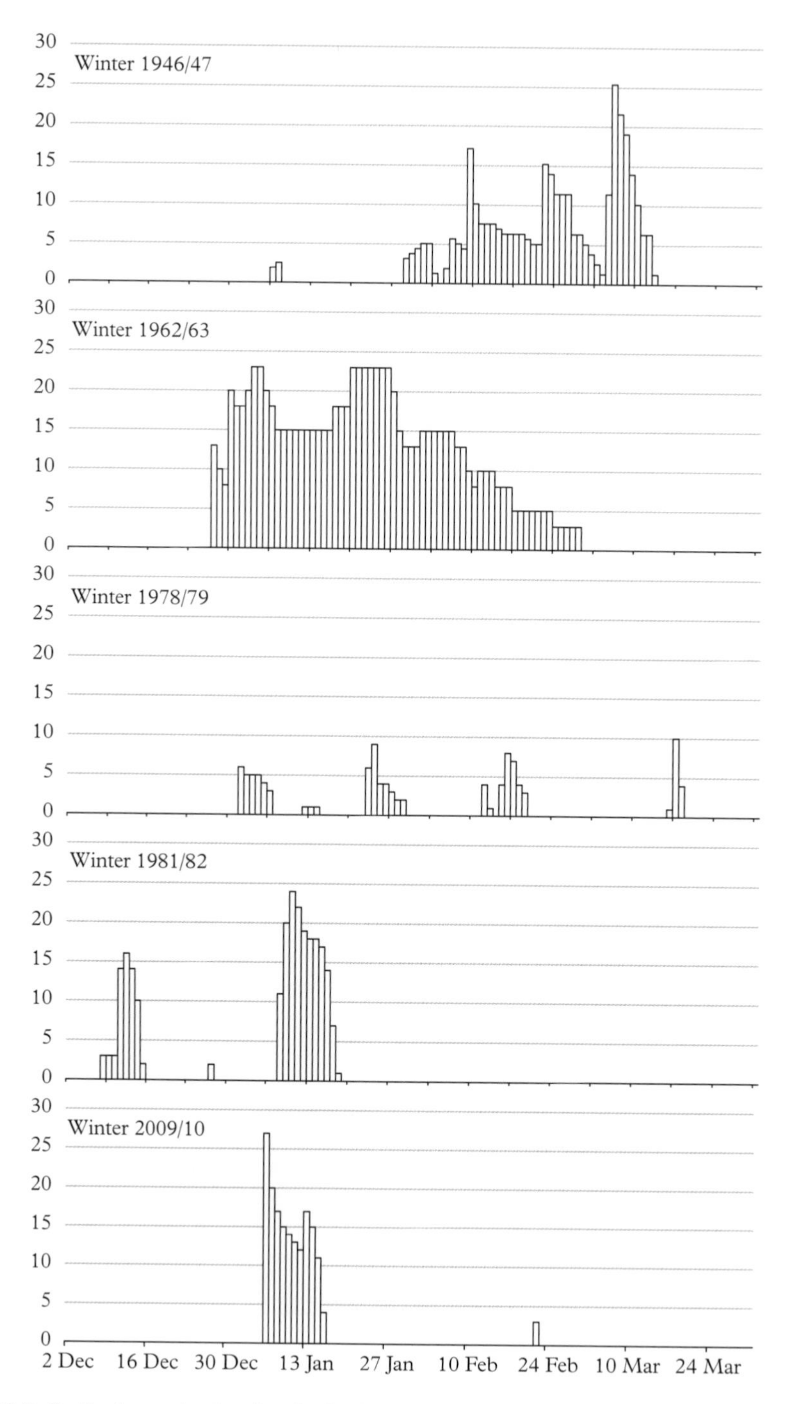


**Figure 20.3** *Daily time series showing the depth of level snow each morning (in centimetres) in Oxford, December to March, for the winters of 1947, 1963, 1979, 1982 and 2010*

## Snowfall and lying snow

Winter 1963 was easily the snowiest winter, with 'snow lying' on 60 mornings, followed by 1947 with 39. Nil snow cover has been recorded in 13 winters since 1926 when these observations began. Figure 20.3 shows the depth of snow (in centimetres) every morning during the winters of 1947, 1963, 1979, 1982 and 2010.

## Sunshine and visibility

Winter is the dullest season of the year, with a sunshine quota of just two and a quarter hours each day; about one in three days will remain sunless. The hours of bright sunshine are of course limited by astronomical day length. Thus, December, which includes the winter solstice, is the dullest of the three winter months (daily average 1 hour 43 minutes), and February, the sunniest (average 2 hours 47 minutes). Altogether, winter has an average sunshine duration of 194 hours. However, this single statistic hides the fact that there has been a significant upward trend in winter sunshine during the period of observation, by linear trendline from just over 150 hours on average in 1881 to almost 200 hours by 2018, an increase of one-third of an hour every year. Both increases in sunshine and decreases in fog frequencies during the winter months are likely to have come about as a lasting benefit of the Clean Air Acts, introduced from 1956; see also the discussion on winter fog, below.

The distribution of winter sunshine hours is somewhat skewed; thus, in the dullest winter (1885), there was less than half the current average figure whereas in the sunniest winter (1952), the excess is only one-third above average. Thus, it is unusual for winter to receive much more than about 225 hours of bright sunshine (a surplus of some

**Table 20.4** *Winter sunshine duration at Oxford, 1881 to 2018, dated by January year*

**Winter mean sunshine duration 194 hours, 2.17 hours per day** (average 1981–2010)

*Possible daylength: 787 hours (leap year 797 hours). Mean sunshine duration as percentage of possible: 24.7*

| Sunniest winters | | | | Dullest winters | | | |
|---|---|---|---|---|---|---|---|
| Duration, hours | Per cent of average | Per cent of possible | Year | Duration, hours | Per cent of average | Per cent of possible | Year |
| 260.6 | 134 | 32.7 | 1952 | 89.5 | 46 | 11.4 | 1885 |
| 258.7 | 133 | 32.9 | 2015 | 98.9 | 51 | 12.4 | 1972 |
| 247.9 | 127 | 31.1 | 2000 | 103.9 | 53 | 12.6 | 2011 |
| 245.1 | 126 | 31.1 | 2018 | 104.3 | 53 | 13.3 | 1897 |
| 243.7 | 125 | 30.6 | 2008 | 110.6 | 57 | 13.9 | 1884 |
| *Last 50 years to 2018* | | | | | | | |
| 258.7 | 133 | 32.9 | 2015 | 98.9 | 51 | 12.4 | 1972 |

30 hours when compared to the average): however, frequent and extensive cloud cover can much more easily lead to deficits of 30 hours or more. With days getting longer as the season evolves, the sunniest days in winter tend to occur towards the end of February. Given the strong upward trend in sunshine hours, the dullest winters mostly fall in the nineteenth century, but there are always exceptions, the winter of 2011 being particularly dull by recent experience. By the same token, the sunniest winters have tended to be more recent, with the winter of 1952 being an exception.

Visibility observations have been made, once daily at 0900 GMT, since 1926. 'Fog' is defined in meteorological terms as visibility less than 1 km: 'thick fog' is visibility less than 200 m. Figure 20.4 shows the annual incidence of fog at Oxford since 1926, including a ten-year running mean for the year ending. There has been a very significant decline since the 1950s, following the various Clean Air Acts which began the demise of coal fire and associated soot emissions. We are uncertain about the observations from the early 2000s, whether this is a real increase or an inadvertent change in observation practice; comparison with the record from the University of Reading [102] suggests the latter. On average, there were 30 incidences of fog each winter at the start of the record; much lower winter totals now seem the norm—for example, six in winter 2017 and only four in winter 2018. This clear downward trend in the incidence of fog in winter may well be partly responsible for the increase in winter sunshine noted above. There are highly significant upward trends for sunshine hours in November, December and January, the period of lowest solar elevation; February has a significant upward trend too.

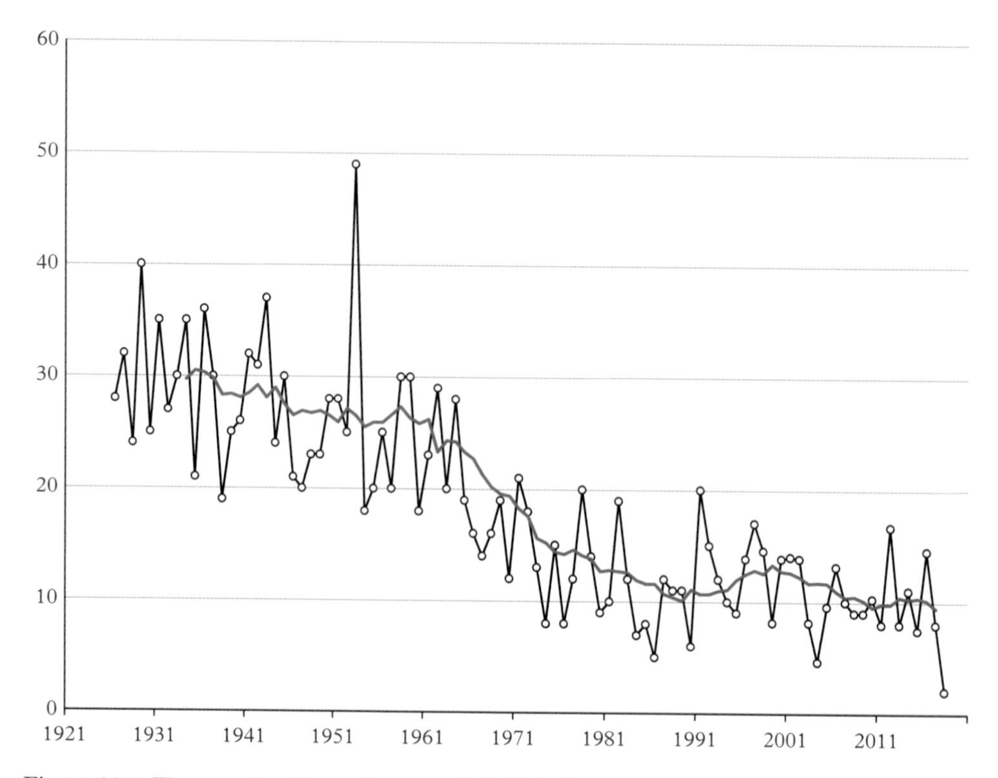


**Figure 20.4** *The annual incidence of fog at Oxford 1926–2018, including a ten-year running mean for the year ending. The decrease in fog frequency since the various Clean Air Acts starting in 1956 is very clear—even a 'foggy' winter in recent decades lies below the minimum for any year from the 1920s to early 1960s. Includes some estimates for recent years based upon the University of Reading fog statistics*

## Oxford temperature, precipitation and sunshine in graphs—Winter

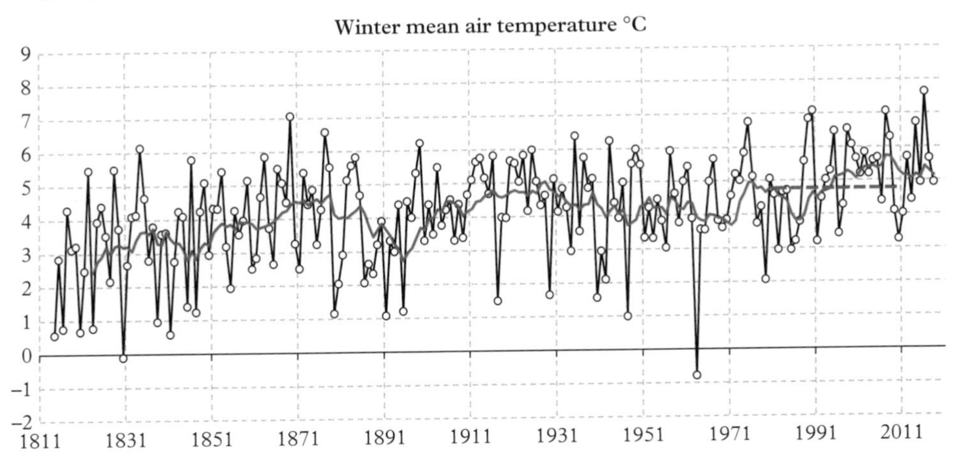


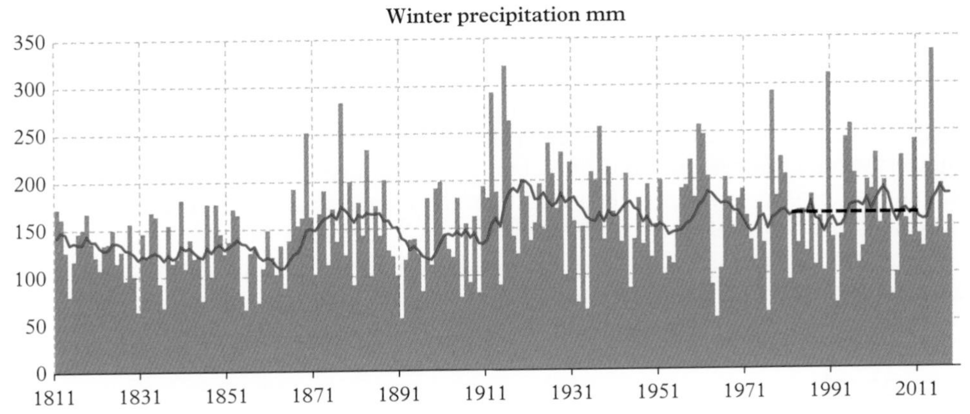


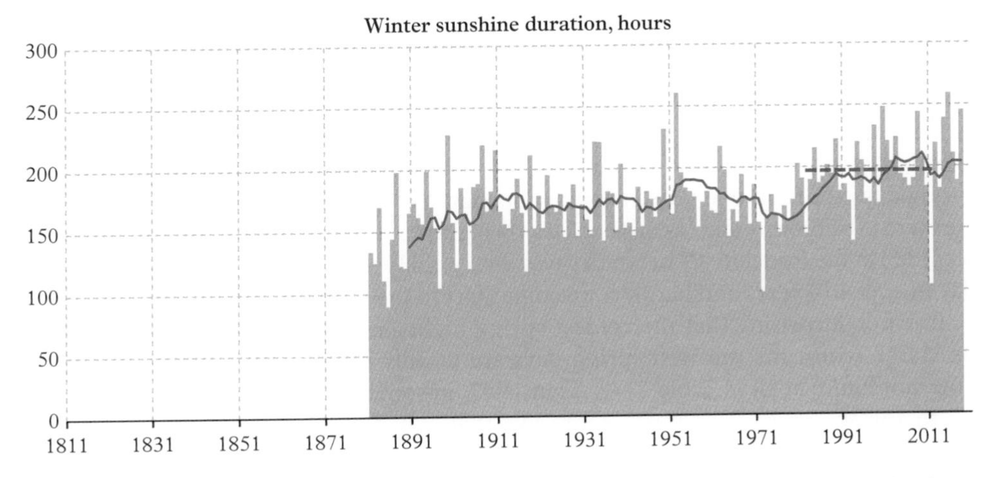


**Figure 20.5** *Seasonal values of (from top) mean temperature (°C, since 1814), total precipitation (mm, since 1811) and sunshine duration (hours, since 1881) for winter at Oxford, dated by January year. The 1981–2010 averages are indicated by the thick blue dashed line, while the ten-year unweighted running mean ending at the year shown is indicated by the red line*

# 21

# Spring

*March, April and May*

Spring often seems disappointing, as winter stumbles towards summer. Expectations in March and especially April can be dashed by spells of cool, even cold, weather; May can be very warm and dry but is equally often cooler and damp. Spring marks the transition from the frosts and snows of winter to the warmth of summer, as the days get longer and the temperature rises—rather unsteadily in most springs. Average temperatures rise from 9 °C by day and 2 °C by night in early March to 19 °C and 9 °C, respectively, by late May.

Despite the association with showers, spring's precipitation average of 154 mm makes it just about the driest season of the year; summer is marginally drier (153 mm). Spring's showery weather is perhaps confirmed by measurable precipitation being expected on 42 days whereas in summer there are 35 'rain days' on average. Snow falls occasionally in spring, most often early in the season: May rarely sees snow but more often the first summer thunderstorms. As daylength increases, monthly sunshine totals increase from an average of 111 hours in March to 193 hours in May.

## Temperature

Since 1815, spring temperatures in Oxford have ranged from −12.0 °C to 30.6 °C. The warmest and coldest spring days and nights are shown in Table 21.1. The coldest spring night remains 13 March 1845, when the temperature fell to −12.0 °C, but more recently −10.8 °C was recorded on 7 March 1947, coming at the end of the very cold winter already discussed. Ironically, May in 1947 was extremely hot and several of the warmest spring days are from that year including the joint record holder, 30 May 1947. Records are there to be broken, and (as this book was being written) on 1 March 2018 the maximum recorded at the Radcliffe Observatory was a very chilly −1.9 °C, Oxford's coldest spring day in over 170 years and hardly a welcome start to the season.

It is not surprising that the coldest spring nights and days come at the beginning of March whilst the warmest spring days are usually towards the end of May; April does not figure at all in Table 21.1. As in 1947, in some years the transition from winter to spring can be very rapid, more like that experienced regularly in continental

*Oxford Weather and Climate since 1767*. Stephen Burt and Tim Burt, Oxford University Press (2019).
 DOI: 10.1093/oso/9780198834632.001.0001

**Table 21.1** *Highest and lowest maximum and minimum temperatures in spring, Radcliffe Observatory Oxford 1815–2018*

| Rank | Hottest days | Coldest days | Warmest nights | Coldest nights |
|---|---|---|---|---|
| 1 | 30.6 °C, 29 May 1944 and 30 May 1947 | *−4.1 °C, 13 Mar 1845* | 16.7 °C, 9 May 1945 | *−12.0 °C, 13 Mar 1845* |
| 2 | 30.1 °C, 31 May 1947 | −1.9 °C, *16 Mar 1845* and 1 Mar 2018 | 15.6 °C, *28 May 1847*, 27 May 1855, 22 May 1922 and 28 May 2001 | −10.8 °C, 7 Mar 1947 |
| 3 | *30.0 °C, 27 May 1841* | −1.1 °C, 6 Mar 1942 | 15.4 °C, 18 May 1868 and 25 May 1922 | −10.6 °C, 10 Mar 1858 |
| 4 | 29.8 °C, 23 May 1922 | −0.9 °C, 4 Mar 1965 | 15.3 °C, 31 May 1895 | −9.7 °C, 12 Mar 1891 |
| 5 | 29.6 °C, 29 May 1947 | *−0.5 °C, 6 Mar 1845* | 15.2 °C, 5 May 1862 | *−8.9 °C, 25 Mar 1850* |
| ***Last 50 years to 2018*** | | | | |
| | 28.9 °C, 24 May 2010 | −1.9 °C, 1 Mar 2018 | 15.6 °C, 28 May 2001 | −6.1 °C, 1 Mar 2018 |

*Temperatures recorded by unscreened thermometers prior to 1853 are shown in italics; these are likely to be a little too cold by modern (screened thermometer) standards in winter and a little too warm in summer—see Appendix 1 for more details. In spring the mean temperature correction for unscreened thermometry varies from +0.2 °C in March to −0.1 °C in May, although this will vary considerably on a day-to-day basis. For March 1845, maximum temperatures have been based on the observed 14h temperature as maximum temperatures are missing for this period (see Appendix 1)*

locations like eastern Canada rather than in the gradually changing oceanic climate of the British Isles. A particularly sharp 'reverse transition' occurred in 1891, when a maximum temperature of 17.4 °C on 28 February was followed within a fortnight by a minimum temperature of -9.7 °C on 12 March.

## Warm and cold springs

As with winter temperatures, the coldest springs are all earlier in the record, with the coldest of all (by more than a degree) in the year that Queen Victoria ascended the throne, 1837 (Table 21.2). Spring 1816 ranks as the equal fourth-coldest spring: 1816 is known as 'the year without a summer' in Europe and North America as a result of the volcanic eruption of Tambora in April 1815 (see Summer, Chapter 22, and the Chronology, Chapter 25), but clearly spring 1816 was also unusually cold too—a year with no spring indeed! As noted in Chapter 13, July 1816 was amongst the coldest on record. The same can be said about each month of spring 1816, with March and May

**Table 21.2** *Spring mean temperatures at Oxford, 1815–2018*

**Spring mean temperature 9.7°C** (average 1981–2010)

| Mildest springs | | | Coldest springs | | |
|---|---|---|---|---|---|
| **Mean temperature, °C** | **Departure from 1981–2010 average degC** | **Year** | **Mean temperature, °C** | **Departure from 1981–2010 average degC** | **Year** |
| 11.9 | +2.2 | 2011 | *5.4* | *−4.3* | *1837* |
| 11.3 | +1.6 | 2017 | 6.6 | *−3.1* | 1879 |
| 10.8 | +1.1 | 1945, 1999 | *6.7* | *−3.0* | *1839* |
| 10.7 | +1.0 | 1893 | 6.8 | *−2.9* | *1816*, 1887 |
| 10.6 | +0.9 | 1952, 1992, 2007, 2009, 2014 | *6.9* | *−2.8* | *1817* |
| *Last 50 years to 2018* | | | | | |
| 11.9 | +2.2 | 2011 | 7.7 | *−2.0* | 2013 |

ranking 16th coldest on record (in 203 years) and April ninth coldest in the same period. The combination of a very cold spring with a very cold summer was bound to be highly detrimental to crop growth: there were widespread crop failures across western Europe. Spring 1817 also appears in Table 21.2, no doubt a continuing effect of the Tambora eruption. Other cold springs include 1839 and 1879: 1879 was especially chilly, and remains the coldest year on record at Oxford, while spring 1839 was fractionally colder than 1816. In terms of growing day degrees (see also Chapter 24), spring 1816 has the third lowest total with only 1860 and 1879 having lower totals, confirming the disastrous start to the growing season of 1816.

Of the warmest springs, most have come this century with just four from the twentieth century and only 1893 from the (late) nineteenth century. As the Second World War came to an end, the very warm spring of 1945 must have been especially welcome.

## Frosts

Based upon the current averaging period 1981–2010, around six air frosts can be expected in spring—four in March, two in April but only one per decade or so in May. Of course, this hides considerable year-to-year variability in the number of air frosts: since 1981, March has seen 10 or more air frosts in four years (including 16 in 2013). There were also six air frosts in April 2013 (none in May), making 2013 the frostiest spring since 1929. In contrast to today, an Oxford spring in the nineteenth century would have expected about twice as many air frosts. This reduction, together with a lengthening of the growing season, has no doubt been welcomed by Oxford gardeners and local farmers.

On average, 28 ground frosts are expected in an Oxford spring (average 1981–2010): fourteen in March, ten in April and the remainder in May. Late ground frosts occur in May in most years, more than ten in four Mays since 1881, very widely spaced in time (1909, 1941, 1984, 1996). Very late ground frosts in June used to occur more often, but more recently only the Junes of 1991 and 2012 have experienced ground frosts.

## Precipitation

Spring's mean precipitation total in Oxford is 154 mm (average 1981–2010), falling on 42 days. Some of this can be expected to fall as snow, on 4–5 days in an average spring. March (48 mm) and April (49 mm) have very similar totals with May slightly wetter (57 mm).

Oxford's driest spring was in 1785 when only 29 mm fell; the wettest spring was in 1862 when 275 mm was recorded, almost ten times as much (Table 21.3). Oxford's wettest spring day was 25 April 1908 when 46.7 mm was recorded, falling almost entirely as snow (see April, Chapter 10, and the Chronology, Chapter 25). Falls in excess of 25 mm occur in spring in around one year in six.

Over the long term, spring precipitation reflects the general pattern seen at Oxford since 1767: after an initially wet period, the 1780s and 1790s were dry; from then on, there has been a gradual increase in spring precipitation, a statistically significant trend over the entire period (correlation $r = 0.208$, probability $p = 0.0009$, number of samples $n = 251$; Figure 21.1). Thus, with the exception of 1990, the driest springs are all from early in the record. The wettest springs tend to be more recent, although well spread across the record.

**Table 21.3** *Spring precipitation at the Radcliffe Observatory, Oxford 1767–2018 (wettest days 1827–2018)*

| **Spring mean precipitation 153.8 mm** (average 1981–2010) | | | | | | | |
|---|---|---|---|---|---|---|---|
| **Wettest springs** | | | **Driest springs** | | | **Wettest days** | |
| **Total fall, mm** | **Per cent of normal** | **Year** | **Total fall, mm** | **Per cent of normal** | **Year** | **Daily fall, mm** | **Date** |
| 275.4 | 179 | 1862 | 28.8 | 19 | 1785 | 46.7 | 25 Apr 1908 |
| 274.0 | 178 | 1979 | 34.0 | 22 | 1893 | 42.1 | 15 Mar 2008 |
| 261.3 | 170 | 1981 | 38.5 | 25 | 1788 | 40.7 | 27 May 2007 |
| 255.6 | 166 | 1932 | 41.9 | 27 | 1803 | 39.5 | 14 May 1985 |
| 246.6 | 160 | 1903 | 47.5 | 31 | 1990 | 38.0 | 11 Apr 2000 |
| ***Last 50 years to 2018*** | | | | | | | |
| 274.0 | 178 | 1979 | 47.5 | 31 | 1990 | 42.1 | 15 Mar 2008 |

## Thunderstorms

Thunder is more likely in May than in March or April, and can be expected on around three days in an average spring. Although about one spring in ten misses out on thunder entirely, in spring 1967 there were 12 days with thunder, and in spring 1983, 10.

## Snowfall and lying snow

Although snowless springs are not uncommon, seven springs since 1908 have seen 10 or more days with snow or sleet observed to fall: 1917 saw 18 days with snow or sleet and 1970 16 days. The very cold spring of 1917 also had seven mornings with lying snow, and there were five such days in spring 1965.

Snow has been observed to fall as late as 16 May in 1935, and on 17 May in 1955. Lying snow on the ground (covering at least half the surface) was observed as late as 26 April in 1981, when it lay 4 cm deep.

# Sunshine

Spring averages almost exactly five hours of sunshine daily, ranging from about 3 hours 35 minutes in March to 6 hours 13 minutes daily in May—indeed, May in Oxford is almost as sunny as each of the three summer months. Around twelve days will remain sunless in a typical spring, half of these in March. The springs of 1889 and 1981 must have been very disappointing in terms of sunshine, well below anything else in the record: 1981 was particularly dismal with the third dullest March (58 hours), the 17th equal dullest April and then easily the dullest May on record, only 98 hours, less than half the average and only one-third of the sunshine in the sunniest May (1989, 301 hours).

**Table 21.4** *Spring sunshine duration at Oxford, 1880–2018*

**Spring mean sunshine duration 465 hours, 5.06 hours per day** (average 1981–2010)

*Possible daylength: 1268 hours. Mean sunshine duration as percentage of possible: 36.7*

| **Sunniest springs** | | | **Dullest springs** | | |
|---|---|---|---|---|---|
| **Duration, hours** | **Per cent of possible** | **Year** | **Duration, hours** | **Per cent of possible** | **Year** |
| 663.5 | 52.3 | 1990 | 266.5 | 21.0 | 1981 |
| 655.9 | 51.7 | 1893 | 293.2 | 23.1 | 1889 |
| 622.3 | 49.1 | 1995 | 346.1 | 27.3 | 1930 |
| 602.5 | 47.5 | 1948 | 356.5 | 28.1 | 1923 |
| 601.5 | 47.4 | 1997 | 356.7 | 28.1 | 1932 |
| *Last 50 years to 2018* | | | | | |
| 663.5 | 52.3 | 1990 | 266.5 | 21.0 | 1981 |

## Oxford temperature, precipitation and sunshine in graphs—Spring

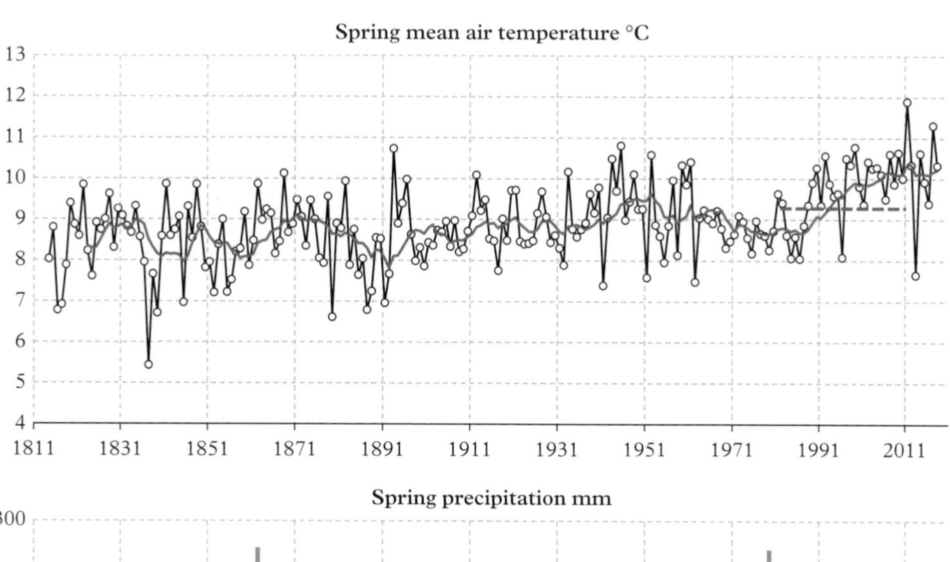


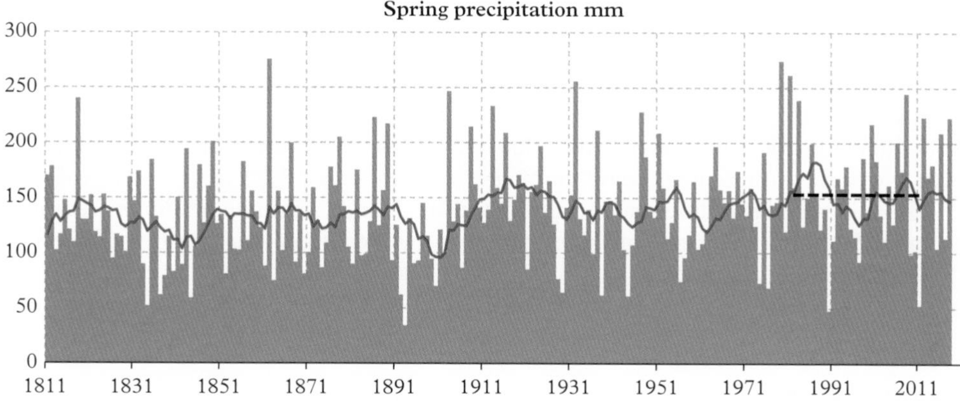


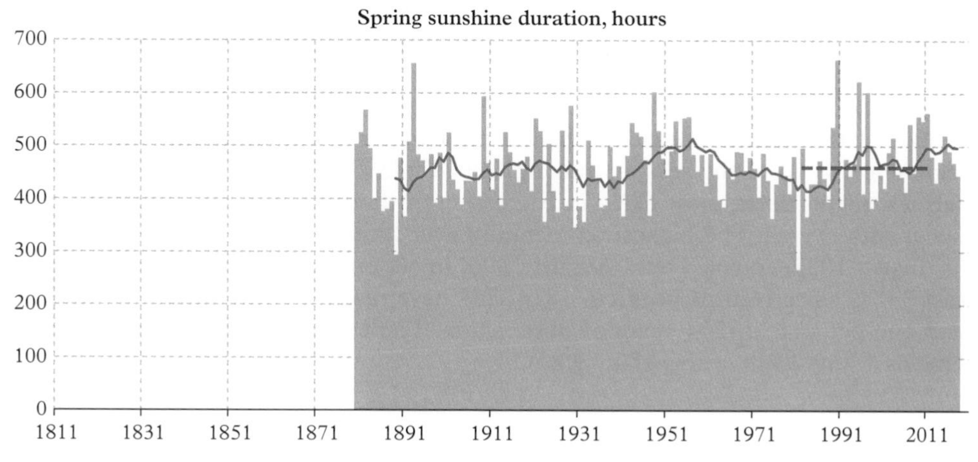


**Figure 21.1** *Seasonal values of (from top) mean temperature (°C, since 1814), total precipitation (mm, since 1811) and sunshine duration (hours, since 1880) for spring at Oxford. The 1981–2010 averages are indicated by the thick blue dashed line, while the ten-year unweighted running mean ending at the year shown is indicated by the red line*

# 22

# Summer

*June, July and August*

Summer is the warmest season of the year, with the mean daily maximum temperature 21.7 °C, and mean daily minimum 12.3 °C. July is, on average, the warmest month but with August close behind. Average daily temperatures reach their annual peak at the end of July—daily maximum temperatures average 23–24 °C, and daily minimum temperatures average 14 °C at this time of year, although, obviously, there is large variability from year to year.

Summer's mean precipitation total is 153 mm, just marginally drier than spring, with August being the wettest of all three summer months—although it is only some 5 mm wetter than the driest, June, on average. Rain falls on about 35 days in an average summer, with five or six of those days being thundery.

There is very little difference in the month-to-month sunshine amounts (seasonal average 594 hours of bright sunshine), although by the end of August the reduction in daylength becomes apparent.

## Temperature

During the period 1815–2018, summer temperatures in Oxford have ranged from 0.2 °C to 35.1 °C. The hottest and coldest summer days and nights are shown in Table 22.1. No air frost has ever been recorded in summer at Oxford, although 26 August 1864 came very close. The coldest summer nights are equally likely at the beginning or end of the season. Judging from Table 22.1, the very hottest days and warmest nights seem most likely in August, even though July is the warmest summer month overall. The highest temperature ever recorded at Oxford is 35.1 °C on 19 August 1932, equalled on 3 August 1990. In all, only seven days have reached at least 34 °C since records began in 1814; 257 have reached or surpassed 30 °C, just over one per year in 204 years of observation. The coldest summer day at Oxford remains 4 June 1909, with a chilly 9.8 °C.

*Oxford Weather and Climate since 1767*. Stephen Burt and Tim Burt, Oxford University Press (2019).
 DOI: 10.1093/oso/9780198834632.001.0001

**Table 22.1** *Highest and lowest maximum and minimum temperatures in summer, 1815–2018*

| Rank | Hottest days | Coldest days | Warmest nights | Coldest nights |
|---|---|---|---|---|
| 1 | 35.1 °C, 19 Aug 1932 and 3 Aug 1990 | 9.8 °C, 4 June 1909 | 21.2 °C, 20 July 2016 | 0.2 °C, 26 Aug 1864 |
| 2 | 34.8 °C, 9 Aug 1911 and 19 July 2006 | 10.0 °C, 3 June 1953 | 20.5 °C, 11 Aug 1997 | 1.4 °C, 24 Aug 1864 and 15 June 1892 |
| 3 | 34.6 °C, 9 Aug 2003 | 10.3 °C, 6 June 1905 | 20.3 °C, 26 Aug 1859 and 4 July 1976 | 1.5 °C, 1 June 1893 |
| 4 | 34.3 °C, 27 June 1976 | 10.4 °C, 14 June 1903 | 20.1 °C, 5 Aug 1975 and 23 Aug 1997 | 1.6 °C, 1 June 1962 |
| 5 | 34.0 °C, 26 June 1976 | 10.6 °C, 2 June 1941 | 20.0 °C, 30 July 1827, 18 Aug 1893 and 20 July 2006 | *1.7 °C, 21 July 1825* |
| *Last 50 years to 2018* | | | | |
| | 35.1 °C, 3 Aug 1990 | 11.2 °C, 1 June 1989 | 21.2 °C, 20 July 2016 | 1.8 °C, 2 June 1991 |

*Temperatures recorded by unscreened thermometers prior to 1853 are shown in italics; these are likely to be a little too cold by modern (screened thermometer) standards in winter and a little too warm in summer—see* Appendix 1 *for more details. In summer the mean temperature correction for unscreened thermometry varies around −0.2 to −0.3 degC, although this will vary considerably on a day-to-day basis*

## Warm and cool summers

All but one of the six warmest summers on Oxford's records have occurred since 1975 (Table 22.2): the exception being 1826, mean temperature 18.7 °C, which until 1976 stood over half a degree clear of the next-hottest (1911, mean 18.1 °C). Until summer 2018, 1826 and 1976 tied as the hottest summers on the long Oxford record, with a mean temperature of 18.7 °C; summer 2018 then took the lead in the table, by just a tenth of a degree in mean temperature, despite August 2018 being only a little warmer than normal and no days of great heat (the hottest day was 26 July, when the temperature reached 32.4 °C; only five days reached or exceeded 30 °C). The synoptic context of summer 2018, noteworthy also for a long drought and an exceptionally sunny two-month spell in midsummer as well as for its high mean temperature, appears in Chapter 4: the summer is compared with others on the Oxford record by means of an index in a later section in this chapter.

The position of summer 1826 as Oxford's equal second-hottest Oxford summer with 1976 is confirmed by its second place in the Central England Temperature (CET)

**Table 22.2** *Summer mean temperatures at Oxford 1814–2018*

| **Summer mean temperature 17.0 °C** (average 1981–2010) | | | | | |
|---|---|---|---|---|---|
| **Hottest summers** | | | **Coolest summers** | | |
| **Mean temperature, °C** | **Departure from 1981–2010 normal degC** | **Year** | **Mean temperature, °C** | **Departure from 1981–2010 normal degC** | **Year** |
| 18.8 | +1.8 | 2018 | 14.2 | −2.8 | *1816*, 1860 |
| 18.7 | +1.7 | *1826*, 1976 | 14.4 | −2.6 | 1879, 1888, 1907, 1920, 1922 |
| 18.6 | +1.6 | 2006 | 14.8 | −2.2 | 1862, 1892, 1903, 1954 |
| 18.4 | +1.4 | 2003 | 14.9 | −2.1 | *1821, 1833,1839, 1841*, 1890, 1891, 1909, 1912 |
| 18.3 | +1.3 | 1983, 1995 | 15.0 | −2.0 | 1894, 1916, 1956, 1977 |
| *Last 50 years to 2018* | | | | | |
| 18.8 | +1.8 | 2018 | 15.0 | −2.0 | 1977 |

summer rankings, beaten there only by 1976 (summer 2018 is ranked fifth in summer mean temperature in the CET series behind 2003, 1995, 1826 and 1976). Of course, there could be a circular argument here as the CET value may well have been unduly influenced by Oxford in 1826, at a time when few temperature records were available. In 1826, the summer started slowly with five out of the first six days of June seeing maxima below 20 °C at the Radcliffe Observatory (Figure 22.1). By the end of the month and into July, there were maxima above 30 °C. June 1826 ends up as the sixth-equal warmest June on record, with July ranking equal 13th warmest and August eighth warmest. Clearly, all three months have to be very warm if the summer is to be record-breaking. Living in London, the diarist William Godwin (married to Mary Wollstonecraft and father of Mary Shelley), not a frequent observer of the weather, noted on 25 June 1826 that there had been hot weather for 17 days and, on 31 July, he noted that the thermometer read '81 and a half' Fahrenheit (27.5 °C), not exceptionally high but clearly worthy of note. The summer of 1826 was therefore sufficiently unusual for even a political philosopher to take note!

The long summer of 1976 is rather clearer in the memory, at least to those as old as your authors: weeks of sunshine, high temperatures and a severe drought. There were 14 days continuously above 30 °C from 25 June to 8 July inclusive (see 'Notable heatwaves and sunny spells' in Chapter 30). The 1976 drought broke at the end of August and was followed by a very dull and wet autumn and winter.

The next four warmest summers are all quite recent: 2006 (mean air temperature 18.6 °C), 2003 (18.4 °C), 1983 and 1995 (both 18.3 °C). Then come two somewhat earlier but very well-known hot summers: 1911 (18.1 °C) and 1947 (18.0 °C). Next comes summer 1975 (17.9 °C), the start of the 1975–76 drought, followed by six summer seasons with an average temperature of 17.8 °C, namely 1818, 1846, 1899, 1933, 1997 and 2017. Summer 1818 must have been welcome indeed after the very poor summers of 1816 and 1817; a similar comment can be made for 1947 following the long, cold winter of that year.

None of the seven coolest summers are more recent than 1922 (Table 22.2). Summer 1816 was the coldest, known as the 'year without a summer', largely due to the effects of the previous year's enormous volcanic eruption at Tambora (see also the Chronology in Chapter 25). Summer 1860 was equally cold; its warmest day came on 5 July, at just 23.1 °C. Next come five summers with an average temperature of 14.4 °C: 1879, 1888, 1907, 1920 and 1922, followed by a group averaging 14.8 °C. Of all the coolest summers listed in Table 22.2, the most recent is 1977 (mean temperature 15.0 °C), and before that 1956. Several come from the period 1816–1841 with another cluster between 1879 and 1916. From the most recent warm period, 1988 is the first cool summer to make an

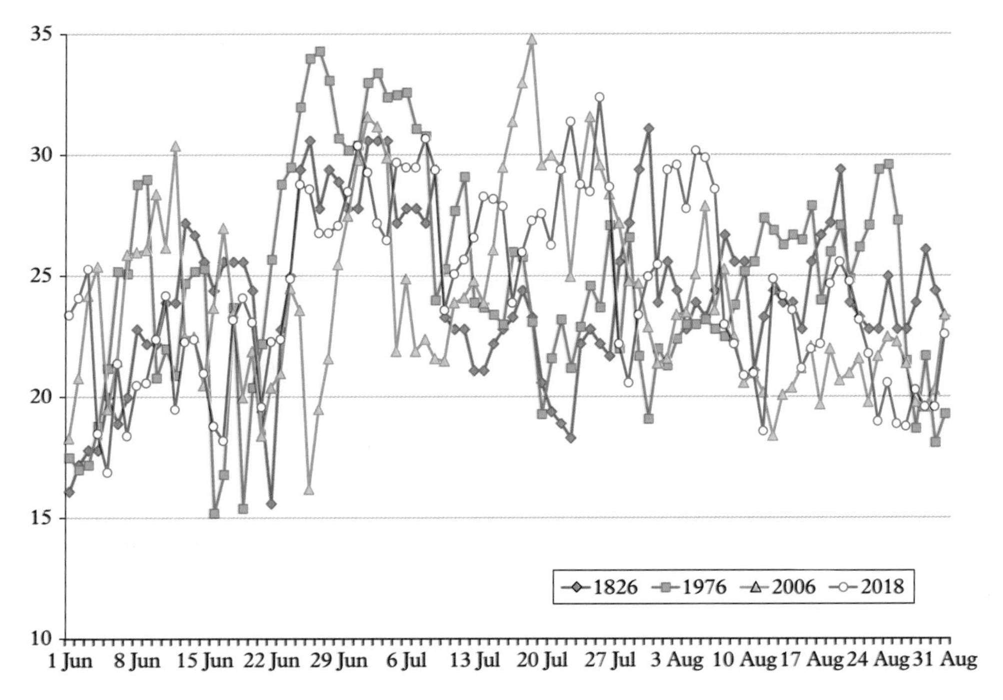


**Figure 22.1** *Daily maximum temperatures in the four hottest summers on Oxford's records—1826, 1976, 2006 and 2018*

appearance, but only ranking 33rd coolest. Of the twenty-first-century summers to 2018, 2011 has been the coolest at 15.5 °C, ranking equal 46th; in some past decades, this would have been regarded as very much an average summer (Figure 22.7).

During the summer months, mean monthly maximum temperatures arguably provide a better indication of the warmth or otherwise of the summer than mean air temperatures, which are derived from the average of the daily mean maximum and minimum temperatures. This is because hot summer months tend to be sunny (less cloud than normal), and reduced cloud amounts (clear nights) often lead to mean minimum temperatures being close to, sometimes even slightly below, normal. In contrast, cool summer months are often much cloudier (and sometimes wetter) than normal; increased cloud cover and higher humidity reduces the daily range in temperature, such that, while mean maximum temperatures may be well below normal, mean minimum temperatures may even be somewhat above, resulting in the mean temperature differing less from normal than might be expected. Table 22.3 therefore shows, for the peak summer months of July and August only, the highest and lowest mean monthly maximum temperatures on record. Since 1815, mean monthly maximum temperatures have exceeded 25.0 °C in fifteen months (eleven Julys and four Augusts), the earliest July 1859, the latest July 2018. The only instance of two consecutive months each surpassing this threshold came in summer 1995 (July 1995 mean maximum 25.1 °C, August 1995 26.4 °C).

At the other end of the scale, six peak summer months have recorded mean maximum temperatures of 18.0 °C or below, the coolest month (August 1912, at 17.1 °C) being

**Table 22.3** *Peak summer (July and August) mean monthly maximum temperatures at Oxford 1815–2018*

| **Mean maximum temperature July 22.7°C, August 22.3 °C** (average 1981–2010) | | | | | |
|---|---|---|---|---|---|
| **Hottest peak summer months** | | | **Coolest peak summer months** | | |
| **Mean max temperature, °C** | **Departure from 1981–2010 normal degC** | **Month/year** | **Mean max temperature, °C** | **Departure from 1981–2010 normal degC** | **Month/ year** |
| 27.4 | +4.7 | July 2018 | 17.1 | −5.2 | Aug 1912 |
| 27.1 | +4.3 | July 2006 | 17.5 | −4.8 | July 1879 |
| 26.8 | +4.1 | July 1983 | 17.9 | −4.4 | Aug 1860 |
| 26.4 | +3.7 | Aug 1995 | 18.0 | −4.7 | July 1888 |
| 26.3 | +3.4 | July 1911 and July 1921 | 18.0 | −4.3 | Aug 1891, Aug 1920 |
| ***Last 50 years to 2018*** | | | | | |
| 27.4 | +4.7 | July 2018 | 18.4 | −3.9 | Aug 1986 |

more than 10 degC cooler than the hottest summer month on record by this measure (July 2018, 27.4 °C). No summer month since August 1920 has recorded a mean monthly maximum of 18.0 °C or below; the coolest summer month by mean maximum temperature within the last 50 years has been August 1986, at 18.4 °C.

### Frosts

No summer has (yet) recorded an air frost in Oxford since records began in 1814, although ground frosts have occurred in all three summer months. Since 1931, there have been a total of 25 ground frosts in June, just one in July (in 1990) and two in August (both in 1964).

## Precipitation

The mean precipitation for summer in Oxford is 153 mm, falling on an average of 35 days. Summer is the driest season at Oxford, marginally drier than spring.

The summer of 1852 saw a reasonable July sandwiched between two poor months. June 1852 remains the wettest June on record (170 mm, 346 per cent of normal), and ranks 21st equal lowest mean air temperature for the month. July was wetter than normal but not excessively so (66 mm, rank 96th) and had the equal 20th highest temperature. August returned to wetter conditions, its total of 104 mm being the 21st highest on record for August; it was slightly (0.9 degC) below the current average temperature for the month. Overall, summer 1852 was the wettest on record (340 mm, Table 22.4), but

**Table 22.4** *Summer precipitation at the Radcliffe Observatory, Oxford 1767 to 2018, wettest days 1827–2018*

| Summer mean precipitation 153 mm (average 1981–2010) | | | | | | | |
|---|---|---|---|---|---|---|---|
| Wettest summers | | | Driest summers | | | Wettest days | |
| Total, mm | Per cent of normal | Year | Total, mm | Per cent of normal | Year | Daily fall, mm | Date |
| 340.3 | 223 | 1852 | 30.0 | 20 | 1818 | 87.9 | 10 July 1968 |
| 332.1 | 217 | 2012 | 47.9 | 31 | 1800 | 81.3 | 22 June 1960 |
| 326.2 | 213 | 1879 | 48.8 | 32 | 1995 | 70.8 | 6 Aug 1922 |
| 314.7 | 206 | 1903 | 55.0 | 36 | 1913 | 67.3 | 27 June 1973 |
| 306.3 | 200 | 1971 | 55.3 | 36 | 1976 | 56.1 | 12 Aug 1957 |
| *Last 50 years to 2018* | | | | | | | |
| 332.2 | 217 | 2012 | 48.8 | 32 | 1995 | 67.3 | 27 June 1973 |

was only slightly cooler than the current average—in fact, it was slightly warmer than the contemporary average of the previous 30 summers. There is a statistical tendency (correlation $r = -0.46$) for wet summers at Oxford to be cooler than dry ones, no doubt controlled by the dominant atmospheric circulation. Summers dominated by cyclonic circulations tend to be cooler and wetter than those dominated by anticyclonic weather, which are more likely to be fine, dry and warm—the exception being warm, thundery summers where rainfall totals can be above normal owing to the occasional torrential downpour. As it happens, both 1852 and 2012 stand somewhat apart from the general trend, being rather warmer than their high rainfalls would suggest.

Summer 2012 was a very wet summer in the middle of an exceptional period of wet weather. Nothing at the start of 2012 could anticipate what was to follow. January through March was very dry, just 80 mm in all, all three months in the lowest 40 monthly totals. In complete contrast, April 2012 then turned out the to be wettest April on record, and the start of easily the wettest April to December period at Oxford with 899 mm falling in just nine months. Even if no rain had fallen in the first three months, it would still have been the sixth wettest year on record; as it was 2012 broke all records with the annual total amounting to 979 mm, the nearest Oxford has yet come to an annual rainfall total of 1000 mm. Summer 2012 ranks as the second wettest summer (332 mm), only a little way behind 1852 (340 mm). June (152 mm) was the second wettest on record, July (101 mm) 28th and August (79 mm) 68th. As we shall see later in this chapter, 2012 was the worst summer since 1977 in terms of overall 'weather'; and yet, the weather was mostly fine and dry for much of the London Olympic and Paralympic Games that year, starting on 27 July.

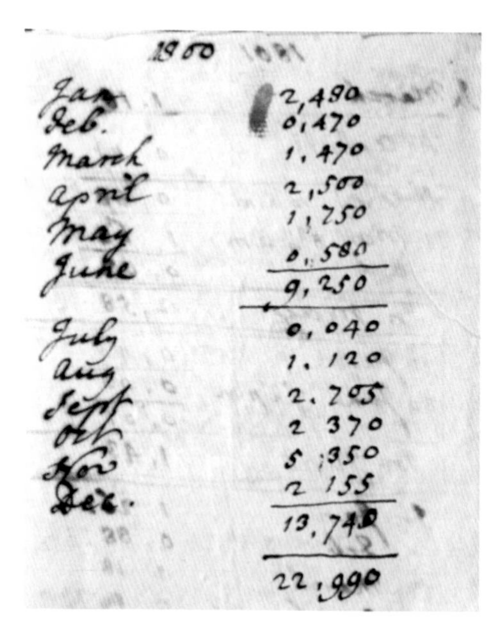

| 1800 | |
|---|---|
| Jan | 2,490 |
| Feb. | 0,470 |
| March | 1,470 |
| April | 2,500 |
| May | 1,750 |
| June | 0,580 |
| | 9,250 |
| July | 0,040 |
| Aug | 1,120 |
| Sept | 2,705 |
| Oct | 2 370 |
| Nov | 5,350 |
| Dec. | 2 155 |
| | 13,740 |
| | 22,990 |

**Figure 22.2** *Monthly rainfall totals (inches: 1 inch = 25.4 mm) for 1800, in Thomas Hornsby's own hand; summer 1800 remains the second-driest on Oxford's records, with just 48 mm of rainfall in three months*

The two driest summers at Oxford come very early in the record: 1800 and 1818. We have Thomas Hornsby's list of monthly rainfall totals for 1800 (Figure 22.2): July 1800 remains the second driest July on record (and the fifth lowest total for any month), Hornsby's recorded total was just 0.040 inch (1.0 mm), a 'corrected' total of just 1.2 mm, beaten only by July 1825 (0.8 mm). Perhaps the dry summer of 1818 reflected the continuing influence of Tambora. There was a memorably hot and dry summer in 1995, with a major drought alleviated only by the fact that winter 1994/95 had been very wet (equal eighth wettest winter on record: 257 mm). Summer 1976 ranks only fifth on the dry summer list

(Table 22.4) while holding the title of the equal hottest summer on record (Table 22.2), until eclipsed by summer 2018.

## Summer vs winter rainfall

Over the 250 years of the Oxford rainfall record, the ratio between winter and summer precipitation has varied markedly. There are, of course, huge variations from year to year, but smoothing out short-term fluctuations by taking a ten-year running mean of each season and then taking the ratio between the two gives us the plot shown in Figure 22.3, where a ratio greater than 1 indicates a period of winters wetter than summers, and vice versa. The ratio is more important than ever to society, because much of southern England's summer water supply depends upon captured winter rainfall, whether stored in surface reservoirs or groundwater (although the former returns to the atmosphere much more rapidly than the latter), while climate change scenarios for southern England generally predict a change towards wetter winters and drier summers—and thus an increase in the ratio. The figure suggests that there is, indeed, a general trend towards winters becoming relatively wetter than summers. For much of the nineteenth century, summers were considerably wetter than winters. This pattern began to reverse at the beginning of the twentieth century, and winters were much wetter than summers in the closing decades of that century. Recent years have seen a sharp reversal of this behaviour, despite several wetter winters. Both seasons show considerable interdecadal variability, but the upward trend in winter precipitation is more monotonic.

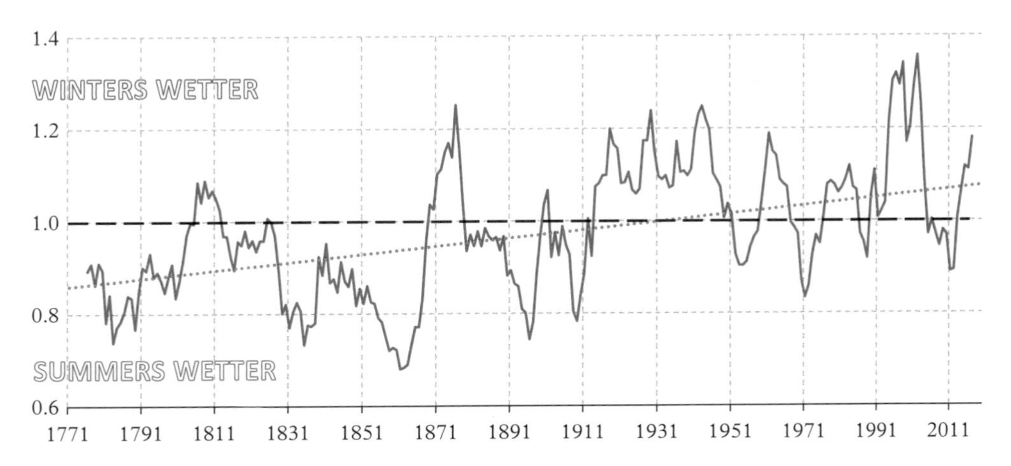


**Figure 22.3** *The long-term variation between summer and winter rainfall at Oxford, expressed here as the ratio between the unweighted ten-year running means for each season (a ratio greater than 1 indicating a period of winters wetter than summers, and vice versa), plotted at the year ending. The average over 1981–2010 is shown by the thicker dashed line, and the least-squares trendline over the entire 250-year period by the thin dotted line*

## Sunshine

Summer is the sunniest season of the year, with each of the three months having a similar average sunshine duration, although with considerable year to year and month-to-month variability. On average, only about four days will remain sunless in summer.

Summer 1888 remains the dullest on record at Oxford, by almost 50 hours or 3 per cent less possible sunshine than the nearest ranked summer, 1912: a daily average of just 4 hours, less than half of that in the sunniest summer, 1989 (8.6 hours per day). Summer 1913 was unusual, in being both very dry (the fourth-driest summer on record, Table 22.4) and very dull—just outside the 'bottom five' in Table 22.5 with 454 hours.

Whilst the dullest summers have all tended to come from earlier in the record, there is no upward trend in summer sunshine. Summer 1977 (eighth dullest, 440 hours) and 1981 (tenth dullest, 452 hours) are the first of the more recent summers to appear near the foot of this table, whilst the very wet summer of 2012 (483 hours) ranks only 29th. At the other end of the rankings, 1989 tops the list with 794 hours; the following summer, 1990, is also highly ranked (34th sunniest, 626 hours). Other pairs of summers near the top of list include 1995 (tenth sunniest, 727 hours) and 1996 (fifth sunniest, 759 hours) and 1975 and 1976 (Table 22.5). Summer 2018 ranked only ninth sunniest, due to a near-normal August.

**Table 22.5** *Summer sunshine duration at Oxford, 1880 to 2018*

**Summer mean sunshine duration 595 hours, 6.46 hours per day** (average 1981–2010)

*Possible daylength: 1449 hours. Mean sunshine duration as percentage of possible: 41.0*

| **Sunniest summers** | | | **Dullest summers** | | |
|---|---|---|---|---|---|
| **Duration, hours** | **Per cent of possible** | **Year** | **Duration, hours** | **Per cent of possible** | **Year** |
| 794.3 | 54.9 | 1989 | 367.5 | 25.3 | 1888 |
| 777.4 | 53.8 | 1899 | 413.8 | 28.6 | 1912 |
| 773.0 | 53.3 | 1976 | 422.1 | 29.2 | 1931 |
| 762.0 | 52.7 | 1975 | 422.8 | 29.2 | 1894 |
| 758.9 | 52.5 | 1996 | 428.5 | 29.6 | 1927 |
| *Last 50 years to 2018* | | | | | |
| 794.3 | 54.9 | 1989 | 440.2 | 30.4 | 1977 |

## Summer weather

Climatologists have spent a good deal of effort in devising indices of summer weather, in an attempt to summarise how 'good' or 'bad' a particular summer season has been. 'Good' and 'bad' are relative terms, of course; a hot, dry summer might be ideal from a holidaymaker or tourist's perspective, but not so marvellous if you are a farmer, or if your water supply becomes rationed owing to falling reservoir levels. The optimum summer weather index of Davis [106] is one of the most frequently quoted. The summer index ($I$) is derived from the formula

$$I = 18T_{max} + 20S_{d} - 0.276R$$

where $T_{max}$ is the mean daily maximum temperature (°C) over the period—usually the three summer months June–July–August but sometimes the 'extended summer' including May and September, $S_d$ is the mean daily sunshine (hours) and $R$ is the total rainfall (mm). Figure 22.4 shows the Davis index for every summer in Oxford since 1880, when sunshine records commenced. There is a gradual upward trend, largely the result of improving temperatures. Most of the years listed in Table 22.6 have already been mentioned. It is no surprise that the hot, dry summer of 1976 comes out on top, with summer 2018 now in second place, followed by 1995 and 1911. Predictably, summer 1888 comes out worst, while no summer since 1954 appears in the 'worst five'. The dramatic contrast between the summers of 1911 (best on record until 1976) and 1912 (no summer has been as poor since) is again very marked. The extremely wet summer of 2012 ranks only 17th worst, the worst since 1977 ($I = 381$, ranked 13th worst since 1880). As noted earlier, 2012 was somewhat warmer than its very high precipitation total would suggest, helping to keep it somewhat higher in the rankings.

## What about the summers of the 1870s?

Records of sunshine at Oxford did not begin until February 1880, but with some reasonable assumptions we can estimate summer sunshine duration for summers since

**Table 22.6** *Optimum Davis summer index values [106] for Oxford, June–July–August 1880–2018*

| Best summer Index | Year | Worst summer Index | Year |
|---|---|---|---|
| 601 | 1976 | 347 | 1888 |
| 582 | 2018 | 355 | 1912 |
| 573 | 1995 | 366 | 1927 |
| 572 | 1911 | 367 | 1954 |
| 566 | 1899 and 1975 | 369 | 1903 |

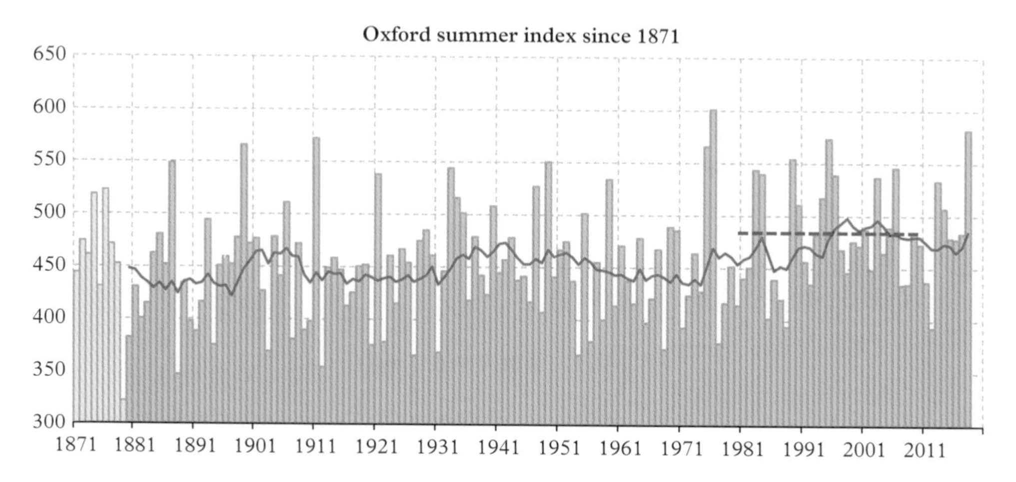


**Figure 22.4** *Values of the Davis Summer Index at Oxford for June–July–August 1880–2018, with a superimposed ten-year running mean ending at the year indicated. The 1981–2010 average (479) is shown by the dashed line*

1871. We have already seen earlier in this chapter how cold and wet the summer of 1879 was in Oxford, and, indeed, most of England, so the exercise has relevance in assessing how summer 1879 might compare with the worst summers in Table 22.6.

The mid-1870s saw the first development prototypes of what became known as the Campbell–Stokes sunshine recorder, the description of which was first published in the Meteorological Society's *Quarterly Journal** by Sir George Stokes in 1880 [107]. Early models were sent for 'field testing' to the two London observatories, Kew and Greenwich, in spring 1876, and monthly sunshine duration totals at Greenwich were published from May 1876 to April 1880 by William Ellis [108]. To the best of our knowledge, the records from Kew pre-1880 have never been published, although they probably exist†. A later paper by Frederick Brodie in 1891 [109] usefully included monthly averages for the Greenwich Observatory record for the period 1877 to 1890. Knowing the totals from 1877 to 1879, the Greenwich average for 1880–1890 could easily be derived and thus compared with Oxford's average over the same period. From this, a factor relating Greenwich sunshine duration to Oxford by month was evaluated and used to estimate monthly sunshine duration at Oxford from the known Greenwich total for each month back to May 1876. The actual monthly totals for Greenwich and the Greenwich:Oxford

* The Meteorological Society became 'Royal' only in 1883.

† A second monthly sunshine record for 1879 was given for Glynde, near Lewes in Sussex, in the printed discussion following Ellis's paper: the observer was the then Speaker of the House of Commons.

**Table 22.7** *Monthly sunshine duration in 1879 (hours) from an early Campbell–Stokes sunshine recorder at Greenwich Observatory in London, together with the average difference between Oxford and Greenwich over 1880–1890 and the derived estimate of monthly sunshine duration at Oxford in 1879 (hours), with percentage of the 1981–2010 normal*

| | Greenwich monthly sunshine duration totals for 1879, h | Oxford average 1880–1890 as percent of Greenwich | Estimated Oxford monthly sunshine duration totals for 1879, h | Per cent of 1981–2010 average sunshine for estimated 1879 total |
|---|---|---|---|---|
| January | 14.8 | 148 | 21.9 | 35 |
| February | 31.7 | 131 | 41.5 | 53 |
| March | 91.0 | 116 | 105.8 | 95 |
| April | 74.6 | 107 | 79.9 | 50 |
| May | 135.6 | 102 | 138.6 | 72 |
| June | 141.9 | 108 | 152.6 | 80 |
| July | 99.3 | 107 | 105.9 | 51 |
| August | 139.1 | 111 | 154.9 | 79 |
| September | 116.5 | 110 | 128.0 | 91 |
| October | 66.7 | 133 | 88.7 | 80 |
| November | 43.3 | 131 | 56.5 | 80 |
| December | 28.4 | 199 | 56.5 | 106 |
| Annual total | 982.9 | 113 | 1130.9 | 72 |

adjustment factor over 1880–1890, together with estimated monthly totals for Oxford for 1879, are listed in Table 22.7. The values thus generated are entirely plausible and, although the value for July is very meagre, four Julys since 1880 have registered lower totals than in 1879. The annual total for 1879 would represent a new minimum extreme on Oxford's record but, even so, it is only 2 per cent below 1888's actual calendar year total.

A second method of estimating summer sunshine duration, rather than individual months, is provided by the fairly good relationship between mean summer maximum temperature and average daily sunshine duration, as might be expected: over the period 1880–2017 (138 years), the correlation $r$ was 0.61. Using regression, we can then estimate summer 1879 sunshine total based on the known average maximum temperature at Oxford. By this method, summer 1879's estimated sunshine total (June to August) comes out as 405 hours, against 413 hours calculated by adjustment from Greenwich data above, a difference of less than 2 per cent. Working this assumed value through the Davis formula, with the other two components known, gives a Davis index for summer 1879 of just 322, which would replace 1888 as the lowest on the record*.

We can use this regression method to estimate summer sunshine totals for the years 1871–1878 to provide an estimate of the summer index for these years. Figure 22.5 updates Figure 22.4 with the Davis summer index for Oxford extended back to 1871, using estimated sunshine totals for the summers of 1871 to 1879. This provides useful context for the 1870s decade, which included, as well as 1879's very low value, two good summers in 1874 and 1876 (index 518 and 523, respectively), values which were exceeded only twice (in the summers of 1887 and 1899) until 1911. Aside from the dreadful summer of 1879, the lowest value in what is often thought of as a succession of poor summers in the 1870s came in 1875 (index 431); lower summer index values than in 1875 occurred in nine of the fifteen summers following 1879, and thus it would appear that the summers of the 1880s and early 1890s were significantly worse than those of the maligned 1870s.

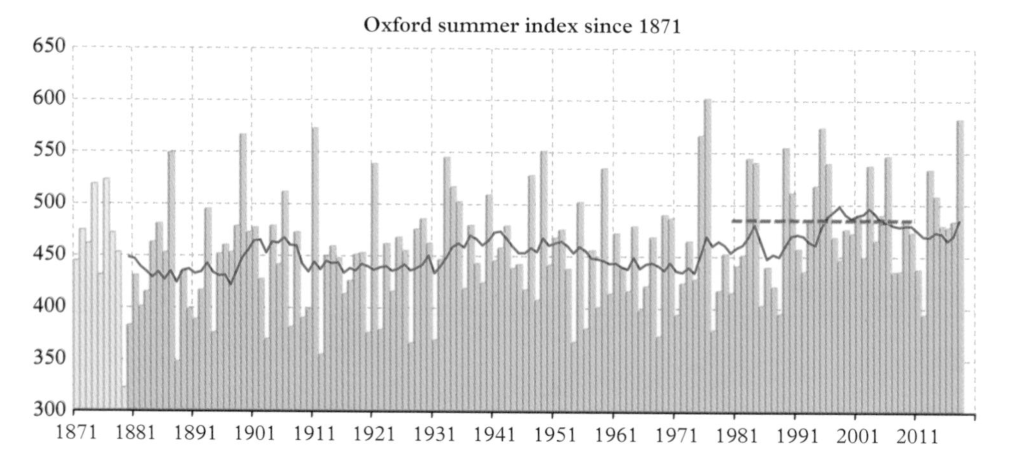


**Figure 22.5** *Values of the Davis Summer Index at Oxford for June–July–August 1871–2018 (using estimated sunshine duration 1871–1879 as described in the text), with a superimposed ten-year running mean ending at the year indicated. The 1981–2010 average (479) is shown by the dashed line plot*

* This figure is quite robust. We know the temperature and rainfall components of the equation, so the estimate only affects one of the three parameters. If we assume that 1879's sunshine was 10 per cent higher or lower than our estimate of 405 hours based upon the Greenwich comparison, this would only change the summer index value by 9 points either way. To equal the 1888 summer index value of 345, the lowest on the 1880–2018 record, the sunshine duration would need to be 27 per cent higher than our estimate or 527 hours in all. When compared against the known sunshine duration totals for summer 1879 at Greenwich (380 hours) and at Glynde, Sussex (371 hours), this appears highly unlikely.

## An 'extended summer' index

As outlined above, the Davis summer index can also be evaluated over the 'extended summer' period May to September. The best and worst 'extended summers' over the period 1880 to 2018 are shown in Table 22.8. Using this measure, years with a fine May (1989) or a fine September (1911, 1959, 1921) are rated higher than in the summer-only Table 22.6, 1911 now topping the table, while 1976 with its dull and wet September drops down to equal fifth position. The years 1903, 1912 and 1927 remain in the 'worst summers' table, but their rankings change; 1968 now tops the table as the worst summer since sunshine records began in 1880, owing to heavy falls of rain in both July and September. The 'extended summer' series 1880–2018 is plotted in Figure 22.6.

**Table 22.8** *Davis summer index values [106] for Oxford for the 'extended summer' May to September 1880-2018*

| Best summer Index | Year | Worst summer Index | Year |
|---|---|---|---|
| 509 | 1911 | 294 | 1968 |
| 505 | 1989 | 301 | 1931 |
| 499 | 1959 | 308 | 1903 |
| 493 | 2018 | 311 | 1912 = 1927 |
| 485 | 1921 and 1976 | 322 | 1894 |

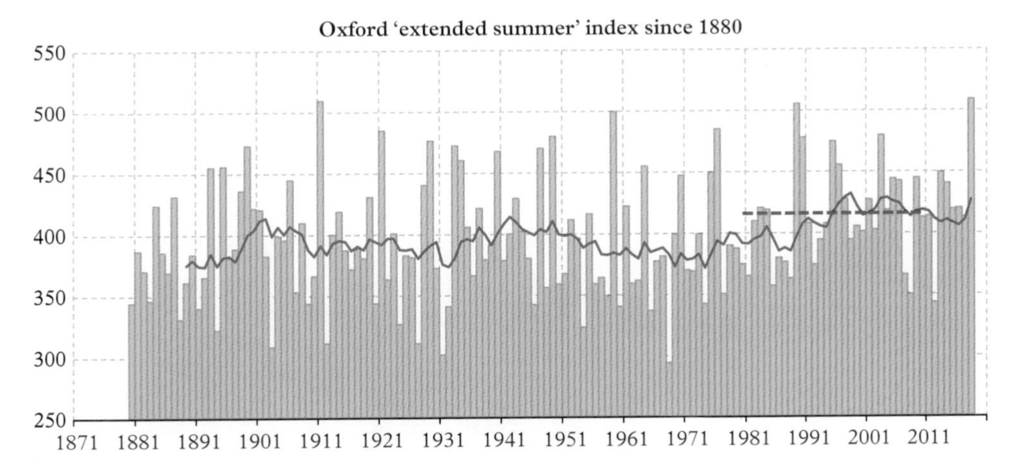


**Figure 22.6** *Values of the Davis Summer Index at Oxford for the 'extended summer' period May to September 1880-2018, with a superimposed ten-year running mean ending at the year indicated. The 1981–2010 average (413) is shown by the dashed line*

# Oxford temperature, precipitation and sunshine in graphs—Summer

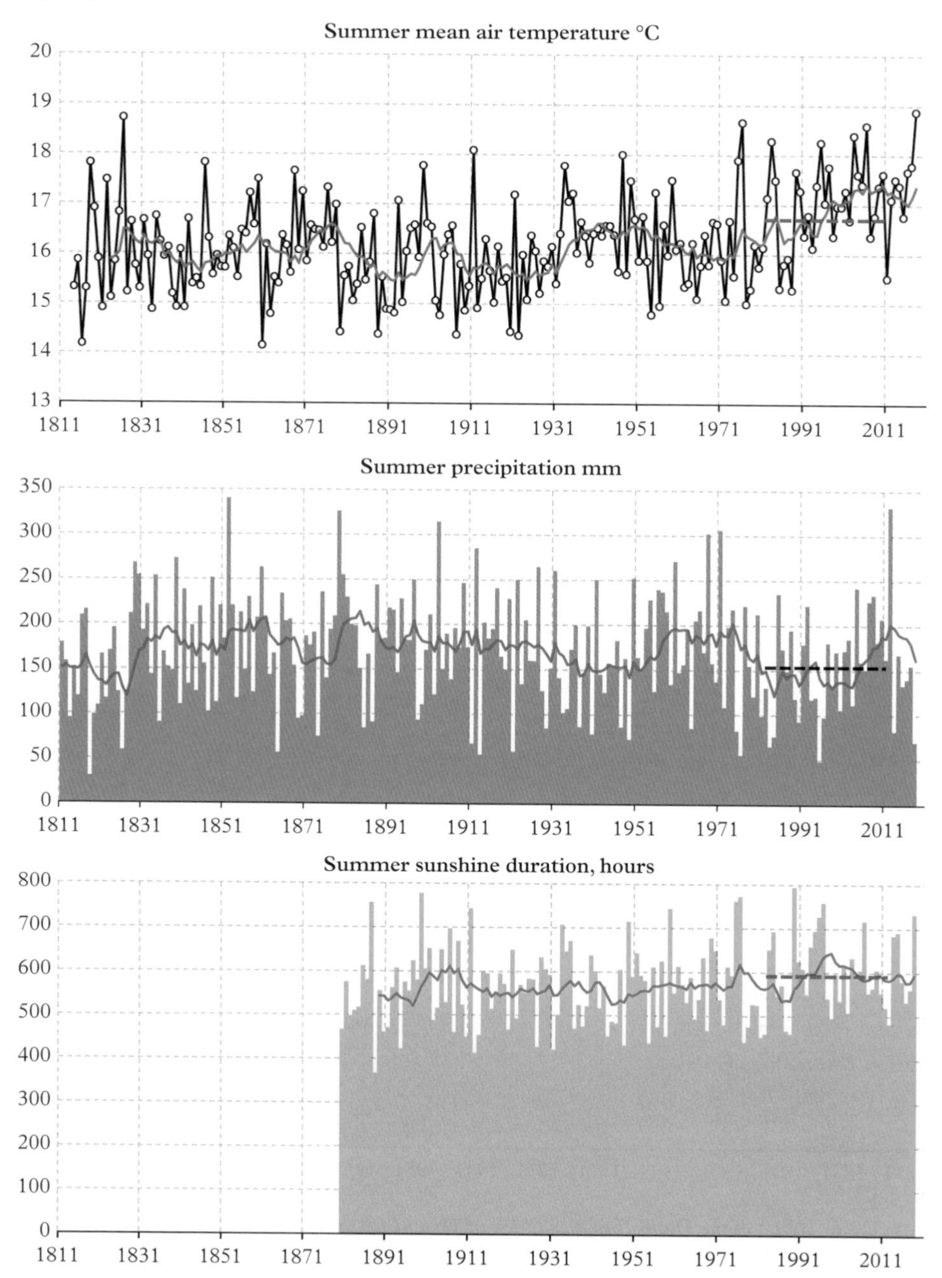


**Figure 22.7** *Seasonal values of (from top) mean temperature (°C, since 1814), total precipitation (mm, since 1811) and sunshine duration (hours, since 1880) for summer at Oxford. The 1981–2010 averages are indicated by the thick blue dashed line, while the ten-year unweighted running mean ending at the year shown is indicated by the red line*

# 23

# Autumn

*September, October and November*

Autumn can be the season of strong winds and storms, of calm and fog, of sunny days and sparkling autumn colours. The first air frosts associated with the shortening days, and often the first snows of the coming winter, will become evident. As daylength decreases, the temperature falls, from a mean air temperature around 16 °C in early September to 6 °C by late November.

Autumn contains the two wettest months of the year (October and November) and is consequently the wettest season of the year, with 191 mm of precipitation on average. Rain typically falls on 43 days (two days fewer than in the rather drier winter), with snow or sleet falling in one autumn in two on average, but rarely before mid-November.

Sunshine can be expected to total 323 hours during autumn, with November receiving a little more than half of the sunshine expected in September.

## Temperature

Oxford's temperatures in autumn have ranged from −9.1 °C to 33.4 °C during the period 1815–2018. The warmest autumn days and coldest autumn nights during this period are shown in Table 23.1.

It is not surprising to find that the warmest days and nights tend to occur in early September, when the weather can often be as good as, if not better than, August. Whilst part of 'autumn', very warm days in early September can hardly be described as 'unseasonably warm'; it is more like the norm for the start of autumn. Hot days can be accompanied by warm nights, of course: night-time minima in excess of 18 °C would be high for summer, let alone early autumn. It is clear that the hot weather of summer 1911 continued into September, since this year appears twice in Table 23.1. Not surprisingly, the coldest autumn nights all occur towards the end of November. Note that 1816 figures once again, a cold end to a very cold year.

*Oxford Weather and Climate since 1767*. Stephen Burt and Tim Burt, Oxford University Press (2019).
 DOI: 10.1093/oso/9780198834632.001.0001

**Table 23.1** *Highest and lowest maximum and minimum temperatures in autumn at Oxford, 1815–2018*

| Rank | Hottest days | Coldest days | Warmest nights | Coldest nights |
|---|---|---|---|---|
| 1 | 33.4 °C, 8 Sep 1911 | −2.7 °C, 28 Nov 1890 | 20.6 °C, 5 Sept 1949 | *−10.1 °C, 17 Nov 1815* |
| 2 | 33.1 °C, 1 and 2 Sep 1906 | −1.9 °C, 2v7 Nov 1915 | 18.4 °C, 8 Sept 1898 | −9.1 °C, 23 Nov 1858 |
| 3 | 32.2 °C, 8 Sep 1898 | −1.7 °C, 23 Nov 1858 | 18.1 °C, 7 Sept 2016 | *−8.9 °C, 24 Nov 1816* |
| 4 | 31.8 °C, 7 Sep 1911 | *−1.6 °C, 17 Nov 1815* | 18.0 °C, 11 Sept 1999 | −8.8 °C, 26 Nov 1923 |
| 5 | 30.0 °C, 4 Sep 1929 | −1.4 °C, 25 Nov 1923 | 17.8 °C, 2 Sept 1939, 3 Sept 1939 and 5 Sept 2005 | −8.4 °C, 24 Nov 1904 and 27 Nov 1923 |
| *Last 50 years to 2018* | | | | |
| | 29.9 °C, 5 Sept 2004 | −1.2 °C, 28 Nov 2010 | 18.1 °C, 7 Sep 2016 | −7.0 °C, 23 Nov 1983 |

*Temperatures recorded by unscreened thermometers prior to 1853 are shown in italics; these are likely to be a little too cold by modern (screened thermometer) standards in winter and a little too warm in summer—see Appendix 1 for more details. In autumn the mean temperature correction for unscreened thermometry varies from zero in September to +0.3 degC in November, although this will vary considerably on a day-to-day basis*

## Warm and cold autumns

By just the smallest margin, autumn has shown a greater rate of warming than any other season: 0.88 degC per century compared to winter (0.86 degC), spring (0.73 degC) and summer (0.53 degC). After some cold autumns at the start of the record, the autumns of the 1820s and early 1830s were quite mild before colder autumns returned in the late 1830s and early 1840s. Since then, there has been a steady rise in autumn mean air temperatures, with just one cooler spell in the late 1910s and early 1920s. The 1990s saw a particularly rapid rate of autumn warming (although 1993 was a notable exception). All five of the warmest autumns are from the twenty-first century, with 2006 0.6 degC warmer than its nearest challenger, 2011. An indication of the importance of autumnal warming is that the growing season for grapes (using a day degree index based on 10 °C) lasted only to 11 October in 1853 but extended to 28 October by 2017 (see also Chapter 24 for a longer discussion on growing seasons).

It is hardly surprising that the coolest autumns come from earlier in the record, none of the ‘coldest five’ being more recent than 1952; other than 1919 and 1952, all the rest were more than one hundred years ago. In relation to the Tambora volcanic eruption (see the Chronology in Chapter 25), it is interesting that autumn 1815 ranks coldest and

Table 23.2 *Autumn mean temperatures at Oxford 1814–2018*

| Autumn mean temperature 11.2 °C (average 1981–2010) | | | | | |
|---|---|---|---|---|---|
| Mildest autumns | | | Coldest autumns | | |
| Mean temperature, °C | Departure from 1981–2010 normal degC | Year | Mean temperature, °C | Departure from 1981–2010 normal degC | Year |
| 13.4 | +2.2 | 2006 | *7.6* | *−3.6* | *1815* |
| 12.8 | +1.6 | 2011 | *7.8* | *−3.4* | *1829* |
| 12.7 | +1.5 | 2014 | 8.0 | −3.2 | *1840,* 1887, 1919 |
| 12.5 | +1.3 | 2009 | 8.2 | −3.0 | 1952 |
| 12.3 | +1.1 | 2005 | 8.5 | −2.7 | 1905 |
| *Last 50 years to 2018* | | | | | |
| 13.4 | +2.2 | 2006 | 9.1 | −2.1 | 1993 |

1816 equal 17th coldest (Table 23.2); November 1815 remains the coldest November on record at Oxford, and November 1816, the equal third coldest.

## Frosts

Over the period 1981–2010, an average of five air frosts was the norm for autumn in Oxford (typically, none in September, one in October and four in November). As expected, there is considerable variability in the frequency of air frosts in autumn, which tends to be determined largely by how cold (or otherwise) November is: cold Novembers can still contribute a relatively large number of air frosts to otherwise warm autumns, for example 11 frosts in autumn 2005, the fifth-mildest autumn on record (Table 23.2). Autumn air frosts have become later and less frequent since the beginning of the record—the first 30 years averaged eight or nine, but since 2000 this has declined to less than half of that number. Autumns without a single air frost are still uncommon: there have been only nine since 1815, the first in 1938, and the latest at the time of writing in 2009.

Although records of grass minimum temperatures began in 1857, a digitised daily record only extends back as far as 1931, and (perhaps surprisingly) there is no overall trend since then. The frostiest autumn in terms of ground frosts since 1931 was 1993, with 31 (21 in November); the least frosty 1984, 2002 and 2009, each with just six. Every year since 1931 has recorded at least two ground frosts by the end of November.

Taken together, the strong warming trend in autumn, together with an extension of the growing season and later air and ground frosts, will have had a major impact on gardeners, especially those with lawns to mow!

## Precipitation

The average autumn precipitation in Oxford is 191 mm, falling on 43 days. Two of the wettest autumns come very early in the record (Table 23.3), followed soon after by two of the driest—the 1780s was overall one of the driest periods on record at Oxford, rivalling the early 1800s. The most recent very wet autumn to appear in the top five is 1960 (349 mm, on 63 rain days), although 1974 (318 mm, 62 rain days) was also very wet. How often a very wet period is immediately followed by a major drought, as here with the 1975/76 drought and more recently, the wet winter of 1994/95 before a very hot, dry summer, is interesting—or is it merely coincidence? It is perhaps surprising that a cluster of memorably wet autumns in recent years do not feature in Table 23.3: 2000 (302 mm, rank 11th wettest), 2002 (268 mm, rank 24th), 2006 (295 mm, rank 15th) and 2012 (265 mm, rank 26th). The 1976 autumn, which seemed excessively wet after the extreme drought that came before it, ranks only 43rd (246 mm) in Oxford.

Autumn precipitation averaged 229 mm from 1767 to 1780 whilst the averages in the 1780s and 1800s were 141 mm and 150 mm, respectively; as noted above, the 1981–2010 average is 191 mm but there has been no consistent upward trend in autumn precipitation, rather a series of inter-decadal fluctuations from wetter to drier periods, very much like the summer record. The 1870s were a notably wet decade across much of the country [79], the 'ruin of British agriculture'. Here, 1875 shows up as one of the wettest autumns, with two of the wettest autumn days. The daily total of 84.8 mm on 6 September 1951 is by far the wettest autumn day at Oxford since 1827.

**Table 23.3** *Autumn precipitation at the Radcliffe Observatory, Oxford 1767–2018, wettest days 1827–2018*

| **Autumn mean precipitation 191 mm** (average 1981–2010) | | | | | | | |
|---|---|---|---|---|---|---|---|
| Wettest autumns | | | Driest autumns | | | Wettest days | |
| Total fall, mm | Per cent of normal | Year | Total fall, mm | Per cent of normal | Year | Daily fall, mm | Date |
| 376.6 | 197 | 1772 | 52.6 | 27 | 1978 | 84.8 | 6 Sep 1951 |
| 361.6 | 189 | 1768 | 58.0 | 30 | 1964 | 48.3 | 9 Oct 1875 |
| 349.5 | 183 | 1960 | 64.9 | 34 | 1788 | 46.4 | 26 Sep 1998 |
| 335.2 | 175 | 1875 | 66.3 | 35 | 1783 | 43.7 | 24 Sep 1915 |
| 325.3 | 170 | 1841 | 70.9 | 37 | 1817 | 43.0 | 19 Oct 1875 |
| *Last 50 years to 2018* | | | | | | | |
| 317.6 | 166 | 1974 | 52.6 | 27 | 1978 | 46.4 | 26 Sep 1998 |

## Sunshine and visibility

Autumn sees a marked decline in daily sunshine totals, with an average duration for the season of 323 hours, the sunniest days usually occurring during the first week of September. The autumns of 1959 and 1971 remain the sunniest on record, the only seasons of that name to receive over 400 hours of sunshine at Oxford since records commenced in 1880. Autumn 1976, despite not being as wet as it felt at the time (see 'Precipitation'), was nevertheless a very dull one, quite unlike the sunny days of the drought summer. Autumn 1945 remains the dullest on record, with just 21 per cent of the possible amount and less than half that of the sunniest autumns; 1946 was almost as dull. Only 1976 (dull) and 1971 (sunny) represent the more recent decades in Table 23.4.

Of course, autumn is Keats' 'season of mists and mellow fruitfulness'. In fact, the incidence of autumnal fog at Oxford has decreased since the Clean Air Acts of the 1950s, falling from an average incidence of 10 mornings in the 1930s and 1940s to 5 in the 1970s and 1980s to only 3 on average each autumn in the 1990s.

**Table 23.4** *Autumn sunshine duration at Oxford 1880–2018*

**Autumn mean sunshine duration 323 hours, 3.55 hours per day** (average 1981–2010)

*Possible daylength: 974 hours. Mean sunshine duration as percentage of possible: 33.2*

| Sunniest autumns | | | Dullest autumns | | |
|---|---|---|---|---|---|
| Duration, hours | Per cent of possible | Year | Duration, hours | Per cent of possible | Year |
| 428.7 | 44.0 | 1959 | 207.6 | 21.3 | 1945 |
| 403.7 | 41.4 | 1971 | 215.9 | 22.2 | 1976 |
| 399.9 | 41.1 | 1964 | 220.0 | 22.6 | 1894 |
| 394.5 | 40.5 | 1928 | 220.8 | 22.7 | 1946 |
| 392.7 | 40.3 | 1921 | 229.3 | 23.5 | 1968 |
| *Last 50 years to 2018* | | | | | |
| 403.7 | 41.4 | 1971 | 215.9 | 22.1 | 1976 |

# Oxford temperature, precipitation and sunshine in graphs—Autumn

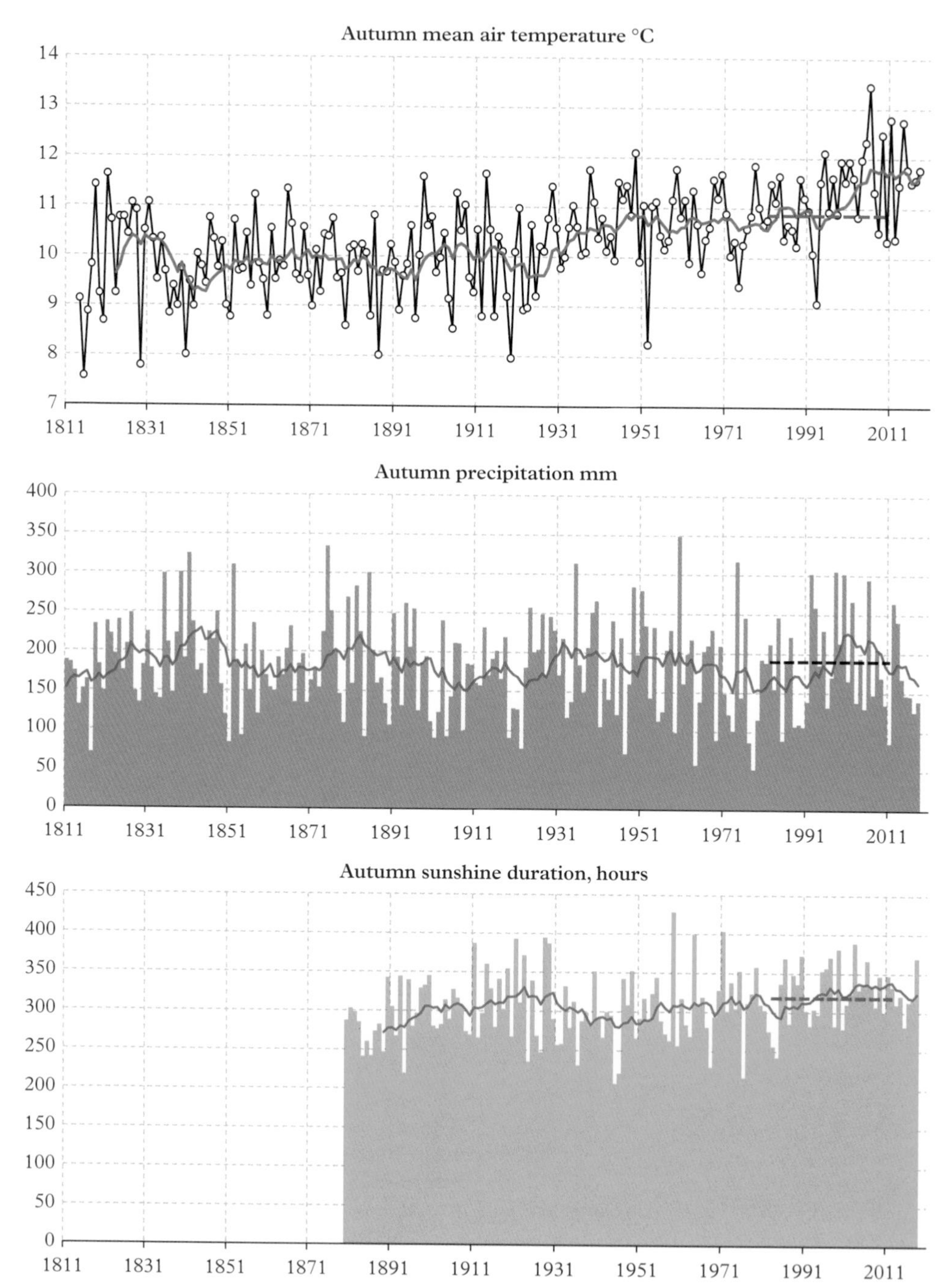


**Figure 23.1** *Seasonal values of (from top) mean temperature (°C, since 1814), total precipitation (mm, since 1811) and sunshine duration (hours, since 1880) for autumn at Oxford. The 1981–2010 averages are indicated by the thick blue dashed line, while the ten-year unweighted running mean ending at the year shown is indicated by the red line*

# Part 4

# Long-term climate change in Oxford

# 24

# Climate change in Oxford

Human influence on the climate system is clear, and anthropogenic emissions of greenhouse gases are now higher than ever. This has led to atmospheric concentrations of greenhouse gases (carbon dioxide, methane, nitrous oxide) that are unprecedented in at least the last 800,000 years. The Inter-Governmental Panel on Climate Change (IPCC) has stated that the effect of these increasing concentrations, together with the impact of other anthropogenic drivers, are extremely likely to have been the dominant cause of the observed warming since the mid-twentieth century [110]. IPCC regards this warming of the climate system as unequivocal; since the 1950s, many of the observed changes are unprecedented over timescales of decades to millennia. Changes in the intensity and/or frequency of many extreme weather and climate events have also been observed since the middle of the twentieth century. Some of these changes have been linked to human influences, including a decrease in low temperature extremes, an increase in high temperature extremes and an increase in the number of heavy precipitation events in a number of regions [110].

An increasing body of observations gives a collective picture of a warming world and other related changes in the climate system. Each of the last three decades has been successively warmer at Earth's surface than any preceding decade since 1850 [110]. The period from 1983 to 2012 was *likely* the warmest 30-year period of the last 1400 years in the Northern Hemisphere. The globally averaged combined land and ocean surface temperature data as calculated by a linear trend show a warming of 1.0 degC in the period 1901 to 2016 [111] and this continues, driven by ongoing emissions of greenhouse gases, the dominant cause of the observed warming since the mid-twentieth century [110, 111].

Of course, given the very long climate records for the Radcliffe Observatory, we should expect to observe other aspects of climate change in addition to recent human-induced global warming. Some of these changes will relate to global drivers such as levels of solar insolation whilst others will reflect regional or local changes such as urbanisation or changing levels of air pollution.

Note that, in this chapter, the Spearman Rank correlation coefficient ($r_s$) is used to avoid the necessity for time series data to be normally distributed. Pearson linear regression and product-moment correlation ($r$) are quoted where we wish to quantify linear trends but without acknowledging any statistical significance. In practice, it makes little

*Oxford Weather and Climate since 1767*. Stephen Burt and Tim Burt, Oxford University Press (2019).
 DOI: 10.1093/oso/9780198834632.001.0001

difference which correlation test is used to assess the reported correlation coefficients or their statistical significance.

## Temperature

### Mean air temperature

Table 24.1 shows decadal mean air temperature (°C) at Oxford, 1814–2018. Figure 24.1 shows annual mean air temperature from 1814 together with a decadal running mean; data for annual mean maximum and annual mean minimum temperatures are also plotted. The 1810s are the coldest decade on record (with the caveat that the first full calendar year of data is 1814). The winter of 1813/14 was very cold and there soon followed the global cooling associated with the eruption of Tambora (see the Chronology in Chapter 25). Whilst there was a general tendency for warming from around 1840 onwards, there was a good deal of variability too, with the 1880s being colder. The 1930s and 1940s were relatively mild but there followed a slightly cooler period before the most dramatic warming from the 1970s onwards. Although there were some cooler years at the end of the 2000s, 2010 especially, the most recent years have again been very warm, with 2014 the warmest year on record (11.8 °C, Table 19.1). Before 1989, only 1921 and 1949 had reached an annual mean air temperature above 11 °C. From 1989 onwards, there have been twenty such years; thirteen of those years have been since 2000. Since 1989, only nine years have had annual mean air temperatures below 11 °C, the coolest of which was 1996, which nevertheless only ranks 93rd equal lowest in a series of 204 years.

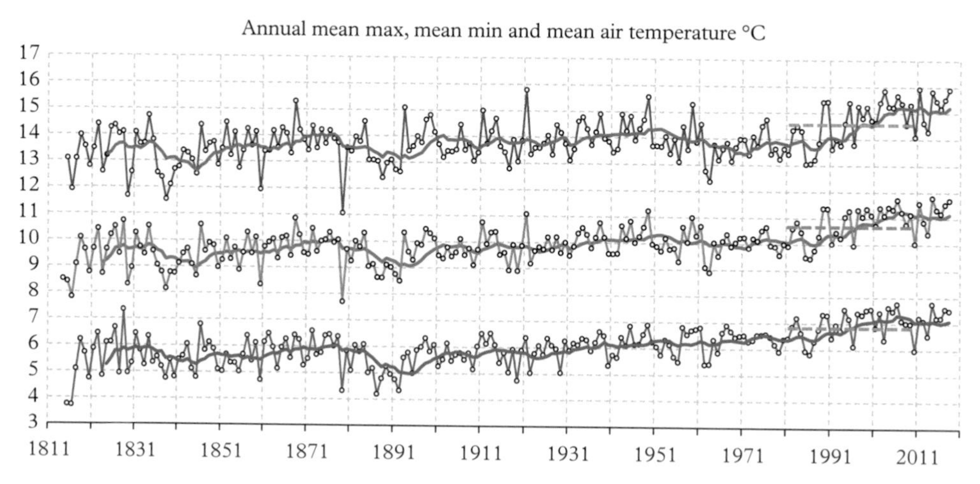


**Figure 24.1** *Mean annual air temperature (°C) at the Radcliffe Observatory, Oxford, 1814–2018, together with a ten-year unweighted running mean plotted at year ending; annual mean maximum and annual mean minimum temperatures are also plotted. Air temperatures before 1853 are from unscreened thermometers*

**Table 24.1** *Decadal mean air temperature (°C) at the Radcliffe Observatory, Oxford, 1814–2018, in date order. Note that the first and last period have fewer than ten years.*

| Decade | Mean annual temperature °C | Decade | Mean annual temperature °C | Decade | Mean annual temperature °C |
|---|---|---|---|---|---|
| 1814–20 | 8.9 | 1881–90 | 9.4 | 1951–60 | 10.0 |
| 1821–30 | 9.7 | 1891–1900 | 9.8 | 1961–70 | 9.9 |
| 1831–40 | 9.3 | 1901–10 | 9.6 | 1971–80 | 10.1 |
| 1841–50 | 9.5 | 1911–20 | 9.8 | 1981–90 | 10.3 |
| 1851–60 | 9.5 | 1921–30 | 9.9 | 1991–2000 | 10.8 |
| 1861–70 | 9.9 | 1931–40 | 10.1 | 2001–10 | 11.1 |
| 1871–80 | 9.7 | 1941–50 | 10.3 | 2011–18 | 11.3 |

Since 1815, the rate of warming in mean air temperature at Oxford has been 0.77 degC per century. Since 1901 the rate of warming has been 1.23 degC per century and, since 1965, this has increased to an extraordinary 2.96 degC per century. Wuebbles *et al* [111] describe the period since 1901 as the warmest in the history of modern civilisation. Globally, the linear regression change over the period 1901–2016 is 1.0 degC [111], just a little less than the change observed at Oxford. Global annual average temperature has increased by 0.65 degC for the period 1986–2016 compared to 1901–1960 [111]; at Oxford the equivalent difference is 0.94 degC (1901–1960 mean = 9.97 °C; 1986–2016 mean = 10.91 °C). The slightly higher rate of warming at Oxford may simply reflect macroclimatic changes in the temperate mid-latitudes of the Northern Hemisphere, but a possible local cause is the increasingly built-up nature of the area immediately adjacent to the Observatory site, both within the grounds of what is now Green Templeton College and immediately to the south-east (see also Chapter 3).

Since mean air temperature is derived from monthly means of daily maximum and daily minimum temperatures, we should expect similar long-term trends to those described above for maximum and minimum temperatures. This is broadly correct, but there are subtle differences, some of which were identified in Chapter 3 when considering the scale and development of Oxford's urban influence over the last two centuries. Since 1815, the average rate of warming for the annual mean maximum air temperature (Tmax) has been 0.73 degC per century and, thus, over the record period of a little over 200 years, an absolute increase of 1.49 degC would be expected. For the annual mean minimum air temperature (Tmin) over the same period, the average rate of warming per century is very similar: 0.78 degC. Since 1901, the century warming rates for Tmax and Tmin are 1.07 degC and 1.41 degC, respectively, but, since 1965, Tmax has shown a greater rate of warming than Tmin: the century warming rates are 3.96 degC and 1.96 degC for Tmax and Tmin, respectively, equivalent to absolute increases over 52 years of 2.06 degC and 1.02 degC.

Figure 24.1 shows that the increase in Tmin since 1853 has been a fairly steady, gradual rise, whereas Tmax has been more variable between warmer and cooler periods. Thus, Tmax has shown the larger absolute increase in the last half-century (see also Chapter 3). This explains why, when comparing the 1901–1960 means with those for 1986–2018, why Tmin shows a slightly larger difference than Tmax (1.06 degC compared to 0.92 degC). For Tmax, the 1901–1960 mean was 13.95 °C whilst for 1986–2016 it was 14.87 °C; for Tmin the respective values are 6.02 °C and 7.08 °C. Figure 24.1 shows that Tmin did not display the same cooling as seen for Tmax in the 1950s and 1960s.

The increases in air temperature at Oxford in the last half-century are remarkable, and the question of what is driving these increases must be addressed. Wuebbles *et al* [111] conclude, based on extensive evidence, that it is extremely likely that human activities, especially emissions of greenhouse gases, are the dominant cause of the observed warming since the mid-twentieth century. For the warming over the last century, there is no convincing alternative explanation supported by the extent of the observational evidence. Radiative forcing due to human activities has become increasingly positive (warming) since about 1870 and has grown at an accelerating rate since about 1970. In contrast, there is high confidence that the contributions of natural radiative forcing (variations in solar radiation, impact of volcanic eruptions) are minor. Natural variability over a timescale of months to years, including El Niño events and other recurring patterns of atmosphere–ocean interaction, is limited to a small fraction of observed climate trends over decades [111]. It must, however, be noted that, whilst anthropogenic climate change is the overriding cause of the long-term temperature trend seen at Oxford, some contribution from the urban heat island effect seems to be part of the explanation too (see also Chapter 3).

Until January 1853, thermometers at the Radcliffe Observatory were unscreened (see Appendix 1) and absolute maximum temperatures are less reliable prior to that date—summer maximum temperatures are probably a little higher than would be recorded in a modern Stevenson screen, owing to the influence of infrared radiation from the surroundings, while winter minimum temperatures would be expected to be somewhat lower as a result of radiative losses to the sky. For this reason, annual absolute maximum and minimum temperatures prior to 1853 are plotted in grey on Figure 24.2. (Figure 24.2 also suggests there is perhaps some evidence for a slightly greater range in annual temperatures prior to the introduction of screened thermometers in 1853.)

As might be expected, both annual maximum and annual minimum temperature series show a significant, gradual upward trend over time, the more so for minima in recent decades. The lowest annual absolute maximum temperature has been 23.1 °C in 1860; in the twentieth century, the lowest annual absolute maximum was 23.9 °C in 1931 but, in the twenty-first century (to 2018), the lowest annual absolute maximum is as high as 28.3 °C in 2008. Whilst the time series for annual absolute minimum temperature shows the same upward trend, it still remains possible to observe extremely low temperatures at Oxford under favourable meteorological conditions: anticyclonic with a well-developed inversion, and often a deep, fresh snow cover. The lowest air temperature yet recorded at Oxford was −17.8 °C on Christmas Eve 1860. The second lowest, −16.6 °C, was recorded as recently as January 1982. The lowest minimum recorded in the twenty-first century to 2018 has been −10.9 °C on 20 December 2010.

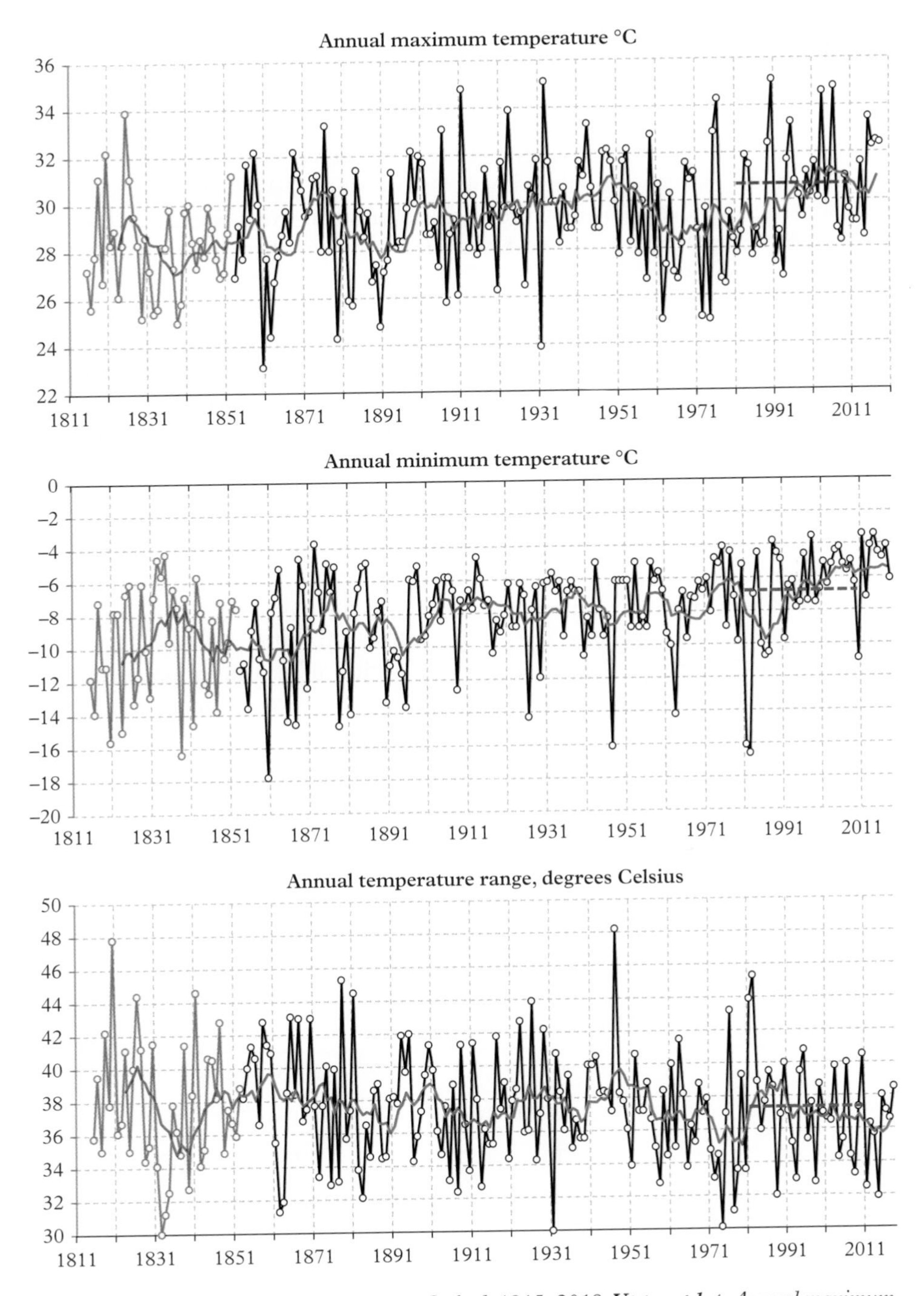


**Figure 24.2** *Annual temperature extremes at Oxford, 1815–2018.* ***Upper plot:*** *Annual maximum temperature, °C.* ***Middle plot:*** *Annual minimum temperature, °C.* ***Lower plot:*** *Annual range in temperature (highest maximum minus lowest minimum) in degC. Ten-year unweighted running means are shown at the year ending in all three plots, together with the 1981–2010 average annual extremes, 30.4 °C and −7.1 °C, respectively, giving a mean annual range in temperature of 37.5 degC. Records prior to the introduction of screened thermometers in 1853 are shown in grey.*

## Temperature-related indices

The incidence of frost is one of the most significant aspects of meteorology for people, both in relation to transport and growing conditions. An 'air frost' is recorded when the air temperature falls below 0 °C. Figure 19.6 (in Calendar Year statistics, Chapter 19) shows the annual frequency of air frosts at Oxford since 1815. There has been a very significant reduction over almost two centuries: over 60 air frosts per year at the start of the record but fewer than 40 each year now; in a few years there have been fewer than 20 air frosts, the lowest total being 14 in 1974.

'Ground frosts' are recorded when the grass minimum temperature falls below 0 °C. Although grass minimum temperature records commenced at the Radcliffe Observatory in 1857, digitised daily records are not available prior to December 1930; a change in the definition of ground frost in 1961 means that tabulated values of ground frost frequency in the Radcliffe Observatory ledgers prior to that date are incompatible with current records. Accordingly, we produced a reanalysis of grass minimum observations from 1931 to date using current definitions. These are discussed within each monthly and seasonal chapter and in the annual chapter.

Climate change, and in particular the rise in temperatures, is reflected in a number of other climatic indices. Whilst the discussion on air and ground frosts was concerned with the number of days that minimum temperatures fell below a given threshold, the number of days with maximum temperature above a given threshold is also of interest. For example, Figure 19.1 in Chapter 19 shows the number of 'hot days' (maximum temperature 25 °C or greater) and 'heatwave days' (30 °C or greater) for every year since 1815. There is, of course, a good deal of inter-year and inter-decadal variability, and the upward trend in mean temperatures is perhaps less immediately obvious presented in this form. Nonetheless, the frequency of 'hot days' and 'heatwave days' have risen substantially over the 200 years of the Radcliffe Observatory record, as can be seen from Table 24.2.

**Table 24.2** *Annual averages of 'hot days' (maximum temperature 25 °C or greater) and 'heatwave days' (30 °C or greater) by decade at the Radcliffe Observatory, Oxford, 1815–2018. Data to 1852 are from unscreened thermometers and are indicated in italics: see Appendix 1 for details.*

| Decade | ≥ 25 °C days | ≥ 30 °C days | Decade | ≥ 25 °C days | ≥ 30 °C days | Decade | ≥ 25 °C days | ≥ 30 °C days |
|---|---|---|---|---|---|---|---|---|
| *1814–20* | *13.7* | *1.2* | 1881–90 | 9.2 | 0.5 | 1951–60 | 13.3 | 1.0 |
| *1821–30* | *13.4* | *1.0* | 1891–1900 | 16.1 | 0.6 | 1961–70 | 8.0 | 0.5 |
| *1831–40* | *5.5* | *0* | 1901–10 | 10.2 | 0.6 | 1971–80 | 13.3 | 2.0 |
| *1841–50* | *9.9* | *0.1* | 1911–20 | 13.3 | 1.3 | 1981–90 | 18.7 | 1.9 |
| 1851–60 | 13.0 | 0.8 | 1921–30 | 13.3 | 1.7 | 1991–2000 | 18.1 | 2.3 |
| 1861–70 | 16.2 | 1.5 | 1931–40 | 18.2 | 1.0 | 2001–10 | 21.4 | 2.5 |
| 1871–80 | 12.8 | 1.4 | 1941–50 | 19.9 | 2.7 | 2011–18 | 22.5 | 2.0 |

The air temperature reached 30 °C only once (on 27 May 1841) in the 25 years between 1827 and 1851; yet, since the 1970s, this figure has been reached on average twice per year. Similarly, with the 25 °C threshold: in the 1830s only about five days per year would reach this level, whereas since 2000 the average has been over 20 such days per annum. Of course, these totals are hugely influenced by the occasional very warm summer, as is evident from Figure 19.1, but a fourfold increase in the number of hot days reflects both more frequent attainment in 'ordinary' summers allied with a decrease in the return period of very warm summers.

Another relevant aspect of increasing warmth is the lengthening of the growing season. The Growing Degree-Day (GDD) is a cumulative heat index that can be used to predict when a crop will reach maturity. Each day's GDD is calculated by subtracting a reference temperature from the daily mean air temperature; values less than zero being ignored. GDDs are then aggregated over months and seasons to provide a measure of the accumulated warmth of the growing season: the daily record of GDD can be used to identify the start and finish of the growing season in each year. In the UK, initial interest was in the growth of grass using a threshold value of 5.5 °C (42 °F). Today, GDDs are available for a wide range of crops.

Annual totals of growing day degrees for the 5.5 °C threshold (GDD5.5) at Oxford have shown a 20 per cent increase since 1853, from a linear regression predicted value then of 1742 to 2084 degree-days in 2016. For annual totals of growing day degrees for the 10 °C threshold (GDD10), the estimated increase is even larger: 29 per cent. Figure 24.3 shows annual GDD10 at Oxford for the entire series from 1815, together with a ten-year running mean. The strong warming in recent decades is clearly evident, with the decadal running mean of annual GDD10 totals increasing from around 850 in the 1960s to well over 1000 now (1981–2010 average 990 degree-days). This is a crucial increase for marginal crops: for example, grapes would have been very difficult to grow commercially for

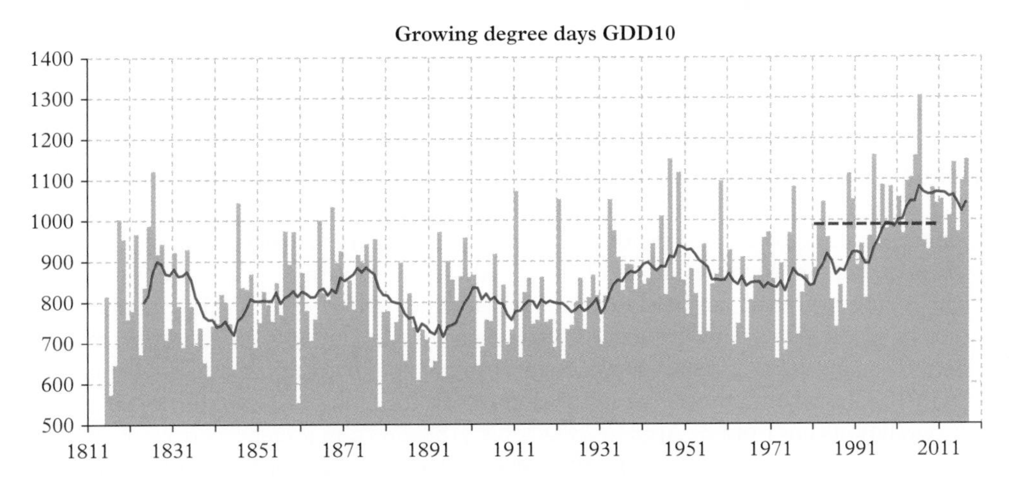


**Figure 24.3** *Annual growing degree-days above 10 °C (GDD10) at Oxford, 1815–2017, together with the ten-year unweighted running mean; the 1981–2010 average (990 degree-days) is shown by the dashed line*

**Figure 24.4** *A modern commercial vineyard near Marlow, Buckinghamshire (Tim Burt)*

most of the observational period but, since about 1990, the running mean has remained above 900 degree-days, and commercially viable vineyards have appeared in the Thames Valley and elsewhere in southern England (Figure 24.4). In some cooler years, these may produce low yields but, generally, production is good, while excellent vintages (quality and quantity) result from warm and sunny summers such as 2018. Of course, the GDD index is based solely on air temperature measured at a single standard meteorological site, and local conditions such as slope aspect, topographic position, soil type and soil wetness can all combine to render a marginal site more productive. The *World Atlas of Wine* [112] reports a GDD10 index of 1000 for the Champagne region of northern France. With global warming, the benefits of a more continental climate are now being experienced in more oceanic locations, such as the Oxford region, where the wine industry is now flourishing.

Figure 24.5 shows an estimate of the length of the growing season at Oxford since 1853. A running mean was used to even out daily fluctuations in mean air temperature in order to identify the period during which, each year, mean air temperature remained above 10 °C. Both linear regression lines are strongly significant and suggest that, on average, in 1853, the continuous growing season was from 6 May to 11 October (158 days), whereas now it runs from 21 April to 28 October (188 days), an average increase of 30 days, a very important extension of the growing season in terms of crop production. Thus, both the length and the intensity of the growing season have increased significantly over the last century and a half.

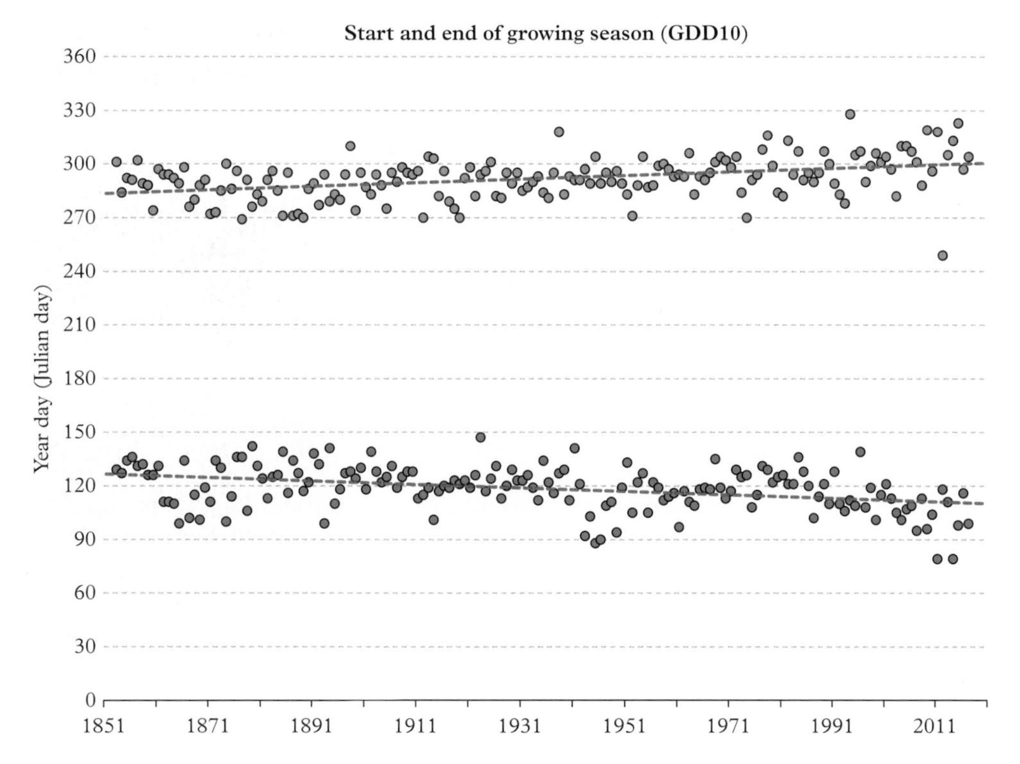


**Figure 24.5** *An indication of the length of the growing season in Oxford every year from 1853 to 2017. The start and end of the growing season (green and orange circles, respectively, for spring and autumn) are defined by GDD10 thresholds and are shown in Julian Days (days from 1 January): day 120 is the end of April and day 270 the end of September. Least-squares regression linear trends over the period are shown by the red dotted lines*

An earlier start to the growing season is, of course, not simply reflected in agricultural crops. More generally, climatic warming has been shown to alter the timing of important developmental or behavioural events in birds, plants, amphibians and insects. The timing of flowering is a key event for plants. It affects their chances of pollination, especially when the pollinator (for example, an insect) is itself seasonal, and determines the timing of seed ripening and dispersal. Flowering time also influences animals for which pollen, nectar, and seeds are important resources, and earlier flowering also implies earlier activity in other processes (leaf expansion, root growth, nutrient uptake) that are important for niche differentiation among coexisting species and so will alter competitive interactions between species. Large changes in flowering date will therefore disrupt ecosystem structure [113]. One of the most important studies of plant flowering was conducted at Chinnor, some 25 km east of Oxford, just below the Chalk escarpment: 557 plant species were observed over a 47-year period by a single observer, R. S. R. Fitter. These data reveal one of the strongest biological signals yet recorded of climatic change. The average first flowering date of 385 British plant species advanced by 4.5 days during the

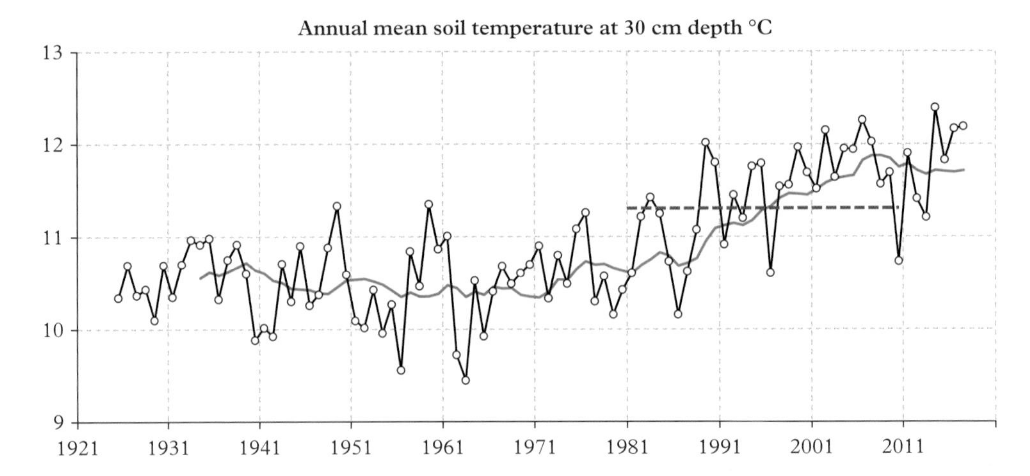


**Figure 24.6** *Annual mean soil temperatures at 30 cm depth (°C) in Oxford since 1925. The ten-year unweighted running mean is shown in read, while the dashed line shows the 1981–2010 average (11.4 °C)*

1990s compared with the previous four decades: 16 per cent of species flowered significantly earlier in the 1990s than previously, with an average advancement of 15 days in a decade. Ten species (3 per cent) flowered significantly later in the 1990s than previously. Fitter and Fitter [113] concluded that these phenological changes, combined with more obvious consequences of climate change such as changes in geographic range, would have profound ecosystem and evolutionary consequences.

Discussion of growing season leads naturally to consideration of soil temperatures. Two Symons-pattern earth thermometers were installed in 1905 at two depths (6½ inches and 3 feet 6 inches: [21]) but it was not until 1925 that measurements started at the standard depths of 1 ft (30 cm) and 4 ft (120 cm). Since 1971, the measurements have been made at 30 and 100 cm depths. Figure 24.6 shows the annual mean soil temperature at 30 cm depth at 0900 GMT since 1925. As expected, there has been a significant increase over the last ninety years, with the record showing a slight cooling in the 1960s and a strong rate of increase since then. Whilst winter soil temperatures are discussed elsewhere, it is worth noting here that the average winter soil temperature at 30 cm depth is now above 5 °C, suggesting that grass growth can continue throughout the winter season in some, if not all, winters.

## Precipitation

The first period of precipitation measurement at the Radcliffe Observatory remains the wettest on record, as Table 24.3 shows, although the first 'decade' consists of only 4 years' records. Given the care and attention that has been paid to the climatic records at the Radcliffe Observatory (see Chapter 2), we can be confident that that the early precipitation record is reliable. The analysis which follows here is in part an update of Burt *et al* [114].

Figure 24.7 shows Oxford's annual rainfall totals from 1767, in millimetres, together with ten-year unweighted running means. The first two calendar years in the Oxford rainfall record were both very wet: 1767 (841 mm) and 1768 (933 mm), the latter still the fourth-wettest year on record. In contrast, the 1780s and the first two decades of the

**Table 24.3.** *Decadal precipitation: average annual totals (mm) at Oxford, 1767– 2018 (first and last periods are less than 10 years long and are shown in italics)*

| Decade | Average annual precipitation mm | Decade | Average annual precipitation mm | Decade | Average annual precipitation mm |
|---|---|---|---|---|---|
| *1767–70* | *777* | 1851–60 | 629 | 1941–50 | 620 |
| 1771–80 | 658 | 1861–70 | 616 | 1951–60 | 710 |
| 1781–90 | 532 | 1871–80 | 700 | 1961–70 | 652 |
| 1791–1800 | 620 | 1881–90 | 643 | 1971–80 | 632 |
| 1801–10 | 546 | 1891–1900 | 609 | 1981–90 | 613 |
| 1811–20 | 579 | 1901–10 | 636 | 1991–2000 | 683 |
| 1821–30 | 612 | 1911–20 | 703 | 2001–10 | 686 |
| 1831–40 | 625 | 1921–30 | 665 | *2011–18* | *666* |
| 1841–50 | 650 | 1931–40 | 655 | | |

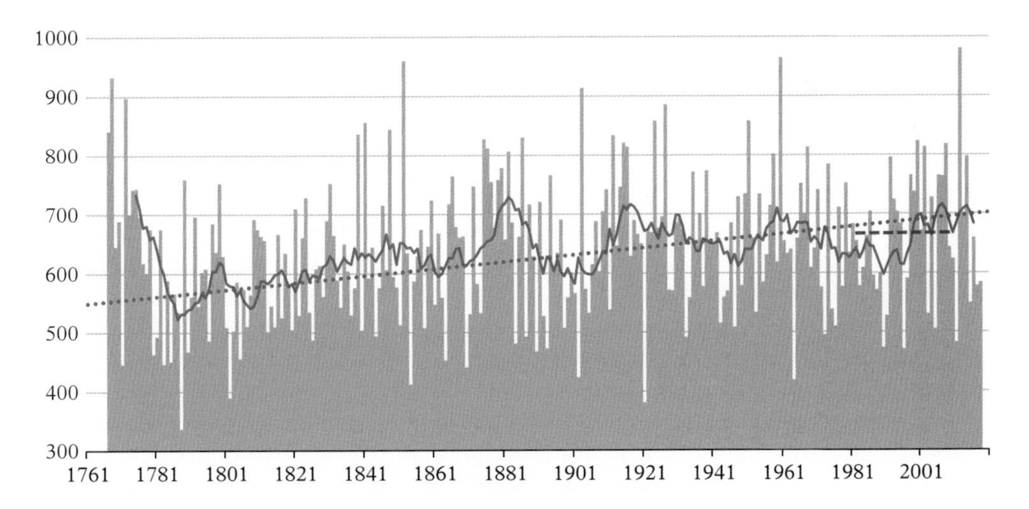


**Figure 24.7** *Annual precipitation total (mm) in Oxford, 1767 to 2018. The red solid line is a ten-year unweighted running mean, the blue dashed line is the least-squares linear regression line over the whole period, and the dark red dashed line represents the 1981–2010 normal fall (660 mm). The extremes on this plot are 1788 (337 mm, 51% of normal) and 2012 (979 mm, 148% of normal). For more information on dry and wet years, see The Calendar Year, Chapter 19*

nineteenth century remain the driest on record. We shall comment on drought in more detail in the next section but suffice it to note here that the late 1780s and the early 1800s are the times of greatest rainfall deficit on record at Oxford.

Despite a good deal of inter-decadal variability, there has been a clear monotonic upward trend in total precipitation ever since this very dry period at the end of the eighteenth and start of the nineteenth centuries. The overall linear trend is highly significant despite the low correlation coefficient with time (i.e. year number) (correlation $r_s = 0.181$, probability $p = 0.0028$, samples $n = 250$). The time series starting in 1780 has an even stronger trend ($r_s = 0.261$, $p = 0.00004$, $n = 237$). As we have already seen, the long-term linear trend in mean air temperature with time (data from 1815) is also very strong ($r_s = 0.424$, $p < 0.00001$, $n = 202$) and, taken together, these two results might suggest a possible long-term temperature-precipitation linkage. Note, however, that the mean air temperature series from 1815 (Figure 24.1) exhibits significant serial correlation, so this makes it harder, on the basis of statistics alone, to argue for a significant, long-term correlation between temperature and rainfall.

In terms of seasons, there are significant upward trends in precipitation in winter ($r_s = 0.311$, $p < 0.0001$) and spring ($r_s = 0.213$, $p = 0.0005$) but not in summer or autumn (see Chapters 20–23 for seasonal plots and commentaries).

## Rainfall frequency

Rainfall totals were not recorded every day in the first part of the 'daily' record from 1827; even including days when precipitation is mentioned but not recorded, the total number of rain days appears low through to at least 1850 (when ground-level measurements began). It is not until 1853 that the published *Radcliffe Results* volumes state that rainfall was measured every day, rather than after a known fall of rain, so there is some doubt about the early observations of measurable precipitation—see also Appendix 1. There is also the possibility of under-catch of snowfall early in the record, and corrections to the record may not have taken this fully into account [98]. Accordingly, Figure 24.8 shows the annual number of rain days (daily totals of at least 0.2 mm) at Oxford only since 1853. As with total rainfall, when considering the entire series (i.e. 1853 to date), there has been a significant upward trend in the number of rain days each year ($r_s = 0.239$, $p = 0.003$, $n = 165$), although almost all of this rise occurred between the 1880s and the 1950s: the trend has, if anything, been slightly downward since the 1950s, as suggested by the dotted red line in Figure 24.8, showing the 30-year running mean against year ending. Winter is the only season with a significant (upward) trend with time for the number of rain days ($r_s = 0.227$, $p = 0.004$) over the entire period since 1853. Mean rainfall per rain day has tended to increase significantly over the last 100 years in south-east England [115] but this trend is not seen at Oxford, perhaps because Oxford is too far west to be strongly influenced by the European mainland. There are no significant trends for any season or for the year as a whole from 1853. For data from 1900, there is only one trend to be noted, for spring, but this is not quite statistically significant ($r_s = 0.183$, $p = 0.055$). However, for data from 1950, spring (alone) shows a highly significant trend in mean rain per rain day ($r_s = 0.313$, $p = 0.011$). This is equivalent

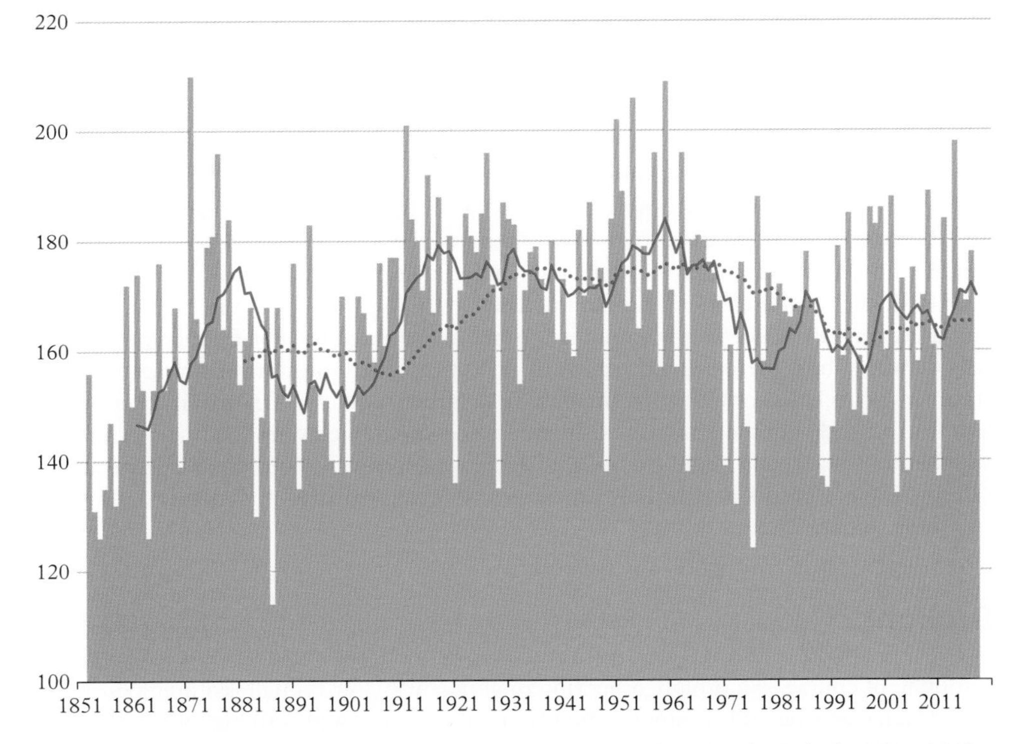


**Figure 24.8** *Annual frequency of rain days (0.2 mm or more precipitation) in Oxford, 1853 to 2018. The red solid line is a ten-year unweighted running mean; the red dotted line is a thirty-year unweighted running mean plotted at year ending. The extremes on this plot are 114 days in 1887 and 210 days in 1872. For more information on rain day extremes, see Table 19.10*

to an increase in mean rainfall per rain day in spring from 4.6 mm/day in 1950 to 6.3 mm/day in 2017, although further research is needed to identify the reasons for any changes in the frequency and intensity of rainfall at Oxford.

As part of a wider study of rainfall intensity in south-east England [113, 115], an hourly rainfall record was produced for 'Oxford' by blending records from RAF Benson (data from 1980), the Radcliffe Meteorological Station (data from 1983 to 1999, with some gaps) and the ECN site at Wytham (data from 1992). More analysis is needed, so only brief mention is made here. There is a weak but statistically insignificant upward trend in *hourly* rainfall intensity in winter (data for 1980–2014, $r = 0.195$, $p = 0.26$) but, as yet, evidence is lacking at the sub-daily timescale to support any conclusions with regard to increasing hourly rainfall frequency and intensity.

There are significant upward trends at Oxford since 1853 for the number of daily totals of at least 1 mm in winter ($r_s = 0.284$, p = 0.0003), in spring ($r_s = 0.225$, p = 0.005), for the year as a whole ($r_s = 0.275$, $p = 0.0005$) and for the number of daily totals exceeding 5 mm in winter ($r_s = 0.276$, $p = 0.0005$), but not for numbers of daily totals of 10 mm or more. In terms of large daily totals, there has been no significant change over the study period at Oxford, for seasonal or annual data. Since the 1850s, the numbers of daily totals of at least 15 mm has averaged six per year, ranging from only one in 1902 to 17 in the record-breaking year of 2012. There is no significant trend in the annual maximum daily series since 1827 (Figure 19.7, Chapter 19).

To summarise long-term trends in the Oxford precipitation data, there is good evidence over 250 years for monotonic linear trends in winter, spring and annual totals. In terms of the frequency of precipitation events (as indicated by the number of rain days) and the number of small daily totals (up to 5 mm), there is some evidence of long-term trends from 1853 from when the daily data are reliable, at least until the 1950s. There is little evidence for increases in rainfall intensity (rainfall per rain day) in recent decades and there is no evidence of any significant trend in the magnitude or frequency of heavy falls of rain. Whilst there is good reason to expect that the long-term tendency for drier summers is likely to continue, the increase in summer rainfall at Oxford in recent years seems to accord with the 2009 UK Climate Projections, which were too uncertain to say whether seasonal totals might increase or decrease. However, there is some expectation of more intense events at sub-daily timescales, showing the need for further analysis of sub-daily rainfall data.

## Drought

Drought, of course, is a prolonged period of abnormally low rainfall, leading to a shortage of water. Spring and summer droughts are often associated with higher than normal temperatures, increasing evaporative losses and adding further to the water deficit. Drought has been quantified in the past in a number of ways. Historically, and with agriculture particularly in mind, the UK Met Office used to focus on runs of days with little or no measurable rainfall: a 'dry spell' was defined as a period of at least 15 consecutive days to none of which is credited 1 mm or more, and an 'absolute drought' as a period of at least 15 consecutive days to none of which is credited 0.2 mm or more (see also 'Droughts and wet spells' in Chapter 28). Today, longer-period totals are more relevant, since water supplies are usually well buffered in the shorter term but can become scarce if the drought is prolonged. Any definition of 'prolonged' will depend on local conditions: in Oxford, water supplies are predominantly from the River Thames which receives much of its flow from groundwater (from the Chalk and Jurassic limestone rocks) with the added protection of the reservoir at Farmoor, close to the Thames just upstream of Oxford.

On 27 May 1992, *The Independent* newspaper claimed that the then current drought was the worst on record. Whatever records they had consulted, it was not the rainfall records of the Radcliffe Meteorological Station! True, in terms of rainfall totals over the previous two years, the total up to and including February 1992 (884 mm) was the lowest observed in the twentieth century, but *The Independent* was clearly unaware of the very severe drought in the early nineteenth century: the 24-month total up to and including October 1803 (800 mm) is easily the lowest on Oxford's record back to 1767. This was soon pointed out in the traditional way—a Letter to the Editor!

Here, we prefer to use accumulated totals rather than specially designed indices such as the Drought Severity Index [116]. In either case, the choice of time period for analysis is crucial: short-period totals indicate many short droughts, whereas much longer periods will identify the severe droughts when water supplies can be seriously stretched. In this sense, the modern drought of 1992 would likely have had much more impact than that of 1803, affecting a much larger urban population entirely dependent on public water supply. Figure 24.9 shows running rainfall totals for 60-month periods expressed in terms of multiples of the standard deviation $\sigma$ of the entire series 1767–2017, 116 mm. This emphasises the very dry periods of the 1780s and 1800s already mentioned above and the monotonic increase in average rainfall since then. In relation to the overall mean, these two droughts are furthest below the mean, the 60-month total for June 1788 being 2.55 standard deviations ($\sigma$) below the mean and the 60-month total for November 1805 being 2.61 $\sigma$ below. In comparison, the 60-month total ending July 1976 is 'only' $-1.89\sigma$ and, for May 1992, $-1.56\sigma$. However, if the Oxford monthly rainfall record is detrended, a somewhat different picture emerges: relevant statistics are now July 1788 $-1.98\sigma$, November 1805 $-2.09\sigma$, July 1976 $-2.09\sigma$ and May 1992 $-2.13\sigma$, the lowest value

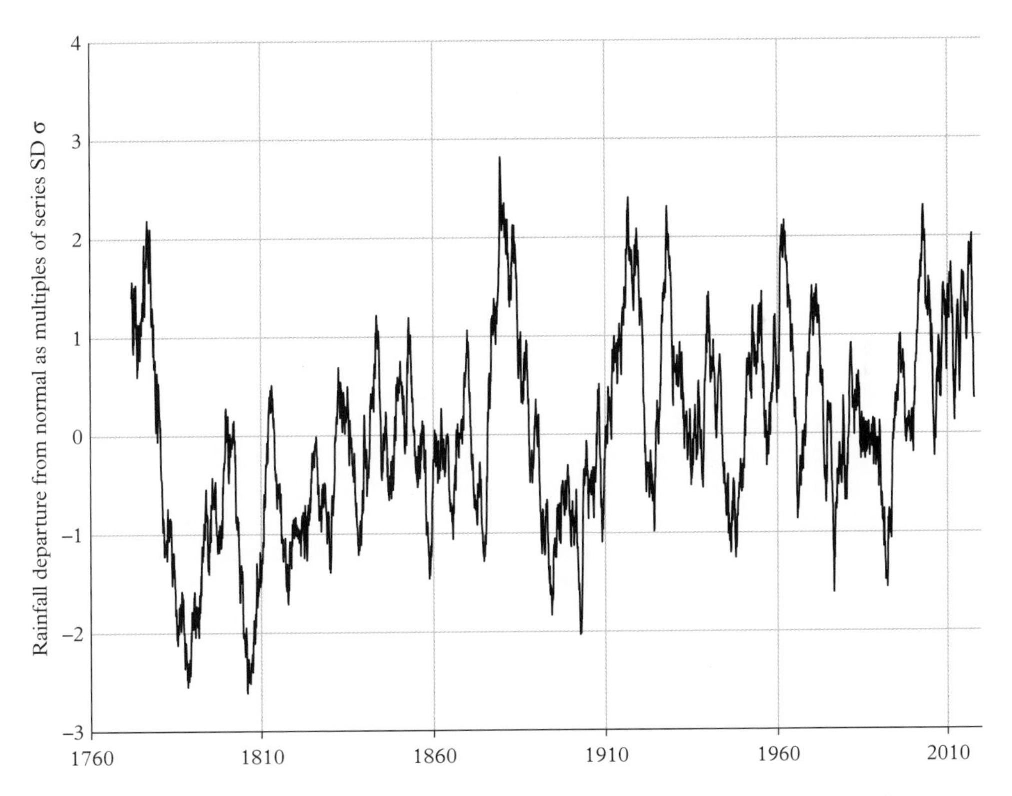


**Figure 24.9** *Five-year rainfall totals at Oxford by month since 1767, plotted in Z-score units (multiples of standard deviation)*

for any 60-month period. Maybe *The Independent* was right after all!—at least in terms of how far the water deficit in 1992 was below 'normal'. On this basis, the drought at the beginning of the twentieth century also enters the frame, with the detrended residual for the 60 months to August 1902 being the fourth lowest at $-2.07\sigma$.

The drought of 1975–76 was quite different. It is generally taken to have begun in Britain in May 1975, after a wetter than average winter, and to have ended at the end of August 1976 [52]. However, rainfall at Oxford in May 1975 was only a little below average, so the drought did not really start in Oxford until June 1975. Summer 1975 was hot and dry, receiving only 54 per cent of normal rainfall (82 mm), ranking 15th driest in the period 1767–2018. Summer 1975 was the ninth warmest summer on record at Oxford (1815–2018). September 1975 was wetter than average: 95 mm, ranking 39th wettest since 1767. There followed an exceptionally dry extended winter: October 1975 to April 1976 was the driest such period at Oxford since records began (132 mm, 33 per cent of average). May 1976 had close to normal rainfall at Oxford, followed by an even drier summer than 1975: summer 1976 received only 55 mm, ranking fifth driest since 1767, just 36 per cent of normal. Summer 1976 was also the hottest on record, surpassing 1826 which was the hottest summer to that date. Altogether, the 1975–76 drought was the driest 15-month spell at Oxford since records began (410 mm, 51 per cent of normal). It ended abruptly with a thunderstorm on 29 August during which 20.8 mm was recorded; other than the day before (0.2 mm), there had been no rain at all since 20 July and that was the wettest day since 13 September 1975. The autumn of 1976 was very wet, September and October especially; the 43rd wettest autumn does not sound exceptional but, after such a drought, it certainly seemed like it was (see also the synoptic contexts of the 1976 drought and recovery in Chapter 4).

Naturally enough, river flow fell to very low levels during the 1975–76 drought. Figure 24.10 shows gauged daily flow ($m^3/s$) for the River Thames at Day's Weir, just downstream of Oxford, for the water years 1975–76 and 1976–77; by convention, a water year (WY) runs from 1 October to 30 September. WY 1976 (October 1975 to September 1976) remains the lowest average daily flow (3.5 $m^3/s$) since records began (1938), equivalent to an annual depth of runoff across the catchment of only 32 mm, just 12 per cent of average. Gregory (in [52]) notes that flow in the River Windrush fell to just 10 per cent of normal for WY 1976, so the Day's Weir record shows that river flows right across the upper Thames basin were at a similarly low level, the lowest anywhere in the country (see the map in [52]). The most obvious point about the river flow data shown on Figure 24.10 is the complete absence of any flood runoff in the 1975/76 winter and, indeed, baseflow hardly increases either. Flow in the summer of 1976 fell to record low levels but recovered remarkably quickly from October onwards, given the very wet autumn. WY 1977 ranks 11th highest on record.

The 1975–76 drought can be ascribed to a lack of cyclonic weather and an abnormal incidence of anticyclonic weather. Atypically high surface pressure occurred over southern Britain (averaging 5 hPa above normal) with much more (dry) anticyclonic weather than is normal (see also the synoptic background in Chapter 4, and Perry in [52]). Not surprisingly, analysis of Oxford droughts in relation to weather types shows a strong positive (negative) correlation with the frequency of anticyclonic (cyclonic) conditions.

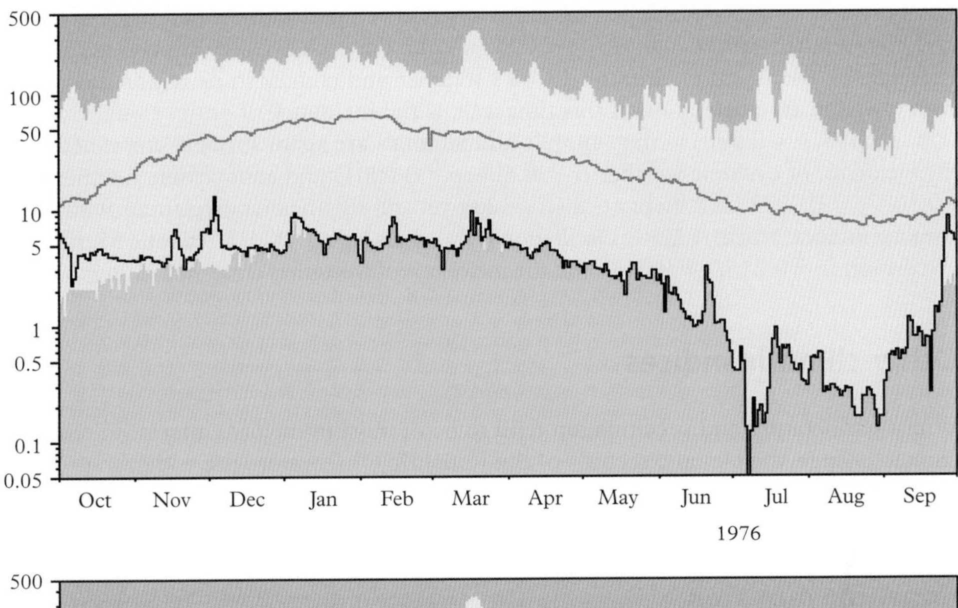


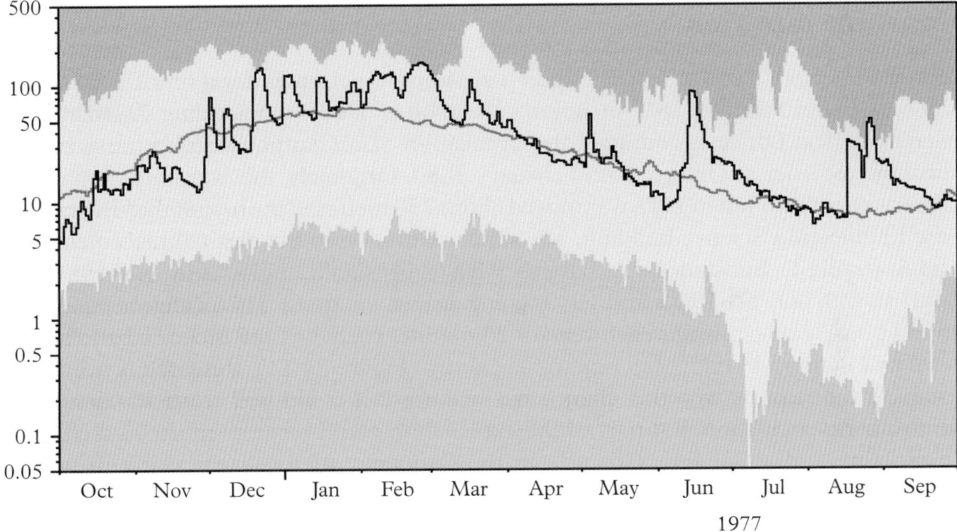


**Figure 24.10** *Thames flood hydrograph for Day's Weir, near Dorchester-on-Thames in Oxford, for water years 1975/76 (October 1975 to September 1976,* **top***) and 1976/77 (October 1976 to September 1977,* **bottom***), showing persistent record low flows from early December 1975 to late September 1976, followed by a very rapid recovery to above-normal flow by early December 1976, and record flows in late February 1977. River flow hydrographs show the daily mean flows (bold black line, units m³/s), together with the long-term daily average flows in grey and maximum and minimum daily flows prior to 2017 (shown by the blue and pink shaded areas, respectively). Daily flows falling outside the maximum/minimum range are indicated where the bold trace enters or delineates the shaded areas. The dashed line represents the period-of-record average daily flow (Data from National Rivers Flow Archive – NRFA courtesy Terry Marsh, CEH Wallingford)*

Rainfall totals over 24-month periods are strongly correlated with the incidence of Lamb cyclonic weather type ($r_s = 0.597, p < 0.0001$); the only other global climate index (see Chapter 4's discussion regarding Oxford's weather and climate in its regional context) with a significant correlation at this timescale is the incidence of anticyclonic weather ($r_s = -0.353, p < 0.0001$). Sixty-month rainfall totals are again strongly correlated with the incidence of cyclonic weather ($r_s = 0.606, p < 0.0001$) and anticyclonic weather ($r_s = -0.307, p = 0.0003$) but there are also weaker yet still significant correlations with other climatic indices: North Atlantic Oscillation ($r_s = -0.282, p = 0.0001$), Atlantic Meridional Oscillation ($r_s = 0.215, p = 0.013$) and the incidence of westerlies ($r_s = -0.191, p = 0.027$).

## Other climatic indices

Whilst temperature and precipitation tend to be of most immediate interest in terms of climate change, the climatic records of the Radcliffe Observatory are wide-ranging and many of them show interesting changes over time, for a variety of reasons, both relating to global climate and more local and regional factors.

### Sunshine duration

Figure 19.3 (in Calendar Year, Chapter 19) shows the annual duration of bright sunshine at Oxford since 1881, in hours. As at other stations, including Durham, the Meteorological Office encouraged the installation of a Campbell–Stokes glass sphere sunshine recorder at Oxford from February 1880. Originally, this was at the top of the Observatory tower but in 1976 was moved about 230 metres to the roof of the University's new Engineering Science building, at a similar elevation to the original placement (see also Appendix 1). Sunshine hours have shown a highly significant increase since 1881 ($p = 0.0014$), with only the 1960s and 1970s going against the trend. This change is equivalent to 137 hours' extra sunshine each year, or 23 minutes each day, and is clearly beneficial to local and national initiatives in renewable energy (Figure 24.11). Research is needed to establish the reasons why this change has occurred: it could well relate to changes in atmospheric circulation, but part of the explanation could relate to more local changes in visibility and air quality. Changes in the frequency of fog, particularly in winter, provide some pointers—see Figure 20.4 in Winter, Chapter 20.

### Evaporation

Burt and Shahgedanova [117] presented a historical series of evaporation losses, both potential and actual, and differences between precipitation and evaporation, for the Radcliffe Meteorological Station since 1815. The Thornthwaite method was used to calculate potential evaporation each month and observations from a nearby station, where the Penman method could be used to calculate daily losses, were used to calibrate the monthly estimates. Whilst no long-term trends were evident, the most recent results from the late 1980s and early 1990s showed potential evaporative losses had been above

**Figure 24.11** *Solar panels on a house in Beechcroft Road, Oxford. Oxford now has over 800 houses with solar panels (Brenda Boardman, pers. comm.), all benefitting from the increased amount of bright sunshine compared to the pre-1950s (John Boardman)*

the long-term average, and differences between precipitation and potential evaporation, an index which shows availability of water for runoff, had noticeably declined as a result. These trends were particularly marked in summer months. The soil moisture deficits observed in summer during the last 20 years of analysis were the largest on record; these moisture deficits persisted into late autumn, delaying the seasonal recovery in river flow. Whilst more recent summers have generally been wetter than those of the late twentieth century, higher temperatures will nevertheless have encouraged greater evaporation, with consequent impact on soil moisture deficits and river flow.

## The importance of maintaining long records

It is hardly surprising that a record as long as that at the Radcliffe Meteorological Station shows clear evidence of climate change, and global warming in particular. The importance of long-term observation of the natural environment has long been recognised, and yet 'monitoring' is often dismissed as low-grade science which can contribute little to our understanding [84]. Many environmental processes vary slowly, over relatively long periods of time, typically much longer than the duration of a research project or period of study for a research degree. Subtle processes are embedded within highly variable systems so that their weak signal cannot be extracted without a long record. In systems jargon, the signal-to-noise ratio is low. Although a clear pattern may eventually be identified, high-frequency variation will obscure such a trend, and short-term study will be unable to recognise it. Such problems are typical of climatic systems: there is no way to distinguish a normal extreme from an entirely new trajectory without long-term studies. By definition, rare events occur infrequently. If one is lucky, rare events may be observed during a short-term study but there is clearly a much better chance of doing so with a long-established measurement programme. Very long records like that at the Radcliffe provide three important benefits: they reveal important patterns for scientists to examine; they are essential for testing hypotheses undreamt of at the time monitoring was set up; and they provide an essential way of discovering whether there are significant changes taking place in the natural environment that may ultimately be harmful to ecosystems and to people themselves [84]. For all these reasons and more, the records of the Radcliffe Meteorological Station are a rare and valuable resource. Whilst short-term studies and computer simulation modelling are important scientific tools, their value would be much less without carefully collected and well-documented measurements made to exacting standards, archived in quality-controlled datasets and maintained over very long periods of time. Thankfully, Oxford University is committed to maintaining meteorological observations at the Radcliffe Observatory as long as there is a need to do so. Put simply, we see that continuing need stretching far into the future.

# Part 5

# Chronology of noteworthy weather events in and around Oxford

# 25

# Chronology

## 7 May 1663 (old style)—Great spring floods

On 7 May 1663, Oxford diarist Anthony à Wood (1632–95) wrote that floodwaters rose '...in the morning in such abundance that it seemed like to the coming in of a tide and by one of the clock it came up to the backside of Merton College within 4 yards; all [Christ] Ch walkes and [Magdalen] walkes were drowned.' Quoted in E. G. W. Bill [118] and the BHS Chronology of British Hydrological Events [119], reference 12226.

> 'There were great rains in the upper Cherwell Valley but none at Oxford such that the Cherwell had a prodigious flood, not only over the meadows but over the raised walks at Magdalen.' ([119], reference 6657)
>
> 'The Cherwell backed up as far as Ivy-Hinksey a mile from its confluence with the Thames; it came up almost to Merton College and the water was level almost with the common way at Magdalen Bridge.' ([119], reference 6658)

*The British Hydrological Society site [119] contains many other references to noteworthy floods and droughts in and around Oxford; only a few are listed here.*

## 31 May 1682 (old style)—'Cloudburst' in Oxford

Robert Harrison, then an undergraduate at Queen's College, described the 'Suddain and Violent tempest' which affected Oxford on 31 May 1682 (New Style or Gregorian Calendar date 10 June) in a 12-page pamphlet, a copy of which is held in the Bodleian Library. The following is part of his account as related by Meaden [120], in the original English:

> 'Upon the 31 of May last. The morning was calm, serene and clear, at ten of the clock an uncouth and intense heat of the sun seem'd to scourge the moistned plains, as if he would have redeem'd in a moment the continued bathings they had been so long plung'd into, by the quick and penetrating stroaks of his redoubled beams. This continued till about a quarter of an hour after noon; when presently a steddy and soft wind (south-west, or south-west by south) seem'd to dislodge from his back, and hoord up several clouds and

*Oxford Weather and Climate since 1767*. Stephen Burt and Tim Burt, Oxford University Press (2019).
 DOI: 10.1093/oso/9780198834632.001.0001

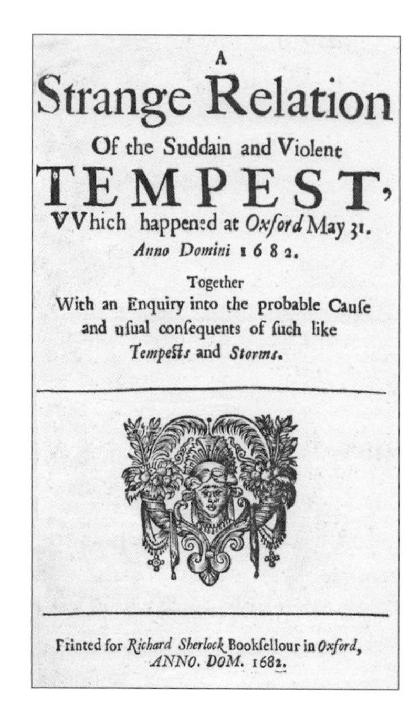
A
Strange Relation
Of the Suddain and Violent
TEMPEST,
VVhich happen'd at *Oxford* May 31.
*Anno Domini* 1 6 8 2.
Together
With an Enquiry into the probable Cause
and usual consequents of such like
*Tempests* and *Storms*.

Printed for *Richard Sherlock* Bookseller in *Oxford*,
*ANNO. DOM.* 1682.

**Figure 25.1** *Cover of Harrison's pamphlet of the Oxford storm of 31 May 1682, now in the Bodleian Library*

vapours not far from our zenith. Having observ'd this, I forthwith erected a scheme, and beheld the position of the heavens, from which I had no sooner learnt the impending event, but immediately the presaged storm came out of hand to attest the veracity of my judments. A miraculous, dismal, and hideous storm followed: First, a cataegis, or rushing, murmur was heard in the upper regions, which anon was felt in the lower. A huge, blustering, and boisterous wind descended with such vehemence and irresistable force, that it proved inimical to antient trees and antiq. edifices; was dangerous, if not destructive, to way-faring men, and out-laying cattle; 'twas thick, and black, and with such violence reflected upon the terraqueous globe, that thereby it wheeled, and contorted it self into such windings, that it hoised up, and as it were absorbed the more subtile, light, pulveriz'd and dry particles of the terrestrial bodies, which had been divided and separated by the preceding ardour of the sun's scorching beames; these opaceous particles were so gross and many, and so variously tossed in our atmosphere, they denyed us our meridian lustre of the sun; for the beams the upper clouds transmitted, they terminated: so that without a contradiction, we might have asserted a solar eclipse, at two signs elongation of the luminaries....And the concomitant thunder was no less astonishing, then the described wind: for besides the frightfull claps at every eruption, the lightening was deadly, ruinous, and powerful; the fulgur quick and crispisulcant, and appeared several times to compose and distort its self to the form and similitude of an oxe's horn, otherwhiles to the likeness of the pyramidical flame of a burning torch, or indeed not much unlike the spiritual cloven tongues...

Lastly, the rain was thick, strong, and ponderous; its fall caus'd the tender scions as it were reverence it, and bow to its presence; several of the drops were extended to the full breadth of a six-penny-piece, which also followed one another so closely that they seemed one continued spout or stream; so that in less then half a quarter of an hour, these pouring cataracts raised the water in a round and uniform vessel of about 4 foot diamiter, near two foot higher then before, without the assistance of any other interfluent rivulet, or commixing water....truth delights in a plain dress, such in reality was this, as I have here described it; whose more immediate and direful effects, are sadly evident in several instances, as in the subversion of bridges, the discussion of walls, and demolition of houses; the despoliation of some trees of the boughs, and others of their fruits, the eradication of newly sown seeds, the exsiccation of those more deeply rooted, and a general either deprivation or depravation of the radical and vividicating moisture of all trees, herbs, fruits and plants: what the wind left, the ruin beat down; and what that spared, the lightening struck: these are tokens too evident.'

In the remainder of the pamphlet, Harrison gives his own theories on clouds and how they came together to produce this storm, finishing by ascribing the ultimate cause of the storm to 'the positions of the planets and the stars, which led him to predict the general weather for the coming summer and autumn' [120].

Attention has focused on this event firstly as what has been interpreted as the first eyewitness description of a tornado in Oxford and, secondly, owing to the prodigious quantity and intensity of the rainfall referred to:

> '...in less then half a quarter of an hour, these pouring cataracts raised the water in a round and uniform vessel of about 4 foot diamiter, near two foot higher then before...'

Taking Harrison's account at face value, Elsom [121] posited 'near two foot' as 21 inches or 530 mm, and 'less than half a quarter of an hour' as 7 minutes, leading to the astonishing rate of almost 90 mm *per minute*. Short of proposing some form of catastrophic and very localised falling to earth of a mass of water suspended within the cloud, perhaps by the sudden collapse of a strong convective updraught, such a rate is simply not credible.

Unusually, we have a second eyewitness account of this storm, by the already-mentioned Oxford diarist Anthony à Wood, as quoted by Elsom [121]:

> 'In the afternoon about 12 and 1 the sky was most prodigiously darkned. A great storme of wind came, which was so circular that it blew all the dust in the street up in the aire that you could not see any houses; afterwards followed a smart shore [shower] of raine. A hurricane; this was never knowne in the memory of man. A prodigious hericane that broke bows and armes of trees; blew of thatch; and did a great deal of harme in the country. A pamphlet of this [by Robert Harrison] I have.'

## November 1768

The year 1768, only the second in the Oxford rainfall record, remains the fourth wettest on record in the entire series to date. It was a very wet autumn, with September ranking second-wettest with 171 mm, October ranking 34th (106 mm) and November ranking 57th (85 mm); not surprisingly, autumn 1768 ranks second, too (362 mm), just behind 1772. Griffiths [122] (as cited by Frank Law in the BHS Chronology of British Hydrological Events [119] [123], record 12128) quotes from the parish records at Whitchurch-on-Thames, between Oxford and Reading:

> 'In the year 1768 November, there was a greater flood than ever was known, the water rose to such a height as to be 2 feet deep [60 cm] in ye barns at Mr Wallis Farm, and at the parsonage, and for the ferry boat to go up to the house opposite Mr Whistlers.'

Although November itself was not excessively wet, the two previous months had been so wet that any significant amount of rain would likely generate a high flood.

## September 1774—The wettest month on Oxford's records

September is not normally a particularly wet month—its current average of 54 mm places it right in the middle of Oxford's ranked monthly mean rainfalls, with five months drier and seven months wetter—and it is rarely (about one year in ten) the wettest month of the year. Starting from such a point makes it all the more difficult to comprehend how extraordinarily wet September 1774 was. Not only does it remain the wettest *September* on Oxford's 250 year record—by a margin of over 50 mm—it is the wettest month *of any name* in that record, and still lies almost 30 mm clear of any other month. When it is considered that there is less than 30 mm separating the remaining months in Oxford's 'top ten' wettest months (Chapter 29, Table 29.7), a little further digging into the circumstances was clearly warranted.

The evidence we have is from Thomas Hornsby's surviving manuscript weather records. These started in 1760 [24] and an almost complete daily record survives from 1767 to 1776, and another from July 1794 to 1805—the records from some of the intervening years are missing and presumed lost. Fortunately, we have a complete daily diary for September 1774 in Thomas Hornsby's own hand: Figure 25.2 shows the record for the period 1–27 September. After the date and (astronomical) time of

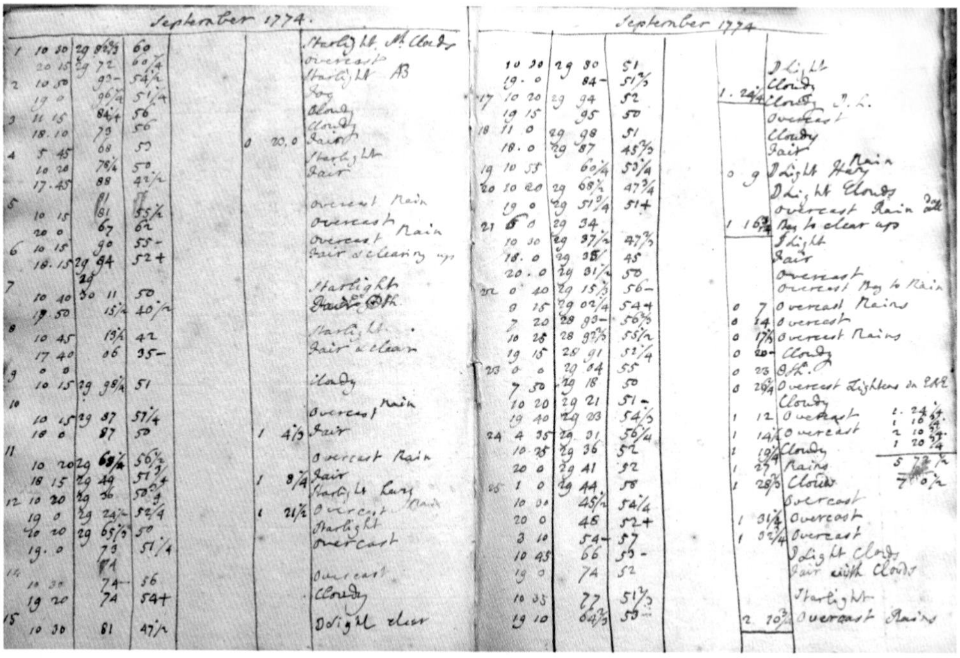

**Figure 25.2** *Thomas Hornsby's manuscript weather records for 1–27 September 1774, from his original ledgers held in the School of Geography, University of Oxford (University of Oxford)*

the observation, the following columns note readings of the barometer (in inches of mercury, inHg), outside air temperature (in °F), rainfall (in inches and 36ths, recorded cumulatively and the receiver then emptied—for the details see [24]), and brief weather notes. For most days, there are two observations daily, the first usually a little after 10 p.m. (10 h on the astronomical clock used, starting at noon), presumably about the start of the evening's astronomical observations, and a second the following morning, usually between about 6 a.m. and 8 a.m. (18 h to 20 h astronomical clock). Some observations were made at other times, and this was particularly the case during the second half of the month when low barometric pressure coincided with large amounts of rainfall; on 22 September six observations were noted, all but one of them including a rainfall total.

The gauge in use at this time was mounted on the east wing terrace of the Observatory and was probably 3 in/76 mm square (replaced four months later by a 12 in/300 mm diameter funnel) with a downpipe leading into the sextant room, where it was measured in a glass measuring cylinder. Such an exposure would have led to an under-catch of rainfall when compared with the modern standard of a 5 in/127 mm diameter raingauge mounted 300 mm above ground level. Later experiments at the Observatory (see Appendix 1) confirmed that the terrace-mounted gauge under-read by about 13 per cent, owing to wind errors, compared to a modern gauge in a standard exposure, and an additional correction of 13 per cent was introduced to all the surviving early records to allow for losses in the downpipe arrangement, and thus Hornsby's monthly total of '7 0½' in shown in the margin of Figure 25.2 (= 7.014 in, or 178.1 mm) was eventually adjusted upwards by just under 26 per cent to 224 mm to enable comparisons and continuity with the long Observatory record made with a raingauge at ground level (from August 1850).

What do we know about that September's weather? Plotting the month's accumulated rainfall from Hornsby's observations (adjusted as described above) against Hornsby's pressure observations in Figure 25.3 (pressure 'as read'—add 5–6 hPa for MSL values), we can infer the passage of a fairly deep depression on 11th/12th, producing about 19 mm rainfall on 11th and a further 12 mm on 13th. In a letter written on 13 September 1774, Gilbert White in Selborne noted [124]: 'Wall-fruit abounds with me this year; but my grapes, that used to be forward and good, are at present backward beyond all precedent: and this is not the worst of the story; for the same ungenial weather, the same cold black solstice, has injured the more necessary fruits of the earth, and discoloured and blighted our wheat. The crop of hops promises to be very large.' There would be beer if not bread after such a wet and poor autumn!

The next spell of wet weather began with a steep and prolonged fall in the barometer 18th–22nd, the lowest barometer reading of the month being 28.91 inHg (979.0 hPa) at 7.15 a.m. on 23 September (approx. 984.4 hPa at MSL). Between 11 p.m. on 19 September and 6 p.m. on 21 September, almost 40 mm fell; Hornsby noted at the latter observation (optimistically, as it turned out) 'Beginning to clear up'. The lack of wind directions makes analysis less certain, but this deluge may have resulted from

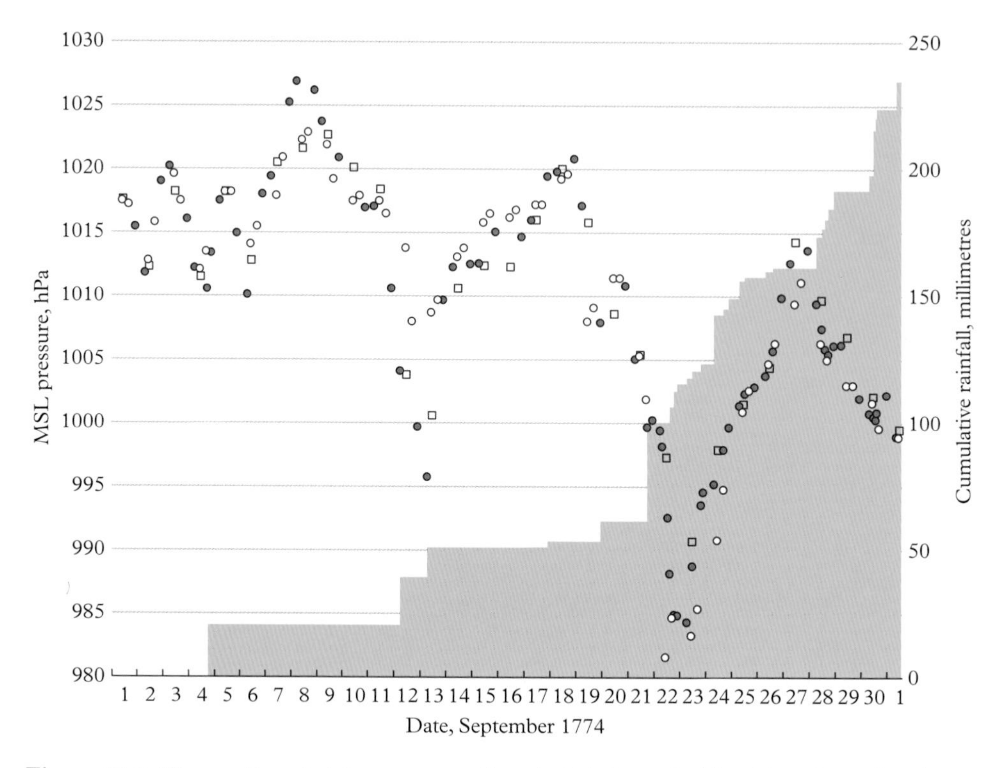


**Figure 25.3** *Thomas Hornsby's barometer readings (red circles, units hPa, left scale—reduced to MSL from observed station level pressure, assuming barometer temperature of 16 °C throughout) together with the monthly accumulated rainfall (orange columns, in mm and right scale, original values adjusted for gauge under-catch as described in the text) for September 1774; dates are shown along the horizontal axis. Two other barometer readings are shown (using approximate MSL corrections)—for London (open squares, after Cornes [62] and pers. comm)—daily averages, shown at a nominal 1200 position for each date: and for Exeter (open circles) from the weather diary of Samuel Milford—daily readings at 0900 and 1500 (courtesy of Ed Hawkins and Harry Coad, University of Reading)*

prolonged cyclonic rainfall to the north of a deep but slow-moving depression over southern England or northern France. Spells of heavy rain continued until the afternoon of 26th, by which time a further 61 mm had fallen, 56 mm in the three days 22–24 September. After a short dry respite, a further 30 mm fell on 27th/28th, and another 37 mm on the last day of the month.

The disruption and flooding from this wet spell, some 173 mm in 10 days, must have been immense and catastrophic. Some contemporary accounts survive—this from the British Hydrological Society online chronology [119], record 6042 (3 October 1774), referring to floods on the River Cherwell:

> 'We hear from Oxford, that the waters are so much out round about that place, that several roads leading to the city are impassable, and a great number of sheep and other cattle have been lost; and the waters are so much out in St Thomas's parish, that the inhabitants are obliged to live up two pair of stairs, and have their provisions brought to them by boats; and the walks belonging to Christ Church College are entirely washed away, and likewise Merton College walks. There is no land to be seen within three or four miles of that place. The damage done to the walks of Christ Church College is computed to be upwards of 200 *l*.'

September 1774 remains the fifth-wettest on record in the long England and Wales monthly rainfall series, which extends back to 1766 (see Chapter 5), acting to confirm Oxford's very high monthly rainfall total (although it should of course be noted that there were very few raingauges operating at the time, and the early years of the England and Wales series necessarily include the Oxford records amongst its constituent sites). In London, the total for the month at Kew (95 mm; [125]) was about twice normal, but not exceptional. Thomas Barker, at Lyndon in Rutland, measured 203 mm of rainfall in September 1774—'I never measured so wet a month before.' [126]. His diary account for the month is as follows (with original spellings retained):

> 'The haytime and beginning of harvest were showery, yet more hindering than hurting; but the latter part of harvest in September was exceeding bad indeed. No grain could be carried for three weeks together; for it rained every day, and in great quantities. I never measured so wet a month before. The wheat and oats were chiefly got in before it, and a great deal of barley; yet, as it was a late harvest, there was a great deal of the barley out, some wheat, and almost all the beans and pease. The wheat through the severe and wet winter was all along thin, and much of it mildewed by the wet towards harvest. The crop of barley was not amiss, if it could have been all well got; but some of it suffered by the wet after it was cut. The beans and pease were a remarkable great crop till harvest; but almost intirely spoiled in it. There was a great deal of winter meat for the cattle this year, plenty of good grabs, a great deal of hay, and fine crops of turneps; but the straw of that corn, which was out in the wet, was spoiled.'

The estimated monthly total for Pode Hole, near Spalding, was 211 mm [127]; there it also remains the wettest September on that long record (from 1726), by over 50 mm.

## January 1776—Thomas Hornsby's wine begins to freeze

The low temperatures and snowfall of January 1776 were closely observed by both Thomas Hornsby in Oxford and Gilbert White in Selborne and, later in the month, in London. Heavy snow, driven by a strong wind, fell on 7th and there was further snow on several days following, particularly on 12th when Hornsby noted 'snow and wind' and White reported 'a prodigious mass overwhelmed all the works of man'. As frequently happens during a prolonged cold spell following a heavy fall of snow in midwinter, the cold intensified when skies cleared overnight and temperatures fell very low indeed [104, 128, 129]. It was common for Hornsby to increase the number of his observations at times of great interest and, instead of the usual three, he made eight observations on both 27 and 28 January and six, five and five on the following three days. The lowest temperature observed was 6 °F (−14.4 °C), at 1915h on 30 January (by the astronomical clock, or 7.15 a.m. clock time on the morning of 31 January). Earlier, on the evening of 26th, Hornsby had noted ruefully that 'wine beg[an] to freeze in my study'—tough times for an Oxford professor indeed (Figure 2.7)! At the time the temperature was recorded as $16\frac{2}{3}$ °F (−8.5 °C). White, who had travelled to South Lambeth on 22 January through 'a sort of Laplandian scene' reported temperatures of 11, 7, 6 and 6 °F (between −12 and −14 °C) for the nights of 28th to 31st at South Lambeth and 7, 6, 10 and 0 °F (−14 to −18 °C) at Selborne on the same nights. The lowest temperatures read by Hornsby on the same nights were 10.5, 14, 6 and 8 °F (−12 to −14 °C). Hornsby's lowest observed temperature of −14.4 °C early on 31 January 1776 still ranks as one of the lowest ever observed at Oxford.

Gilbert White gathered together the various passages from his journal about the very cold January 1776 into a single essay, which has been reprinted several times including in *Weather* [130]. After a very wet start to the month, it snowed from the 7th to the 12th. White was 'obliged to be much abroad' on the 14th and had '…never before or since encountered such rugged Siberian weather. Many of the narrow roads were now filled above the tops of the hedges…' In relation to Oxford, White's closest reference is to a group of people stranded in Marlborough on their return to London from celebrating the Queen's birthday at Bath:

> 'From the 14th, the snow continued to increase and began to stop the road waggons and coaches, which could no longer keep on their regular stages—and especially on the western roads, where the fall appears to have been deeper than in the south…Many carriages of persons who got in their way to town from Bath as far as Marlborough, after strange embarrassments here met with a ne plus ultra. The ladies fretted and offered large rewards to labourers if they would shovel them a track to London. But the relentless heaps of snows were too bulky to be removed.'

The very cold weather persisted to the end of the month, with some very low temperatures as noted above. Then, on 1 February, 'without any apparent cause', a rapid thaw took place. Amongst the bird populations killed by the cold, White especially lamented the partridges, so 'thinned' that few remained to breed the following year.

## 14 February 1795

From *The Gentleman's Magazine*, March 1795, [119] reference 13038:

> 'Oxford: We have not experienced so great a flood at this place and in its neighbourhood for 22 years. The waters both in the Isis and the Cherwell were swelled to an alarming height; many of the roads were so much inundated as to render them in many places dangerous, and in some impassable. In St Thomas's parish in this city in particular, a great many houses were mid-leg deep in water, and in some much higher; so that they passed from house to house in boats, and inhabited the upper rooms.'

This was almost certainly a snow-melt flood following on from the intensely cold January of that year, for the total precipitation in Oxford that month was only 57 mm, little more than the current monthly average.

## 29 January 1809

This event is included as one of the largest floods on record in much of the Thames Valley: the rainfall total at the Radcliffe Observatory for the month was 105 mm. From *The Oxford Guide* of 1818 (p 94): '*...the raised gravel walk in [Christ Church] Meadow was completely inundated by a sudden flood, great part of it was washed away, and the repairing it attended with great expense to the college*'. Griffiths [122] (BHS Chronology, record 12129) again quotes from the parish records at Whitchurch-on-Thames, between Oxford and Reading:

> 'In January 1809 there was a very high flood. The water reached up the street beyond Mr Simeon's back gates out of which it flowed with great rapidity. Part of the south and east wall in Mr Simeon's kitchen garden was beaten down, and it stood 3 ft 8 inches [112 cm] in his offices. The height of the water at The Parsonage is marked by a notch in the upper stone step: January 29, 1809. No service in church on account of the flood.'

A later entry (BHS entry 12130) adds: '21 April 1809—A very deep snow and another high flood in consequence of it.'

## 14 November 1813

After various false starts and breaks in the record, this day marks the start of the continuous daily weather record at the Radcliffe Observatory (Figure 25.4) and it is therefore fitting to be included in this chronology. It was a cool November day with outside ('without') temperatures recorded of 40 °F and 42 °F (4.4–5.6 °C). Notably, the barometer was falling quickly, while a west wind brought half an inch of rain (13 mm), a significant fall at Oxford, and 'very dark' conditions.

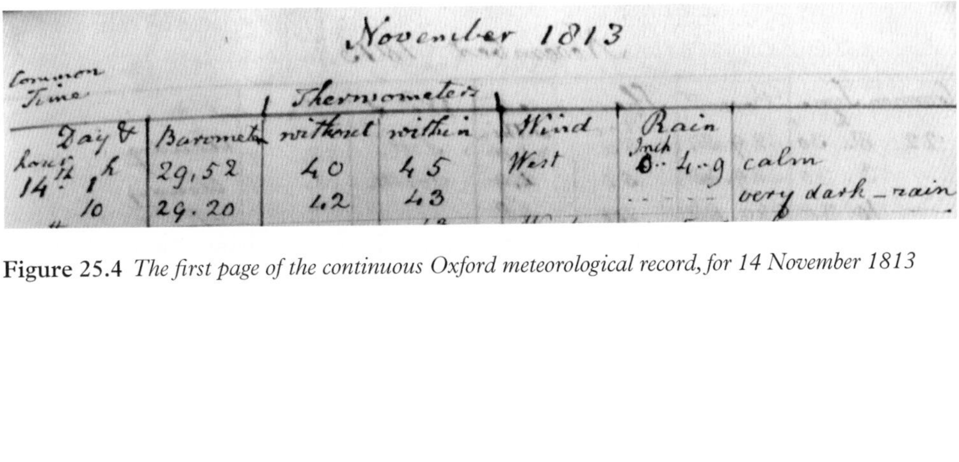

**Figure 25.4** *The first page of the continuous Oxford meteorological record, for 14 November 1813*

## 1816—Indonesian volcano perturbs Oxford's weather

The April 1815 eruption of Tambora volcano (Sumbawa island, Indonesia) resulted in profound climatic perturbations. This was probably the largest caldera-forming eruption of the last few centuries [131]. Anomalously cold weather is known to have hit eastern North America and Western Europe, but the eruption had global climatic impact [131–135]. The eruption injected ~60 Mt of sulphur into the stratosphere, six times the amount of the 1991 Pinatubo eruption, forming a sulphate aerosol veil in the stratosphere, causing a major disturbance of global climate. The year 1816 came to be known as 'the year without a summer' [134, 136]. Post [137] characterised the period 1816–19 as the last great subsistence crisis to affect the western world: 1816–17 witnessed the worst famine in over a century. The United Kingdom, still recovering from the long effects of the Napoleonic War, suffered badly: crop failures were widespread and the famine was followed by a major typhus epidemic affecting almost every town and village in England [137]. At the same time, the economy was stagnating, as over 400,000 men from the armed services re-entered the labour market and the country had failed to recapture European markets. Mass reaction to the dire circumstances included food riots, looting of granaries and arson (see [138] for a detailed account).

Summer 1816 saw a weak Azores high and a strong Icelandic low [131]. Britain and the rest of western Europe were affected by anomalous northerly and north-westerly airflow, bringing cooler temperatures [135]. This was one of the coldest summers over much of western Europe from central Scandinavia to the Mediterranean. Veale and Endfield [138] provide a detailed review of the UK's weather in 1816 and the surrounding period; they summarise 1816 itself as cold, very wet and sunless with frequent strong winds and storms. Unusually cold weather continued well into the spring, with snow falling across the Midlands and Wales on 12 May; no snow is mentioned in the Radcliffe ledger for that day, but it was clearly cloudy and windy with 'flying showers' and a maximum temperature of only 8 °C, decidedly chilly for mid-May.

The records of the Radcliffe Observatory show the climatic impact of Tambora very clearly. The year 1816 remains the second-coldest on record at Oxford (mean air temperature 7.8 °C), beaten only by 1879 (7.7 °C), while the summer of 1816 is the equal-coldest on record (14.2 °C) with 1860. Spring 1816 remains the equal fourth-coldest on record and spring 1817 the fifth coldest. The significance of the cold 1816 spring may have been overlooked in the past: if the start of the growing season is significantly delayed, then, in combination with a very cold summer, crop failures can only be expected. Using a crude estimate of growing season at Oxford based on a 5.5 °C threshold, 1815 is estimated to have had 235 growing degree days but 1816 only 202, with a recovery to 240 growing degree days in 1817; 1822 a rather warmer year had 301 growing degree days. In comparison, the average year over the period 1981–2010 accumulates around 286 growing degree days.

Up to and including January 1817, 20 consecutive months experienced below-average temperatures; this is not a record at Oxford (incredibly, 45 consecutive months between and including July 1836 and March 1840 were all below the current average temperature

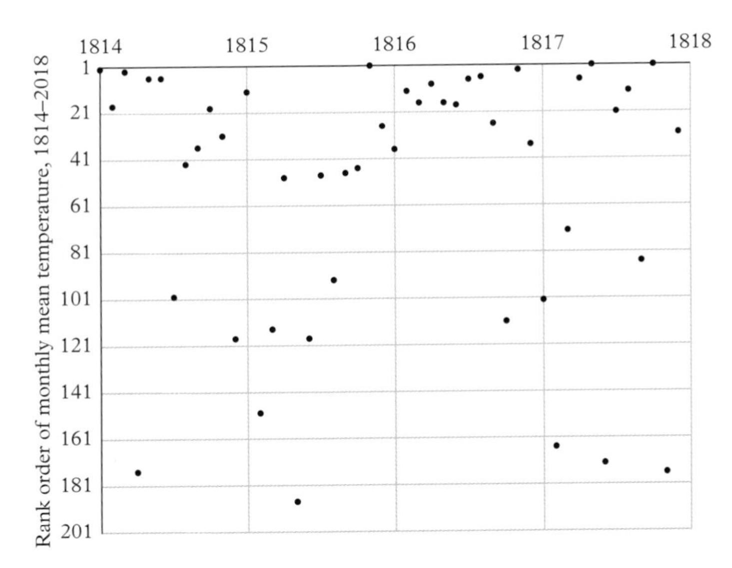


**Figure 25.5** *Monthly temperature rankings in 204 years for the period 1814 to 1817 at Oxford; low numbers indicate the coldest months of that name on the record*

for each month) but a calamitous deficit under the circumstances. The year 1816 was regarded as wet by those who experienced the very cold weather; summer was wet (209 mm fell at the Radcliffe Observatory, well above the current average of 153 mm, the 65th wettest summer to 2018) but it was probably the combined frequency and persistence of cold, rainy days that persisted in the memory. Spring 1816 (121 mm) was drier than average, while summer 1817, another relatively cold one (15.0 °C), was even slightly wetter than 1816 (215 mm). The 1816/17 winter was milder, with an enhanced North Atlantic Oscillation, enhanced westerlies and a strengthened polar vortex [131]. Figure 22.2 shows the rank of each month in the Oxford temperature series from 1814 to 1817 (1 = coldest, $n$ = 204). Other than January 1814, 1814 and 1815 show a mix of rankings, whereas the second half of 1815 and all of 1816 have rankings below 50, many in the bottom 20, a clear and significant anomaly compared to what might be expected in a normal year. The year 1817 shows more variability but still two notably cold months—May ranking coldest on record, October second-coldest.

Finally, note that, in relation to the Central England Temperature (CET) record since 1659, 1816 is the ninth coldest year on record, and only 1879 has been colder since. Of the other seven years colder than 1816, six were in the seventeenth century and only 1740 from the eighteenth century, a notably harsh winter followed by a very cold spring. Summer 1816 is the third-coolest in the CET record (13.4 °C), behind only 1695 (13.2 °C) and 1725 (13.1 °C). Figure 25.6 shows CET mean summer temperatures since 1659 together with an unweighted thirty-year running mean. The 1810s stand out as a cool decade; this was likely to have been a cooler period anyway [131, 138], made significantly worse by the aftermath of Tambora. No Central England summer has come within a degree of being so cold in recent decades. Note that 1814 was also very cold, with the

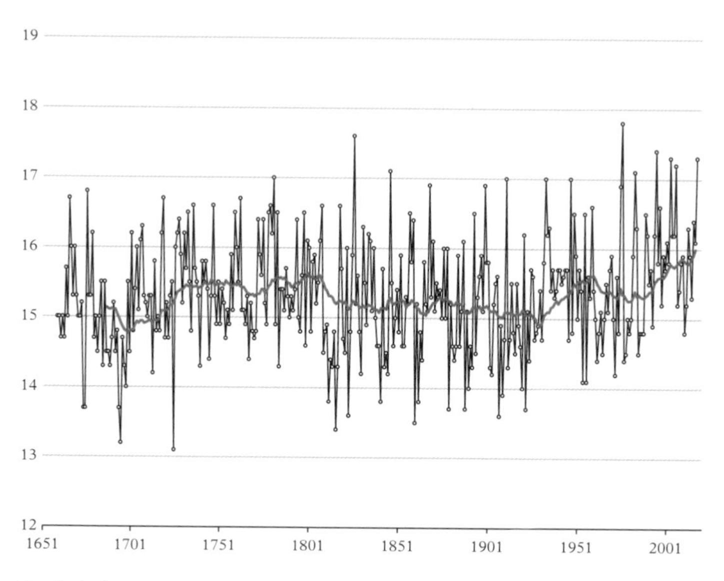


**Figure 25.6** *Manley's Central England Temperature (CET) mean summer (June–July–August) temperatures °C 1659–2018, together with a 30-year unweighted running mean (Met Office Hadley Centre)*

second-coldest January at Oxford on record (see above). Raible *et al* [131] mention evidence for an unknown eruption in 1808/09, followed by a period of negative temperature trend.

## 19 November 1850

The limited nature of the early observations is well illustrated by the 'great storm' of Tuesday 19 November 1850. Elsewhere, the part-constructed Crystal Palace was severely tested by high winds but survived [139], whilst on the west coast of Ireland 98 people lost their lives when the passenger vessel *Edmond* was wrecked in Kilkee Bay (Wikipedia entry). At Oxford, winds were gentler: nothing greater than Beaufort Force 4 or 5 was noted in the ledger and there is no mention of high winds in the Remarks. Perhaps Oxford missed the worst of the weather, but an anemograph record would have been interesting nevertheless. It is notable that Hubert Lamb did not include this event in his list of the greatest storms to hit the British Isles.

## 11 November 1852

November 1852 was a very wet month at the end of a very wet year, the third-wettest year on record at Oxford since 1767 and the second-wettest November. A total of 32.3 mm was recorded for 11 November 1852 with four daily totals later in the month exceeding 12 mm. By December, Christ Church Meadows were so badly flooded that it was possible to sail small boats across (see Figure 17.1 in November's Chapter) [140]. In Reading, the Thames was highest on 17 November [102], the so-called Duke of Wellington's flood (his funeral was held in London the following day, when the flood peak was close to its highest). A report stated: 'No parallel flood has occurred since 1841, and none exceeding it except 1809.' (There is a brief note on the 1809 flood above.) The River Kennet was at its highest on 28 November. Given the very wet period that had gone before, June and August in particular, it is little surprise that heavy and prolonged rainfall in November produced major and protracted flooding.

Elsom [140] goes on to report that in November 1852 the railway line south of the Abingdon Road railway bridge was flooded. Ballast supporting the railway sleepers was washed away and this led to a railway engine coming off the rails. Following this flood, the Great Western Railway company decided to raise the rails by 35 cm. However, history repeated itself on 15 November 1875, when both lines of rails, still held together by the fastenings of the sleepers, were swept bodily away from their position. Once more the railway lines were raised, only to find in November 1894 the rail services had to be suspended for seven days because the ballast was washed from beneath the rails. The November 1894 flood is reported below.

## The very cold Christmas of 1860 in Oxford

The lowest air temperature yet recorded in Oxford was −17.8 °C, on Christmas Eve 1860. We know from the published records that the thermometers were read to a precision of 0.1 degrees Fahrenheit (and any required thermometer calibration corrections were applied before publication), and the minimum that morning was 0.0 °F.

The first half of December 1860 was uneventful, even mild, the temperature reaching 11.7 °C on 6th. Conditions became colder from mid-month, with a 10-hour snowfall noted on 19th, and several days failing to reach 0 °C between 20th and 28th, during which it would be reasonable to assume the persistence of a snow cover, although routine daily records of snow cover/snow depth were not kept at that time. Conditions on 23 December were noted as 'Fair, with persistent fog', the maximum temperature reaching only −1.1 °C. The temperature fell rapidly overnight to the minimum of −17.8 °C, although by 10 a.m. on 24th the temperature had risen to −13.3 °C. The weather on 24th was noted as 'Fine till sunset—immediately after which a thick fog arose'.

There are surviving manuscript notes in the original observation registers which confirm the veracity of the reading. Under the initials *JL* (John Lucas, the experienced Assistant at the Observatory at the time) the following comment was entered:

> 'In looking at the minimum readings Mr Main thinks it should be 5.0° [°F] instead of 0.5°. The readings as entered are correct as I was especially careful in reading and examined the readings after entering.'

Clearly, the experienced Mr Lucas was taken at his word, for 0.0 °F was duly entered as the day's minimum temperature (once the 0.5 degrees Fahrenheit calibration correction to the minimum thermometer was included) and subsequently published. The weather continued very cold. Four days later came another very severe frost, the temperature falling to −16.7 °C (1.9 °F) on the morning of 28 December, before milder conditions returned as southerly winds set in on 29th, accompanied by heavy rain. By 31 December the temperature had risen to 7.5 °C, and a historic cold snap was over.

## 4 July 1862—*Alice in Wonderland* Day in Oxford

The story of *Alice in Wonderland* was first told on a boating expedition from Oxford to Godstow, on the River Isis, on this date [141]. Present were Lewis Carroll (real name Charles Lutwidge Dodgson), his friend Robinson Duckworth, who did the rowing with him, and of course the real Alice, Alice Liddell and her two sisters, the children of the Dean of Christ Church.

They left Oxford after lunch and did not get back until after eight that evening. Alice Liddell and Robinson Duckworth left details of the afternoon's weather—Alice spoke of the 'burning sun' and 'that blazing summer afternoon with the heat haze shimmering over the meadows where the party landed to shelter for a while in the shade cast by the haycocks near Godstow', while Robinson referred to 'that beautiful summer afternoon…described in the introductory verses to the story' and ever since immortalised in this perennial favourite children's book.

Sunshine records at the Radcliffe Observatory did not begin until almost 20 years after this particular boat trip, but we do know that the morning started cloudy (10/10 cloud at the morning observation) and that the afternoon turned out fine, with light south-southwesterly winds and a maximum temperature of 19.9 °C. The sunshine probably turned hazy later in the afternoon and evening, as high cloud thickened in advance of a warm front which brought rain after 2 a.m. next morning; 4.2 mm had fallen by 8 a.m.

## 1879

The year 1879 was the coldest on Oxford's records: within Manley's Central England Temperature series [35], only 1695 and 1740 are colder. If 1695 was not a surprise, being part of a cold decade, 1740 certainly was, two degrees colder than the year before. The year 1879 was equally a surprise, 1.6 degC colder than the year before in Oxford, where mean temperatures were below the current normal for every month from July 1878 to January 1880 inclusive (19 consecutive months). January 1879 remains the tenth coldest January since 1814 (mean temperature -0.7 °C) and the winter of 1878/79 (which includes December 1878, of course) ranks equal eleventh coldest. There followed the second-coldest spring, equal-second-coldest summer and eighth equal-coldest autumn. To round off the year, December 1879 remains the sixth coldest December on record, even colder than the previous December (which ranks equal tenth coldest): the air temperature fell as low as −11 °C on both 6 and 7 December. In some years, a poor winter is followed by some good weather later in the year—1947 is perhaps the best example—but there was no such respite in 1879. The 1870s were a wet decade and 1879 was fully a part of that pattern, the 38th wettest year since 1767 with 758 mm of rainfall and even wetter than 1872 (747 mm), a notably wet year right across the British Isles [79] and only exceeded in the 1870s by 1875 and 1876. Cold and wet must have made for a very depressing year. Elsewhere, the best known weather-related event in 1879 was the collapse of the Tay bridge near Dundee on 28 December 1879 during a severe gale [142], evidently a 'bomb' depression as we would now understand the term, judging from the Meteorological Office's *Daily Weather Report*.

## November 1894—One of the greatest Thames floods

By late autumn, soil moisture deficits have usually been restored and heavy rain makes flooding likely. This was very much the case in November 1894, when 83 mm fell in the four days commencing on 11 November (Figure 25.7). This spell included two of the five wettest November days in almost 200 years of records, only two days apart (see November, Table 17.3), namely 33.3 mm on 12 November and 37.9 mm on 14th (the latter still the wettest November day on record at Oxford). For the Thames catchment as a whole, accumulated rainfall totals for mid-November 1894 rank as the highest in the Thames daily catchment series over spans of 2 to 7 days. This is not quite the case for Oxford, where July 1968 and July 1901 come at the top of the list of accumulated totals; nevertheless, the 3-day total for 12–14 November 1894 ranks ninth on Oxford's records: it is no surprise that the flood response was very large, given the season. The sustained heavy rainfall generated the greatest recorded flood on the River Thames since records began in 1883; the highest recorded mean daily flow for the Thames at Teddington remains that for 18 November 1894 at 800 cubic metres per second ($m^3/s$) (see Figure 25.8) [97].

**Figure 25.7** *Flooding on the railway near Oxford during the great flood of November 1894 (© Oxfordshire History Centre HT 1796)*

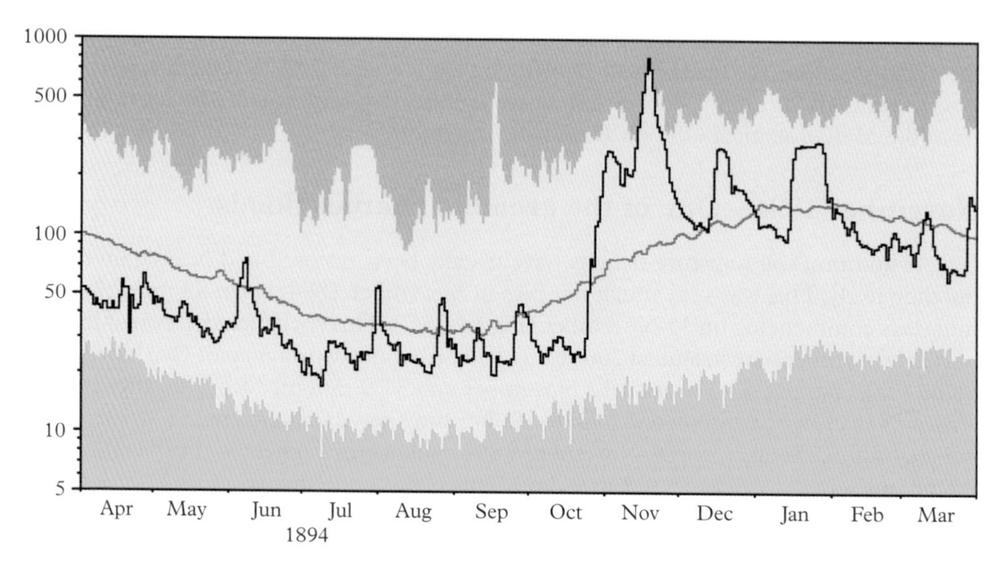


**Figure 25.8** *Thames flood hydrograph for Teddington showing the record Thames flood of November 1894. River flow hydrographs show the daily mean flows (bold black line, units m³/s), together with the long-term daily average flows in grey and maximum and minimum daily flows since the commencement of the record (1883), shown by the blue and pink shaded areas, respectively. Daily flows falling outside the maximum/minimum range are indicated where the bold trace enters or delineates the shaded areas. The dashed line represents the period-of-record average daily flow: the featured flows are 'naturalised' (adjusted to account for the major abstractions upstream to meet London's water supply needs—a relatively modest adjustment in 1894 but, in some recent dry years, the abstractions account for more than 40% of the flow (Data from National Rivers Flow Archive – NRFA courtesy Terry Marsh, CEH Wallingford)*

## 24–25 April 1908—Oxford's heaviest snowfall of the twentieth century

Oxford and much of the south of England were affected by an exceptional and unseasonal deep snowfall in Easter week, in late April 1908. The preceding weeks had been cool but dry, with winds frequently northerly or north-easterly, although the temperature had reached 16.3 °C on 9 April. The weather turned colder from 23rd, with the Radcliffe Observatory records noting 'Rain 8 a.m. to noon, sleet to 2.30 p.m. then snow' on 23rd. Snow fell during the afternoon and night and at 8 a.m. on 24 April the mean depth was noted as 9 cm. Aside from a few shaded areas, this melted away during the afternoon of 24 April; the maximum temperature that day reached 5.3 °C. There were occasional snow showers the following day (25th) and the ground was just covered with snow once more at midnight. The snowfall resulted from the passage of a small depression from North Wales to the English Channel on 24 April, after which the cyclonic centre made an abrupt turn to the left and crossed south-east England during 25 April [143, 144]. In Oxford, snow fell throughout the night and the following

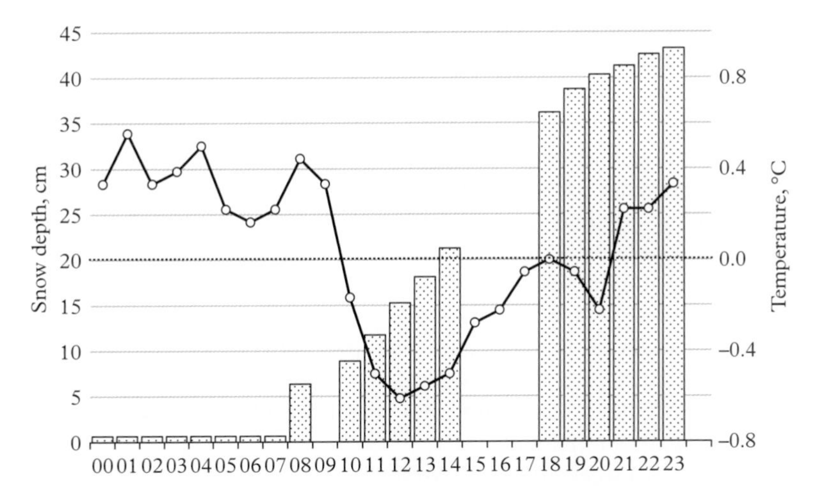


**Figure 25.9** *Hourly temperatures (solid line, right axis, °C) and snow depths (columns, left axis, cm) measured at the Radcliffe Observatory, Oxford, on 25 April 1908, the greatest snowstorm of the twentieth century in Oxford*

day, ceasing only after 24 hours, by which time it lay 43 cm deep at the Radcliffe Observatory; the total precipitation (melted snowfall rainfall equivalent) on 25 April was 46.7 mm, which remains to this day the greatest daily 'rainfall' measured in April in Oxford. Corroboration of this astonishing fall is given by the reported depth of snow at another location in Banbury Road, Oxford, of 43–46 cm [143]. The maximum temperature on 25 April reached only 1.3 °C—the hourly temperatures and snow depths are shown in Figure 25.9. A selection of contemporary photographs can be found in Chapter 10, April.

The Radcliffe observers made careful notes of the snowfall as it happened. The previous day (Friday 24 April), there were occasional snow showers between 1530 and 1900. Snow recommenced about 2230, and was falling lightly at 2300 and midnight, when the ground was 'just covered'. Snow was still falling at 0400 h on Saturday 25 April, and thereafter snow fell continuously throughout the day, heavily at times, until about 2230. The depth of snow was measured almost every hour as shown in Figure 25.9; at 1000 h the depth was 9 cm, at noon 15 cm, at 1400 21 cm, at 1800 36 cm and at 2100 41 cm. This was no gentle fall, either—throughout the 24 hours the wind averaged 6.9 metres per second (13.4 knots); at 1700 the wind was north-westerly, hourly mean 13 metres per second (25 knots). At 2130 the snow lay 42 cm deep, and snow continued to fall lightly for another hour, after which the snow reached its greatest depth of 43 cm. The observation registers record that in many places the snow lay almost 50 cm deep, with drifts over 60 cm in the field south of the Observatory building.

At 0800 h Sunday 26 April, the snow had settled somewhat, to a mean depth of 41 cm, and a 'rapid thaw' began after sunrise.

This was deepest snowfall in Oxford since 13–14 February 1888, when the depth reached 61 cm, and has not been exceeded since. The observer's notes recall 'Much damage was done to the laburnum, box and other trees in the shrubbery around the south grounds of the Observatory, whilst the south cedar had a large limb stripped off'. The rapid thaw continued throughout Sunday 26 April, which was sunny (10.3 hours of sunshine) and much milder (maximum 9.6 °C), with the wind now in the west-south-west. Tremendous flooding resulted on the Thames in the days following as the deep snow quickly thawed; at Maidenhead the Thames reached its highest level since June 1903 on 30 April. The Observatory notes record that, at noon 28 April, 'Port Meadow has become very much flooded since yesterday', while at 6 p.m. '...floods rising rapidly'. At noon on 29 April 'The floodwater on Port Meadow has risen considerably and extends, with the exception of a few small patches, to beyond Fiddlers Island'. At 6 p.m. that day 'The path leading from Medley to Tumbling Bay and from Osney Lock to Folly Bridge is covered with floodwater in many places.'

By 1 May, less than a week after this extraordinary snowstorm, the weather was 'very fine and warm', and the temperature had risen to 23.6 °C. On the following day, another fine, very warm and sunny day, '...a punt containing four undergraduates was upset in the flooded Cherwell by being driven against a pier while attempting to get through Magdalen Bridge, and one of the occupants was drowned.'

## 9 June 1910—Record-breaking hail and rain at Wheatley

The period 5–10 June 1910 was one of a series of remarkable thunderstorms day after day, chiefly in the south of England [92], covered in detail in *British Rainfall 1910*. As with most thundery activity, some places had little or no rainfall while others just a few kilometres distant suffered torrential rain and destructive hailstorms. In Oxfordshire, both 7 and 9 June saw storms of remarkable ferocity. On the first of these dates, heavy showers accompanied by thunder and lightning took place at frequent intervals from before daylight until after dark, the heaviest of the rainfall occurring in a period of 4–5 hours during the evening; at Churchill School (5 km south-west of Chipping Norton) 108 mm of rain was recorded. Nearby at Swerford, the observer reported seven separate thunderstorms (three directly overhead) during this extraordinary day, culminating in a prolonged and alarming storm from 2045 to 2330h; thunder was heard during 12 hours of the day. There were many damaging lightning strikes in Oxfordshire: Deddington parish church was struck, with damage to the pinnacles, and houses were struck in Kingham, Lyneham, Churchill, Deddington, Hook Norton, South Newington, Ducklington (Witney) and Boars Hill (Oxford). At Chipping Norton two hayricks were struck and set on fire, while farm livestock were killed by lightning at Great Rollright, Hook Norton and Salford.

Even worse was to come on 9 June, in severe thunderstorms that occurred between Reading and Oxford in the early afternoon. Much of what follows is taken from an anniversary account of the storms by Jonathan Webb [92] published in *Weather*. At Wheatley the deluge commenced at 1242h, while at nearby Waterstock the storm began at 1300h with a *hurricane* (sic) followed by the hailstorm which lasted from 1315 to 1415h; rain continued until 1530h though most of it fell before 1500h. At Pyrton Hill, pioneering meteorologist W. H. Dines wrote '...thunder was first heard soon after noon and was practically continuous from 1230 to 1600h, very severe from 1400 to 1500h. Nearby, three trees were struck by lightning.' At the Radcliffe Observatory, heavy thunderstorms were reported from 1200 to 1600h and 15.5 mm of rain fell*. Just east of Oxford, however, the rainfall was much heavier. There were at the time two raingauges in Wheatley; the gauge at School House recorded 132 mm (including a carefully measured 110 mm in one hour, which remains a British record to this day), while another at Holton Cottage received 127 mm. Although both gauges support each other's enormous totals, there are caveats to both readings since the former was from a non-standard gauge and the latter from a gauge in an unorthodox exposure. A gauge in Waterstock recorded 100 mm before becoming choked by hail, and the true fall was probably similar to those in Wheatley; the force of the floodwater here was sufficient to uproot a large elm tree. The most destructive hail was at Waterstock (Figure 25.10). Hailstones the size of marbles fell in Wheatley but were the size of walnuts (about 32 mm diameter) at Waterstock, where hail wrecked glasshouses and frames. Here hailstones lay to a depth of 10 cm

**Figure 25.10** *Hail at Waterstock, just east of Wheatley, on 9 June 1910.* ***Left:*** *Hail on a lawn at Waterstock at 4.15 p.m., an hour after the fall had ceased.* ***Right:*** *Hail in Waterstock village street at 6.15 p.m., three hours after the fall ceased (Photographs by 'Miss Ashurst', from* British Rainfall 1910*)*

* A very thundery and wet four-day spell in Oxford; three days in four recorded 'thunder heard' with 15 mm or more of rainfall on each. The four-day total was 48 mm, about the normal for the entire month of June.

during the storm, with drifts of hailstones more than a metre deep piled up by floodwater. Plants were reduced to bare stalks with the complete devastation of vegetation and glass remarked upon by a letter in *The Times*.

The flood and damage at Wheatley was described in the *Henley Chronicle* of 17 June 1910. Floodwater flowed 45–90 cm deep in most streets of the village, the water carrying earth and debris washed down from the surrounding small hills—as well as many household goods, live ducks and geese. The main street became a raging torrent within 30 minutes with many houses flooded to depths of 45–90 cm; one woman was rescued from drowning. In Crown Street, where water was 180 cm deep, tenants in one dwelling escaped via a bedroom window. A wall was washed away at the west end of the village, a 180 cm gulley was scoured in the meadow turf and a hayrick weighing nearly two tonnes was bodily moved by the force of the floodwater. The village Post Office was inundated, cutting communications. The *Oxford Journal Illustrated* published photographs showing the peak flood-tide marks in Crown Street (with a resident pointing to one at 180 cm), the displaced hayrick and damage to roads which had been under 30–60 cm of debris. Numerous incidents of lightning damage were reported, some involving serious injuries: a woman was struck in Wheatley and lost the use of her arm for several days, while in Farthinghoe, north Oxfordshire, a man was struck in a horse-driven cart: he was then thrown out of it as the startled horse fled.

## 1911

At the time, this was the warmest year on record at Oxford, equally as warm as 1868 (mean air temperature 10.8 °C). It was the hottest summer, and the first time that any summer had exceeded 18 °C, since 1826. Like 1838 and 1953, there was a Coronation Day (King George V and Queen Mary), on 22 June, but most of the very warm days came after this, with maximum temperatures always above 20 °C from 3 July to 22 August. The maximum temperature reached 34.8 °C on 9 August, at that time the highest temperature ever recorded at Oxford (surpassing 33.9 °C on 19 July 1825), and only twice surpassed since. The record stood until 19 August 1932, when 35 °C was reached for the first time; since then, only 3 August 1990 has equalled this extreme. As is typical of a warm year at Oxford, 1911 was also a dry year, a total of only 538 mm, which ranks 55th driest since 1767.

## 1921

The record-equalling mean air temperature of 1911 stood for only ten years, when 1921 became the warmest year on record at Oxford with a mean air temperature of 11.1 °C, the first time 11 °C had been surpassed. Since 1921, 22 more years have reached or exceeded 11 °C but, 1949 apart, all have occurred since 1989; indeed, 14 are in the present millennium, including the warmest seven years. Calendar year 1921 was also very dry over most of England and Wales and remains the driest on record at most sites with less than 200 years' record: at Oxford just 381 mm fell. Only calendar year 1788 has been drier (337 mm), although only 297 mm fell in the 12 months commencing 20 September 1975 (Table 28.3).

## 19 August 1932—Oxford's hottest day

The summer of 1932 was altogether unexceptional—indeed, July was dull and wet—apart from a very short but intense heatwave 18–20 August [145]. There had been a week-long hot spell earlier in the month culminating on the 11th (30.3 °C at the Radcliffe Observatory), followed by much cooler weather from the 13th to the 15th—Oxford's maximum temperature was just 18.7 °C on 15th (see Figure 14.1). By the 16th, high pressure extended from the Atlantic across southern parts of the British Isles to the Low Countries and, over the next few days, a slight northward movement of a depression over Spain and the Bay of Biscay established a warm southerly gradient over England and Wales. In already warm air, abundant sunshine (undoubtedly aided by anticyclonic subsidence) raised temperatures in the south-east of England to the highest since August 1911, although in some places those were surpassed. The hottest day in Oxford was the 19th, which reached 35.1 °C—just topping 34.8 °C recorded on 9 August 1911—a value which was equalled in August 1990 but has yet to be exceeded. In Essex and in London, 36 °C was reached. Figure 25.11 shows the highest temperatures recorded during August 1932.

The break in the short, sharp heatwave came with the incursion of much cooler air from the north-west and was marked by spectacular thunderstorms and locally intense rainfall. At Oxford 51 mm fell during violent thunderstorms on 20th/21st, while the maximum temperature of 35.1 °C on 19th was followed by 25.5 °C on 20th (after an overnight minimum of 17.8 °C with heavy thunderstorms) and just 18.2 °C on 22nd and 23rd.

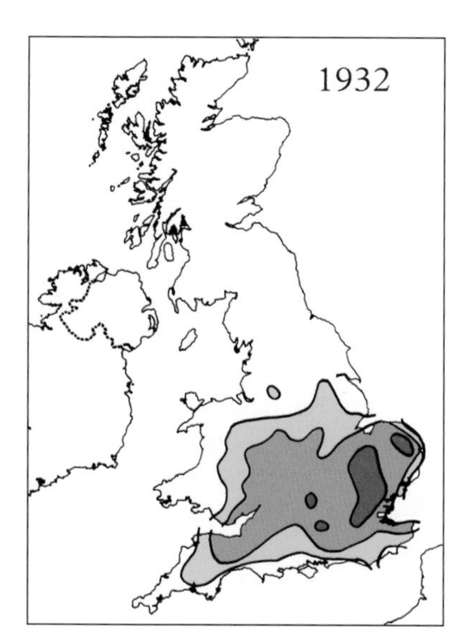


**Figure 25.11** *Maximum temperatures in August 1932, mapped from the* Monthly Weather Report *of the Met Office. For clarity, only temperatures above 30 °C are shown. Isotherms are at 30 °C (yellow), 32 °C (orange) and 35 °C (red). Figure 10c from Burt 2004 [145] (Stephen Burt)*

## March 1947

As noted in Chapter 9, March 1947 saw probably the most severe flooding of the Thames Valley in the twentieth century, a result of the sudden thaw of melting snow coinciding with still-frozen impermeable soil conditions. The Thames flood of 1894 has already been discussed above, the largest recorded flood at the Kingston gauging station (NRFA 39001); this was simply the result of heavy and persistent rain. The Day's Weir gauging station just downstream of Oxford was only opened in 1938 so we have no direct comparison with Kingston for 1894. However, we can compare flood responses at the two sites in March 1947 for this rain-on-snow flood event. At Kingston, the second-highest flood peak ever recorded (709 cubic metres per second, $m^3/s$) was on 20 March 1947. The highest daily flow ever recorded for the Thames at Day's Weir (NRFA Station 39002) was 349 $m^3/s$ on both 17 and 19 March 1947, with 346 $m^3/s$ on 18 March. The eight highest daily flows yet recorded at Day's Weir are all in the period 14–21 March 1947 (see Figure 9.2 in March, Chapter 9). The impact of flooding was made much worse by coming at the end of a prolonged cold and snowy winter during a time of post-war austerity, although, as noted elsewhere, a very warm summer in 1947 provided a welcome counterbalance to the winter cold and floods.

## 2 June 1953

The Coronation of Queen Elizabeth II took place on 2 June 1953. News reports always comment on the disappointing weather that day. At Oxford, the maximum temperature was only 12.2 °C, the equal 35th coldest June day on record since 1853 ($n = 4950$). Still, it was rather better than the following day when the maximum only reached 10.0 °C, the second-coldest June day on record; only 4 June 1909 has been colder at 9.8 °C. On the 2nd, 2.3 mm rain underlined the lacklustre weather of Coronation Day. By contrast, Queen Victoria's Coronation Day on 28 June 1838 was completely dry, with a maximum temperature of 19.2 °C (Figure 2.9).

## 16 October 1966

A strong tornado (TORRO scale F4–5) affected parts of Oxford on Sunday 16 October 1966. A full account of this event, which saw five houses damaged beyond repair, was given by Terence Meaden in the *Journal of Meteorology* [146]. The first damage was at Coppock Close in Headington Quarry, where a chimney stack was blown down and tiles torn from the roof. The tornado also caused damage in Pitts Road, where it lifted the roof from a garage and threw a summerhouse to the bottom of the garden. It then crossed the west side of the Headington roundabout, levelled a fence at the Fox and Hounds and raced through the Barton Estate towards Stow Farm. Several brick houses had their roofs damaged, and then the tornado struck the 20-year-old prefabricated houses in Mather Road and Stowford Road, where two were destroyed and four severely damaged and later demolished, leaving five families homeless. The length of the damage track was about 1.2 km, but the width only about 50 m. The storm was only 4 km from the Radcliffe Observatory, where there was a sharp change of wind direction from east to south-west coincident with the onset of heavy rainfall; 12.7 mm fell at the site.

## 10 July 1968

This was, and remains, the wettest day in almost 200 years of daily rainfall records in Oxford; 87.9 mm fell at the Radcliffe Observatory in the 24 hours commencing 0900 GMT on 10 July. The high value is confirmed by other records in the Oxford area—101.6 mm fell at Farmoor, 93 mm in Norham Road and 87.6 mm at Osney Lock. Somewhat larger totals were recorded further north and west in the Cotswolds (101.6 mm at Chipping Camden and 122 mm at Sudeley Castle). In Oxford, the Radcliffe Meteorological Station weather diary noted that the day was 'overcast throughout with heavy showers alternating with longer spells of continuous heavy rain; thunderstorm with heavy rain at 1350; heavy rain, easing off to moderate, continuing into the early morning of 11 July'. Figure 25.12 shows the hydrograph for Day's Weir for Water Year 1968: only 2007 has since seen higher flows in July. The peak daily gauged flow on 14 July (189 m³/s) ranks 84th highest since the record commenced in October 1938 ($n$ = 28,490) and was the highest for the calendar year. Aside from July 2007, this is the only occasion that a summer flow has provided the annual peak.

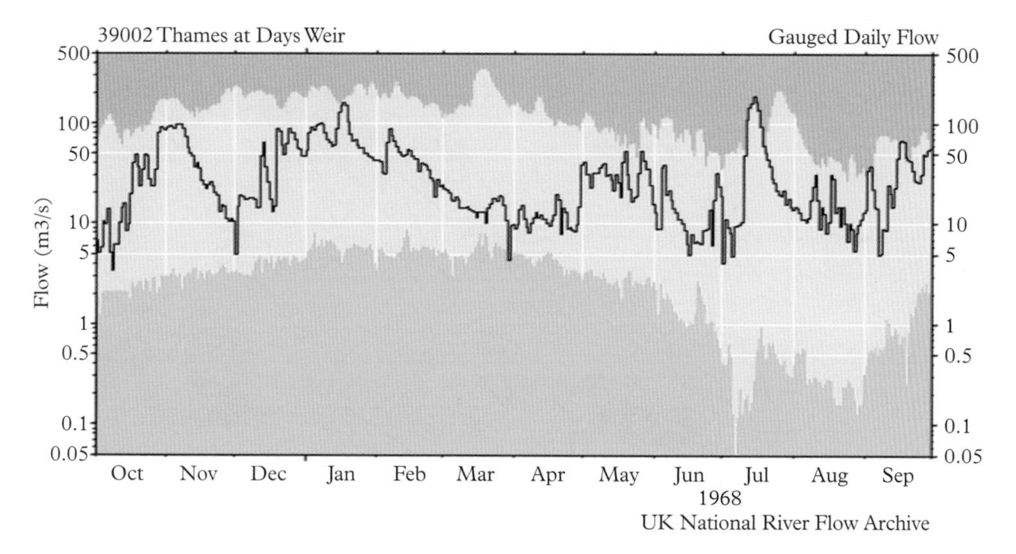


**Figure 25.12** *Thames flood hydrograph for Day's Weir, near Dorchester-on-Thames in Oxfordshire, for Water Year 1968, showing the greatest recorded summer flood on the Thames following the heavy rainfall on 10 July, the wettest day yet recorded at the Radcliffe Observatory. River flow hydrographs show the daily mean flows (bold black line, units m³/s), together with the long-term daily average flows in grey and maximum and minimum daily flows on record since 1938 (shown by the blue and pink shaded areas, respectively). Daily flows falling outside the maximum/minimum range are indicated where the bold trace enters or delineates the shaded areas. The dashed line represents the period-of-record average daily flow (Data from National Rivers Flow Archive – NRFA courtesy Terry Marsh, CEH Wallingford)*

*British Rainfall 1968* (p 160) summarised the synoptic situation as follows:

> 'Winds NE light to moderate becoming N to NW; depression, frontal, thundery; warm-sector depression occluding slowly and moving east-north-east over southern part of England.'

The Met Office issued the following forecast with its *Daily Weather Report* at noon on 10 July:

> 'High pressure has persisted to the north-west of the British Isles, and during the day a ridge extended down the North Sea to Germany. The warm front over France, with Mediterranean air to the south of it, has moved north to the English Channel associated with a thundery low over Biscay. The high pressures will continue to maintain mostly dry

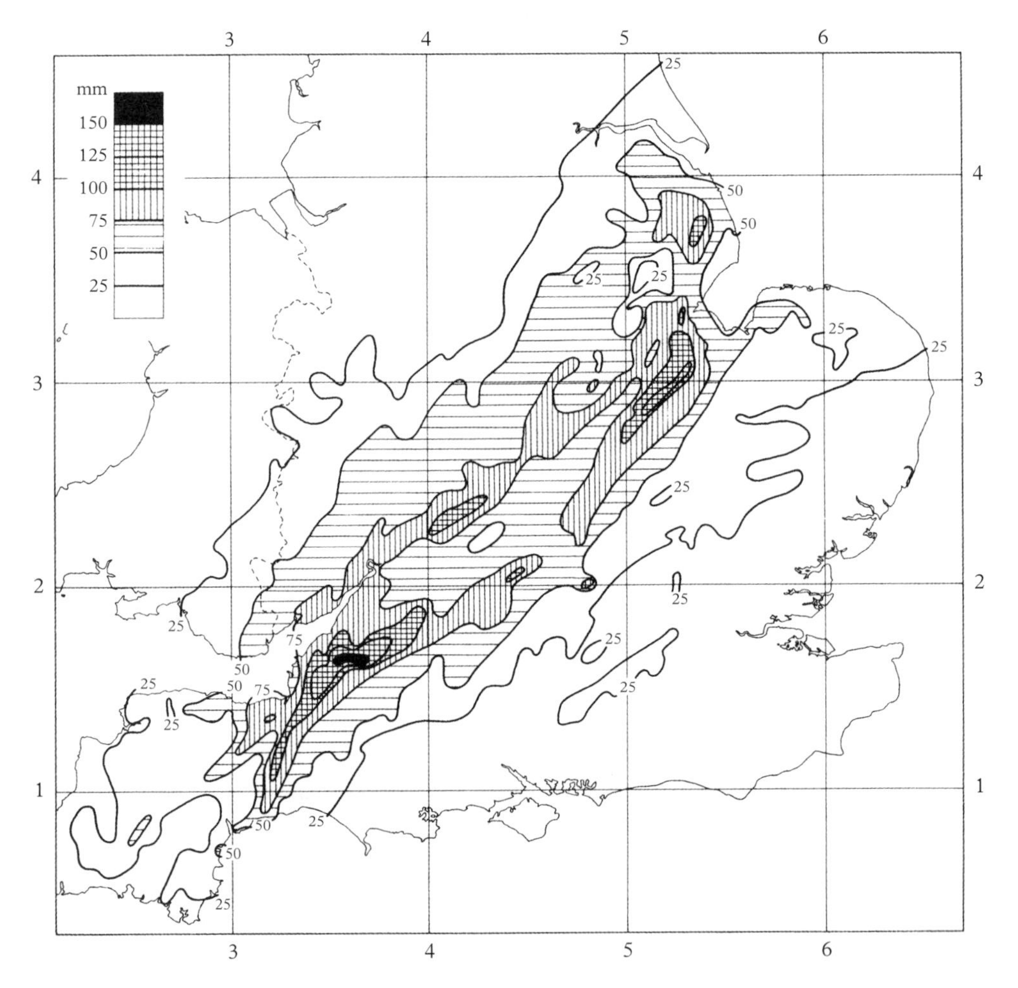


**Figure 25.13** *Rainfall (mm) for the 24 hours commencing 0900 GMT on 10 July 1968, Oxford's wettest day on record (from* British Rainfall 1968; *Crown Copyright)*

weather in the northwest, but a complex low-pressure system will spread thundery outbreaks to much of England and Wales.'

The forecast does not quite anticipate the severity of the storms, which proved much greater than expected, and the records show that this might be accounted for by factors hardly predictable at the time [147]. More recent experience coupled with much more sophisticated forecasting capability would now likely identify a rapidly deepening and occluding depression in the Bay of Biscay, moving quickly north-east over southern and eastern England. The storm was driven by a jet stream that had intensified from a normal zonal track above the British Isles to a more southwest-northeast track. (This rapid evolution is not dissimilar to the 'Great Storm' of 16 October 1987, where wind, rather than rain, was the problem).

Figure 25.13 clearly portrays the track of the depression from Lyme Bay to the Wash. The heaviest rainfall occurred just to the lee of the Mendip Hills, where several sites recorded daily totals in excess of 140 mm: continuous heavy rain was interspersed with violent downpours, with rainfall intensities reaching 50 mm per hour at times. The highest daily total, 173 mm, was recorded at an automatic raingauge at Chew Stoke [148]. Not surprisingly, there was major flooding in Bath and Bristol, and at Pensford the A37 bridge was washed away. Seven people died and thousands had to leave their homes as flood water up to 3 m deep flowed through the streets. On Mendip itself, caves were flooded while the extensive dry valley system functioned as a normal river channel network, causing extensive erosion: water flowed strongly down Cheddar Gorge at least a metre deep.

## 22–26 August 1986

Hurricane *Charley* was the third tropical storm and second hurricane of the 1986 season [149]. *Charley* formed as a subtropical low on 13 August along the Florida panhandle. After moving off the coast of South Carolina, the system transitioned into a tropical cyclone and intensified into a tropical storm on 15 August. *Charley* later attained hurricane status before moving across eastern North Carolina. It gradually weakened over the north Atlantic Ocean before transitioning into an extratropical cyclone on 20 August. *Charley*'s remnants remained identifiable for over a week after crossing the British Isles, finally dissipating on 30 August [149]. As an extratropical cyclone, *Charley* brought heavy rainfall and strong winds to the British Isles, causing at least 11 deaths. In eastern Ireland, Phoenix Park, Dublin, recorded its highest 24-hour rainfall total (85.1 mm) in a record stretching back to 1885, and there was widespread flooding in Dublin as a result. In the UK, the storm also caused rivers to flood, with severe floods in Cumbria and Gloucestershire, and brought down trees and power lines.

Before *Charley* arrived, another cyclone had passed over, with its centre (then 996 hPa) at approximately 51.5°N 11°W at 0600 GMT on Friday 22 August, the start of the Bank Holiday weekend in the UK. Thunderstorms and heavy rain occurred over southern Britain [150]. By 2400 GMT, the centre (now 1004 hPa) had reached 49°N 6°W with its by then occluded front over East Anglia; 63.3 mm of rain was recorded at Bushey Heath in Hertfordshire. At Oxford, 25.7 mm fell on 22 August, the 115th wettest summer day at Oxford. Ex-hurricane *Charley* arrived on Bank Holiday Monday. At 0600 GMT the pressure was 990 hPa at 49°N 12°W, the centre reaching 52°N 3°W by 0600 on

26 August and deepening to 981 hPa. Many places in England and Wales experienced more than 12 hours of continuous, heavy rain and this was reflected in the totals, with a maximum 24-hour total at Aber College Farm, Gwynedd, of 134.9 mm. The highest hourly amount was 38.2 mm at Preston, Lancashire, early on 26 August [150]. At Oxford, 29.3 mm was recorded on 25 August, the 75th wettest summer day at Oxford.

Oxford's four-day total of 55.4 mm is large, if not exceptional, ranking 138th highest in nearly 70,000 possible summer four-day totals. On closer inspection, it seems that having two daily totals over 25 mm in a single summer happens on average about one year in seven: there have been 27 summers with at least two such falls since 1827 ($n = 192$). Nevertheless, having two totals in excess of 25 mm within four days is very unusual, made worse by it being a holiday weekend. Five years have had three such totals in a single summer, the most recent being 1971. Three of these years come very early in the record: 1829, 1830 and 1834. In 1834, the three falls occurred between 18 and 29 July: the 12 day total of 135.6 mm is the third highest on record, marginally below the totals for 19 and 20 June 1903.

## The 'Great Storm'—16 October 1987

The 'Great Storm' of 16 October 1987 caused extensive damage in south-east England [151]. Oxford was just too far north to experience the worst of the storm, but the meteorological record is still of great interest. The rapidly deepening depression started life as a small wave on a cold front just off the north-west corner of Spain. A very strong jet stream was responsible for the formation of the depression, its rapid deepening over the Bay of Biscay and its swift passage north-east. At 1200 GMT on 15 October, the depression lay over the Bay of Biscay, with a central pressure of 970 hPa; 12 hours later, it had deepened to 953 hPa and lay in the English Channel between Brittany and Cornwall [151].

The approach of the warm front was accompanied by the usual light wind and rain. As the warm front passed over Oxford just before midnight, there was a sharp rise in temperature of 7 degC and a burst of heavy rain. The warm sector was associated with strong winds—mean wind speed exceeded 20 knots (10 m/s), while the strongest gust reached 59 knots (30.3 m/s) (see Figure 25.14). There was a sharp change of wind direction as the warm front passed over but no rain at that time. The passage of the cold front at 0230 GMT was marked by a fall in temperature, more rain, a drop in wind speed and a further shift in wind direction. The strongest winds came at the passage of the occlusion—mean speeds reached gale force with a peak gust of 62 knots (31.9 m/s) just before 0600 GMT.

The strongest winds and most extensive damage occurred to the south of a line approximately from Dorset to the Wash. Over much of south-east England and in the London area, this was the strongest gale for at least a hundred years, possibly since Defoe's 'Great Storm' of 1703 [152, 153]; 18 people lost their lives that night. Gusts reached 90 knots (46 m/s) along parts of the south coast, with the highest recorded gust of 98 knots (50.4 m/s) at Shoreham-by-Sea [151]. Both hourly mean wind speeds and gusts exceeded the 100 year recurrence interval at many places along the south coast as well as at inland locations such as Heathrow, Gatwick and central London. The event was also the first recognised occurrence of the 'sting jet' phenomenon [154].

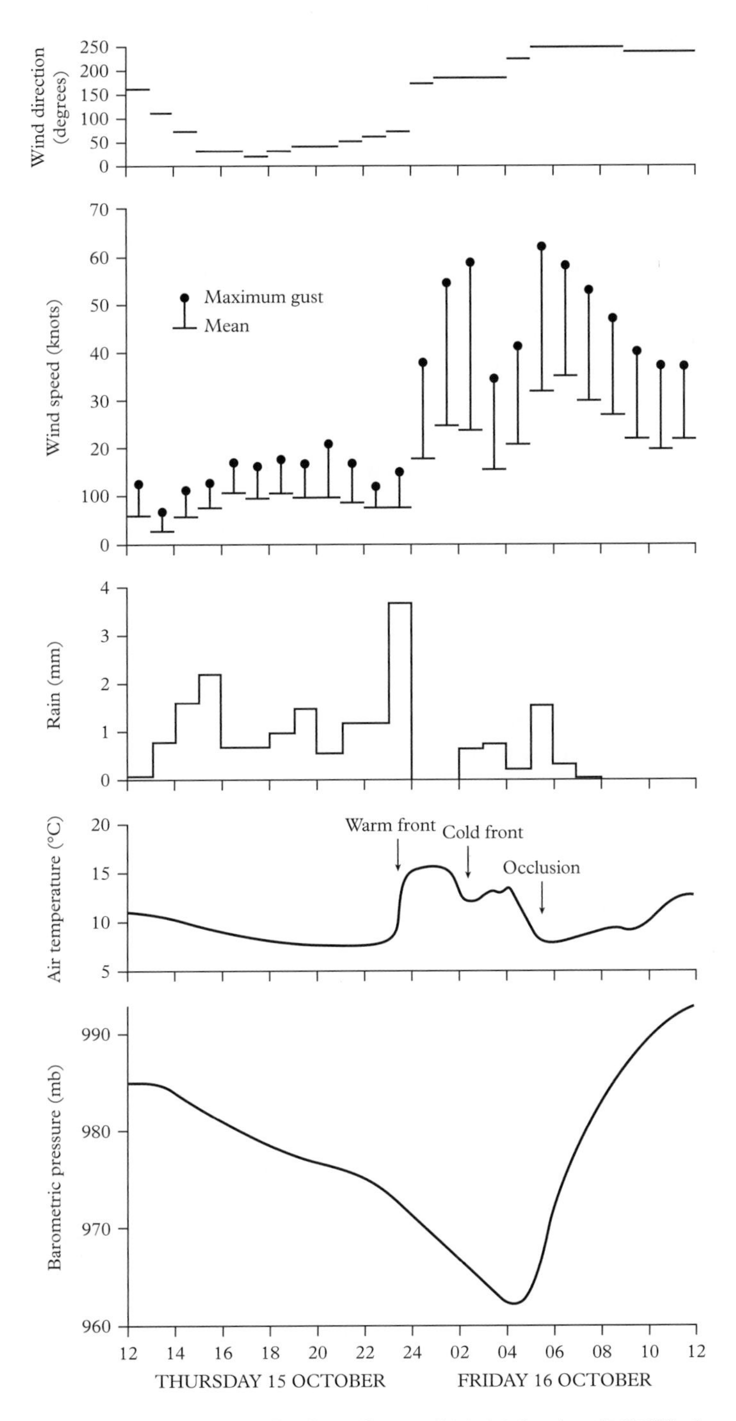


**Figure 25.14** *Oxford weather data for the Great Storm of 15–16 October 1987. Wind speeds are those recorded at 45 m above ground level (Chris Orton).*

There was considerable criticism in the media of the weather forecasts at the time. BBC weatherman Michael Fish stated in an earlier television forecast that there would not be a 'hurricane' and technically he was correct, as a true 'hurricane' is a particular type of circular storm that only occurs over tropical or subtropical oceans. Nevertheless, his comment perhaps implied some complacency on the part of the Met Office. Weather forecasting models at the time were of insufficient resolution to provide confidence in the development and path of the storm; today, forecasting models operate at much higher resolution (in both space and time) and the depth and track of the depression would be much more accurately forecast, together with additional emphasis being given to possible uncertainties in the forecast. The event led directly to the current 'weather warnings' system and indirectly to the naming of potentially damaging storms as a means of increasing public awareness.

## Summer heatwave—24 July 1989

Peter Venters did the weather readings that morning, under the watchful gaze of an *Oxford Mail* photographer (Figure 25.15). It was a hot summer and, of course, the local newspaper took a keen interest. Memories of 1976 were still fresh in the mind and comparisons were inevitable. It was also the time when the issue of global warming was beginning to be taken seriously, although as Tim Burt pointed out in the report, it was still too early for evidence to be conclusive. As is customary, the maximum thermometer records yesterday's maximum temperature so Peter would have recorded 31.3 °C for the 23rd, the third day in a row that temperatures had topped 30 °C. On 21 July the maximum was 31.0 °C and on 22 July it was 32.4 °C, the hottest day that summer. It was back over 30 °C on 25th, but only reached 29 °C on the day of the newspaper report. That summer remains one of the best summers on record at Oxford, still the sunniest on Oxford's long record and the 18th equal warmest since 1815; altogether 40 days reached 25 °C in 1989—four in May, nine in June, fifteen in July, nine in August and three in September, the latest on 21 September. Not unexpectedly, it was a dry summer too, but not exceptionally so—the total of 119 mm ranked only 51st driest since 1767 by 2018.

LUCY TENNYSON reports on why the Big Heat of '89 isn't such hot stuff, after all

**We're heating up under pressure**

Surprising

Peter Venters measuring rainfall outside Green College

Detect

Frost

Drastic

Concerned

Unpredictable

**Figure 25.15** Oxford Mail *headline for 24 July 1989*

## December 1989

December 1989 was the third-wettest December on Oxford's record, with 141 mm of precipitation. The month was odd in that the first ten days of the month, and the last six, were completely dry: the entire month's rainfall fell in 15 consecutive days from 11 December—including the wettest fortnight in

Oxford since June 1903 (see Table 28.3). Of these fifteen days, six received 12 mm or more of rainfall. Had this rate persisted for just another seven days, the month would have surpassed 200 mm for the first time since September 1774. Local streams and rivers were very low at the beginning of the month, following the warm, dry and sunny summer and a dry autumn, yet by Christmas floods were raging at near-record levels across southern England [155].

## Exceptional storms January–February 1990

On 25 January 1990, exceptional gales caused havoc across southern Britain, a little over two years after the 'Great Storm' of 16 October 1987. The deep depression which caused all the damage began life off the east coast of North America. Under the influence of a powerful jet stream, it crossed the Atlantic in less than two days, deepening rapidly as it travelled. In mid-Atlantic the pressure at its centre was 992 hPa, but only a day later its centre lay over southern Scotland at a pressure of 953 hPa. Depressions that deepen by more than 24 hPa in 24 hours are now known as 'bombs'.

The arrival of the warm front was heralded by southerly winds and light rain. The warm front passed Oxford at about 0500 GMT, temperatures rose a few degrees and the winds shifted to the south-west. The cold front passed Oxford at about 0900. Wind speeds increased, and the wind became gustier; there were some sharp showers of rain. Like the 1987 storm, the highest winds were associated with the occluded front which had 'bent back' around the centre of the depression. The highest wind speeds occurred between 1300 and 1530 GMT with several gusts reaching 74 knots (38 m/s) at 45 m above ground level. The wind veered to the west and barometric pressure at Oxford reached its lowest point: 964 hPa. The sustained impact of high wind speeds over two hours was particularly important in relation to the large amount of damage.

The gusts during this storm were not the highest ever recorded at Oxford: in March 1987 there was a gust of 76 knots (39 m/s), but this was an isolated event and there was generally little damage in the Oxford area. The 25 January 1990 storm seems more comparable to that of 2 January 1976 referenced in January's Chapter 7, but, as noted above, the anemometer then in use could not record gusts, though at nearby RAF Abingdon a gust of 75 knots (38.6 m/s) was recorded on that occasion.

The storm affected most of southern Britain and adjacent coastal waters. Forty-seven people died and at least 3 million trees were blown down, possibly as many as 5 million. Many of the fatalities were caused by trees falling on to cars. The greater damage and death toll compared to 1987 related to the wider area affected and its daytime occurrence. A succession of strong gusts was thought to be significant in the weakening of buildings and other structures. Restoration of power lines was hampered by inclement January weather.

There was little rainfall in the storm itself, but January 1990 rainfall as a whole was above average and more heavy rainfall at the start of February led to significant flooding, adding to the difficulties of clearing up the storm damage (Figure 25.16).

**Figure 25.16** *Flooding and windthrown trees on the River Thames at Clifton Hampden, early February 1990 (Tim Burt)*

## Oxford's hottest day equalled—3 August 1990

After a spell of three weeks with little or no rainfall across southern England, the advection of very warm continental air around an anticyclone centred near Denmark during the first four days of August 1990 resulted in a short but intense hot spell. The intensity and spatial extent of high temperatures were without precedent within the previous 150 years [145, 156] and new absolute maximum temperatures were established in England (37.1 °C at Cheltenham, Gloucestershire, on 3 August) and Wales (35.2 °C at Hawarden Bridge, Clwyd, on 2 August), while the existing record for Scotland was closely approached. Minimum temperatures were also exceptional, again establishing new record levels for the British Isles. In Oxford, four consecutive days surpassed 30 °C, with the maximum temperature of 35.1 °C on 3rd equalling the all-time Oxford record set in August 1932. (Since 1990, this value has been closely approached twice—namely 34.6 °C on 9 August 2003 and 34.8 °C on 19 July 2006—but not yet exceeded.) Figure 25.17 shows the highest temperatures recorded during the 1990 heatwave.

## Solar eclipse—11 August 1999

There was a near-total eclipse of the Sun at Oxford on this date: at Wytham, this led to a drop in average hourly air temperature of 1.5 degC, from 16.6 °C for the hour commencing 0900 GMT (10 a.m. clock time) to 15.1 °C during the following hour

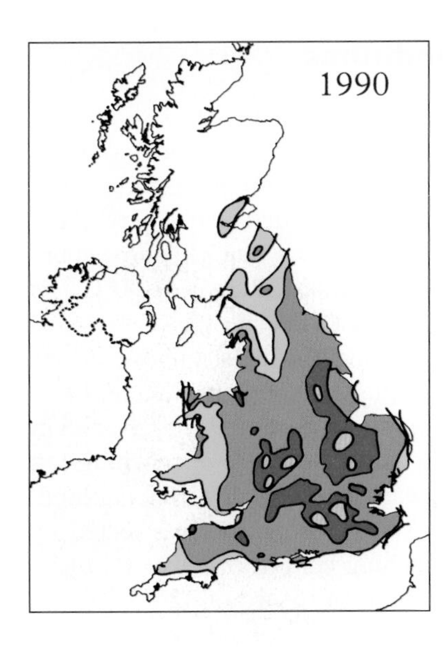


**Figure 25.17** *Maximum temperatures during August 1990, from [145, 156]. For clarity, only temperatures above 30 °C are shown. Isotherms are at 30 (yellow), 32 (orange), 35 (red) and 36 °C (pale orange) (Stephen Burt)*

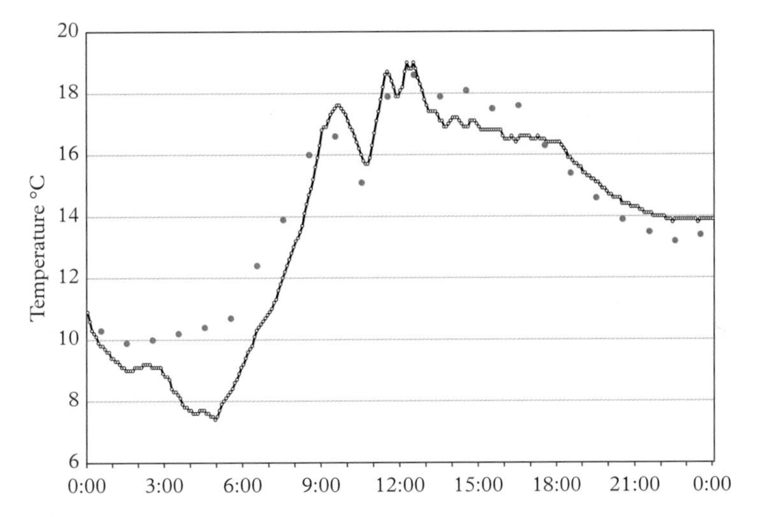


**Figure 25.18** *Air temperatures (°C) during the near-total solar eclipse on 11 August 1999. The black line is from the automatic weather station at Stratfield Mortimer, south-west of Reading (5-minute data points); the red circles are from Wytham Woods, west of Oxford (hourly averages, plotted on the half hour). Time is in GMT (Wytham data from Environmental Change Network, with permission)*

(Figure 25.18, red solid circles, plotted at the midpoint of each hour). The five-minute temperature record from an automatic weather station near Reading is also shown (black line and open circles), where the fall in temperature was slightly greater at 2.0 degC.

## August 2003—Near-record heatwave conditions

As high pressure built during the first few days of August 2003, the unsettled conditions prevailing at the end of July gave way to sunny, warm weather across England and Wales and temperatures began to climb into what became a prolonged and exceptional heatwave [145]. A maximum temperature of 25 °C or more was recorded somewhere in the United Kingdom every day 1–18 August inclusive, and 30 °C was reached somewhere in the United Kingdom for the ten consecutive days 3–12 August. In Oxford, 27 °C was reached or exceeded on 11 consecutive days 3–13 August. The heatwave saw two peaks—on 5/6 August and 9/10 August. Very broadly speaking, the highest temperatures were reached in the north and west between 5 and 8 August, in southern Scotland and all other parts of England apart from East Anglia and the south-east on 9 August, and in the east and south of England on 10 August. After the early clearance of morning fog and low cloud, Saturday 9 August became another fine and very hot day over England, Wales and most of Scotland, although Ireland and western Scotland were affected by cloud and patchy light rain and drizzle from an encroaching cold front. Over southern Scotland, 30 °C was reached quite widely with a new Scottish record of 32.9 °C being attained at Greycrook in Borders Region. Almost everywhere in England reached 30 °C except for the north-east, the south-west and coastal areas, while 34 °C was reached widely in the Midlands and south-east. For much of central and northern England (including Oxford), Wales and the southern part of Scotland, this was the hottest day of the August 2003 heatwave, and 36.4 °C was reached at Enfield in north London. At the Radcliffe Observatory, the day's maximum was 34.6 °C, which remains the third-hottest day on Oxford's record back to 1815. Figure 25.19 shows the distribution of maximum temperatures on 9 August 2003.

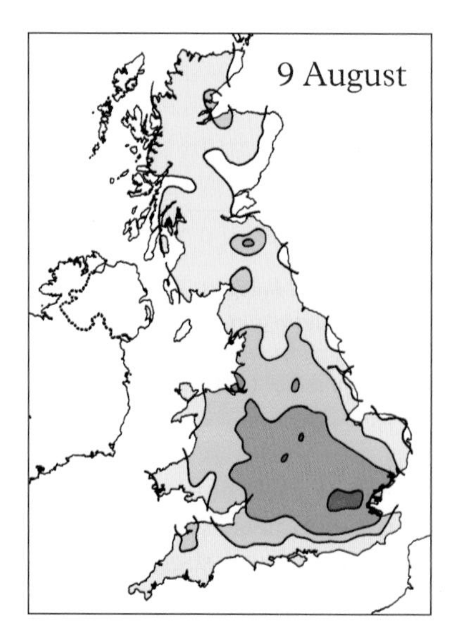


**Figure 25.19** *Maximum temperatures on 9 August 2003, from [145], Figure 5. For clarity, only temperatures above 25 °C are shown. Isotherms are at 25 °C (pale yellow), 30 °C (yellow), 32 °C (orange) and 35 °C (red). From [145, 156] (Stephen Burt)*

## July 2006—Another exceptional heatwave

Sustained warmth and prolonged sunshine resulted in July 2006 being one of the hottest and sunniest months on record over much of the UK [157]: with a mean Central England Temperature (CET) of 19.7 °C it ranks ahead of July 1983 (CET mean temperature 19.5 °C), August 1995 (19.2 °C) and July 2018 (19.1 °C) on the series which commenced in 1659. The heatwave resulted from anomalous high pressure over northern Europe, with persistent southerly winds over the UK. The heat and sunshine placed strains on water and energy utilities, road and rail transport and the health and fire services but benefitted the tourist industry, food and drink retailers and, to some extent, farmers and growers. Anticyclonic weather became firmly established over the UK and the North Sea from the 13th, advecting increasingly warm air from the east or south-east with little or no cloud and almost unbroken sunshine for over a week; Oxford recorded an average of almost 13 hours sunshine daily from 11th to 19th. Temperatures first reached 30 °C over much of England and Wales on 16th and exceeded 35 °C in places on the 19th (Figure 25.20) when new UK and Welsh temperature records for July were set. The highest temperature recorded was 36.5 °C at Wisley in Surrey on the 19th; Oxford reached 34.8 °C, its second-highest temperature on record, equalling the value attained on 9 August 1911 (Table 29.1). Figure 25.20 shows the maximum temperatures recorded on this date.

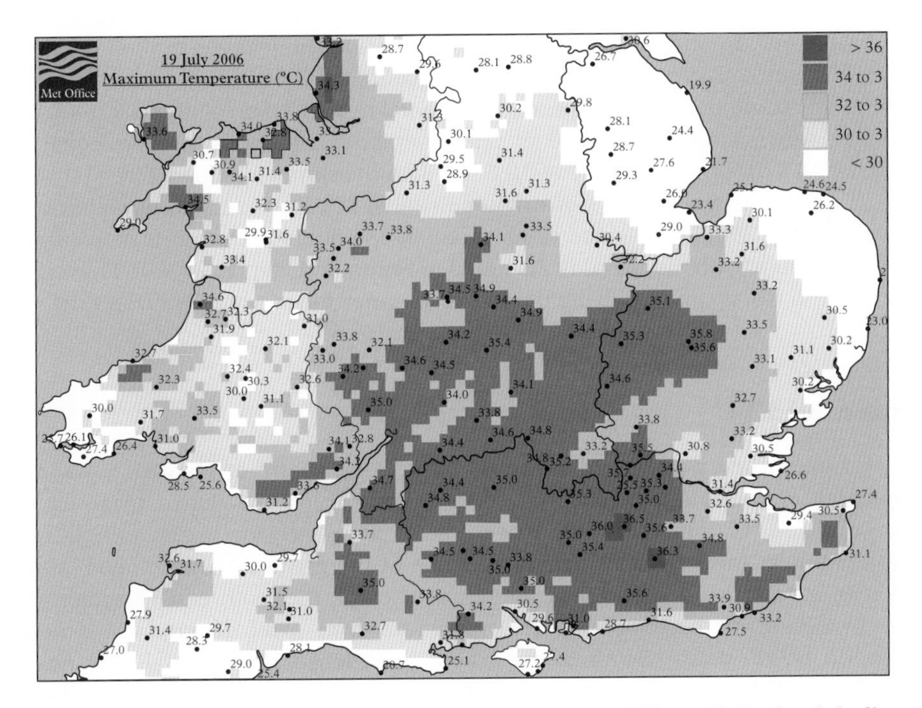


**Figure 25.20** *Maximum temperatures on 18 July 2006, from [157], Figure 1b. Dark red shading corresponds to temperatures above 34 °C (Crown Copyright)*

## 8 May 2012—Tornado near Kidlington

A small tornado swept through Oxfordshire during the afternoon and was seen (and photographed) in several places, including Bicester, Eynsham, Kidlington and South Leigh. The parent supercell thunderstorm developed over northern Wiltshire, moving across Oxfordshire and into part of Buckinghamshire, eventually dying out before it reached Cambridgeshire. It caused only minor damage to trees and roof tiles, although large hailstones were also reported during the storm. At the Radcliffe Observatory site, 5.4 mm of rainfall was recorded.

## 2012

This was the wettest year on record at Oxford since records began in 1767. The total of 979.5 mm is remarkable, the closest an Oxford annual rainfall total has yet come to 1000 mm. At 148 per cent of the 1981–2010 average, 2012 exceeded the previous record holders (all above 900 mm): 1960, 1852, 1768 and 1903. In simple terms, an annual rainfall total above 900 mm happens on average once every fifty years at Oxford. What is even more remarkable is that 2012 started with a winter drought. The total for the first three months was only 80 mm, the 32nd lowest total for the first quarter in 251 years. The total for the next nine months was 899 mm, a total that in itself would have exceeded all but the top five annual totals. The summer was the second-wettest on record (332 mm), beaten only by summer 1852 (340 mm). The autumn was wet but not exceptionally so (265 mm, ranking 26th wettest) but, of course, it followed an exceptionally wet summer. The combined summer plus autumn total ranks fourth highest. Despite all the rain, 2012 was not a particularly cool year: overall, in 204 years, it ranks equal 25th warmest with a mean temperature of 10.8 °C, with the summer being equal 41st warmest (17.1 °C). In terms of hours of bright sunshine, 2012 was very disappointing with summer ranking 29th lowest since 1880, a total of 483 hours (81 per cent of average) or equivalent to just 5¼ hours per day. As noted in Summer, Chapter 22, 2012 ranks overall as the 18th worst summer using the Davis Summer Index since 1880, and the worst since 1977.

## Winter 2013/14

This was one of the wettest and stormiest winters on record at Oxford. It was not a cold winter, quite the opposite: the seventh-mildest on record, with a mean air temperature of 6.7 °C. Thus, the floods were generated solely by excess rainfall rather than the rain-on-snow that has generated some of the greatest winter floods, 1947 in particular. It was also a very windy winter: at Oxford, the tenth equal windiest since 1881. Elsewhere in the country there was widespread wind damage, most notably at Dawlish on the south Devon coast, where a section of the sea wall collapsed on 5 February 2014 under ferocious attack from massive waves, leaving the railway to Cornwall suspended in mid-air.

**Figure 25.21** *Oxford flooding in January 2014—Oxford's wettest January, and its wettest winter, in 250 years of record.* ***Top:*** *Botley Road at Lamarsh Road, 9 January 2014 (Jane Buekett).* ***Middle:*** *Water on the line—trains cancelled at Oxford Station, 9 January 2014 (Jane Buekett).* ***Bottom:*** *Flooding on Christ Church meadow on 11 January 2014 (Stephen Burt)*

The wettest winter on record, 2013/14, was exceptional across wide areas of the country. Residents of homes on the Somerset Levels were evacuated amid fears that flood defences could be overwhelmed. Serious flooding once again affected parts of west Oxford; many houses affected in November 2012 were inundated again. The Oxford winter rainfall total, the highest on record (335 mm, more than twice normal) was the result of exceptional rainfall in all three winter months: December (98 mm) ranked equal 27th wettest since 1767, and February (90 mm) 11th wettest, while January (147 mm) established a new extreme for the month. In Oxford the first extreme flood came on 9 January, the peak flow at Day's Weir (near Dorchester-on-Thames) being the sixth highest flood peak (237 $m^3/s$) since records commenced in 1938 (Figures 25.21 and 25.22). Two further floods followed in February, another very large one on 10th (227 $m^3/s$, ranking ninth highest at Day's Weir) and a lesser flood on 21st (150 $m^3/s$, ranking 57th).

As a result of the impact of the 2007, 2012 and 2014 floods, an Oxford Flood Alleviation Scheme is planned that will involve channelling water round the west of the city, partly using the Seacourt stream. The design includes a two-stage channel, with a narrow permanently wet section flanked by a much wider area of lowered ground. The secondary channel will be grassed and normally dry. The aim is to enable flood water to pass more quickly and safely through the city, rather than accumulating in the floodplain as happens currently. Two large culverts under the southern bypass are another critical part of the design. The Scheme includes an ambitious environmental plan which aims to deliver improvements in biodiversity and species habitats in the project area. The Scheme was subject to planning permission at the time of writing.

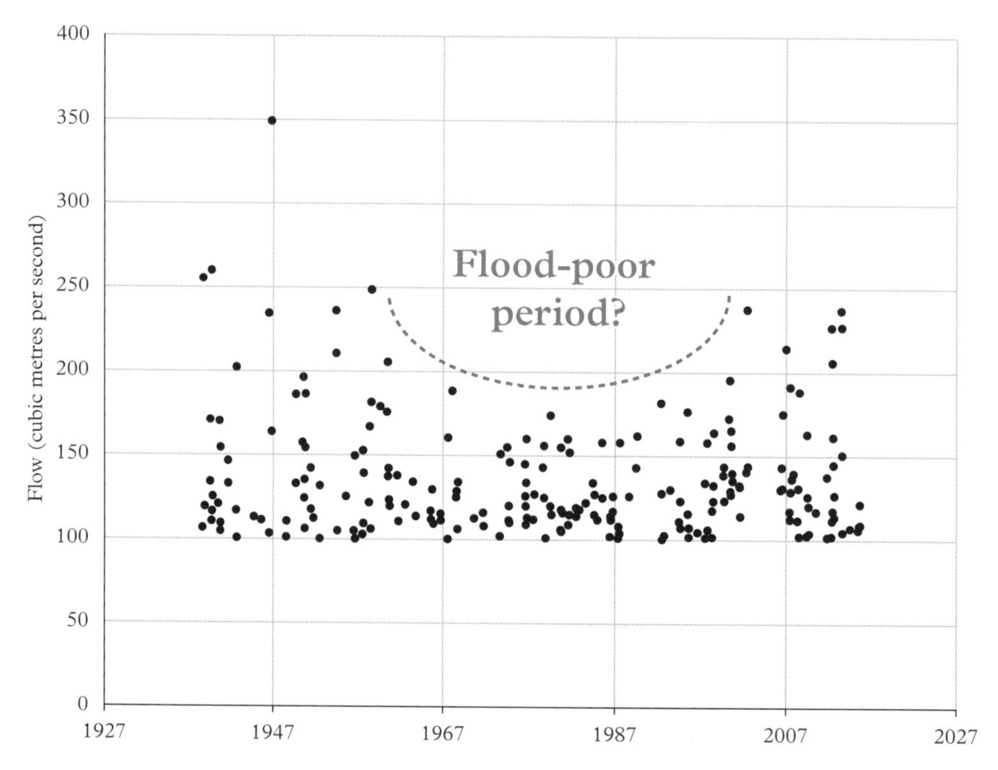


**Figure 25.22** *Flood peaks above 100 m³/s at Day's Weir by year, 1938–2016. Data from NRFA (Station 39002) with permission*

**Figure 25.23** *There have been plenty of floods in the last three decades following a 'flood-poor' period before that. Here, Tim Burt drives through a flood near Wytham in the early 1990s (Elizabeth Burt)*

Figure 25.22 shows flood peaks at Day's Weir on the Thames below Oxford from 1938 to 2016. There is a noticeable lack of high peaks in the 1970s and 1980s. By contrast, there were more high peaks in the early record and some indication of a return to more frequent high floods since the 1990s. It has been argued [158, 159] that much of the UK's flood engineering has been based on conditions in the 'flood-poor' period from the 1960s to the 1980s; more recent floods have caused surprise and havoc by being larger than expected.

## 2018

The year 2018 has been remarkable, featuring record or near-record extremes in almost every month:

- January: the third-mildest January day on record, 14.6 °C on 23rd;
- February: the second-sunniest February on record, 117.6 hours;
- March: the coldest March day since 1845 on 1st, maximum −1.9 °C, with two notable snowstorms during the month;
- April: the second-hottest April day on record, 26.9 °C on 19th, and the warmest April night on record, 13.8 °C on 22nd;
- May: the fifth-wettest May day on record on 31st, with 36.4 mm, Oxford's wettest day for 6 years;
- June: the driest June since 1925, with just 2.5 mm of rainfall;
- June–July: only 1.4 mm of rainfall in 48 days 2 June to 19 July, the driest 45-day spell on record;
- June–July: the sunniest 21-day, 30-day and 60-day periods on record—with 332.5 hours of sunshine (an average of 11.1 hours per day) for the 30 days commencing 18 June;
- July: the second-hottest month on Oxfords records at 21.05 °C, less than a tenth of a degree behind the hottest to date, July 1983; and
- Summer: the hottest on Oxford's records, the highest frequency of hot days (25 °C or more), the equal-highest frequency of warm nights (15 °C or more), and the second-best summer index (behind only 1976) since at least 1880.
- October: the warmest October *night* on record (17.0 °C on 13th), followed only two weeks later by the coldest October *day* in more than 40 years (6.6 °C on 27th).

For the year as a whole, there were 48 days which reached or surpassed 25 °C, surpassing the previous highest (46 days, in 1976). The year also equalled (with 1983) the greatest number of warm nights (minimum air temperature 15 °C or more) with 34.

The year 2018 became the third-warmest on Oxford's records. The mean temperature of 11.70 °C in 2018 has been surpassed only by 2014 (11.79 °C) and 2006 (11.71 °C), and all five of the warmest years on Oxford's records have occurred since 2006. The annual sunshine duration, 1791.7 hours, was also the highest for any year since 1995

(1880 hours) and the tenth sunniest year on record. Compared to normal, 2018 had more than twice as many days with 12 hours or more of sunshine; in all, there were 38 such days, eclipsing the previous highest (36 days, in 1990).

The year presented a considerable challenge to your authors, for each of these records or near-records necessitated rewriting part, or in July's case most of, the existing material for that month's chapter as the year has gone on, to ensure this book is as up to date as possible when it goes to print.

# Part 6

# Oxford weather averages and extremes

# 26

# Warmest, driest, sunniest,...

The following analysis is based upon daily statistics for the Radcliffe Observatory, Oxford over the standard 30-year period 1981–2010, and excludes 29 February. Different analysis periods would probably show slightly different results.

The most reliably warm week of the year in Oxford is the final week of July; this also starts out as one of the least rainy periods of the year, although the highest frequency of thunder is most likely around the same time. Mid-July and mid-August are the most reliably sunny times of the year, and amongst the driest too. Early to mid-February tends to be the coldest time of the year, and the most likely to see snowfall, while late December is the least sunny. See also 'annual cycle' plots, Figure 6.1.

| | | DAILY VALUES | | WEEKLY VALUES | |
|---|---|---|---|---|---|
| | | Value | Date | Value | Week comm. |
| WARMEST | Highest mean daily maximum temperature | 23.8 °C | 28 July | 23.4 °C | 25 July |
| | Highest mean daily minimum temperature | 14.1 °C | 31 July | 13.7 °C | 26 July |
| | Highest mean daily temperature | 18.9 °C | 29 July | 18.5 °C | 26 July |
| COLDEST | Lowest mean daily maximum temperature | 6.6 °C | 28 Dec | 7.1 °C | 25 Dec |
| | Lowest mean daily minimum temperature | 0.3 °C | 14 Feb | 1.0 °C | 14 Feb |
| | Lowest mean daily temperature | 3.7 °C | 14 Feb | 4.2 °C | 14 Feb |

*(continued)*

*Oxford Weather and Climate since 1767*. Stephen Burt and Tim Burt, Oxford University Press (2019).
 DOI: 10.1093/oso/9780198834632.001.0001

| | | DAILY VALUES | | WEEKLY VALUES | |
|---|---|---|---|---|---|
| | | Value | Date | Value | Week comm. |
| **WETTEST** | Greatest mean daily rainfall | 5.4 mm | 9 Aug | 2.87 mm/day | 19 Oct |
| | Greatest frequency of rain days | 22 days in 30 years (73%) | 8 Nov | 2.7 days/wk | 3 Jan |
| **DRIEST** | Lowest mean daily rainfall | 0.36 mm<br>0.43 mm | 28 Aug<br>2 May | 0.84 mm/day | 26 Aug |
| | Lowest frequency of rain days | 6 days in 30 years (20%) | 15 and 28 Aug | 1.33 days/wk<br>1.35 days/wk | 12 June<br>26 Aug |
| **SUNNIEST** | Greatest mean daily sunshine duration | 8.35 hours | 31 May | 6.98 hours/day<br>6.97 hours/day | 15 Aug<br>15 July |
| **DULLEST** | Lowest mean daily sunshine duration | 0.76 hours | 30 Nov | 1.51 hours/day | 20 Dec |

# 27

# Earliest and latest dates

## Temperature

Air temperatures were recorded by unscreened thermometers prior to January 1853—see Appendix 1 for more details. These are likely to be a little too cold by modern (screened thermometer) standards in winter and a little too warm in summer. For this reason, they are shown in *italics* in the tables following, together with the nearest extreme recorded since January 1853. The mean temperature correction for unscreened thermometry varies throughout the year, from +0.3 degC in December and January to −0.3 degC in July, although this will vary considerably on a day-to-day basis. Temperatures were recorded in a Stevenson screen from December 1878.

For parts of the record, the maximum minimum temperatures were read to, or are known only to, a precision of 1 degree Fahrenheit. For these, the Celsius conversion assumes the exact °F value, although the actual temperature could lie within ± 0.3 degC of that value.

**Table 27.1** *Earliest threshold maximum temperatures by date at the Radcliffe Observatory, Oxford, April 1815 to December 2018*

| Maximum temperature | Earliest date | Value, °C | Latest date | Value, °C |
|---|---|---|---|---|
| ≤ 0 °C | 15 Nov 1965 | 0.0 | 24 March 1879 | −0.2 |
| | *16 Nov 1841* | *−0.9* | | |
| | *17 Nov 1815* | *−2.6* | | |
| | 24 Nov 1936 | −0.6 | | |
| > 20 °C | 9 March 1948 | 21.1 | 31 Oct 2014 | 21.6 |
| 21 °C (70 °F) | 9 March 1948 | 21.1 | 31 Oct 2014 | 21.6 |
| 25 °C | 3 April 1946 | 25.0 | 9 Oct 1921 | 25.7 |
| 27 °C (80 °F) | 23 April 2011 | 27.6 | 5 Oct 1921 | 27.2 |

(*continued*)

*Oxford Weather and Climate since 1767*. Stephen Burt and Tim Burt, Oxford University Press (2019).
 DOI: 10.1093/oso/9780198834632.001.0001

Table 27.1 *Continued*

| Maximum temperature | Earliest date | Value, °C | Latest date | Value, °C |
|---|---|---|---|---|
| 30 °C | *27 May 1841* | *30.0* | 8 Sept 1898 | 32.2 |
| | 29 May 1944 | 30.6 | 8 Sept 1911 | 33.4 |
| 32 °C (90 °F) | 2 June 1947 | 32.2 | 8 Sept 1898 | 32.2 |
| | | | 8 Sept 1911 | 33.4 |
| 35 °C | 3 Aug 1990 | 35.1 | 19 Aug 1932 | 35.1 |

Table 27.2 *Earliest threshold low minimum temperatures by date at the Radcliffe Observatory, Oxford, April 1815 to December 2018*

| Threshold temperature | Earliest date | Value, °C | Latest date | Value, °C |
|---|---|---|---|---|
| Dates within the winter season, starting 1 July | | | | |
| 5 °C | *2 July 1837* | *4.4* | 30 June 1892 | 4.6 |
| | *4 July 1821* | *4.4* | | |
| | 4 July 1965 | 4.4 | | |
| 0 °C | *23 Sept 1845* | *−1.2* | *27 May 1821* | *0.0* |
| | *28 Sept 1851* | *−3.3* | 27 May 1914 | −0.1 |
| | 29 Sept 1919 | −0.4 | | |
| −5.°C | 24 Oct 1859 | −5.7 | *11 April 1817* | *−5.6* |
| | | | *6 April 1851* | *−5.6* |
| | | | *5 April 1830* | *−5.0* |
| | | | 31 March 1955 | −5.6 |
| −10.°C | 6 Dec 1879 | −11.3 | *13 March 1845* | *−12.0* |
| | | | 7 March 1947 | −11.1 |
| −15.°C | 13 Dec 1981 | −16.1 | 25 Feb 1947 | −16.1 |

Table 27.3 *Earliest threshold high minimum temperatures by date at the Radcliffe Observatory, Oxford, April 1815 to December 2018*

| Threshold temperature | Earliest date | Value, °C | Latest date | Value, °C |
|---|---|---|---|---|
| Dates within the calendar year | | | | |
| 15 °C | 5 May 1862 | 15.2 | 22 Nov 1947 | 15.0 |
| | 9 May 1945 | 16.7 | | |
| 20 °C | 1 July 1968 | 20.0 | 5 Sept 1949 | 20.6 |

## Snowfall

Table 27.4 *Snow Lying at 0900 GMT, period January 1959 to December 2018*

| Depth threshold | Earliest date | Latest date |
|---|---|---|
| > *50% cover,* > 0 cm | 20 Nov 2008, depth 1 cm | 26 April 1981, depth 4 cm |
| ≥ 5 cm | 8 December 1967, 8 cm<br>9 December 1990, 5 cm | 15 April 1966, 5 cm |
| ≥ 10 cm | 11 December 1981, 14 cm | 17 March 1979, 10 cm |
| ≥ 15 cm | 12 December 1981, 16 cm | 12 March 1996, 15 cm |
| ≥ 20 cm | 30 December 1962, 20 cm | 5 March 1970, 20 cm |

The greatest snow depth known from the Oxford record was on 13 and 14 February 1888, when the depth was reported as 61 cm (see Chapter 8). In 1908, snow lay 43 cm deep at 2130h 25 April and 41 cm deep at 0800h 26 April (see Chapter 10 and Chronology, Chapter 25).

## Sunshine

Table 27.5 *Earliest sunshine duration thresholds by date at the Radcliffe Observatory, Oxford, January 1921 to December 2018*

| Daily duration, hours | Earliest date | Duration, hours | Latest date | Duration, hours |
|---|---|---|---|---|
| ≥ 8.0 | 25 Jan 1986 | 8.5 | 21 Nov 1996 | 8.0 |
| 9.0 | 16 Feb 1985<br>16 Feb 2008 | 9.2<br>9.0 | 28 Oct 1987<br>28 Oct 1997 | 9.1<br>9.4 |
| 10.0 | 5 March 1961†† | 10.0 | 10 Oct 1988<br>10 Oct 1997 | 10.0<br>10.0 |
| 11.0 | 7 March 2016 | 11.0 | 18 Sept 1971* | 11.2 |
| 12.0 | 27 March 1996 | 12.4 | 4 Sept 1997 | 12.0 |
| 13.0 | 16 April 1942 | 13.0 | 27 Aug 2001 | 13.1 |
| 14.0 | 3 May 1953 | 14.4 | 7 Aug 1988 | 14.1 |
| 15.0 | 24 May 1977† | 15.1 | 22 July 1995 | 15.0 |

* One year between 1880 and 1920 had 11.1 h sunshine on 25 September
† Equalled on 24 May with 15.15 h on that date in one year between 1880 and 1920
†† 10.3 hours was recorded on 27 February 2019

# 28

# Droughts and wet spells

Droughts can take many forms. Meteorologically, two definitions are still sometimes used:

*Absolute drought:* A period of at least 15 consecutive days, during which no day receives as much as 0.2 mm of precipitation.

*Partial drought:* A period of at least 29 consecutive days, whose mean daily precipitation does not exceed 0.2 mm.

These definitions were introduced by G. J. Symons in *British Rainfall 1887* (itself a notable drought year) but ceased to be used officially over 50 years ago; however, they remain useful ways of summarising short-term rainfall deficits, particularly in central southern England. Periods of drought are also usually easier to define in hindsight.

Various categories of drought affect water consumers differently, depending upon their water usage requirements. Drought impacts are more keenly felt in summer when temperatures are higher and the rate of evaporation of water from the soil is higher; for a farmer, the development of a drought occurs more rapidly and might be said to occur when the surface soil layers become too dry for crops to grow. In winter the soil can stay quite moist even during a 15-day absolute drought, due to minimal evaporation of soil moisture, particularly in cold weather. The water supply industry is more likely to be affected by a dry winter, however, when underground (groundwater) water levels may be insufficiently replenished by winter rainfall—evaporation exceeds rainfall in most summers in southern England. A major cause of drought in Britain is the persistence of 'blocking' anticyclones, and the consequent displacement of mid-latitude depressions northwards or southwards of the British Isles.

*Oxford Weather and Climate since 1767*. Stephen Burt and Tim Burt, Oxford University Press (2019).
 DOI: 10.1093/oso/9780198834632.001.0001

## Absolute droughts

In Oxford, an absolute drought lasting 15 days or more can be expected, on average, almost every year. Table 28.1 shows the longest absolute droughts (25 days or more without measurable rainfall) recorded at the Radcliffe Observatory, Oxford, since the start of daily rainfall recordings in 1827. The average recurrence interval of 25-day droughts has been about 7 years, but intervals vary widely, from consecutive years (1844 and 1845, 1935 and 1936, 1940 and 1941) to a gap of 36 years (1854 to 1880). Such droughts tend to occur in the summer rather than the winter half of the year, but it is interesting to note that the longest partial drought began towards the end of winter in 1893. Table 28.2 shows the longest partial droughts (45 days or more during which the mean daily rainfall was 0.2 mm or less) recorded at the Radcliffe Observatory, Oxford, since the start of daily rainfall recordings in 1827. Table 28.3 shows the frequency by starting month of both absolute and partial droughts since 1827. Both categories of drought are most likely to start in late winter or early spring, with a secondary frequency peak in high summer.

## Partial droughts

A partial drought of 29 days or more can be expected, on average, slightly less than once per year (176 such events in 191 years 1827–2018), although some years have seen two (1844, 1929, 2003) while sometimes many years pass without any (none between 1854 and 1887, 1899 and 1921, for example). Prolonged partial droughts are much rarer—only six have lasted longer than 60 days in that period. The longest and most intense partial drought on Oxford's records took place during the spring of 1893, when just 4.3 mm fell in the 75 days 2 March to 15 May. Measurable rain fell on only 6 days during this 75-day spell, the wettest day receiving just 1.4 mm. Excluding minor falls at the beginning of this long drought, only 2.7 mm fell in the 72 days commencing 5 March—a daily average fall of just 0.04 mm. Perhaps surprisingly, the notable drought years of 1921, 1959 and 1976 barely feature in the partial droughts table.

Table 28.4 shows both the driest and wettest spells of particular durations recorded between 1827 and 2018. Notable wet spells have included those of late 1929 to January 1930, the very wet winters of 2000/01 (this latter event part of a wet spell lasting over a year and ending in May 2001) and more recently 2013/14, and the prolonged wet conditions during the First World War, from December 1914 to early 1916.

Notable dry periods include early 1975 to September 1976, the dry year of 1990, and particularly dry spells during 1938, 1976 and 1990, along with spring 1929 and the prolonged fine summers of 1959 and 2018. Note that 1929 is the only year to appear in both tables—the driest 70 day spell from February to April that year received only 15 mm of precipitation, whereas the 70 day spell commencing in November received 353 mm—more than 20 times as much rainfall as in the earlier period.

**Table 28.1** *Periods of absolute drought in Oxford lasting at least 25 days, in date order, January 1827 to December 2018. Dates are first and last days with precipitation 0.1 mm or nil. Ranked droughts 1–5 by duration are indicated in each column.*

| **ABSOLUTE DROUGHTS ≥ 25 days** | | | |
|---|---|---|---|
| **Year** | **Start and end date** | **Duration, days** | **Rank 1-5** |
| 1827 | 1-27 Feb | 27 | |
| 1829 | 28 Feb—28 March | 29 | |
| 1835 | 13 July—6 Aug | 25 | |
| 1840 | 19 Feb—14 March | 25 | |
| 1844 | 14 April—10 May | 27 | |
| 1845 | 12 Oct—5 Nov | 25 | |
| 1849 | 2-27 March | 26 | |
| 1852 | 3-29 March | 27 | |
| 1854 | 21 March—15 April | 26 | |
| 1880 | 8 Aug—3 Sept | 27 | |
| 1887 | 4 June—3 July | 30 | |
| 1891 | 2 Feb—6 March | 33 | 4 |
| 1893 | 17 March—16 April | 31 | |
| 1911 | 3-28 July | 26 | |
| 1929 | 1-28 Sept | 28 | |
| 1935 | 21 July—17 Aug | 28 | |
| 1936 | 27 April—21 May | 25 | |
| 1940 | 11 Aug—8 Sept | 29 | |
| 1941 | 10 June—10 July | 31 | |
| 1947 | 5 Aug—10 Sept | 37 | 3 |
| 1949 | 4 June—3 July | 30 | |
| 1955 | 4 July—1 Aug | 29 | |
| 1959 | 14 Aug—20 Sept | 38 | 2 |
| 1972 | 9 Aug—7 Sept | 30 | |
| **1976** | **19 July—27 Aug** | **40** | **1** |
| 1979 | 25 June—27 July | 33 | 4 |
| 1997 | 28 March—24 April | 28 | |

(*continued*)

Table 28.1 *Continued*

| **ABSOLUTE DROUGHTS ≥ 25 days** | | | |
|---|---|---|---|
| **Year** | **Start and end date** | **Duration, days** | **Rank 1-5** |
| 2003 | 2-27 Aug | 26 | |
| 2011 | 31 March—27 April | 28 | |
| 2018 | 18 June—19 July | 32 | 5 |

Table 28.2 *Periods of partial drought in Oxford lasting at least 45 days, in date order, January 1827 to December 2018. Dates are first and last days with mean precipitation over the period below 0.2 mm. Ranked droughts 1–5 by duration are indicated in each column.*

| **PARTIAL DROUGHTS ≥ 45 days** | | | | | |
|---|---|---|---|---|---|
| **Year** | **Start and end date** | **Duration (days)** | **Period total rainfall (mm)** | **mm/day** | **Rank 1-5** |
| 1827 | 12 Jan—27 Feb | 47 | 4.4 | 0.09 | |
| 1829 | 25 Apr—17 June | 54 | 5.8 | 0.11 | |
| 1834 | 13 Feb—29 Mar | 45 | 4.7 | 0.10 | |
| 1835 | 6 July—23 Aug | 49 | 4.8 | 0.10 | |
| 1844 | 27 March—5 June | 71 | 12.7 | 0.18 | 2 |
| 1844 | 15 Nov—29 Dec | 45 | 6.9 | 0.15 | |
| 1850 | 15 Feb—1 Apr | 46 | 8.1 | 0.18 | |
| 1852 | 10 Feb—29 Mar | 49 | 8.7 | 0.18 | |
| 1854 | 24 Feb—20 Apr | 56 | 10.6 | 0.19 | |
| 1887 | 4 June—23 July | 50 | 6.7 | 0.13 | |
| 1893 | **2 Mar—15 May** | **75** | **4.3** | **0.06** | 1 |
| 1895 | 29 Apr—25 June | 58 | 7.0 | 0.12 | |
| 1896 | 17 Apr—3 June | 48 | 7.0 | 0.15 | |
| 1898 | 8 Aug—28 Sept | 52 | 9.6 | 0.18 | |
| 1899 | 16 Feb—3 Apr | 47 | 7.6 | 0.16 | |
| 1921 | 4 June—22 July | 49 | 9.6 | 0.20 | |

(*continued*)

Table 28.2 *Continued*

**PARTIAL DROUGHTS ≥ 45 days**

| Year | Start and end date | Duration (days) | Period total rainfall (mm) | mm/day | Rank 1-5 |
|---|---|---|---|---|---|
| 1929 | 10 Feb—31 Mar | 50 | 8.4 | 0.17 | |
| 1929 | 6 Aug—28 Sept | 54 | 6.8 | 0.13 | |
| 1932 | 17 Jan—20 March | 64 | 7.7 | 0.12 | 4 |
| 1938 | 1 March—1 May | 62 | 11.6 | 0.19 | 5 |
| 1940 | 28 July—11 Sept | 46 | 2.3 | 0.05 | |
| 1955 | 4 July—3 Sept | 62 | 11.8 | 0.19 | 5 |
| 1959 | 14 Aug—9 Oct | 57 | 5.4 | 0.09 | |
| 1969 | 16 Sept—2 Nov | 48 | 6.6 | 0.14 | |
| 1976 | 14 July—28 Aug | 46 | 8.3 | 0.18 | |
| 1978 | 28 Sept—11 Nov | 45 | 6.4 | 0.14 | |
| 1980 | 2 Apr—18 May | 47 | 5.4 | 0.11 | |
| 1984 | 26 March—9 May | 45 | 7.2 | 0.16 | |
| 1993 | 28 Jan—20 March | 52 | 8.6 | 0.17 | |
| 1995 | 15 July—28 Aug | 45 | 7.8 | 0.17 | |
| 1997 | 27 Feb—24 Apr | 57 | 9.2 | 0.16 | |
| 2003 | 8 March—23 Apr | 47 | 6.6 | 0.14 | |
| 2003 | 2 Aug—21 Sept | 51 | 8.4 | 0.16 | |
| 2007 | 22 March—5 May | 45 | 3.7 | 0.08 | |
| 2011 | 1 March—5 May | 66 | 9.9 | 0.15 | 3 |
| 2017 | 23 March-10 May | 49 | 7.0 | 0.14 | |
| 2018 | 2 June—26 July | 55* | 8.9 | 0.16 | |

* The 48 days ended 19 July 2018 received only 1.4 mm, 0.03 mm/day

**Table 28.3** *Frequency of absolute droughts lasting at least 25 days, and partial droughts lasting at least 45 days, by starting date (month). Radcliffe Observatory, Oxford, January 1827 to December 2018*

| | Jan | Feb | Mar | April | May | June | July | Aug | Sept | Oct | Nov | Dec |
|---|---|---|---|---|---|---|---|---|---|---|---|---|
| Absolute droughts ≥ 25 days | 0 | 4 | 6 | 2 | 0 | 5 | 5 | 6 | 1 | 1 | 0 | 0 |
| Partial droughts ≥ 45 days | 3 | 7 | 8 | 4 | 0 | 3 | 5 | 4 | 2 | 0 | 1 | 0 |

**Table 28.4** *Precipitation depth-duration extremes at the Radcliffe Observatory, Oxford, January 1827 to December 2018.*

| | Wettest spells | | |
|---|---|---|---|
| **Period length** | **Amount mm** | **mm/day** | **(Start and) end dates** |
| 1 day | 87.9 | 87.9 | 10 July 1968 |
| 2 days | 98.1 | 49.1 | 9–10 July 1968 |
| 3 days | 104.5 | 34.8 | 8–10 July 1968 |
| 4 days | 109.8 | 27.5 | 24–27 July 1901 |
| 5 days | 111.2 | 22.2 | 24–28 July 1901 |
| 7 days | 118.0 | 16.9 | 8–14 July 1968 |
| | 115.9 | 16.5 | 9–15 June 1903 |
| 10 days | 127.6 | 12.8 | 7–16 July 1968 |
| | 127.4 | 12.7 | 10–19 June 1903 |
| 14 days *(2 weeks)* | 145.2 | 10.4 | 18–31 July 1834 |
| | 141.8 | 10.1 | 8–19 June 1903 |
| | 140.2 | 10.0 | 11–24 Dec 1989 |
| 21 days *(3 weeks)* | 179.7 | 8.5 | 30 Oct—19 Nov 1940 |
| | 178.0 | 8.5 | 25 Oct—14 Nov 1894 |
| 28 days *(4 weeks)* | 200.7 | 7.2 | 15 Nov—12 Dec 1929 |
| | 198.1 | 7.1 | 24 Oct—20 Nov 1894 |

| | Wettest spells | | | Driest spells | | |
|---|---|---|---|---|---|---|
| **Period length** | **Amount mm** | **mm/ day** | **Start and end dates** | **Amount mm** | **mm / day** | **Start and end dates** |
| 35 days *(5 weeks)* | 230.7 | 6.6 | 9 Oct—12 Nov 1875 | nil | nil | 1947, 1959 and 1976 (see Absolute Droughts) |
| | 230.1 | 6.6 | 5 Nov—9 Dec 1929 | | | |

*(continued)*

**Table 28.4** *Continued*

| | Wettest spells | | | Driest spells | | |
|---|---|---|---|---|---|---|
| Period length | Amount mm | mm/ day | Start and end dates | Amount mm | mm / day | Start and end dates |
| 40 days | 251.1 | 6.3 | 9 Oct—17 Nov 1875 | 0.1 | trace | 19 July—27 Aug 1976 |
| | 244.8 | 6.1 | 5 Nov—14 Dec 1929 | 0.5 | < 0.1 | 4 June—13 July 1949 |
| 45 days | 274.2 | 6.1 | 1 Oct—14 Nov 1875 | 1.3 | < 0.1 | 5 June—19 July 2018 |
| | 254.9 | 5.7 | 15 Nov—29 Dec 1929 | 1.5 | < 0.1 | 17 Mar—30 Apr 1893 |
| 50 days | 282.4 | 5.6 | 1 Oct—19 Nov 1875 | 1.9 | < 0.1 | 17 Mar—15 May 1893 |
| | 272.1 | 5.4 | 9 Nov—28 Dec 1929 | 2.8 | < 0.1 | 6 Mar—24 Apr 1997 |
| 60 days *(2 months)* | 323.6 | 5.4 | 19 Sept—17 Nov 1875 | 2.1 | < 0.1 | 17 Mar—15 May 1893 |
| | 303.0 | 5.1 | 25 Apr—19 June 1903 | 7.7 | 0.1 | 17 Jan—16 Mar 1932 |
| 75 days | 336.9 | 4.5 | 23 Apr—6 July 2012 | 4.3 | < 0.1 | 2 Mar—15 May 1893 |
| | 336.1 | 4.5 | 19 Sept—2 Dec 1875 | 17.2 | 0.2 | 23 Mar—5 June 1844 |
| 90 days *(3 months)* | 414.3 | 4.6 | 16 Apr—14 July 2012 | 24.2 | 0.3 | 18 Mar—15 June 1844 |
| | 375.0 | 4.2 | 5 Nov 1929—2 Feb 1930 | 25.2 | 0.3 | 5 Mar—15 May 1893 |
| 120 days *(4 months)* | 482.6 | 4.0 | 11 Oct 2013—7 Feb 2014 | 45.0 | 0.4 | 4 Mar—1 July 1893 |
| | 471.8 | 3.9 | 10 Apr—7 Aug 2012 | 50.9 | 0.4 | 15 July—11 Nov 1964 |
| 180 days *(6 months)* | 668.0 | 3.7 | 22 June—18 Dec 1960 | 99.7 | 0.5 | 1 Mar—27 Aug 1976 |
| | 651.0 | 3.6 | 3 June—29 Nov 1852 | 114.9 | 0.6 | 3 Feb—31 July 1921 |
| 365 days *(1 year)* | 1089.3 | 3.0 | 16 Apr 2012—15 Apr 2013 | 297.4 | 0.8 | 20 Sept 1975—18 Sept 1976 |
| | 1027.4 | 2.8 | 11 May 1960—10 May 1961 | 340.8 | 0.9 | 14 Jan 1921—13 Jan 1922 |

Where precipitation amounts are identical in adjacent periods, only the earliest period is shown. All are based on 'rain day' (0900–0900 GMT) daily totals. Only one instance is given for each unique (non-overlapping) spell, even though the spell may include other greater extremes than any second-shown occurrence. The period totals are exact day totals, the monthly equivalents shown are an approximation to the number of days shown.

# 29

# Oxford's 'top ten' extremes

## Temperature

Air temperatures were recorded by unscreened thermometers prior to January 1853—see Appendix 1 for more details. As noted above (Chapter 27), these are likely to be a little too cold by modern (screened thermometer) standards in winter and a little too warm in summer. For this reason, they are shown in *italics* in the tables following, together with the nearest extreme recorded since January 1853. The mean temperature correction for unscreened thermometry varies throughout the year, from +0.3 degC in December and January to −0.3 degC in July, although this will vary considerably on a day-to-day basis. Temperatures were recorded in a Stevenson screen from December 1878.

For parts of the record, the maximum minimum temperatures were read to, or are known only to, a precision of 1 degree Fahrenheit. For these, the Celsius conversion assumes the exact °F value, although the actual temperature could lie within ± 0.3 degC of that value.

*In the following tables, ' = ' indicates that the value on the previous line was equalled on one or more subsequent occasions.*

*Oxford Weather and Climate since 1767.* Stephen Burt and Tim Burt, Oxford University Press (2019).
 DOI: 10.1093/oso/9780198834632.001.0001

**Table 29.1** *Hottest and coldest days at the Radcliffe Observatory, Oxford, by daily maximum temperature, period April 1815 to December 2018*

| | Hottest days | | Coldest days | |
|---|---|---|---|---|
| Rank | Maximum temperature, °C | Date | Maximum temperature, °C | Date |
| 1 | 35.1 | 19 Aug 1932<br>= 3 Aug 1990 | *−9.6* | *8 Jan 1841* |
| 2 | 34.8 | 9 Aug 1911<br>= 19 July 2006 | *−7.9* | *2 Feb 1830* |
| 3 | 34.6 | 9 Aug 2003 | *−7.8* | *19 Jan 1823* |
| 4 | 34.3 | 27 June 1976 | *−7.4* | *20 Jan 1838* |
| 5 | 34.0 | 26 June 1976 | −7.1 | *15 Jan 1838*<br>*= 19 Jan 1838*<br>= 13 Jan 1982 |
| 6 | 33.9 | *19 July 1825*<br>= 12 July 1923 | *−6.9* | *7 Jan 1841* |
| 7 | 33.7 | 13 July 1923 | −6.7 | 26 Jan 1881<br>= 22 Dec 1890<br>= 23 Jan 1963<br>= 24 Jan 1963 |
| 8 | 33.5 | 1 July 2015 | *−6.5* | *31 Jan 1830* |
| 9 | 33.4 | 8 Sept 1911<br>= 3 July 1976 | −6.4 | *28 Jan 1848*<br>= 14 Jan 1982 |
| 10 | 33.3 | 13 Aug 1876<br>= 31 July 1943<br>= 2 Aug 1995 | −6.2 | *3 Feb 1830*<br>= 12 Jan 1987 |
| *Last 50 years to 2018* | | | | |
| | 35.1 | 3 Aug 1990 | −7.1 | 13 Jan 1982 |

**Table 29.2** *Warmest and coldest nights at the Radcliffe Observatory, Oxford, by daily minimum temperature, period April 1815 to December 2018*

| | Warmest nights | | Coldest nights | | |
|---|---|---|---|---|---|
| Rank | Minimum temperature, °C | Date | Minimum temperature, °C | Date | Grass minimum temperature*, °C |
| 1 | 21.2 | 20 July 2016 | −17.8 | 24 Dec 1860 | −17.8 |
| 2 | 20.6 | 5 Sept 1949 | −16.7 | 28 Dec 1860 | −13.3 |
| 3 | 20.5 | 11 Aug 1997 | −16.6 | 14 Jan 1982 | −22.5 |
| 4 | 20.3 | 26 Aug 1859<br>= 4 July 1976 | *−16.4* | *20 Jan 1838* | — |
| 5 | 20.1 | 5 Aug 1975<br><br>= 23 Aug 1997 | −16.1 | 24 Feb 1947<br>= 25 Feb 1947<br>= 13 Dec 1981 | −17.9<br>−22.2<br>−20.1 |
| 6 | 20.0 | *2 Sept 1825*<br>= *30 July 1827*<br>= 18 Aug 1893<br>= 1 July 1968<br>= 20 July 2006 | *−15.6* | *15 Jan 1820* | — |
| 7 | 19.7 | *21 June 1829*<br>= *13 Aug 1829*<br>= 29 July 2001 | −15.5 | 15 Jan 1982 | −18.4 |
| 8 | 19.6 | 22 July 1989<br>= 4 Aug 1994<br>= 24 Aug 1997<br>= 9 Aug 2004<br>= 27 June 2011 | *−15.0* | *19 Jan 1823* | — |
| ***Last 50 years to 2018*** | | | | | |
| | 21.2 | 20 July 2016 | −16.6 | 14 Jan 1982 | −22.5 |

* Records of grass minimum temperature commenced in 1857. Note that the nominal 'grass minimum temperature' in very cold weather is more than likely to be above a snow surface rather than grass. Where the 'grass/snow' minimum is shown as higher than the air minimum, the grass thermometer may have been buried in snow and read higher than the actual minimum temperature as a result.

Other low grass minimum temperatures

| Date | *Air minimum, °C* | Grass or snow minimum, °C |
|---|---|---|
| 5 Jan 1894 | *−11.6* | −16.4 |
| 8 Feb 1895 | *−13.6* | −17.8 |
| 7 Feb 1917 | *−10.4* | −19.3 |
| 9 Jan 1918 | *−8.4* | −16.3 |
| 29 Jan 1947 | *−12.4* | −18.3 |
| 30 Jan 1947 | *−11.2* | −18.8 |
| 7 March 1947 | *−10.8* | −19.1 |

**Table 29.3** *Warmest and coldest days at the Radcliffe Observatory, Oxford, by mean daily temperature, period April 1815 to December 2018*

| Rank | Hottest days<br>Mean daily temperature, °C | Date | Minimum temperature, °C | Maximum temperature, °C |
|---|---|---|---|---|
| 1 | 26.5 | 13 July 1923 | 19.2 | 33.7 |
| 2 | 26.3 | 4 July 1976 | 20.3 | 32.4 |
| 3 | 26.1 | 26 June 1976 | 18.2 | 34.0 |
| 4 | 26.1 | 27 June 1976 | 17.8 | 34.3 |
| 5 | 26.0 | 22 July 1989 | 19.6 | 32.4 |
| 6 | 25.9 | 2 Aug 1995 | 18.6 | 33.3 |
| 7 | 25.9 | *18 July 1825* | *18.9* | *32.8* |
| | | *= 19 July 1825* | *17.8* | *33.9* |
| | | = 28 June 1976 | 18.7 | 33.1 |
| 8 | 25.8 | 1 July 1968 | 20.0 | 31.6 |
| 9 | 25.7 | 18 Aug 1893 | 20.0 | 31.3 |
| | | = 3 Aug 1990 | 16.4 | 35.1 |
| 10 | 25.6 | 12 July 1923 | 17.3 | 33.9 |
| | | = 19 Aug 1932 | 16.1 | 35.1 |
| ***Last 50 years to 2018*** | | | | |
| | 26.3 | 4 July 1976 | 20.3 | 32.4 |

| Rank | Coldest days<br>Mean daily temperature, °C | Date | Minimum temperature, °C | Maximum temperature, °C |
|---|---|---|---|---|
| 1 | *−12.1* | *8 Jan 1841* | *−14.6* | *−9.6* |
| 2 | *−11.9* | *20 Jan 1838* | *−16.4* | *−7.4* |
| 3 | −11.5 | 14 Jan 1982 | −16.6 | −6.4 |
| 4 | *−11.4* | *19 Jan 1823* | *−15.0* | *−7.8* |
| 5 | *−10.7* | *15 Jan 1838* | *−14.3* | *−7.1* |
| 6 | *−10.7* | *7 Jan 1841* | *−14.4* | *−6.9* |
| 7 | −10.6 | 23 Jan 1963 | −14.4 | −6.7 |
| 8 | −10.3 | 24 Jan 1963 | −13.9 | −6.7 |
| 9 = | −10.0 | 24 Dec 1860 | −17.8 | −2.2 |
| | | = 22 Dec 1890 | −13.3 | −6.7 |
| | | = 4 Jan 1867 | −14.6 | −5.3 |
| ***Last 50 years to 2018*** | | | | |
| | −11.5 | 14 Jan 1982 | −16.6 | −6.4 |

By standard meteorological convention, the 'mean daily temperature' is defined as the average of the daily maximum and minimum temperatures

**Table 29.4a** *Hottest calendar months at the Radcliffe Observatory, Oxford, by mean daily temperature, period April 1815 to December 2018*

*By standard meteorological convention, the 'mean daily temperature' is defined as the average of the daily maximum and minimum temperatures: the monthly mean temperature is taken as the average of all available mean daily temperatures within that month.*

| Rank | Hottest months<br>Mean temperature, °C | Departure from 1981–2010 normal degC | Month and year | Mean daily max °C | Mean daily min °C |
|---|---|---|---|---|---|
| 1 | 21.14 | +3.2 | July 1983 | 26.8 | 15.4 |
| 2 | 21.05 | +3.1 | July 2018 | 27.4 | 14.7 |
| 3 | 21.02 | +3.1 | July 2006 | 27.1 | 14.9 |
| 4 | 20.3 | +2.7 | Aug 1997 | 24.9 | 15.7 |
| 5 | 20.1 | +2.5 | Aug 1995 | 26.4 | 13.9 |
| 6 | 19.7 | +1.8 | July 1921 | 26.1 | 13.3 |
| | | +1.8 | = July 2013 | 25.6 | 13.9 |
| 7 | 19.6 | +1.7 | July 1911 | 26.1 | 13.1 |
| | | +2.0 | = Aug 1975 | 25.1 | 14.1 |
| | | +1.7 | = July 1976 | 25.9 | 13.2 |
| 8 | 19.5 | +1.6 | July 1989 | 24.9 | 14.0 |
| | | +1.6 | July 1995 | 25.1 | 13.9 |
| | | +1.9 | Aug 2003 | 25.5 | 13.5 |
| *Last 50 years to 2018* | | | | | |
| | 21.14 | +3.2 | July 1983 | 26.8 | 15.4 |

**Table 29.4b** *Coldest calendar months at the Radcliffe Observatory, Oxford, by mean daily temperature, period April 1815 to December 2018*

| Rank | Coldest months<br>Mean temperature, °C | Departure from 1981–2010 normal degC | Month and year | Mean daily max °C | Mean daily min °C |
|---|---|---|---|---|---|
| 1 | −3.0 | −7.9 | Jan 1963 | −0.3 | −5.8 |
| 2 | *−2.6* | *−7.5* | *Jan 1814* | *N/A* | *N/A* |
| 3 | *−2.4* | *−7.2* | *Jan 1838* | *−0.2* | *−4.6* |
| 4 | −2.2 | −7.1 | Feb 1947 | −0.1 | −4.3 |
| 5 | −1.9 | −6.9 | Dec 1890 | 0.2 | −4.0 |
| 6 | −1.8 | −6.7 | Feb 1895 | 1.8 | −5.3 |
| 7 | −1.5 | −6.4 | Feb 1986 | 1.0 | −4.0 |
| 8 | −1.4 | −6.3 | Feb 1855 | 1.7 | −4.5 |
| 9 | *−1.3* | *−6.1* | *Jan 1830* | *0.7* | *−3.2* |
| 10 | −1.3 | −6.2 | Jan 1940 | 1.9 | −4.6 |
| *Last 50 years to 2018* | | | | | |
| | −1.5 | −6.4 | Feb 1986 | 1.0 | −4.0 |

Records prior to 1852 have had mean temperatures adjusted for exposure and as a result will differ from the mean of the daily maximum and minimum temperature by the value of the monthly correction given in Table A1.1

**Table 29.5** *Warmest and coldest calendar months at the Radcliffe Observatory, Oxford, by departure from the 1981-2010 monthly mean temperature, period April 1815 to December 2018*

| | Warmest months | | | Coldest months | | |
|---|---|---|---|---|---|---|
| Rank | Mean temperature, °C | Departure from 1981–2010 monthly normal degC | Month and year | Mean temperature, °C | Departure from 1981–2010 monthly normal degC | Month and year |
| 1 | 10.9 | +5.9 | Dec 2015 | −3.0 | −7.9 | Jan 1963 |
| 2 | 9.0 | +4.0 | Dec 1852 | *−2.6* | *−7.5* | *Jan 1814* |
| 3 | 13.3 | +4.0 | April 2011 | *−2.4* | *−7.2* | *Jan 1838* |
| 4 | 8.3 | +3.4 | Feb 1990 | −2.2 | −7.1 | Feb 1947 |
| 5 | 21.1 | +3.2 | July 1983 | −1.9 | −6.9 | Dec 1890 |
| 6 | 8.2 | +3.2 | Dec 1934 | −1.8 | −6.7 | Feb 1895 |
| 7 | 8.2 | +3.2 | Dec 1974 | −1.5 | −6.4 | Feb 1986 |
| 8 | 21.0 | +3.1 | July 2018 | −1.4 | −6.3 | Feb 1855 |
| 9 | 21.0 | +3.1 | July 2006 | −1.3 | −6.2 | Jan 1940 |
| 10 | 10.6 | +3.0 | Nov 1994 | *−1.3* | *−6.1* | *Jan 1830* |
| ***Last 50 years to 2018*** | | | | | | |
| | 10.9 | +5.9 | Dec 2015 | −1.5 | −6.4 | Feb 1986 |

**Table 29.6** *Greatest daily ranges in temperature at the Radcliffe Observatory, Oxford, period April 1815 to December 2018*

| Rank | Greatest daily ranges<br>Daily range, degC | Date | Minimum temperature, °C | Maximum temperature, °C |
|---|---|---|---|---|
| 1 | 22.5 | 23 Sept 1895 | 2.8 | 25.3 |
| 2 | 22.5 | 8 Sept 1911 | 10.9 | 33.4 |
| 3 | 21.8 | 24 April 1855 | −2.1 | 19.7 |
| 4 | 21.7 | 7 Sept 1911 | 10.1 | 31.8 |
| 5 | *21.6* | *24 June 1822* | *6.7* | *28.3* |
| 6 | 21.4 | 11 March 1929 | −2.2 | 19.2 |
| | | = 29 March 1965 | 0.7 | 22.1 |
| 7 | 21.3 | 9 April 1909 | 0.7 | 22.0 |
| 8 | 21.2 | 28 March 1907 | −0.9 | 20.3 |
| | | = 25 March 1953 | −0.6 | 20.6 |
| | | = 15 July 1990 | 9.3 | 30.5 |
| 9 | 21.1 | 18 March 1929 | −4.6 | 16.5 |
| | | = 2 July 1976 | 11.9 | 33.0 |
| 10 | 21.0 | 30 Aug 1906 | 7.2 | 28.2 |
| | | = 30 June 1995 | 11.2 | 32.2 |
| ***Last 50 years to 2018*** | | | | |
| | 21.2 | 15 July 1990 | 9.3 | 30.5 |

*The daily temperature range is the difference between the maximum and minimum temperatures over the period 0900-0900 GMT.*

## Precipitation

See also Table 28.4 for details of other periods from 1 day to 1 year.

Since November 1940, the most recent entry in the 'wettest months' table (Table 29.7) as we write in early 2019, the wettest months have been October 1949 (162.3 mm), November 1970 (150.3 mm), September 1974 (156.1 mm), June 2012 (151.7 mm) and January 2014 (146.9 mm).

For a heavy fall of rain to appear in Table 29.8, the timing of heavy rain needs to somewhat fortuitously fall within the 0900–0900 GMT standard 'rain day'. There are other examples in the record where the fall has been split across two rain days and therefore does not appear in the table—a relatively recent example being 20 July 2007, when 58.8 mm fell, almost all of it within the 'civil day' (see Chapter 13, July for hourly data for another site in the Oxford area on this date).

It is perhaps noteworthy that eight of the ten highest daily falls (Table 29.8) have occurred since 1950, although as at 2018 no fall in excess of 60 mm has been recorded since 1973.

**Table 29.7** *Wettest and driest calendar months at the Radcliffe Observatory, Oxford, by total precipitation, period January 1767 to December 2018*

| | **Wettest months** | **Driest months** | | | | |
|---|---|---|---|---|---|---|
| **Rank** | **Total precipitation, mm** | **Per cent of monthly 1981–2010 normal** | **Month and year** | **Total precipitation, mm** | **Per cent of monthly 1981–2010 normal** | **Month and year** |
| 1 | 223.9 | 413 | Sept 1774 | 0.5 | 1 | April 1817, April 2011 |
| 2 | 192.4 | 288 | Nov 1770 | 0.6 | 1 | April 1912 |
| 3 | 189.0 | 267 | Oct 1875 | 0.8 | 2 | July 1825 |
| 4 | 175.7 | 264 | Nov 1852 | 1.2 | 3 | July 1800 |
| 5 | 175.5 | 264 | Nov 1940 | 1.4 | 3 | March 1929 |
| 6 | 174.8 | 359 | July 1834 | 1.5 | 3 | May 1795, April 1893 |
| 7 | 174.0 | 261 | Nov 1772 | 1.6 | 3 | April 1984 |
| 8 | 170.9 | 315 | Sept 1768 | 1.7 | 4 | Dec 1780 = June 1925 |
| 9 | 170.4 | 351 | June 1852 | 1.8 | 3 | Aug 1940 = April 2007 |
| 10 | 165.9 | 249 | Nov 1929 | 1.9 | 4 | March 1781 |
| *Last 50 years to 2018* | | | | | | |
| | 156.1 | 310 | Sept 1974 | 0.5 | 1 | April 2011 |

**Table 29.8** *Wettest days at the Radcliffe Observatory, Oxford, by daily precipitation total 0900-0900 GMT, period January 1827 to December 2018*

| Rank | Wettest days<br>Total precipitation, mm | Date |
|---|---|---|
| 1 | 87.9 | 10 July 1968 |
| 2 | 84.8 | 6 Sept 1951 |
| 3 | 81.3 | 22 June 1960 |
| 4 | 70.8 | 6 Aug 1922 |
| 5 | 67.3 | 27 June 1973 |
| 6 | 56.1 | 12 Aug 1957 |
| 7 | 53.6 | 6 Aug 1962 |
| 8 | 51.1 | 14 June 1903 |
| 9 | 51.1 | 25 Aug 2010 |
| 10 | 50.0 | 8 July 2004 |

## Greatest snow depths

Cold winters tend to produce a large number of days with similar snow depths, often a slow reduction after one or two major snowfalls. Table 29.9 shows the list of absolute ranked snow depths since January 1959, which is dominated by the three largest snowfalls in Oxford within the last 60 years—in the winters of 1962/63, 1981/82 and 2009/10.

The right half of Table 29.9 shows the greatest single snow depth for every winter since 1959/60, dated by the January, where there was at least one morning with 10 cm or more snow cover. The winter of 1981/82 has two entries, because the two large snowfalls of that winter were almost a month apart and were separated by a mild spell at New Year which melted all the existing snow cover. After 1982, it was to be another 25 years before Oxford saw 10 cm of lying snow.

Table 29.10 provides a measure of the 'snowiest' winter months in Oxford during the past 60 years, using an accumulated snow depth index. This index is derived simply by accumulating all of the snow depths at 0900 during the month; thus, a month with three mornings with snow cover (50 per cent cover or more), each of 2 cm, will accumulate 6 cm. This does not imply that the snow depth reached this level—it is only an integration of the recorded snow depths during the month. High values can arise from a few days with deep snow, or a long period with relatively small but persistent snow cover.

Winter 1963 is the only winter in which all three months feature in the table (winter total 868 cm). See also the plot of daily snow depths in the winters of 1947, 1963, 1979 and 2010 in Winter, Chapter 20.

**Table 29.9** *Greatest snow depths (centimetres) at 0900 GMT at the Radcliffe Observatory, Oxford, period January 1959 to December 2018*

| Greatest snow depths | | | Greatest snow depths each winter *Depths of 10 cm or more only* | | |
|---|---|---|---|---|---|
| Rank | Snow depth, cm | | Winter (January year) | Snow depth, cm | Date |
| 1 | 27 | 6 Jan 2010 | 1960 | 13 | 15 Jan |
| 2 | 24 | 10 Jan 1982 | 1962 | 15 | 1 Jan |
| 3 | 23 | 3–4 Jan 1963 | 1963 | 23 | 3–4 Jan and 20–26 Jan |
| | | and 20–26 Jan 1963 | 1965 | 10 | 5 March |
| 4 | 20 | 30 Dec 1962 | 1968 | 10 | 9 and 13 Jan |
| | | 2 and 5 Jan 1963 | 1969 | 10 | 20 Feb |
| | | 27 Jan 1963 | 1970 | 20 | 5 March |
| | | 5 March 1970 | 1979 | 10 | 17 March |
| | | 9 Jan 1982 | 1982 | 16 | 12 Dec 1981 |
| | | 7 Jan 2010 | | 24 | 10 Jan 1982 |
| | | | 2007 | 11 | 8 Feb |
| | | | 2009 | 10 | 6 Feb |
| | | | 2010 | 27 | 6 Jan |
| | | | 2011 | 18 | 19 and 20 Dec 2010 |

In earlier years, we know the snow lay 61 cm deep on 13 and 14 February 1888, and 43 cm deep at 2230h on 25 April 1908; see also Table 27.4 for 'earliest and latest' dates with different depth thresholds.

**Table 29.10** *Greatest accumulated snow depth index (centimetre-days) at the Radcliffe Observatory, Oxford, period January 1959 to December 2018*

| Month | Accumulated snow depth (cm) |
|---|---|
| January 1963 | 561 |
| February 1963 | 238 |
| January 1982 | 171 |
| January 2010 | 165 |
| December 2010 | 149 |
| March 1970 | 71 |
| December 1962 | 69 |
| December 1981 | 67 |
| February 1985 | 66 |
| January 1979 | 61 |

## Sunshine

**Table 29.11** *Sunniest and dullest calendar months by total sunshine duration at the Radcliffe Observatory, Oxford, period February 1880 to December 2018*

| Rank | Sunniest months | | | | Dullest months | | | |
|---|---|---|---|---|---|---|---|---|
| | Total sunshine, hours | Per cent of 1981–2010 monthly normal | Per cent of possible | Month and year | Total sunshine, hours | Per cent of 1981–2010 monthly normal | Per cent of possible | Month and year |
| 1 | 310.45 | 150 | 62.1 | July 1911 | 5.0 | 9 | 2.0 | Dec 1890 |
| 2 | 303.7 | 147 | 60.8 | July 2006 | 14.9 | 24 | 5.7 | Jan 1885 |
| 3 | 301.0 | 158 | 60.4 | June 1975 | 18.2 | 34 | 7.5 | Dec 1956 |
| 4 | 300.8 | 156 | 61.9 | May 1989 | 20.4 | 38 | 8.4 | Dec 2010 |
| 5 | 297.3 | 144 | 59.5 | July 2013 | 22.8 | 29 | 8.1 | Feb 1940 |
| 6 | 297.1 | 156 | 59.7 | June 1957 | 23.2 | 43 | 9.5 | Dec 1884 |
| 7 | 293.6 | 152 | 60.4 | May 1909 | 23.3 | 43 | 9.5 | Dec 1930 |
| 8 | 290.7 | 152 | 58.4 | June 1996 | 24.2 | 39 | 9.3 | Jan 1917 |
| 9 | 285.1 | 145 | 63.2 | Aug 1995 | 25.6 | 47 | 10.5 | Dec 1927 |
| 10 | 285.0 | 148 | 58.6 | May 1990 | 26.3 | 33 | 9.3 | Feb 1947 |
| *Last 50 years to 2018* | | | | | | | | |
| | 303.7 | 147 | 60.8 | July 2006 | 20.4 | 38 | 8.4 | Dec 2010 |

# 30

# Notable heatwaves and cold spells, sunny and dull periods, in Oxford since 1815

## Prolonged heatwaves

Table 30.1 lists all spells of at least three consecutive days reaching or surpassing 30.0 °C at the Radcliffe Observatory since 1814. To and including summer 2018, there have been 24 such spells in just over 200 years—although almost half of those (11 to summer 2018) have occurred since 1975. Only 1876 and 2003 recorded more than one spell in the year—in 2003 just two slightly less hot days separated the spells shown in Table 30.1.

The word 'unprecedented' is often used, and abused, when referring to extreme weather events. However, one event on Oxford's long weather record does deserve the term without reservation or hesitation: the heatwave of June–July 1976. This event is unlike any other on the 200+ year record, before or since, and truly deserves to be identified as 'unprecedented'.

**Table 30.1** *The longest heatwaves (consecutive days reaching at least 30 °C) on Oxford's records, between December 1813 and December 2018. The longest spell is highlighted in bold.*

| Year | Dates in spell | Spell length | Highest max reached, °C | Hottest day in spell |
|---|---|---|---|---|
| *1825* | *17–19 July* | *3 days* | *33.9* | *19 July* |
| *1826* | *2–4 July* | *3* | *30.6* | *2,3 and 4 July* |
| 1856 | 1–3 August | 3 | 31.7 | 2 August |
| 1876 | 14–16 July | 3 | 31.8 | 16 July |
| 1876 | 13–17 August | 5 | 33.3 | 13 August |
| 1901 | 18–20 July | 3 | 31.7 | 19 July |
| 1906 | 31 August—2 September | 3 | 33.1 | 1 September |

(*continued*)

*Oxford Weather and Climate since 1767*. Stephen Burt and Tim Burt, Oxford University Press (2019).
 DOI: 10.1093/oso/9780198834632.001.0001

Table 30.1 *Continued*

| Year | Dates in spell | Spell length | Highest max reached, °C | Hottest day in spell |
|---|---|---|---|---|
| 1911 | 12–14 August | 3 | 32.5 | 12 August |
| 1921 | 9–11 July | 3 | 31.7 | 10 July |
| 1923 | 11–13 July | 3 | 33.9 | 12 July |
| 1930 | 27–29 August | 3 | 31.8 | 28 August |
| 1947 | 16–18 August | 3 | 31.1 | 16 August |
| 1948 | 26–30 July | 5 | 32.2 | 29 July |
| 1975 | 2–4 August | 3 | 32.9 | 4 August |
| **1976** | **25 June—8 July** | **14** | **34.3** | **27 June** |
| 1983 | 12–16 July | 5 | 31.9 | 14 July |
| 1989 | 21–23 July | 3 | 32.4 | 22 July |
| 1990 | 1–4 August | 4 | 35.1 | 3 August |
| 1995 | 29 July—3 August | 6 | 33.3 | 2 August |
| 2003 | 4–6 August | 4 | 31.5 | 5 August |
| 2003 | 9–11 August | 3 | 34.6 | 9 August |
| 2006 | 17–19 July | 3 | 34.8 | 19 July |
| 2009 | 29 June—2 July | 4 | 31.0 | 1 July |
| 2017 | 19–21 June | 3 | 32.5 | 21 June |

*Temperatures recorded by unscreened thermometers prior to 1853 are shown in italics; these are likely to be a little too warm by modern (screened thermometer) standards in summer—see* Appendix 1 *for more details*

## Prolonged cold spells

Table 30.2 lists all spells of at least eight consecutive days when the maximum temperature did not exceed 0.0 °C at the Radcliffe Observatory since April 1815. To and including 2018, there have been 13 such spells in just over 200 years—although the thermometer exposure in use up to 1853 would have resulted in somewhat lower maximum temperatures than a standard Stevenson screen-based record today, and this is reflected in the Table by showing pre-1853 values in *italics*.

Tables 30.2 and 30.3 show the coldest spells of various lengths recorded at the Radcliffe Observatory since April 1815. These are arranged in date order with the lowest mean temperature in each spell length shown in **bold.** See also Table 29.4b, 'Coldest calendar months by mean temperature'.

**Table 30.2** *The longest spells of sub-freezing days (consecutive days not exceeding 0.0 °C) on Oxford's records, April 1815 to December 2018. The longest spell is highlighted in bold.*

| Year | Dates in spell | Spell length | Lowest max, °C and date | Coldest night, °C and date |
|---|---|---|---|---|
| *1820/1* | *24 Dec 1820–5 Jan 1821* | *13 days* | *−3.9, 29 and 30 Dec* | *−8.3, 31 Dec* |
| ***1823*** | ***11–26 Jan*** | ***17*** | ***−7.8, 19 Jan*** | ***−15.0, 19 Jan*** |
| *1826* | *9–16 Jan 1826* | *9* | *−4.4, 16 Jan* | *−13.3, 16 Jan* |
| *1829* | *16–24 Jan* | *9* | *−5.6, 23 Jan* | *−10.0, 20 Jan* |
| *1829/30* | *23 Dec 1829–2 Jan 1830* | *11* | *−4.2, 29 Dec* | *−8.8, 24 Dec* |
| *1838* | *8–20 Jan* | *13* | *−7.4, 20 Jan* | *−16.4, 20 Jan* |
| *1841* | *1–10 Feb* | *10* | *−5.6, 3 Feb* | *−10.7, 3 Feb* |
| *1844* | *6–14 Dec* | *9* | *−3.7, 13 Dec* | *−6.7, 6 Dec* |
| 1881 | 15–22 Jan | 8 | −3.4, 16 Jan | −14.0, 22 Jan |
| 1890 | 13–22 Dec | 10 | −6.7, 22 Dec | −13.3, 22 Dec |
| 1947 | 11–24 Feb | 14 | −2.9, 17 Feb | −15.7, 24 Feb |
| 1963 | 17–24 Jan | 8 | −6.7, 23 and 24 Jan | −14.2, 23 Jan |
| 1982 | 7–14 Jan | 8 | −7.1, 13 Jan | −16.6, 14 Jan |

*Since 1982, the longest unbroken spell below 0 °C has been the six days 17–22 Dec 2010: maximum −4.4 °C on 19 December, minimum −10.9 °C on 20 December.*

**Table 30.3** *The coldest 7, 14 and 21 day periods on Oxford's records, April 1815 to December 2018, in chronological order. In the case of multiple candidates within a particular range of dates, only the lowest mean temperature is quoted. The coldest spell in each duration category is highlighted in bold.*

| Dates | Mean temperature |
|---|---|
| ***7-day spells*** | |
| *19–25 January 1823* | *−6.0 °C* |
| *31 Jan—6 Feb 1830* | *−7.0* |
| ***14–20 January 1838*** | ***−7.8*** |
| 20–26 January 1881 | −6.4 |
| 19–25 January 1963 | −6.5 |
| 9–15 January 1982 | −6.4 |
| *Since 1982, the coldest 7 day spell has been 11–17 January 1987, mean temperature −4.2 °C* | |

(*continued*)

Table 30.3 *Continued*

| Dates | Mean temperature |
|---|---|
| ***14-day spells*** | |
| *3–16 January 1830* | *−4.5 °C* |
| *12–25 January 1823* | *−5.0* |
| ***8–21 January 1838*** | ***−6.3*** |
| 8–21 February 1855 | −4.4 |
| 21 Dec 1870–3 Jan 1871 | −4.3 |
| 13–26 January 1881 | −5.8 |
| 5–18 February 1895 | −4.6 |
| 11–24 January 1963 | −5.3 |
| *Within the last 50 years to 2018, the coldest 14 day spell has been 7–20 January 1987, mean temperature −2.4 °C* | |
| ***21-day spells*** | |
| *28 Dec 1819–17 Jan 1820* | *−4.0 °C* |
| *7–27 January 1823* | *−3.6* |
| *17 Jan—6 Feb 1830* | *−3.4* |
| ***8–28 January 1838*** | ***−4.7*** |
| 20 Dec 1860–9 Jan 1861 | −3.1 |
| 7–27 January 1881 | −3.8 |
| 11–31 December 1890 | −3.4 |
| 26 Jan—15 Feb 1895 | −3.9 |
| 6–26 February 1947 | −3.0 |
| 7–27 January 1963 | −4.4 |
| *Within the last 50 years to 2018, the coldest 21 day spell has been 7–27 February 1986, mean temperature −2.3 °C* | |

## Prolonged sunny weather

Table 30.4 lists the sunniest spells of various lengths recorded at the Radcliffe Observatory since January 1921 (although sunshine records commenced in February 1880, only a few months yet have digitised daily data prior to January 1921). These are arranged in date order with the highest sunshine duration in each spell length shown in **bold.** See also Table 29.11, 'Sunniest and dullest calendar months'. As with the heatwave extremes in Table 30.1, the summer of 1976 has a 'clean sweep' in the sunniest 10- to 21-day periods, although, over the longer durations, 1989, 1990 and 2018 edge ahead.

**Table 30.4** *The sunniest periods of 7–60 consecutive days on Oxford's records, January 1921 to December 2018, in chronological order (includes some specifically-digitised earlier sunny months such as July 1911). In the case of multiple candidates within a particular range of dates, only the greatest period total is included*. The sunniest spell in each duration category is highlighted in bold.*

| Dates | Total sunshine, hours | Average per day, hours |
|---|---|---|
| *7-day spells* | | |
| 4–10 July 1934 | 100.2 | 14.31 |
| 3–9 June 1939 | 101.5 | 14.50 |
| 2–8 June 1940 | 101.5 | 14.50 |
| 11–17 June 1957 | 101.1 | 14.44 |
| 26 June—2 July 1976 | 100.7 | 14.39 |
| **12–18 June 1996** | **102.3** | **14.61** |
| *10-day spells* | | |
| 2–11 July 1934 | 139.4 | 13.94 |
| 1–10 June 1939 | 134.0 | 13.40 |
| 12–21 June 1957 | 137.2 | 13.72 |
| 30 May—8 June 1962 | 130.1 | 13.01 |
| 6–15 June 1969 | 132.3 | 13.23 |
| **23 June—2 July 1976** | **142.1** | **14.21** |
| 9–18 May 1980 | 132.4 | 13.24 |
| 24 May—2 June 1977 | 137.9 | 13.79 |
| 24 June—3 July 2018 | 131.1 | 13.11 |
| *14-day spells* | | |
| 28 June—11 July 1934 | 170.8 | 12.20 |
| 28 May—10 June 1939 | 171.1 | 12.22 |
| 10–23 June 1957 | 170.0 | 12.14 |
| **24 June—7 July 1976** | **183.1** | **13.08** |
| 11–24 July 1990 | 171.9 | 12.28 |
| 4–17 June 1996 | 170.3 | 12.16 |
| 6–19 July 2013 | 178.5 | 12.75 |
| 21 June—4 July 2018 | 180.5 | 12.89 |

(*continued*)

Table 30.4 *Continued*

| Dates | Total sunshine, hours | Average per day, hours |
|---|---|---|
| *21-day spells* | | |
| 23 May—12 June 1939 | 238.5 | 11.36 |
| 20 June—10 July 1941 | 237.7 | 11.32 |
| 11 June—1 July 1957 | 236.6 | 11.27 |
| 6–26 June 1975 | 238.0 | 11.33 |
| 22 June—12 July 1976 | 251.6 | 11.98 |
| 30 May—19 June 1996 | 240.4 | 11.45 |
| 5–25 July 2013 | 236.2 | 11.25 |
| **21 June—11 July 2018** | **262.0** | **12.48** |
| *30-day spells* | | |
| 2–31 July 1911 | 307.3 | 10.24 |
| 15 June—14 July 1941 | 320.3 | 10.68 |
| 25 May—23 June 1957 | 307.6 | 10.25 |
| 6 June—5 July 1975 | 306.8 | 10.23 |
| 23 June—22 July 1976 | 304.6 | 10.15 |
| 11 July—9 August 1990 | 323.0 | 10.77 |
| 22 July—20 August 1995 | 322.7 | 10.76 |
| 5 July—3 August 2013 | 310.3 | 10.34 |
| **18 June—17 July 2018** | **332.5** | **11.08** |
| *60-day spells* | | |
| 17 June—15 August 1949 | 543.0 | 9.05 |
| 6 June—4 August 1975 | 553.2 | 9.22 |
| 24 June—22 August 1976 | 546.2 | 9.10 |
| 11 June—9 August 1989 | 560.7 | 9.35 |
| 22 June—20 August 1995 | 557.7 | 9.29 |
| 30 May—28 July 1996 | 545.8 | 9.10 |
| 2 June—31 July 2006 | 549.3 | 9.15 |
| **10 June—8 August 2018** | **573.3** | **9.55** |

* Only the summer of 1989 has two overlapping periods separated by more than a few days; the 60 days 10 July to 7 September 1989 received 550.0 h sunshine but overlaps the 11 June to 9 August period in that year (which was the sunnier, 560.7 h). The 90 days 10 June to 7 September 1989 received 808.8 h sunshine, a daily average of a tiny fraction under 9 hours.

## Prolonged dull weather

Table 30.5 lists the dullest 30-day and 60-day spells recorded at the Radcliffe Observatory since January 1921 (although sunshine records commenced in February 1880, only a few months have digitised daily data prior to January 1921). These are arranged in date order with the lowest sunshine duration in each spell length shown in bold. See also Table 29.11, 'Sunniest and dullest calendar months'.

Since January 1921, the longest consecutive runs of sunless days have been the 15 days 26 January to 9 February 1940 and 14 days 14–27 January 1987.

**Table 30.5** *The dullest 30 day and 60 day periods on Oxford's records, January 1921 to December 2018, in chronological order. In the case of multiple candidates within a particular range of dates, only the lowest period total is included. The dullest spell in each duration category is highlighted in bold.*

| Dates | Total sunshine, hours | Average per day, hours |
|---|---|---|
| *30-day spells* | | |
| 24 Jan—22 Feb 1947 | 12.3 | 0.41 |
| 20 Nov—19 Dec 1953 | 12.9 | 0.43 |
| 20 Jan—18 Feb 1966 | 14.2 | 0.47 |
| 20 Jan—18 Feb 1993 | 14.4 | 0.48 |
| **9 Dec 2010–7 Jan 2011** | **11.8** | **0.39** |
| *60-day spells* | | |
| 24 Dec 1934–22 Jan 1935 | 59.5 | 0.99 |
| **26 Nov 1968–24 Jan 1969** | **57.4** | **0.96** |
| 27 Nov 1995–25 Jan 1996 | 59.7 | 0.99 |
| 16 Nov 2010–14 Jan 2011* | 58.4 | 0.97 |

* Includes 13 days' data from CEH Wallingford

# Appendices

# Appendix 1
# Metadata: Meteorological observations at the Radcliffe observatory, Oxford, 1772 to date

This appendix sets out the main details of the instruments used and their exposures, together with any corrections which may have been applied to enhance homogeneity of the older series with current observations. For instrumental details, we have made extensive use of the monograph prepared by Gordon Wallace in 1997 [21], to which the reader is referred for additional details over and above those given here; our intention is to provide a brief, stand-alone reference account based upon Wallace and relevant contemporary sources, amplifying detail where necessary and bringing the account up to date at the time of writing (autumn 2018).

## Temperature

### Sources of data

Outside air temperature observations from January 1811 exist in the original Radcliffe Observatory registers, held in the School of Geography and the Environment at the University of Oxford. The records from 1811 to 1853 consist mostly of three temperature observations per day, with daily Six's maximum–minimum temperatures from April 1815. The latter were photographed and digitised by the authors: more details on thermometry used, exposure and precision of readings are given below.

The main source for the majority of the daily record of maximum and minimum temperatures was the Met Office/BADC digital record for the Radcliffe Observatory, which commences on 1 January 1854: this has only a few short gaps which were completed by reference to the original manuscript records or, more recently, estimated from neighbouring sites. This dataset appears to have been assembled from the digitisation of published records from the Radcliffe Observatory (the meteorological *Radcliffe Results* volumes, for example, [30]) from 1854 to 1930 and digitised climatological returns from the Radcliffe Observatory since 1930. This source, provided under Open Government Licence from the Met Office, was used for data to 31 December 2016: at the time, the dataset had been quality-control checked up to the end of 2016*. Data were provided in degrees Celsius to 1 decimal place, although from 1 December 1930 until Celsius scale thermometers

* For some inexplicable reason, the maximum temperatures on the Met Office/BADC dataset are held against the day of reading rather than being thrown back to the previous day, per normal climatological practice. The corrected dates have been used throughout this publication, in both daily values and extremes and monthly/seasonal and annual means and extremes.

replaced Fahrenheit units in January 1972 most are conversions of values originally digitised to the nearest degree Fahrenheit (0.55 degC). All daily maximum and minimum temperatures for the period 1961–70 were taken from a University of Oxford daily dataset, where they were held to a precision of 0.1 degF before conversion to degC, although there are a small number of days in this 10-year period where differences from the Met Office dataset exceed 1 degC (23 maxima, 13 minima)—the reason for these differences is presently unknown and being investigated: where values differ, the University of Oxford daily dataset has been used in preference. Published daily values of maximum and minimum temperature to 0.1 degF are available in the *Radcliffe Results* volumes from 1853 to 1935 inclusive: some months have already been re-digitised to this precision from these sources or the original manuscript registers. The entire series is gradually being re-digitised to 0.1 degF precision from published records or original manuscript ledgers: minor variations in daily values and monthly means from previously published values to 1 degF precision can be expected as this work continues after the publication of this volume.

From January 2017, temperature observations published monthly by the Radcliffe Meteorological Station have been used in this publication, with occasional short gaps completed by reference to neighbouring stations as necessary.

## Thermometry and thermometer exposure at the Radcliffe Observatory since 1811

Aside from the Hornsby records referred to in Chapter 2, the earliest temperature and pressure records from the Observatory date from January 1811 (Figure A1.1). Observations were made three or four times daily for just over six months: the records cease after 25 July 1811. They resumed in a similar fashion, now mostly with two observations daily, between 20 October 1811

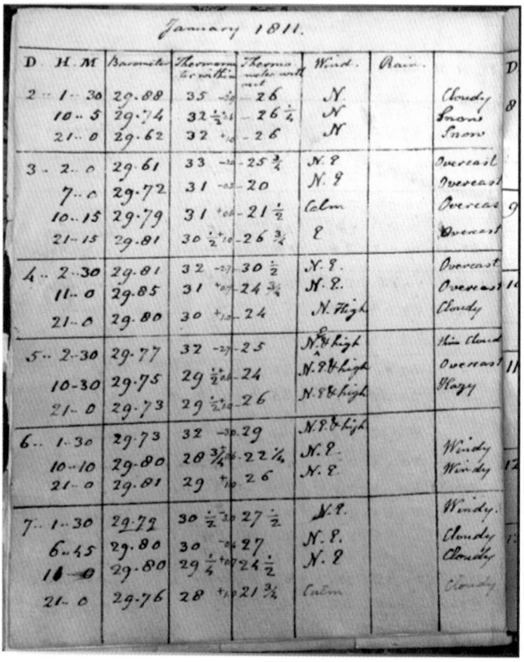

January 1811.

| D | H.M | Barometer | Thermometer within | Thermometer outside | Wind | Rain | |
|---|---|---|---|---|---|---|---|
| 2 | 1..30 | 29.88 | 35 | 26 | N | | Cloudy |
| | 10..5 | 29.74 | 32½ | 26¼ | N | | Snow |
| | 21..0 | 29.62 | 32 | 26 | N | | Snow |
| 3 | 2..0 | 29.61 | 33 | 25¾ | N.E | | Overcast |
| | 7..0 | 29.72 | 31 | 20 | N.E | | Overcast |
| | 10..15 | 29.79 | 31 | 21½ | Calm | | Overcast |
| | 21..15 | 29.81 | 30½ | 26¾ | E | | Overcast |
| 4 | 2..30 | 29.81 | 32 | 30½ | N.E. | | Overcast |
| | 11..0 | 29.85 | 31 | 24¾ | N.E. | | Overcast |
| | 21..0 | 29.80 | 30 | 24 | N. High | | Cloudy |
| 5 | 2..30 | 29.77 | 32 | 25 | N.E. high | | Thin Cloud |
| | 10-30 | 29.75 | 29½ | 24 | N.E. & high | | Overcast |
| | 21-0 | 29.73 | 29½ | 26 | N.E. & high | | Hazy |
| 6 | 1..30 | 29.73 | 32 | 29 | N.E. & high | | Windy |
| | 10..10 | 29.80 | 28¾ | 22¼ | N.E. | | Windy |
| | 21..0 | 29.81 | 29 | 26 | N.E. | | |
| 7 | 1..30 | 29.72 | 30½ | 27½ | N.E. | | Windy |
| | 6..45 | 29.80 | 30 | 27 | N.E. | | Cloudy |
| | 10..0 | 29.80 | 29½ | 24½ | N.E. | | Cloudy |
| | 21..0 | 29.76 | 28 | 21¾ | Calm | | Cloudy |

**Figure A1.1** *The first post-Hornsby meteorological records from the Radcliffe Observatory, for 2–7 January 1811 (Tim Burt)* and 2 March 1812. There is then an 18-month gap until observations restarted on 14 November 1813, resuming three observations per day most days. As described elsewhere in this book, this observation record has continued without significant interruption to the present day.

## Early air temperature records

The earliest air temperature records are simply headed 'Thermometer without' (the 'Thermometer within' column evident in Figure A1.1 referred to the temperature of the barometer, required to correct for thermal expansion of the mercury column). As was probably the case since Hornsby's day, the 'thermometer without' was an unshielded mercury thermometer fixed to the north wall of the Observatory on its eastern wing, outside the window of the Transit room, probably at or close to position 2 on Figure A1.2.

It is of interest to note that at least one of Hornsby's original thermometers, made by London instrument-maker John Bird* about 1760, survives in the Museum of the History of Science

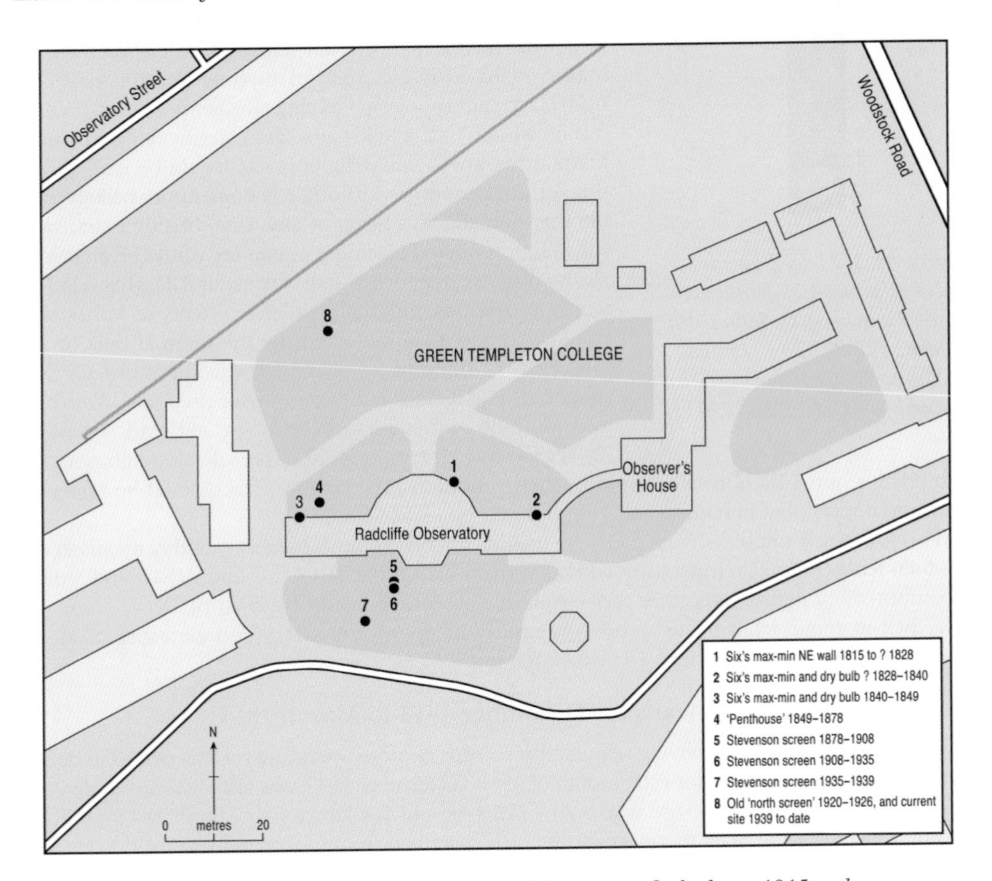


**Figure A1.2** *Thermometry positions at the Radcliffe Observatory Oxford, pre-1815 to date (Chris Orton)*

* John Bird died in 1776. In his will he left a bequest to the Observatory, which provided a small additional observer stipend until 1914 (Ivor Guest, 1991: *Dr John Radcliffe and his Trust.* London: The Radcliffe Trust, p. 241).

**Figure A1.3** *Metal-framed mercury thermometer, graduated in degrees Fahrenheit, by John Bird, London, c. 1760: 426 × 45 mm (© Museum of the History of Science, University of Oxford: Inv. 29413)*

in Oxford (Figure A1.3), to whom it was presented by the Radcliffe Trustees in 1933. Six Bird thermometers were ordered by Hornsby for the Radcliffe Observatory, four with wooden scales and two with metal scales: a thermometer with a metal scale would be more weather-resistant. However, this is not the thermometer used for exterior temperature records throughout Hornsby's time; because the scale reads up to boiling point, it was most likely a general-purpose laboratory thermometer, or used in teaching. It is very likely, though, that the Bird thermometer(s) in daily use would have closely resembled this instrument.

An unshielded thermometer such as this would be likely to read somewhat lower than true air temperature during the hours of darkness, owing to radiative cooling to the sky, and somewhat higher than true air temperature during daylight hours, owing to interception of thermal infrared radiation from surrounding objects. Cooling effects would predominate during winter, owing to lower solar angles by day and longer hours of darkness, while the opposite would be true during the summer months (although it does appear likely that the thermometer was never at any time directly exposed to sunshine). Radiative warming or cooling would be greater in clear/sunny weather with light winds, and least in cloudy, windy conditions, although radiative effects of either sign would be offset to some extent by the thermal bulk of the Observatory building upon which the thermometer was mounted. The unshielded thermometer itself would also act as a wet bulb in precipitation or in fog and read somewhat lower than true air temperature as a result, although, with the high relative humidity normally found in such conditions, the cooling effect would be relatively small and intermittent in nature.

Two previously unpublished records are included in this volume, namely monthly means of the dry-bulb temperature ('temperature without') from December 1813 to March 1815 and a daily maximum–minimum temperature series from a Six's thermometer from April 1815. These are described in some detail below: a brief summary of the thermometry and exposure of all air temperature measurements since 1811 then follows.

## Monthly temperature means for December 1813 to March 1815

The authors digitised the surviving manuscript records of air temperature for this period to derive monthly mean temperatures for each month of 1814 (December 1813 was also included to derive a mean temperature for the severe winter of 1813/14), and for January to March 1815 until the commencement of the Six's thermometer record, described in the following section. There were normally three observations per day, typically at 0830 to 0900 clock time, 1230 to 1300 and 2200 to 2230. The mean temperature of each of the three observation hours was calculated for each month and the 'raw' monthly mean calculated as the average of the three (a manuscript note added retrospectively to the January 1851 observation registers confirms that this was the same method used by Knox-Shaw and Balk [30] when they prepared the published monthly temperature means for the Radcliffe Observatory from 1815). A monthly correction was then added to the 'raw' mean

**Table A1.1** *Monthly temperature corrections applied by Knox-Shaw and Balk to correct for radiative effects on the mean monthly temperature derived from the average of thrice-daily unscreened dry bulb temperatures at the Radcliffe Observatory, 1815–1849.*

| | Jan | Feb | Mar | Apr | May | June | July | Aug | Sept | Oct | Nov | Dec |
|---|---|---|---|---|---|---|---|---|---|---|---|---|
| Correction to mean temperature from three hourly means, degF | +0.5 | +0.4 | +0.3 | +0.1 | −0.2 | −0.3 | −0.5 | −0.3 | 0.0 | +0.3 | +0.5 | +0.6 |
| Conversion to degC | +0.28 | +0.22 | +0.17 | +0.06 | −0.11 | −0.17 | −0.28 | −0.17 | 0.00 | +0.17 | +0.28 | +0.33 |

to make allowance for radiative effects upon the thermometer, positive in winter and negative in summer. The derivation of this correction was not explicitly stated by Knox-Shaw and Balk but was presumably based upon overlapping measurements: the corrections applied are given in Table A1.1, obtained from manuscript entries in James Balk's handwriting in the 1849 registers prior to the introduction of 'screened' external thermometers (see below). The final 'corrected' monthly mean was the sum of the 'raw' mean and the monthly radiative correction. For this exercise, monthly mean temperatures in °F were evaluated to 2 decimal places and the correction added at that point, prior to conversion to °C, to avoid rounding errors. The resulting monthly and annual air temperature means for 1814 can be found in Appendix 6.

## A daily series of maximum and minimum temperatures from April 1815

During the course of the research leading to this book, the authors came across a previously unknown manuscript ledger containing daily readings of two Six's thermometers [9] at the Observatory. The record commenced on 9 April 1815, and manuscript entries in the ledger continue until early January 1828. Daily wind directions are also included. Readings of a Six's thermometer first appear in the main Observatory meteorological ledgers in January 1828 and, from comparison of the three-week overlap record between the two registers, it became clear that this was the same instrument.

We know that one of two Six's thermometers (the one still in use until in 1862) was made by Troughton & Simms*; presumably, both were, but we have only brief details regarding their exposure. The opening page of the record describes two such thermometers initially, one placed

* Edward Troughton (1753–1835) was a highly regarded maker of mathematical, optical and surveying instruments, who traded at The Orrery, 136 Fleet Street, London between 1804 and 1826, in partnership with his brother John Troughton (c. 1739–1807) between 1788 and 1804. The manufacturer of the instrument is first identified in a manuscript note in the August 1827 register (as 'Mr Troughton's thermometer') but not in the *Results* until 1862, when it was identified as having been made by 'Troughton and Simms'. The partnership with William Simms and successors was set up in 1826 so, if the instrument in use in 1862 was really from 'Troughton and Simms', it cannot have been the original instrument in use from 1815. It is possible, of course, that this was an assumption or mistake on the part of the collator of the 1862 inventory, as other Observatory instruments by Troughton (singular) are mentioned in the 1853 *Results*. (Biographical details courtesy of the London Science Museum website.)

on the east side of the Observatory tower, the other on the west side. The only other note on exposure within the manuscript record occurs on 29 October 1817, when it is stated that '...the thermometers were removed from the Tower to the NE and NW windows of the Section Room [in the centre of the ground floor of the Observatory].' The north-east window is position 1 on Figure A1.2. This location appears to be unlikely as a permanent site, for here the thermometer(s) would have caught early morning or late evening sunshine at midsummer—a risk the (astronomical) Observatory would, of course, have been well aware of (and there is no evidence of this in the temperature records themselves). Minimum temperatures on the 'west' instrument appear to have been erratic from the start and are missing for most of the record. From February 1820, only the readings from the 'east' instrument are given. The records are in degrees Fahrenheit, mostly to a precision of 1 degF, although a few days have ½ °F entries but, after 1828, readings are increasingly given to one decimal place.

It is very likely that this thermometer was unscreened, as was the dry-bulb thermometer then in use, and that similar radiative corrections should be applied (see above). Surviving records noted in Wallace confirm that, until 1840, the dry-bulb thermometer was exposed outside the Transit room on the east side of the Observatory; after 1840 it was moved to the west side of the Observatory (Figure A1.2, positions 2 and 3). It would seem reasonable to assume that both thermometers were mounted roughly at eye height above ground floor level, for observer convenience if for no other reason. Further, as the Six's thermometer and the dry-bulb one were both used as a mutual calibration check, they may well have been mounted adjacent to each other: perhaps the Six's thermometer was moved to the dry-bulb position between 1815 and 1828, when it was first included in the standard Observatory meteorological registers. Indeed, the 1853 *Meteorological Observations* from the Observatory state that the two thermometers were mounted next to each other, and this had presumably been the case since at least 1840 when both were moved to the west wing. In 1853 (and presumably before then), the thermometers were mounted at a height of 6 feet (1.8 m) above ground level.

There are records of the Six's maximum and minimum temperatures in the Observatory ledgers until 1862, after which the thermometer was moved to the top of the Observatory tower. (Records from the tower were made daily and published in the regular Observatory meteorological records until December 1908, by which time the original thermometer had been in daily use for 93 years). Other than occasional short gaps, there are four periods up to May 1842 when both maximum and minimum temperatures are missing from the record of the Six's max–min thermometer: these are 21 September 1815 to 9 February 1816, 8 November to 16 December 1818, 28 April to 20 August 1827, and 20 December 1839 to 22 February 1840. For these gaps, the occasional shorter gap, and also for 1 January to 9 April 1815, in order to complete the record for 1815, maximum and minimum temperatures were estimated from the three 'spot' temperature observations made daily, using the procedure described below:

- Daily maximum temperatures were estimated using the daily 1400h temperature observation (1230h 1815–1818) plus an empirical correction, derived as follows:
  - The difference between daily 1400h temperatures and the daily maximum from the Six's thermometer was evaluated for a period of 3 years (1850–52) immediately following the gap, and monthly averages taken.
  - As a sense check, a comparison was also made of current monthly means of daily 1400 GMT temperatures compared with daily maximum temperatures recorded over a 17-year period at an observatory located near Reading, and the two were found to be very similar in both absolute values and their seasonal variation.

**Table A1.2** *The (smoothed) values of the empirical monthly correction applied to the observed Radcliffe Observatory 1400h air temperature to provide an estimate of the daily maximum temperature where the latter was missing from the record in the 1840s; units degC. All corrections are positive.*

| | Jan | Feb | Mar | Apr | May | June | July | Aug | Sept | Oct | Nov | Dec |
|---|---|---|---|---|---|---|---|---|---|---|---|---|
| Corr'n to 14h temp, degC | 0.8 | 0.7 | 0.9 | 1.0 | 1.1 | 1.2 | 1.2 | 1.1 | 1.0 | 1.0 | 1.0 | 0.9 |

- The derived corrections to the 1400h temperature were then smoothed slightly (shown in Table A1.2) and then added to each daily 1400h temperature to derive an estimated daily maximum temperature for the missing data period. Each daily value was rounded to one decimal place in degrees Celsius, and all values thus generated flagged as such within the working database.

It is accepted that, in reality, the correction factor would vary somewhat from day to day, but this method was chosen in preference to simple substitution of the daily 1400h temperature as the day's maximum temperature, which would otherwise result in daily values, and monthly means of both maximum temperature and mean daily temperature, being slightly too low for the period where the Six's thermometer maximum temperature was missing.

For the period when minimum temperatures were missing, the situation was a little more complicated, since the 10 p.m. outside temperature reading is likely to bear a less close relation to the eventual minimum temperature (usually) later in the night. In this case, regression analysis was used to estimate the missing minimum temperature values (samples $n = 180$, regression $r = 0.96$).

For a much longer period (May 1842 to November 1848), maximum temperatures are missing from the Six's thermometer record (minimum temperatures were recorded as normal). For this period, and again for the occasional short gap, maximum temperatures were estimated from the daily 1400h temperature using the method above.

The daily Six's maximum and minimum temperatures (and many of the sub-daily temperature records where necessary) were digitised by the authors, and now form a near-complete daily record back to 1815. Prior to publication as part of this volume, the earliest published daily temperature records from the Radcliffe Observatory were for 1853. The early records form a valuable daily series now extending back over 200 years, but it must be accepted that, although these records were made with exceptional care and diligence throughout, the instruments and exposure practices of the time differ from modern standards. Accordingly, appropriate care should be taken when making comparisons with temperature records made prior to the introduction of a standard Stevenson screen thermometer enclosure at the Observatory in December 1878. More details are set out in the section 'Restating the monthly mean temperatures'.

## The 'penthouse screen'—From August 1849

From 8 August 1849, a second dry-bulb thermometer was exposed in a screened wooden 'penthouse' originally located about 2 m north of the west side of the Observatory (position 4 on Figure A1.2). The 1858 *Radcliffe Observations* [160] give this brief description:

> 'The penthouse is so constructed of open work as to allow a perfectly free passage of air, and, at the same time, to produce a good protection to the instruments from storms and rain.'

**Figure A1.4** *The 'Kew screen', an early thermometer shelter from 1853, probably similar in design to the 'penthouse' erected at the Radcliffe Observatory, Oxford, in 1849 (from the Frontispiece to* Symons's Meteorological Magazine 1869)

More exact details of this penthouse structure have not survived, but it is not unreasonable to assume it was similar to—and apparently pre-dated—a passively ventilated thermometer shelter first erected at Kew Observatory in late 1853. This thermometer shelter, which became known as the 'Kew screen' [9] or 'Welsh's stand', is shown in Figure A1.4. The thermometer bulbs within the Oxford 'penthouse screen' screen were stated in 1849 as being about 1.8 m above ground, later reduced (or possibly corrected) to about 1.25–1.5 m above ground—very similar to modern standard screen heights.

The penthouse screen was presumably something of an experiment which proved its worth, for the exterior north-wall dry-bulb thermometer was discontinued after March 1852, while in January 1853 the Six's thermometer was relocated into the penthouse alongside the dry- and wet-bulb thermometers. Many years later, James Balk examined the 2½ years overlap between unscreened and screened dry-bulb temperatures, and proposed corrections to the 1815 to 1848 mean temperatures to make them retrospectively comparable with the screened records from 1849 onwards; these were the monthly mean temperature series published in Knox-Shaw and Balk [30] and referred to subsequently.

Separate maximum and minimum thermometers by Negretti and Zambra were installed in the penthouse screen in January 1857, although, as stated above, the Six's max–min thermometer remained in the penthouse until 1862. The penthouse was moved about 3 m further north from the west wall on 2 March 1868, as there were concerns that the proximity of the Observatory building was affecting the records (Figure A1.2, position 4), although no sensible difference was subsequently detected. According to Wallace, on 24 October 1868, the penthouse was destroyed by a severe gale (and several thermometers broken), although this is not mentioned in that year's published *Radcliffe Meteorological Observations*. It is assumed that the structure was quickly rebuilt in the same or similar location, for there was no break in the maximum and minimum thermometer readings. 'Additional screening' was added to the penthouse structure in June 1870. Its position in 1876 is marked on the Ordnance Survey map of that date (Figure A1.5).

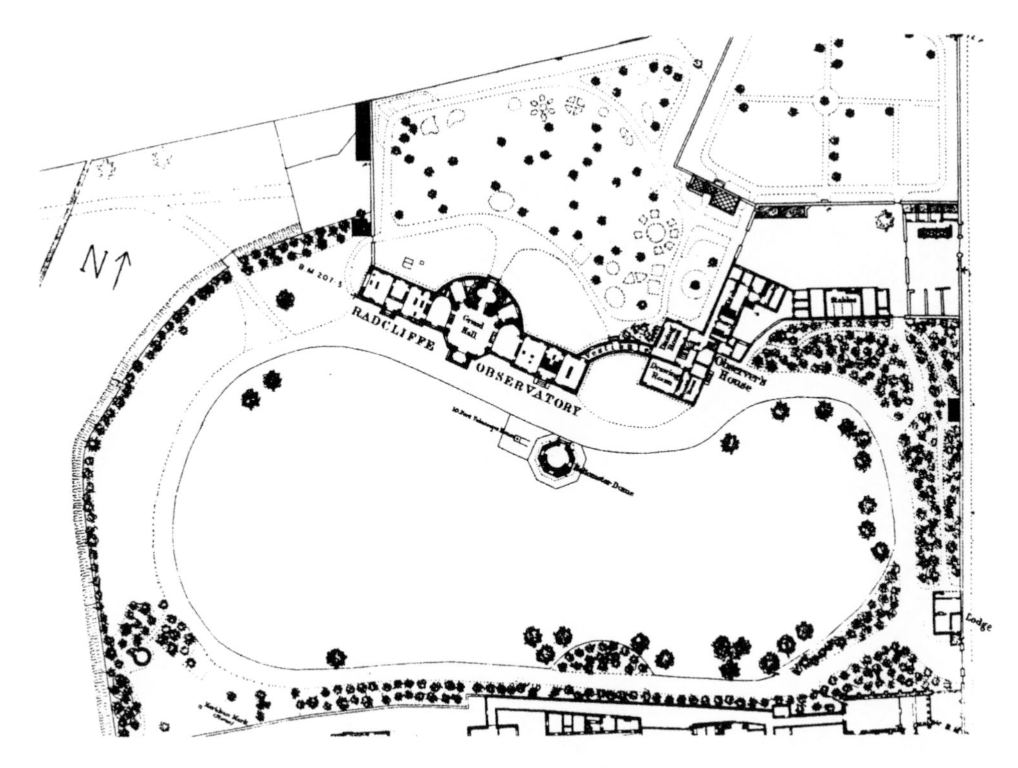


**Figure A1.5** *Detail of the Observatory site from the First Edition Ordnance Survey map of 1876, 1:500 scale. The square structure just north of the west wing of the Observatory is believed to be the penthouse screen, in use between 1849 and 1878 [22]*

## Stevenson screen—from December 1878

On 2 December 1878, the thermometers (dry and wet bulb, maximum and minimum) were relocated to a double-louvred Stevenson screen within a railed enclosure on the Observatory's south lawn, about 10 m south of the transit circle room (Figure A1.2, position 5, and Figure A1.6); the penthouse readings from the north lawn were then discontinued. The thermometers were initially at 1.5 m above ground level and were subsequently lowered to 1.2 m in September 1885. In late November 1908, the screen was moved 1.5 m further south (Figure A1.2, position 6), where it remained until 1935, apart from a relocation to an enclosure in the north lawn (almost certainly the current site, Figure A1.2 position 8) from 1920 to 1926 before moving back to the previous site on the south lawn. On 1 January 1923, the existing Stevenson screen was replaced with a slightly larger model 'with slightly better ventilation' [161]. The existing thermometer screen remained in place throughout 1923, and comparisons were made during that year between carefully calibrated thermometers in both screens, illustrating the extreme care the Observatory took when it came to documenting metadata and instrumental or process changes. Results suggested that the diurnal range was slightly increased in the newer screen when compared with the older, by about 0.3 degC (annual means of +0.10 degC for maximum temperatures, and −0.16 degC for

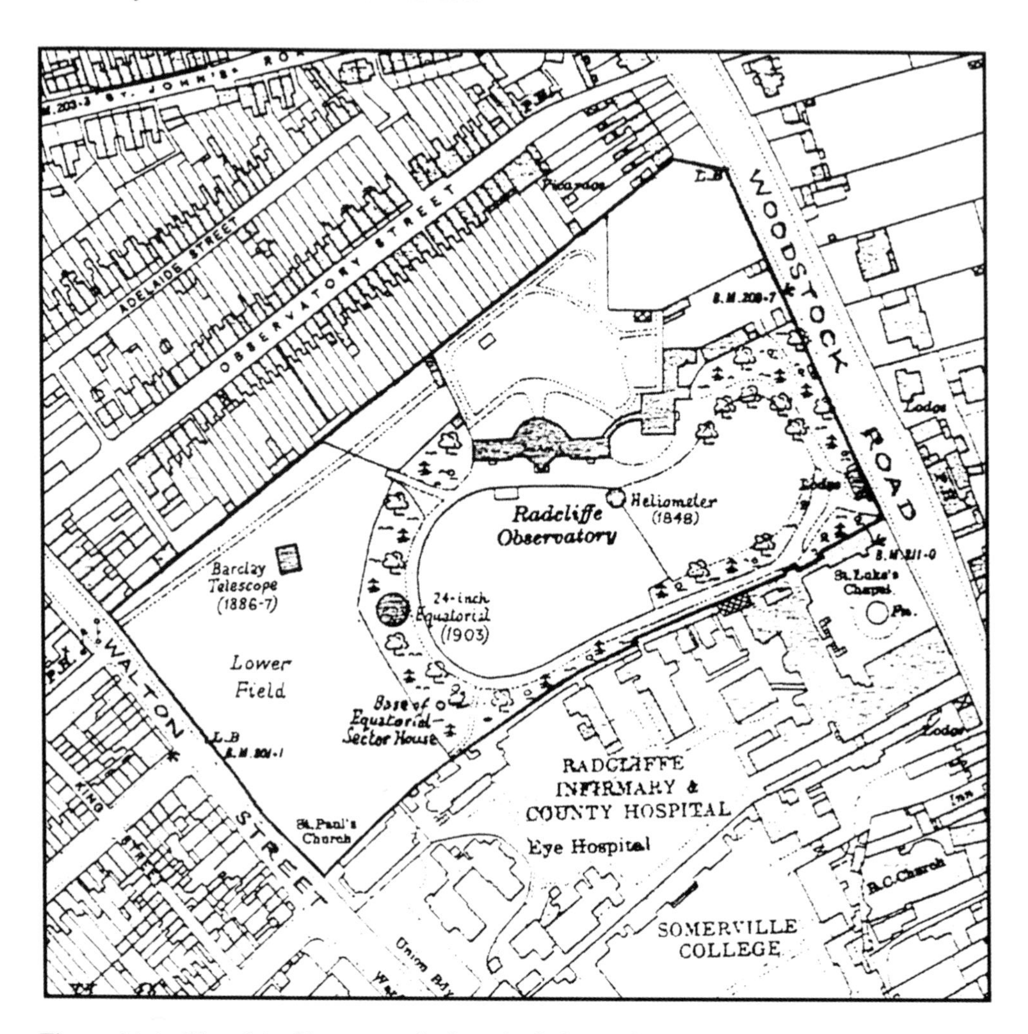


**Figure A1.6** *Plan of the Observatory site from the Ordnance Survey map of 1887. The enclosure housing the Stevenson screen is marked just south-west of the centre of the Observatory tower, on the front lawn, and the raingauge enclosure is shown in its current site on the north lawn: see also Figure A1.7 (Radcliffe Trust)*

minimum temperatures, Table A1.3), although this would of course almost cancel out when calculating daily mean temperatures. These minor differences have been used to correct monthly means of maximum and minimum temperature during the period December 1878 to December 1922, based upon a slightly smoothed monthly correction (Table A1.3, final two rows). It should be appreciated that these are monthly means: daily variations might be expected to vary between near-zero in cloudy, windy conditions to twice these values or greater on sunny days or clear nights with little wind and, accordingly, there would be little merit in adjusting daily values by these average amounts. The monthly means in Appendix 6 therefore differ from the arithmetic average

**Table A1.3** *Monthly temperature differences observed between 'old' (1878–1922) and 'new' (1923 onwards) Stevenson screens at the Radcliffe Observatory, based upon a comparison of observations in both screens during 1923 (from the Radcliffe Results volume for 1921–25 [161], page viii). These minor differences have been applied to monthly means for the period 1878 to 1922 during the preparation of the current volume. All maximum temperature differences are positive, and all minimum temperature differences negative. A positive difference implies that the new (1923) screen recorded higher values than the old (1878) screen, and vice versa, and monthly values have been corrected accordingly in Appendix 6*

| | Jan | Feb | Mar | Apr | May | June | July | Aug | Sept | Oct | Nov | Dec |
|---|---|---|---|---|---|---|---|---|---|---|---|---|
| Max temp difference, degF (+) | 0.06 | 0.30 | 0.23 | 0.11 | 0.32 | 0.21 | 0.35 | 0.23 | 0.19 | 0.24 | 0.14 | 0.35 |
| Min temp difference, degF (−) | 0.29 | 0.20 | 0.17 | 0.17 | 0.28 | 0.29 | 0.35 | 0.36 | 0.37 | 0.28 | 0.31 | 0.26 |
| Max temp difference, degC (+) | 0.03 | 0.17 | 0.13 | 0.06 | 0.18 | 0.12 | 0.19 | 0.13 | 0.11 | 0.13 | 0.08 | 0.19 |
| Min temp difference, degC (−) | 0.16 | 0.11 | 0.09 | 0.09 | 0.16 | 0.16 | 0.19 | 0.20 | 0.21 | 0.16 | 0.17 | 0.14 |
| **Smoothed max temp difference applied, degC (+)** | **0.11** | **0.12** | **0.13** | **0.13** | **0.14** | **0.15** | **0.16** | **0.15** | **0.13** | **0.12** | **0.11** | **0.10** |
| **Smoothed min temp difference applied, degC (−)** | **0.14** | **0.12** | **0.10** | **0.11** | **0.14** | **0.17** | **0.19** | **0.20** | **0.19** | **0.18** | **0.16** | **0.16** |

of the daily values by these adjustments and (between 1915 and 1924) for additional minor adjustments owing to the variation in terminal hours (see 'Terminal hours').

Until December 1924, the observational programme continued at three observations per day, one of which was the morning climatological observation at or close to 0900 GMT. From 1 January 1925, the observational routine was curtailed to a single observation at 0900 GMT daily, and this has continued without significant interruption to the current day.

The 1908–35 south-lawn site (Figure A1.2, position 6) was described in the 1931–35 *Radcliffe Meteorological Observations* as being '...in a railed enclosure 38 feet [11.6 m] south of the Observatory building, on the lawn near the broad gravel drive' and '...on the edge of the lawn near the building, a few yards [metres] to the west of the main porch.' With careful examination, two screens can be seen in the Observatory's *Country Life* photograph of 1930 (Figure 2.10, p. 19);

an enlarged image is shown in Figure A1.7. The second screen contained a bimetallic thermograph. On 1 January 1935, the enclosure was moved 18 m to the south-south-west to a position nearly in the middle of the front lawn (Figure A1.2, position 7). The enclosure was widened to 6 m and now also contained the grass minimum thermometers and the standard and the Beckley raingauges, which were previously in the garden on the north side of the Observatory. (The January 1935 site was illustrated in the Frontispiece in the 1931–35 *Radcliffe Meteorological Observations* volume, but the quality of the image is too poor to reproduce it here.)

In November 1930, at the suggestion of the Meteorological Office, a second Stevenson screen was erected on the north side of the Observatory near the raingauges (at or very close to what is now the current site) to ascertain what effect, if any, new buildings about to be erected for the Radcliffe Infirmary on the southern part of the old Observatory grounds might have on the temperatures in the Stevenson screen on the south lawn. The new buildings were begun in June 1931 and completed by the end of 1932; their nearest point was 32 m from the middle of the new enclosure. The comparison showed that they made no appreciable difference. The same screen remained on the north lawn until the end of 1936, and the readings were used to assess the effect, if any, of the January 1935 change in the position of the screens on the south lawn. Both mean maximum and mean minimum for the comparable period in the '1935' enclosure were within 0.2 degC of those in the north lawn, to all intents and purposes within the calibration error of the thermometers used, and from this it was assumed that the impact of the change in the position of the instruments on the south lawn was negligible.

This comparison proved opportune, for shortly after the outbreak of war in September 1939 the instrument enclosure was again relocated to the north lawn, because of the expected expansion

**Figure A1.7** *The two screens on the south side of the Observatory, 25 February 1930. This is an enlargement from the view shown on Figure 2.10 (Courtesy Country Life Picture Library)*

of the Radcliffe Infirmary under wartime conditions. This site is position 8 on Figure A1.2, and the instrument enclosure has remained in this location to the present day. The thermometers (maximum, minimum, dry bulb and wet bulb) have been enclosed within a standard pattern Stevenson screen throughout; the existing wooden screen was replaced by a modern aluminium and plastic 'Metspec' screen on 15 February 2012 (Figure 2.12, page 21).

In October 2017, as part of a national programme to remove mercury-based thermometry from its observing network, the Met Office withdrew the mercury thermometers, replacing them with electrical resistance sensors connected to a Vaisala logger. Initial installation was unsatisfactory and, unfortunately, 11 days' records were lost as a result. The logger was subsequently replaced on two further occasions in late 2017. At the time of writing, the limited battery life of this logger continues to result in loss of record for the occasional day, although estimates based upon regional comparisons are used to complete the record. The lack of any overlap period with existing thermometry is extremely regrettable and poses a serious risk to the homogeneity of the long Oxford temperature record, particularly as there was additional uncertainty with regard to the accuracy of the remaining spirit minimum thermometer at the site from May to August 2018. At the time of writing, insufficient data has accumulated to assess any lasting impact based on comparisons with nearby stations, although this exercise in itself may prove futile, as all sites within the Met Office climatological network were migrated to the new system within weeks of each other and thus few 'undisturbed' records exist with which to compare. The manner and lack of forethought regarding potential homogeneity impacts of introducing a completely different sensor system at such long-period sites is itself an exercise in how *not* to undertake such changes to a climatological network.

## Terminal hours

The hour of reading and resetting the maximum and minimum thermometers is occasionally specified in the early records as 8 a.m. clock time, with only minor variations. It is possible that, until about 1839, maximum temperatures were not 'thrown back' as per current standard, but more detailed investigation is required to confirm or deny this point. From 1853 the hour of reading and resetting has been assumed to be the same as the raingauges, viz. 10 a.m., although this is not stated explicitly in the *Results* volumes until 1876 (Table A1.4). From 1881 the terminal hours were subsequently confirmed by Knox–Shaw and Balk in a note in *Meteorological Magazine* in 1933 [162], and these are set out in Table A1.4.

From 1889 to 1924, maximum and minimum thermometers were read and reset at 2000 or 2100 clock time (see Table A1.4). Between 1915 and 1924, they were also read at 0900 and the (mostly very minor) differences between the 2100–2100 and 0900–0900 extremes published as individual monthly corrections to the monthly mean maximum and minimum temperature by Knox–Shaw and Balk in 1933 (op. cit.). The authors recommended applying the average monthly corrections over 1915–24 to the monthly means of the evening-read maximum and minimum temperatures for the period 1889–1924. Further minor monthly corrections to the monthly mean minimum for the period June 1880 to December 1888, when the thermometer was reset at 1200 clock time, were suggested, in order to, as the authors stated, 'reduce the whole series…to a common standard, even though this can only be done in a statistical sense'. All of these terminal hour corrections have been applied to the relevant monthly mean maximum and monthly mean minimum for this period, as set out below, in order to homogenise the record for this period to the same standard as the rest of the record as far as is statistically feasible. No corrections have been applied to daily values, and as a result the monthly means of the daily maximum and minimum temperatures differ very slightly from the corrected means of those elements.

**Table A1.4** *Hours of reading (clock time or GMT) and period of record of the maximum and minimum thermometers at the Radcliffe Observatory. From 1876 to 1880, this has been taken from the introduction in the relevant Radcliffe Results annual or collected years volumes. Between 1881 and 1924, the terminal hours were explicitly stated by Knox-Shaw and Balk [162]; the times of reading are also given in the collected Radcliffe Results volumes, identified as 'Table II' from the 1911–15 volume*

| Date | Times of reading (clock time) | Terminal hour max (24 h ending at) | Terminal hour min (24 h ending at) |
|---|---|---|---|
| Jan 1876 to May 1880 | 0800, 1800 | 0800, thrown back | 0800, same day |
| June 1880 to Dec 1886 | 0800, 2000 | 0800, thrown back | 1200, same day |
| 1887–88 | 0800, 1200, 2000 | | |
| 1889 | 1000, 1200, 2000 | 2000, same day | 2000, same day |
| 1890–1914 | 0800, 2000 | | |
| 1915–24 | 0900, 2100 *(No allowance seems to have been made for Summer Time, first introduced in 1916)* | 2100, same day | 2100, same day |
| 1925 to date | 0900 GMT | 0900, thrown back | 0900, same day |

In the absence of any information to the contrary, it has been assumed that the thermometers were reset only once per day, at the times shown in Table A1.4, and in all cases represent a 24-hour record period.

## Restating the monthly mean temperatures

In 1932, Knox-Shaw and Balk [30] published comprehensive tables of the monthly mean temperature at the Radcliffe Observatory since 1815. These were based upon sub-daily observations, usually three times daily, of a calibrated dry-bulb thermometer, painstakingly adjusted on a monthly basis for diurnal factors (time of observation) and exposure (no screen/screen).

In January 1925, when the Observatory reverted to a once-daily (0900 GMT) climatological observation, sub-daily temperatures were no longer available to calculate the daily mean temperature on the same basis as had been employed at the Observatory for many years. To provide consistency with previously derived and published tables of monthly mean temperature extending back to 1815, from 1925 a correction was applied to the mean temperature derived from the average of the monthly mean maximum and monthly mean minimum temperatures. This correction (which was always negative) was derived from a multi-year comparison between the two, although the exact period of comparison is not known. The monthly values of this correction are given in Table A1.5. This correction was still being applied by the Radcliffe Meteorological Station observers in 2018. We have taken the decision *not* to apply this correction to the results in this volume, in order to allow comparison with most other stations where no such correction is applied.

It is modern UK climatological practice to take the mean temperature of a month as the average of the monthly mean maximum and monthly mean minimum temperatures, without site-specific

**Table A1.5** *Monthly temperature corrections (degC, negative) applied by the Radcliffe Observatory to adjust the mean temperature derived from the average of the monthly mean maximum and monthly mean minimum temperatures to a notional 24-hour mean temperature comparable with previously published monthly mean temperatures for the Observatory from 1815. These corrections are* **not** *applied in this volume.*

| | Jan | Feb | Mar | Apr | May | June | July | Aug | Sept | Oct | Nov | Dec |
|---|---|---|---|---|---|---|---|---|---|---|---|---|
| Correction to ½ (max+min) mean temperature, degC | 0.2 | 0.3 | 0.4 | 0.4 | 0.3 | 0.2 | 0.3 | 0.4 | 0.4 | 0.3 | 0.2 | 0.1 |

adjustments. Accordingly, we have recalculated the Radcliffe's historical temperature records in a manner consistent with modern practice, avoiding the need for this empirical correction to a notional 24-hour mean. This has been facilitated by the fortuitous discovery of the early record of daily maximum and minimum temperatures from a Six's thermometer dating back to April 1815.

We have taken the unusual step of restating the monthly mean temperatures for the Radcliffe Observatory for two reasons:

- Firstly, because the climatological standard for the UK (and many other countries) is to define the mean daily temperature simply as the mean of the day's maximum and minimum temperatures, without correction or adjustment. Changing the calculation method to this basis enables the Oxford records to be standardised in this important respect, and thus made easier to compare with other sites on a like-for-like basis.
- Secondly, the discovery of the daily maximum and minimum temperatures for the period 1815 to 1827 gave us the opportunity to extend a daily temperature record for the Observatory back to 1815; previously, daily values had only been available from 1853. In doing so, it seemed advantageous to utilise a single and consistent approach to calculating daily and monthly mean temperatures, viz. the mean of daily maximum and minimum temperatures (the monthly figure being the mean of the available daily observations). The methods adopted by Knox-Shaw and Balk in 1932 made perfect sense at the time, when the majority of the record (110 years, 1815–1924) was available as multiple sub-daily observations. Since that time, however, the record length for which daily maximum and minimum temperatures are available has increased from 72 years (i.e. 1853 to 1924) to 203 years (1815 to 2018), whereas no additional sub-daily observations have been made (other than occasionally by the on-site automatic weather station) since December 1924.
- Thirdly, the opportunity has been taken to include the various minor corrections to monthly mean, maximum or minimum temperatures published by the Radcliffe Observatory, adjusting for variations in screen type (1878–1923) and terminal hours (1915–24), as detailed in this appendix, to produce an internally consistent 200-year monthly time series of maximum, minimum and mean temperatures as far as it is possible and realistic to do so.

The restated monthly mean temperatures up to the end of 2018 are given in Appendix 6 and were published on the University of Oxford, School of Geography and the Environment website at https://www.geog.ox.ac.uk/research/climate/rms to coincide with publication of this volume; online sources will be updated regularly to maintain the series.

The calculation basis for the new series varies slightly during the period of record, as stated below:

For 1814, the monthly and annual mean air temperatures are derived using the methods set out in 'Monthly temperature means for December 1813 to March 1815'.

- Between 1815 and 1852, the monthly and annual mean air temperatures are the average of the mean daily maximum temperature and mean daily minimum temperature from the unscreened Six's thermometer, with the monthly radiative correction per Table A1.1 applied on a monthly basis. (Daily figures are normally quoted as recorded in the original registers, without radiative correction, as such corrections would in reality vary significantly from day to day, and different empirical corrections for maximum and minimum temperatures would be preferable—for which no overlapping observations or correction tables exist). Where data from the Six's thermometer is missing, estimates have been made using the sub-daily temperature observations as described previously.
- Between January 1853 and November 1878, the monthly and annual mean air temperatures are the average of the mean daily maximum temperature and mean daily minimum temperature from 'penthouse' screened thermometers. Where data are missing, estimates have again been made using the sub-daily temperature observations.
- Since December 1878, the monthly and annual mean air temperatures are the average of the mean daily maximum temperature and mean daily minimum temperature from thermometers mounted in a standard Stevenson screen. Where data are missing, estimates have been made using the thrice-daily manual temperature records up to 1924, and from surviving thermograph records or neighbouring sites since 1925. Between December 1878 and December 1922 inclusive, the monthly and annual mean maximum and minimum air temperatures have been corrected slightly to conform to the 'better ventilated' Stevenson screen introduced in January 1923 (see above), although the effect on monthly *mean* temperatures is negligible (less than 0.05 degC). Between June 1880 and December 1914, minor corrections to monthly mean maximum and minimum temperatures have also been included to compensate for different terminal hours, as set out above.
- From January 1915 to December 1924, mean monthly maximum and minimum temperatures have been individually corrected by month from dual observations to reflect a different terminal hour from the 'morning-to-morning' standard of most of the rest of the record. These corrections are very slight—only a few mean minimum temperatures in winter months exceed 0.5 degC—and reflect a balance of positive and negative corrections. The original corrections by month are not reproduced here but are given in Knox-Shaw and Balk (1933), page 297 [162].

It is as well to state explicitly that the tables of monthly and annual mean temperature presented in this book can be expected to differ slightly from those previously published by the Radcliffe Observatory (for example [30]), and as used by Gordon Manley [35] and others, for the reasons stated above, and that, between 1815 and 1852, and 1878 and 1924, the monthly mean temperature will differ slightly from the average of the monthly mean maximum and monthly mean minimum temperatures because of the minor corrections applied as stated above.

### Have any corrections for urbanisation been applied?

No, the only corrections applied are those listed above for exposure differences to current standards. Details of the effect of urbanisation on the Oxford record are given in Chapter 3. Our best estimate of this is 0.20 degC, and this has increased by 0.1 degC in the last 40 years.

### Is the entire Radcliffe Observatory temperature record completely homogeneous?

In a word—no, although the results presented in this book and online are as close to homogeneity as we can achieve with minor adjustments to surviving records for varying instruments and exposures, and disregarding the cumulative effects of the progressive urbanisation of Oxford (see Chapter 3 for a complete discussion). As stated previously, the instruments and exposure of the early Radcliffe records were significantly different to modern standards, particularly before 1853, and, although corrections have been applied in an attempt to render the entire series statistically homogeneous, appropriate allowance should be made when making comparisons with more recent temperature records made to modern standards.

## Precipitation

### Sources of data

The Radcliffe Meteorological Station is undoubtedly one of the best documented meteorological sites in the world, with very detailed records of instrumentation, sites and exposure, and any corrections made to original readings [26]; Wallace [21] provides a detailed history of all instrumentation and observing procedures up to 1995. The site has the longest continuous series of rainfall records in the United Kingdom at one site, daily records being continuous from April 1814. Before this time, the records are fragmentary and incomplete, although a monthly rainfall record has been reconstructed back to 1767 [24] based mainly upon Thomas Hornsby's surviving records. Even after 1814, however, rainfall readings were irregular and, even after 1827, when the daily record seems reliable, the number of observed 'rain days' still appears too low. It was not until 1853 that the observational records confirm that the raingauge was read every day, and we assume that, prior to this, some multi-day accumulations are included in the record. Accordingly, in this volume, we regard only the daily rainfall record from 1853 as fully comparable with the standards of modern observation.

### Monthly totals 1767–1814

Monthly rainfall totals for the early years are from Craddock and Craddock [24], converted to millimetres. These are mostly from the surviving records of Thomas Hornsby in Oxford, with gaps filled or estimated from other sites, notably nearby Shirburn Castle. The series was constructed so as to be homogeneous with the Radcliffe Observatory record from April 1814, within the limits of the available data and the statistical techniques employed.

### Raingauges and their exposure at the Radcliffe Observatory since 1806

A number of storage raingauges have been in use at various times, but it is only necessary to have recourse to three of them to obtain a complete daily series from 1815 [30] [98]:

- From 1806 to 1850, the only raingauge was on the east wing parapet of the Observatory, a 305 mm (12 inch) diameter orifice at a height of 6.7 m (22 feet) above the ground (position 1 on Figure A1.8). Such an exposed location would be avoided today, but roof-mounted gauges were common in the early nineteenth century.

Remarkably, Hornsby's original raingauge funnel, made by London opticians P. and J. Dollond in 1774, survives (together with Dollond's handwritten invoice) in the Museum of the History of Science in Oxford collections (inventory number 83577), to whom it was presented by the

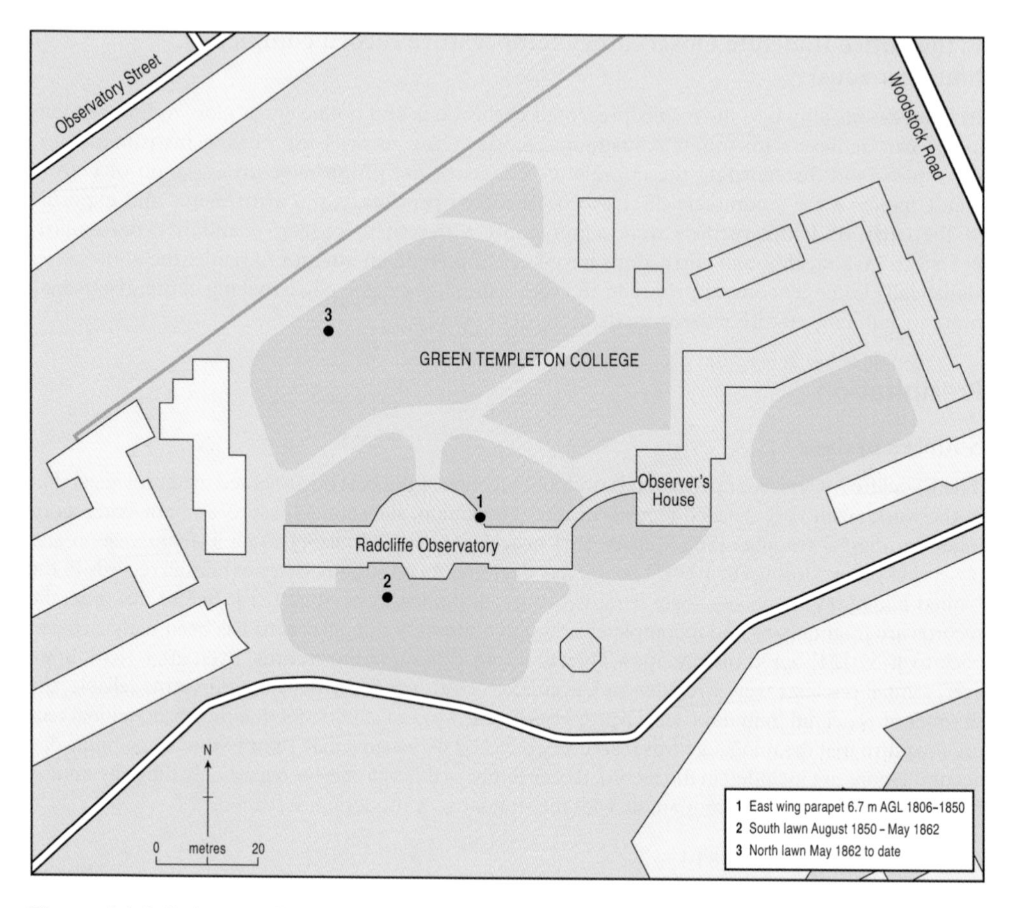


**Figure A1.8** *Rain gauge locations at the Radcliffe Observatory, Oxford, 1806 to date (Chris Orton)*

Radcliffe Trustees around 1933. This was installed on the parapet of the Radcliffe Observatory in 1775 and was used by Hornsby until his death in 1810 and thence by the Observatory from 1814 to 1879, and again from 1923–30 to provide an overlap and thus conversion factor with contemporary raingauges mounted near ground level [21] [22].

- A near ground-level gauge on the south lawn, of 254 mm (10 inch) orifice diameter, with its rim 280 mm (11 inches) above the ground was first recorded on 22 August 1850 (position 2 on Figure A1.7). In May 1862, the ground gauge was moved to the north lawn to improve exposure (somewhere near position 3 on Figure A1.8).
- In 1877, a new 'Glaisher' gauge of 203 mm (8 inch) diameter with its rim 280 mm (11 inches) above ground was installed close to the old gauge on the north lawn; this gauge provides the common standard to which all other records have been reduced [30]. In order to improve exposure, the gauge was raised to 510 mm (20 inches) above the ground in 1887, where it remained until 1935, when it was lowered to the standard 300 mm (12 inches) above the ground when both Glaisher and Beckley gauges were moved to the position on the front (south) lawn. Both gauges were relocated to the north lawn once more in September 1939 when the enclosure was migrated from the south-lawn site.

- The Beckley gauge was discontinued in April 1963, being replaced by a new Dines tilting siphon rain recorder. At the same time, the 8" Glaisher gauge was replaced by a standard copper 'five-inch' (127 mm diameter funnel) UK Met Office gauge [21], its rim remaining at 300 mm above ground (the current gauge can be seen in Figure 2.12, page 21). Both Beckley and Glaisher gauges had been in daily use since 1877.
- The 1963 5" gauge has been replaced at least twice since, by an identical unit in the early to mid-1990s, subsequently by a stainless-steel unit of otherwise identical construction prior to 2014. The latter remains in daily use as we write.
- The Dines tilting siphon rain recorder within the enclosure has not been in use for some years; it is hoped to commence hourly rainfall records using an AWS logger and a tipping-bucket raingauge during 2019.

## Conversion factors to bring earlier observations to a common standard

Given the use of different gauges at varying degrees of exposure, it is not surprising that there has been much effort to apply conversion factors to bring the readings of the various gauges to a common standard—the 203 mm/8" Glaisher gauge of 1877.

- The first corrections applied were for the period March 1830 to March 1835 inclusive: all measurements were adjusted by a factor of 1.108026. This is not mentioned either by Craddock and Smith [26] or by Wallace [21]. There is a note dated 21 October 1922 at the start of the 1831 ledger, written and signed by James Balk, but no explanation of the reason for the adjustment is given; most likely, the glass measuring cylinder was incorrectly calibrated. This correction has been applied to the original observations.
- The original 305 mm/12 inch roof gauge was re-installed between 1923 and 1930, probably at the instigation of James Balk, to obtain comparison of its results with the (by now standard exposure) low-level gauge near ground level. Knox-Shaw and Balk [30] found that the readings of the roof gauge needed to be multiplied by a factor, the mean of which is 1.130, to adjust them to the standard of the north-lawn ground gauge. Since the conversion factor varied more by month than by the rainfall total, individual factors were applied for each month ([30], p. 95). These correction factors have been applied to all readings from 1827 through to 21 August 1850; they range between 1.15 in December and January and 1.115 in August, the variation presumably reflecting the greater loss of catch with elevation in the windier winter months.
- Craddock and Craddock [24] applied the single correction factor of 1.130 for the years 1775–76 and 1795–1805, arguing that a single value for all months was as accurate as the data would justify. For the start of the record, 1767–74, their conversion factor was 1.3223.
- From 22 August 1850, ground-level observations have been used, reduced to the common standard using the factors derived by Knox-Shaw and Balk [30], namely 0.962 until 19 May 1862 (correcting from the south-lawn gauge to the 203 mm/8" Glaisher gauge) and 0.98 from then until 1877 (correcting from the 254 mm/10" north-lawn gauge to the 203 mm/8" Glaisher gauge). Both these corrections are applied equally across the year.
- Having applied the conversion factors derived by Knox-Shaw and Balk [30], the homogeneity of the record from 1829 was checked against the England and Wales (E&W) annual precipitation record [163] using double-mass curves [164]. A break of slope around 1862 was found and so analysis of covariance was applied to test its significance [164], comparing 1829–1861 with 1862–1894. The break in the double-mass curve is

significant at the 1 per cent level and was very probably caused by a change in the relation between Oxford and the E&W series and not by the vagaries of sampling, namely which combination of gauges was used to produce the E&W series. All daily rainfall observations to the end of April 1862 have therefore been multiplied by a factor of 0.953 (applying the Searcy and Hardison [164] formula on p. 38). It seems that the Knox-Shaw and Balk conversion factors therefore overestimated Oxford rainfall totals to 1862, compared to all subsequent measurements made using the north-lawn gauges. This may be because their re-installed roof gauge was at a height of 7.3 m (24 feet), not at the original 6.7 m (22 feet) ([21], p. 47).

In presenting monthly data back to 1767, the 0.953 correction factor has been applied to all values before February 1827; data for 1767 to 1814 are derived from Craddock and Craddock [24].

Table A1.6 summarises the sources of record and corrections used. The corrected series differs from the Met Office 'historic station data' for Oxford. Monthly totals in the period 1853–1862 are 91.7 per cent (0.962 × 0.953) of the Met Office figures. For the period 1862–1877, the new series is 0.98 of the Met Office record, presumably because the Met Office did not apply the Knox-Shaw and Balk [30] corrections to the original ground-level observations.

## Daily readings

Whilst daily rainfall records appear largely reliable from February 1827, with almost all references to rain or showers being accompanied by a measured total, the number of rain days seems too low until the 1850s, and it is only from the 1853 published observations that reference is clearly made to the raingauge being read every day. We therefore regard the daily rainfall record as reliable from 1853; our time series of rain days and rain per rain day therefore generally start in 1853. The one doubt is the recording of snow. It is not clear when snowfall was first counted as 'precipitation' and included in the rainfall totals; certainly, there are many references in the ledgers to 'snow' for which no rainfall total is given, suggesting that the snow was not melted so that an equivalent depth of water could be noted, although it should be noted that light falls of snow often leave little or no evidence in the raingauge bottle.

The daily dataset from 1827 is available at https://www.geog.ox.ac.uk/research/climate/rms. Prior to metrication in 1971, daily values were converted from inches, and rounded to the nearest 0.1 mm (the vast majority of the Radcliffe Observatory rainfall record was recorded in inches to three places of decimals). Monthly totals are the sum of the daily values: owing to rounding errors, the values presented in this volume and online may differ slightly from previous compilations or previously published summaries.

Finally, at the time Burt and Howden [98] were unable to find the daily rainfall data for the first six months of 1828. The ledger has since been found and so a continuous daily record is now available from 1827.

## The use of self-recording gauges

Self-recording raingauges were first used at Oxford in 1856 (a continuous record exists from 1877 until the mid-1990s) but, other than daily totals, these records have not been digitised and no use has been made of them here. The Beckley self-recording gauge installed in 1877 (funnel diameter 11.2 inches or 284 mm, rim height 28 inches or 710 mm in 1899) was used to form the published record from 1882 to 1924, the period of record being the civil day (midnight to midnight) rather than

**Table A1.6** *Summary table showing the rainfall data available making up the daily Radcliffe Observatory record, 1767 to date*

| Rain gauge and site | Source of data | Period of use | Period used for our Radcliffe series | Adjustment factor to adjust to current records |
|---|---|---|---|---|
| Various incl Thomas Hornsby records | Craddock & Craddock [24] | 1767–1814 | Monthly 1767–1774 | 1.3223 |
| 12" east wing gauge (6.7 m AGL) | | 1775–1814 | Monthly 1775–1814 with some estimates | 1.130 1775–76 and 1795–1805 |
| | Original MS ledgers | 1814–1850 Aug | Monthly 1814–1827 Daily Feb 1827 – Aug 1850 | 1.130, per 1923–30 overlap |
| 12" east wing gauge (7.3 m AGL) | Original MS ledgers | 1923–30 | To derive overlap factor only | 1.130, varies slightly by month |
| 10" gauge, south lawn, 280 mm AGL | Original MS ledgers | Aug 1850 – May 1862 | Aug 1850 – May 1862 | 0.917 (0.962 × 0.953) |
| 10" gauge, north lawn, 280 mm AGL | Original MS ledgers | May 1862 to 1877 | 1862–1877 | 0.98 |
| 11.2" Beckley gauge | Published data | 1880–1963 | 1880–1920 | |
| 8" Glaisher gauge, north lawn (south lawn 1935–39) 280 mm AGL 1877–1887, 510 mm 1887–1935, 300 mm 1936–1963 | MetO daily database except 1921–24, taken from MS ledgers | 1877–1963 | 1877–1879, 1921–1963 | 1.000 |
| 5" MetO gauge, current enclosure | Current observational record | 1963 to date | 1963 to date | 1.000 |

morning to morning [21]. For the period 1921–24, entries from a manuscript ledger giving the 0900–0900 GMT totals from the Glaisher 8" gauge (inches to 3 decimal places, converted to millimetres to 1 decimal place) were used in place of the published (Beckley gauge, midnight to midnight) record. Over these four years, the difference between the Beckley and Glaisher gauges was less than 1 per cent so, aside from terminal hour differences (see below), the series has been taken to be homogeneous.

Records from a 0.2 mm tipping-bucket raingauge, part of an automatic weather station system, have been made erratically since about 1993. It is hoped that records from this AWS will soon be resumed to provide backup for the manually read temperature and precipitation observations.

### Terminal hours

The raingauges have been read during the morning for most of the period of record. For the early years, the hour of reading was typically 8 or 9 a.m. Oxford time (5 minutes behind GMT, which was not brought into use in the Observatory until 1889). From the 1853 *Results* publication onwards, the hour of reading is given as 10 a.m., and the readings were 'thrown back' to the previous day as remains standard practice today. From 1 June 1880 to 31 December 1888, the gauges were read daily at noon Oxford time; the record refers to the 24 hours commencing noon on the start date (i.e. it is thrown back from the day of reading). From 1 January 1889, the record of the Beckley self-recording gauge was used to derive a 'civil day' rainfall record (midnight to midnight GMT) and this measure was the published record until 31 December 1924. From 1 January 1925, the published record was from the eight-inch gauge read daily at 0900 GMT, and the 'standard' rain day (the 24 hours starting at 0900 GMT) was adopted; this remains the case today. For this publication, daily 0900–0900 GMT rainfall totals from the eight-inch gauge have been extended back to 1 January 1921, transcribed from a manuscript ledger within the School of Geography archive and converted from thousandths of inches to millimetres, as stated above.

## Sunshine

### The sunshine recorder and its exposure at the Radcliffe Observatory

A Campbell–Stokes sunshine recorder [165] was installed on the roof of the Observatory tower in February 1880, from where it had an unbroken horizon. It remained here until late 1976, when it was transferred to the roof of the Engineering Science building, some 230 metres east-south-east of the Observatory site and with a similarly unobstructed viewpoint*. The sunshine cards are changed daily, shortly after the 0900 GMT observation at the Green–Templeton site (Figure A1.9).

### Sources of data

'Sunshine' refers to the duration of bright sunshine, as measured from the burn length on the daily card from the Campbell–Stokes sunshine recorder and tabulated on a 'civil day' (midnight to midnight) basis†. From the commencement of the record in February 1880 until 1935, daily and monthly totals (to 0.05 h precision) were published in the *Radcliffe Results* volumes (for example, [30]), and this source provided monthly sunshine totals to 1930 inclusive, and daily totals from January 1921 (and a few earlier months: the daily record is steadily being digitised to 0.05 h precision from these sources or the original manuscript registers, and it is hoped that the entire daily series will soon be available in digital format).

For the period 1 January 1931 to 31 December 2016, the majority of the daily sunshine record was taken from the Met Office/BADC digital record for the Radcliffe Observatory, quality-control checked by the Met Office up to the end of 2016. This dataset has occasional gaps, mostly since 1998 (the longest single gap being 143 days from April to August 2012; this and the other gaps were filled with the record from the Centre for Ecology and Hydrology site at Wallingford, 22 km south-east of the Radcliffe Observatory). Data were provided in hours and tenths.

* It is of interest to note that the original 1880 instrument remained in daily use until the 1976 relocation, when it was retired after 96 years' service, and was subsequently acquired by the Met Office instrument museum.

† Until 1888, sunshine records were tabulated on a noon-to-noon basis and entered and published against the day the card was changed. The difference is almost immaterial for monthly totals but, once daily records from this period are digitised, it is hoped that true 'civil day' records can be reconstituted from the surviving hourly data noted in the manuscript ledgers.

**Figure A1.9** *DPhil student Amy Creese changing the sunshine recorder in its current position on the roof of the Engineering Science building, May 2018 (Stephen Burt)*

From January 2017, daily sunshine durations have been taken from the Radcliffe Meteorological Station monthly summary.

## The Radcliffe Observatory sunshine series 1880 to date

Daily sunshine data for 1880 have not yet been traced, but daily records were published from 1881 to 1935 in the *Radcliffe Results* series (for example [30]). Monthly sunshine totals from February 1880 to date are available on the University of Oxford, School of Geography and the Environment website at https://www.geog.ox.ac.uk/research/climate/rms, and are updated regularly.

## Terminal hours

From the start of the record in February 1880 until 31 December 1888, the sunshine cards were changed at noon, and the published sunshine duration was for the 24 hours ending at noon on the date the card was changed. From 1 January 1889 to the last published *Results* records in 1935, the sunshine record refers to the civil day, midnight to midnight GMT. Since January 1925, the sunshine card has been changed during or shortly after the 0900 GMT observation, and the records tabulated according to the civil day date by adding together the pre-0900 and post-0900 portions of the burn for the date in question.

## Is the Radcliffe Observatory sunshine record completely homogeneous?

Within the limits of instrument performance and minor analysis variances between different observers over the near 140 year record—yes. There has been an increase in average annual sunshine duration, particularly since about 1980 (see also Chapter 19 and Figure 19.3), most likely due to changes in atmospheric clarity and pollution levels, particularly in the winter months, superimposed on background changes resulting from regional changes in atmospheric circulation patterns.

## Wind

Wind-speed records have been kept at the Radcliffe Observatory Oxford since 1856. In January 1880, a Beckley–Robinson anemograph was installed on the roof of the Observatory at 34 m above ground level, replaced in July 1893 by a Munro anemograph at the same height above ground level; daily and monthly mean wind speeds from this instrument were published up to 1935 and are available in manuscript up to 1970. This instrument was unable to resolve gust wind speeds and was dismantled and replaced by a Dines Pressure Tube Anemometer mounted on the roof of the nearby Engineering Science building, at 45 m above ground level, in 1976. (At the time of writing, this instrument has been out of order since 2014, pending replacement by an automated logger-based system.) Because the various anemometers have been mounted at heights well above the standard 10 m above ground level, measured wind speeds are about 20 per cent higher than a conventional exposure.

## Dataset online availability

To coincide with the publication of this volume, the entire Radcliffe Meteorological Station dataset will be made available online through the website of the School of Geography, University of Oxford at https://www.geog.ox.ac.uk/research/climate/rms.

There are two datasets:

- The *monthly* dataset contains monthly, annual and seasonal totals or means for the elements listed above, with corrections and adjustments applied as set out in this appendix.
- The *daily* dataset holds daily records of maximum and minimum temperatures (°C), mean daily temperature (average of the maximum and minimum temperature, °C), grass minimum temperature (°C), precipitation (mm) and sunshine (hours), together with various derived values such as rain days (0.2 mm or more of precipitation) and air frosts (screen minimum temperature below 0 °C). The dataset starts in December 1813; the period of record of each element is given in Chapter 2, and daily records prior to the start of records are shown as −999. Other than the daily rainfall record, the corrections and adjustments detailed in this appendix have *not* been applied to daily records.

Each dataset will be regularly updated with current records.

The datasets used to prepare this book are still being worked on, primarily to extend the period of digitised record but also to correct any remaining errors found. It is for this reason that minor discrepancies may occasionally be evident between the values given in this work and the online records; the latter should be regarded as the most up-to-date version. There is a facility on the website to report incorrect values as they are found, in order continually to improve the quality of the Radcliffe Meteorological Station datasets.

# Appendix 2
# The Radcliffe observers

The Radcliffe Observatory was based in Oxford for 163 years. During this time, the Observer was in charge and just eight men served in that role; their contributions are briefly summarised below (see [22] for more details). The astronomical observatory moved to South Africa in 1935, and since then the Radcliffe Meteorological Station has been run by a Director, of whom there have so far been five. The Directors have been assisted by a large number of 'observers', including those specifically employed to do the day-to-day work, plus many deputies providing cover on Sundays and public holidays. Occasionally, the Directors themselves have even been known to make the morning observations, Christmas Day being a favourite choice! Of all the meteorological observers, James Balk deserves special mention, providing the necessary continuity after the last Radcliffe Observer in Oxford, Harold Knox-Shaw, moved with the astronomical observatory to South Africa.

## The Radcliffe Observers

### Thomas Hornsby (1733–1810), first Radcliffe Observer 1772–1810

A full account of Thomas Hornsby's life and principal accomplishments is given in Chapter 2. Of course, his single greatest achievement lay in persuading the Radcliffe Trustees to build their fine Observatory, but he had more than a passing interest in meteorology and today he is best remembered for starting meteorological observations at the Observatory. His earliest surviving weather records, observations of rainfall, were kept regularly in 1760 but had petered out by March 1761. Thereafter, he made numerous, if irregular, observations of rainfall, temperature, winds and cloud cover from 1767 until 1804; his weather journal is only complete for about a third of this period. Craddock and Craddock [24] note that the observations of 1767 show all the signs of a fresh start, and this marks the point from which the long, homogenous monthly rainfall series for Oxford commences. Hornsby did not publish his meteorological observations, but they were carefully made with sound instruments and have allowed others to reconstruct aspects of the climate record during that period, including the rainfall record and temperatures during the severe winters of 1767 and 1776.

### Abram Robertson (1751–1826), second Radcliffe Observer 1810–1826

Robertson employed and supervised an assistant in making some astronomical observations and made three main contributions to the Observatory's meteorological work: firstly, he purchased for the Observatory a number of good quality meteorological instruments and introduced from 1816 more systematic observations of pressure and temperature; secondly, he completed the task of obtaining the freehold over the land on which the new Observatory had been built; lastly, in 1822, he persuaded the *Edinburgh Philosophical Journal* to publish the Radcliffe meteorological results for 1816–21, the first meteorological results of the Radcliffe Observatory to be published ([22], pp. 106–7).

## Stephen Peter Rigaud (1774–1839), third Radcliffe Observer 1827–1839

Rigaud took scientific interest in weather as a continuous natural phenomenon, rather than regarding the weather as merely an aid or hindrance to astronomical observation. He was also the 'King's Observer' to George III, and was a key influence in the establishment of the King's Observatory, later Kew Observatory, in 1769 [29]. He recognised the interdependence of the various meteorological elements and, starting in 1828, organised the taking daily of three complete sets of eye observations of all the main meteorological elements at the Radcliffe Observatory. Starting in June 1838, Rigaud brought into use as his main barometer the splendid Newman Standard No. 1220 that still graces the downstairs Common Room of Green Templeton College (Figure A2.1, [22], pp. 107–8).

**Figure A2.1** *The Newman Standard Barometer no. 1220 in the Common Room of Green Templeton College in the Radcliffe Observatory building in Oxford, photographed in April 2018. This instrument has been in daily use since it was installed in June 1838 (Stephen Burt)*

## Manuel John Johnson (1805–1859), fourth Radcliffe Observer 1839–1859

Johnson was responsible for two innovations that gave long-term value to the Radcliffe Observatory's meteorological observations. First, he consolidated the systematic observations and, from 1847, he had these published together with the astronomical observations in the Observatory's annual *Results* volumes. Johnson's second important contribution to meteorology was to introduce several self-recording instruments at the Radcliffe Observatory: barograph, thermograph and hygrograph in 1854, anemograph and pluviometer (or hyetograph) in 1856 ([22], pp. 108–10).

## Robert Main (1808–1878), fifth Radcliffe Observer 1860–1878

Despite Main carrying out his duties faithfully and ensuring that the observations were made and published, the *Introductions to the Meteorological Results* in the annual Radcliffe volumes suddenly become perfunctory following Main's appointment in 1860. Nevertheless, he seems to have taken an interest in the quality of the observations and it was he who had the raingauge moved from the south lawn to a more exposed position on the north lawn in 1862. He also introduced a maximum and minimum thermometer at the top of the tower to compare with those at ground level. In 1872, the Radcliffe Observatory became a 'reporting station' to the Meteorological Office and from 1873 to 1916 its daily observations were telegraphed to the Meteorological Office to contribute to the preparation of synoptic charts and reports ([22], pp. 110–11).

## Edward James Stone (1831–1897), sixth Radcliffe Observer 1879–1897

Stone alerted the Trustees in his first annual report to the fact that, over the previous 30 years, the equipment of the Radcliffe Observatory had fallen behind more advanced equipment introduced in observatories elsewhere. Although the Trustees could not afford expensive new astronomical instruments, improvements were made almost immediately in the meteorological instrumentation, with little expense to the Trustees. In 1880 a self-registering photographic barograph, a dry-bulb thermograph and a wet-bulb thermograph were installed, comparable to the 1854 set of Radcliffe instruments but this time borrowed from the Meteorological Office and 'of the usual Kew type' ([22], pp. 112–3).

## Arthur Alcock Rambaut (1859–1923), seventh Radcliffe Observer 1897–1923

Having previously been the Royal Astronomer for Ireland, Rambaut's main achievement was persuading the Trustees to renew obsolete astronomical instruments, but he did not neglect the meteorological instruments and introduced platinum-resistance soil thermometers in 1898. In 1908 he rationalised the location of the meteorological instruments within a new enclosure on the south lawn. In 1923 a Stevenson screen of a modern standard pattern was erected in the south lawn enclosure, close to and in place of the existing double-louvred Stevenson 'stand'. Through no fault of Rambaut, the status and importance of the Radcliffe Observatory as a source of meteorological observations was already declining in relative terms. More and more meteorological stations were making detailed observations, and the Radcliffe Observatory lost its status as a 'reporting station' to the Meteorological Office in 1916 ([22], pp. 113–5).

## Harold Knox-Shaw (1885–1970), eighth and last Radcliffe Observer 1924–1935

Previously, Knox-Shaw had been Director of the Meteorological Service of Egypt and the Sudan. Despite all the pressures on him as a consequence of the decision to move the astronomical observatory to South Africa, Knox-Shaw was fully aware of the long-term value for climatic studies of the Radcliffe Observatory's long continuous meteorological observations. He recognised also that there had been a wide range of instruments in use over the previous century and that some of these instruments had been exposed in different locations and with different screening at different times, so leading to doubts about their homogeneity and the reliability of results. With the assistance of his indefatigable assistant James Balk, Knox-Shaw examined closely all the earlier records. Where it seemed possible to make reliable reductions to a common standard, this was done. If it became necessary to re-establish instruments in order to provide calibration with current instruments, then this was done, as with the roof-mounted raingauge. In the appendix to Volume LV for the *Results of Meteorological Observations made at the Radcliffe Observatory, Oxford, in the Five Years 1926–1930* [30], revised tables were published of all the main weather elements from 1881 to 1930, while monthly mean dry-bulb temperatures and monthly rainfall totals were tabulated from 1815 ([22], p. 115).

*The 'Radcliffe Observers' of the Radcliffe Observatory up to 1935 were first and foremost professional astronomers and, from the earliest published records in 1840 until 1892, the annual Radcliffe Results volumes mostly consist of astronomical observations, although some include details of both astronomical and meteorological instruments and staff members. After 1892, the meteorological observations were published in their own volume, but usually combined for several years and thus published some years in arrears. Details of the early* meteorological *observers are quite sketchy, but two of the longer-serving observers are referred to in the Results:*

**George Green** was described as 'the journeyman of the establishment'. He joined the Radcliffe Observatory as a boy assistant in the 1840s and is first mentioned in the staff lists in 1854. He was responsible for most of the meteorological observations from at least 1854 until he left the Observatory in August 1861.

**John Lucas** was first mentioned in the 1859–60 *Radcliffe Results*, but was taken on as the (astronomical) assistant at the Observatory under Johnson in 1840, deputising for the meteorological observations as occasion demanded (it was John Lucas who recorded the lowest-ever temperature at the Observatory on Christmas Eve 1860, of -17.8 °C). When George Green departed in August 1861, John Lucas assumed responsibility for the meteorological observations. He was promoted to the office of first assistant in June 1869 and remained in charge of meteorology until his retirement on £100 per annum pension in June 1878, after 38 years' service at the Observatory. He died in 1883.

## The Directors of the Radcliffe Meteorological Station (RMS)

### Wilfred George Kendrew (1936–1950)

In 1935, the astronomical work of the Radcliffe Observatory was transferred to South Africa. Responsibility for the supervision of weather observations was taken over by the School of Geography, and W. G. Kendrew (1884–1962) served as Director of the Radcliffe Meteorological Station from 1936 until his retirement in 1950, although James Balk did much of the work when Kendrew was absent on war service, weather forecasting in the Indian Ocean for the Royal Navy [22]. Kendrew was a University Lecturer in Climatology, best known for *The Climates of the Continents*. A full record of his career was given by Joan Kenworthy [166].

### C. G. (Gordon) Smith (1950–1985)

A former student of Kendrew's, Gordon Smith (1921–1999) was a University Lecturer in Climatology (from 1949) and Fellow of Keble College (from 1957), posts he held until his retirement in 1983. A native of Gloucestershire, with the outbreak of war in 1939 he went from there straight into the army to serve initially in the Middle East. It was this period of his life that stimulated his later interest in the politics and hydrology of the region.

Gordon Smith will perhaps be best remembered for his meteorological work. He held the post of Director of the Radcliffe Meteorological Observatory from 1950 to 1985, carrying on after retirement until his successor had settled into his new life in Oxford and was able properly to take over running the RMS, a role to which Gordon Smith gave the highest attention. Gordon was meticulous in his scrutiny of the observations and always keen to point out dubious data to his successor! He was responsible for moving the anemometer and sunshine recorder to the roof of the Engineering building in 1976 and generally for ensuring that the Radcliffe Meteorological Station survived intact, in the face of possible closure. Once the anemometer had been removed from the Tower of the Winds, Gordon ensured that the wooden structure that held the instruments was removed so that the building could be returned to its former glory [28].

Gordon Smith was instrumental in the analysis for, and publication of many papers on, the Oxford dataset, including his papers in *Weather* on the Station's main weather records and the periodic fluctuations in rainfall. Several journals, including *Weather* and *Meteorological Magazine*, carried his papers on interesting meteorological events occurring in both contemporary (e.g. the

**Figure A2.2** *The last four Directors have all been Fellows of Keble College as well as members of the School of Geography and the Environment. This snowy picture of Keble chapel was taken on 29 January 2004 (Faye McLeod)*

gale of 2 January 1976) and historical times (e.g. his analysis of very cold winters in the RMS record).

## Tim Burt (1986–1996)

Tim Burt took over the RMS from Gordon Smith in 1986 and ran the Station for 10 years. During that time, the work on digitising the records was completed, a job started by Gordon Smith and Basil Gomez. Tim was responsible for the installation of an automatic weather station (AWS) in the early 1990s and, through his work for the UK Environmental Change Network, helped with the installation of an AWS at Wytham Woods. Having moved to Durham, Tim took over responsibility for the Durham Observatory weather station in 2001, with not as long a record as at Oxford, but still one of the longest continuous records available in the UK (from 1850: [78] [80]). Notwithstanding retirement to Devon, Tim continues to write monthly and annual weather reports about the weather at Durham.

## David N. Collins (1996–1999)

David Collins (1949–2016) had only a brief tenure at the RMS. He came to Oxford from the University of Manchester but soon returned to that city with a Chair at the University of Salford. His research interests were in glacial hydrology and climatology.

### Richard Washington (1999–present)

Richard Washington is Professor of Climate Science at the School of Geography and the Environment and, like all the Directors from Gordon Smith onwards, a Fellow of Keble College. He specialises in African climate science and runs the African Climate research group. Richard's research is concerned with African climate systems, including climate change, the mechanisms leading to floods and drought, and the way these characteristics of climate are represented in climate models. He has also worked extensively on aerosols, particularly so on dust storms in the central Sahara and in southern Africa. He has run major observational campaigns involving both ground observations and instrumented aircraft in North and southern Africa. In recent years, the Radcliffe Meteorological Observers have been Richard's own DPhil students.

## The Meteorological Observers

### James Balk (1903–1954)

James Balk (1889–1956) is a key figure in the history of the Radcliffe Meteorological Station. Balk joined the Observatory from Oxford Technical College in April 1903, aged 14. It was James Balk who took on the Herculean task—well before computers and spreadsheets—of manually organising or reorganising the Observatory's meteorological archives and, from there, comparing and adjusting to common standards the whole of the records. His distinctive handwriting, signature red ink, and initialled and dated comments appear throughout the early records, carefully assessing, adjusting and correcting errors where needed. These eventually appeared in the appendix to Volume LV for the *Results of Meteorological Observations* [30] and his neatly written logbooks and registers survive today in the University's School of Geography, where both authors have examined them carefully and have been extremely grateful for his unsung efforts in maintaining and archiving the series.

Without James Balk, it is likely that many of the early Observatory records would have been lost or discarded when the astronomical observatory moved to Pretoria in 1935. It seems self-evident that Balk, by then an Observatory employee of more than 25 years, was a strong advocate of maintaining the meteorological record at the Observatory site once the astronomical functions emigrated to the clearer skies of South Africa; we owe to him the survival of the meteorological record past 1935 at a time when it seemed very likely that the Station would be closed. He remained the principal Observer after 1935, in the employ of the University of Oxford, until his retirement in 1954 after 51 years at the Observatory (less his military service in the First World War). He died in 1956.

*As part of the 200-year anniversary celebrations, Ian Curtis from the School of Geography and the Environment collated memories from some of those who were employed as meteorological observer, of which some excerpts are reproduced here.*

### Derek Winstanley (1966–69)

*Then*: DPhil student in climatology. *Now*: Retired Chief of the Illinois State Water Survey, Champaign, Illinois

'Serving as Radcliffe Meteorological Observer was an important step in my life. I had previously set up a little observation station in our back yard at home in Wigan and specialized in Climatology while earning my BA at the School of Geography and at Hertford College. Under Gordon Smith's tutelage, I earned my DPhil in Climatology at the School of Geography and it was he who invited me to serve as Radcliffe Meteorological Observer, from 1966 through 1969, if

I remember correctly. At that time, I lived in Kidlington and rode my bicycle to arrive promptly at the Observatory by 9 a.m. come rain or shine. As well as dedication to accurate meteorological observations, I learned a deep commitment to punctuality.'

### Guy Robinson (1977–78)

*Then*: DPhil student and Acting Demonstrator at the School of Geography. *Now*: Adjunct Professor of Geography, University of Adelaide

'The "delights" [of being the Met Observer] were many and varied—from riding up to the top of the paternoster lift in the Thom Building to changing the card in the Campbell–Stokes sunshine recorder on the roof, to changing the paper and ink (Oh, that damned ink!) in an anemograph connected to a cup anemometer, to phoning the *Oxford Mail* to deliver the previous day's weather readings, to taking the daily readings of temperature and precipitation under the stern gaze of the Tower of the Winds. So, each morning I would drive down the Woodstock Road to the Observatory, with its Stevenson screen, and then to the Thom Building to ride the paternoster before proceeding to my small office at the School of Geography—the only office in the building from which there was no proper view of the sky. After the phone call to the *Mail* I would work on the archives, constructing moving means and searching for unusual weather occurrences. This analysis would then be passed to Gordon Smith to form the basis of several of his papers on Oxford's weather, including one published in *Weather* in 1979 [25] and a School of Geography monograph on *The Gale of 2 January 1976*, also published in 1979 [91].'

### Neil Calton (1979–81)

*Then*: MSc student. *Now*: Systems Analyst at the Rutherford Appleton Laboratory, based in Oxford

'I took on the role of Radcliffe Meteorological Observer while a member of Linacre College studying for an MSc under my tutor Gordon Smith. I was grateful to take up the opportunity to be the official observer to help supplement my grant as it allowed me to stay longer in Oxford and extend my research. Luckily, I did not live too far from the Observatory and managed to keep the early morning deadline required to take the temperature and rainfall readings there and the wind speed measurements at the nearby Engineering building. My only gap in these duties being when I sprained my ankle and Mr Smith kindly stepped in until I was mobile enough to continue making the daily journey.

I found the job particularly apposite as my research topic required me to make extensive use of the weather records from the Observatory. I was attempting to find a correlation between the width of tree rings from oaks in Wytham Woods and the climate of the Oxford region. So much of my time was spent typing into the computer the figures from the meteorological records and producing lots of graphs.'

### Basil Gomez (1982–87)

*Then*: Lecturer in Physical Geography at Jesus and St Catherine's Colleges. *Now*: Consultant and member of the council which shapes the scope, vision and strategic direction of the National Science Foundation's 'EarthCube' cyberinfrastructure programme, based in Honolulu

'For a young researcher returning to the UK from a post-doctoral position in Uppsala, Gordon Smith's offer of the post of Radcliffe Meteorological Observer was a welcome bright spot in the bleak academic landscape created by Margaret Thatcher's cuts in higher education spending. In 1982, electronic record keeping was in its infancy, but a concerted effort was already being made to convert the Station's entire daily rainfall and (max, min) temperature records to digital format. This involved many visits to the Oxford University Computing Service and much use of the VAX minicomputers in Banbury Road (which each had 8 MB of main memory!). The process was finally completed before Gordon Smith retired and Tim Burt took over as Director of the Radcliffe Meteorological Station in 1986.

Come rain or shine, winter or summer, my daily visits to the instrument enclosure (in the then Green College grounds), the magnificent Observatory building and the roof of the Thom Building, afforded a year-round perspective of Oxford and the surrounding landscape that few people have the privilege of experiencing. Indeed, it was the view from the from the roof of the Thom Building that inspired Gordon Smith and I to examine, in two short papers published in *Weather*, changes in fog frequency and visibility documented in the Station's records [167].'

## Peter Venters (1987–90)

*Then:* DPhil student, Department of Atmospheric Physics. *Now:* Managing Director, Sea Pebble Ltd

'Looking back now from three decades on, being Radcliffe Met Observer was, of course, about reliability and repeatability, temporarily taking on responsibility for a regular and unbroken chain of accurate measurements, then stretching back for over a century and a half. But now, for me, it is less analytical and more impressionistic memories which endure.

Memories of chill winter 9 o'clocks in the walled quiet of Green College's immaculately trimmed garden, crunching over the frosted lawn to the instrument enclosure. Or warmer summer mornings, with an extra hour of BST in hand, and perhaps time for a brief conversation with the College's gardener, friendly, wise and splendidly moustached.

The instruments themselves—all brass and glass, paper and clockwork, mercury and muslin—with the air of museum pieces but still tracking the modest chaos of the Oxford climate with enduring precision. Most venerable of all, secure in the carpeted silence of the old Observatory building, the barometer with its pool of mercury and Vernier scale, all imperial units and squinting in the gloom of an overcast (eight oktas) autumn morning.

And by contrast, a jog up eight flights—or a ponderously clunking lift journey—to the top of Engineering, for the sunshine and wind measurements and the best view in Oxford. Then with charred sunshine cards, damp rainfall charts and flimsy, inky anemometer traces, back to the monastic gloom of the School of Geography, to record and report and—most challenging of all—write up the monthly summary. How, oh how, to make the 34th driest May since records began sound interesting, or the 15th windiest?

When I was Observer, climate change was at the very beginning of its long, painful transition from academic study to public awareness. Now, as the world slowly turns to face the music, the value of reliable, long-term climatological data sets has never been greater. Those morning measurements in the tranquillity of Green College garden have their part to play in the science that may yet stave off disaster, and I'm glad to have been involved all those years ago.'

## Rosemary Munro (1989–91)

*Then*: DPhil student. *Now*: Atmospheric Composition Competence Area Manager, EUMETSAT, based in Germany

'The clearest memories I have of my days as the Radcliffe Meteorological Observer were of the beautiful environment of Green College and in particular the immaculate wisteria which covered the buildings around the quadrangle where the Radcliffe Observatory is located. In spring they were glorious in purple and in the winter the beautifully pruned stems were stunningly architectural. Of course, that wasn't the only thing I remember and enjoyed. In particular I remember being extremely grateful for the help from the very kind and competent technician, with an immaculate white beard, who more than once had to retrieve the 1 metre thermometer from the bottom of its tube when it became detached from the chain. I also very clearly remember the eight floors to the top of the Engineering Science building where the sunshine recorder (a Campbell–Stokes instrument—a lovely little glass sphere in a brass mount, used to focus the sun when it is out and burn a very controlled hole in a cardboard measuring strip) and anemograph resided. Every morning I set off up the stairs, resolutely ignoring the elevator which might be slightly less easy today, to admire the view and take the observations. I particularly liked the little piece of card in the sunshine recorder with the burnt holes indicating the sunshine hours although I recall interpretation was sometimes difficult.

Last but not least, as a student in the Department of Atmospheric, Oceanic and Planetary Physics, I was always very envious of the building occupied by the School of Geography and the Environment where the Meteorological Observer's office resided. It had a beautiful wooden staircase and the walls were covered in large photos from exotic locations where staff members and students had been on expeditions. I have to say it always made me wonder if atmospheric physics was the right choice! I did however consider myself very lucky to have the position of Meteorological Observer, which at that time was being passed from one DPhil student in the Department of Atmospheric, Oceanic and Planetary Physics to the next, as it was the perfect way to support one's studies and to start the day.'

### Maria Shahgedanova (1991–97)

*Then*: DPhil student and part-time Departmental lecturer. *Now*: Associate Professor in Climate Science at the University of Reading

'Running the Radcliffe Weather Station taught me to appreciate continuity and consistency of observations and the importance of maintaining quality of observation over long period of time. One does feel responsibility if the station had been going on for 200 years before you! I worked with climate data before and after this job but what I still value about it that it taught me about the basics and standards of meteorological observations and how to operate manage different types of equipment. An AWS was established on site when I was the Observer and I had to learn how to run this one too. Evidently this was useful—I am running a network of high-altitude AWS across the mountains of Asia now!'

### Norman Cheung (1998–2000)

*Then*: DPhil student in Typhoon Climatology. *Now*: Senior CAT Analyst at Scor Reinsurance Company (Asia) Limited

'I remember the heavy fog in the morning while I was riding the bicycle to Green College. I remember gazing over the white snow-covered spikes from the top of the Engineering building in the morning. How would I forget the time when I got wet by the splash from the wheels of my bike on riding to changing the rain bucket! How could I let myself forget the best time of my life! I am so proud of being an Oxonian. I even feel so privileged to ever be an Oxford Radcliffe Station observer. This is what I call history!'

### Helen Bray (now Wain) (2000–03)

*Then*: DPhil student. *Now*: Police Sergeant at West Mercia Police

'Whilst I was the Observer, for a short period the weekday readings were done by Steve, the School of Geography porter. He showed me how all the instruments worked when I started. I recall it very well because as I bent down to read the soil thermometers on my first ever day my trousers split. Steve affected not to have noticed but when I returned to the SoG later on, he just said, "Changed your trousers, then?!"

Dave Banfield, the lab technician at the time, also did some of the readings, to cover if I was away. I quite often had friends tag along on the weekends as the views from the Engineering building are so lovely. One of those friends is now my husband!'

### Gillian Kay (2003–07)

*Then*: DPhil student. *Now*: Climate Scientist at the Met Office, based in Exeter

'The station was in the beautiful gardens of Green College, and I loved the fact that going there every day, I was so aware of how the gardens changed throughout the year, and the measurements I was taking reflected the passing of the seasons and changes I saw. I knew which weekend of the year saw the purples of the wisteria and the Californian lilac blooming together around the gate.

My favourite instrument was the barometer, because it was such an old instrument. Generations of Observers had been twiddling the same knobs as I was twiddling! And during the holidays, when the Observatory was locked, I had to collect the giant old key from the lodge to gain entry. It gave me a real sense of the history of what I was doing. The wind instruments were a bit of a rite of passage – filling the pens with ink using the syringe was nerve-racking: so often I applied that little bit too much pressure and psshhht! I got a spray of blue ink in the face. The mark of the Observer! And what a view from the top of the Engineering building! It was a privilege.

Finally, I enjoyed putting my measurements in the context of the historical observations. The old Observer books were an absolute treasure trove, and testament to the care that has been taken over the years, and the importance attached to this work that forms the foundation of weather and climate science today.'

### Helen Pearce (2008–12)

*Then*: DPhil student. *Now*: Senior Science Programmes Officer, Natural Environment Research Council

'I have many memories from the time I spent observing, with particular recollections of the stillness of Oxford at 9 a.m. on Christmas Day, with only a few dedicated runners for company on the streets. I also remember a number of weekends walking up the nine flights of stairs to the Campbell–Stokes sunshine recorder on the roof of the engineering department, not always trusting the unreliable lifts when the building was deserted—being the observer was good for fitness, in addition to the obvious benefits for timekeeping and reliability!

I now work for the Natural Environment Research Council, in a role that allows me to combine my continued interest in environmental science, and climate variability and change in particular, with gaining new knowledge and perspectives on the research and science landscapes. My time as the meteorological observer and the rich history of the Radcliffe Meteorological Station always serves as a good talking point, and I hope the tradition of taking daily manual measurements at the station continues for many years to come.'

### Ian Ashpole (2012–15)

*Then*: DPhil student. *Now*: Postdoc in Halifax (Nova Scotia), using satellite data to identify trends in, and sources of, air pollution over Canadian cities

'I loved the sense that I was part of a tradition stretching back hundreds of years. I felt great responsibility and pride in maintaining this tradition, and often found myself lost in thought, imagining how the rest of the day would have unfolded for my nineteenth-century counterparts after they had finished taking the observations. How different their lives must have been, but how amazing that we shared the same morning routine. I felt that this connected us, in a way.

Moreover, being custodian of the position gave me something to say when asked about my work. For some reason people would always glaze over when I told them that I studied dust; mention the fact that I was responsible for maintaining the longest continuous set of weather records in the UK and, all of a sudden, they were interested. I think I even enlightened some people about the difference between weather and climate and maybe, maybe, even got them to take the notion of climate change seriously by showing them a quick time series plot of mean annual temperature since records began. Maybe. I lost count of the number of times I took friends, family, and friends of friends with me to do the measurements just because they were interested. No doubt part of the seduction was the promise of getting the best view of Oxford, from the top of the Thom building. Not something that the majority of past observers were lucky enough to enjoy, I often found clarity of thought while contemplating that breath-taking vista. I won't lie—I also liked the media attention!'

## Amy Creese (2015–19)

DPhil student, African climate science

'I love the sense of history you get when taking measurements and checking records in the RMS data. During my time as observer, RMS reached its 200th anniversary of daily readings, and so every record can now be put into context with the phrase '...in over 200 years!' As my DPhil focuses on African climate science, I've enjoyed the opportunity to look at UK weather trends, which is about as far away from my usual expertise as you can get whilst still falling under the "weather and climate" umbrella. As well as its serious aspects the job has lots of fun elements too. There has always been a bit of a competitive streak between me and recent fellow observers; whenever there's a particularly big storm, those of us on the WhatsApp chat make bets as to how much

**Figure A2.3** *For many years, schoolchildren have been employed to take the Sunday readings, giving the observer a day off...here Emma Burt changes the chart on the Dines tilting siphon rain gauge in the early 1990s (Tim Burt)*

rainfall there will have been, and curse inwardly if someone else has beat our personal best! Heat records don't induce quite the same excitement, as they speak of a worrying long-term trend in our climate, but it is still fascinating to calculate how much warmer it has been than the long-term average, as is often the case. The summer of 2018 was one of the warmest on record, and certainly in my lifetime, and there is a sense of responsibility in collecting and communicating this kind of data. Again, on a lighter note, being an RMS observer always provides an interesting topic of conversation and has also given me several opportunities to speak on the local news. A highlight has been talking as an expert on a BBC4 documentary about the science of temperature, though filming the segment did involve walking to the top of the Observatory tower spiral staircase three times in a row—a feat I'm sure all previous observers can agree is quite impressive!'

**Figure A2.4** *Gordon Smith (see page 374) on active service in Egypt (School of Geography and Environment, University of Oxford)*

# Appendix 3
# Climatological averages and extremes for Oxford, 1981–2010

# OXFORD

**Radcliffe Meteorological Station**

*Averages for period*
**1981–2010**

*Extremes for period*
**1981–2018**

**Site details**

| *Record began* | *Lat* | *Long* | *NGR* | *Altitude AMSL* |
|---|---|---|---|---|
| 1772 | 51.76N | 1.26W | SP (42) 509 072 | 63 m |
| | | *Terminal hour* | *Temperature* | 09–09 GMT |
| | | 0900 GMT | *Rainfall* | 09–09 GMT |

| **TEMPERATURE °C** | **Jan** | **Feb** | **Mar** | **Apr** | **May** | **June** | **July** | **Aug** | **Sept** | **Oct** | **Nov** | **Dec** | **Annual** | **Yrs** |
|---|---|---|---|---|---|---|---|---|---|---|---|---|---|---|
| Mean daily maximum | 7.6 | 8.0 | 10.9 | 13.6 | 17.0 | 20.2 | 22.7 | 22.3 | 19.1 | 14.8 | 10.5 | 7.7 | 14.6 | *30* |
| Mean daily minimum | 2.1 | 1.8 | 3.7 | 5.0 | 7.9 | 10.9 | 13.0 | 12.9 | 10.7 | 7.8 | 4.6 | 2.3 | 6.9 | *30* |
| Mean temperature | 4.9 | 4.9 | 7.3 | 9.3 | 12.5 | 15.6 | 17.9 | 17.6 | 14.9 | 11.3 | 7.5 | 5.0 | 10.7 | *30* |
| Highest maximum | 15.9 | 18.5 | 21.3 | 27.6 | 28.9 | 32.5 | 34.8 | 35.1 | 29.9 | 29.1 | 18.0 | 15.9 | 35.1 | *38* |
| Year | 2016 | 1998 | 2012 | 2011 | 2010 | 2017 | 2006 | 1990 | 2004 | 2011 | 2005 | 2015 | 3 Aug 1990 | |
| Lowest minimum | −16.6 | −10.7 | −6.1 | −2.9 | −0.2 | 1.8 | 5.6 | 4.5 | 1.2 | −3.9 | −7.0 | −16.1 | −16.6 | *38* |
| Year | 1982 | 1986 | 2018 | 1996 | 1997 | 1991 | 2015 | 2003 | 2003 | 1997 | 1983 | 1981 | 14 Jan 1982 | |
| Highest minimum | 12.1 | 12.0 | 12.8 | 13.8 | 15.6 | 19.6 | 21.2 | 20.5 | 18.1 | 17.0 | 14.7 | 12.5 | 21.2 | *38* |
| Year | 2008 | 2004 | 1998 | 2018 | 2001 | 2011 | 2016 | 1997 | 2016 | 2018 | 1996 | 1985 | 20 Jul 2016 | |
| Lowest maximum | −7.1 | −3.6 | −1.9 | 4.5 | 8.3 | 11.2 | 15.3 | 14.3 | 9.8 | 6.6 | −1.2 | −4.4 | −7.1 | *38* |
| Year | 1982 | 1991 | 2018 | 2013 | 1996 | 1989 | 1987 | 1986 | 1993 | 2018 | 2010 | 2010 | 13 Jan 1982 | |
| Air frosts | 8.5 | 9.1 | 3.8 | 1.7 | 0.1 | 0 | 0 | 0 | 0 | 0.9 | 4.0 | 9.1 | 37.1 | *30* |
| Ground frosts | 17.3 | 16.2 | 14.4 | 10.2 | 3.4 | 0.3 | 0.03 | 0 | 0.8 | 4.9 | 11.7 | 16.7 | 95.9 | *30* |

| **PRECIPITATION mm** | **Jan** | **Feb** | **Mar** | **Apr** | **May** | **June** | **July** | **Aug** | **Sept** | **Oct** | **Nov** | **Dec** | **Annual** | **Yrs** |
|---|---|---|---|---|---|---|---|---|---|---|---|---|---|---|
| Monthly mean | 56.9 | 42.8 | 47.7 | 48.9 | 57.2 | 49.2 | 48.8 | 54.8 | 54.2 | 70.0 | 66.7 | 63.1 | 660.3 | *30* |
| Days ≥ 0.2 mm | 16.2 | 13.0 | 15.1 | 13.4 | 13.5 | 11.7 | 11.4 | 11.9 | 12.4 | 15.3 | 15.6 | 15.5 | 164.9 | *30* |
| Days ≥ 1.0 mm | 11.4 | 9.0 | 10.1 | 9.0 | 9.7 | 8.2 | 7.8 | 8.2 | 9.2 | 10.9 | 11.3 | 10.7 | 115.5 | *30* |
| Wettest day | 21.5 | 25.9 | 42.1 | 38.0 | 40.7 | 27.9 | 50.0 | 51.1 | 46.4 | 40.6 | 34.9 | 38.3 | 51.1 | *38* |
| Year | 2003 | 2009 | 2008 | 2000 | 2007 | 2008 | 2004 | 2010 | 1998 | 2007 | 2008 | 1985 | 25 Aug 2010 | |
| Wettest month | 146.9 | 94.2 | 129.7 | 143.0 | 135.2 | 151.7 | 110.2 | 146.2 | 119.4 | 138.8 | 133.4 | 141.5 | 979.5 | *38* |
| Year | 2014 | 1990 | 1981 | 2012 | 2007 | 2012 | 2007 | 2010 | 1998 | 1987 | 2002 | 1989 | 2012 | |
| Driest month | 9.9 | 7.1 | 9.2 | 0.5 | 8.6 | 2.5 | 3.6 | 3.0 | 4.1 | 17.1 | 20.0 | 13.4 | 470.8 | *38* |
| Year | 1987 | 1993 | 1997 | 2011 | 1991 | 2006 | 2016 | 2003 | 2014 | 2017 | 1990 | 1988 | 1996 | |

| **SUNSHINE hours** | Jan | Feb | Mar | Apr | May | June | July | Aug | Sept | Oct | Nov | Dec | Annual | Yrs |
|---|---|---|---|---|---|---|---|---|---|---|---|---|---|---|
| Monthly mean | 62.3 | 78.9 | 111.2 | 160.9 | 192.9 | 191.0 | 207.0 | 196.5 | 140.4 | 111.3 | 71.0 | 53.2 | 1576.5 | *30* |
| *Daily mean* | *2.01* | *2.79* | *3.59* | *5.36* | *6.22* | *6.37* | *6.68* | *6.34* | *4.68* | *3.59* | *2.37* | *1.72* | *4.32* | *30* |
| *Daylight hours* | *260.8* | *282.4* | *369.4* | *417.7* | *486.0* | *498.0* | *499.9* | *451.4* | *378.8* | *329.7* | *265.3* | *244.2* | *4483.7* | |
| *% possible* | *23.9* | *27.9* | *30.1* | *38.5* | *39.7* | *38.3* | *41.4* | *43.5* | *37.1* | *33.8* | *26.8* | *21.8* | *35.2* | |
| Days nil sunshine | 11.6 | 7.6 | 6.2 | 3.3 | 2.7 | 2.0 | 0.9 | 1.2 | 2.6 | 5.5 | 8.8 | 13.0 | 65.3 | *30* |
| Sunniest day | 8.5 | 9.7 | 12.4 | 13.9 | 15.5 | 15.5 | 15.4 | 14.1 | 12.6 | 10.4 | 8.9 | 7.5 | 15.5 | *38* |
| Year | 1986 | 1996 | 1996 | 1990 | 1985 | Vs | 1994 | Vs | 2002 | 1994 | 1990 | 2016 | Vs | |
| Sunniest month | 87.7 | 124.1 | 198.7 | 236.8 | 300.8 | 290.7 | 303.7 | 285.1 | 175.3 | 153.8 | 104.1 | 96.9 | 1880.3 | *38* |
| Year | 2003 | 2008 | 1995 | 1984 | 1989 | 1996 | 2006 | 1995 | 2003 | 1997 | 1989 | 2014 | 1995 | |
| Dullest month | 29.3 | 33.8 | 48.8 | 105.8 | 98.3 | 100.4 | 129.0 | 135.3 | 92.4 | 66.3 | 33.8 | 20.4 | 1204.7 | *38* |
| Year | 1996 | 1982 | 1984 | 1998 | 1981 | 2016 | 1981 | 2015 | 1983 | 1982 | 2015 | 2010 | 1981 | |
| **DAYS WITH…** | **Jan** | **Feb** | **Mar** | **Apr** | **May** | **June** | **July** | **Aug** | **Sept** | **Oct** | **Nov** | **Dec** | **Annual** | **Yrs** |
| Snow/sleet falling* | 4.1 | 3.7 | 3.4 | 1.3 | 0.2 | 0 | 0 | 0 | 0 | 0 | 0.8 | 2.6 | 16.1 | *30* |
| Snow lying 0900 GMT | 2.1 | 2.5 | 0.3 | 0.1 | 0 | 0 | 0 | 0 | 0 | 0 | 0.2 | 1.5 | 6.7 | *30* |
| Thunder heard* | 0.2 | 0.1 | 0.4 | 0.6 | 1.8 | 1.9 | 2.1 | 1.7 | 0.7 | 0.3 | 0.1 | 0.1 | 10.0 | *30* |

***Site and instrument metadata***

*A well-maintained meteorological enclosure in a walled college garden. Screen and raingauge have been in this enclosure since September 1939; sunshine from nearby Engineering Building, since September 1976. *Averages for period 1961–90 only*

# Appendix 4
# **Climatological averages and extremes for Oxford, since records began**

# OXFORD

**Radcliffe Meteorological Station**

*Averages and extremes for entire available record*

**Site details**

| *Record began* | *Lat* | *Long* | *NGR* | *Altitude AMSL* |
|---|---|---|---|---|
| 1772 | 51.76N | 1.26W | SP (42) 509 072 | 63 m |
| | | Terminal hour 0900 GMT | *Temperature* | 09–09 GMT since 1925 |
| | | | *Rainfall* | 09–09 GMT since 1921 |

| **TEMPERATURE °C** | Jan | Feb | Mar | Apr | May | June | July | Aug | Sept | Oct | Nov | Dec | Annual | Yrs |
|---|---|---|---|---|---|---|---|---|---|---|---|---|---|---|
| ***Period 1815–2018*** | | | | | | | | | | | | | | |
| Mean daily maximum | 6.4 | 7.3 | 9.9 | 13.1 | 16.8 | 20.1 | 21.8 | 21.2 | 18.4 | 14.1 | 9.6 | 7.2 | 13.9 | *204* |
| Mean daily minimum | 1.1 | 1.3 | 2.3 | 4.2 | 7.1 | 10.3 | 12.1 | 11.9 | 9.8 | 6.8 | 3.5 | 1.8 | 6.0 | *204* |
| Mean temperature | 3.8 | 4.3 | 6.1 | 8.7 | 12.0 | 15.2 | 17.0 | 16.6 | 14.1 | 10.4 | 6.5 | 4.5 | 10.0 | *204* |
| Highest maximum | 15.9 | 18.5 | 22.1 | 27.6 | 30.6 | 34.3 | 34.8 | 35.1 | 33.4 | 29.1 | 18.9 | 15.9 | 35.1 | *204* |
| Year | 2016 | 1998 | 1965 | 2011 | 1947 | 1976 | 2006 | 1990 | 1911 | 2011 | 1946 | 2015 | 3 Aug 1990 | *204* |
| Lowest minimum | −16.6 | −16.1 | −12.0 | −5.6 | −2.6 | 1.4 | 2.4 | 0.2 | −3.3 | −5.7 | −10.1 | −17.8 | −17.8 | *204* |
| Year | 1982 | 1947 | 1845 | 1817 | 1831 | 1892 | 1863 | 1864 | 1851 | 1859 | 1815 | 1860 | 24 Dec 1860 | *204* |
| Lowest grass minimum | −22.5 | −22.2 | −19.1 | −9.5 | −8.3 | −3.6 | −0.1 | −0.1 | −4.4 | −10.0 | −11.0 | −20.1 | −22.5 | *88* |
| Year *since 1931* | 1982 | 1947 | 1947 | 1990 | 1944 | 1991 | 1990 | 1964 | 1943 | 1997 | 2010 | 1981 | 14 Jan 1982 | *88* |
| Highest minimum | 12.1 | 12.0 | 12.8 | 13.8 | 16.7 | 19.6 | 21.2 | 20.5 | 20.6 | 17.0 | 15.0 | 12.5 | 21.2 | *204* |
| Year | 2008 | 2004 | 1998 | 2018 | 1945 | 2011 | 2016 | 1997 | 1949 | 2018 | 1947 | 1985 | 20 Jul 2016 | *204* |
| Lowest maximum | −9.6 | −7.9 | −4.1 | 0.8 | 5.4 | 9.8 | 11.1 | 12.1 | 7.9 | 2.3 | −2.7 | −6.7 | −9.6 | *204* |
| Year | 1841 | 1830 | 1845 | 1830 | 1866 | 1909 | 1821 | 1879 | 1918 | 1836 | 1890 | 1890 | 8 Jan 1841 | *204* |
| Air frosts | 11.1 | 9.6 | 7.5 | 3.0 | 0.4 | 0 | 0 | 0 | 0.0 | 1.3 | 5.5 | 9.2 | 47.5 | *204* |
| Ground frost *since 1931* | 17.5 | 16.2 | 15.0 | 9.4 | 3.4 | 0.3 | 0.0 | 0.0 | 0.8 | 5.2 | 11.2 | 16.4 | 95.4 | *88* |

| **PRECIPITATION** *mm* | Jan | Feb | Mar | Apr | May | June | July | Aug | Sept | Oct | Nov | Dec | Annual | Yrs |
|---|---|---|---|---|---|---|---|---|---|---|---|---|---|---|
| ***Period 1767–2018*** | | | | | | | | | | | | | | |
| Monthly mean | 52.4 | 41.1 | 41.2 | 43.4 | 51.4 | 52.6 | 58.9 | 57.9 | 58.6 | 64.4 | 62.0 | 55.7 | 639.6 | *252* |
| Days ≥ 0.2 mm *since 1853* | 16.4 | 13.2 | 13.7 | 13.1 | 13.3 | 11.7 | 12.4 | 13.0 | 12.4 | 15.0 | 15.3 | 16.0 | 165.6 | *166* |

| | | | | | | | | | | | | | | |
|---|---|---|---|---|---|---|---|---|---|---|---|---|---|---|
| Days ≥ 1.0 mm *since 1853* | 11.1 | 8.8 | 9.0 | 8.9 | 9.5 | 8.5 | 8.8 | 9.3 | 8.7 | 10.6 | 10.2 | 10.8 | 114.3 | *166* |
| Wettest day *since 1827* | 36.1 | 40.8 | 42.1 | 46.7 | 40.7 | 81.3 | 87.9 | 70.8 | 84.8 | 48.3 | 37.9 | 38.3 | 87.9 | *192* |
| Year | 1958 | 1888 | 2008 | 1908 | 2007 | 1960 | 1968 | 1922 | 1951 | 1875 | 1894 | 1985 | 10 Jul 1968 | |
| Wettest month | 146.9 | 120.3 | 132.9 | 143.0 | 154.9 | 170.4 | 174.8 | 146.2 | 223.9 | 189.0 | 192.4 | 147.9 | 979.5 | *252* |
| Year | 2014 | 1937 | 1947 | 2012 | 1773 | 1852 | 1834 | 2010 | 1774 | 1875 | 1770 | 1914 | 2012 | |
| Driest month | 8.1 | 2.2 | 1.4 | 0.5 | 1.5 | 1.7 | 0.8 | 1.8 | 2.6 | 4.1 | 4.8 | 1.7 | 336.7 | *252* |
| Year | 1825 | 1959 | 1929 | 2011 | 1795 | 1925 | 1825 | 1940 | 1929 | 1809 | 1945 | 1780 | 1788 | |

| **SUNSHINE hours** | **Jan** | **Feb** | **Mar** | **Apr** | **May** | **June** | **July** | **Aug** | **Sept** | **Oct** | **Nov** | **Dec** | **Annual** | **Yrs** |
|---|---|---|---|---|---|---|---|---|---|---|---|---|---|---|
| *Period 1881–2018* | | | | | | | | | | | | | | |
| Monthly mean | 55.6 | 71.2 | 114.2 | 153.8 | 192.2 | 197.0 | 194.4 | 180.3 | 139.4 | 103.5 | 65.5 | 49.7 | 1516.7 | *138* |
| *Daily mean* | *1.79* | *2.52* | *3.68* | *5.13* | *6.20* | *6.57* | *6.27* | *5.82* | *4.64* | *3.34* | *2.18* | *1.60* | *4.15* | *138* |
| Days nil sunshine *1921–2018* | 11.5 | 8.1 | 5.8 | 3.0 | 2.4 | 1.8 | 1.2 | 1.6 | 2.4 | 5.3 | 9.4 | 12.4 | 65.0 | *98* |
| Sunniest month | 94.1 | 124.1 | 198.7 | 252.8 | 300.8 | 301.0 | 310.5 | 285.1 | 222.6 | 159.7 | 107.9 | 96.9 | 1880.3 | *138* |
| Year | 1952 | 2008 | 1995 | 1893 | 1989 | 1975 | 1911 | 1995 | 1911 | 1921 | 1923 | 2014 | 1995 | |
| Dullest month | 14.9 | 22.8 | 48.8 | 81.5 | 98.3 | 100.4 | 95.3 | 97.7 | 64.9 | 43.6 | 29.7 | 5.0 | 1154.5 | *138* |
| Year | 1885 | 1940 | 1984 | 1966 | 1981 | 2016 | 1888 | 1912 | 1945 | 1976 | 1888 | 1890 | 1888 | |
| Sunniest day *since 1921* | 8.5 | 9.7 | 12.4 | 13.9 | 15.5 | 15.9 | 15.5 | 14.1 | 12.6 | 10.5 | 8.9 | 7.5 | 15.9 | *97* |
| Year | 1986 | 1996 | 1996 | 1990 | 1985 | 1921 | 1934 | 1989 | 2002 | 1928 | 1990 | 2016 | 7 Jun 1921 | |

| **DAYS WITH...** | **Jan** | **Feb** | **Mar** | **Apr** | **May** | **June** | **July** | **Aug** | **Sept** | **Oct** | **Nov** | **Dec** | **Annual** | **Yrs** |
|---|---|---|---|---|---|---|---|---|---|---|---|---|---|---|
| Snow/sleet falling [A] | 5.0 | 4.8 | 3.2 | 1.2 | 0.2 | 0 | 0 | 0 | 0 | 0.1 | 0.7 | 2.7 | 17.8 | *62* |
| Snow lying 0900 GMT [B] | 3.5 | 3.0 | 0.9 | 0.1 | 0 | 0 | 0 | 0 | 0 | 0 | 0.2 | 1.8 | 9.5 | *93* |
| Thunder heard [C] | 0.1 | 0.1 | 0.2 | 0.8 | 2.4 | 2.4 | 2.8 | 2.2 | 1.1 | 0.5 | 0.2 | 0.1 | 13.0 | *54* |

***Site and instrument metadata***

See Appendix 1 for instrument and site details. *A well-maintained meteorological enclosure in a walled college garden. Screen and raingauge have been in this enclosure since September 1939; sunshine from nearby Engineering Building, since September 1976.*

Where an extreme has occurred more than once, only the most recent year is shown. Annual values may differ slightly from the sum of the monthly totals owing to rounding errors. 'Days with' period of averages [A] 1926–86, [B] 1926–2018, [C] 1931–84

# Appendix 5
# Monthly rainfall totals for Oxford, 1767–1814

These are from Craddock and Craddock [24] updated by Burt and Howden [98]. Units: millimetres

**Table A5.1** *Monthly precipitation totals representative of the Radcliffe Observatory, Oxford, 1767–1813. Units: millimetres. See Appendix 1 for sources.*

| Year | Jan | Feb | Mar | Apr | May | June | July | Aug | Sept | Oct | Nov | Dec | Annual | % 1981–2010 average |
|---|---|---|---|---|---|---|---|---|---|---|---|---|---|---|
| 1767 | 68.7 | 88.1 | 54.9 | 17.7 | 100.0 | 74.6 | 124.9 | 93.2 | 70.2 | 113.8 | 27.6 | 7.7 | 841.4 | 127.4 |
| 1768 | 68.3 | 62.2 | 4.1 | 79.6 | 39.2 | 131.9 | 76.7 | 39.9 | 170.9 | 105.8 | 85.0 | 69.0 | 932.7 | 141.2 |
| 1769 | 34.6 | 52.3 | 29.5 | 19.6 | 57.1 | 88.8 | 47.7 | 58.1 | 123.5 | 13.8 | 54.7 | 65.6 | 645.3 | 97.7 |
| 1770 | 20.1 | 28.8 | 60.3 | 24.2 | 60.0 | 94.2 | 19.1 | 26.1 | 64.9 | 24.2 | 192.4 | 74.6 | 688.9 | 104.3 |
| 1771 | 13.8 | 22.8 | 37.0 | 24.2 | 21.8 | 65.8 | 28.8 | 62.9 | 22.0 | 105.1 | 13.3 | 28.3 | 445.9 | 67.5 |
| 1772 | 97.6 | 89.6 | 74.6 | 39.7 | 45.3 | 39.0 | 31.0 | 72.1 | 123.9 | 86.9 | 165.8 | 32.2 | 897.6 | 135.9 |
| 1773 | 31.0 | 15.7 | 11.1 | 46.7 | 154.9 | 25.9 | 31.2 | 80.8 | 80.1 | 44.3 | 125.6 | 52.0 | 699.6 | 105.9 |
| 1774 | 88.8 | 64.9 | 88.6 | 15.7 | 11.1 | 40.7 | 44.1 | 67.5 | 223.9 | 11.1 | 40.9 | 45.0 | 742.4 | 112.4 |
| 1775 | 61.5 | 36.6 | 60.5 | 67.8 | 34.4 | 69.5 | 35.3 | 119.3 | 126.8 | 67.3 | 62.2 | 2.4 | 743.6 | 112.6 |
| 1776 | 92.0 | 79.2 | 57.1 | 8.0 | 24.4 | 94.6 | 64.4 | 104.8 | 48.9 | 21.8 | 54.2 | 47.0 | 696.4 | 105.5 |
| 1777 | 29.0 | 36.1 | 42.6 | 46.7 | 109.2 | 71.2 | 80.6 | 36.6 | 17.9 | 91.3 | 34.6 | 22.0 | 617.7 | 93.6 |
| 1778 | 62.9 | 19.1 | 26.9 | 20.1 | 44.3 | 47.2 | 101.4 | 2.4 | 30.5 | 70.0 | 99.0 | 76.7 | 600.6 | 91.0 |
| 1779 | 10.4 | 11.1 | 13.8 | 45.5 | 54.9 | 63.2 | 117.6 | 34.9 | 59.3 | 75.5 | 68.0 | 120.5 | 674.9 | 102.2 |
| 1780 | 26.4 | 15.0 | 26.1 | 56.9 | 32.4 | 34.1 | 49.1 | 28.1 | 69.0 | 67.5 | 56.6 | 1.7 | 463.1 | 70.1 |
| 1781 | 39.2 | 41.4 | 1.9 | 26.9 | 44.1 | 79.6 | 42.1 | 32.9 | 63.2 | 7.5 | 69.7 | 44.3 | 492.8 | 74.6 |
| 1782 | 48.9 | 14.8 | 64.6 | 60.5 | 89.1 | 32.0 | 121.3 | 103.6 | 59.3 | 27.8 | 34.6 | 18.6 | 675.1 | 102.2 |
| 1783 | 37.5 | 74.8 | 31.5 | 13.6 | 55.7 | 41.4 | 55.2 | 48.4 | 14.3 | 15.2 | 36.8 | 22.3 | 446.6 | 67.6 |
| 1784 | 35.8 | 15.5 | 46.5 | 76.5 | 54.5 | 57.9 | 54.7 | 56.6 | 34.9 | 47.2 | 63.9 | 45.0 | 588.9 | 89.2 |
| 1785 | 36.1 | 22.8 | 5.6 | 6.3 | 16.9 | 35.8 | 31.2 | 51.8 | 103.1 | 56.6 | 42.1 | 42.6 | 451.0 | 68.3 |
| 1786 | 72.9 | 15.0 | 31.0 | 20.8 | 59.1 | 31.0 | 12.6 | 42.4 | 76.2 | 97.8 | 54.2 | 53.5 | 566.4 | 85.8 |
| 1787 | 7.3 | 43.8 | 61.5 | 19.4 | 35.8 | 25.2 | 126.4 | 26.1 | 33.4 | 57.6 | 34.9 | 62.9 | 534.2 | 80.9 |
| 1788 | 16.2 | 52.5 | 18.6 | 10.2 | 9.7 | 56.6 | 18.9 | 81.6 | 48.7 | 4.1 | 12.1 | 7.5 | 336.7 | 51.0 |
| 1789 | 64.1 | 49.6 | 44.1 | 32.2 | 63.4 | 121.5 | 107.7 | 31.0 | 75.3 | 90.8 | 37.3 | 42.4 | 759.3 | 115.0 |

| Year | Jan | Feb | Mar | Apr | May | June | July | Aug | Sept | Oct | Nov | Dec | Annual | % 1981–2010 average |
|---|---|---|---|---|---|---|---|---|---|---|---|---|---|---|
| 1790 | 33.4 | 4.6 | 9.7 | 45.5 | 64.6 | 28.1 | 53.3 | 53.0 | 15.7 | 14.8 | 84.2 | 60.8 | 467.7 | 70.8 |
| 1791 | 78.4 | 37.5 | 9.9 | 36.6 | 22.5 | 14.5 | 68.3 | 35.1 | 17.2 | 77.9 | 109.9 | 53.3 | 561.1 | 85.0 |
| 1792 | 72.9 | 12.8 | 43.3 | 72.4 | 46.5 | 66.6 | 71.2 | 79.4 | 90.3 | 86.4 | 21.1 | 33.6 | 696.4 | 105.5 |
| 1793 | 56.2 | 29.8 | 40.2 | 47.2 | 42.1 | 15.0 | 58.3 | 39.5 | 107.0 | 22.5 | 40.5 | 46.0 | 544.2 | 82.4 |
| 1794 | 12.1 | 24.2 | 31.2 | 50.3 | 51.1 | 14.8 | 65.1 | 45.3 | 65.4 | 76.2 | 116.4 | 50.1 | 602.3 | 91.2 |
| 1795 | 12.3 | 57.1 | 50.8 | 43.6 | 1.5 | 133.9 | 50.6 | 54.0 | 5.8 | 103.6 | 47.2 | 47.7 | 608.1 | 92.1 |
| 1796 | 73.1 | 35.1 | 11.4 | 5.6 | 57.4 | 20.6 | 84.0 | 42.4 | 29.0 | 52.3 | 17.4 | 57.6 | 485.8 | 73.6 |
| 1797 | 36.3 | 7.7 | 43.8 | 36.3 | 32.0 | 103.1 | 62.2 | 68.0 | 139.2 | 23.2 | 52.3 | 80.4 | 684.6 | 103.7 |
| 1798 | 52.3 | 2.5 | 11.1 | 41.4 | 44.1 | 27.8 | 115.7 | 49.1 | 57.1 | 100.5 | 106.5 | 28.1 | 636.3 | 96.4 |
| 1799 | 44.1 | 66.6 | 25.9 | 69.7 | 52.3 | 13.3 | 85.7 | 84.5 | 153.2 | 84.5 | 54.7 | 17.7 | 752.1 | 113.9 |
| 1800 | 67.8 | 12.8 | 40.2 | 68.5 | 47.9 | 16.0 | 1.2 | 30.7 | 74.1 | 64.9 | 146.4 | 58.8 | 629.4 | 95.3 |
| 1801 | 34.4 | 10.4 | 32.2 | 5.1 | 70.7 | 39.2 | 54.9 | 51.8 | 42.4 | 46.5 | 71.4 | 49.6 | 508.6 | 77.0 |
| 1802 | 8.7 | 49.6 | 6.5 | 6.5 | 42.8 | 43.6 | 81.8 | 6.8 | 6.3 | 39.2 | 40.9 | 56.6 | 389.5 | 59.0 |
| 1803 | 50.1 | 39.2 | 5.8 | 16.7 | 19.4 | 68.7 | 42.6 | 18.4 | 21.8 | 6.8 | 107.5 | 105.3 | 502.3 | 76.1 |
| 1804 | 70.7 | 16.2 | 77.5 | 50.6 | 52.8 | 14.5 | 86.2 | 50.8 | 8.0 | 73.1 | 65.1 | 20.6 | 586.0 | 88.8 |
| 1805 | 51.1 | 25.2 | 15.2 | 40.7 | 37.5 | 43.6 | 36.1 | 50.6 | 41.9 | 44.3 | 21.3 | 48.2 | 455.6 | 69.0 |
| 1806 | 82.1 | 33.9 | 42.8 | 22.0 | 24.9 | 17.7 | 87.4 | 45.0 | 37.3 | 24.7 | 69.0 | 86.4 | 573.2 | 86.8 |
| 1807 | 30.0 | 29.8 | 11.4 | 9.2 | 81.6 | 30.3 | 43.3 | 38.5 | 59.8 | 45.7 | 98.0 | 32.9 | 510.5 | 77.3 |
| 1808 | 24.0 | 22.8 | 8.5 | 69.5 | 48.7 | 31.7 | 74.3 | 51.3 | 61.2 | 85.0 | 55.9 | 31.7 | 564.5 | 85.5 |
| 1809 | 104.8 | 63.2 | 30.5 | 85.0 | 31.7 | 35.8 | 77.5 | 77.9 | 78.2 | 4.1 | 36.1 | 67.1 | 691.8 | 104.8 |
| 1810 | 15.7 | 36.8 | 55.4 | 54.5 | 44.3 | 26.9 | 84.5 | 62.2 | 28.8 | 51.6 | 131.0 | 82.8 | 674.4 | 102.1 |
| 1811 | 37.5 | 51.6 | 34.1 | 42.8 | 92.7 | 50.6 | 80.4 | 47.7 | 45.0 | 82.5 | 60.8 | 37.5 | 663.2 | 100.4 |
| 1812 | 39.0 | 85.0 | 69.0 | 40.4 | 68.5 | 74.3 | 49.1 | 34.4 | 23.0 | 105.8 | 56.9 | 10.9 | 656.2 | 99.4 |
| 1813 | 28.1 | 87.4 | 18.2 | 32.0 | 51.8 | 27.4 | 46.2 | 21.3 | 36.6 | 111.3 | 26.6 | 14.8 | 501.6 | 76.0 |

# Appendix 6
# Monthly and annual summaries of Oxford's weather by year, 1813–2018

This Appendix contains monthly and annual summaries of temperature, rainfall and sunshine in Oxford for each month and year from 1813 to 2018 for which data are available. Details of the layout and content of these tables can be found in Table A6.1.

Each table is headed with the year followed by monthly totals or means for various elements as detailed in Table A6.1. Where no data are available (for example, prior to the start of sunshine records in February 1880) the relevant columns are greyed out. Monthly rainfall records are available from 1767 [24] [98], but daily rainfall totals are not available until January 1827. Sunshine data commenced in February 1880, but daily data have not yet been digitised prior to January 1921, although this process is ongoing and daily records will be added online in due course.

Temperatures were recorded by unscreened thermometers prior to January 1853. These are likely to be a little too cold by modern (screened thermometer) standards in winter and a little too warm in summer – see Appendix 1 for more details. The mean temperature correction for unscreened thermometry varies seasonally from +0.3 degC in December and January to −0.3 degC in July (Table A1.1), although this will vary considerably on a day-to-day basis. The *mean* temperatures in these tables have been corrected accordingly for the period to December 1852 inclusive, but daily values of maximum and minimum temperature, and the monthly means of those daily values, have *not* been corrected in this manner.

For all months and years *except* 1813 to 1852 inclusive, the mean monthly temperature is the average of the mean daily maximum temperature and the mean daily minimum temperature. Between December 1813 and March 1815 the mean temperature is based upon two or three observations daily, corrected for diurnal variation; from April 1815 to December 1852 it is the average of the mean daily maximum temperature and the mean daily minimum temperature, then corrected as per Table A1.1. Between December 1878 and December 1922 inclusive minor corrections have been made to the calculated monthly mean maximum and minimum temperatures to reflect the change of screen in January 1923, as set out in Appendix 1, while from 1881 to 1924 minor adjustments to the calculated monthly mean maximum and minimum temperatures aim to homogenise terminal hours, as detailed in Appendix 1. Again, daily values have *not* been corrected in this manner, and as a result the arithmetic mean of the daily maximum and minimum temperatures differs slightly from the adjusted monthly means given in the tables which follow.

The period of the 'daily' values (the 'terminal hour') varies slightly across the dataset: details where known, and any corrections applied, are given by element in Appendix 1. Since 1925, daily temperature and precipitation measurements refer to the period 0900 to 0900 GMT; since 1888, sunshine duration refers to the 'civil day', midnight to midnight.

Monthly and annual totals and means presented here and elsewhere in this volume may disagree slightly owing to rounding errors. Continuing work on digitising the Oxford dataset (and re-digitising to greater precision, for instance to 0.1 degF temperatures from 1 degF) may lead to slight inconsistencies over time between values in these tables and the daily, monthly and annual values published online.

For annual values, the same terminology and terms apply, except that the totals and means relate to the calendar year, and the date of each of the extremes is replaced by the number of the month of occurrence (1 = January, 2 = February, etc.: reference back to the month line will give the date of the event within the month, from which the date of the annual extreme can easily be determined). In the case of two or more months having the same extreme value, only the first month value is shown.

**Table A6.1** *Explanation of column headings in the monthly and annual tables*

| Column abbreviation | Value | Unit |
|---|---|---|
| *Month* | Month name | |
| *Mean max* | Mean daily maximum temperature | °C |
| *Mean min* | Mean daily minimum temperature | °C |
| *Mean temp* | Mean daily temperature (average of mean daily max and mean daily min) | °C |
| *Anom* | Difference ('anomaly') of the mean daily temperature above from the 1981–2010 normal (Appendix 3); negative value indicates a value below the current normal | degC |
| *Highest max* | The highest daily maximum temperature during the month (warmest day) | °C |
| *Date* | The date of the highest daily maximum during the month. Where the extreme occurred on more than one date during the month, only the first date is shown. (This is more likely during the early part of the record, when temperatures were read only to 1 degF resolution.) | |
| *Lowest min* | The lowest daily minimum temperature during the month (coldest night) | °C |
| *Date* | The date of the lowest daily minimum during the month. Where the extreme occurred on more than one date during the month, only the first date is shown. | |
| *Lowest max* | The lowest daily maximum temperature during the month (coldest day) | °C |
| *Date* | The date of the lowest daily maximum during the month. Where the extreme occurred on more than one date during the month, only the first date is shown. | |
| *Highest min* | The highest daily minimum temperature during the month (warmest night) | °C |
| *Date* | The date of the highest daily minimum during the month. Where the extreme occurred on more than one date during the month, only the first date is shown. | |
| *Air frost* | Count of the number of days in the month when the minimum temperature fell to −0.1 °C or below | Count |
| *Days ≥ 25 °C* | Count of the number of days in the month when the maximum temperature reached 25.0 °C or above ('hot days') | Count |
| *Total pptn* | Total precipitation during the month; from 1827, this is the sum of the daily values | mm |
| *Anom* | Total precipitation during the month as a percentage of the 1981–2010 normal (Appendix 3); > 100 indicates a wetter month than the current normal | % |
| *Rain days* | Count of the number of days in the month when 0.2 mm or more of rainfall was recorded; only available from 1827, when the daily record starts. See comments in Appendix 1 regarding the reliability of the rain day count in the early years of the record | Count |
| *Wettest day* | The amount of precipitation on the wettest day in the month | mm |
| *Date* | The date of the wettest day during the month. Where the extreme occurred on more than one date during the month, only the first date is shown. | |
| *Total sunshine* | Total duration of bright sunshine during the month; from 1921, this is the sum of the daily values | hours |
| *Anom* | Total sunshine duration during the month as a percentage of the 1981–2010 normal (Appendix 3); > 100 indicates a sunnier month than the current normal | % |
| *Sunniest day* | The duration of bright sunshine on the sunniest day in the month | hours |
| *Date* | The date of the sunniest day during the month. Where the extreme occurred on more than one date during the month, only the first date is shown. | |

## OXFORD 1813

| Month | Mean max | Mean min | Mean temp | Anom | Highest max | Date | Lowest min | Date | Lowest max | Date | Highest min | Date | Air frost | Days ≥ 25 °C | Total pptn | Anom | Rain days | Wettest day | Date | Total sunshine | Anom | Sunniest day | Date |
|---|---|---|---|---|---|---|---|---|---|---|---|---|---|---|---|---|---|---|---|---|---|---|---|
| Jan | | | | | | | | | | | | | | | 28.1 | *49* | | | | | | | |
| Feb | | | | | | | | | | | | | | | 87.4 | *204* | | | | | | | |
| Mar | | | | | | | | | | | | | | | 18.2 | *38* | | | | | | | |
| Apr | | | | | | | | | | | | | | | 32.0 | *65* | | | | | | | |
| May | | | | | | | | | | | | | | | 51.8 | *91* | | | | | | | |
| June | | | | | | | | | | | | | | | 27.4 | *56* | | | | | | | |
| July | | | | | | | | | | | | | | | 46.2 | *95* | | | | | | | |
| Aug | | | | | | | | | | | | | | | 21.3 | *39* | | | | | | | |
| Sep | | | | | | | | | | | | | | | 36.6 | *67* | | | | | | | |
| Oct | | | | | | | | | | | | | | | 111.3 | *159* | | | | | | | |
| Nov | | | | | | | | | | | | | | | 26.6 | *40* | | | | | | | |
| Dec | | | 2.6 | *−2.4* | | | | | | | | | | | 14.8 | *23* | | | | | | | |
| **1813** | | | | | | | | | | | | | | | 501.6 | *76* | | | | | | | |

## OXFORD 1814

| Month | Mean max | Mean min | Mean temp | Anom | Highest max | Date | Lowest min | Date | Lowest max | Date | Highest min | Date | Air frost | Days ≥ 25 °C | Total pptn | Anom | Rain days | Wettest day | Date | Total sunshine | Anom | Sunniest day | Date |
|---|---|---|---|---|---|---|---|---|---|---|---|---|---|---|---|---|---|---|---|---|---|---|---|
| Jan | | | −2.6 | *−7.4* | | | | | | | | | | | 57.6 | *101* | | | | | | | |
| Feb | | | 1.7 | *−3.2* | | | | | | | | | | | 7.5 | *18* | | | | | | | |
| Mar | | | 3.2 | *−4.1* | | | | | | | | | | | 30.7 | *64* | | | | | | | |
| Apr | | | 10.6 | *+1.3* | | | | | | | | | | | 60.5 | *124* | | | | | | | |
| May | | | 10.3 | *−2.1* | | | | | | | | | | | 25.2 | *44* | | | | | | | |
| June | | | 13.0 | *−2.6* | | | | | | | | | | | 56.2 | *114* | | | | | | | |
| July | | | 17.1 | *−0.8* | | | | | | | | | | | 64.6 | *132* | | | | | | | |
| Aug | | | 15.8 | *−1.7* | | | | | | | | | | | 30.7 | *56* | | | | | | | |
| Sep | | | 13.2 | *−1.7* | | | | | | | | | | | 38.5 | *71* | | | | | | | |
| Oct | | | 8.8 | *−2.5* | | | | | | | | | | | 56.4 | *81* | | | | | | | |
| Nov | | | 5.4 | *−2.2* | | | | | | | | | | | 36.3 | *54* | | | | | | | |
| Dec | | | 5.5 | *+0.5* | | | | | | | | | | | 80.4 | *127* | | | | | | | |
| **1814** | | | 8.5 | *−2.2* | | | | | | | | | | | 544.6 | *82* | | | | | | | |

## OXFORD 1815

| Month | Mean max | Mean min | Mean temp | Anom | Highest max | Date | Lowest min | Date | Lowest max | Date | Highest min | Date | Air frost | Days ≥ 25 °C | Total pptn | Anom | Rain days | Wettest day | Date | Total sunshine | Anom | Sunniest day | Date |
|---|---|---|---|---|---|---|---|---|---|---|---|---|---|---|---|---|---|---|---|---|---|---|---|
| Jan | 2.6 | −5.6 | −1.2 | −6.3 | 7.7 | 10 | −11.8 | 24 | −3.5 | 24 | −0.4 | 10 | 31 | 0 | 18.7 | 33 | | | | | | | |
| Feb | 8.5 | 0.5 | 4.7 | −0.4 | 12.6 | 16 | −3.8 | 27 | 4.2 | 27 | 4.7 | 16 | 9 | 0 | 17.9 | 42 | | | | | | | |
| Mar | 9.2 | 1.0 | 5.3 | −2.2 | 13.4 | 8 | −5.6 | 11 | 2.7 | 11 | 5.3 | 8 | 11 | 0 | 55.9 | 117 | | | | | | | |
| Apr | 11.5 | 3.5 | 7.6 | −1.8 | 17.8 | 13 | −3.1 | 16 | 5.6 | 14 | 9.7 | 12 | 3 | 0 | 32.9 | 67 | | | | | | | |
| May | 18.8 | 8.7 | 13.7 | +1.3 | 25.6 | 28 | 5.6 | 17 | 15.0 | 13 | 12.2 | 29 | 0 | 1 | 58.8 | 103 | | | | | | | |
| June | 20.9 | 9.8 | 15.2 | −0.2 | 23.9 | 28 | 3.9 | 2 | 17.8 | 12 | 13.3 | 20 | 0 | 0 | 55.2 | 112 | | | | | | | |
| July | 21.2 | 10.5 | 15.6 | −2.0 | 25.6 | 12 | 5.3 | 28 | 17.2 | 2 | 16.1 | 15 | 0 | 3 | 20.8 | 43 | | | | | | | |
| Aug | 21.6 | 11.1 | 16.2 | −1.2 | 25.3 | 4 | 5.6 | 8 | 18.6 | 6 | 16.7 | 23 | 0 | 1 | 43.1 | 79 | | | | | | | |
| Sep | 20.1 | 7.4 | 13.7 | −1.1 | 27.2 | 14 | 1.3 | 22 | 14.4 | 30 | 15.6 | 16 | 0 | 4 | 71.9 | 133 | | | | | | | |
| Oct | 12.2 | 4.1 | 8.3 | −3.1 | 18.0 | 20 | −1.5 | 27 | 6.8 | 27 | 9.9 | 20 | 2 | 0 | 61.5 | 88 | | | | | | | |
| Nov | 5.0 | −3.3 | 1.1 | −6.7 | 13.5 | 10 | −10.1 | 17 | −1.6 | 17 | 5.3 | 10 | 21 | 0 | 18.9 | 28 | | | | | | | |
| Dec | 5.1 | −3.2 | 1.3 | −4.1 | 12.8 | 30 | −11.8 | 9 | −3.4 | 9 | 4.7 | 30 | 21 | 0 | 53.9 | 85 | | | | | | | |
| **1815** | 13.0 | 3.7 | 8.5 | −2.4 | 27.2 | 9 | −11.8 | 1 | −3.5 | 1 | 16.7 | 8 | 98 | 9 | 509.4 | 77 | | | | | | | |

## OXFORD 1816

| Month | Mean max | Mean min | Mean temp | Anom | Highest max | Date | Lowest min | Date | Lowest max | Date | Highest min | Date | Air frost | Days ≥ 25 °C | Total pptn | Anom | Rain days | Wettest day | Date | Total sunshine | Anom | Sunniest day | Date |
|---|---|---|---|---|---|---|---|---|---|---|---|---|---|---|---|---|---|---|---|---|---|---|---|
| Jan | 4.8 | −3.3 | 1.0 | −4.1 | 10.5 | 9 | −10.7 | 31 | −2.4 | 31 | 2.5 | 9 | 28 | 0 | 57.8 | 102 | | | | | | | |
| Feb | 4.4 | −3.3 | 0.8 | −4.3 | 11.1 | 24 | −13.9 | 10 | −5.3 | 9 | 4.4 | 20 | 19 | 0 | 34.4 | 80 | | | | | | | |
| Mar | 7.0 | 0.4 | 3.9 | −3.6 | 12.2 | 12 | −5.0 | 1 | 2.8 | 9 | 7.8 | 12 | 12 | 0 | 32.9 | 69 | | | | | | | |
| Apr | 11.3 | 1.5 | 6.5 | −2.9 | 20.0 | 28 | −3.9 | 5 | 2.8 | 14 | 10.0 | 29 | 9 | 0 | 50.8 | 104 | | | | | | | |
| May | 14.9 | 5.5 | 10.1 | −2.3 | 21.1 | 17 | −1.1 | 13 | 7.2 | 12 | 11.1 | 29 | 1 | 0 | 37.1 | 65 | | | | | | | |
| June | 18.6 | 8.4 | 13.4 | −2.0 | 25.6 | 29 | 3.3 | 6 | 11.1 | 6 | 12.2 | 26 | 0 | 1 | 42.6 | 87 | | | | | | | |
| July | 19.3 | 10.0 | 14.4 | −3.2 | 23.3 | 20 | 5.0 | 31 | 16.1 | 18 | 17.2 | 21 | 0 | 0 | 104.5 | 214 | | | | | | | |
| Aug | 18.9 | 9.8 | 14.2 | −3.3 | 21.1 | 4 | 5.6 | 21 | 12.2 | 31 | 13.3 | 8 | 0 | 0 | 61.8 | 113 | | | | | | | |
| Sep | 17.0 | 8.5 | 12.7 | −2.1 | 22.8 | 15 | 1.7 | 3 | 11.7 | 1 | 13.3 | 16 | 0 | 0 | 46.0 | 85 | | | | | | | |
| Oct | 14.2 | 7.1 | 10.8 | −0.6 | 20.6 | 6 | −0.6 | 23 | 8.9 | 25 | 12.8 | 5 | 1 | 0 | 71.4 | 102 | | | | | | | |
| Nov | 6.5 | −0.1 | 3.5 | −4.4 | 12.8 | 13 | −8.9 | 24 | 0.0 | 23 | 6.7 | 6 | 11 | 0 | 46.5 | 70 | | | | | | | |
| Dec | 5.6 | −0.4 | 3.0 | −2.4 | 10.0 | 25 | −10.0 | 22 | −2.8 | 21 | 3.3 | 24 | 13 | 0 | 80.6 | 128 | | | | | | | |
| **1816** | 11.9 | 3.7 | 7.9 | −3.0 | 25.6 | 6 | −13.9 | 2 | −5.3 | 2 | 17.2 | 7 | 94 | 1 | 666.4 | 101 | | | | | | | |

## OXFORD 1817

| Month | Mean max | Mean min | Mean temp | Anom | Highest max | Date | Lowest min | Date | Lowest max | Date | Highest min | Date | Air frost | Days ≥ 25 °C | Total pptn | Anom | Rain days | Wettest day | Date | Total sunshine | Anom | Sunniest day | Date |
|---|---|---|---|---|---|---|---|---|---|---|---|---|---|---|---|---|---|---|---|---|---|---|---|
| Jan | 6.9 | 1.3 | 4.4 | *−0.8* | 12.2 | 4 | −5.6 | 16 | 0.0 | 10 | 7.8 | 25 | 10 | 0 | 47.9 | *84* | | | | | | | |
| Feb | 9.1 | 3.2 | 6.4 | *+1.3* | 12.2 | 18 | −1.1 | 12 | 4.4 | 11 | 6.7 | 18 | 3 | 0 | 21.3 | *50* | | | | | | | |
| Mar | 9.6 | 1.3 | 5.6 | *−1.8* | 13.9 | 13 | −7.2 | 22 | 0.6 | 20 | 7.8 | 13 | 11 | 0 | 34.1 | *72* | | | | | | | |
| Apr | 11.6 | 1.1 | 6.4 | *−2.9* | 17.2 | 3 | −5.6 | 11 | 3.9 | 10 | 7.2 | 15 | 11 | 0 | 0.5 | *1* | | | | | | | |
| May | 13.6 | 4.3 | 8.9 | *−3.5* | 18.9 | 17 | 0.6 | 7 | 7.2 | 20 | 7.8 | 4 | 0 | 0 | 74.8 | *131* | | | | | | | |
| June | 21.3 | 11.0 | 16.0 | *+0.6* | 27.8 | 19 | 4.4 | 1 | 14.4 | 9 | 16.7 | 22 | 0 | 8 | 55.5 | *113* | | | | | | | |
| July | 19.9 | 10.6 | 15.0 | *−2.6* | 23.3 | 9 | 5.6 | 20 | 16.7 | 1 | 15.0 | 21 | 0 | 0 | 69.2 | *142* | | | | | | | |
| Aug | 19.1 | 9.9 | 14.3 | *−3.1* | 22.8 | 6 | 3.9 | 23 | 15.6 | 21 | 13.9 | 8 | 0 | 0 | 90.8 | *166* | | | | | | | |
| Sep | 18.3 | 9.4 | 13.9 | *−1.0* | 25.6 | 8 | 3.3 | 29 | 12.2 | 27 | 15.0 | 16 | 0 | 2 | 8.5 | *16* | | | | | | | |
| Oct | 11.0 | 3.7 | 7.5 | *−3.9* | 17.5 | 8 | −1.7 | 3 | 4.1 | 29 | 14.1 | 8 | 4 | 0 | 17.6 | *25* | | | | | | | |
| Nov | 11.4 | 5.1 | 8.5 | *+0.7* | 14.4 | 5 | −1.1 | 20 | 7.2 | 23 | 11.1 | 18 | 2 | 0 | 44.8 | *67* | | | | | | | |
| Dec | 4.9 | 0.1 | 2.8 | *−2.5* | 12.8 | 1 | −7.2 | 24 | −1.1 | 30 | 7.8 | 2 | 14 | 0 | 59.8 | *95* | | | | | | | |
| **1817** | 13.1 | 5.1 | 9.1 | *−1.7* | 27.8 | 6 | −7.2 | 3 | −1.1 | 12 | 16.7 | 6 | 55 | 10 | 524.8 | *79* | | | | | | | |

## OXFORD 1818

| Month | Mean max | Mean min | Mean temp | Anom | Highest max | Date | Lowest min | Date | Lowest max | Date | Highest min | Date | Air frost | Days ≥ 25 °C | Total pptn | Anom | Rain days | Wettest day | Date | Total sunshine | Anom | Sunniest day | Date |
|---|---|---|---|---|---|---|---|---|---|---|---|---|---|---|---|---|---|---|---|---|---|---|---|
| Jan | 6.3 | 2.8 | 4.8 | *−0.3* | 11.1 | 15 | −4.4 | 1 | −2.8 | 1 | 8.9 | 16 | 3 | 0 | 66.3 | *117* | | | | | | | |
| Feb | 4.7 | 0.0 | 2.5 | *−2.6* | 10.0 | 27 | −11.1 | 4 | −1.1 | 11 | 6.7 | 18 | 16 | 0 | 41.2 | *96* | | | | | | | |
| Mar | 7.8 | 1.9 | 5.0 | *−2.4* | 10.6 | 19 | −1.7 | 10 | 4.4 | 10 | 6.1 | 5 | 5 | 0 | 69.0 | *145* | | | | | | | |
| Apr | 10.8 | 4.4 | 7.7 | *−1.7* | 19.4 | 26 | −1.7 | 1 | 5.6 | 7 | 11.7 | 27 | 2 | 0 | 107.2 | *219* | | | | | | | |
| May | 15.9 | 6.5 | 11.1 | *−1.3* | 22.2 | 31 | 3.3 | 30 | 11.7 | 1 | 8.9 | 4 | 0 | 0 | 63.7 | *111* | | | | | | | |
| June | 23.6 | 12.1 | 17.7 | *+2.2* | 29.4 | 12 | 8.9 | 7 | 17.8 | 19 | 15.6 | 17 | 0 | 12 | 5.3 | *11* | | | | | | | |
| July | 24.7 | 13.7 | 19.0 | *+1.4* | 31.1 | 24 | 9.4 | 3 | 19.4 | 27 | 18.9 | 24 | 0 | 15 | 8.7 | *18* | | | | | | | |
| Aug | 21.7 | 11.0 | 16.2 | *−1.2* | 31.1 | 5 | 5.6 | 24 | 17.2 | 14 | 19.4 | 6 | 0 | 4 | 16.0 | *29* | | | | | | | |
| Sep | 18.2 | 9.9 | 14.1 | *−0.8* | 23.9 | 4 | 4.4 | 11 | 12.2 | 10 | 17.2 | 5 | 0 | 0 | 91.8 | *169* | | | | | | | |
| Oct | 16.2 | 9.1 | 12.8 | *+1.4* | 19.4 | 2 | 2.2 | 8 | 11.1 | 24 | 13.9 | 4 | 0 | 0 | 46.2 | *66* | | | | | | | |
| Nov | 11.4 | 3.7 | 7.8 | *+0.0* | 15.7 | 17 | −2.1 | 22 | 6.2 | 21 | 9.4 | 1 | 8 | 0 | 96.3 | *144* | | | | | | | |
| Dec | 5.6 | −1.2 | 2.6 | *−2.8* | 13.4 | 1 | −6.1 | 25 | 0.7 | 17 | 7.2 | 18 | 22 | 0 | 23.0 | *36* | | | | | | | |
| **1818** | 13.9 | 6.2 | 10.1 | *−0.7* | 31.1 | 7 | −11.1 | 2 | −2.8 | 1 | 19.4 | 8 | 56 | 31 | 634.7 | *96* | | | | | | | |

## OXFORD 1819

| Month | Mean max | Mean min | Mean temp | Anom | Highest max | Date | Lowest min | Date | Lowest max | Date | Highest min | Date | Air frost | Days ≥ 25 °C | Total pptn | Anom | Rain days | Wettest day | Date | Total sunshine | Anom | Sunniest day | Date |
|---|---|---|---|---|---|---|---|---|---|---|---|---|---|---|---|---|---|---|---|---|---|---|---|
| Jan | 6.9 | 0.5 | 4.0 | *−1.2* | 11.7 | 14 | −5.0 | 1 | 1.1 | 1 | 3.9 | 10 | 8 | 0 | 44.8 | *79* | | | | | | | |
| Feb | 6.3 | 1.1 | 4.0 | *−1.2* | 10.6 | 9 | −5.0 | 25 | 1.7 | 3 | 6.1 | 10 | 10 | 0 | 67.8 | *158* | | | | | | | |
| Mar | 9.1 | 3.4 | 6.4 | *−1.0* | 15.6 | 31 | 0.0 | 1 | 3.9 | 3 | 10.0 | 31 | 0 | 0 | 23.9 | *50* | | | | | | | |
| Apr | 13.4 | 4.8 | 9.2 | *−0.2* | 18.9 | 7 | −2.2 | 28 | 7.8 | 24 | 10.0 | 1 | 2 | 0 | 31.9 | *65* | | | | | | | |
| May | 17.9 | 7.8 | 12.7 | *+0.4* | 23.3 | 9 | 1.7 | 30 | 11.1 | 20 | 14.4 | 6 | 0 | 0 | 74.0 | *129* | | | | | | | |
| June | 20.0 | 9.1 | 14.4 | *−1.0* | 22.2 | 4 | 5.6 | 5 | 16.1 | 16 | 13.3 | 25 | 0 | 0 | 50.8 | *103* | | | | | | | |
| July | 22.7 | 12.4 | 17.3 | *−0.3* | 26.7 | 24 | 7.8 | 1 | 17.2 | 15 | 16.7 | 5 | 0 | 5 | 32.7 | *67* | | | | | | | |
| Aug | 23.3 | 13.9 | 18.4 | *+1.0* | 26.1 | 9 | 10.6 | 31 | 16.7 | 31 | 16.7 | 9 | 0 | 10 | 15.2 | *28* | | | | | | | |
| Sep | 19.1 | 9.5 | 14.3 | *−0.6* | 25.0 | 9 | 2.8 | 20 | 13.9 | 16 | 16.1 | 8 | 0 | 2 | 94.2 | *174* | | | | | | | |
| Oct | 13.1 | 6.0 | 9.7 | *−1.7* | 21.7 | 1 | −4.4 | 27 | 5.0 | 21 | 15.6 | 1 | 5 | 0 | 46.5 | *66* | | | | | | | |
| Nov | 6.7 | 1.0 | 4.1 | *−3.7* | 11.7 | 5 | −6.1 | 24 | 1.7 | 26 | 8.9 | 30 | 12 | 0 | 42.4 | *64* | | | | | | | |
| Dec | 3.5 | −1.5 | 1.4 | *−4.0* | 12.2 | 20 | −11.1 | 11 | −3.9 | 13 | 10.6 | 20 | 20 | 0 | 55.0 | *87* | | | | | | | |
| **1819** | 13.5 | 5.7 | 9.7 | *−1.2* | 26.7 | 7 | −11.1 | 12 | −3.9 | 12 | 16.7 | 7 | 57 | 17 | 579.2 | *88* | | | | | | | |

## OXFORD 1820

| Month | Mean max | Mean min | Mean temp | Anom | Highest max | Date | Lowest min | Date | Lowest max | Date | Highest min | Date | Air frost | Days ≥ 25 °C | Total pptn | Anom | Rain days | Wettest day | Date | Total sunshine | Anom | Sunniest day | Date |
|---|---|---|---|---|---|---|---|---|---|---|---|---|---|---|---|---|---|---|---|---|---|---|---|
| Jan | 2.0 | −4.3 | −0.9 | *−6.0* | 9.4 | 27 | −15.6 | 15 | −3.9 | 12 | 6.1 | 27 | 23 | 0 | 52.3 | *92* | | | | | | | |
| Feb | 4.9 | −0.6 | 2.3 | *−2.8* | 11.7 | 7 | −8.3 | 18 | 0.6 | 3 | 6.7 | 8 | 16 | 0 | 13.8 | *32* | | | | | | | |
| Mar | 8.4 | 0.9 | 4.8 | *−2.6* | 16.1 | 29 | −6.1 | 7 | 0.6 | 3 | 10.6 | 31 | 12 | 0 | 7.1 | *15* | | | | | | | |
| Apr | 14.9 | 4.0 | 9.5 | *+0.2* | 19.4 | 19 | −1.1 | 10 | 10.0 | 9 | 9.4 | 19 | 4 | 0 | 58.1 | *119* | | | | | | | |
| May | 17.8 | 7.2 | 12.4 | *+0.0* | 26.1 | 22 | −1.1 | 5 | 11.7 | 4 | 11.7 | 31 | 1 | 1 | 64.6 | *113* | | | | | | | |
| June | 20.4 | 9.9 | 15.0 | *−0.4* | 32.2 | 25 | 3.9 | 11 | 13.3 | 11 | 17.2 | 28 | 0 | 7 | 22.5 | *46* | | | | | | | |
| July | 21.6 | 11.4 | 16.3 | *−1.3* | 28.3 | 31 | 7.2 | 2 | 15.6 | 2 | 16.7 | 17 | 0 | 6 | 46.2 | *95* | | | | | | | |
| Aug | 20.7 | 11.2 | 15.8 | *−1.6* | 24.4 | 15 | 4.4 | 24 | 15.0 | 21 | 17.2 | 1 | 0 | 0 | 40.2 | *73* | | | | | | | |
| Sep | 18.1 | 7.5 | 12.8 | *−2.1* | 23.9 | 12 | 1.7 | 20 | 11.1 | 26 | 14.4 | 11 | 0 | 0 | 49.4 | *91* | | | | | | | |
| Oct | 11.6 | 5.2 | 8.6 | *−2.9* | 17.8 | 15 | 1.1 | 4 | 7.2 | 17 | 9.4 | 1 | 0 | 0 | 55.0 | *79* | | | | | | | |
| Nov | 7.6 | 2.1 | 5.1 | *−2.7* | 13.9 | 8 | −2.8 | 4 | 2.8 | 14 | 8.3 | 8 | 9 | 0 | 45.3 | *68* | | | | | | | |
| Dec | 5.3 | 2.0 | 4.0 | *−1.4* | 11.7 | 4 | −8.3 | 31 | −3.9 | 29 | 10.0 | 8 | 12 | 0 | 50.1 | *79* | | | | | | | |
| **1820** | 12.8 | 4.7 | 8.8 | *−2.0* | 32.2 | 6 | −15.6 | 1 | −3.9 | 1 | 17.2 | 6 | 77 | 14 | 504.6 | *76* | | | | | | | |

## OXFORD 1821

| Month | Mean max | Mean min | Mean temp | Anom | Highest max | Date | Lowest min | Date | Lowest max | Date | Highest min | Date | Air frost | Days ≥ 25 °C | Total pptn | Anom | Rain days | Wettest day | Date | Total sunshine | Anom | Sunniest day | Date |
|---|---|---|---|---|---|---|---|---|---|---|---|---|---|---|---|---|---|---|---|---|---|---|---|
| Jan | 4.8 | 0.4 | 2.9 | *−2.3* | 10.6 | 18 | −6.7 | 4 | −3.3 | 2 | 7.8 | 19 | 11 | 0 | 50.1 | *88* | | | | | | | |
| Feb | 4.0 | −1.7 | 1.4 | *−3.7* | 11.1 | 1 | −7.8 | 27 | −0.6 | 16 | 7.8 | 1 | 22 | 0 | 7.2 | *17* | | | | | | | |
| Mar | 9.3 | 2.0 | 5.8 | *−1.6* | 12.8 | 10 | −3.3 | 17 | 1.7 | 5 | 8.3 | 10 | 7 | 0 | 47.0 | *99* | | | | | | | |
| Apr | 14.5 | 5.3 | 10.0 | *+0.6* | 22.8 | 25 | −0.6 | 6 | 9.4 | 5 | 13.3 | 26 | 1 | 0 | 71.2 | *146* | | | | | | | |
| May | 15.5 | 5.0 | 10.2 | *−2.2* | 21.1 | 3 | 0.0 | 22 | 10.6 | 23 | 12.2 | 12 | 0 | 0 | 33.6 | *59* | | | | | | | |
| June | 17.0 | 7.3 | 12.0 | *−3.4* | 25.0 | 30 | 3.3 | 24 | 12.2 | 8 | 11.7 | 5 | 0 | 1 | 36.3 | *74* | | | | | | | |
| July | 20.1 | 10.4 | 15.0 | *−2.6* | 25.6 | 19 | 4.4 | 4 | 11.1 | 2 | 15.0 | 1 | 0 | 1 | 42.6 | *87* | | | | | | | |
| Aug | 22.5 | 12.1 | 17.1 | *−0.3* | 28.3 | 24 | 8.3 | 12 | 12.8 | 28 | 16.1 | 16 | 0 | 8 | 86.9 | *159* | | | | | | | |
| Sep | 19.4 | 12.4 | 15.9 | *+1.0* | 22.8 | 3 | 6.1 | 30 | 13.9 | 29 | 17.2 | 18 | 0 | 0 | 96.5 | *178* | | | | | | | |
| Oct | 14.3 | 6.9 | 10.8 | *−0.7* | 19.4 | 3 | 1.7 | 23 | 10.6 | 21 | 14.4 | 4 | 0 | 0 | 54.2 | *77* | | | | | | | |
| Nov | 10.9 | 5.9 | 8.7 | *+0.9* | 16.1 | 2 | −1.1 | 5 | 6.1 | 4 | 13.9 | 2 | 3 | 0 | 87.1 | *131* | | | | | | | |
| Dec | 8.8 | 3.8 | 6.6 | *+1.2* | 14.4 | 10 | −1.1 | 26 | 4.4 | 25 | 9.4 | 9 | 2 | 0 | 96.8 | *153* | | | | | | | |
| **1821** | 13.4 | 5.8 | 9.7 | *−1.1* | 28.3 | 8 | −7.8 | 2 | −3.3 | 1 | 17.2 | 9 | 46 | 10 | 709.7 | *107* | | | | | | | |

## OXFORD 1822

| Month | Mean max | Mean min | Mean temp | Anom | Highest max | Date | Lowest min | Date | Lowest max | Date | Highest min | Date | Air frost | Days ≥ 25 °C | Total pptn | Anom | Rain days | Wettest day | Date | Total sunshine | Anom | Sunniest day | Date |
|---|---|---|---|---|---|---|---|---|---|---|---|---|---|---|---|---|---|---|---|---|---|---|---|
| Jan | 6.3 | 2.0 | 4.4 | *−0.7* | 10.0 | 24 | −1.7 | 6 | 1.7 | 5 | 6.1 | 20 | 8 | 0 | 13.3 | *23* | | | | | | | |
| Feb | 9.2 | 2.8 | 6.2 | *+1.1* | 11.7 | 15 | −3.9 | 28 | 5.0 | 12 | 7.8 | 10 | 5 | 0 | 23.4 | *55* | | | | | | | |
| Mar | 11.9 | 4.2 | 8.2 | *+0.8* | 16.7 | 28 | −3.9 | 1 | 5.6 | 31 | 9.4 | 10 | 3 | 0 | 18.9 | *40* | | | | | | | |
| Apr | 12.6 | 4.1 | 8.4 | *−0.9* | 21.1 | 29 | −0.6 | 10 | 6.7 | 10 | 7.8 | 15 | 2 | 0 | 57.7 | *118* | | | | | | | |
| May | 18.4 | 7.8 | 13.0 | *+0.7* | 26.1 | 20 | 2.2 | 9 | 10.6 | 12 | 12.8 | 28 | 0 | 1 | 42.1 | *74* | | | | | | | |
| June | 24.5 | 12.0 | 18.1 | *+2.6* | 28.3 | 4 | 5.6 | 13 | 18.3 | 12 | 16.7 | 1 | 0 | 18 | 8.0 | *16* | | | | | | | |
| July | 22.8 | 12.4 | 17.3 | *−0.3* | 25.6 | 5 | 8.3 | 3 | 19.4 | 12 | 16.1 | 24 | 0 | 2 | 105.5 | *216* | | | | | | | |
| Aug | 21.8 | 11.4 | 16.4 | *−1.0* | 28.9 | 21 | 6.7 | 3 | 18.3 | 4 | 17.8 | 11 | 0 | 5 | 3.1 | *6* | | | | | | | |
| Sep | 17.3 | 9.2 | 13.2 | *−1.7* | 23.3 | 17 | 2.2 | 14 | 12.8 | 24 | 16.1 | 6 | 0 | 0 | 52.8 | *97* | | | | | | | |
| Oct | 14.2 | 7.5 | 11.0 | *−0.4* | 18.9 | 2 | 0.0 | 18 | 8.9 | 14 | 12.8 | 13 | 0 | 0 | 70.7 | *101* | | | | | | | |
| Nov | 10.5 | 5.6 | 8.3 | *+0.5* | 15.0 | 1 | −1.1 | 9 | 5.0 | 29 | 12.8 | 2 | 1 | 0 | 98.7 | *148* | | | | | | | |
| Dec | 2.9 | −1.9 | 0.8 | *−4.5* | 9.4 | 1 | −7.8 | 30 | −1.7 | 30 | 4.4 | 6 | 18 | 0 | 34.6 | *55* | | | | | | | |
| **1822** | 14.4 | 6.4 | 10.5 | *−0.3* | 28.9 | 8 | −7.8 | 12 | −1.7 | 12 | 17.8 | 8 | 37 | 26 | 528.9 | *80* | | | | | | | |

## OXFORD 1823

| Month | Mean max | Mean min | Mean temp | Anom | Highest max | Date | Lowest min | Date | Lowest max | Date | Highest min | Date | Air frost | Days ≥ 25 °C | Total pptn | Anom | Rain days | Wettest day | Date | Total sunshine | Anom | Sunniest day | Date |
|---|---|---|---|---|---|---|---|---|---|---|---|---|---|---|---|---|---|---|---|---|---|---|---|
| Jan | 1.1 | −3.4 | −0.9 | *−6.0* | 9.4 | 29 | −15.0 | 19 | −7.8 | 19 | 5.6 | 4 | 21 | 0 | 34.9 | *61* | | | | | | | |
| Feb | 5.5 | 0.4 | 3.2 | *−1.9* | 11.1 | 11 | −7.8 | 7 | −0.6 | 5 | 6.1 | 12 | 10 | 0 | 65.6 | *153* | | | | | | | |
| Mar | 8.2 | 1.3 | 4.9 | *−2.5* | 14.4 | 31 | −3.3 | 7 | 2.2 | 19 | 6.7 | 14 | 9 | 0 | 28.3 | *59* | | | | | | | |
| Apr | 11.6 | 2.8 | 7.3 | *−2.1* | 17.8 | 30 | −2.2 | 12 | 6.7 | 8 | 10.0 | 2 | 6 | 0 | 50.1 | *102* | | | | | | | |
| May | 17.9 | 7.6 | 12.6 | *+0.3* | 23.3 | 6 | 2.8 | 5 | 12.8 | 14 | 13.3 | 31 | 0 | 0 | 35.5 | *62* | | | | | | | |
| June | 19.2 | 8.0 | 13.4 | *−2.0* | 24.4 | 1 | 4.4 | 24 | 12.2 | 22 | 13.9 | 14 | 0 | 0 | 30.2 | *61* | | | | | | | |
| July | 20.2 | 10.7 | 15.2 | *−2.4* | 26.1 | 5 | 6.7 | 9 | 15.0 | 8 | 16.1 | 20 | 0 | 1 | 60.0 | *123* | | | | | | | |
| Aug | 20.8 | 11.8 | 16.1 | *−1.3* | 25.0 | 25 | 6.7 | 6 | 16.7 | 9 | 18.3 | 26 | 0 | 1 | 79.7 | *145* | | | | | | | |
| Sep | 18.0 | 7.5 | 12.7 | *−2.2* | 26.1 | 5 | 0.6 | 29 | 12.2 | 30 | 15.6 | 14 | 0 | 1 | 62.7 | *116* | | | | | | | |
| Oct | 12.2 | 4.7 | 8.6 | *−2.8* | 16.7 | 6 | 0.0 | 31 | 2.8 | 31 | 13.9 | 6 | 0 | 0 | 94.2 | *135* | | | | | | | |
| Nov | 8.8 | 4.3 | 6.9 | *−1.0* | 13.3 | 30 | −3.3 | 13 | 3.3 | 12 | 10.0 | 7 | 7 | 0 | 39.7 | *60* | | | | | | | |
| Dec | 7.1 | 2.1 | 4.9 | *−0.4* | 12.2 | 3 | −4.4 | 20 | 2.2 | 19 | 8.9 | 1 | 5 | 0 | 79.9 | *127* | | | | | | | |
| **1823** | 12.5 | 4.8 | 8.7 | *−2.1* | 26.1 | 7 | −15.0 | 1 | −7.8 | 1 | 18.3 | 8 | 58 | 3 | 660.8 | *100* | | | | | | | |

## OXFORD 1824

| Month | Mean max | Mean min | Mean temp | Anom | Highest max | Date | Lowest min | Date | Lowest max | Date | Highest min | Date | Air frost | Days ≥ 25 °C | Total pptn | Anom | Rain days | Wettest day | Date | Total sunshine | Anom | Sunniest day | Date |
|---|---|---|---|---|---|---|---|---|---|---|---|---|---|---|---|---|---|---|---|---|---|---|---|
| Jan | 5.4 | 0.9 | 3.4 | *−1.7* | 11.1 | 25 | −6.7 | 14 | −1.7 | 14 | 9.4 | 26 | 10 | 0 | 17.9 | *32* | | | | | | | |
| Feb | 6.7 | 1.5 | 4.3 | *−0.8* | 12.2 | 8 | −3.3 | 16 | 2.2 | 25 | 8.9 | 9 | 7 | 0 | 52.0 | *122* | | | | | | | |
| Mar | 7.6 | 1.7 | 4.9 | *−2.6* | 13.3 | 19 | −4.4 | 31 | 2.2 | 3 | 7.8 | 7 | 6 | 0 | 75.5 | *158* | | | | | | | |
| Apr | 11.3 | 3.2 | 7.3 | *−2.0* | 18.3 | 29 | −2.8 | 3 | 6.1 | 10 | 11.7 | 22 | 10 | 0 | 44.3 | *91* | | | | | | | |
| May | 14.9 | 6.8 | 10.8 | *−1.6* | 23.3 | 27 | −0.6 | 21 | 6.1 | 14 | 11.7 | 30 | 1 | 0 | 32.9 | *57* | | | | | | | |
| June | 18.6 | 9.6 | 13.9 | *−1.5* | 24.4 | 29 | 5.0 | 13 | 12.8 | 24 | 12.8 | 28 | 0 | 0 | 79.2 | *161* | | | | | | | |
| July | 21.7 | 11.6 | 16.4 | *−1.2* | 25.6 | 12 | 7.2 | 29 | 17.2 | 3 | 15.6 | 9 | 0 | 3 | 36.6 | *75* | | | | | | | |
| Aug | 21.1 | 12.4 | 16.6 | *−0.8* | 26.1 | 29 | 6.7 | 24 | 19.4 | 7 | 17.8 | 14 | 0 | 1 | 79.4 | *145* | | | | | | | |
| Sep | 18.9 | 10.5 | 14.7 | *−0.2* | 28.3 | 2 | 0.6 | 28 | 10.0 | 27 | 16.7 | 4 | 0 | 3 | 82.5 | *152* | | | | | | | |
| Oct | 13.4 | 7.4 | 10.6 | *−0.9* | 18.3 | 6 | −1.1 | 16 | 6.7 | 15 | 15.6 | 1 | 4 | 0 | 70.7 | *101* | | | | | | | |
| Nov | 10.1 | 4.3 | 7.5 | *−0.3* | 14.4 | 1 | −3.3 | 6 | 6.1 | 26 | 13.3 | 2 | 2 | 0 | 86.9 | *130* | | | | | | | |
| Dec | 8.1 | 2.8 | 5.8 | *+0.4* | 11.1 | 25 | −4.4 | 6 | 1.1 | 4 | 9.4 | 22 | 9 | 0 | 70.4 | *112* | | | | | | | |
| **1824** | 13.2 | 6.1 | 9.7 | *−1.1* | 28.3 | 9 | −6.7 | 1 | −1.7 | 1 | 17.8 | 8 | 49 | 7 | 728.4 | *110* | | | | | | | |

## OXFORD 1825

| Month | Mean max | Mean min | Mean temp | Anom | Highest max | Date | Lowest min | Date | Lowest max | Date | Highest min | Date | Air frost | Days ≥ 25 °C | Total pptn | Anom | Rain days | Wettest day | Date | Total sunshine | Anom | Sunniest day | Date |
|---|---|---|---|---|---|---|---|---|---|---|---|---|---|---|---|---|---|---|---|---|---|---|---|
| Jan | 6.3 | 1.7 | 4.3 | *−0.9* | 13.6 | 2 | −2.8 | 6 | 2.2 | 5 | 10.1 | 2 | 8 | 0 | 5.1 | *9* | | | | | | | |
| Feb | 6.6 | 0.9 | 4.0 | *−1.2* | 10.0 | 2 | −5.0 | 5 | 1.7 | 4 | 7.2 | 1 | 11 | 0 | 39.2 | *92* | | | | | | | |
| Mar | 8.5 | 1.0 | 4.9 | *−2.5* | 13.9 | 9 | −5.6 | 17 | 1.7 | 16 | 9.4 | 11 | 13 | 0 | 28.1 | *59* | | | | | | | |
| Apr | 15.5 | 4.0 | 9.8 | *+0.5* | 18.3 | 26 | −2.8 | 2 | 11.7 | 1 | 9.4 | 23 | 5 | 0 | 39.7 | *81* | | | | | | | |
| May | 17.1 | 7.4 | 12.1 | *−0.2* | 25.6 | 23 | 2.8 | 15 | 11.7 | 26 | 13.3 | 7 | 0 | 2 | 69.5 | *121* | | | | | | | |
| June | 21.0 | 9.3 | 15.0 | *−0.4* | 25.6 | 11 | 3.3 | 6 | 15.6 | 4 | 14.4 | 13 | 0 | 3 | 42.9 | *87* | | | | | | | |
| July | 24.0 | 11.7 | 17.6 | *−0.0* | 33.9 | 19 | 6.7 | 25 | 15.6 | 5 | 18.9 | 18 | 0 | 13 | 0.8 | *2* | | | | | | | |
| Aug | 22.2 | 12.7 | 17.3 | *−0.2* | 30.6 | 1 | 6.1 | 20 | 17.2 | 28 | 18.9 | 2 | 0 | 4 | 80.3 | *147* | | | | | | | |
| Sep | 19.9 | 12.0 | 15.9 | *+1.1* | 23.3 | 1 | 5.6 | 6 | 14.4 | 30 | 17.2 | 2 | 0 | 0 | 68.5 | *126* | | | | | | | |
| Oct | 14.3 | 7.9 | 11.3 | *−0.2* | 20.6 | 2 | 0.0 | 22 | 5.6 | 20 | 15.0 | 2 | 0 | 0 | 57.8 | *83* | | | | | | | |
| Nov | 8.2 | 2.4 | 5.6 | *−2.2* | 14.4 | 1 | −4.4 | 13 | 2.8 | 12 | 8.9 | 3 | 6 | 0 | 49.7 | *74* | | | | | | | |
| Dec | 7.0 | 1.9 | 4.8 | *−0.6* | 11.1 | 15 | −6.1 | 31 | 0.0 | 27 | 7.8 | 17 | 9 | 0 | 52.5 | *83* | | | | | | | |
| **1825** | 14.2 | 6.1 | 10.2 | *−0.6* | 33.9 | 7 | −6.1 | 12 | 0.0 | 12 | 18.9 | 7 | 52 | 22 | 534.1 | *81* | | | | | | | |

## OXFORD 1826

| Month | Mean max | Mean min | Mean temp | Anom | Highest max | Date | Lowest min | Date | Lowest max | Date | Highest min | Date | Air frost | Days ≥ 25 °C | Total pptn | Anom | Rain days | Wettest day | Date | Total sunshine | Anom | Sunniest day | Date |
|---|---|---|---|---|---|---|---|---|---|---|---|---|---|---|---|---|---|---|---|---|---|---|---|
| Jan | 2.1 | −2.6 | 0.0 | *−5.1* | 8.3 | 31 | −13.3 | 16 | −4.4 | 16 | 3.9 | 2 | 20 | 0 | 28.1 | *49* | | | | | | | |
| Feb | 9.4 | 3.4 | 6.6 | *+1.5* | 12.2 | 6 | −3.9 | 10 | 5.0 | 9 | 8.3 | 3 | 4 | 0 | 46.2 | *108* | | | | | | | |
| Mar | 9.2 | 2.5 | 6.0 | *−1.4* | 18.9 | 9 | −3.3 | 18 | 3.9 | 26 | 12.2 | 10 | 6 | 0 | 38.0 | *80* | | | | | | | |
| Apr | 13.8 | 5.1 | 9.5 | *+0.2* | 18.9 | 21 | −2.8 | 1 | 7.2 | 28 | 11.1 | 9 | 3 | 0 | 25.6 | *52* | | | | | | | |
| May | 15.7 | 6.2 | 10.8 | *−1.6* | 23.9 | 18 | −1.1 | 2 | 8.3 | 4 | 11.7 | 20 | 1 | 0 | 31.3 | *55* | | | | | | | |
| June | 23.5 | 11.9 | 17.5 | *+2.1* | 30.6 | 26 | 7.2 | 17 | 15.6 | 22 | 16.1 | 29 | 0 | 13 | 10.4 | *21* | | | | | | | |
| July | 24.7 | 13.8 | 19.0 | *+1.4* | 31.1 | 31 | 4.4 | 13 | 18.3 | 23 | 17.8 | 1 | 0 | 13 | 20.3 | *42* | | | | | | | |
| Aug | 24.4 | 14.0 | 19.0 | *+1.6* | 29.4 | 20 | 8.3 | 12 | 21.1 | 11 | 18.9 | 24 | 0 | 10 | 28.3 | *52* | | | | | | | |
| Sep | 18.8 | 10.8 | 14.8 | *−0.1* | 22.8 | 3 | 2.8 | 23 | 14.4 | 7 | 16.1 | 30 | 0 | 0 | 107.2 | *198* | | | | | | | |
| Oct | 15.2 | 8.6 | 12.1 | *+0.7* | 19.4 | 21 | 0.3 | 6 | 10.6 | 31 | 15.0 | 13 | 0 | 0 | 69.2 | *99* | | | | | | | |
| Nov | 7.2 | 1.9 | 4.8 | *−3.0* | 12.2 | 11 | −5.6 | 26 | 1.7 | 25 | 8.3 | 11 | 10 | 0 | 33.2 | *50* | | | | | | | |
| Dec | 7.9 | 3.9 | 6.2 | *+0.9* | 12.2 | 6 | −2.2 | 22 | 3.3 | 5 | 9.4 | 11 | 4 | 0 | 49.4 | *78* | | | | | | | |
| **1826** | 14.3 | 6.6 | 10.5 | *−0.3* | 31.1 | 7 | −13.3 | 1 | −4.4 | 1 | 18.9 | 8 | 48 | 36 | 487.2 | *74* | | | | | | | |

## OXFORD 1827

| Month | Mean max | Mean min | Mean temp | Anom | Highest max | Date | Lowest min | Date | Lowest max | Date | Highest min | Date | Air frost | Days ≥ 25 °C | Total pptn | Anom | Rain days | Wettest day | Date | Total sunshine | Anom | Sunniest day | Date |
|---|---|---|---|---|---|---|---|---|---|---|---|---|---|---|---|---|---|---|---|---|---|---|---|
| Jan | 3.3 | −1.6 | 1.1 | *−4.0* | 10.0 | 14 | −10.6 | 5 | −2.8 | 22 | 5.6 | 1 | 16 | 0 | 27.0 | *47* | 4 | 12.9 | 11 | | | | |
| Feb | 3.0 | −3.5 | 0.0 | *−5.2* | 11.7 | 27 | −11.7 | 17 | −1.7 | 17 | 7.8 | 27 | 24 | 0 | 20.0 | *47* | 1 | 20.0 | 28 | | | | |
| Mar | 9.7 | 2.9 | 6.5 | *−1.0* | 13.9 | 24 | −4.4 | 10 | 3.9 | 9 | 8.3 | 21 | 5 | 0 | 55.4 | *116* | 6 | 16.8 | 6 | | | | |
| Apr | 14.1 | 4.2 | 9.2 | *−0.1* | 25.4 | 30 | −2.2 | 26 | 6.7 | 22 | 9.4 | 6 | 2 | 1 | 20.2 | *41* | 6 | 5.0 | 13 | | | | |
| May | 18.9 | 4.3 | 11.5 | *−0.9* | 24.2 | 19 | −3.4 | 8 | 11.7 | 8 | 13.3 | 21 | 5 | 0 | 40.9 | *72* | 7 | 15.9 | 7 | | | | |
| June | 21.2 | 6.5 | 13.7 | *−1.7* | 26.2 | 29 | 0.4 | 8 | 14.5 | 4 | 13.0 | 20 | 0 | 2 | 40.9 | *83* | 8 | 18.6 | 7 | | | | |
| July | 25.0 | 9.0 | 16.7 | *−0.9* | 29.5 | 16 | 4.7 | 6 | 21.8 | 1 | 16.2 | 30 | 0 | 14 | 37.9 | *78* | 9 | 15.1 | 2 | | | | |
| Aug | 21.1 | 8.6 | 14.7 | *−2.8* | 28.9 | 2 | 3.7 | 8 | 15.6 | 26 | 13.0 | 15 | 0 | 3 | 45.6 | *83* | 9 | 25.5 | 17 | | | | |
| Sep | 18.2 | 10.9 | 14.6 | *−0.3* | 21.7 | 11 | 6.9 | 23 | 15.0 | 19 | 16.7 | 11 | 0 | 0 | 88.3 | *163* | 9 | 24.1 | 30 | | | | |
| Oct | 14.2 | 8.9 | 11.7 | *+0.2* | 17.8 | 1 | 0.6 | 29 | 8.3 | 28 | 14.1 | 21 | 0 | 0 | 115.2 | *165* | 9 | 38.7 | 9 | | | | |
| Nov | 9.5 | 4.6 | 7.4 | *−0.5* | 15.6 | 13 | −5.6 | 22 | 0.6 | 23 | 10.6 | 6 | 4 | 0 | 45.3 | *68* | 6 | 18.5 | 29 | | | | |
| Dec | 9.3 | 3.8 | 6.9 | *+1.6* | 12.9 | 8 | −4.4 | 30 | 1.1 | 29 | 9.8 | 19 | 4 | 0 | 71.6 | *114* | 12 | 16.9 | 16 | | | | |
| 1827 | 14.0 | 4.9 | 9.5 | *−1.3* | 29.5 | 7 | −11.7 | 2 | −2.8 | 1 | 16.7 | 9 | 60 | 20 | 608.3 | *92* | 86 | 38.7 | 10 | | | | |

## OXFORD 1828

| Month | Mean max | Mean min | Mean temp | Anom | Highest max | Date | Lowest min | Date | Lowest max | Date | Highest min | Date | Air frost | Days ≥ 25 °C | Total pptn | Anom | Rain days | Wettest day | Date | Total sunshine | Anom | Sunniest day | Date |
|---|---|---|---|---|---|---|---|---|---|---|---|---|---|---|---|---|---|---|---|---|---|---|---|
| Jan | 7.1 | 2.5 | 5.1 | *−0.1* | 12.8 | 18 | −5.0 | 10 | −1.1 | 9 | 8.6 | 20 | 6 | 0 | 65.0 | *114* | 8 | 24.5 | 13 | | | | |
| Feb | 7.7 | 2.6 | 5.3 | *+0.2* | 15.2 | 29 | −3.9 | 16 | −0.6 | 12 | 10.6 | 26 | 8 | 0 | 19.8 | *46* | 5 | 7.5 | 14 | | | | |
| Mar | 10.2 | 3.0 | 6.7 | *−0.7* | 16.5 | 15 | −4.4 | 7 | 2.8 | 6 | 8.9 | 9 | 8 | 0 | 9.7 | *20* | 6 | 4.2 | 27 | | | | |
| Apr | 13.2 | 5.0 | 9.2 | *−0.2* | 21.7 | 29 | −1.7 | 3 | 7.2 | 4 | 11.1 | 30 | 3 | 0 | 60.1 | *123* | 9 | 11.9 | 16 | | | | |
| May | 18.1 | 8.3 | 13.1 | *+0.7* | 22.4 | 30 | 2.8 | 7 | 7.8 | 19 | 14.1 | 28 | 0 | 0 | 44.6 | *78* | 10 | 11.4 | 25 | | | | |
| June | 21.5 | 11.9 | 16.5 | *+1.1* | 28.3 | 25 | 8.3 | 5 | 15.6 | 7 | 15.9 | 27 | 0 | 6 | 44.0 | *89* | 9 | 12.8 | 17 | | | | |
| July | 21.0 | 13.8 | 17.1 | *−0.5* | 25.6 | 3 | 6.7 | 30 | 14.2 | 31 | 18.3 | 3 | 0 | 2 | 95.1 | *195* | 14 | 28.8 | 9 | | | | |
| Aug | 19.4 | 12.1 | 15.6 | *−1.8* | 22.8 | 24 | 7.8 | 15 | 13.3 | 14 | 16.1 | 30 | 0 | 0 | 72.2 | *132* | 12 | 16.9 | 14 | | | | |
| Sep | 18.5 | 11.3 | 14.9 | *+0.1* | 23.3 | 8 | 3.9 | 16 | 11.7 | 14 | 16.7 | 9 | 0 | 0 | 65.9 | *122* | 8 | 17.3 | 10 | | | | |
| Oct | 13.5 | 7.3 | 10.6 | *−0.9* | 17.2 | 12 | 0.4 | 20 | 8.1 | 30 | 12.8 | 6 | 0 | 0 | 55.0 | *79* | 6 | 19.3 | 6 | | | | |
| Nov | 9.7 | 5.1 | 7.7 | *−0.2* | 13.9 | 28 | −6.1 | 12 | 2.2 | 8 | 10.0 | 22 | 4 | 0 | 28.6 | *43* | 8 | 6.5 | 15 | | | | |
| Dec | 9.1 | 5.0 | 7.4 | *+2.0* | 12.8 | 17 | −1.6 | 2 | 3.9 | 29 | 10.3 | 21 | 4 | 0 | 54.1 | *86* | 9 | 16.9 | 8 | | | | |
| 1828 | 14.1 | 7.3 | 10.8 | *−0.0* | 28.3 | 6 | −6.1 | 11 | −1.1 | 1 | 18.3 | 7 | 33 | 8 | 613.9 | *93* | 104 | 28.8 | 7 | | | | |

## OXFORD 1829

| Month | Mean max | Mean min | Mean temp | Anom | Highest max | Date | Lowest min | Date | Lowest max | Date | Highest min | Date | Air frost | Days ≥ 25 °C | Total pptn | Anom | Rain days | Wettest day | Date | Total sunshine | Anom | Sunniest day | Date |
|---|---|---|---|---|---|---|---|---|---|---|---|---|---|---|---|---|---|---|---|---|---|---|---|
| Jan | 1.8 | −2.3 | 0.0 | *−5.1* | 7.9 | 1 | −10.1 | 25 | −5.6 | 23 | 3.4 | 31 | 21 | 0 | 15.2 | *27* | 6 | 5.2 | 3 | | | | |
| Feb | 6.7 | 2.1 | 4.6 | *−0.5* | 10.6 | 15 | −7.4 | 2 | 0.3 | 2 | 6.1 | 15 | 5 | 0 | 31.5 | *74* | 8 | 14.2 | 27 | | | | |
| Mar | 7.8 | 0.4 | 4.3 | *−3.1* | 15.0 | 20 | −6.3 | 25 | 2.3 | 1 | 9.6 | 20 | 13 | 0 | 5.5 | *12* | 1 | 5.4 | 29 | | | | |
| Apr | 12.0 | 4.2 | 8.1 | *−1.2* | 19.2 | 4 | −2.1 | 2 | 7.6 | 29 | 8.8 | 26 | 2 | 0 | 91.4 | *187* | 18 | 18.1 | 16 | | | | |
| May | 17.6 | 7.8 | 12.6 | *+0.3* | 21.8 | 23 | 5.0 | 4 | 14.0 | 1 | 11.6 | 6 | 0 | 0 | 3.9 | *7* | 2 | 3.1 | 25 | | | | |
| June | 20.2 | 10.8 | 15.3 | *−0.1* | 25.2 | 13 | 3.1 | 7 | 15.1 | 6 | 15.0 | 24 | 0 | 1 | 77.1 | *157* | 11 | 28.8 | 26 | | | | |
| July | 20.0 | 12.5 | 15.9 | *−1.7* | 23.5 | 24 | 7.7 | 27 | 16.8 | 26 | 17.1 | 25 | 0 | 0 | 97.8 | *200* | 17 | 35.2 | 18 | | | | |
| Aug | 18.7 | 12.4 | 15.4 | *−2.0* | 24.4 | 9 | 7.0 | 17 | 13.1 | 15 | 19.7 | 13 | 0 | 0 | 93.0 | *170* | 13 | 25.9 | 15 | | | | |
| Sep | 15.7 | 8.2 | 11.9 | *−3.0* | 19.2 | 10 | 2.5 | 26 | 11.5 | 29 | 15.9 | 2 | 0 | 0 | 68.9 | *127* | 13 | 11.6 | 10 | | | | |
| Oct | 11.5 | 4.8 | 8.3 | *−3.1* | 16.5 | 19 | −1.7 | 24 | 3.8 | 7 | 11.7 | 20 | 7 | 0 | 39.5 | *57* | 8 | 10.4 | 7 | | | | |
| Nov | 6.2 | 0.5 | 3.6 | *−4.2* | 13.1 | 12 | −7.1 | 21 | −0.2 | 19 | 8.8 | 13 | 13 | 0 | 26.2 | *39* | 7 | 9.0 | 11 | | | | |
| Dec | 1.7 | −2.1 | 0.1 | *−5.2* | 8.2 | 5 | −8.8 | 24 | −4.2 | 29 | 4.9 | 5 | 23 | 0 | 11.2 | *18* | 4 | 6.2 | 21 | | | | |
| **1829** | 11.7 | 4.9 | 8.4 | *−2.5* | 25.2 | 6 | −10.1 | 1 | −5.6 | 1 | 19.7 | 8 | 84 | 1 | 561.2 | *85* | 108 | 35.2 | 7 | | | | |

## OXFORD 1830

| Month | Mean max | Mean min | Mean temp | Anom | Highest max | Date | Lowest min | Date | Lowest max | Date | Highest min | Date | Air frost | Days ≥ 25 °C | Total pptn | Anom | Rain days | Wettest day | Date | Total sunshine | Anom | Sunniest day | Date |
|---|---|---|---|---|---|---|---|---|---|---|---|---|---|---|---|---|---|---|---|---|---|---|---|
| Jan | 0.7 | −3.2 | −1.0 | *−6.1* | 5.5 | 7 | −10.0 | 18 | −6.5 | 31 | 0.4 | 5 | 30 | 0 | 21.5 | *38* | 5 | 9.4 | 20 | | | | |
| Feb | 3.9 | −1.6 | 1.4 | *−3.7* | 12.8 | 25 | −12.3 | 6 | −7.9 | 2 | 7.5 | 25 | 18 | 0 | 31.0 | *73* | 8 | 11.6 | 9 | | | | |
| Mar | 11.8 | 2.9 | 7.5 | *+0.1* | 18.2 | 27 | −2.8 | 7 | 5.3 | 16 | 6.8 | 18 | 4 | 0 | 9.1 | *19* | 6 | 4.8 | 9 | | | | |
| Apr | 13.0 | 4.3 | 8.7 | *−0.6* | 22.1 | 30 | −5.0 | 5 | 0.8 | 2 | 9.6 | 23 | 6 | 0 | 67.3 | *138* | 12 | 19.4 | 2 | | | | |
| May | 16.7 | 6.7 | 11.6 | *−0.7* | 23.6 | 6 | 1.7 | 3 | 10.3 | 11 | 11.4 | 20 | 0 | 0 | 91.9 | *161* | 11 | 34.4 | 23 | | | | |
| June | 17.5 | 9.1 | 13.1 | *−2.3* | 23.5 | 27 | 3.6 | 5 | 12.6 | 17 | 15.3 | 26 | 0 | 0 | 86.6 | *176* | 13 | 33.6 | 3 | | | | |
| July | 21.6 | 13.4 | 17.3 | *−0.3* | 28.6 | 30 | 8.2 | 13 | 17.1 | 9 | 19.3 | 26 | 0 | 6 | 72.8 | *149* | 7 | 27.1 | 3 | | | | |
| Aug | 19.3 | 10.8 | 14.9 | *−2.5* | 24.4 | 4 | 6.1 | 31 | 16.1 | 18 | 14.5 | 5 | 0 | 0 | 94.9 | *173* | 16 | 33.2 | 9 | | | | |
| Sep | 16.4 | 9.6 | 13.0 | *−1.9* | 19.4 | 2 | 5.6 | 30 | 12.8 | 30 | 12.2 | 16 | 0 | 0 | 109.8 | *203* | 16 | 17.6 | 12 | | | | |
| Oct | 14.7 | 7.3 | 11.2 | *−0.3* | 20.0 | 21 | 0.3 | 17 | 10.0 | 30 | 13.3 | 3 | 0 | 0 | 20.7 | *30* | 9 | 7.4 | 25 | | | | |
| Nov | 10.2 | 4.9 | 7.8 | *+0.0* | 14.9 | 3 | −1.6 | 25 | 4.4 | 27 | 11.3 | 6 | 2 | 0 | 52.4 | *78* | 10 | 17.5 | 6 | | | | |
| Dec | 4.2 | −0.9 | 2.0 | *−3.4* | 9.8 | 6 | −12.9 | 25 | −5.2 | 24 | 7.7 | 7 | 17 | 0 | 31.4 | *50* | 7 | 11.0 | 9 | | | | |
| **1830** | 12.5 | 5.3 | 9.0 | *−1.8* | 28.6 | 7 | −12.9 | 12 | −7.9 | 2 | 19.3 | 7 | 77 | 6 | 689.5 | *104* | 120 | 34.4 | 5 | | | | |

## OXFORD 1831

| Month | Mean max | Mean min | Mean temp | Anom | Highest max | Date | Lowest min | Date | Lowest max | Date | Highest min | Date | Air frost | Days ≥ 25 °C | Total pptn | Anom | Rain days | Wettest day | Date | Total sunshine | Anom | Sunniest day | Date |
|---|---|---|---|---|---|---|---|---|---|---|---|---|---|---|---|---|---|---|---|---|---|---|---|
| Jan | 3.7 | −1.2 | 1.5 | *−3.6* | 9.9 | 22 | −6.9 | 8 | −0.2 | 25 | 4.6 | 22 | 20 | 0 | 57.7 | *101* | 11 | 23.6 | 21 | | | | |
| Feb | 8.4 | 1.7 | 5.3 | *+0.2* | 15.2 | 10 | −5.8 | 3 | 2.1 | 1 | 8.4 | 9 | 7 | 0 | 56.7 | *132* | 12 | 20.0 | 4 | | | | |
| Mar | 10.8 | 3.2 | 7.2 | *−0.2* | 14.7 | 28 | −2.4 | 24 | 3.5 | 24 | 9.2 | 17 | 4 | 0 | 51.4 | *108* | 15 | 13.0 | 25 | | | | |
| Apr | 13.7 | 5.2 | 9.5 | *+0.1* | 18.2 | 12 | −0.6 | 18 | 7.7 | 2 | 8.5 | 25 | 1 | 0 | 38.8 | *79* | 8 | 10.9 | 13 | | | | |
| May | 15.9 | 5.7 | 10.7 | *−1.7* | 21.4 | 23 | −2.6 | 7 | 7.6 | 6 | 12.2 | 24 | 1 | 0 | 56.3 | *99* | 6 | 30.6 | 2 | | | | |
| June | 20.2 | 10.3 | 15.1 | *−0.3* | 23.9 | 23 | 5.6 | 2 | 15.2 | 6 | 16.6 | 26 | 0 | 0 | 37.0 | *75* | 8 | 13.4 | 27 | | | | |
| July | 22.1 | 12.2 | 16.9 | *−0.7* | 27.2 | 30 | 9.3 | 25 | 17.9 | 13 | 16.3 | 28 | 0 | 4 | 107.4 | *220* | 11 | 20.6 | 10 | | | | |
| Aug | 21.9 | 13.3 | 17.4 | *+0.0* | 25.3 | 9 | 9.7 | 26 | 17.9 | 19 | 16.7 | 6 | 0 | 1 | 48.2 | *88* | 9 | 14.4 | 2 | | | | |
| Sep | 18.2 | 10.0 | 14.1 | *−0.8* | 21.5 | 4 | 5.3 | 23 | 12.8 | 1 | 14.2 | 29 | 0 | 0 | 93.2 | *172* | 14 | 28.9 | 5 | | | | |
| Oct | 16.3 | 9.6 | 13.1 | *+1.7* | 19.9 | 1 | 3.2 | 30 | 12.1 | 30 | 14.0 | 1 | 0 | 0 | 76.2 | *109* | 14 | 18.3 | 27 | | | | |
| Nov | 9.4 | 3.0 | 6.5 | *−1.4* | 14.2 | 23 | −5.2 | 29 | 3.7 | 29 | 10.2 | 23 | 10 | 0 | 55.8 | *84* | 11 | 20.2 | 1 | | | | |
| Dec | 8.2 | 3.8 | 6.3 | *+1.0* | 12.4 | 12 | −2.7 | 25 | 2.6 | 25 | 9.1 | 13 | 3 | 0 | 73.3 | *116* | 13 | 23.5 | 9 | | | | |
| **1831** | 14.1 | 6.4 | 10.3 | *−0.5* | 27.2 | 7 | −6.9 | 1 | −0.2 | 1 | 16.7 | 8 | 46 | 5 | 752.1 | *114* | 132 | 30.6 | 5 | | | | |

## OXFORD 1832

| Month | Mean max | Mean min | Mean temp | Anom | Highest max | Date | Lowest min | Date | Lowest max | Date | Highest min | Date | Air frost | Days ≥ 25 °C | Total pptn | Anom | Rain days | Wettest day | Date | Total sunshine | Anom | Sunniest day | Date |
|---|---|---|---|---|---|---|---|---|---|---|---|---|---|---|---|---|---|---|---|---|---|---|---|
| Jan | 5.2 | 0.7 | 3.2 | *−1.9* | 9.6 | 11 | −4.6 | 4 | 1.6 | 3 | 5.1 | 23 | 14 | 0 | 37.5 | *66* | 6 | 15.5 | 27 | | | | |
| Feb | 6.1 | 0.5 | 3.5 | *−1.6* | 11.9 | 5 | −4.3 | 24 | 1.9 | 29 | 7.2 | 6 | 14 | 0 | 9.1 | *21* | 4 | 5.5 | 6 | | | | |
| Mar | 9.7 | 2.1 | 6.0 | *−1.4* | 13.6 | 21 | −3.1 | 10 | 5.6 | 8 | 7.9 | 17 | 8 | 0 | 47.5 | *100* | 8 | 21.0 | 14 | | | | |
| Apr | 13.4 | 4.4 | 9.0 | *−0.4* | 19.3 | 5 | 0.7 | 28 | 7.1 | 24 | 11.3 | 2 | 0 | 0 | 57.6 | *118* | 8 | 17.7 | 15 | | | | |
| May | 16.7 | 6.7 | 11.6 | *−0.8* | 22.7 | 28 | 1.6 | 10 | 10.3 | 12 | 11.9 | 28 | 0 | 0 | 68.4 | *120* | 8 | 19.1 | 30 | | | | |
| June | 20.2 | 11.1 | 15.5 | *+0.1* | 23.3 | 12 | 5.0 | 2 | 16.4 | 1 | 14.3 | 12 | 0 | 0 | 72.1 | *146* | 11 | 21.1 | 21 | | | | |
| July | 21.1 | 11.3 | 15.9 | *−1.7* | 24.6 | 17 | 6.0 | 21 | 16.7 | 22 | 14.4 | 13 | 0 | 0 | 38.3 | *78* | 6 | 17.1 | 12 | | | | |
| Aug | 20.8 | 11.2 | 15.8 | *−1.6* | 25.4 | 9 | 5.9 | 27 | 15.0 | 29 | 17.7 | 22 | 0 | 2 | 110.8 | *202* | 11 | 36.8 | 21 | | | | |
| Sep | 18.5 | 8.7 | 13.6 | *−1.3* | 21.1 | 29 | 3.4 | 20 | 15.3 | 14 | 12.9 | 30 | 0 | 0 | 24.9 | *46* | 4 | 17.8 | 1 | | | | |
| Oct | 14.6 | 7.2 | 11.1 | *−0.4* | 19.6 | 11 | 1.2 | 20 | 10.7 | 8 | 13.5 | 12 | 0 | 0 | 79.8 | *114* | 9 | 29.0 | 4 | | | | |
| Nov | 9.3 | 3.7 | 6.8 | *−1.0* | 15.3 | 1 | −1.2 | 17 | 4.4 | 9 | 8.7 | 3 | 1 | 0 | 73.9 | *111* | 8 | 23.6 | 14 | | | | |
| Dec | 7.9 | 2.1 | 5.3 | *−0.0* | 12.9 | 1 | −1.6 | 28 | 2.8 | 7 | 9.4 | 2 | 6 | 0 | 44.5 | *70* | 8 | 10.8 | 15 | | | | |
| **1832** | 13.6 | 5.8 | 9.8 | *−1.0* | 25.4 | 8 | −4.6 | 1 | 1.6 | 1 | 17.7 | 8 | 43 | 2 | 664.4 | *101* | 91 | 36.8 | 8 | | | | |

## OXFORD 1833

| Month | Mean max | Mean min | Mean temp | Anom | Highest max | Date | Lowest min | Date | Lowest max | Date | Highest min | Date | Air frost | Days ≥ 25 °C | Total pptn | Anom | Rain days | Wettest day | Date | Total sunshine | Anom | Sunniest day | Date |
|---|---|---|---|---|---|---|---|---|---|---|---|---|---|---|---|---|---|---|---|---|---|---|---|
| Jan | 3.8 | −0.6 | 1.9 | *−3.3* | 8.3 | 2 | −5.6 | 24 | 0.0 | 10 | 2.7 | 14 | 15 | 0 | 17.3 | *30* | 2 | 9.5 | 13 | | | | |
| Feb | 8.9 | 2.8 | 6.1 | *+1.0* | 12.5 | 10 | −1.0 | 23 | 5.1 | 23 | 8.8 | 5 | 4 | 0 | 106.3 | *248* | 12 | 17.3 | 10 | | | | |
| Mar | 7.3 | 0.0 | 3.8 | *−3.6* | 12.3 | 29 | −3.2 | 13 | 3.4 | 9 | 6.3 | 3 | 17 | 0 | 26.8 | *56* | 6 | 16.3 | 22 | | | | |
| Apr | 12.9 | 3.4 | 8.2 | *−1.1* | 17.6 | 23 | −0.6 | 17 | 9.2 | 16 | 7.7 | 27 | 3 | 0 | 55.3 | *113* | 13 | 18.7 | 24 | | | | |
| May | 20.2 | 8.2 | 14.1 | *+1.8* | 25.6 | 15 | 0.6 | 1 | 13.1 | 1 | 11.4 | 26 | 0 | 2 | 7.4 | *13* | 2 | 4.0 | 1 | | | | |
| June | 20.6 | 9.7 | 14.9 | *−0.5* | 25.4 | 1 | 5.7 | 25 | 15.5 | 23 | 13.8 | 8 | 0 | 1 | 61.3 | *125* | 11 | 12.6 | 26 | | | | |
| July | 20.3 | 10.6 | 15.2 | *−2.4* | 24.4 | 27 | 5.6 | 2 | 15.2 | 12 | 15.7 | 29 | 0 | 0 | 53.2 | *109* | 10 | 13.6 | 8 | | | | |
| Aug | 19.3 | 8.9 | 13.9 | *−3.5* | 21.6 | 9 | 4.1 | 7 | 12.2 | 31 | 13.5 | 21 | 0 | 0 | 29.0 | *53* | 7 | 15.6 | 30 | | | | |
| Sep | 16.6 | 7.3 | 11.9 | *−2.9* | 18.9 | 23 | 1.0 | 1 | 13.3 | 1 | 12.8 | 11 | 0 | 0 | 31.2 | *58* | 8 | 8.8 | 28 | | | | |
| Oct | 14.7 | 6.0 | 10.5 | *−1.0* | 17.0 | 26 | 1.2 | 13 | 11.6 | 19 | 11.1 | 25 | 0 | 0 | 65.4 | *93* | 14 | 15.9 | 22 | | | | |
| Nov | 9.6 | 3.1 | 6.6 | *−1.2* | 15.6 | 1 | −3.4 | 9 | 4.2 | 26 | 8.3 | 3 | 4 | 0 | 49.6 | *74* | 8 | 18.2 | 7 | | | | |
| Dec | 9.8 | 3.6 | 7.0 | *+1.7* | 13.3 | 9 | −1.4 | 26 | 4.7 | 12 | 10.1 | 18 | 2 | 0 | 86.5 | *137* | 16 | 22.5 | 20 | | | | |
| 1833 | 13.7 | 5.2 | 9.5 | *−1.3* | 25.6 | 5 | −5.6 | 1 | 0.0 | 1 | 15.7 | 7 | 45 | 3 | 589.3 | *89* | 109 | 22.5 | 12 | | | | |

## OXFORD 1834

| Month | Mean max | Mean min | Mean temp | Anom | Highest max | Date | Lowest min | Date | Lowest max | Date | Highest min | Date | Air frost | Days ≥ 25 °C | Total pptn | Anom | Rain days | Wettest day | Date | Total sunshine | Anom | Sunniest day | Date |
|---|---|---|---|---|---|---|---|---|---|---|---|---|---|---|---|---|---|---|---|---|---|---|---|
| Jan | 9.5 | 3.7 | 6.9 | *+1.8* | 13.1 | 23 | −2.9 | 2 | 3.5 | 29 | 10.2 | 28 | 5 | 0 | 64.7 | *114* | 12 | 18.3 | 16 | | | | |
| Feb | 8.7 | 1.5 | 5.3 | *+0.2* | 14.2 | 27 | −4.1 | 13 | 2.7 | 9 | 6.9 | 28 | 8 | 0 | 12.3 | *29* | 9 | 2.7 | 8 | | | | |
| Mar | 11.2 | 2.3 | 6.9 | *−0.5* | 15.4 | 4 | −4.0 | 19 | 6.8 | 18 | 8.8 | 5 | 8 | 0 | 10.3 | *22* | 3 | 8.7 | 30 | | | | |
| Apr | 12.7 | 2.9 | 7.9 | *−1.5* | 18.4 | 27 | −2.7 | 14 | 7.8 | 9 | 10.4 | 28 | 5 | 0 | 14.6 | *30* | 5 | 7.2 | 29 | | | | |
| May | 18.8 | 8.0 | 13.3 | *+0.9* | 22.3 | 4 | 3.1 | 11 | 14.9 | 25 | 14.3 | 5 | 0 | 0 | 26.7 | *47* | 8 | 14.8 | 13 | | | | |
| June | 21.4 | 10.4 | 15.7 | *+0.3* | 28.2 | 21 | 5.7 | 12 | 16.7 | 5 | 15.9 | 21 | 0 | 3 | 27.8 | *56* | 11 | 7.9 | 12 | | | | |
| July | 22.2 | 13.5 | 17.5 | *−0.1* | 27.3 | 17 | 10.1 | 27 | 15.7 | 19 | 19.1 | 18 | 0 | 1 | 174.8 | *358* | 13 | 42.4 | 29 | | | | |
| Aug | 21.3 | 11.7 | 16.3 | *−1.1* | 24.7 | 4 | 4.5 | 28 | 17.5 | 25 | 16.2 | 1 | 0 | 0 | 51.3 | *94* | 12 | 12.7 | 24 | | | | |
| Sep | 19.0 | 10.1 | 14.5 | *−0.3* | 23.2 | 17 | 4.0 | 15 | 15.1 | 11 | 14.5 | 5 | 0 | 0 | 81.4 | *150* | 6 | 34.6 | 11 | | | | |
| Oct | 14.5 | 6.1 | 10.5 | *−1.0* | 20.9 | 5 | 0.6 | 26 | 5.7 | 24 | 12.3 | 9 | 0 | 0 | 24.9 | *36* | 5 | 18.4 | 14 | | | | |
| Nov | 9.2 | 3.3 | 6.6 | *−1.3* | 16.0 | 5 | −1.7 | 13 | 4.2 | 21 | 11.2 | 5 | 5 | 0 | 33.1 | *50* | 6 | 10.9 | 9 | | | | |
| Dec | 7.7 | 2.2 | 5.3 | *−0.1* | 13.2 | 7 | −4.3 | 24 | 3.5 | 11 | 9.6 | 31 | 7 | 0 | 20.1 | *32* | 8 | 7.8 | 1 | | | | |
| 1834 | 14.7 | 6.3 | 10.6 | *−0.2* | 28.2 | 6 | −4.3 | 12 | 2.7 | 2 | 19.1 | 7 | 38 | 4 | 542.1 | *82* | 98 | 42.4 | 7 | | | | |

## OXFORD 1835

| Month | Mean max | Mean min | Mean temp | Anom | Highest max | Date | Lowest min | Date | Lowest max | Date | Highest min | Date | Air frost | Days ≥ 25 °C | Total pptn | Anom | Rain days | Wettest day | Date | Total sunshine | Anom | Sunniest day | Date |
|---|---|---|---|---|---|---|---|---|---|---|---|---|---|---|---|---|---|---|---|---|---|---|---|
| Jan | 6.2 | 0.5 | 3.6 | *−1.5* | 11.1 | 12 | −8.2 | 21 | −1.7 | 7 | 7.2 | 16 | 13 | 0 | 19.8 | *35* | 5 | 6.9 | 18 | | | | |
| Feb | 9.1 | 2.2 | 5.9 | *+0.8* | 12.7 | 14 | −4.2 | 11 | 3.5 | 10 | 8.3 | 2 | 6 | 0 | 53.1 | *124* | 12 | 9.9 | 19 | | | | |
| Mar | 9.6 | 1.5 | 5.7 | *−1.7* | 14.1 | 21 | −2.6 | 30 | 3.3 | 1 | 6.8 | 6 | 11 | 0 | 75.3 | *158* | 14 | 14.7 | 11 | | | | |
| Apr | 13.1 | 4.1 | 8.7 | *−0.7* | 19.1 | 2 | −3.8 | 17 | 7.2 | 17 | 9.8 | 3 | 6 | 0 | 38.7 | *79* | 6 | 26.1 | 29 | | | | |
| May | 16.6 | 6.6 | 11.5 | *−0.9* | 21.2 | 19 | 1.6 | 5 | 11.5 | 14 | 13.5 | 24 | 0 | 0 | 70.0 | *122* | 11 | 21.0 | 13 | | | | |
| June | 20.3 | 9.4 | 14.7 | *−0.7* | 27.4 | 11 | 3.1 | 28 | 12.7 | 25 | 15.0 | 17 | 0 | 5 | 59.4 | *121* | 8 | 23.6 | 26 | | | | |
| July | 22.8 | 11.4 | 16.8 | *−0.8* | 27.3 | 28 | 6.3 | 31 | 18.8 | 13 | 19.5 | 26 | 0 | 7 | 17.2 | *35* | 3 | 13.2 | 5 | | | | |
| Aug | 22.6 | 10.9 | 16.6 | *−0.8* | 28.2 | 11 | 5.7 | 9 | 17.6 | 26 | 15.9 | 22 | 0 | 4 | 13.8 | *25* | 3 | 12.4 | 24 | | | | |
| Sep | 18.3 | 8.8 | 13.6 | *−1.3* | 23.2 | 3 | 3.8 | 18 | 12.7 | 28 | 14.0 | 15 | 0 | 0 | 98.7 | *182* | 15 | 18.9 | 18 | | | | |
| Oct | 13.1 | 5.1 | 9.3 | *−2.2* | 16.9 | 6 | −0.7 | 20 | 8.3 | 30 | 10.5 | 2 | 2 | 0 | 134.2 | *192* | 15 | 30.8 | 1 | | | | |
| Nov | 9.3 | 3.5 | 6.7 | *−1.1* | 13.3 | 26 | −5.8 | 12 | 4.8 | 9 | 10.2 | 26 | 5 | 0 | 66.3 | *99* | 10 | 20.2 | 3 | | | | |
| Dec | 4.5 | −0.1 | 2.6 | *−2.8* | 12.2 | 1 | −9.6 | 25 | −3.7 | 25 | 8.3 | 3 | 12 | 0 | 3.9 | *6* | 5 | 1.1 | 1 | | | | |
| 1835 | 13.8 | 5.3 | 9.6 | *−1.2* | 28.2 | 8 | −9.6 | 12 | −3.7 | 12 | 19.5 | 7 | 55 | 16 | 650.2 | *98* | 107 | 30.8 | 10 | | | | |

## OXFORD 1836

| Month | Mean max | Mean min | Mean temp | Anom | Highest max | Date | Lowest min | Date | Lowest max | Date | Highest min | Date | Air frost | Days ≥ 25 °C | Total pptn | Anom | Rain days | Wettest day | Date | Total sunshine | Anom | Sunniest day | Date |
|---|---|---|---|---|---|---|---|---|---|---|---|---|---|---|---|---|---|---|---|---|---|---|---|
| Jan | 5.7 | 0.6 | 3.4 | *−1.7* | 11.3 | 23 | −6.4 | 13 | −1.4 | 2 | 8.0 | 5 | 13 | 0 | 35.3 | *62* | 11 | 9.7 | 15 | | | | |
| Feb | 5.6 | 0.3 | 3.2 | *−1.9* | 9.8 | 9 | −6.0 | 20 | 1.2 | 26 | 7.1 | 10 | 12 | 0 | 28.5 | *67* | 11 | 12.0 | 2 | | | | |
| Mar | 9.3 | 3.0 | 6.3 | *−1.1* | 17.2 | 19 | −2.3 | 27 | 4.4 | 28 | 8.6 | 18 | 3 | 0 | 62.3 | *131* | 15 | 15.7 | 15 | | | | |
| Apr | 10.6 | 3.6 | 7.2 | *−2.2* | 14.4 | 16 | −3.3 | 9 | 5.6 | 2 | 9.7 | 18 | 4 | 0 | 55.5 | *114* | 9 | 15.5 | 6 | | | | |
| May | 15.6 | 5.5 | 10.5 | *−1.9* | 21.7 | 18 | 0.3 | 1 | 8.9 | 2 | 9.8 | 19 | 0 | 0 | 14.7 | *26* | 3 | 10.4 | 5 | | | | |
| June | 20.3 | 11.4 | 15.6 | *+0.2* | 25.8 | 15 | 7.1 | 2 | 14.8 | 5 | 14.7 | 23 | 0 | 1 | 51.0 | *103* | 13 | 11.3 | 24 | | | | |
| July | 21.0 | 12.3 | 16.4 | *−1.2* | 29.8 | 5 | 7.2 | 21 | 13.7 | 20 | 17.2 | 5 | 0 | 5 | 58.9 | *121* | 9 | 11.4 | 6 | | | | |
| Aug | 20.1 | 10.6 | 15.2 | *−2.2* | 24.3 | 14 | 5.3 | 25 | 16.6 | 24 | 14.9 | 17 | 0 | 0 | 58.9 | *107* | 7 | 23.7 | 24 | | | | |
| Sep | 15.7 | 8.8 | 12.2 | *−2.6* | 20.5 | 26 | 4.3 | 22 | 11.1 | 30 | 14.3 | 26 | 0 | 0 | 53.0 | *98* | 11 | 11.3 | 4 | | | | |
| Oct | 12.0 | 5.6 | 8.9 | *−2.5* | 16.6 | 17 | −4.1 | 31 | 2.3 | 29 | 12.3 | 18 | 3 | 0 | 73.1 | *104* | 12 | 13.9 | 1 | | | | |
| Nov | 8.2 | 2.9 | 5.8 | *−2.0* | 13.3 | 13 | −3.1 | 8 | 2.2 | 25 | 12.2 | 13 | 6 | 0 | 85.8 | *129* | 16 | 18.1 | 28 | | | | |
| Dec | 6.0 | 2.0 | 4.3 | *−1.0* | 12.1 | 4 | −5.5 | 25 | −1.7 | 25 | 9.5 | 4 | 9 | 0 | 27.0 | *43* | 7 | 9.9 | 7 | | | | |
| 1836 | 12.5 | 5.5 | 9.0 | *−1.7* | 29.8 | 7 | −6.4 | 1 | −1.7 | 12 | 17.2 | 7 | 50 | 6 | 604.1 | *91* | 124 | 23.7 | 8 | | | | |

## OXFORD 1837

| Month | Mean max | Mean min | Mean temp | Anom | Highest max | Date | Lowest min | Date | Lowest max | Date | Highest min | Date | Air frost | Days ≥ 25 °C | Total pptn | Anom | Rain days | Wettest day | Date | Total sunshine | Anom | Sunniest day | Date |
|---|---|---|---|---|---|---|---|---|---|---|---|---|---|---|---|---|---|---|---|---|---|---|---|
| Jan | 4.7 | 0.5 | 2.9 | *−2.2* | 10.9 | 22 | −7.5 | 2 | 0.0 | 11 | 6.8 | 23 | 13 | 0 | 65.9 | *116* | 13 | 17.8 | 26 | | | | |
| Feb | 7.4 | 2.1 | 5.0 | *−0.2* | 12.2 | 16 | −3.2 | 26 | 3.3 | 4 | 7.4 | 11 | 4 | 0 | 61.2 | *143* | 16 | 15.1 | 11 | | | | |
| Mar | 5.2 | −1.2 | 2.2 | *−5.3* | 9.1 | 9 | −7.1 | 24 | 0.5 | 24 | 2.9 | 10 | 20 | 0 | 13.0 | *27* | 6 | 7.2 | 22 | | | | |
| Apr | 8.4 | 0.8 | 4.6 | *−4.7* | 14.1 | 26 | −4.1 | 12 | 2.8 | 16 | 8.5 | 30 | 15 | 0 | 33.2 | *68* | 13 | 9.4 | 25 | | | | |
| May | 14.3 | 5.1 | 9.6 | *−2.8* | 19.4 | 29 | 0.2 | 10 | 8.8 | 10 | 10.9 | 29 | 0 | 0 | 15.3 | *27* | 9 | 5.9 | 8 | | | | |
| June | 20.2 | 10.2 | 15.0 | *−0.4* | 24.4 | 16 | 2.9 | 8 | 14.4 | 7 | 14.8 | 14 | 0 | 0 | 36.2 | *74* | 10 | 6.6 | 18 | | | | |
| July | 21.9 | 12.1 | 16.8 | *−0.8* | 27.3 | 27 | 4.4 | 2 | 18.1 | 1 | 16.1 | 26 | 0 | 2 | 31.2 | *64* | 8 | 20.7 | 29 | | | | |
| Aug | 20.4 | 11.8 | 15.9 | *−1.5* | 26.0 | 17 | 5.6 | 27 | 14.8 | 27 | 17.7 | 20 | 0 | 2 | 84.9 | *155* | 11 | 19.3 | 2 | | | | |
| Sep | 16.8 | 9.0 | 12.9 | *−1.9* | 20.7 | 20 | 3.2 | 6 | 13.3 | 3 | 16.1 | 18 | 0 | 0 | 63.8 | *118* | 9 | 24.2 | 9 | | | | |
| Oct | 14.0 | 6.8 | 10.6 | *−0.9* | 20.2 | 3 | 0.0 | 15 | 8.2 | 25 | 15.1 | 4 | 0 | 0 | 36.0 | *51* | 10 | 12.3 | 6 | | | | |
| Nov | 8.0 | 1.7 | 5.2 | *−2.7* | 12.9 | 1 | −5.1 | 18 | 3.1 | 17 | 10.1 | 23 | 9 | 0 | 48.3 | *72* | 14 | 18.1 | 1 | | | | |
| Dec | 6.7 | 3.1 | 5.2 | *−0.1* | 12.3 | 19 | −4.9 | 14 | 0.6 | 4 | 10.2 | 25 | 8 | 0 | 40.3 | *64* | 14 | 8.5 | 19 | | | | |
| 1837 | 12.3 | 5.2 | 8.8 | *−2.0* | 27.3 | 7 | −7.5 | 1 | 0.0 | 1 | 17.7 | 8 | 69 | 4 | 529.4 | *80* | 133 | 24.2 | 9 | | | | |

## OXFORD 1838

| Month | Mean max | Mean min | Mean temp | Anom | Highest max | Date | Lowest min | Date | Lowest max | Date | Highest min | Date | Air frost | Days ≥ 25 °C | Total pptn | Anom | Rain days | Wettest day | Date | Total sunshine | Anom | Sunniest day | Date |
|---|---|---|---|---|---|---|---|---|---|---|---|---|---|---|---|---|---|---|---|---|---|---|---|
| Jan | −0.2 | −4.6 | −2.1 | *−7.2* | 8.8 | 1 | −16.4 | 20 | −7.4 | 20 | 4.9 | 1 | 25 | 0 | 7.0 | *12* | 4 | 3.6 | 2 | | | | |
| Feb | 2.4 | −1.8 | 0.5 | *−4.6* | 8.4 | 25 | −7.8 | 13 | −0.2 | 15 | 3.8 | 8 | 19 | 0 | 67.2 | *157* | 10 | 15.6 | 27 | | | | |
| Mar | 8.5 | 2.1 | 5.5 | *−2.0* | 15.1 | 29 | −2.4 | 23 | 3.3 | 23 | 8.2 | 14 | 5 | 0 | 37.1 | *78* | 9 | 18.2 | 2 | | | | |
| Apr | 10.2 | 3.4 | 6.8 | *−2.5* | 16.4 | 11 | −3.3 | 2 | 4.3 | 1 | 9.2 | 7 | 5 | 0 | 18.0 | *37* | 7 | 6.3 | 7 | | | | |
| May | 16.4 | 5.4 | 10.8 | *−1.6* | 22.4 | 31 | −2.2 | 13 | 11.7 | 14 | 11.3 | 3 | 2 | 0 | 23.6 | *41* | 7 | 6.5 | 28 | | | | |
| June | 19.6 | 10.2 | 14.7 | *−0.7* | 23.6 | 17 | 3.9 | 9 | 16.2 | 11 | 15.1 | 18 | 0 | 0 | 72.0 | *146* | 16 | 17.5 | 26 | | | | |
| July | 19.9 | 11.2 | 15.3 | *−2.3* | 25.0 | 13 | 4.3 | 17 | 15.2 | 22 | 16.5 | 12 | 0 | 1 | 39.5 | *81* | 11 | 17.6 | 6 | | | | |
| Aug | 19.2 | 11.0 | 14.9 | *−2.5* | 23.8 | 27 | 3.8 | 30 | 15.8 | 29 | 15.3 | 12 | 0 | 0 | 36.5 | *67* | 12 | 10.5 | 1 | | | | |
| Sep | 16.2 | 8.6 | 12.4 | *−2.4* | 20.5 | 5 | 1.1 | 11 | 10.2 | 24 | 14.1 | 15 | 0 | 0 | 96.7 | *178* | 8 | 22.9 | 21 | | | | |
| Oct | 12.9 | 7.3 | 10.3 | *−1.1* | 16.3 | 20 | −0.9 | 14 | 5.9 | 13 | 12.5 | 22 | 1 | 0 | 53.3 | *76* | 8 | 22.5 | 27 | | | | |
| Nov | 6.9 | 1.9 | 4.7 | *−3.1* | 14.2 | 7 | −3.9 | 25 | 1.1 | 26 | 7.1 | 30 | 8 | 0 | 73.7 | *110* | 12 | 18.3 | 19 | | | | |
| Dec | 5.7 | 1.7 | 4.0 | *−1.3* | 12.1 | 1 | −4.1 | 26 | 1.2 | 9 | 9.0 | 2 | 10 | 0 | 51.4 | *82* | 9 | 11.7 | 23 | | | | |
| 1838 | 11.5 | 4.7 | 8.1 | *−2.6* | 25.0 | 7 | −16.4 | 1 | −7.4 | 1 | 16.5 | 7 | 75 | 1 | 576.0 | *87* | 113 | 22.9 | 9 | | | | |

## OXFORD 1839

| Month | Mean max | Mean min | Mean temp | Anom | Highest max | Date | Lowest min | Date | Lowest max | Date | Highest min | Date | Air frost | Days ≥ 25 °C | Total pptn | Anom | Rain days | Wettest day | Date | Total sunshine | Anom | Sunniest day | Date |
|---|---|---|---|---|---|---|---|---|---|---|---|---|---|---|---|---|---|---|---|---|---|---|---|
| Jan | 5.4 | 1.1 | 3.5 | *−1.6* | 10.5 | 6 | −4.9 | 30 | 1.1 | 9 | 8.2 | 21 | 13 | 0 | 30.5 | *54* | 10 | 9.2 | 21 | | | | |
| Feb | 6.6 | 1.0 | 4.0 | *−1.1* | 11.8 | 9 | −6.9 | 19 | 1.8 | 18 | 9.1 | 9 | 10 | 0 | 36.8 | *86* | 11 | 9.8 | 22 | | | | |
| Mar | 7.0 | 1.6 | 4.4 | *−3.0* | 12.9 | 27 | −5.6 | 9 | 0.7 | 6 | 7.1 | 15 | 10 | 0 | 45.4 | *95* | 10 | 21.5 | 15 | | | | |
| Apr | 9.3 | 2.7 | 6.0 | *−3.3* | 18.5 | 30 | −3.8 | 7 | 1.0 | 3 | 8.9 | 23 | 8 | 0 | 42.4 | *87* | 6 | 10.0 | 18 | | | | |
| May | 15.0 | 4.8 | 9.8 | *−2.6* | 22.2 | 31 | −1.9 | 16 | 7.1 | 14 | 10.3 | 21 | 2 | 0 | 26.8 | *47* | 4 | 20.3 | 8 | | | | |
| June | 18.8 | 10.0 | 14.2 | *−1.2* | 25.8 | 18 | 5.3 | 2 | 11.6 | 14 | 15.2 | 21 | 0 | 2 | 97.4 | *198* | 16 | 18.3 | 14 | | | | |
| July | 19.7 | 11.3 | 15.2 | *−2.4* | 23.9 | 4 | 5.8 | 1 | 15.4 | 19 | 16.1 | 18 | 0 | 0 | 110.3 | *226* | 15 | 19.1 | 30 | | | | |
| Aug | 19.0 | 10.8 | 14.7 | *−2.7* | 22.8 | 2 | 4.6 | 21 | 13.6 | 20 | 14.4 | 30 | 0 | 0 | 65.5 | *120* | 11 | 20.4 | 17 | | | | |
| Sep | 16.6 | 9.5 | 13.1 | *−1.8* | 20.8 | 9 | 4.8 | 23 | 13.6 | 22 | 15.1 | 11 | 0 | 0 | 75.5 | *139* | 16 | 33.5 | 14 | | | | |
| Oct | 12.5 | 5.9 | 9.4 | *−2.1* | 18.3 | 11 | 1.4 | 19 | 3.3 | 31 | 10.8 | 22 | 0 | 0 | 86.8 | *124* | 7 | 24.4 | 3 | | | | |
| Nov | 9.2 | 4.8 | 7.3 | *−0.6* | 12.7 | 17 | −2.2 | 26 | 2.1 | 27 | 10.2 | 17 | 3 | 0 | 138.4 | *207* | 20 | 20.2 | 2 | | | | |
| Dec | 5.8 | 1.9 | 4.2 | *−1.2* | 12.6 | 20 | −3.8 | 30 | 1.4 | 6 | 9.5 | 24 | 12 | 0 | 80.4 | *127* | 13 | 15.1 | 19 | | | | |
| **1839** | 12.1 | 5.4 | 8.8 | *−2.0* | 25.8 | 6 | −6.9 | 2 | 0.7 | 3 | 16.1 | 7 | 58 | 2 | 836.3 | *127* | 139 | 33.5 | 9 | | | | |

## OXFORD 1840

| Month | Mean max | Mean min | Mean temp | Anom | Highest max | Date | Lowest min | Date | Lowest max | Date | Highest min | Date | Air frost | Days ≥ 25 °C | Total pptn | Anom | Rain days | Wettest day | Date | Total sunshine | Anom | Sunniest day | Date |
|---|---|---|---|---|---|---|---|---|---|---|---|---|---|---|---|---|---|---|---|---|---|---|---|
| Jan | 6.3 | 1.1 | 4.0 | *−1.1* | 11.8 | 19 | −8.7 | 8 | −1.9 | 7 | 8.0 | 24 | 10 | 0 | 64.1 | *113* | 15 | 11.3 | 24 | | | | |
| Feb | 6.1 | 0.5 | 3.5 | *−1.6* | 11.9 | 7 | −6.2 | 23 | 0.5 | 21 | 5.2 | 12 | 13 | 0 | 36.5 | *85* | 9 | 8.9 | 16 | | | | |
| Mar | 7.1 | −0.1 | 3.7 | *−3.7* | 11.3 | 10 | −5.4 | 25 | 2.1 | 1 | 5.7 | 13 | 17 | 0 | 8.6 | *18* | 2 | 7.5 | 31 | | | | |
| Apr | 14.8 | 4.2 | 9.6 | *+0.2* | 23.6 | 28 | −0.4 | 8 | 6.6 | 8 | 9.4 | 23 | 2 | 0 | 16.0 | *33* | 3 | 9.2 | 1 | | | | |
| May | 17.3 | 8.2 | 12.6 | *+0.3* | 28.4 | 31 | 2.9 | 19 | 10.0 | 19 | 12.9 | 24 | 0 | 1 | 58.0 | *101* | 9 | 19.5 | 11 | | | | |
| June | 21.6 | 10.7 | 16.0 | *+0.6* | 29.7 | 1 | 6.8 | 2 | 16.1 | 5 | 15.2 | 11 | 0 | 4 | 25.6 | *52* | 8 | 8.2 | 6 | | | | |
| July | 19.7 | 10.8 | 15.0 | *−2.6* | 26.9 | 15 | 5.0 | 13 | 16.4 | 3 | 15.0 | 29 | 0 | 3 | 35.1 | *72* | 13 | 8.9 | 26 | | | | |
| Aug | 21.5 | 12.1 | 16.6 | *−0.8* | 26.6 | 6 | 8.6 | 24 | 14.7 | 17 | 15.0 | 6 | 0 | 4 | 49.1 | *90* | 7 | 21.9 | 17 | | | | |
| Sep | 15.4 | 7.1 | 11.2 | *−3.6* | 23.9 | 1 | 0.8 | 17 | 10.4 | 22 | 15.0 | 1 | 0 | 0 | 75.4 | *139* | 12 | 16.4 | 24 | | | | |
| Oct | 11.6 | 3.2 | 7.5 | *−3.9* | 14.6 | 16 | −1.7 | 8 | 8.3 | 28 | 9.8 | 18 | 6 | 0 | 24.7 | *35* | 7 | 6.1 | 26 | | | | |
| Nov | 8.6 | 2.3 | 5.7 | *−2.1* | 14.5 | 16 | −6.1 | 27 | 2.5 | 27 | 7.8 | 1 | 8 | 0 | 91.4 | *137* | 11 | 21.2 | 6 | | | | |
| Dec | 2.1 | −2.5 | 0.2 | *−5.2* | 11.2 | 1 | −7.6 | 17 | −2.2 | 15 | 3.1 | 1 | 23 | 0 | 19.1 | *30* | 2 | 13.6 | 8 | | | | |
| **1840** | 12.7 | 4.8 | 8.7 | *−2.0* | 29.7 | 6 | −8.7 | 1 | −2.2 | 12 | 15.2 | 6 | 79 | 12 | 503.8 | *76* | 98 | 21.9 | 8 | | | | |

## OXFORD 1841

| Month | Mean max | Mean min | Mean temp | Anom | Highest max | Date | Lowest min | Date | Lowest max | Date | Highest min | Date | Air frost | Days ≥ 25 °C | Total pptn | Anom | Rain days | Wettest day | Date | Total sunshine | Anom | Sunniest day | Date |
|---|---|---|---|---|---|---|---|---|---|---|---|---|---|---|---|---|---|---|---|---|---|---|---|
| Jan | 3.0 | −2.0 | 0.8 | *−4.4* | 10.7 | 17 | −14.6 | 8 | −9.6 | 8 | 5.9 | 17 | 18 | 0 | 63.2 | *111* | 9 | 17.6 | 14 | | | | |
| Feb | 3.6 | −0.7 | 1.6 | *−3.5* | 10.6 | 14 | −10.7 | 3 | −5.6 | 3 | 5.6 | 13 | 13 | 0 | 26.3 | *62* | 8 | 7.5 | 15 | | | | |
| Mar | 11.8 | 3.3 | 7.7 | *+0.3* | 15.4 | 16 | −1.1 | 1 | 3.9 | 1 | 8.7 | 21 | 4 | 0 | 51.2 | *107* | 12 | 10.8 | 2 | | | | |
| Apr | 12.3 | 3.7 | 8.0 | *−1.3* | 19.9 | 27 | −0.1 | 15 | 8.3 | 11 | 10.9 | 25 | 1 | 0 | 53.6 | *110* | 13 | 19.9 | 24 | | | | |
| May | 19.2 | 8.9 | 13.9 | *+1.6* | 30.0 | 27 | 5.0 | 13 | 11.9 | 3 | 15.0 | 27 | 0 | 5 | 45.0 | *79* | 13 | 16.2 | 7 | | | | |
| June | 19.8 | 9.0 | 14.2 | *−1.2* | 26.7 | 1 | 4.0 | 15 | 12.4 | 7 | 12.6 | 24 | 0 | 1 | 71.2 | *145* | 9 | 22.8 | 25 | | | | |
| July | 19.5 | 10.7 | 14.8 | *−2.8* | 26.1 | 4 | 7.7 | 13 | 15.3 | 31 | 14.8 | 2 | 0 | 3 | 66.8 | *137* | 10 | 14.9 | 10 | | | | |
| Aug | 19.4 | 11.2 | 15.1 | *−2.3* | 23.1 | 20 | 4.6 | 31 | 16.4 | 24 | 15.0 | 26 | 0 | 0 | 100.1 | *183* | 13 | 17.3 | 4 | | | | |
| Sep | 17.9 | 10.5 | 14.2 | *−0.7* | 25.0 | 13 | 2.2 | 6 | 11.1 | 4 | 16.0 | 13 | 0 | 1 | 137.9 | *255* | 16 | 29.7 | 30 | | | | |
| Oct | 12.0 | 6.4 | 9.4 | *−2.1* | 16.7 | 2 | 0.0 | 21 | 7.8 | 21 | 10.9 | 14 | 0 | 0 | 92.7 | *132* | 20 | 17.9 | 19 | | | | |
| Nov | 8.0 | 2.2 | 5.4 | *−2.4* | 13.3 | 29 | −8.2 | 16 | −0.9 | 16 | 9.7 | 21 | 9 | 0 | 94.6 | *142* | 16 | 22.7 | 29 | | | | |
| Dec | 6.2 | 2.2 | 4.5 | *−0.9* | 10.4 | 12 | −5.4 | 18 | −1.7 | 18 | 8.4 | 12 | 8 | 0 | 53.1 | *84* | 20 | 9.0 | 6 | | | | |
| **1841** | 12.7 | 5.4 | 9.1 | *−1.7* | 30.0 | 5 | −14.6 | 1 | −9.6 | 1 | 16.0 | 9 | 53 | 10 | 856.0 | *130* | 159 | 29.7 | 9 | | | | |

## OXFORD 1842

| Month | Mean max | Mean min | Mean temp | Anom | Highest max | Date | Lowest min | Date | Lowest max | Date | Highest min | Date | Air frost | Days ≥ 25 °C | Total pptn | Anom | Rain days | Wettest day | Date | Total sunshine | Anom | Sunniest day | Date |
|---|---|---|---|---|---|---|---|---|---|---|---|---|---|---|---|---|---|---|---|---|---|---|---|
| Jan | 1.7 | −1.9 | 0.2 | *−5.0* | 7.2 | 31 | −5.7 | 23 | −3.3 | 9 | 2.7 | 26 | 27 | 0 | 41.3 | *73* | 8 | 16.5 | 27 | | | | |
| Feb | 7.0 | 1.4 | 4.4 | *−0.7* | 11.4 | 12 | −2.9 | 6 | −0.4 | 6 | 8.4 | 11 | 9 | 0 | 44.6 | *104* | 10 | 11.5 | 24 | | | | |
| Mar | 10.0 | 3.7 | 7.0 | *−0.4* | 14.0 | 30 | −1.1 | 5 | 5.6 | 23 | 8.6 | 15 | 3 | 0 | 44.0 | *92* | 14 | 17.7 | 10 | | | | |
| Apr | 12.4 | 2.5 | 7.5 | *−1.9* | 22.2 | 30 | −2.7 | 4 | 5.4 | 13 | 7.4 | 24 | 6 | 0 | 7.4 | *15* | 6 | 2.5 | 26 | | | | |
| May | 16.6 | 6.4 | 11.4 | *−0.9* | 20.4 | 31 | 1.0 | 9 | 8.7 | 9 | 10.5 | 27 | 0 | 0 | 37.8 | *66* | 13 | 7.5 | 26 | | | | |
| June | 21.8 | 11.3 | 16.4 | *+1.0* | 27.3 | 12 | 6.1 | 2 | 16.3 | 30 | 15.0 | 13 | 0 | 4 | 42.9 | *87* | 7 | 22.0 | 22 | | | | |
| July | 20.1 | 10.6 | 15.1 | *−2.5* | 23.9 | 18 | 5.5 | 22 | 15.6 | 7 | 13.9 | 4 | 0 | 0 | 58.9 | *121* | 11 | 32.3 | 1 | | | | |
| Aug | 23.1 | 13.1 | 18.0 | *+0.5* | 28.4 | 18 | 7.3 | 30 | 17.6 | 25 | 16.7 | 18 | 0 | 11 | 31.1 | *57* | 7 | 17.6 | 11 | | | | |
| Sep | 17.7 | 9.7 | 13.7 | *−1.2* | 24.3 | 2 | 4.1 | 21 | 11.8 | 27 | 15.1 | 2 | 0 | 0 | 67.7 | *125* | 12 | 22.0 | 2 | | | | |
| Oct | 11.8 | 3.0 | 7.6 | *−3.9* | 16.2 | 7 | −4.9 | 20 | 6.8 | 26 | 9.5 | 8 | 9 | 0 | 14.6 | *21* | 5 | 6.3 | 23 | | | | |
| Nov | 8.7 | 3.1 | 6.2 | *−1.7* | 12.1 | 11 | −1.1 | 4 | 3.4 | 22 | 7.7 | 12 | 3 | 0 | 155.5 | *233* | 15 | 27.0 | 14 | | | | |
| Dec | 9.2 | 4.1 | 7.0 | *+1.6* | 13.7 | 13 | −1.6 | 24 | 4.0 | 10 | 9.6 | 30 | 4 | 0 | 46.5 | *74* | 9 | 25.8 | 27 | | | | |
| **1842** | 13.4 | 5.6 | 9.5 | *−1.3* | 28.4 | 8 | −5.7 | 1 | −3.3 | 1 | 16.7 | 8 | 61 | 15 | 592.3 | *90* | 117 | 32.3 | 7 | | | | |

## OXFORD 1843

| Month | Mean max | Mean min | Mean temp | Anom | Highest max | Date | Lowest min | Date | Lowest max | Date | Highest min | Date | Air frost | Days ≥ 25 °C | Total pptn | Anom | Rain days | Wettest day | Date | Total sunshine | Anom | Sunniest day | Date |
|---|---|---|---|---|---|---|---|---|---|---|---|---|---|---|---|---|---|---|---|---|---|---|---|
| Jan | 6.7 | 1.9 | 4.6 | *−0.6* | 12.6 | 28 | −6.1 | 2 | 0.5 | 3 | 9.4 | 27 | 11 | 0 | 31.7 | *56* | 10 | 11.6 | 13 | | | | |
| Feb | 4.2 | −0.6 | 2.0 | *−3.1* | 10.7 | 1 | −7.8 | 15 | −1.1 | 15 | 7.2 | 1 | 12 | 0 | 41.0 | *96* | 12 | 8.9 | 20 | | | | |
| Mar | 9.9 | 2.4 | 6.3 | *−1.1* | 16.4 | 20 | −5.1 | 4 | 4.0 | 3 | 9.7 | 22 | 9 | 0 | 22.7 | *48* | 6 | 10.8 | 22 | | | | |
| Apr | 13.3 | 4.8 | 9.1 | *−0.3* | 18.8 | 18 | −3.3 | 11 | 6.6 | 12 | 10.3 | 6 | 4 | 0 | 51.5 | *105* | 12 | 11.1 | 2 | | | | |
| May | 15.0 | 7.1 | 11.0 | *−1.4* | 19.5 | 31 | 1.4 | 7 | 9.0 | 8 | 12.8 | 31 | 0 | 0 | 119.6 | *209* | 20 | 30.5 | 24 | | | | |
| June | 17.5 | 9.4 | 13.3 | *−2.2* | 21.4 | 16 | 5.1 | 20 | 13.6 | 12 | 13.6 | 1 | 0 | 0 | 62.6 | *127* | 13 | 11.4 | 7 | | | | |
| July | 20.4 | 11.7 | 15.8 | *−1.8* | 27.3 | 5 | 6.7 | 19 | 14.5 | 11 | 15.1 | 16 | 0 | 1 | 55.0 | *113* | 15 | 13.0 | 9 | | | | |
| Aug | 20.9 | 12.5 | 16.5 | *−0.9* | 27.2 | 19 | 7.2 | 26 | 14.7 | 22 | 18.3 | 17 | 0 | 3 | 80.5 | *147* | 10 | 17.4 | 24 | | | | |
| Sep | 19.9 | 10.5 | 15.2 | *+0.4* | 24.0 | 17 | 2.2 | 26 | 10.9 | 27 | 15.0 | 18 | 0 | 0 | 27.3 | *50* | 5 | 13.3 | 20 | | | | |
| Oct | 12.7 | 4.9 | 9.0 | *−2.5* | 20.4 | 5 | −3.9 | 18 | 6.5 | 31 | 13.4 | 4 | 5 | 0 | 96.0 | *137* | 15 | 20.3 | 11 | | | | |
| Nov | 9.2 | 3.0 | 6.4 | *−1.5* | 13.2 | 26 | −2.8 | 12 | 3.9 | 15 | 10.6 | 26 | 8 | 0 | 52.1 | *78* | 16 | 9.2 | 7 | | | | |
| Dec | 9.2 | 4.7 | 7.3 | *+1.9* | 12.8 | 23 | −1.7 | 1 | 3.5 | 12 | 10.1 | 23 | 3 | 0 | 4.7 | *7* | 6 | 2.4 | 10 | | | | |
| 1843 | 13.2 | 6.0 | 9.6 | *−1.1* | 27.3 | 7 | −7.8 | 2 | −1.1 | 2 | 18.3 | 8 | 52 | 4 | 644.7 | *98* | 140 | 30.5 | 5 | | | | |

## OXFORD 1844

| Month | Mean max | Mean min | Mean temp | Anom | Highest max | Date | Lowest min | Date | Lowest max | Date | Highest min | Date | Air frost | Days ≥ 25 °C | Total pptn | Anom | Rain days | Wettest day | Date | Total sunshine | Anom | Sunniest day | Date |
|---|---|---|---|---|---|---|---|---|---|---|---|---|---|---|---|---|---|---|---|---|---|---|---|
| Jan | 6.7 | 0.8 | 4.0 | *−1.1* | 11.6 | 4 | −12.1 | 2 | 0.6 | 2 | 7.2 | 5 | 9 | 0 | 40.6 | *71* | 11 | 10.1 | 4 | | | | |
| Feb | 4.8 | −1.5 | 1.8 | *−3.3* | 9.8 | 25 | −9.4 | 5 | 0.6 | 12 | 4.2 | 29 | 22 | 0 | 76.4 | *179* | 15 | 15.8 | 24 | | | | |
| Mar | 9.2 | 1.6 | 5.6 | *−1.9* | 14.7 | 27 | −4.8 | 5 | 4.2 | 4 | 7.4 | 26 | 6 | 0 | 46.0 | *97* | 14 | 14.7 | 11 | | | | |
| Apr | 16.3 | 4.5 | 10.5 | *+1.1* | 21.0 | 26 | −0.9 | 7 | 12.4 | 5 | 10.1 | 20 | 1 | 0 | 6.0 | *12* | 3 | 4.0 | 13 | | | | |
| May | 16.4 | 6.3 | 11.3 | *−1.1* | 22.1 | 14 | 0.1 | 17 | 9.4 | 17 | 11.3 | 11 | 0 | 0 | 6.7 | *12* | 3 | 2.9 | 22 | | | | |
| June | 21.1 | 10.8 | 15.8 | *+0.4* | 28.5 | 23 | 4.4 | 2 | 14.4 | 2 | 17.2 | 23 | 0 | 3 | 17.9 | *36* | 5 | 6.6 | 19 | | | | |
| July | 20.9 | 11.9 | 16.1 | *−1.5* | 28.4 | 25 | 7.2 | 16 | 15.3 | 19 | 17.2 | 23 | 0 | 3 | 59.5 | *122* | 10 | 18.0 | 14 | | | | |
| Aug | 18.7 | 9.6 | 14.0 | *−3.4* | 21.5 | 5 | 5.0 | 28 | 13.0 | 14 | 14.2 | 16 | 0 | 0 | 47.2 | *86* | 9 | 13.0 | 6 | | | | |
| Sep | 18.1 | 9.5 | 13.8 | *−1.1* | 23.4 | 2 | 0.1 | 29 | 13.1 | 18 | 15.3 | 16 | 0 | 0 | 41.4 | *76* | 10 | 23.1 | 18 | | | | |
| Oct | 13.2 | 5.9 | 9.7 | *−1.7* | 18.6 | 3 | −1.6 | 22 | 8.9 | 30 | 12.7 | 1 | 2 | 0 | 75.1 | *107* | 13 | 18.8 | 16 | | | | |
| Nov | 8.6 | 3.5 | 6.4 | *−1.5* | 14.6 | 16 | −3.3 | 25 | 3.3 | 26 | 10.1 | 16 | 7 | 0 | 67.0 | *100* | 13 | 16.2 | 13 | | | | |
| Dec | 2.4 | −1.6 | 0.7 | *−4.6* | 8.7 | 29 | −6.7 | 6 | −2.8 | 13 | 4.3 | 28 | 21 | 0 | 9.9 | *16* | 4 | 5.5 | 30 | | | | |
| 1844 | 13.0 | 5.1 | 9.1 | *−1.7* | 28.5 | 6 | −12.1 | 1 | −2.8 | 12 | 17.2 | 6 | 68 | 6 | 493.5 | *75* | 110 | 23.1 | 9 | | | | |

## OXFORD 1845

| Month | Mean max | Mean min | Mean temp | Anom | Highest max | Date | Lowest min | Date | Lowest max | Date | Highest min | Date | Air frost | Days ≥ 25 °C | Total pptn | Anom | Rain days | Wettest day | Date | Total sunshine | Anom | Sunniest day | Date |
|---|---|---|---|---|---|---|---|---|---|---|---|---|---|---|---|---|---|---|---|---|---|---|---|
| Jan | 6.1 | 1.0 | 3.8 | *−1.3* | 10.2 | 11 | −6.1 | 29 | 0.8 | 31 | 6.3 | 5 | 11 | 0 | 36.9 | *65* | 12 | 8.9 | 19 | | | | |
| Feb | 3.6 | −3.1 | 0.5 | *−4.6* | 8.6 | 26 | −12.7 | 11 | −2.2 | 12 | 2.2 | 27 | 22 | 0 | 28.2 | *66* | 5 | 10.8 | 14 | | | | |
| Mar | 5.6 | −1.5 | 2.3 | *−5.2* | 13.9 | 27 | −12.0 | 13 | −4.1 | 13 | 8.4 | 27 | 17 | 0 | 31.0 | *65* | 7 | 13.1 | 24 | | | | |
| Apr | 13.8 | 3.6 | 8.8 | *−0.6* | 19.9 | 23 | −2.4 | 6 | 6.3 | 11 | 11.1 | 25 | 3 | 0 | 26.5 | *54* | 8 | 14.3 | 26 | | | | |
| May | 14.1 | 6.1 | 10.0 | *−2.4* | 18.3 | 31 | −0.1 | 5 | 9.2 | 24 | 11.1 | 15 | 1 | 0 | 48.1 | *84* | 15 | 12.7 | 25 | | | | |
| June | 20.7 | 11.3 | 15.8 | *+0.4* | 27.8 | 13 | 5.4 | 28 | 14.7 | 8 | 16.2 | 13 | 0 | 5 | 68.6 | *139* | 13 | 13.6 | 18 | | | | |
| July | 19.7 | 11.4 | 15.3 | *−2.3* | 26.8 | 6 | 6.7 | 11 | 14.0 | 30 | 15.1 | 6 | 0 | 2 | 77.6 | *159* | 16 | 15.0 | 31 | | | | |
| Aug | 18.6 | 10.4 | 14.3 | *−3.1* | 22.2 | 31 | 6.3 | 21 | 12.8 | 19 | 13.4 | 4 | 0 | 0 | 73.0 | *133* | 16 | 25.1 | 19 | | | | |
| Sep | 16.2 | 7.2 | 11.7 | *−3.2* | 20.8 | 12 | −1.2 | 23 | 11.7 | 23 | 13.1 | 17 | 1 | 0 | 59.5 | *110* | 10 | 20.7 | 22 | | | | |
| Oct | 13.7 | 5.7 | 9.9 | *−1.6* | 17.9 | 3 | −2.8 | 25 | 9.3 | 26 | 14.4 | 2 | 3 | 0 | 34.1 | *49* | 7 | 10.4 | 7 | | | | |
| Nov | 10.3 | 3.7 | 7.3 | *−0.6* | 18.5 | 7 | −5.3 | 3 | 4.4 | 24 | 10.6 | 7 | 7 | 0 | 51.6 | *77* | 13 | 13.7 | 30 | | | | |
| Dec | 7.3 | 1.2 | 4.6 | *−0.8* | 11.6 | 30 | −3.7 | 12 | 1.2 | 13 | 6.1 | 16 | 8 | 0 | 40.8 | *65* | 15 | 10.4 | 5 | | | | |
| **1845** | 12.5 | 4.8 | 8.6 | *−2.1* | 27.8 | 6 | −12.7 | 2 | −4.1 | 3 | 16.2 | 6 | 73 | 7 | 575.9 | *87* | 137 | 25.1 | 8 | | | | |

## OXFORD 1846

| Month | Mean max | Mean min | Mean temp | Anom | Highest max | Date | Lowest min | Date | Lowest max | Date | Highest min | Date | Air frost | Days ≥ 25 °C | Total pptn | Anom | Rain days | Wettest day | Date | Total sunshine | Anom | Sunniest day | Date |
|---|---|---|---|---|---|---|---|---|---|---|---|---|---|---|---|---|---|---|---|---|---|---|---|
| Jan | 8.9 | 4.2 | 6.8 | *+1.7* | 15.1 | 31 | −2.2 | 2 | 2.4 | 12 | 8.3 | 25 | 3 | 0 | 107.0 | *188* | 15 | 17.6 | 20 | | | | |
| Feb | 9.4 | 3.7 | 6.8 | *+1.6* | 14.5 | 26 | −3.5 | 10 | 3.5 | 10 | 10.7 | 23 | 4 | 0 | 28.8 | *67* | 8 | 11.6 | 2 | | | | |
| Mar | 10.2 | 2.5 | 6.5 | *−0.9* | 13.3 | 31 | −5.1 | 20 | 5.1 | 20 | 9.4 | 14 | 6 | 0 | 44.9 | *94* | 14 | 9.5 | 5 | | | | |
| Apr | 12.0 | 4.9 | 8.5 | *−0.9* | 16.6 | 12 | −0.6 | 28 | 6.6 | 7 | 9.6 | 15 | 2 | 0 | 100.0 | *204* | 15 | 19.5 | 6 | | | | |
| May | 18.2 | 8.1 | 13.0 | *+0.7* | 24.0 | 31 | 1.7 | 15 | 13.7 | 18 | 12.2 | 22 | 0 | 0 | 34.4 | *60* | 10 | 10.5 | 18 | | | | |
| June | 24.3 | 12.7 | 18.3 | *+2.9* | 29.9 | 21 | 8.3 | 2 | 14.4 | 23 | 16.1 | 15 | 0 | 14 | 24.7 | *50* | 6 | 17.5 | 23 | | | | |
| July | 22.4 | 13.3 | 17.5 | *−0.1* | 29.3 | 30 | 8.8 | 26 | 15.6 | 24 | 17.2 | 31 | 0 | 7 | 52.4 | *107* | 12 | 9.6 | 25 | | | | |
| Aug | 21.5 | 12.9 | 17.0 | *−0.4* | 25.9 | 2 | 8.3 | 24 | 17.2 | 20 | 17.0 | 6 | 0 | 3 | 78.6 | *143* | 12 | 25.5 | 3 | | | | |
| Sep | 20.0 | 10.6 | 15.3 | *+0.4* | 24.3 | 6 | 4.4 | 29 | 15.7 | 30 | 14.4 | 5 | 0 | 0 | 45.7 | *84* | 5 | 21.4 | 23 | | | | |
| Oct | 13.1 | 6.7 | 10.1 | *−1.4* | 17.6 | 2 | 1.2 | 31 | 7.4 | 28 | 13.3 | 1 | 0 | 0 | 145.2 | *208* | 23 | 16.0 | 18 | | | | |
| Nov | 9.7 | 4.6 | 7.4 | *−0.4* | 15.9 | 5 | −5.6 | 29 | 1.5 | 30 | 10.2 | 3 | 4 | 0 | 34.4 | *52* | 10 | 8.0 | 26 | | | | |
| Dec | 2.8 | −2.5 | 0.4 | *−4.9* | 9.8 | 21 | −8.3 | 14 | −1.0 | 14 | 5.6 | 20 | 24 | 0 | 19.1 | *30* | 8 | 5.8 | 23 | | | | |
| **1846** | 14.4 | 6.8 | 10.6 | *−0.2* | 29.9 | 6 | −8.3 | 12 | −1.0 | 12 | 17.2 | 7 | 43 | 24 | 715.3 | *108* | 138 | 25.5 | 8 | | | | |

## OXFORD 1847

| Month | Mean max | Mean min | Mean temp | Anom | Highest max | Date | Lowest min | Date | Lowest max | Date | Highest min | Date | Air frost | Days ≥ 25 °C | Total pptn | Anom | Rain days | Wettest day | Date | Total sunshine | Anom | Sunniest day | Date |
|---|---|---|---|---|---|---|---|---|---|---|---|---|---|---|---|---|---|---|---|---|---|---|---|
| Jan | 3.8 | −0.2 | 2.1 | *−3.1* | 9.6 | 27 | −4.4 | 1 | −2.0 | 17 | 5.0 | 7 | 18 | 0 | 42.6 | *75* | 9 | 11.6 | 28 | | | | |
| Feb | 4.8 | −1.3 | 2.0 | *−3.1* | 12.7 | 17 | −13.8 | 11 | −3.5 | 12 | 8.1 | 17 | 17 | 0 | 38.5 | *90* | 7 | 20.0 | 14 | | | | |
| Mar | 9.5 | 1.3 | 5.6 | *−1.8* | 16.2 | 26 | −7.8 | 10 | 1.3 | 11 | 7.8 | 16 | 12 | 0 | 23.9 | *50* | 6 | 11.6 | 29 | | | | |
| Apr | 11.4 | 2.5 | 7.0 | *−2.3* | 15.7 | 25 | −4.2 | 16 | 5.2 | 2 | 8.3 | 11 | 8 | 0 | 25.6 | *52* | 7 | 7.6 | 12 | | | | |
| May | 18.2 | 8.4 | 13.2 | *+0.8* | 26.7 | 23 | 1.4 | 2 | 10.8 | 1 | 15.6 | 28 | 0 | 3 | 77.4 | *135* | 14 | 16.7 | 16 | | | | |
| June | 18.3 | 10.1 | 14.0 | *−1.4* | 24.0 | 3 | 4.3 | 6 | 13.4 | 15 | 13.4 | 29 | 0 | 0 | 54.5 | *111* | 12 | 15.7 | 17 | | | | |
| July | 23.6 | 13.2 | 18.2 | *+0.6* | 29.0 | 13 | 8.9 | 23 | 17.4 | 2 | 17.6 | 10 | 0 | 8 | 23.5 | *48* | 7 | 6.8 | 7 | | | | |
| Aug | 21.1 | 11.4 | 16.1 | *−1.3* | 27.3 | 1 | 5.0 | 24 | 16.1 | 16 | 16.9 | 11 | 0 | 3 | 24.1 | *44* | 9 | 7.6 | 29 | | | | |
| Sep | 16.0 | 7.4 | 11.7 | *−3.2* | 20.3 | 10 | 1.1 | 27 | 10.4 | 18 | 12.5 | 12 | 0 | 0 | 58.3 | *108* | 11 | 17.9 | 19 | | | | |
| Oct | 14.6 | 8.6 | 11.8 | *+0.3* | 20.2 | 11 | 0.8 | 25 | 10.1 | 24 | 12.2 | 9 | 0 | 0 | 106.9 | *153* | 12 | 32.6 | 10 | | | | |
| Nov | 10.8 | 4.7 | 8.0 | *+0.2* | 16.6 | 1 | −3.2 | 18 | 4.9 | 19 | 11.1 | 7 | 3 | 0 | 50.5 | *76* | 13 | 13.0 | 29 | | | | |
| Dec | 7.5 | 3.1 | 5.6 | *+0.3* | 14.2 | 3 | −1.7 | 26 | 1.0 | 21 | 11.1 | 9 | 6 | 0 | 80.6 | *128* | 14 | 14.7 | 30 | | | | |
| 1847 | 13.3 | 5.8 | 9.5 | *−1.2* | 29.0 | 7 | −13.8 | 2 | −3.5 | 2 | 17.6 | 7 | 64 | 14 | 606.4 | *92* | 121 | 32.6 | 10 | | | | |

## OXFORD 1848

| Month | Mean max | Mean min | Mean temp | Anom | Highest max | Date | Lowest min | Date | Lowest max | Date | Highest min | Date | Air frost | Days ≥ 25 °C | Total pptn | Anom | Rain days | Wettest day | Date | Total sunshine | Anom | Sunniest day | Date |
|---|---|---|---|---|---|---|---|---|---|---|---|---|---|---|---|---|---|---|---|---|---|---|---|
| Jan | 3.3 | −1.1 | 1.4 | *−3.8* | 11.5 | 3 | −7.2 | 27 | −5.6 | 28 | 6.4 | 3 | 20 | 0 | 24.9 | *44* | 7 | 6.3 | 18 | | | | |
| Feb | 9.3 | 3.4 | 6.6 | *+1.4* | 12.7 | 6 | −4.7 | 17 | 2.9 | 1 | 9.2 | 13 | 5 | 0 | 71.3 | *167* | 14 | 14.2 | 26 | | | | |
| Mar | 9.5 | 3.1 | 6.4 | *−1.0* | 19.4 | 31 | −0.6 | 19 | 4.0 | 17 | 7.2 | 22 | 3 | 0 | 73.6 | *154* | 19 | 15.5 | 21 | | | | |
| Apr | 13.2 | 4.5 | 8.9 | *−0.4* | 21.8 | 4 | −1.0 | 9 | 7.4 | 9 | 8.9 | 21 | 2 | 0 | 64.2 | *131* | 12 | 11.3 | 16 | | | | |
| May | 20.6 | 8.3 | 14.3 | *+2.0* | 26.1 | 12 | 1.7 | 1 | 13.5 | 19 | 12.7 | 25 | 0 | 3 | 22.3 | *39* | 4 | 17.0 | 18 | | | | |
| June | 19.4 | 10.9 | 15.0 | *−0.4* | 25.2 | 22 | 6.1 | 30 | 14.3 | 1 | 14.0 | 25 | 0 | 1 | 91.5 | *186* | 19 | 12.7 | 18 | | | | |
| July | 21.3 | 12.2 | 16.4 | *−1.1* | 27.7 | 6 | 6.7 | 1 | 12.1 | 1 | 17.8 | 6 | 0 | 2 | 41.6 | *85* | 11 | 8.8 | 9 | | | | |
| Aug | 19.0 | 10.7 | 14.7 | *−2.8* | 22.2 | 16 | 7.3 | 24 | 13.6 | 24 | 17.2 | 27 | 0 | 0 | 118.5 | *216* | 18 | 19.0 | 9 | | | | |
| Sep | 17.8 | 8.8 | 13.3 | *−1.5* | 24.8 | 5 | 1.8 | 18 | 13.2 | 29 | 13.6 | 9 | 0 | 0 | 82.0 | *151* | 10 | 31.3 | 30 | | | | |
| Oct | 13.5 | 7.1 | 10.4 | *−1.0* | 21.6 | 7 | 0.1 | 17 | 4.1 | 18 | 14.9 | 4 | 0 | 0 | 146.2 | *209* | 14 | 33.4 | 1 | | | | |
| Nov | 8.9 | 2.6 | 6.0 | *−1.8* | 13.3 | 29 | −3.3 | 4 | 2.9 | 4 | 8.3 | 28 | 7 | 0 | 22.3 | *33* | 8 | 10.3 | 1 | | | | |
| Dec | 8.3 | 3.0 | 6.0 | *+0.7* | 14.0 | 8 | −5.8 | 23 | −0.2 | 21 | 9.4 | 7 | 7 | 0 | 85.7 | *136* | 15 | 21.3 | 28 | | | | |
| 1848 | 13.7 | 6.1 | 9.9 | *-0.9* | 27.7 | 7 | −7.2 | 1 | −5.6 | 1 | 17.8 | 7 | 44 | 6 | 844.0 | *128* | 151 | 33.4 | 10 | | | | |

## OXFORD 1849

| Month | Mean max | Mean min | Mean temp | Anom | Highest max | Date | Lowest min | Date | Lowest max | Date | Highest min | Date | Air frost | Days ≥ 25 °C | Total pptn | Anom | Rain days | Wettest day | Date | Total sunshine | Anom | Sunniest day | Date |
|---|---|---|---|---|---|---|---|---|---|---|---|---|---|---|---|---|---|---|---|---|---|---|---|
| Jan | 6.8 | 1.5 | 4.4 | *−0.7* | 12.4 | 17 | −6.9 | 2 | −2.6 | 2 | 10.0 | 13 | 12 | 0 | 39.3 | *69* | 12 | 22.3 | 8 | | | | |
| Feb | 9.2 | 1.7 | 5.7 | *+0.5* | 13.4 | 22 | −4.6 | 12 | 5.7 | 13 | 6.7 | 3 | 8 | 0 | 20.0 | *47* | 8 | 9.2 | 21 | | | | |
| Mar | 10.4 | 2.4 | 6.6 | *−0.9* | 14.2 | 12 | −4.3 | 8 | 5.1 | 9 | 8.1 | 14 | 7 | 0 | 36.8 | *77* | 4 | 25.2 | 1 | | | | |
| Apr | 11.4 | 2.4 | 7.0 | *−2.4* | 17.4 | 30 | −3.9 | 20 | 4.1 | 13 | 8.8 | 30 | 8 | 0 | 79.3 | *162* | 19 | 13.0 | 23 | | | | |
| May | 17.6 | 8.6 | 13.0 | *+0.7* | 24.9 | 31 | 0.6 | 11 | 11.1 | 8 | 15.0 | 27 | 0 | 0 | 84.5 | *148* | 12 | 18.7 | 29 | | | | |
| June | 20.1 | 9.5 | 14.6 | *−0.8* | 25.6 | 4 | 3.9 | 10 | 10.5 | 10 | 16.1 | 4 | 0 | 2 | 27.9 | *57* | 7 | 15.9 | 5 | | | | |
| July | 21.4 | 11.1 | 16.0 | *−1.6* | 26.9 | 7 | 8.1 | 11 | 16.1 | 25 | 15.0 | 28 | 0 | 2 | 56.7 | *116* | 9 | 15.6 | 24 | | | | |
| Aug | 21.6 | 12.0 | 16.6 | *−0.8* | 26.5 | 7 | 6.3 | 3 | 17.0 | 17 | 16.2 | 30 | 0 | 2 | 28.2 | *51* | 7 | 14.2 | 10 | | | | |
| Sep | 18.3 | 10.1 | 14.2 | *−0.7* | 23.0 | 5 | 2.7 | 18 | 13.5 | 12 | 16.1 | 3 | 0 | 0 | 83.7 | *154* | 13 | 20.7 | 30 | | | | |
| Oct | 13.6 | 8.3 | 11.1 | *−0.3* | 18.6 | 28 | 0.8 | 15 | 9.2 | 14 | 16.1 | 28 | 0 | 0 | 39.3 | *56* | 12 | 13.2 | 7 | | | | |
| Nov | 8.7 | 2.6 | 6.0 | *−1.9* | 13.9 | 8 | −6.8 | 27 | −1.1 | 27 | 11.1 | 8 | 7 | 0 | 35.5 | *53* | 10 | 12.9 | 30 | | | | |
| Dec | 5.3 | 0.0 | 2.9 | *−2.4* | 12.7 | 15 | −10.6 | 28 | −2.2 | 28 | 7.8 | 14 | 18 | 0 | 63.5 | *101* | 16 | 15.4 | 3 | | | | |
| **1849** | 13.7 | 5.9 | 9.8 | *−1.0* | 26.9 | 7 | −10.6 | 12 | −2.6 | 1 | 16.2 | 8 | 60 | 6 | 594.5 | *90* | 129 | 25.2 | 3 | | | | |

## OXFORD 1850

| Month | Mean max | Mean min | Mean temp | Anom | Highest max | Date | Lowest min | Date | Lowest max | Date | Highest min | Date | Air frost | Days ≥ 25 °C | Total pptn | Anom | Rain days | Wettest day | Date | Total sunshine | Anom | Sunniest day | Date |
|---|---|---|---|---|---|---|---|---|---|---|---|---|---|---|---|---|---|---|---|---|---|---|---|
| Jan | 2.5 | −2.5 | 0.3 | *−4.8* | 9.9 | 25 | −9.6 | 7 | −3.3 | 6 | 7.2 | 25 | 27 | 0 | 27.7 | *49* | 5 | 15.7 | 19 | | | | |
| Feb | 9.2 | 3.2 | 6.4 | *+1.3* | 12.4 | 1 | −3.4 | 13 | 4.4 | 13 | 8.2 | 1 | 3 | 0 | 32.8 | *77* | 11 | 12.0 | 1 | | | | |
| Mar | 8.0 | −1.1 | 3.6 | *−3.9* | 13.2 | 31 | −8.9 | 25 | 1.8 | 26 | 7.1 | 31 | 19 | 0 | 4.6 | *10* | 3 | 3.4 | 27 | | | | |
| Apr | 13.7 | 5.2 | 9.5 | *+0.2* | 17.4 | 7 | 0.0 | 28 | 10.3 | 25 | 10.0 | 3 | 0 | 0 | 63.0 | *129* | 14 | 12.2 | 4 | | | | |
| May | 14.9 | 6.3 | 10.5 | *−1.9* | 21.4 | 30 | −1.7 | 2 | 7.2 | 6 | 12.7 | 31 | 2 | 0 | 58.6 | *102* | 12 | 16.2 | 25 | | | | |
| June | 20.7 | 10.4 | 15.4 | *−0.0* | 27.1 | 25 | 2.8 | 15 | 14.4 | 15 | 15.6 | 24 | 0 | 4 | 52.5 | *107* | 10 | 27.1 | 27 | | | | |
| July | 20.9 | 12.4 | 16.3 | *−1.3* | 26.7 | 23 | 7.8 | 9 | 15.0 | 8 | 16.6 | 17 | 0 | 3 | 137.1 | *281* | 14 | 34.8 | 19 | | | | |
| Aug | 19.6 | 10.4 | 14.8 | *−2.6* | 23.6 | 3 | 2.5 | 21 | 14.7 | 21 | 15.4 | 1 | 0 | 0 | 30.7 | *56* | 12 | 5.0 | 25 | | | | |
| Sep | 16.7 | 7.7 | 12.2 | *−2.7* | 21.2 | 2 | 1.7 | 4 | 13.3 | 8 | 13.8 | 1 | 0 | 0 | 39.1 | *72* | 9 | 12.7 | 30 | | | | |
| Oct | 11.3 | 4.2 | 7.9 | *−3.5* | 16.6 | 18 | −0.4 | 26 | 6.6 | 24 | 10.1 | 18 | 1 | 0 | 37.5 | *54* | 14 | 7.3 | 28 | | | | |
| Nov | 10.2 | 3.9 | 7.3 | *−0.5* | 15.9 | 2 | −4.9 | 28 | 1.1 | 30 | 11.9 | 1 | 6 | 0 | 43.4 | *65* | 14 | 10.0 | 19 | | | | |
| Dec | 6.6 | 1.3 | 4.3 | *−1.1* | 12.2 | 5 | −5.0 | 9 | 0.7 | 9 | 10.0 | 31 | 11 | 0 | 50.1 | *79* | 13 | 14.0 | 15 | | | | |
| **1850** | 12.9 | 5.1 | 9.0 | *−1.8* | 27.1 | 6 | −9.6 | 1 | −3.3 | 1 | 16.6 | 7 | 69 | 7 | 576.9 | *87* | 131 | 34.8 | 7 | | | | |

## OXFORD 1851

| Month | Mean max | Mean min | Mean temp | Anom | Highest max | Date | Lowest min | Date | Lowest max | Date | Highest min | Date | Air frost | Days ≥ 25 °C | Total pptn | Anom | Rain days | Wettest day | Date | Total sunshine | Anom | Sunniest day | Date |
|---|---|---|---|---|---|---|---|---|---|---|---|---|---|---|---|---|---|---|---|---|---|---|---|
| Jan | 7.9 | 2.6 | 5.5 | *+0.4* | 12.8 | 1 | −3.2 | 23 | 3.2 | 24 | 11.5 | 1 | 4 | 0 | 67.7 | *119* | 17 | 12.7 | 18 | | | | |
| Feb | 7.7 | −0.1 | 4.0 | *−1.1* | 12.6 | 18 | −6.1 | 16 | 3.3 | 16 | 8.7 | 18 | 15 | 0 | 18.9 | *44* | 7 | 5.8 | 6 | | | | |
| Mar | 9.6 | 1.3 | 5.7 | *−1.8* | 13.3 | 27 | −3.9 | 6 | 5.0 | 1 | 7.5 | 25 | 11 | 0 | 82.0 | *172* | 19 | 10.2 | 18 | | | | |
| Apr | 12.7 | 2.0 | 7.4 | *−2.0* | 20.0 | 20 | −5.0 | 6 | 8.3 | 6 | 9.7 | 20 | 11 | 0 | 29.8 | *61* | 10 | 7.8 | 17 | | | | |
| May | 16.0 | 6.1 | 10.9 | *−1.4* | 21.2 | 29 | 0.0 | 3 | 7.8 | 4 | 11.9 | 17 | 0 | 0 | 22.9 | *40* | 10 | 6.6 | 19 | | | | |
| June | 20.4 | 10.0 | 15.1 | *−0.4* | 28.8 | 27 | 5.0 | 3 | 13.7 | 10 | 14.5 | 12 | 0 | 6 | 39.6 | *80* | 10 | 16.8 | 6 | | | | |
| July | 20.0 | 10.8 | 15.1 | *−2.5* | 24.2 | 2 | 5.7 | 10 | 15.3 | 24 | 16.7 | 31 | 0 | 0 | 89.7 | *184* | 15 | 17.0 | 24 | | | | |
| Aug | 21.0 | 12.0 | 16.4 | *−1.1* | 27.0 | 13 | 5.2 | 30 | 13.3 | 29 | 16.8 | 21 | 0 | 4 | 54.2 | *99* | 8 | 26.0 | 28 | | | | |
| Sep | 17.9 | 7.1 | 12.5 | *−2.4* | 22.6 | 3 | −3.3 | 28 | 12.1 | 26 | 15.7 | 2 | 1 | 0 | 8.8 | *16* | 5 | 4.0 | 26 | | | | |
| Oct | 15.3 | 7.0 | 11.4 | *−0.1* | 21.4 | 12 | 0.0 | 16 | 8.1 | 31 | 13.5 | 19 | 0 | 0 | 59.7 | *85* | 13 | 17.3 | 15 | | | | |
| Nov | 6.0 | −0.7 | 2.9 | *−4.9* | 10.1 | 1 | −7.1 | 18 | 1.2 | 29 | 5.0 | 8 | 19 | 0 | 15.7 | *24* | 7 | 4.8 | 24 | | | | |
| Dec | 6.7 | 1.6 | 4.5 | *−0.8* | 12.7 | 9 | −6.2 | 26 | 0.0 | 12 | 8.1 | 20 | 11 | 0 | 23.6 | *37* | 9 | 9.0 | 22 | | | | |
| 1851 | 13.4 | 5.0 | 9.2 | *−1.5* | 28.8 | 6 | −7.1 | 11 | 0.0 | 12 | 16.8 | 8 | 72 | 10 | 512.6 | *78* | 130 | 26.0 | 8 | | | | |

## OXFORD 1852

| Month | Mean max | Mean min | Mean temp | Anom | Highest max | Date | Lowest min | Date | Lowest max | Date | Highest min | Date | Air frost | Days ≥ 25 °C | Total pptn | Anom | Rain days | Wettest day | Date | Total sunshine | Anom | Sunniest day | Date |
|---|---|---|---|---|---|---|---|---|---|---|---|---|---|---|---|---|---|---|---|---|---|---|---|
| Jan | 8.3 | 0.9 | 4.9 | *−0.2* | 13.1 | 15 | −3.4 | 4 | 1.9 | 1 | 6.4 | 11 | 9 | 0 | 123.7 | *217* | 18 | 32.6 | 13 | | | | |
| Feb | 7.4 | 0.9 | 4.4 | *−0.7* | 13.1 | 1 | −6.0 | 20 | 1.9 | 20 | 7.3 | 4 | 12 | 0 | 24.1 | *56* | 12 | 6.1 | 9 | | | | |
| Mar | 9.2 | −0.8 | 4.4 | *−3.1* | 18.2 | 22 | −7.6 | 4 | 4.0 | 5 | 6.4 | 29 | 18 | 0 | 14.7 | *31* | 3 | 11.8 | 30 | | | | |
| Apr | 13.4 | 0.7 | 7.1 | *−2.2* | 20.6 | 14 | −4.7 | 20 | 8.7 | 4 | 8.3 | 29 | 15 | 0 | 14.6 | *30* | 3 | 11.5 | 29 | | | | |
| May | 15.4 | 5.3 | 10.2 | *−2.1* | 21.3 | 18 | −2.3 | 2 | 9.9 | 2 | 11.1 | 17 | 3 | 0 | 51.3 | *90* | 12 | 17.7 | 27 | | | | |
| June | 18.4 | 9.0 | 13.5 | *−1.9* | 21.2 | 23 | 3.4 | 11 | 13.2 | 11 | 12.8 | 28 | 0 | 0 | 170.4 | *346* | 21 | 39.3 | 10 | | | | |
| July | 24.8 | 12.6 | 18.4 | *+0.8* | 31.2 | 5 | 9.1 | 1 | 19.9 | 25 | 17.3 | 16 | 0 | 12 | 65.7 | *134* | 6 | 26.8 | 26 | | | | |
| Aug | 21.3 | 12.0 | 16.5 | *−0.9* | 25.0 | 1 | 8.8 | 9 | 17.5 | 12 | 15.2 | 16 | 0 | 1 | 104.2 | *190* | 16 | 19.3 | 14 | | | | |
| Sep | 18.5 | 9.2 | 13.8 | *−1.1* | 23.4 | 4 | 1.1 | 16 | 13.9 | 30 | 14.9 | 8 | 0 | 0 | 71.3 | *132* | 11 | 15.9 | 19 | | | | |
| Oct | 12.7 | 5.0 | 9.0 | *−2.4* | 16.2 | 22 | −0.6 | 8 | 8.8 | 27 | 11.1 | 21 | 1 | 0 | 63.7 | *91* | 15 | 17.4 | 5 | | | | |
| Nov | 12.4 | 6.6 | 9.8 | *+2.0* | 18.3 | 8 | −3.1 | 30 | 4.4 | 29 | 14.4 | 1 | 2 | 0 | 175.7 | *263* | 23 | 32.3 | 12 | | | | |
| Dec | 11.6 | 6.4 | 9.4 | *+4.0* | 14.7 | 20 | 0.3 | 28 | 8.3 | 1 | 11.7 | 4 | 0 | 0 | 80.7 | *128* | 20 | 15.4 | 27 | | | | |
| 1852 | 14.5 | 5.7 | 10.1 | *−0.7* | 31.2 | 7 | −7.6 | 3 | 1.9 | 1 | 17.3 | 7 | 60 | 13 | 960.1 | *145* | 160 | 39.3 | 6 | | | | |

## OXFORD 1853

| Month | Mean max | Mean min | Mean temp | Anom | Highest max | Date | Lowest min | Date | Lowest max | Date | Highest min | Date | Air frost | Days ≥ 25 °C | Total pptn | Anom | Rain days | Wettest day | Date | Total sunshine | Anom | Sunniest day | Date |
|---|---|---|---|---|---|---|---|---|---|---|---|---|---|---|---|---|---|---|---|---|---|---|---|
| Jan | 8.8 | 3.0 | 5.9 | *+1.1* | 12.3 | 7 | −3.3 | 31 | 4.4 | 26 | 7.8 | 1 | 2 | 0 | 57.6 | *101* | 18 | 8.6 | 12 | | | | |
| Feb | 3.9 | −1.2 | 1.3 | *−3.6* | 6.5 | 2 | −6.6 | 13 | 0.4 | 11 | 2.3 | 6 | 17 | 0 | 26.9 | *63* | 7 | 11.2 | 8 | | | | |
| Mar | 8.2 | −0.2 | 4.0 | *−3.2* | 14.3 | 13 | −6.1 | 25 | 2.2 | 17 | 7.9 | 31 | 18 | 0 | 23.7 | *50* | 10 | 5.0 | 5 | | | | |
| Apr | 13.2 | 5.0 | 9.1 | *−0.2* | 18.3 | 18 | 0.5 | 27 | 8.2 | 25 | 10.6 | 4 | 0 | 0 | 55.1 | *113* | 16 | 9.3 | 4 | | | | |
| May | 17.5 | 6.6 | 12.1 | *−0.4* | 25.0 | 26 | 0.4 | 10 | 7.8 | 7 | 10.7 | 30 | 0 | 1 | 54.5 | *95* | 11 | 20.3 | 4 | | | | |
| June | 21.1 | 11.1 | 16.1 | *+0.5* | 26.9 | 24 | 4.1 | 3 | 15.0 | 2 | 14.9 | 25 | 0 | 2 | 75.2 | *153* | 15 | 16.5 | 14 | | | | |
| July | 21.2 | 12.2 | 16.7 | *−1.2* | 26.1 | 7 | 9.0 | 1 | 17.7 | 2 | 16.4 | 7 | 0 | 2 | 79.0 | *162* | 17 | 42.3 | 14 | | | | |
| Aug | 20.3 | 10.8 | 15.5 | *−2.1* | 24.9 | 19 | 6.3 | 17 | 16.6 | 17 | 15.4 | 19 | 0 | 0 | 66.3 | *121* | 12 | 19.1 | 26 | | | | |
| Sep | 17.3 | 8.4 | 12.8 | *−2.0* | 21.1 | 1 | 3.9 | 13 | 12.9 | 2 | 12.7 | 11 | 0 | 0 | 47.0 | *87* | 11 | 10.3 | 13 | | | | |
| Oct | 14.0 | 7.3 | 10.7 | *−0.6* | 17.3 | 26 | 0.1 | 2 | 10.4 | 16 | 12.8 | 21 | 0 | 0 | 93.8 | *134* | 21 | 12.8 | 5 | | | | |
| Nov | 8.9 | 2.3 | 5.6 | *−2.0* | 15.0 | 2 | −5.7 | 21 | 3.1 | 23 | 9.9 | 30 | 10 | 0 | 45.5 | *68* | 9 | 16.3 | 26 | | | | |
| Dec | 3.7 | −1.3 | 1.2 | *−3.8* | 10.4 | 1 | −11.3 | 28 | −0.3 | 28 | 4.8 | 6 | 19 | 0 | 9.8 | *16* | 9 | 4.0 | 16 | | | | |
| 1853 | 13.2 | 5.3 | 9.2 | *−1.5* | 26.9 | 6 | −11.3 | 12 | −0.3 | 12 | 16.4 | 7 | 66 | 5 | 634.3 | *96* | 156 | 42.3 | 7 | | | | |

## OXFORD 1854

| Month | Mean max | Mean min | Mean temp | Anom | Highest max | Date | Lowest min | Date | Lowest max | Date | Highest min | Date | Air frost | Days ≥ 25 °C | Total pptn | Anom | Rain days | Wettest day | Date | Total sunshine | Anom | Sunniest day | Date |
|---|---|---|---|---|---|---|---|---|---|---|---|---|---|---|---|---|---|---|---|---|---|---|---|
| Jan | 6.7 | 1.5 | 4.1 | *−0.7* | 12.9 | 29 | −10.9 | 2 | 0.3 | 3 | 10.3 | 29 | 11 | 0 | 50.0 | *88* | 15 | 23.6 | 8 | | | | |
| Feb | 8.0 | 0.6 | 4.3 | *−0.6* | 12.1 | 6 | −5.2 | 3 | 1.9 | 3 | 9.2 | 6 | 12 | 0 | 20.7 | *48* | 9 | 8.4 | 2 | | | | |
| Mar | 11.2 | 2.2 | 6.7 | *−0.5* | 16.0 | 13 | −4.3 | 2 | 6.4 | 20 | 10.0 | 8 | 8 | 0 | 9.7 | *20* | 5 | 3.8 | 19 | | | | |
| Apr | 15.0 | 3.5 | 9.3 | *−0.0* | 22.2 | 20 | −3.6 | 24 | 7.8 | 23 | 11.2 | 20 | 2 | 0 | 18.2 | *37* | 5 | 11.1 | 22 | | | | |
| May | 16.2 | 5.7 | 11.0 | *−1.5* | 20.6 | 31 | −0.9 | 16 | 13.1 | 16 | 10.7 | 20 | 1 | 0 | 75.3 | *132* | 19 | 16.8 | 24 | | | | |
| June | 18.6 | 9.7 | 14.1 | *−1.4* | 26.9 | 25 | 4.2 | 7 | 13.2 | 2 | 16.5 | 25 | 0 | 1 | 41.4 | *84* | 9 | 13.7 | 13 | | | | |
| July | 21.7 | 11.5 | 16.6 | *−1.3* | 29.1 | 24 | 6.4 | 28 | 16.0 | 12 | 17.6 | 24 | 0 | 5 | 34.1 | *70* | 12 | 7.7 | 11 | | | | |
| Aug | 20.9 | 10.7 | 15.8 | *−1.8* | 25.6 | 13 | 5.0 | 25 | 13.3 | 4 | 17.1 | 27 | 0 | 5 | 41.6 | *76* | 10 | 16.9 | 1 | | | | |
| Sep | 20.2 | 8.4 | 14.3 | *−0.6* | 25.0 | 4 | 3.3 | 26 | 15.8 | 22 | 16.1 | 15 | 0 | 1 | 10.4 | *19* | 7 | 4.0 | 15 | | | | |
| Oct | 13.9 | 5.6 | 9.8 | *−1.5* | 20.3 | 5 | −1.1 | 12 | 7.4 | 25 | 12.1 | 4 | 2 | 0 | 52.9 | *76* | 15 | 10.6 | 26 | | | | |
| Nov | 8.2 | 2.0 | 5.1 | *−2.4* | 14.1 | 2 | −5.6 | 26 | 2.4 | 26 | 8.9 | 4 | 5 | 0 | 29.7 | *45* | 11 | 10.2 | 29 | | | | |
| Dec | 8.3 | 2.1 | 5.2 | *+0.2* | 12.0 | 14 | −4.3 | 10 | 2.9 | 28 | 8.8 | 14 | 8 | 0 | 28.1 | *45* | 14 | 9.0 | 20 | | | | |
| 1854 | 14.1 | 5.3 | 9.7 | *−1.0* | 29.1 | 7 | −10.9 | 1 | 0.3 | 1 | 17.6 | 7 | 49 | 12 | 412.2 | *62* | 131 | 23.6 | 1 | | | | |

## OXFORD 1855

| Month | Mean max | Mean min | Mean temp | Anom | Highest max | Date | Lowest min | Date | Lowest max | Date | Highest min | Date | Air frost | Days ≥ 25 °C | Total pptn | Anom | Rain days | Wettest day | Date | Total sunshine | Anom | Sunniest day | Date |
|---|---|---|---|---|---|---|---|---|---|---|---|---|---|---|---|---|---|---|---|---|---|---|---|
| Jan | 4.4 | −0.3 | 2.0 | *−2.8* | 10.8 | 2 | −6.9 | 31 | −0.9 | 17 | 7.6 | 4 | 18 | 0 | 6.3 | *11* | 4 | 4.2 | 25 | | | | |
| Feb | 1.7 | −4.5 | −1.4 | *−6.3* | 9.8 | 28 | −13.6 | 16 | −2.8 | 18 | 4.8 | 28 | 20 | 0 | 30.9 | *72* | 6 | 14.2 | 5 | | | | |
| Mar | 7.7 | 0.1 | 3.9 | *−3.4* | 13.6 | 20 | −5.3 | 26 | 1.9 | 22 | 4.8 | 2 | 15 | 0 | 51.3 | *108* | 13 | 12.1 | 29 | | | | |
| Apr | 13.0 | 2.6 | 7.8 | *−1.4* | 19.7 | 24 | −4.0 | 1 | 8.2 | 2 | 8.6 | 13 | 9 | 0 | 9.7 | *20* | 7 | 6.1 | 4 | | | | |
| May | 15.0 | 4.9 | 10.0 | *−2.5* | 27.3 | 26 | −2.4 | 5 | 8.4 | 13 | 15.6 | 27 | 3 | 1 | 41.3 | *72* | 14 | 11.3 | 14 | | | | |
| June | 19.4 | 9.3 | 14.4 | *−1.2* | 27.7 | 6 | 3.0 | 18 | 12.8 | 1 | 16.7 | 30 | 0 | 2 | 55.0 | *112* | 11 | 18.7 | 1 | | | | |
| July | 22.8 | 13.4 | 18.1 | *+0.2* | 26.8 | 23 | 8.3 | 5 | 18.3 | 26 | 17.6 | 23 | 0 | 3 | 118.5 | *243* | 14 | 38.4 | 27 | | | | |
| Aug | 21.7 | 12.2 | 17.0 | *−0.6* | 25.1 | 17 | 8.9 | 31 | 19.4 | 8 | 17.2 | 12 | 0 | 1 | 39.0 | *71* | 11 | 10.8 | 20 | | | | |
| Sep | 19.0 | 10.4 | 14.7 | *−0.2* | 22.6 | 23 | 2.8 | 8 | 14.6 | 14 | 14.9 | 24 | 0 | 0 | 54.7 | *101* | 5 | 18.0 | 29 | | | | |
| Oct | 14.0 | 8.2 | 11.1 | *−0.2* | 17.9 | 2 | 1.3 | 29 | 7.6 | 31 | 15.1 | 1 | 0 | 0 | 131.2 | *187* | 20 | 19.3 | 17 | | | | |
| Nov | 7.8 | 3.4 | 5.6 | *−2.0* | 13.3 | 6 | −4.0 | 15 | 4.2 | 21 | 9.4 | 7 | 4 | 0 | 23.1 | *35* | 11 | 12.4 | 1 | | | | |
| Dec | 5.6 | −0.1 | 2.8 | *−2.2* | 10.8 | 28 | −9.7 | 22 | −3.8 | 21 | 5.7 | 28 | 13 | 0 | 25.8 | *41* | 10 | 10.1 | 24 | | | | |
| 1855 | 12.7 | 5.0 | 8.8 | *−1.9* | 27.7 | 6 | −13.6 | 2 | −3.8 | 12 | 17.6 | 7 | 82 | 7 | 586.9 | *89* | 126 | 38.4 | 7 | | | | |

## OXFORD 1856

| Month | Mean max | Mean min | Mean temp | Anom | Highest max | Date | Lowest min | Date | Lowest max | Date | Highest min | Date | Air frost | Days ≥ 25 °C | Total pptn | Anom | Rain days | Wettest day | Date | Total sunshine | Anom | Sunniest day | Date |
|---|---|---|---|---|---|---|---|---|---|---|---|---|---|---|---|---|---|---|---|---|---|---|---|
| Jan | 6.8 | 2.0 | 4.4 | *−0.4* | 11.9 | 24 | −5.9 | 14 | 1.8 | 11 | 9.2 | 23 | 10 | 0 | 68.5 | *120* | 19 | 10.8 | 20 | | | | |
| Feb | 8.2 | 3.0 | 5.6 | *+0.7* | 13.7 | 9 | −6.2 | 3 | 1.1 | 19 | 8.7 | 6 | 8 | 0 | 30.6 | *72* | 6 | 7.1 | 14 | | | | |
| Mar | 7.9 | 0.6 | 4.3 | *−3.0* | 13.3 | 31 | −5.6 | 31 | 4.4 | 4 | 5.6 | 19 | 11 | 0 | 21.9 | *46* | 3 | 12.3 | 17 | | | | |
| Apr | 12.8 | 4.1 | 8.5 | *−0.8* | 21.7 | 25 | −2.8 | 1 | 7.8 | 27 | 8.6 | 26 | 2 | 0 | 59.5 | *122* | 11 | 14.3 | 27 | | | | |
| May | 13.9 | 5.7 | 9.8 | *−2.6* | 20.3 | 21 | −1.1 | 5 | 6.7 | 1 | 11.1 | 22 | 2 | 0 | 100.9 | *176* | 17 | 30.7 | 27 | | | | |
| June | 20.2 | 9.5 | 14.8 | *−0.8* | 27.8 | 27 | 5.0 | 2 | 15.8 | 1 | 16.3 | 27 | 0 | 2 | 59.4 | *121* | 8 | 29.5 | 20 | | | | |
| July | 22.2 | 11.3 | 16.8 | *−1.1* | 29.4 | 31 | 5.6 | 2 | 14.9 | 8 | 16.4 | 20 | 0 | 5 | 14.5 | *30* | 6 | 3.8 | 9 | | | | |
| Aug | 22.7 | 12.5 | 17.6 | *+0.0* | 31.7 | 2 | 5.6 | 23 | 16.1 | 18 | 15.8 | 2 | 0 | 12 | 75.0 | *137* | 13 | 20.3 | 18 | | | | |
| Sep | 16.8 | 8.5 | 12.6 | *−2.3* | 21.7 | 10 | 4.4 | 21 | 12.1 | 28 | 13.3 | 11 | 0 | 0 | 73.6 | *136* | 12 | 15.9 | 28 | | | | |
| Oct | 13.9 | 6.9 | 10.4 | *−0.9* | 18.3 | 4 | −1.4 | 29 | 9.4 | 29 | 13.6 | 4 | 1 | 0 | 55.3 | *79* | 14 | 13.3 | 4 | | | | |
| Nov | 8.1 | 2.3 | 5.2 | *−2.3* | 13.3 | 24 | −3.9 | 29 | 1.4 | 30 | 10.0 | 24 | 8 | 0 | 22.2 | *33* | 11 | 10.5 | 26 | | | | |
| Dec | 7.3 | 1.4 | 4.3 | *−0.7* | 15.0 | 6 | −8.9 | 2 | 0.6 | 27 | 11.9 | 8 | 12 | 0 | 47.7 | *76* | 15 | 7.2 | 3 | | | | |
| 1856 | 13.4 | 5.7 | 9.5 | *−1.2* | 31.7 | 8 | −8.9 | 12 | 0.6 | 12 | 16.4 | 7 | 54 | 19 | 629.0 | *95* | 135 | 30.7 | 5 | | | | |

## OXFORD 1857

| Month | Mean max | Mean min | Mean temp | Anom | Highest max | Date | Lowest min | Date | Lowest max | Date | Highest min | Date | Air frost | Days ≥ 25 °C | Total pptn | Anom | Rain days | Wettest day | Date | Total sunshine | Anom | Sunniest day | Date |
|---|---|---|---|---|---|---|---|---|---|---|---|---|---|---|---|---|---|---|---|---|---|---|---|
| Jan | 5.1 | −0.6 | 2.3 | *−2.6* | 10.3 | 1 | −7.2 | 29 | −0.6 | 29 | 5.8 | 1 | 17 | 0 | 74.0 | *130* | 18 | 21.2 | 9 | | | | |
| Feb | 7.6 | 0.4 | 4.0 | *−0.9* | 12.2 | 28 | −5.6 | 3 | −0.8 | 3 | 5.0 | 21 | 11 | 0 | 7.5 | *18* | 8 | 2.3 | 18 | | | | |
| Mar | 9.1 | 1.8 | 5.4 | *−1.8* | 16.1 | 18 | −3.3 | 9 | 3.9 | 10 | 6.1 | 29 | 10 | 0 | 35.9 | *75* | 15 | 6.8 | 29 | | | | |
| Apr | 11.9 | 3.6 | 7.8 | *−1.5* | 18.9 | 18 | −2.5 | 28 | 5.0 | 27 | 10.8 | 5 | 3 | 0 | 47.0 | *96* | 13 | 17.7 | 4 | | | | |
| May | 16.5 | 6.3 | 11.4 | *−1.1* | 22.8 | 14 | −0.8 | 4 | 9.4 | 5 | 11.7 | 15 | 4 | 0 | 27.8 | *49* | 7 | 9.3 | 10 | | | | |
| June | 22.0 | 10.6 | 16.3 | *+0.7* | 29.4 | 28 | 3.6 | 13 | 17.2 | 9 | 15.6 | 19 | 0 | 9 | 70.6 | *143* | 11 | 20.0 | 30 | | | | |
| July | 21.9 | 12.7 | 17.3 | *−0.6* | 26.7 | 14 | 7.5 | 7 | 16.1 | 1 | 16.4 | 22 | 0 | 4 | 73.8 | *151* | 7 | 41.7 | 27 | | | | |
| Aug | 22.8 | 13.2 | 18.0 | *+0.4* | 28.3 | 24 | 8.9 | 27 | 15.0 | 7 | 16.4 | 23 | 0 | 7 | 85.7 | *156* | 10 | 22.1 | 14 | | | | |
| Sep | 19.0 | 11.2 | 15.1 | *+0.2* | 24.4 | 17 | 6.1 | 28 | 14.4 | 3 | 16.1 | 15 | 0 | 0 | 90.7 | *168* | 17 | 25.6 | 2 | | | | |
| Oct | 14.6 | 7.8 | 11.2 | *−0.1* | 18.9 | 1 | 1.1 | 30 | 8.9 | 22 | 13.6 | 12 | 0 | 0 | 106.6 | *152* | 17 | 35.8 | 22 | | | | |
| Nov | 10.2 | 4.5 | 7.4 | *−0.2* | 16.1 | 3 | −1.1 | 12 | 4.4 | 25 | 11.7 | 5 | 2 | 0 | 39.0 | *59* | 18 | 19.1 | 3 | | | | |
| Dec | 9.6 | 4.5 | 7.1 | *+2.0* | 13.3 | 2 | −0.6 | 30 | 3.6 | 31 | 10.0 | 22 | 3 | 0 | 15.4 | *24* | 6 | 6.5 | 3 | | | | |
| 1857 | 14.2 | 6.3 | 10.3 | *−0.5* | 29.4 | 6 | −7.2 | 1 | −0.8 | 2 | 16.4 | 7 | 50 | 20 | 674.1 | *102* | 147 | 41.7 | 7 | | | | |

## OXFORD 1858

| Month | Mean max | Mean min | Mean temp | Anom | Highest max | Date | Lowest min | Date | Lowest max | Date | Highest min | Date | Air frost | Days ≥ 25 °C | Total pptn | Anom | Rain days | Wettest day | Date | Total sunshine | Anom | Sunniest day | Date |
|---|---|---|---|---|---|---|---|---|---|---|---|---|---|---|---|---|---|---|---|---|---|---|---|
| Jan | 6.4 | −0.1 | 3.1 | *−1.7* | 11.7 | 8 | −6.4 | 23 | −2.2 | 5 | 8.9 | 8 | 20 | 0 | 15.9 | *28* | 7 | 5.6 | 30 | | | | |
| Feb | 4.7 | −1.2 | 1.7 | *−3.2* | 10.8 | 5 | −5.6 | 18 | 0.6 | 11 | 3.1 | 5 | 18 | 0 | 41.2 | *96* | 8 | 11.2 | 3 | | | | |
| Mar | 9.4 | 1.1 | 5.2 | *−2.0* | 19.2 | 24 | −10.6 | 10 | 0.0 | 1 | 8.9 | 30 | 12 | 0 | 18.7 | *39* | 10 | 4.9 | 31 | | | | |
| Apr | 13.1 | 4.0 | 8.5 | *−0.8* | 21.7 | 22 | −3.9 | 1 | 5.6 | 6 | 11.9 | 21 | 2 | 0 | 82.6 | *169* | 12 | 25.6 | 16 | | | | |
| May | 15.9 | 6.2 | 11.1 | *−1.4* | 26.1 | 31 | 0.8 | 7 | 9.4 | 1 | 14.4 | 31 | 0 | 1 | 54.5 | *95* | 14 | 18.6 | 24 | | | | |
| June | 23.6 | 11.9 | 17.7 | *+2.2* | 32.2 | 15 | 6.7 | 27 | 18.3 | 17 | 17.5 | 15 | 0 | 11 | 37.0 | *75* | 4 | 13.9 | 16 | | | | |
| July | 20.8 | 10.6 | 15.7 | *−2.2* | 28.9 | 15 | 5.8 | 29 | 16.4 | 3 | 16.1 | 15 | 0 | 3 | 53.4 | *109* | 13 | 18.6 | 27 | | | | |
| Aug | 21.4 | 11.2 | 16.3 | *−1.3* | 27.2 | 11 | 6.1 | 6 | 16.7 | 14 | 15.8 | 10 | 0 | 3 | 33.4 | *61* | 9 | 12.6 | 14 | | | | |
| Sep | 19.5 | 11.2 | 15.3 | *+0.5* | 26.1 | 12 | 5.0 | 18 | 15.8 | 19 | 15.6 | 3 | 0 | 2 | 58.1 | *107* | 11 | 13.3 | 4 | | | | |
| Oct | 13.4 | 6.9 | 10.2 | *−1.1* | 17.5 | 4 | 0.8 | 8 | 7.8 | 29 | 12.8 | 14 | 0 | 0 | 46.2 | *66* | 14 | 8.6 | 10 | | | | |
| Nov | 7.4 | 0.8 | 4.1 | *−3.5* | 12.5 | 26 | −9.1 | 23 | −1.7 | 23 | 8.6 | 26 | 12 | 0 | 16.4 | *25* | 9 | 4.7 | 25 | | | | |
| Dec | 7.2 | 2.3 | 4.8 | *−0.3* | 12.2 | 21 | −2.8 | 5 | 2.2 | 14 | 6.7 | 1 | 2 | 0 | 50.0 | *79* | 21 | 11.2 | 18 | | | | |
| 1858 | 13.6 | 5.4 | 9.5 | *−1.3* | 32.2 | 6 | −10.6 | 3 | −2.2 | 1 | 17.5 | 6 | 66 | 20 | 507.5 | *77* | 132 | 25.6 | 4 | | | | |

## OXFORD 1859

| Month | Mean max | Mean min | Mean temp | Anom | Highest max | Date | Lowest min | Date | Lowest max | Date | Highest min | Date | Air frost | Days ≥ 25 °C | Total pptn | Anom | Rain days | Wettest day | Date | Total sunshine | Anom | Sunniest day | Date |
|---|---|---|---|---|---|---|---|---|---|---|---|---|---|---|---|---|---|---|---|---|---|---|---|
| Jan | 7.5 | 1.8 | 4.6 | *−0.2* | 11.9 | 17 | −2.2 | 8 | 1.4 | 3 | 6.7 | 17 | 9 | 0 | 19.1 | *34* | 15 | 6.5 | 23 | | | | |
| Feb | 9.4 | 2.7 | 6.1 | *+1.2* | 13.3 | 16 | −1.4 | 27 | 5.6 | 7 | 7.2 | 15 | 5 | 0 | 39.5 | *92* | 15 | 11.9 | 5 | | | | |
| Mar | 11.2 | 4.7 | 8.0 | *+0.7* | 17.8 | 4 | −5.6 | 31 | 4.2 | 31 | 10.6 | 12 | 4 | 0 | 35.0 | *73* | 13 | 10.3 | 30 | | | | |
| Apr | 12.3 | 3.5 | 7.9 | *−1.4* | 23.6 | 7 | −2.8 | 19 | 7.2 | 16 | 10.6 | 8 | 8 | 0 | 58.6 | *120* | 15 | 9.1 | 24 | | | | |
| May | 16.6 | 6.7 | 11.7 | *−0.8* | 22.2 | 30 | 2.8 | 2 | 10.3 | 1 | 11.9 | 28 | 0 | 0 | 43.4 | *76* | 10 | 7.9 | 7 | | | | |
| June | 20.7 | 11.1 | 15.9 | *+0.4* | 25.0 | 25 | 5.8 | 20 | 16.1 | 28 | 16.9 | 25 | 0 | 2 | 52.6 | *107* | 9 | 13.0 | 12 | | | | |
| July | 25.1 | 13.7 | 19.4 | *+1.5* | 30.0 | 12 | 7.5 | 24 | 19.4 | 1 | 17.8 | 18 | 0 | 16 | 65.3 | *134* | 8 | 27.5 | 20 | | | | |
| Aug | 21.8 | 12.5 | 17.2 | *−0.4* | 27.2 | 20 | 8.3 | 10 | 15.8 | 31 | 20.3 | 26 | 0 | 6 | 89.1 | *163* | 10 | 26.3 | 8 | | | | |
| Sep | 17.4 | 9.4 | 13.4 | *−1.5* | 21.9 | 24 | 3.9 | 17 | 13.3 | 13 | 15.6 | 24 | 0 | 0 | 82.3 | *152* | 15 | 18.4 | 13 | | | | |
| Oct | 13.7 | 6.6 | 10.1 | *−1.1* | 22.2 | 4 | −5.7 | 24 | 5.0 | 21 | 14.7 | 2 | 6 | 0 | 69.9 | *100* | 12 | 26.6 | 25 | | | | |
| Nov | 8.5 | 1.5 | 5.0 | *−2.5* | 14.7 | 6 | −6.4 | 14 | 3.1 | 30 | 10.3 | 6 | 8 | 0 | 48.4 | *73* | 12 | 11.2 | 29 | | | | |
| Dec | 5.0 | −1.1 | 1.9 | *−3.1* | 13.6 | 31 | −11.4 | 17 | −5.4 | 17 | 11.4 | 31 | 14 | 0 | 42.7 | *68* | 10 | 8.6 | 23 | | | | |
| **1859** | 14.1 | 6.1 | 10.1 | *−0.6* | 30.0 | 7 | −11.4 | 12 | −5.4 | 12 | 20.3 | 8 | 54 | 24 | 645.9 | *98* | 144 | 27.5 | 7 | | | | |

## OXFORD 1860

| Month | Mean max | Mean min | Mean temp | Anom | Highest max | Date | Lowest min | Date | Lowest max | Date | Highest min | Date | Air frost | Days ≥ 25 °C | Total pptn | Anom | Rain days | Wettest day | Date | Total sunshine | Anom | Sunniest day | Date |
|---|---|---|---|---|---|---|---|---|---|---|---|---|---|---|---|---|---|---|---|---|---|---|---|
| Jan | 6.8 | 0.8 | 3.8 | *−1.0* | 12.8 | 1 | −3.3 | 25 | 1.7 | 10 | 6.4 | 1 | 12 | 0 | 75.9 | *133* | 19 | 9.5 | 19 | | | | |
| Feb | 5.2 | −1.6 | 1.8 | *−3.1* | 10.6 | 27 | −8.1 | 13 | −0.8 | 13 | 3.3 | 26 | 20 | 0 | 29.9 | *70* | 6 | 9.3 | 11 | | | | |
| Mar | 8.4 | 1.9 | 5.2 | *−2.1* | 13.3 | 17 | −5.6 | 10 | 2.2 | 9 | 7.2 | 30 | 8 | 0 | 44.4 | *93* | 18 | 6.3 | 20 | | | | |
| Apr | 10.6 | 1.7 | 6.1 | *−3.2* | 17.2 | 30 | −2.5 | 21 | 6.9 | 19 | 6.1 | 7 | 5 | 0 | 13.6 | *28* | 8 | 4.4 | 23 | | | | |
| May | 17.0 | 7.7 | 12.4 | *−0.1* | 22.8 | 21 | 0.3 | 6 | 11.7 | 5 | 12.2 | 10 | 0 | 0 | 65.0 | *114* | 15 | 12.6 | 12 | | | | |
| June | 17.0 | 9.5 | 13.2 | *−2.4* | 20.6 | 24 | 5.8 | 7 | 13.3 | 9 | 13.1 | 23 | 0 | 0 | 119.0 | *242* | 26 | 21.9 | 2 | | | | |
| July | 19.3 | 10.3 | 14.8 | *−3.1* | 23.1 | 5 | 7.2 | 4 | 15.3 | 10 | 12.8 | 5 | 0 | 0 | 55.7 | *114* | 10 | 14.2 | 23 | | | | |
| Aug | 17.9 | 10.9 | 14.4 | *−3.2* | 20.3 | 16 | 6.7 | 31 | 15.3 | 6 | 14.2 | 3 | 0 | 0 | 88.6 | *162* | 19 | 14.4 | 10 | | | | |
| Sep | 15.6 | 7.0 | 11.3 | *−3.6* | 19.4 | 8 | 0.6 | 11 | 11.7 | 28 | 12.2 | 5 | 0 | 0 | 58.7 | *108* | 12 | 22.8 | 24 | | | | |
| Oct | 13.4 | 7.3 | 10.3 | *−1.0* | 17.2 | 6 | −0.4 | 11 | 7.8 | 11 | 12.8 | 30 | 1 | 0 | 36.6 | *52* | 15 | 7.2 | 10 | | | | |
| Nov | 7.4 | 2.1 | 4.8 | *−2.8* | 11.7 | 14 | −0.3 | 3 | 4.4 | 25 | 5.3 | 29 | 2 | 0 | 75.3 | *113* | 13 | 14.9 | 16 | | | | |
| Dec | 4.4 | −1.4 | 1.5 | *−3.6* | 11.7 | 6 | −17.8 | 24 | −2.8 | 28 | 7.5 | 6 | 13 | 0 | 61.0 | *97* | 11 | 21.9 | 31 | | | | |
| **1860** | 11.9 | 4.7 | 8.3 | *−2.4* | 23.1 | 7 | −17.8 | 12 | −2.8 | 12 | 14.2 | 8 | 61 | 0 | 723.6 | *110* | 172 | 22.8 | 9 | | | | |

## OXFORD 1861

| Month | Mean max | Mean min | Mean temp | Anom | Highest max | Date | Lowest min | Date | Lowest max | Date | Highest min | Date | Air frost | Days ≥ 25 °C | Total pptn | Anom | Rain days | Wettest day | Date | Total sunshine | Anom | Sunniest day | Date |
|---|---|---|---|---|---|---|---|---|---|---|---|---|---|---|---|---|---|---|---|---|---|---|---|
| Jan | 3.8 | −0.7 | 1.5 | *−3.3* | 11.4 | 25 | −7.8 | 7 | −1.4 | 6 | 6.8 | 26 | 16 | 0 | 15.4 | *27* | 2 | 11.2 | 25 | | | | |
| Feb | 7.9 | 3.1 | 5.5 | *+0.6* | 12.1 | 17 | −5.2 | 13 | 1.6 | 12 | 7.2 | 16 | 4 | 0 | 44.0 | *103* | 15 | 9.1 | 9 | | | | |
| Mar | 10.1 | 3.4 | 6.7 | *−0.5* | 13.9 | 23 | −1.1 | 13 | 6.8 | 17 | 7.2 | 7 | 1 | 0 | 39.5 | *83* | 17 | 8.6 | 27 | | | | |
| Apr | 11.8 | 3.0 | 7.4 | *−1.9* | 17.8 | 17 | −1.2 | 5 | 7.1 | 27 | 7.2 | 30 | 5 | 0 | 16.2 | *33* | 5 | 12.6 | 27 | | | | |
| May | 16.1 | 6.5 | 11.3 | *−1.2* | 23.7 | 23 | −2.2 | 8 | 7.6 | 11 | 12.7 | 22 | 1 | 0 | 32.0 | *56* | 8 | 12.6 | 10 | | | | |
| June | 20.0 | 11.2 | 15.6 | *−0.0* | 26.5 | 15 | 7.4 | 1 | 13.8 | 7 | 15.9 | 21 | 0 | 3 | 73.2 | *149* | 18 | 20.7 | 21 | | | | |
| July | 19.7 | 12.0 | 15.9 | *−2.0* | 22.5 | 20 | 8.1 | 6 | 16.9 | 4 | 15.2 | 19 | 0 | 0 | 119.9 | *246* | 22 | 47.8 | 25 | | | | |
| Aug | 21.6 | 12.7 | 17.1 | *−0.5* | 27.7 | 12 | 5.7 | 31 | 17.9 | 24 | 17.5 | 14 | 0 | 1 | 13.8 | *25* | 6 | 4.2 | 15 | | | | |
| Sep | 17.9 | 9.9 | 13.9 | *−1.0* | 24.0 | 1 | 3.9 | 26 | 14.2 | 15 | 16.7 | 5 | 0 | 0 | 45.5 | *84* | 16 | 10.4 | 28 | | | | |
| Oct | 16.2 | 9.5 | 12.8 | *+1.5* | 21.8 | 8 | 2.7 | 16 | 10.7 | 29 | 15.7 | 13 | 0 | 0 | 37.4 | *53* | 13 | 16.8 | 24 | | | | |
| Nov | 8.4 | 1.6 | 5.0 | *−2.6* | 13.7 | 26 | −4.8 | 17 | 2.0 | 17 | 11.7 | 29 | 10 | 0 | 71.9 | *108* | 16 | 33.7 | 13 | | | | |
| Dec | 7.0 | 1.5 | 4.2 | *−0.8* | 11.7 | 12 | −6.2 | 29 | −1.7 | 29 | 9.6 | 12 | 12 | 0 | 38.4 | *61* | 12 | 12.8 | 6 | | | | |
| **1861** | 13.4 | 6.1 | 9.7 | *−1.0* | 27.7 | 8 | −7.8 | 1 | −1.7 | 12 | 17.5 | 8 | 49 | 4 | 547.2 | *83* | 150 | 47.8 | 7 | | | | |

## OXFORD 1862

| Month | Mean max | Mean min | Mean temp | Anom | Highest max | Date | Lowest min | Date | Lowest max | Date | Highest min | Date | Air frost | Days ≥ 25 °C | Total pptn | Anom | Rain days | Wettest day | Date | Total sunshine | Anom | Sunniest day | Date |
|---|---|---|---|---|---|---|---|---|---|---|---|---|---|---|---|---|---|---|---|---|---|---|---|
| Jan | 6.6 | 2.2 | 4.4 | *−0.4* | 12.1 | 31 | −5.6 | 18 | −1.7 | 19 | 10.3 | 31 | 8 | 0 | 57.0 | *100* | 23 | 7.4 | 13 | | | | |
| Feb | 7.3 | 3.4 | 5.3 | *+0.4* | 12.3 | 20 | −6.1 | 8 | 0.3 | 8 | 9.4 | 2 | 4 | 0 | 6.7 | *16* | 9 | 1.6 | 19 | | | | |
| Mar | 9.4 | 3.9 | 6.6 | *−0.6* | 15.8 | 24 | −6.2 | 3 | 1.1 | 2 | 9.1 | 6 | 4 | 0 | 127.3 | *267* | 17 | 26.6 | 16 | | | | |
| Apr | 12.9 | 6.2 | 9.6 | *+0.3* | 20.6 | 25 | −1.1 | 11 | 5.7 | 12 | 10.6 | 19 | 4 | 0 | 52.8 | *108* | 10 | 21.0 | 9 | | | | |
| May | 17.5 | 9.2 | 13.4 | *+0.9* | 23.3 | 19 | 3.8 | 2 | 11.6 | 15 | 15.2 | 5 | 0 | 0 | 95.2 | *167* | 17 | 26.7 | 7 | | | | |
| June | 18.1 | 9.6 | 13.9 | *−1.7* | 23.2 | 2 | 5.4 | 9 | 14.5 | 20 | 13.5 | 6 | 0 | 0 | 55.7 | *113* | 16 | 13.9 | 12 | | | | |
| July | 19.9 | 10.6 | 15.3 | *−2.6* | 24.4 | 26 | 6.2 | 10 | 16.2 | 3 | 14.5 | 24 | 0 | 0 | 43.9 | *90* | 16 | 11.0 | 11 | | | | |
| Aug | 20.0 | 10.5 | 15.3 | *−2.3* | 22.9 | 3 | 4.6 | 23 | 15.4 | 17 | 14.1 | 1 | 0 | 0 | 43.8 | *80* | 9 | 15.4 | 16 | | | | |
| Sep | 18.2 | 9.9 | 14.0 | *−0.8* | 21.2 | 8 | 4.1 | 17 | 14.9 | 22 | 15.0 | 28 | 0 | 0 | 54.0 | *100* | 11 | 18.6 | 29 | | | | |
| Oct | 14.5 | 7.6 | 11.0 | *−0.2* | 19.7 | 4 | −1.9 | 29 | 8.7 | 28 | 15.7 | 2 | 1 | 0 | 71.6 | *102* | 17 | 13.7 | 18 | | | | |
| Nov | 6.7 | 0.4 | 3.6 | *−4.0* | 13.8 | 4 | −6.9 | 12 | 2.7 | 13 | 9.8 | 1 | 14 | 0 | 24.8 | *37* | 16 | 5.2 | 6 | | | | |
| Dec | 9.2 | 3.8 | 6.5 | *+1.5* | 12.8 | 6 | 0.3 | 21 | 4.6 | 21 | 9.8 | 5 | 0 | 0 | 35.3 | *56* | 13 | 8.2 | 9 | | | | |
| **1862** | 13.4 | 6.4 | 9.9 | *−0.8* | 24.4 | 7 | −6.9 | 11 | −1.7 | 1 | 15.7 | 10 | 35 | 0 | 668.1 | *101* | 174 | 26.7 | 5 | | | | |

## OXFORD 1863

| Month | Mean max | Mean min | Mean temp | Anom | Highest max | Date | Lowest min | Date | Lowest max | Date | Highest min | Date | Air frost | Days ≥ 25 °C | Total pptn | Anom | Rain days | Wettest day | Date | Total sunshine | Anom | Sunniest day | Date |
|---|---|---|---|---|---|---|---|---|---|---|---|---|---|---|---|---|---|---|---|---|---|---|---|
| Jan | 7.7 | 2.5 | 5.1 | *+0.3* | 12.0 | 1 | −3.4 | 11 | 3.2 | 16 | 8.8 | 22 | 2 | 0 | 80.5 | *141* | 22 | 14.4 | 6 | | | | |
| Feb | 9.1 | 2.7 | 5.9 | *+1.0* | 11.8 | 27 | −5.2 | 16 | 6.6 | 14 | 8.1 | 6 | 7 | 0 | 16.8 | *39* | 10 | 10.2 | 18 | | | | |
| Mar | 10.8 | 1.9 | 6.3 | *−0.9* | 15.4 | 3 | −4.4 | 17 | 4.3 | 11 | 8.8 | 28 | 9 | 0 | 17.0 | *36* | 7 | 6.5 | 7 | | | | |
| Apr | 14.2 | 4.7 | 9.4 | *+0.1* | 18.2 | 26 | −1.8 | 30 | 10.1 | 30 | 9.8 | 21 | 2 | 0 | 34.8 | *71* | 12 | 10.7 | 6 | | | | |
| May | 16.4 | 6.0 | 11.2 | *−1.3* | 23.6 | 29 | 0.5 | 22 | 9.4 | 19 | 12.9 | 30 | 0 | 0 | 22.7 | *40* | 9 | 6.0 | 12 | | | | |
| June | 19.3 | 9.6 | 14.4 | *−1.1* | 24.0 | 3 | 4.5 | 1 | 15.0 | 5 | 14.7 | 20 | 0 | 0 | 84.9 | *172* | 15 | 15.2 | 10 | | | | |
| July | 21.7 | 9.5 | 15.6 | *−2.2* | 26.7 | 11 | 2.4 | 18 | 16.3 | 21 | 14.7 | 12 | 0 | 6 | 16.7 | *34* | 3 | 11.7 | 21 | | | | |
| Aug | 21.2 | 11.9 | 16.5 | *−1.1* | 25.4 | 9 | 5.9 | 20 | 15.9 | 20 | 17.3 | 7 | 0 | 2 | 65.8 | *120* | 16 | 17.4 | 4 | | | | |
| Sep | 15.8 | 7.6 | 11.7 | *−3.2* | 19.1 | 19 | 1.3 | 29 | 13.1 | 28 | 11.3 | 1 | 0 | 0 | 68.9 | *127* | 17 | 16.7 | 4 | | | | |
| Oct | 14.1 | 7.0 | 10.6 | *−0.7* | 17.8 | 8 | −0.6 | 23 | 9.2 | 31 | 15.1 | 3 | 2 | 0 | 74.0 | *106* | 18 | 17.2 | 7 | | | | |
| Nov | 10.3 | 4.6 | 7.5 | *−0.1* | 13.9 | 25 | −2.7 | 28 | 5.2 | 29 | 11.9 | 4 | 6 | 0 | 50.1 | *75* | 16 | 11.0 | 1 | | | | |
| Dec | 9.1 | 3.3 | 6.2 | *+1.2* | 12.2 | 12 | −3.2 | 30 | 4.6 | 31 | 8.2 | 7 | 6 | 0 | 27.1 | *43* | 8 | 9.5 | 2 | | | | |
| **1863** | 14.1 | 5.9 | 10.0 | *−0.7* | 26.7 | 7 | −5.2 | 2 | 3.2 | 1 | 17.3 | 8 | 34 | 8 | 559.1 | *85* | 153 | 17.4 | 8 | | | | |

## OXFORD 1864

| Month | Mean max | Mean min | Mean temp | Anom | Highest max | Date | Lowest min | Date | Lowest max | Date | Highest min | Date | Air frost | Days ≥ 25 °C | Total pptn | Anom | Rain days | Wettest day | Date | Total sunshine | Anom | Sunniest day | Date |
|---|---|---|---|---|---|---|---|---|---|---|---|---|---|---|---|---|---|---|---|---|---|---|---|
| Jan | 4.8 | −0.2 | 2.3 | *−2.6* | 12.1 | 22 | −10.7 | 6 | −3.8 | 6 | 7.9 | 22 | 13 | 0 | 24.5 | *43* | 14 | 4.7 | 19 | | | | |
| Feb | 5.5 | 0.0 | 2.7 | *−2.2* | 13.4 | 13 | −7.9 | 9 | −0.7 | 20 | 8.2 | 2 | 16 | 0 | 36.2 | *84* | 11 | 11.5 | 11 | | | | |
| Mar | 9.5 | 1.4 | 5.5 | *−1.8* | 16.0 | 19 | −4.1 | 9 | 2.3 | 9 | 7.6 | 13 | 8 | 0 | 61.4 | *129* | 13 | 18.6 | 5 | | | | |
| Apr | 14.2 | 4.3 | 9.2 | *−0.1* | 21.3 | 20 | 0.4 | 1 | 5.4 | 5 | 9.7 | 10 | 0 | 0 | 40.6 | *83* | 7 | 13.9 | 16 | | | | |
| May | 18.2 | 7.9 | 13.0 | *+0.6* | 27.5 | 19 | 0.4 | 29 | 10.8 | 31 | 12.5 | 21 | 0 | 4 | 53.7 | *94* | 12 | 11.5 | 6 | | | | |
| June | 19.7 | 9.8 | 14.7 | *−0.8* | 23.1 | 7 | 3.0 | 3 | 15.1 | 3 | 14.1 | 17 | 0 | 0 | 25.1 | *51* | 11 | 7.9 | 22 | | | | |
| July | 22.7 | 10.3 | 16.5 | *−1.3* | 27.8 | 17 | 6.3 | 8 | 15.6 | 3 | 15.3 | 18 | 0 | 5 | 11.8 | *24* | 7 | 5.2 | 2 | | | | |
| Aug | 21.0 | 8.8 | 14.9 | *−2.7* | 27.8 | 5 | 0.2 | 26 | 16.2 | 23 | 16.7 | 30 | 0 | 7 | 19.4 | *35* | 8 | 7.0 | 8 | | | | |
| Sep | 18.3 | 9.4 | 13.8 | *−1.1* | 22.6 | 9 | 4.8 | 30 | 14.7 | 30 | 17.6 | 7 | 0 | 0 | 73.4 | *136* | 17 | 24.1 | 15 | | | | |
| Oct | 13.6 | 6.7 | 10.1 | *−1.1* | 18.2 | 19 | 1.5 | 30 | 9.9 | 30 | 11.3 | 25 | 0 | 0 | 39.0 | *56* | 8 | 18.6 | 26 | | | | |
| Nov | 8.9 | 2.0 | 5.5 | *−2.1* | 12.7 | 28 | −3.8 | 6 | 4.4 | 11 | 8.2 | 21 | 8 | 0 | 58.2 | *87* | 12 | 17.4 | 23 | | | | |
| Dec | 5.6 | 1.3 | 3.5 | *−1.6* | 11.6 | 5 | −6.4 | 17 | −0.1 | 17 | 8.2 | 4 | 13 | 0 | 9.3 | *15* | 6 | 4.7 | 7 | | | | |
| **1864** | 13.5 | 5.1 | 9.3 | *−1.4* | 27.8 | 7 | −10.7 | 1 | −3.8 | 1 | 17.6 | 9 | 58 | 16 | 452.6 | *69* | 126 | 24.1 | 9 | | | | |

## OXFORD 1865

| Month | Mean max | Mean min | Mean temp | Anom | Highest max | Date | Lowest min | Date | Lowest max | Date | Highest min | Date | Air frost | Days ≥ 25 °C | Total pptn | Anom | Rain days | Wettest day | Date | Total sunshine | Anom | Sunniest day | Date |
|---|---|---|---|---|---|---|---|---|---|---|---|---|---|---|---|---|---|---|---|---|---|---|---|
| Jan | 4.7 | −0.8 | 1.9 | *−2.9* | 10.5 | 10 | −14.4 | 28 | −0.1 | 28 | 4.7 | 9 | 14 | 0 | 76.5 | *135* | 15 | 20.4 | 26 | | | | |
| Feb | 5.0 | 0.2 | 2.6 | *−2.3* | 10.8 | 23 | −7.3 | 14 | −1.7 | 13 | 7.1 | 23 | 13 | 0 | 51.8 | *121* | 16 | 16.7 | 16 | | | | |
| Mar | 6.0 | −0.2 | 2.9 | *−4.4* | 11.3 | 31 | −5.1 | 20 | 2.7 | 20 | 5.7 | 31 | 13 | 0 | 24.4 | *51* | 9 | 10.0 | 4 | | | | |
| Apr | 17.2 | 5.4 | 11.3 | *+2.0* | 23.6 | 27 | −1.7 | 1 | 11.1 | 19 | 12.7 | 17 | 3 | 0 | 22.8 | *47* | 7 | 11.2 | 14 | | | | |
| May | 18.4 | 8.1 | 13.2 | *+0.8* | 25.8 | 21 | 3.1 | 12 | 9.6 | 10 | 13.7 | 27 | 0 | 1 | 54.7 | *96* | 16 | 8.2 | 8 | | | | |
| June | 21.9 | 10.8 | 16.3 | *+0.7* | 28.7 | 21 | 5.8 | 19 | 14.1 | 30 | 14.4 | 27 | 0 | 7 | 87.3 | *177* | 6 | 35.9 | 1 | | | | |
| July | 22.3 | 12.4 | 17.4 | *−0.5* | 26.6 | 4 | 8.1 | 1 | 14.6 | 31 | 17.0 | 15 | 0 | 7 | 69.7 | *143* | 13 | 11.2 | 5 | | | | |
| Aug | 20.0 | 10.8 | 15.4 | *−2.2* | 23.1 | 9 | 5.4 | 4 | 15.1 | 3 | 14.8 | 10 | 0 | 0 | 77.4 | *141* | 16 | 16.5 | 6 | | | | |
| Sep | 22.3 | 10.9 | 16.6 | *+1.7* | 26.9 | 15 | 5.2 | 27 | 17.0 | 21 | 17.7 | 10 | 0 | 4 | 4.6 | *9* | 4 | 3.2 | 20 | | | | |
| Oct | 14.6 | 6.1 | 10.4 | *−0.9* | 21.7 | 2 | 0.5 | 31 | 7.7 | 19 | 13.1 | 10 | 0 | 0 | 134.9 | *193* | 24 | 26.9 | 23 | | | | |
| Nov | 10.2 | 4.0 | 7.1 | *−0.4* | 13.4 | 17 | −2.3 | 1 | 7.7 | 13 | 11.1 | 19 | 6 | 0 | 64.4 | *96* | 16 | 22.0 | 19 | | | | |
| Dec | 8.4 | 3.4 | 5.9 | *+0.9* | 11.7 | 6 | −1.7 | 15 | 4.4 | 15 | 10.7 | 6 | 4 | 0 | 48.8 | *77* | 11 | 13.4 | 28 | | | | |
| **1865** | 14.2 | 5.9 | 10.1 | *−0.7* | 28.7 | 6 | −14.4 | 1 | −1.7 | 2 | 17.7 | 9 | 53 | 19 | 717.5 | *109* | 153 | 35.9 | 6 | | | | |

## OXFORD 1866

| Month | Mean max | Mean min | Mean temp | Anom | Highest max | Date | Lowest min | Date | Lowest max | Date | Highest min | Date | Air frost | Days ≥ 25 °C | Total pptn | Anom | Rain days | Wettest day | Date | Total sunshine | Anom | Sunniest day | Date |
|---|---|---|---|---|---|---|---|---|---|---|---|---|---|---|---|---|---|---|---|---|---|---|---|
| Jan | 8.6 | 3.4 | 6.0 | *+1.1* | 11.9 | 21 | −4.5 | 12 | 2.3 | 11 | 8.8 | 21 | 5 | 0 | 71.6 | *126* | 22 | 13.4 | 10 | | | | |
| Feb | 7.9 | 1.3 | 4.6 | *−0.3* | 12.7 | 1 | −8.7 | 28 | 1.1 | 28 | 10.0 | 6 | 11 | 0 | 71.7 | *168* | 16 | 10.0 | 1 | | | | |
| Mar | 8.4 | 1.8 | 5.1 | *−2.2* | 16.1 | 30 | −5.4 | 2 | 1.1 | 1 | 8.9 | 28 | 9 | 0 | 41.8 | *88* | 15 | 6.2 | 23 | | | | |
| Apr | 13.6 | 4.7 | 9.2 | *−0.1* | 23.9 | 27 | −0.6 | 1 | 7.8 | 4 | 12.2 | 27 | 3 | 0 | 51.2 | *105* | 14 | 24.1 | 28 | | | | |
| May | 15.7 | 4.8 | 10.2 | *−2.2* | 21.5 | 19 | −1.7 | 3 | 5.4 | 1 | 10.7 | 31 | 2 | 0 | 41.7 | *73* | 7 | 16.5 | 31 | | | | |
| June | 21.9 | 11.1 | 16.5 | *+0.9* | 28.1 | 27 | 5.8 | 16 | 15.8 | 17 | 16.0 | 9 | 0 | 7 | 80.1 | *163* | 15 | 20.7 | 4 | | | | |
| July | 21.7 | 11.2 | 16.4 | *−1.4* | 29.7 | 12 | 7.4 | 22 | 16.6 | 24 | 15.2 | 9 | 0 | 7 | 49.4 | *101* | 11 | 19.7 | 28 | | | | |
| Aug | 19.8 | 11.3 | 15.6 | *−2.0* | 23.7 | 26 | 5.7 | 18 | 15.8 | 29 | 14.8 | 23 | 0 | 0 | 73.6 | *134* | 15 | 16.5 | 28 | | | | |
| Sep | 17.0 | 10.3 | 13.7 | *−1.2* | 19.8 | 4 | 4.5 | 24 | 13.4 | 22 | 16.2 | 4 | 0 | 0 | 142.5 | *263* | 23 | 21.7 | 23 | | | | |
| Oct | 14.2 | 8.0 | 11.1 | *−0.2* | 19.4 | 3 | 0.4 | 15 | 8.8 | 25 | 13.6 | 3 | 0 | 0 | 52.3 | *75* | 11 | 12.9 | 22 | | | | |
| Nov | 10.7 | 3.7 | 7.2 | *−0.4* | 16.1 | 2 | −2.4 | 30 | 3.6 | 19 | 10.6 | 2 | 7 | 0 | 38.0 | *57* | 9 | 12.0 | 12 | | | | |
| Dec | 9.2 | 3.3 | 6.2 | *+1.2* | 14.2 | 6 | −3.4 | 19 | 1.1 | 1 | 10.1 | 4 | 7 | 0 | 50.6 | *80* | 18 | 14.4 | 5 | | | | |
| **1866** | 14.1 | 6.2 | 10.1 | *−0.6* | 29.7 | 7 | −8.7 | 2 | 1.1 | 2 | 16.2 | 9 | 44 | 14 | 764.6 | *116* | 176 | 24.1 | 4 | | | | |

## OXFORD 1867

| Month | Mean max | Mean min | Mean temp | Anom | Highest max | Date | Lowest min | Date | Lowest max | Date | Highest min | Date | Air frost | Days ≥ 25 °C | Total pptn | Anom | Rain days | Wettest day | Date | Total sunshine | Anom | Sunniest day | Date |
|---|---|---|---|---|---|---|---|---|---|---|---|---|---|---|---|---|---|---|---|---|---|---|---|
| Jan | 4.6 | −1.1 | 1.7 | *−3.1* | 13.3 | 27 | −14.6 | 4 | −5.3 | 4 | 9.9 | 27 | 18 | 0 | 64.5 | *113* | 14 | 18.4 | 30 | | | | |
| Feb | 10.1 | 4.4 | 7.2 | *+2.3* | 12.6 | 16 | −2.9 | 28 | 5.3 | 27 | 8.4 | 16 | 1 | 0 | 38.5 | *90* | 12 | 9.2 | 5 | | | | |
| Mar | 6.0 | 0.5 | 3.2 | *−4.0* | 13.8 | 24 | −7.3 | 15 | 1.0 | 13 | 8.7 | 25 | 16 | 0 | 71.9 | *151* | 18 | 17.0 | 19 | | | | |
| Apr | 14.0 | 6.6 | 10.3 | *+1.0* | 17.7 | 23 | 0.6 | 11 | 8.4 | 25 | 10.9 | 18 | 0 | 0 | 64.8 | *132* | 18 | 11.2 | 24 | | | | |
| May | 16.8 | 6.8 | 11.8 | *−0.6* | 26.6 | 6 | −1.0 | 23 | 9.2 | 22 | 13.7 | 6 | 3 | 2 | 62.6 | *110* | 10 | 26.9 | 20 | | | | |
| June | 20.0 | 9.6 | 14.8 | *−0.8* | 26.0 | 11 | 5.1 | 28 | 15.2 | 15 | 14.3 | 27 | 0 | 3 | 48.1 | *98* | 8 | 16.7 | 2 | | | | |
| July | 20.3 | 10.5 | 15.4 | *−2.4* | 25.7 | 1 | 5.6 | 27 | 14.1 | 26 | 14.4 | 20 | 0 | 1 | 97.7 | *200* | 17 | 22.6 | 25 | | | | |
| Aug | 21.5 | 11.7 | 16.6 | *−1.0* | 28.4 | 13 | 7.0 | 27 | 15.7 | 1 | 17.1 | 19 | 0 | 4 | 59.6 | *109* | 9 | 20.2 | 15 | | | | |
| Sep | 18.4 | 9.6 | 14.0 | *−0.9* | 23.9 | 1 | 1.8 | 24 | 14.2 | 17 | 15.5 | 3 | 0 | 0 | 40.8 | *75* | 10 | 17.2 | 9 | | | | |
| Oct | 13.5 | 6.3 | 9.9 | *−1.4* | 18.7 | 22 | −0.9 | 10 | 8.7 | 4 | 12.9 | 22 | 1 | 0 | 71.1 | *102* | 21 | 15.2 | 11 | | | | |
| Nov | 8.3 | 1.6 | 5.0 | *−2.6* | 16.0 | 1 | −2.8 | 27 | 4.5 | 11 | 9.6 | 14 | 10 | 0 | 23.9 | *36* | 6 | 18.6 | 30 | | | | |
| Dec | 6.2 | 0.1 | 3.1 | *−1.9* | 13.3 | 14 | −6.2 | 8 | −0.6 | 31 | 9.9 | 15 | 18 | 0 | 35.0 | *55* | 10 | 8.7 | 1 | | | | |
| 1867 | 13.3 | 5.6 | 9.4 | *−1.3* | 28.4 | 8 | −14.6 | 1 | −5.3 | 1 | 17.1 | 8 | 67 | 10 | 678.5 | *103* | 153 | 26.9 | 5 | | | | |

## OXFORD 1868

| Month | Mean max | Mean min | Mean temp | Anom | Highest max | Date | Lowest min | Date | Lowest max | Date | Highest min | Date | Air frost | Days ≥ 25 °C | Total pptn | Anom | Rain days | Wettest day | Date | Total sunshine | Anom | Sunniest day | Date |
|---|---|---|---|---|---|---|---|---|---|---|---|---|---|---|---|---|---|---|---|---|---|---|---|
| Jan | 5.9 | 1.7 | 3.8 | *−1.1* | 12.3 | 31 | −4.2 | 2 | −0.6 | 9 | 10.8 | 31 | 11 | 0 | 85.5 | *150* | 23 | 14.9 | 18 | | | | |
| Feb | 9.8 | 3.3 | 6.6 | *+1.7* | 15.0 | 25 | −2.3 | 8 | 5.6 | 8 | 9.3 | 24 | 5 | 0 | 41.5 | *97* | 11 | 18.1 | 29 | | | | |
| Mar | 11.1 | 2.9 | 7.0 | *−0.3* | 15.0 | 26 | −3.9 | 24 | 5.2 | 24 | 9.8 | 4 | 6 | 0 | 37.9 | *80* | 16 | 12.2 | 7 | | | | |
| Apr | 14.1 | 4.5 | 9.3 | *+0.0* | 18.2 | 30 | −4.6 | 12 | 7.7 | 12 | 10.7 | 15 | 5 | 0 | 38.9 | *80* | 11 | 11.5 | 20 | | | | |
| May | 20.4 | 7.8 | 14.1 | *+1.6* | 29.1 | 19 | 0.7 | 7 | 14.7 | 6 | 15.4 | 18 | 0 | 4 | 14.6 | *26* | 9 | 6.0 | 22 | | | | |
| June | 23.6 | 9.4 | 16.5 | *+0.9* | 29.2 | 20 | 4.0 | 7 | 17.1 | 8 | 15.7 | 27 | 0 | 9 | 23.1 | *47* | 4 | 13.4 | 21 | | | | |
| July | 25.8 | 12.7 | 19.3 | *+1.4* | 32.2 | 22 | 7.3 | 31 | 19.7 | 29 | 18.0 | 21 | 0 | 19 | 46.6 | *95* | 6 | 33.3 | 11 | | | | |
| Aug | 22.1 | 12.5 | 17.3 | *−0.3* | 31.4 | 5 | 8.6 | 25 | 16.7 | 19 | 17.9 | 6 | 0 | 6 | 84.1 | *153* | 16 | 20.2 | 18 | | | | |
| Sep | 19.7 | 9.4 | 14.5 | *−0.3* | 29.7 | 7 | 3.3 | 12 | 15.3 | 14 | 14.8 | 7 | 0 | 5 | 99.3 | *183* | 11 | 20.4 | 29 | | | | |
| Oct | 12.6 | 5.0 | 8.8 | *−2.5* | 15.9 | 13 | −1.9 | 19 | 8.2 | 20 | 11.6 | 31 | 3 | 0 | 55.9 | *80* | 12 | 15.4 | 3 | | | | |
| Nov | 7.9 | 2.5 | 5.2 | *−2.3* | 15.4 | 1 | −3.5 | 5 | 4.3 | 6 | 11.0 | 3 | 7 | 0 | 30.1 | *45* | 13 | 7.9 | 21 | | | | |
| Dec | 10.4 | 5.2 | 7.8 | *+2.8* | 14.1 | 5 | −2.0 | 31 | 4.4 | 30 | 10.6 | 4 | 2 | 0 | 103.2 | *164* | 25 | 17.2 | 28 | | | | |
| 1868 | 15.3 | 6.4 | 10.8 | *+0.1* | 32.2 | 7 | −4.6 | 4 | −0.6 | 1 | 18.0 | 7 | 39 | 43 | 660.5 | *100* | 157 | 33.3 | 7 | | | | |

## OXFORD 1869

| Month | Mean max | Mean min | Mean temp | Anom | Highest max | Date | Lowest min | Date | Lowest max | Date | Highest min | Date | Air frost | Days ≥ 25 °C | Total pptn | Anom | Rain days | Wettest day | Date | Total sunshine | Anom | Sunniest day | Date |
|---|---|---|---|---|---|---|---|---|---|---|---|---|---|---|---|---|---|---|---|---|---|---|---|
| Jan | 8.2 | 3.0 | 5.6 | *+0.8* | 12.8 | 31 | −3.4 | 23 | 0.6 | 24 | 8.1 | 8 | 5 | 0 | 94.4 | *166* | 16 | 20.7 | 28 | | | | |
| Feb | 10.5 | 5.1 | 7.8 | *+2.9* | 14.7 | 5 | −0.9 | 23 | 3.7 | 22 | 10.0 | 7 | 1 | 0 | 54.1 | *126* | 16 | 16.2 | 11 | | | | |
| Mar | 7.1 | 1.1 | 4.1 | *−3.2* | 12.7 | 5 | −3.1 | 6 | 4.0 | 15 | 4.7 | 5 | 10 | 0 | 32.4 | *68* | 14 | 17.9 | 19 | | | | |
| Apr | 16.0 | 6.0 | 11.0 | *+1.7* | 25.2 | 14 | 1.7 | 3 | 8.6 | 3 | 11.1 | 11 | 0 | 1 | 29.2 | *60* | 12 | 7.4 | 23 | | | | |
| May | 15.5 | 6.4 | 11.0 | *−1.5* | 20.2 | 26 | 0.4 | 1 | 9.3 | 28 | 10.7 | 6 | 0 | 0 | 74.4 | *130* | 20 | 18.6 | 3 | | | | |
| June | 18.9 | 8.8 | 13.9 | *−1.7* | 29.6 | 7 | 4.3 | 10 | 13.7 | 17 | 12.3 | 6 | 0 | 2 | 35.5 | *72* | 10 | 18.4 | 13 | | | | |
| July | 24.1 | 12.4 | 18.3 | *+0.4* | 31.3 | 22 | 8.2 | 13 | 15.6 | 2 | 17.0 | 17 | 0 | 13 | 26.5 | *54* | 10 | 10.7 | 29 | | | | |
| Aug | 21.5 | 10.6 | 16.1 | *−1.5* | 29.7 | 27 | 2.7 | 30 | 16.2 | 29 | 15.4 | 8 | 0 | 5 | 32.9 | *60* | 8 | 11.5 | 2 | | | | |
| Sep | 19.0 | 11.3 | 15.1 | *+0.3* | 24.2 | 5 | 4.4 | 3 | 15.3 | 19 | 15.9 | 4 | 0 | 0 | 109.9 | *203* | 16 | 20.7 | 30 | | | | |
| Oct | 13.9 | 6.1 | 10.0 | *−1.2* | 22.4 | 8 | −2.1 | 27 | 3.5 | 27 | 13.8 | 12 | 4 | 0 | 27.5 | *39* | 12 | 6.5 | 17 | | | | |
| Nov | 10.0 | 3.1 | 6.6 | *−1.0* | 15.0 | 14 | −2.6 | 11 | 2.7 | 30 | 10.4 | 14 | 8 | 0 | 60.1 | *90* | 15 | 26.9 | 27 | | | | |
| Dec | 5.6 | 0.9 | 3.3 | *−1.8* | 13.8 | 18 | −6.2 | 28 | −0.7 | 26 | 6.1 | 19 | 10 | 0 | 85.8 | *136* | 19 | 22.0 | 16 | | | | |
| **1869** | 14.2 | 6.3 | 10.2 | *−0.5* | 31.3 | 7 | −6.2 | 12 | −0.7 | 12 | 17.0 | 7 | 38 | 21 | 662.6 | *100* | 168 | 26.9 | 11 | | | | |

## OXFORD 1870

| Month | Mean max | Mean min | Mean temp | Anom | Highest max | Date | Lowest min | Date | Lowest max | Date | Highest min | Date | Air frost | Days ≥ 25 °C | Total pptn | Anom | Rain days | Wettest day | Date | Total sunshine | Anom | Sunniest day | Date |
|---|---|---|---|---|---|---|---|---|---|---|---|---|---|---|---|---|---|---|---|---|---|---|---|
| Jan | 5.9 | 1.3 | 3.6 | *−1.2* | 11.1 | 7 | −8.1 | 27 | 0.6 | 20 | 6.9 | 4 | 10 | 0 | 29.6 | *52* | 18 | 6.2 | 8 | | | | |
| Feb | 5.1 | 0.8 | 2.9 | *−2.0* | 12.2 | 28 | −6.5 | 11 | −2.4 | 12 | 8.0 | 28 | 13 | 0 | 47.7 | *111* | 14 | 16.2 | 6 | | | | |
| Mar | 8.3 | 1.5 | 4.9 | *−2.4* | 14.3 | 17 | −4.8 | 13 | 3.2 | 13 | 9.6 | 1 | 13 | 0 | 41.7 | *88* | 13 | 14.7 | 1 | | | | |
| Apr | 15.6 | 3.7 | 9.6 | *+0.3* | 24.3 | 20 | −3.4 | 4 | 8.4 | 28 | 10.1 | 19 | 4 | 0 | 13.2 | *27* | 6 | 7.0 | 9 | | | | |
| May | 18.0 | 6.3 | 12.2 | *−0.3* | 26.8 | 21 | −1.7 | 3 | 9.7 | 2 | 12.9 | 20 | 2 | 1 | 25.8 | *45* | 8 | 11.0 | 11 | | | | |
| June | 22.3 | 11.2 | 16.8 | *+1.2* | 30.1 | 16 | 6.5 | 27 | 17.6 | 24 | 15.9 | 21 | 0 | 6 | 17.0 | *34* | 6 | 8.7 | 16 | | | | |
| July | 24.9 | 13.0 | 19.0 | *+1.1* | 30.6 | 25 | 6.0 | 1 | 18.4 | 29 | 16.9 | 8 | 0 | 15 | 22.1 | *45* | 8 | 6.0 | 31 | | | | |
| Aug | 21.4 | 10.5 | 16.0 | *−1.6* | 27.2 | 1 | 4.2 | 20 | 17.2 | 29 | 16.3 | 1 | 0 | 5 | 58.6 | *107* | 8 | 21.4 | 22 | | | | |
| Sep | 18.3 | 8.2 | 13.2 | *−1.6* | 20.8 | 27 | 1.3 | 14 | 15.3 | 13 | 13.9 | 1 | 0 | 0 | 33.1 | *61* | 10 | 7.7 | 2 | | | | |
| Oct | 14.1 | 6.7 | 10.4 | *−0.9* | 19.9 | 2 | 0.0 | 14 | 10.4 | 26 | 13.1 | 7 | 0 | 0 | 73.1 | *104* | 21 | 11.2 | 19 | | | | |
| Nov | 8.3 | 2.1 | 5.2 | *−2.3* | 13.2 | 24 | −3.9 | 16 | 3.4 | 10 | 9.1 | 24 | 10 | 0 | 29.0 | *43* | 14 | 8.9 | 22 | | | | |
| Dec | 3.2 | −2.3 | 0.4 | *−4.6* | 14.3 | 14 | −12.4 | 24 | −4.6 | 31 | 7.4 | 13 | 19 | 0 | 49.9 | *79* | 13 | 14.9 | 13 | | | | |
| **1870** | 13.8 | 5.2 | 9.5 | *−1.2* | 30.6 | 7 | −12.4 | 12 | −4.6 | 12 | 16.9 | 7 | 71 | 27 | 440.9 | *67* | 139 | 21.4 | 8 | | | | |

## OXFORD 1871

| Month | Mean max | Mean min | Mean temp | Anom | Highest max | Date | Lowest min | Date | Lowest max | Date | Highest min | Date | Air frost | Days ≥ 25 °C | Total pptn | Anom | Rain days | Wettest day | Date | Total sunshine | Anom | Sunniest day | Date |
|---|---|---|---|---|---|---|---|---|---|---|---|---|---|---|---|---|---|---|---|---|---|---|---|
| Jan | 2.7 | −1.2 | 0.7 | *−4.1* | 8.0 | 6 | −8.2 | 12 | −2.0 | 1 | 3.2 | 6 | 20 | 0 | 27.6 | *49* | 14 | 10.7 | 15 | | | | |
| Feb | 9.0 | 3.8 | 6.4 | *+1.5* | 13.2 | 27 | −2.8 | 11 | 0.9 | 1 | 8.8 | 27 | 4 | 0 | 24.6 | *57* | 14 | 6.8 | 9 | | | | |
| Mar | 11.9 | 3.2 | 7.5 | *+0.3* | 18.7 | 26 | −2.2 | 1 | 4.8 | 15 | 8.3 | 11 | 4 | 0 | 23.8 | *50* | 10 | 8.4 | 15 | | | | |
| Apr | 13.8 | 5.3 | 9.5 | *+0.2* | 18.1 | 29 | −2.9 | 6 | 9.1 | 6 | 10.3 | 18 | 4 | 0 | 65.8 | *134* | 16 | 19.9 | 18 | | | | |
| May | 17.0 | 5.8 | 11.4 | *−1.1* | 26.9 | 24 | 1.4 | 13 | 10.2 | 10 | 12.6 | 24 | 0 | 1 | 10.8 | *19* | 4 | 7.0 | 25 | | | | |
| June | 17.5 | 8.9 | 13.2 | *−2.4* | 22.9 | 16 | 2.7 | 4 | 12.6 | 4 | 15.5 | 14 | 0 | 0 | 79.7 | *162* | 13 | 26.2 | 14 | | | | |
| July | 21.1 | 11.9 | 16.5 | *−1.4* | 26.8 | 17 | 7.1 | 30 | 16.8 | 2 | 16.4 | 13 | 0 | 2 | 94.4 | *193* | 21 | 30.9 | 10 | | | | |
| Aug | 23.8 | 12.0 | 17.9 | *+0.3* | 29.5 | 12 | 7.4 | 27 | 18.8 | 4 | 17.1 | 13 | 0 | 12 | 12.7 | *23* | 6 | 7.0 | 17 | | | | |
| Sep | 17.9 | 9.6 | 13.7 | *−1.2* | 25.2 | 1 | 2.3 | 22 | 10.5 | 29 | 15.3 | 10 | 0 | 1 | 116.4 | *215* | 12 | 25.2 | 29 | | | | |
| Oct | 14.0 | 6.0 | 10.0 | *−1.3* | 18.1 | 18 | −1.7 | 12 | 11.2 | 30 | 13.9 | 18 | 2 | 0 | 29.8 | *43* | 13 | 6.2 | 21 | | | | |
| Nov | 6.3 | 0.2 | 3.3 | *−4.3* | 11.8 | 15 | −7.2 | 18 | 3.0 | 17 | 7.0 | 1 | 14 | 0 | 18.0 | *27* | 9 | 8.9 | 14 | | | | |
| Dec | 6.0 | 0.9 | 3.5 | *−1.6* | 9.8 | 18 | −5.6 | 4 | −1.0 | 8 | 8.2 | 18 | 12 | 0 | 26.9 | *43* | 12 | 10.7 | 19 | | | | |
| 1871 | 13.4 | 5.5 | 9.5 | *−1.3* | 29.5 | 8 | −8.2 | 1 | −2.0 | 1 | 17.1 | 8 | 60 | 16 | 530.5 | *80* | 144 | 30.9 | 7 | | | | |

## OXFORD 1872

| Month | Mean max | Mean min | Mean temp | Anom | Highest max | Date | Lowest min | Date | Lowest max | Date | Highest min | Date | Air frost | Days ≥ 25 °C | Total pptn | Anom | Rain days | Wettest day | Date | Total sunshine | Anom | Sunniest day | Date |
|---|---|---|---|---|---|---|---|---|---|---|---|---|---|---|---|---|---|---|---|---|---|---|---|
| Jan | 7.8 | 2.6 | 5.2 | *+0.4* | 11.7 | 4 | −1.2 | 9 | 3.8 | 15 | 7.3 | 29 | 5 | 0 | 101.6 | *179* | 23 | 16.2 | 23 | | | | |
| Feb | 10.1 | 4.7 | 7.4 | *+2.5* | 13.3 | 24 | 1.1 | 19 | 6.4 | 15 | 10.0 | 29 | 0 | 0 | 37.8 | *88* | 17 | 8.9 | 5 | | | | |
| Mar | 11.2 | 3.6 | 7.4 | *+0.1* | 15.4 | 4 | −3.7 | 25 | 3.3 | 21 | 11.1 | 28 | 8 | 0 | 48.1 | *101* | 15 | 11.7 | 28 | | | | |
| Apr | 13.9 | 4.4 | 9.2 | *−0.1* | 19.8 | 12 | −1.9 | 19 | 8.6 | 3 | 10.2 | 7 | 2 | 0 | 47.0 | *96* | 14 | 10.5 | 27 | | | | |
| May | 15.2 | 5.9 | 10.6 | *−1.9* | 22.0 | 27 | −0.6 | 19 | 8.4 | 18 | 11.2 | 27 | 1 | 0 | 63.9 | *112* | 19 | 12.9 | 7 | | | | |
| June | 20.4 | 10.2 | 15.3 | *−0.3* | 29.7 | 18 | 3.2 | 6 | 14.1 | 8 | 15.3 | 17 | 0 | 5 | 72.6 | *147* | 17 | 9.7 | 11 | | | | |
| July | 23.7 | 12.9 | 18.3 | *+0.4* | 29.6 | 25 | 7.3 | 30 | 17.8 | 15 | 18.4 | 25 | 0 | 13 | 74.4 | *152* | 12 | 17.6 | 24 | | | | |
| Aug | 21.0 | 11.3 | 16.1 | *−1.4* | 26.3 | 17 | 6.0 | 27 | 17.0 | 30 | 15.1 | 21 | 0 | 2 | 29.5 | *54* | 14 | 7.7 | 7 | | | | |
| Sep | 18.3 | 10.0 | 14.2 | *−0.7* | 25.6 | 3 | 1.4 | 21 | 11.4 | 22 | 16.5 | 11 | 0 | 1 | 25.0 | *46* | 14 | 3.7 | 1 | | | | |
| Oct | 12.6 | 5.2 | 8.9 | *−2.4* | 18.6 | 2 | −0.8 | 13 | 8.2 | 15 | 12.4 | 1 | 1 | 0 | 72.9 | *104* | 22 | 10.0 | 24 | | | | |
| Nov | 10.0 | 4.6 | 7.3 | *−0.2* | 16.7 | 6 | −0.9 | 17 | 4.3 | 13 | 10.2 | 5 | 1 | 0 | 79.2 | *119* | 21 | 10.5 | 30 | | | | |
| Dec | 7.9 | 3.5 | 5.7 | *+0.7* | 12.7 | 22 | −3.7 | 11 | 3.4 | 12 | 8.6 | 27 | 6 | 0 | 95.0 | *150* | 22 | 28.9 | 16 | | | | |
| 1872 | 14.4 | 6.6 | 10.5 | *−0.3* | 29.7 | 6 | −3.7 | 3 | 3.3 | 3 | 18.4 | 7 | 24 | 21 | 747.1 | *113* | 210 | 28.9 | 12 | | | | |

## OXFORD 1873

| Month | Mean max | Mean min | Mean temp | Anom | Highest max | Date | Lowest min | Date | Lowest max | Date | Highest min | Date | Air frost | Days ≥ 25 °C | Total pptn | Anom | Rain days | Wettest day | Date | Total sunshine | Anom | Sunniest day | Date |
|---|---|---|---|---|---|---|---|---|---|---|---|---|---|---|---|---|---|---|---|---|---|---|---|
| Jan | 7.8 | 3.2 | 5.5 | *+0.7* | 12.2 | 10 | −4.3 | 28 | 1.6 | 29 | 9.9 | 13 | 9 | 0 | 55.2 | *97* | 18 | 10.7 | 18 | | | | |
| Feb | 4.2 | 0.0 | 2.1 | *−2.8* | 9.9 | 26 | −3.9 | 22 | −0.1 | 2 | 3.8 | 26 | 13 | 0 | 40.1 | *94* | 13 | 8.9 | 25 | | | | |
| Mar | 9.7 | 2.2 | 6.0 | *−1.3* | 17.6 | 30 | −2.7 | 12 | 3.9 | 14 | 5.6 | 3 | 3 | 0 | 58.9 | *124* | 19 | 14.9 | 16 | | | | |
| Apr | 12.8 | 3.9 | 8.3 | *−0.9* | 21.5 | 15 | −1.1 | 24 | 7.6 | 24 | 11.7 | 30 | 3 | 0 | 12.0 | *24* | 11 | 3.2 | 17 | | | | |
| May | 16.1 | 5.4 | 10.7 | *−1.7* | 19.7 | 12 | −0.9 | 19 | 9.2 | 18 | 11.5 | 10 | 1 | 0 | 58.1 | *102* | 15 | 12.2 | 7 | | | | |
| June | 20.0 | 10.3 | 15.2 | *−0.4* | 26.0 | 29 | 5.4 | 3 | 14.0 | 6 | 16.0 | 21 | 0 | 3 | 69.3 | *141* | 12 | 30.9 | 29 | | | | |
| July | 22.6 | 11.8 | 17.2 | *−0.7* | 31.1 | 22 | 7.6 | 18 | 17.3 | 13 | 17.5 | 22 | 0 | 6 | 56.2 | *115* | 14 | 23.9 | 13 | | | | |
| Aug | 21.6 | 12.6 | 17.1 | *−0.5* | 27.2 | 7 | 8.3 | 28 | 16.5 | 29 | 16.6 | 12 | 0 | 3 | 65.5 | *119* | 18 | 12.4 | 24 | | | | |
| Sep | 16.9 | 7.8 | 12.3 | *−2.5* | 20.2 | 27 | 1.6 | 28 | 14.4 | 29 | 13.2 | 20 | 0 | 0 | 45.0 | *83* | 13 | 9.2 | 10 | | | | |
| Oct | 12.9 | 4.7 | 8.8 | *−2.5* | 22.1 | 3 | −5.0 | 28 | 5.7 | 29 | 14.9 | 10 | 6 | 0 | 67.2 | *96* | 13 | 15.2 | 12 | | | | |
| Nov | 9.6 | 3.8 | 6.7 | *−0.8* | 14.9 | 22 | −2.3 | 3 | 6.6 | 16 | 8.0 | 28 | 4 | 0 | 42.8 | *64* | 13 | 15.2 | 5 | | | | |
| Dec | 7.3 | 2.1 | 4.7 | *−0.3* | 14.4 | 16 | −6.6 | 9 | −2.8 | 11 | 8.9 | 2 | 10 | 0 | 12.1 | *19* | 7 | 4.5 | 26 | | | | |
| 1873 | 13.5 | 5.6 | 9.6 | *−1.2* | 31.1 | 7 | −6.6 | 12 | −2.8 | 12 | 17.5 | 7 | 49 | 12 | 582.2 | *88* | 166 | 30.9 | 6 | | | | |

## OXFORD 1874

| Month | Mean max | Mean min | Mean temp | Anom | Highest max | Date | Lowest min | Date | Lowest max | Date | Highest min | Date | Air frost | Days ≥ 25 °C | Total pptn | Anom | Rain days | Wettest day | Date | Total sunshine | Anom | Sunniest day | Date |
|---|---|---|---|---|---|---|---|---|---|---|---|---|---|---|---|---|---|---|---|---|---|---|---|
| Jan | 8.6 | 3.0 | 5.8 | *+0.9* | 12.2 | 20 | −1.2 | 5 | 3.9 | 5 | 7.8 | 14 | 6 | 0 | 57.6 | *101* | 19 | 9.7 | 23 | | | | |
| Feb | 7.1 | 1.0 | 4.1 | *−0.8* | 12.4 | 14 | −5.0 | 10 | −0.1 | 6 | 8.2 | 13 | 12 | 0 | 42.3 | *99* | 15 | 13.9 | 26 | | | | |
| Mar | 11.0 | 3.1 | 7.0 | *−0.2* | 17.6 | 27 | −5.1 | 10 | 2.1 | 11 | 9.4 | 28 | 8 | 0 | 15.5 | *32* | 12 | 3.2 | 8 | | | | |
| Apr | 15.7 | 5.4 | 10.5 | *+1.2* | 24.2 | 27 | −0.3 | 10 | 10.0 | 4 | 11.9 | 24 | 1 | 0 | 31.9 | *65* | 10 | 8.7 | 9 | | | | |
| May | 15.7 | 5.9 | 10.8 | *−1.6* | 22.4 | 27 | −0.8 | 10 | 9.8 | 4 | 13.2 | 31 | 1 | 0 | 38.9 | *68* | 9 | 25.7 | 22 | | | | |
| June | 20.4 | 8.9 | 14.7 | *−0.9* | 27.1 | 9 | 3.1 | 12 | 13.6 | 18 | 15.3 | 30 | 0 | 4 | 17.2 | *35* | 8 | 4.2 | 23 | | | | |
| July | 25.3 | 11.8 | 18.5 | *+0.7* | 31.2 | 9 | 6.4 | 5 | 20.9 | 22 | 15.3 | 9 | 0 | 15 | 12.5 | *26* | 10 | 4.2 | 25 | | | | |
| Aug | 21.2 | 11.2 | 16.2 | *−1.4* | 26.3 | 20 | 7.2 | 21 | 16.8 | 5 | 15.9 | 31 | 0 | 2 | 45.0 | *82* | 16 | 15.4 | 7 | | | | |
| Sep | 18.6 | 10.4 | 14.5 | *−0.4* | 22.8 | 25 | 6.6 | 12 | 15.1 | 17 | 13.7 | 20 | 0 | 0 | 84.0 | *155* | 17 | 22.1 | 3 | | | | |
| Oct | 14.7 | 7.8 | 11.2 | *−0.1* | 17.7 | 1 | 1.5 | 5 | 10.4 | 23 | 12.6 | 25 | 0 | 0 | 78.4 | *112* | 18 | 22.1 | 6 | | | | |
| Nov | 8.7 | 2.5 | 5.6 | *−1.9* | 13.9 | 9 | −4.5 | 23 | 0.4 | 26 | 10.3 | 4 | 9 | 0 | 63.3 | *95* | 14 | 23.1 | 28 | | | | |
| Dec | 3.5 | −2.2 | 0.6 | *−4.4* | 11.6 | 6 | −8.9 | 31 | −3.4 | 22 | 5.6 | 5 | 22 | 0 | 45.9 | *73* | 10 | 11.0 | 8 | | | | |
| 1874 | 14.2 | 5.7 | 10.0 | *−0.8* | 31.2 | 7 | −8.9 | 12 | −3.4 | 12 | 15.9 | 8 | 59 | 21 | 532.5 | *81* | 158 | 25.7 | 5 | | | | |

## OXFORD 1875

| Month | Mean max | Mean min | Mean temp | Anom | Highest max | Date | Lowest min | Date | Lowest max | Date | Highest min | Date | Air frost | Days ≥ 25 °C | Total pptn | Anom | Rain days | Wettest day | Date | Total sunshine | Anom | Sunniest day | Date |
|---|---|---|---|---|---|---|---|---|---|---|---|---|---|---|---|---|---|---|---|---|---|---|---|
| Jan | 9.3 | 4.9 | 7.1 | *+2.3* | 12.8 | 18 | −1.2 | 1 | 5.1 | 30 | 9.7 | 28 | 2 | 0 | 88.6 | *156* | 24 | 11.7 | 24 | | | | |
| Feb | 4.4 | −0.4 | 2.0 | *−2.9* | 12.7 | 14 | −3.9 | 8 | −0.6 | 28 | 5.3 | 13 | 20 | 0 | 36.6 | *85* | 13 | 8.2 | 6 | | | | |
| Mar | 8.6 | 2.4 | 5.5 | *−1.8* | 15.1 | 8 | −2.3 | 17 | 0.8 | 1 | 10.2 | 7 | 9 | 0 | 27.9 | *59* | 10 | 16.2 | 6 | | | | |
| Apr | 13.4 | 3.5 | 8.4 | *−0.9* | 21.6 | 20 | −1.2 | 13 | 6.6 | 9 | 8.9 | 30 | 3 | 0 | 36.8 | *75* | 10 | 10.0 | 6 | | | | |
| May | 18.3 | 7.8 | 13.1 | *+0.6* | 26.9 | 15 | 3.8 | 2 | 13.1 | 29 | 11.7 | 12 | 0 | 1 | 44.2 | *77* | 14 | 11.2 | 28 | | | | |
| June | 20.6 | 10.5 | 15.6 | *−0.0* | 27.2 | 3 | 4.9 | 16 | 16.4 | 13 | 14.6 | 30 | 0 | 2 | 74.3 | *151* | 18 | 9.2 | 10 | | | | |
| July | 19.9 | 11.1 | 15.5 | *−2.4* | 24.6 | 29 | 7.2 | 12 | 15.1 | 14 | 13.9 | 17 | 0 | 0 | 116.8 | *239* | 18 | 42.5 | 14 | | | | |
| Aug | 21.9 | 12.8 | 17.3 | *−0.3* | 28.0 | 16 | 7.0 | 31 | 18.0 | 28 | 17.1 | 8 | 0 | 2 | 45.0 | *82* | 10 | 21.4 | 12 | | | | |
| Sep | 20.1 | 11.4 | 15.7 | *+0.9* | 24.7 | 18 | 6.0 | 10 | 14.1 | 23 | 15.2 | 1 | 0 | 0 | 50.9 | *94* | 12 | 17.6 | 21 | | | | |
| Oct | 12.6 | 6.4 | 9.5 | *−1.8* | 19.9 | 5 | 1.4 | 12 | 6.5 | 30 | 14.2 | 4 | 0 | 0 | 189.0 | *270* | 18 | 48.3 | 9 | | | | |
| Nov | 8.4 | 3.6 | 6.0 | *−1.6* | 15.3 | 4 | −2.2 | 30 | 2.1 | 29 | 12.1 | 5 | 5 | 0 | 95.3 | *143* | 20 | 20.2 | 10 | | | | |
| Dec | 6.2 | 2.4 | 4.3 | *−0.7* | 12.7 | 21 | −4.9 | 5 | 0.4 | 5 | 6.7 | 23 | 9 | 0 | 22.1 | *35* | 12 | 8.4 | 2 | | | | |
| 1875 | 13.6 | 6.4 | 10.0 | *−0.7* | 28.0 | 8 | −4.9 | 12 | −0.6 | 2 | 17.1 | 8 | 48 | 5 | 827.4 | *125* | 179 | 48.3 | 10 | | | | |

## OXFORD 1876

| Month | Mean max | Mean min | Mean temp | Anom | Highest max | Date | Lowest min | Date | Lowest max | Date | Highest min | Date | Air frost | Days ≥ 25 °C | Total pptn | Anom | Rain days | Wettest day | Date | Total sunshine | Anom | Sunniest day | Date |
|---|---|---|---|---|---|---|---|---|---|---|---|---|---|---|---|---|---|---|---|---|---|---|---|
| Jan | 5.5 | 0.5 | 3.0 | *−1.9* | 11.8 | 3 | −5.2 | 15 | −1.5 | 8 | 6.7 | 19 | 14 | 0 | 43.0 | *76* | 11 | 16.5 | 21 | | | | |
| Feb | 8.1 | 3.0 | 5.5 | *+0.6* | 13.8 | 29 | −6.6 | 10 | −1.8 | 11 | 10.1 | 28 | 5 | 0 | 71.3 | *167* | 19 | 9.7 | 15 | | | | |
| Mar | 8.6 | 1.9 | 5.2 | *−2.0* | 16.1 | 31 | −3.3 | 21 | 2.7 | 21 | 7.2 | 2 | 8 | 0 | 85.7 | *180* | 22 | 12.0 | 27 | | | | |
| Apr | 13.0 | 4.8 | 8.9 | *−0.4* | 19.4 | 8 | −0.6 | 1 | 2.6 | 13 | 10.8 | 5 | 2 | 0 | 71.8 | *147* | 19 | 14.9 | 20 | | | | |
| May | 15.4 | 4.7 | 10.1 | *−2.4* | 21.1 | 21 | −0.5 | 2 | 9.6 | 2 | 10.6 | 26 | 1 | 0 | 20.0 | *35* | 7 | 9.7 | 24 | | | | |
| June | 20.8 | 10.0 | 15.4 | *−0.2* | 29.6 | 21 | 3.2 | 10 | 15.8 | 10 | 14.3 | 30 | 0 | 3 | 51.7 | *105* | 11 | 12.2 | 15 | | | | |
| July | 24.5 | 13.1 | 18.8 | *+0.9* | 31.8 | 16 | 6.4 | 11 | 17.8 | 11 | 15.9 | 4 | 0 | 11 | 21.4 | *44* | 10 | 8.4 | 6 | | | | |
| Aug | 23.8 | 11.8 | 17.8 | *+0.2* | 33.3 | 13 | 5.2 | 24 | 16.0 | 31 | 18.6 | 17 | 0 | 10 | 66.9 | *122* | 9 | 19.4 | 19 | | | | |
| Sep | 17.4 | 9.8 | 13.6 | *−1.3* | 20.9 | 4 | 5.6 | 20 | 13.3 | 12 | 14.6 | 4 | 0 | 0 | 145.6 | *269* | 20 | 22.6 | 5 | | | | |
| Oct | 14.8 | 8.7 | 11.8 | *+0.5* | 20.9 | 7 | 0.0 | 31 | 6.4 | 31 | 15.6 | 8 | 0 | 0 | 25.4 | *36* | 10 | 6.5 | 6 | | | | |
| Nov | 9.7 | 4.1 | 6.9 | *−0.6* | 16.1 | 14 | −1.4 | 7 | 3.3 | 9 | 10.7 | 14 | 5 | 0 | 81.5 | *122* | 17 | 16.2 | 12 | | | | |
| Dec | 8.7 | 4.9 | 6.8 | *+1.8* | 13.8 | 1 | −2.2 | 22 | 1.2 | 25 | 10.5 | 2 | 4 | 0 | 127.6 | *202* | 26 | 14.4 | 2 | | | | |
| 1876 | 14.2 | 6.4 | 10.3 | *−0.4* | 33.3 | 8 | −6.6 | 2 | −1.8 | 2 | 18.6 | 8 | 39 | 24 | 812.0 | *123* | 181 | 22.6 | 9 | | | | |

## OXFORD 1877

| Month | Mean max | Mean min | Mean temp | Anom | Highest max | Date | Lowest min | Date | Lowest max | Date | Highest min | Date | Air frost | Days ≥ 25 °C | Total pptn | Anom | Rain days | Wettest day | Date | Total sunshine | Anom | Sunniest day | Date |
|---|---|---|---|---|---|---|---|---|---|---|---|---|---|---|---|---|---|---|---|---|---|---|---|
| Jan | 9.3 | 3.2 | 6.3 | *+1.4* | 13.4 | 19 | −1.5 | 11 | 3.6 | 12 | 8.1 | 16 | 3 | 0 | 116.1 | *204* | 26 | 14.2 | 3 | | | | |
| Feb | 9.4 | 4.0 | 6.7 | *+1.8* | 12.8 | 7 | −5.1 | 28 | 2.6 | 28 | 10.1 | 6 | 3 | 0 | 39.1 | *91* | 14 | 10.2 | 13 | | | | |
| Mar | 9.0 | 2.2 | 5.6 | *−1.7* | 13.2 | 28 | −4.1 | 21 | 4.7 | 9 | 8.6 | 15 | 9 | 0 | 49.5 | *104* | 17 | 8.7 | 23 | | | | |
| Apr | 12.1 | 4.3 | 8.2 | *−1.1* | 16.3 | 22 | 0.6 | 30 | 6.6 | 18 | 9.5 | 3 | 0 | 0 | 53.9 | *110* | 16 | 8.7 | 6 | | | | |
| May | 14.5 | 5.5 | 10.0 | *−2.5* | 19.2 | 26 | −2.4 | 4 | 8.5 | 1 | 11.2 | 31 | 4 | 0 | 57.2 | *100* | 17 | 12.7 | 16 | | | | |
| June | 22.2 | 10.5 | 16.4 | *+0.8* | 28.0 | 18 | 5.1 | 23 | 15.8 | 1 | 15.3 | 11 | 0 | 8 | 25.0 | *51* | 8 | 8.2 | 21 | | | | |
| July | 21.1 | 11.1 | 16.1 | *−1.8* | 27.8 | 31 | 5.6 | 7 | 17.6 | 17 | 15.3 | 28 | 0 | 3 | 83.7 | *171* | 15 | 28.9 | 14 | | | | |
| Aug | 20.5 | 11.9 | 16.2 | *−1.4* | 25.5 | 20 | 4.2 | 23 | 16.7 | 25 | 17.2 | 20 | 0 | 1 | 85.6 | *156* | 20 | 11.7 | 18 | | | | |
| Sep | 15.9 | 7.6 | 11.7 | *−3.1* | 19.8 | 14 | 1.0 | 24 | 12.1 | 3 | 15.5 | 13 | 0 | 0 | 62.2 | *115* | 9 | 33.6 | 3 | | | | |
| Oct | 13.7 | 5.2 | 9.4 | *−1.9* | 19.1 | 14 | −1.7 | 17 | 7.7 | 16 | 13.1 | 13 | 2 | 0 | 42.2 | *60* | 15 | 6.5 | 28 | | | | |
| Nov | 10.8 | 4.3 | 7.5 | *+0.0* | 15.1 | 6 | 0.2 | 18 | 5.0 | 24 | 12.6 | 15 | 0 | 0 | 96.2 | *144* | 21 | 24.4 | 11 | | | | |
| Dec | 7.8 | 2.3 | 5.1 | *+0.0* | 12.2 | 6 | −2.1 | 24 | 2.8 | 27 | 7.4 | 29 | 7 | 0 | 43.7 | *69* | 18 | 12.9 | 5 | | | | |
| 1877 | 13.9 | 6.0 | 9.9 | *−0.8* | 28.0 | 6 | −5.1 | 2 | 2.6 | 2 | 17.2 | 8 | 28 | 12 | 754.5 | *114* | 196 | 33.6 | 9 | | | | |

## OXFORD 1878

| Month | Mean max | Mean min | Mean temp | Anom | Highest max | Date | Lowest min | Date | Lowest max | Date | Highest min | Date | Air frost | Days ≥ 25 °C | Total pptn | Anom | Rain days | Wettest day | Date | Total sunshine | Anom | Sunniest day | Date |
|---|---|---|---|---|---|---|---|---|---|---|---|---|---|---|---|---|---|---|---|---|---|---|---|
| Jan | 7.5 | 2.6 | 5.1 | *+0.2* | 13.6 | 21 | −3.8 | 30 | 2.7 | 11 | 11.7 | 21 | 8 | 0 | 44.3 | *78* | 18 | 13.0 | 3 | | | | |
| Feb | 8.8 | 4.2 | 6.5 | *+1.6* | 14.4 | 17 | −3.8 | 7 | 2.1 | 6 | 10.6 | 28 | 4 | 0 | 34.2 | *80* | 14 | 11.2 | 13 | | | | |
| Mar | 9.5 | 2.2 | 5.8 | *−1.4* | 14.5 | 1 | −3.3 | 25 | 4.2 | 29 | 8.8 | 1 | 9 | 0 | 19.5 | *41* | 9 | 8.1 | 31 | | | | |
| Apr | 14.1 | 5.5 | 9.8 | *+0.5* | 20.0 | 30 | −2.0 | 5 | 7.6 | 3 | 11.8 | 30 | 3 | 0 | 58.4 | *119* | 11 | 19.8 | 10 | | | | |
| May | 17.7 | 8.4 | 13.1 | *+0.6* | 22.0 | 10 | 2.8 | 20 | 13.7 | 21 | 13.3 | 17 | 0 | 0 | 126.7 | *222* | 23 | 20.1 | 7 | | | | |
| June | 21.1 | 11.1 | 16.1 | *+0.5* | 30.6 | 26 | 4.6 | 14 | 13.9 | 14 | 16.7 | 26 | 0 | 8 | 49.0 | *100* | 12 | 10.2 | 16 | | | | |
| July | 22.2 | 12.6 | 17.4 | *−0.5* | 29.1 | 19 | 8.2 | 3 | 17.7 | 1 | 16.2 | 23 | 0 | 6 | 29.1 | *60* | 7 | 16.8 | 24 | | | | |
| Aug | 21.7 | 13.1 | 17.4 | *−0.2* | 24.9 | 5 | 9.9 | 16 | 18.4 | 21 | 16.4 | 9 | 0 | 0 | 130.7 | *239* | 19 | 24.9 | 4 | | | | |
| Sep | 18.1 | 9.4 | 13.8 | *−1.1* | 22.9 | 5 | 0.7 | 23 | 13.2 | 23 | 15.5 | 17 | 0 | 0 | 34.4 | *64* | 10 | 9.4 | 19 | | | | |
| Oct | 14.1 | 7.4 | 10.7 | *−0.5* | 21.1 | 5 | 0.1 | 30 | 6.4 | 31 | 14.1 | 5 | 0 | 0 | 54.1 | *77* | 15 | 13.5 | 23 | | | | |
| Nov | 6.6 | 2.3 | 4.5 | *−3.1* | 11.1 | 10 | −0.9 | 12 | 2.8 | 23 | 9.9 | 10 | 5 | 0 | 58.1 | *87* | 16 | 11.2 | 27 | | | | |
| Dec | 3.3 | −2.2 | 0.5 | *−4.5* | 12.1 | 29 | −14.7 | 24 | −2.6 | 11 | 9.7 | 30 | 20 | 0 | 38.6 | *61* | 10 | 10.2 | 25 | | | | |
| 1878 | 13.7 | 6.4 | 10.1 | *−0.7* | 30.6 | 6 | −14.7 | 12 | −2.6 | 12 | 16.7 | 6 | 49 | 14 | 677.1 | *103* | 164 | 24.9 | 8 | | | | |

## OXFORD 1879

| Month | Mean max | Mean min | Mean temp | Anom | Highest max | Date | Lowest min | Date | Lowest max | Date | Highest min | Date | Air frost | Days ≥ 25 °C | Total pptn | Anom | Rain days | Wettest day | Date | Total sunshine | Anom | Sunniest day | Date |
|---|---|---|---|---|---|---|---|---|---|---|---|---|---|---|---|---|---|---|---|---|---|---|---|
| Jan | 1.6 | −3.0 | −0.7 | *−5.5* | 8.2 | 14 | −8.6 | 12 | −2.8 | 10 | 4.0 | 13 | 27 | 0 | 76.5 | *135* | 9 | 24.9 | 1 | | | | |
| Feb | 6.2 | 1.0 | 3.6 | *−1.3* | 12.1 | 9 | −4.7 | 24 | −0.1 | 22 | 7.4 | 9 | 9 | 0 | 84.6 | *198* | 22 | 9.9 | 8 | | | | |
| Mar | 8.7 | 1.0 | 4.9 | *−2.4* | 16.4 | 19 | −2.1 | 25 | −0.2 | 24 | 5.6 | 30 | 10 | 0 | 22.3 | *47* | 13 | 5.1 | 30 | | | | |
| Apr | 9.6 | 2.3 | 5.9 | *−3.4* | 14.5 | 26 | −2.7 | 18 | 2.1 | 12 | 6.5 | 6 | 7 | 0 | 62.1 | *127* | 16 | 10.9 | 23 | | | | |
| May | 13.0 | 5.1 | 9.1 | *−3.4* | 18.2 | 5 | −1.4 | 9 | 8.1 | 7 | 10.6 | 20 | 3 | 0 | 57.6 | *101* | 15 | 23.9 | 28 | | | | |
| June | 17.2 | 10.0 | 13.6 | *−2.0* | 20.1 | 28 | 4.7 | 3 | 12.0 | 3 | 14.1 | 27 | 0 | 0 | 110.7 | *225* | 27 | 12.7 | 30 | | | | |
| July | 17.7 | 11.2 | 14.5 | *−3.4* | 24.0 | 29 | 8.8 | 5 | 14.8 | 3 | 15.8 | 31 | 0 | 0 | 86.8 | *178* | 19 | 17.5 | 19 | | | | |
| Aug | 18.8 | 11.6 | 15.2 | *−2.4* | 24.3 | 12 | 4.9 | 31 | 12.1 | 17 | 15.4 | 14 | 0 | 0 | 128.7 | *235* | 17 | 32.8 | 19 | | | | |
| Sep | 16.7 | 9.3 | 13.0 | *−1.9* | 20.0 | 3 | 4.6 | 24 | 13.7 | 24 | 13.2 | 17 | 0 | 0 | 73.6 | *136* | 15 | 14.2 | 28 | | | | |
| Oct | 12.3 | 5.8 | 9.1 | *−2.2* | 18.4 | 4 | 0.4 | 15 | 7.7 | 15 | 11.2 | 23 | 0 | 0 | 18.6 | *27* | 10 | 4.6 | 21 | | | | |
| Nov | 6.8 | 0.7 | 3.8 | *−3.8* | 11.9 | 18 | −5.5 | 14 | 1.2 | 22 | 7.3 | 9 | 13 | 0 | 17.4 | *26* | 10 | 6.1 | 21 | | | | |
| Dec | 3.8 | −3.3 | 0.3 | *−4.8* | 11.8 | 28 | −11.4 | 7 | −1.4 | 17 | 7.0 | 31 | 24 | 0 | 18.7 | *30* | 11 | 4.3 | 30 | | | | |
| 1879 | 11.1 | 4.3 | 7.7 | *−3.0* | 24.3 | 8 | −11.4 | 12 | −2.8 | 1 | 15.8 | 7 | 93 | 0 | 757.6 | *115* | 184 | 32.8 | 8 | | | | |

## OXFORD 1880

| Month | Mean max | Mean min | Mean temp | Anom | Highest max | Date | Lowest min | Date | Lowest max | Date | Highest min | Date | Air frost | Days ≥ 25 °C | Total pptn | Anom | Rain days | Wettest day | Date | Total sunshine | Anom | Sunniest day | Date |
|---|---|---|---|---|---|---|---|---|---|---|---|---|---|---|---|---|---|---|---|---|---|---|---|
| Jan | 3.0 | −2.7 | 0.2 | *−4.7* | 11.9 | 1 | −9.0 | 20 | −4.2 | 28 | 9.1 | 1 | 22 | 0 | 11.1 | *20* | 6 | 7.4 | 16 | | | | |
| Feb | 9.1 | 2.3 | 5.7 | *+0.8* | 12.7 | 20 | −3.9 | 1 | 4.6 | 24 | 8.1 | 19 | 5 | 0 | 60.9 | *142* | 22 | 15.2 | 16 | 66.0 | *84* | | |
| Mar | 11.8 | 2.4 | 7.1 | *−0.2* | 16.8 | 25 | −2.8 | 19 | 6.3 | 22 | 9.2 | 4 | 7 | 0 | 35.3 | *74* | 5 | 12.7 | 2 | 155.8 | *140* | | |
| Apr | 12.3 | 4.6 | 8.4 | *−0.9* | 17.7 | 19 | −0.3 | 30 | 8.2 | 14 | 11.1 | 18 | 1 | 0 | 48.7 | *100* | 19 | 13.2 | 14 | 140.7 | *87* | | |
| May | 16.1 | 6.3 | 11.2 | *−1.3* | 22.3 | 21 | 1.8 | 8 | 9.4 | 10 | 13.1 | 26 | 0 | 0 | 21.1 | *37* | 6 | 10.2 | 27 | 206.0 | *107* | | |
| June | 18.1 | 9.8 | 14.0 | *−1.6* | 22.9 | 19 | 4.9 | 5 | 12.1 | 3 | 14.9 | 28 | 0 | 0 | 87.6 | *178* | 16 | 28.4 | 25 | 139.8 | *73* | | |
| July | 19.8 | 12.0 | 15.9 | *−1.9* | 22.1 | 26 | 8.3 | 31 | 15.8 | 14 | 15.1 | 28 | 0 | 0 | 155.1 | *318* | 24 | 46.0 | 14 | 184.6 | *89* | | |
| Aug | 20.6 | 12.9 | 16.8 | *−0.8* | 24.2 | 11 | 9.4 | 3 | 16.7 | 8 | 17.3 | 11 | 0 | 0 | 12.7 | *23* | 4 | 5.8 | 6 | 142.1 | *72* | | |
| Sep | 19.9 | 11.3 | 15.6 | *+0.7* | 28.4 | 4 | 6.1 | 29 | 14.8 | 30 | 17.5 | 5 | 0 | 3 | 88.6 | *164* | 13 | 25.4 | 11 | 129.0 | *92* | | |
| Oct | 12.3 | 4.7 | 8.5 | *−2.8* | 17.9 | 2 | −1.4 | 24 | 6.1 | 21 | 11.7 | 1 | 3 | 0 | 133.5 | *191* | 17 | 27.7 | 6 | 86.0 | *77* | | |
| Nov | 9.8 | 2.9 | 6.4 | *−1.2* | 14.1 | 14 | −6.2 | 22 | 0.9 | 22 | 12.4 | 14 | 9 | 0 | 47.8 | *72* | 13 | 15.5 | 15 | 72.6 | *102* | | |
| Dec | 9.5 | 3.4 | 6.4 | *+1.4* | 13.1 | 11 | −1.4 | 3 | 2.7 | 18 | 9.5 | 23 | 5 | 0 | 75.7 | *120* | 17 | 11.4 | 21 | 40.6 | *76* | | |
| 1880 | 13.5 | 5.8 | 9.7 | *−1.1* | 28.4 | 9 | −9.0 | 1 | −4.2 | 1 | 17.5 | 9 | 52 | 3 | 778.1 | *118* | 162 | 46.0 | 7 | *Incomplete* | | | |

## OXFORD 1881

| Month | Mean max | Mean min | Mean temp | Anom | Highest max | Date | Lowest min | Date | Lowest max | Date | Highest min | Date | Air frost | Days ≥ 25 °C | Total pptn | Anom | Rain days | Wettest day | Date | Total sunshine | Anom | Sunniest day | Date |
|---|---|---|---|---|---|---|---|---|---|---|---|---|---|---|---|---|---|---|---|---|---|---|---|
| Jan | 1.8 | −4.6 | −1.4 | *−6.2* | 9.6 | 31 | −14.0 | 22 | −6.7 | 26 | 4.3 | 3 | 21 | 0 | 31.9 | *56* | 13 | 6.9 | 18 | 49.9 | *80* | | |
| Feb | 6.5 | 0.9 | 3.7 | *−1.2* | 11.1 | 4 | −3.6 | 28 | 1.5 | 22 | 8.1 | 4 | 10 | 0 | 70.2 | *164* | 17 | 19.9 | 9 | 43.0 | *54* | | |
| Mar | 10.4 | 1.8 | 6.1 | *−1.1* | 15.1 | 17 | −4.8 | 1 | 4.1 | 1 | 10.4 | 7 | 11 | 0 | 37.7 | *79* | 12 | 11.4 | 5 | 142.6 | *128* | | |
| Apr | 13.0 | 2.8 | 7.9 | *−1.4* | 17.2 | 14 | −2.3 | 10 | 6.4 | 20 | 9.7 | 29 | 7 | 0 | 19.9 | *41* | 8 | 9.6 | 11 | 153.0 | *95* | | |
| May | 17.8 | 6.7 | 12.3 | *−0.2* | 23.7 | 31 | −0.8 | 11 | 9.7 | 4 | 12.8 | 27 | 1 | 0 | 31.9 | *56* | 10 | 13.4 | 19 | 229.0 | *119* | | |
| June | 19.3 | 10.1 | 14.7 | *−0.9* | 25.8 | 2 | 3.6 | 9 | 11.9 | 8 | 14.9 | 18 | 0 | 3 | 48.0 | *97* | 15 | 19.7 | 5 | 192.7 | *101* | | |
| July | 22.9 | 12.3 | 17.6 | *−0.3* | 30.5 | 6 | 6.1 | 28 | 15.8 | 7 | 16.1 | 19 | 0 | 7 | 63.7 | *130* | 12 | 27.3 | 5 | 235.2 | *114* | | |
| Aug | 19.3 | 10.6 | 15.0 | *−2.6* | 26.6 | 6 | 5.7 | 28 | 14.3 | 13 | 15.8 | 4 | 0 | 2 | 118.2 | *216* | 17 | 13.5 | 23 | 147.7 | *75* | | |
| Sep | 17.7 | 9.3 | 13.5 | *−1.4* | 23.6 | 19 | 2.7 | 29 | 12.7 | 1 | 15.3 | 21 | 0 | 0 | 35.5 | *66* | 14 | 17.8 | 24 | 111.9 | *80* | | |
| Oct | 12.1 | 3.7 | 7.9 | *−3.4* | 17.4 | 1 | −3.6 | 31 | 7.1 | 31 | 11.1 | 11 | 5 | 0 | 42.8 | *61* | 9 | 13.6 | 13 | 127.0 | *114* | | |
| Nov | 12.6 | 6.1 | 9.4 | *+1.8* | 15.3 | 5 | 0.8 | 2 | 5.3 | 2 | 12.8 | 5 | 0 | 0 | 82.0 | *123* | 16 | 22.8 | 24 | 64.1 | *90* | | |
| Dec | 7.3 | 1.3 | 4.3 | *−0.7* | 12.2 | 3 | −5.9 | 24 | 1.6 | 10 | 7.2 | 1 | 7 | 0 | 75.5 | *120* | 11 | 24.5 | 17 | 50.7 | *95* | | |
| 1881 | 13.4 | 5.1 | 9.3 | *−1.5* | 30.5 | 7 | −14.0 | 1 | −6.7 | 1 | 16.1 | 7 | 62 | 12 | 657.3 | *100* | 154 | 27.3 | 7 | 1546.8 | *98* | | |

## OXFORD 1882

| Month | Mean max | Mean min | Mean temp | Anom | Highest max | Date | Lowest min | Date | Lowest max | Date | Highest min | Date | Air frost | Days ≥ 25 °C | Total pptn | Anom | Rain days | Wettest day | Date | Total sunshine | Anom | Sunniest day | Date |
|---|---|---|---|---|---|---|---|---|---|---|---|---|---|---|---|---|---|---|---|---|---|---|---|
| Jan | 7.7 | 2.2 | 5.0 | *+0.1* | 12.4 | 11 | −4.3 | 24 | 1.7 | 17 | 7.2 | 10 | 6 | 0 | 28.4 | *50* | 8 | 11.9 | 8 | 31.9 | *51* | | |
| Feb | 9.4 | 2.9 | 6.2 | *+1.3* | 13.2 | 26 | −3.8 | 3 | 5.1 | 6 | 10.4 | 25 | 3 | 0 | 38.8 | *91* | 8 | 14.7 | 28 | 42.1 | *53* | | |
| Mar | 12.7 | 3.7 | 8.2 | *+0.9* | 17.8 | 18 | −1.5 | 22 | 7.8 | 22 | 10.0 | 9 | 3 | 0 | 28.2 | *59* | 9 | 12.9 | 25 | 135.1 | *121* | | |
| Apr | 13.7 | 4.8 | 9.3 | *−0.0* | 16.6 | 6 | −0.6 | 15 | 9.6 | 5 | 10.4 | 19 | 1 | 0 | 99.7 | *204* | 16 | 20.3 | 25 | 172.3 | *107* | | |
| May | 17.6 | 7.2 | 12.4 | *−0.1* | 20.2 | 30 | 2.1 | 16 | 13.7 | 15 | 11.7 | 10 | 0 | 0 | 47.0 | *82* | 8 | 11.5 | 25 | 260.1 | *135* | | |
| June | 17.6 | 9.5 | 13.6 | *−2.0* | 21.6 | 27 | 5.1 | 16 | 13.6 | 12 | 13.1 | 28 | 0 | 0 | 83.8 | *170* | 17 | 17.8 | 21 | 150.7 | *79* | | |
| July | 20.2 | 11.7 | 16.0 | *−1.9* | 23.5 | 3 | 8.7 | 26 | 17.2 | 6 | 15.4 | 14 | 0 | 0 | 82.2 | *168* | 19 | 21.0 | 11 | 196.2 | *95* | | |
| Aug | 20.0 | 11.3 | 15.7 | *−1.9* | 25.9 | 6 | 7.9 | 30 | 14.8 | 24 | 15.4 | 13 | 0 | 2 | 33.8 | *62* | 11 | 9.9 | 31 | 150.7 | *77* | | |
| Sep | 17.4 | 8.1 | 12.8 | *−2.1* | 19.7 | 3 | 2.1 | 15 | 13.1 | 29 | 14.8 | 1 | 0 | 0 | 51.4 | *95* | 12 | 14.0 | 19 | 145.4 | *104* | | |
| Oct | 13.7 | 6.6 | 10.2 | *−1.1* | 19.5 | 1 | −0.1 | 25 | 8.4 | 16 | 12.1 | 5 | 1 | 0 | 147.9 | *211* | 19 | 37.1 | 23 | 70.5 | *63* | | |
| Nov | 9.6 | 2.7 | 6.2 | *−1.4* | 15.4 | 5 | −5.2 | 17 | 3.6 | 16 | 10.3 | 22 | 3 | 0 | 84.8 | *127* | 18 | 14.8 | 6 | 82.5 | *116* | | |
| Dec | 7.4 | 1.7 | 4.5 | *−0.5* | 12.7 | 27 | −7.9 | 11 | −1.3 | 11 | 11.4 | 27 | 9 | 0 | 80.3 | *127* | 17 | 17.9 | 29 | 27.0 | *51* | | |
| 1882 | 14.0 | 6.0 | 10.0 | *−0.7* | 25.9 | 8 | −7.9 | 12 | −1.3 | 12 | 15.4 | 7 | 26 | 2 | 806.3 | *122* | 162 | 37.1 | 10 | 1464.5 | *93* | | |

## OXFORD 1883

| Month | Mean max | Mean min | Mean temp | Anom | Highest max | Date | Lowest min | Date | Lowest max | Date | Highest min | Date | Air frost | Days ≥ 25 °C | Total pptn | Anom | Rain days | Wettest day | Date | Total sunshine | Anom | Sunniest day | Date |
|---|---|---|---|---|---|---|---|---|---|---|---|---|---|---|---|---|---|---|---|---|---|---|---|
| Jan | 8.2 | 2.6 | 5.4 | *+0.6* | 12.8 | 1 | −1.9 | 30 | 3.6 | 31 | 8.7 | 17 | 3 | 0 | 57.8 | *102* | 21 | 8.6 | 24 | 59.9 | *96* | | |
| Feb | 10.1 | 3.4 | 6.7 | *+1.8* | 12.4 | 22 | 0.3 | 1 | 7.7 | 6 | 9.7 | 21 | 0 | 0 | 95.1 | *222* | 16 | 21.2 | 10 | 83.2 | *105* | | |
| Mar | 7.3 | −1.6 | 2.9 | *−4.4* | 13.4 | 5 | −6.4 | 23 | 2.6 | 9 | 6.6 | 29 | 22 | 0 | 24.0 | *50* | 9 | 12.2 | 19 | 157.5 | *142* | | |
| Apr | 14.0 | 4.2 | 9.1 | *−0.2* | 18.6 | 5 | −1.2 | 7 | 8.7 | 23 | 10.4 | 3 | 3 | 0 | 25.7 | *53* | 9 | 9.4 | 27 | 165.4 | *103* | | |
| May | 16.4 | 6.9 | 11.7 | *−0.8* | 22.4 | 24 | −0.9 | 3 | 7.7 | 9 | 11.6 | 25 | 1 | 0 | 47.9 | *84* | 14 | 10.3 | 9 | 171.5 | *89* | | |
| June | 19.7 | 9.9 | 14.8 | *−0.8* | 25.6 | 29 | 4.9 | 16 | 13.3 | 15 | 15.8 | 28 | 0 | 1 | 91.7 | *186* | 16 | 26.2 | 20 | 195.6 | *102* | | |
| July | 19.0 | 11.0 | 15.0 | *−2.9* | 25.7 | 2 | 5.9 | 15 | 15.2 | 21 | 16.1 | 2 | 0 | 1 | 87.9 | *180* | 20 | 43.9 | 2 | 144.8 | *70* | | |
| Aug | 21.1 | 11.6 | 16.4 | *−1.2* | 24.7 | 13 | 7.6 | 24 | 17.1 | 10 | 15.8 | 13 | 0 | 0 | 18.7 | *34* | 10 | 5.2 | 31 | 168.7 | *86* | | |
| Sep | 17.6 | 10.2 | 13.9 | *−1.0* | 23.3 | 17 | 4.8 | 8 | 10.7 | 30 | 14.0 | 19 | 0 | 0 | 98.4 | *182* | 16 | 30.3 | 29 | 111.5 | *79* | | |
| Oct | 13.9 | 6.9 | 10.4 | *−0.9* | 17.6 | 27 | 3.7 | 1 | 10.1 | 31 | 12.6 | 25 | 0 | 0 | 48.2 | *69* | 11 | 12.0 | 15 | 91.5 | *82* | | |
| Nov | 10.3 | 2.5 | 6.4 | *−1.1* | 14.0 | 28 | −2.8 | 12 | 5.6 | 14 | 8.1 | 27 | 4 | 0 | 79.4 | *119* | 16 | 16.1 | 5 | 82.3 | *116* | | |
| Dec | 7.4 | 1.9 | 4.7 | *−0.3* | 12.6 | 3 | −2.5 | 6 | 2.8 | 31 | 10.6 | 13 | 5 | 0 | 11.4 | *18* | 10 | 4.4 | 10 | 30.3 | *57* | | |
| 1883 | 13.8 | 5.8 | 9.8 | *−1.0* | 25.7 | 7 | −6.4 | 3 | 2.6 | 3 | 16.1 | 7 | 38 | 2 | 686.2 | *104* | 168 | 43.9 | 7 | 1462.2 | *93* | | |

## OXFORD 1884

| Month | Mean max | Mean min | Mean temp | Anom | Highest max | Date | Lowest min | Date | Lowest max | Date | Highest min | Date | Air frost | Days ≥ 25 °C | Total pptn | Anom | Rain days | Wettest day | Date | Total sunshine | Anom | Sunniest day | Date |
|---|---|---|---|---|---|---|---|---|---|---|---|---|---|---|---|---|---|---|---|---|---|---|---|
| Jan | 9.5 | 4.5 | 7.0 | *+2.2* | 12.6 | 23 | 0.7 | 1 | 3.6 | 1 | 10.1 | 22 | 0 | 0 | 55.7 | *98* | 18 | 9.6 | 26 | 27.2 | *44* | | |
| Feb | 9.0 | 2.5 | 5.8 | *+0.9* | 12.3 | 12 | −1.8 | 2 | 5.1 | 7 | 6.9 | 12 | 4 | 0 | 32.2 | *75* | 11 | 12.4 | 21 | 53.1 | *67* | | |
| Mar | 11.0 | 2.7 | 6.8 | *−0.4* | 19.4 | 16 | −2.1 | 2 | 6.4 | 27 | 9.1 | 17 | 4 | 0 | 39.3 | *82* | 8 | 14.8 | 3 | 98.1 | *88* | | |
| Apr | 11.8 | 2.8 | 7.3 | *−2.0* | 17.8 | 2 | −3.1 | 22 | 6.8 | 17 | 9.4 | 1 | 6 | 0 | 40.2 | *82* | 10 | 14.2 | 6 | 102.7 | *64* | | |
| May | 17.7 | 6.5 | 12.1 | *−0.3* | 25.8 | 24 | 2.2 | 5 | 11.4 | 28 | 12.4 | 15 | 0 | 1 | 19.4 | *34* | 8 | 6.7 | 12 | 197.7 | *102* | | |
| June | 19.6 | 9.7 | 14.7 | *−0.9* | 26.6 | 28 | 7.2 | 2 | 14.5 | 9 | 13.8 | 24 | 0 | 3 | 54.6 | *111* | 7 | 18.3 | 5 | 157.1 | *82* | | |
| July | 21.7 | 12.2 | 17.0 | *−0.9* | 27.3 | 3 | 6.4 | 25 | 16.8 | 25 | 15.7 | 30 | 0 | 7 | 57.5 | *118* | 21 | 7.6 | 26 | 137.1 | *66* | | |
| Aug | 24.0 | 11.8 | 17.9 | *+0.3* | 31.4 | 11 | 7.1 | 25 | 13.8 | 26 | 17.3 | 11 | 0 | 13 | 38.7 | *71* | 8 | 14.5 | 31 | 222.3 | *113* | | |
| Sep | 19.8 | 10.7 | 15.3 | *+0.4* | 27.0 | 17 | 1.2 | 29 | 15.9 | 22 | 16.4 | 15 | 0 | 3 | 31.2 | *58* | 9 | 10.4 | 3 | 120.5 | *86* | | |
| Oct | 13.8 | 5.5 | 9.6 | *−1.6* | 17.4 | 16 | 1.2 | 28 | 7.2 | 10 | 11.8 | 16 | 0 | 0 | 19.4 | *28* | 5 | 10.5 | 9 | 84.5 | *76* | | |
| Nov | 8.9 | 1.8 | 5.3 | *−2.2* | 15.6 | 2 | −3.4 | 29 | 3.2 | 24 | 10.3 | 1 | 10 | 0 | 41.4 | *62* | 9 | 16.3 | 30 | 36.5 | *51* | | |
| Dec | 7.5 | 2.1 | 4.8 | *−0.2* | 12.6 | 7 | −5.1 | 30 | 1.7 | 26 | 9.7 | 13 | 4 | 0 | 51.3 | *81* | 16 | 9.2 | 5 | 23.2 | *44* | | |
| 1884 | 14.6 | 6.1 | 10.3 | *−0.4* | 31.4 | 8 | −5.1 | 12 | 1.7 | 12 | 17.3 | 8 | 28 | 27 | 480.9 | *73* | 130 | 18.3 | 6 | 1260.0 | *80* | | |

## OXFORD 1885

| Month | Mean max | Mean min | Mean temp | Anom | Highest max | Date | Lowest min | Date | Lowest max | Date | Highest min | Date | Air frost | Days ≥ 25 °C | Total pptn | Anom | Rain days | Wettest day | Date | Total sunshine | Anom | Sunniest day | Date |
|---|---|---|---|---|---|---|---|---|---|---|---|---|---|---|---|---|---|---|---|---|---|---|---|
| Jan | 5.1 | 0.3 | 2.7 | *−2.2* | 10.9 | 28 | −3.9 | 21 | −0.1 | 20 | 8.4 | 28 | 14 | 0 | 49.4 | *87* | 13 | 7.9 | 8 | 14.9 | *24* | | |
| Feb | 10.0 | 3.2 | 6.6 | *+1.7* | 13.8 | 11 | −4.5 | 20 | 3.6 | 20 | 8.5 | 15 | 4 | 0 | 73.3 | *171* | 16 | 14.5 | 26 | 51.4 | *65* | | |
| Mar | 9.5 | 0.9 | 5.2 | *−2.1* | 14.6 | 20 | −2.7 | 7 | 4.0 | 6 | 5.0 | 26 | 10 | 0 | 29.8 | *63* | 7 | 15.5 | 21 | 116.9 | *105* | | |
| Apr | 13.0 | 3.3 | 8.2 | *−1.1* | 20.8 | 21 | −3.1 | 4 | 7.0 | 8 | 9.3 | 24 | 4 | 0 | 45.0 | *92* | 12 | 10.9 | 1 | 149.7 | *93* | | |
| May | 13.9 | 5.2 | 9.5 | *−2.9* | 18.4 | 27 | 0.2 | 11 | 8.6 | 5 | 11.7 | 27 | 0 | 0 | 53.5 | *94* | 17 | 13.9 | 21 | 180.2 | *93* | | |
| June | 20.0 | 9.6 | 14.8 | *−0.8* | 26.9 | 4 | 4.2 | 10 | 13.8 | 25 | 14.7 | 4 | 0 | 2 | 39.9 | *81* | 11 | 8.1 | 8 | 220.2 | *115* | | |
| July | 22.8 | 11.7 | 17.3 | *−0.6* | 29.7 | 25 | 6.8 | 8 | 17.3 | 16 | 15.6 | 18 | 0 | 5 | 3.1 | *6* | 2 | 2.9 | 18 | 237.3 | *115* | | |
| Aug | 19.0 | 9.7 | 14.4 | *−3.2* | 24.3 | 17 | 5.1 | 13 | 14.1 | 30 | 15.0 | 9 | 0 | 0 | 41.7 | *76* | 9 | 9.6 | 6 | 154.8 | *79* | | |
| Sep | 17.2 | 8.4 | 12.8 | *−2.1* | 23.0 | 15 | 0.1 | 27 | 9.9 | 26 | 14.1 | 14 | 0 | 0 | 111.9 | *207* | 17 | 35.0 | 10 | 125.5 | *89* | | |
| Oct | 11.3 | 4.3 | 7.8 | *−3.5* | 16.0 | 8 | 0.1 | 29 | 7.4 | 30 | 7.9 | 15 | 0 | 0 | 98.4 | *141* | 18 | 19.7 | 30 | 96.5 | *87* | | |
| Nov | 8.5 | 3.2 | 5.8 | *−1.7* | 14.1 | 27 | −3.1 | 17 | 4.4 | 18 | 9.3 | 29 | 5 | 0 | 91.0 | *136* | 16 | 19.7 | 3 | 38.6 | *54* | | |
| Dec | 6.6 | 0.5 | 3.6 | *−1.5* | 10.3 | 15 | −4.9 | 10 | 0.7 | 8 | 7.3 | 31 | 11 | 0 | 24.8 | *39* | 10 | 6.9 | 28 | 48.7 | *92* | | |
| 1885 | 13.1 | 5.0 | 9.1 | *−1.7* | 29.7 | 7 | −4.9 | 12 | −0.1 | 1 | 15.6 | 7 | 48 | 7 | 661.8 | *100* | 148 | 35.0 | 9 | 1434.7 | *91* | | |

## OXFORD 1886

| Month | Mean max | Mean min | Mean temp | Anom | Highest max | Date | Lowest min | Date | Lowest max | Date | Highest min | Date | Air frost | Days ≥ 25 °C | Total pptn | Anom | Rain days | Wettest day | Date | Total sunshine | Anom | Sunniest day | Date |
|---|---|---|---|---|---|---|---|---|---|---|---|---|---|---|---|---|---|---|---|---|---|---|---|
| Jan | 4.6 | −1.1 | 1.8 | *−3.1* | 10.8 | 3 | −10.0 | 7 | −0.1 | 6 | 7.4 | 1 | 15 | 0 | 97.6 | *172* | 21 | 19.4 | 30 | 56.8 | *91* | | |
| Feb | 3.4 | −1.7 | 0.9 | *−4.0* | 9.2 | 13 | −4.8 | 9 | −0.1 | 19 | 4.8 | 12 | 20 | 0 | 19.9 | *46* | 7 | 6.9 | 2 | 38.7 | *49* | | |
| Mar | 8.0 | 1.1 | 4.6 | *−2.7* | 17.1 | 24 | −6.6 | 6 | 1.4 | 12 | 9.4 | 26 | 17 | 0 | 41.8 | *88* | 14 | 10.1 | 28 | 76.8 | *69* | | |
| Apr | 12.9 | 3.5 | 8.2 | *−1.1* | 20.4 | 24 | −1.3 | 11 | 6.4 | 10 | 8.8 | 23 | 4 | 0 | 59.0 | *121* | 15 | 21.8 | 28 | 141.9 | *88* | | |
| May | 15.9 | 6.8 | 11.4 | *−1.1* | 22.5 | 7 | 0.9 | 3 | 8.6 | 13 | 11.8 | 7 | 0 | 0 | 122.3 | *214* | 16 | 36.9 | 12 | 157.2 | *81* | | |
| June | 18.7 | 9.3 | 14.0 | *−1.6* | 24.3 | 29 | 4.2 | 4 | 14.4 | 20 | 14.4 | 29 | 0 | 0 | 33.6 | *68* | 10 | 14.4 | 2 | 196.6 | *103* | | |
| July | 22.0 | 11.8 | 16.9 | *−1.0* | 28.6 | 4 | 6.1 | 27 | 15.0 | 27 | 16.6 | 3 | 0 | 6 | 93.5 | *191* | 13 | 46.3 | 25 | 203.5 | *98* | | |
| Aug | 21.3 | 11.9 | 16.6 | *−1.0* | 28.3 | 30 | 6.1 | 2 | 16.8 | 2 | 16.1 | 6 | 0 | 3 | 39.7 | *72* | 8 | 14.9 | 9 | 180.2 | *92* | | |
| Sep | 19.1 | 10.1 | 14.6 | *−0.3* | 23.3 | 4 | 5.1 | 15 | 13.1 | 24 | 16.1 | 12 | 0 | 0 | 49.7 | *92* | 11 | 12.0 | 2 | 130.5 | *93* | | |
| Oct | 15.2 | 8.1 | 11.6 | *+0.4* | 24.3 | 5 | 2.3 | 21 | 8.1 | 26 | 12.9 | 7 | 0 | 0 | 88.1 | *126* | 20 | 11.1 | 15 | 64.8 | *58* | | |
| Nov | 9.6 | 3.0 | 6.3 | *−1.3* | 14.8 | 1 | −2.4 | 23 | 3.2 | 22 | 9.1 | 19 | 5 | 0 | 64.3 | *96* | 14 | 14.1 | 5 | 47.0 | *66* | | |
| Dec | 5.7 | −1.1 | 2.3 | *−2.7* | 12.0 | 5 | −9.9 | 31 | −1.9 | 31 | 5.8 | 12 | 13 | 0 | 120.8 | *191* | 19 | 33.0 | 26 | 74.1 | *139* | | |
| 1886 | 13.1 | 5.2 | 9.1 | *−1.6* | 28.6 | 7 | −10.0 | 1 | −1.9 | 12 | 16.6 | 7 | 74 | 9 | 830.3 | *126* | 168 | 46.3 | 7 | 1368.1 | *87* | | |

## OXFORD 1887

| Month | Mean max | Mean min | Mean temp | Anom | Highest max | Date | Lowest min | Date | Lowest max | Date | Highest min | Date | Air frost | Days ≥ 25 °C | Total pptn | Anom | Rain days | Wettest day | Date | Total sunshine | Anom | Sunniest day | Date |
|---|---|---|---|---|---|---|---|---|---|---|---|---|---|---|---|---|---|---|---|---|---|---|---|
| Jan | 4.7 | −1.1 | 1.8 | *−3.1* | 10.7 | 31 | −9.4 | 1 | −2.8 | 1 | 6.7 | 28 | 15 | 0 | 65.3 | *115* | 12 | 33.0 | 3 | 30.1 | *48* | | |
| Feb | 7.7 | −0.1 | 3.8 | *−1.1* | 13.3 | 5 | −6.2 | 16 | 0.6 | 16 | 9.3 | 3 | 17 | 0 | 15.2 | *36* | 5 | 6.6 | 2 | 94.0 | *119* | | |
| Mar | 7.5 | −0.4 | 3.5 | *−3.7* | 13.3 | 29 | −8.1 | 16 | 1.8 | 15 | 7.0 | 26 | 13 | 0 | 45.7 | *96* | 12 | 8.7 | 14 | 90.6 | *81* | | |
| Apr | 11.6 | 1.6 | 6.6 | *−2.7* | 18.9 | 19 | −2.9 | 16 | 7.4 | 13 | 7.7 | 21 | 5 | 0 | 29.1 | *59* | 7 | 11.5 | 26 | 158.9 | *99* | | |
| May | 14.5 | 5.9 | 10.2 | *−2.2* | 19.6 | 8 | 1.7 | 13 | 10.1 | 20 | 10.3 | 11 | 0 | 0 | 49.9 | *87* | 15 | 14.4 | 31 | 130.7 | *68* | | |
| June | 21.8 | 10.0 | 15.9 | *+0.3* | 27.7 | 15 | 6.2 | 20 | 11.3 | 2 | 13.7 | 29 | 0 | 6 | 32.8 | *67* | 2 | 25.8 | 2 | 252.4 | *132* | | |
| July | 24.2 | 12.2 | 18.2 | *+0.3* | 29.6 | 3 | 6.5 | 17 | 19.2 | 17 | 17.0 | 11 | 0 | 7 | 18.0 | *37* | 5 | 6.3 | 24 | 261.5 | *126* | | |
| Aug | 22.3 | 10.4 | 16.3 | *−1.2* | 28.9 | 6 | 4.5 | 14 | 17.3 | 13 | 14.7 | 28 | 0 | 6 | 40.5 | *74* | 7 | 17.5 | 17 | 242.3 | *123* | | |
| Sep | 16.4 | 7.8 | 12.1 | *−2.7* | 19.2 | 4 | 0.9 | 27 | 13.3 | 28 | 13.1 | 1 | 0 | 0 | 61.2 | *113* | 14 | 13.7 | 16 | 107.9 | *77* | | |
| Oct | 11.6 | 2.9 | 7.3 | *−4.0* | 15.7 | 7 | −3.3 | 25 | 6.3 | 24 | 9.2 | 2 | 6 | 0 | 53.7 | *77* | 8 | 23.3 | 29 | 115.0 | *103* | | |
| Nov | 7.9 | 1.4 | 4.6 | *−2.9* | 11.9 | 3 | −7.5 | 16 | 1.3 | 16 | 8.4 | 26 | 8 | 0 | 46.0 | *69* | 15 | 8.9 | 9 | 50.9 | *72* | | |
| Dec | 5.9 | −0.4 | 2.7 | *−2.3* | 11.3 | 8 | −5.6 | 28 | 1.2 | 28 | 6.4 | 8 | 17 | 0 | 35.3 | *56* | 12 | 8.6 | 12 | 46.4 | *87* | | |
| 1887 | 13.0 | 4.2 | 8.6 | *−2.1* | 29.6 | 7 | −9.4 | 1 | −2.8 | 1 | 17.0 | 7 | 81 | 19 | 492.7 | *75* | 114 | 33.0 | 1 | 1580.7 | *100* | | |

## OXFORD 1888

| Month | Mean max | Mean min | Mean temp | Anom | Highest max | Date | Lowest min | Date | Lowest max | Date | Highest min | Date | Air frost | Days ≥ 25 °C | Total pptn | Anom | Rain days | Wettest day | Date | Total sunshine | Anom | Sunniest day | Date |
|---|---|---|---|---|---|---|---|---|---|---|---|---|---|---|---|---|---|---|---|---|---|---|---|
| Jan | 5.6 | 0.1 | 2.8 | *−2.0* | 11.9 | 23 | −6.3 | 29 | 0.3 | 13 | 7.1 | 7 | 17 | 0 | 19.8 | *35* | 12 | 6.7 | 20 | 37.7 | *61* | | |
| Feb | 4.2 | −1.2 | 1.5 | *−3.4* | 10.3 | 6 | −7.8 | 24 | −1.1 | 23 | 7.1 | 8 | 20 | 0 | 71.7 | *167* | 14 | 40.8 | 13 | 38.1 | *48* | | |
| Mar | 7.0 | 0.6 | 3.8 | *−3.5* | 11.9 | 9 | −6.3 | 1 | 1.1 | 19 | 8.8 | 8 | 15 | 0 | 77.6 | *163* | 18 | 10.9 | 25 | 74.3 | *67* | | |
| Apr | 10.4 | 2.7 | 6.6 | *−2.7* | 15.6 | 30 | −3.4 | 5 | 5.2 | 8 | 8.2 | 30 | 8 | 0 | 39.5 | *81* | 16 | 6.0 | 15 | 100.7 | *63* | | |
| May | 16.5 | 6.2 | 11.4 | *−1.1* | 23.2 | 19 | 2.1 | 10 | 10.8 | 2 | 14.4 | 18 | 0 | 0 | 39.4 | *69* | 9 | 8.1 | 16 | 219.7 | *114* | | |
| June | 18.4 | 9.5 | 13.9 | *−1.7* | 24.9 | 25 | 3.9 | 17 | 12.9 | 5 | 14.3 | 24 | 0 | 0 | 77.0 | *156* | 16 | 26.6 | 21 | 127.2 | *67* | | |
| July | 18.2 | 10.7 | 14.5 | *−3.4* | 21.8 | 19 | 5.1 | 10 | 12.4 | 11 | 13.9 | 17 | 0 | 0 | 119.6 | *245* | 23 | 19.4 | 15 | 95.4 | *46* | | |
| Aug | 19.2 | 10.3 | 14.8 | *−2.8* | 26.7 | 10 | 4.4 | 18 | 13.5 | 16 | 15.9 | 10 | 0 | 3 | 47.6 | *87* | 12 | 15.7 | 28 | 145.0 | *74* | | |
| Sep | 17.7 | 8.6 | 13.1 | *−1.7* | 21.3 | 15 | 2.2 | 30 | 11.2 | 30 | 13.9 | 4 | 0 | 0 | 33.0 | *61* | 12 | 10.4 | 23 | 132.5 | *94* | | |
| Oct | 13.1 | 3.1 | 8.1 | *−3.2* | 19.5 | 27 | −1.7 | 2 | 9.2 | 6 | 14.8 | 26 | 9 | 0 | 19.5 | *28* | 7 | 8.9 | 28 | 120.5 | *108* | | |
| Nov | 10.6 | 5.2 | 7.9 | *+0.4* | 14.7 | 16 | 0.4 | 27 | 3.4 | 6 | 12.2 | 15 | 0 | 0 | 114.8 | *172* | 17 | 22.5 | 2 | 29.7 | *42* | | |
| Dec | 7.7 | 1.7 | 4.7 | *−0.3* | 14.2 | 3 | −4.9 | 31 | 0.0 | 10 | 10.8 | 5 | 11 | 0 | 56.8 | *90* | 12 | 13.7 | 24 | 33.7 | *63* | | |
| 1888 | 12.4 | 4.8 | 8.6 | *−2.1* | 26.7 | 8 | −7.8 | 2 | −1.1 | 2 | 15.9 | 8 | 80 | 3 | 716.3 | *108* | 168 | 40.8 | 2 | 1154.5 | *73* | | |

## OXFORD 1889

| Month | Mean max | Mean min | Mean temp | Anom | Highest max | Date | Lowest min | Date | Lowest max | Date | Highest min | Date | Air frost | Days ≥ 25 °C | Total pptn | Anom | Rain days | Wettest day | Date | Total sunshine | Anom | Sunniest day | Date |
|---|---|---|---|---|---|---|---|---|---|---|---|---|---|---|---|---|---|---|---|---|---|---|---|
| Jan | 5.2 | −0.4 | 2.4 | *−2.5* | 10.5 | 18 | −6.8 | 5 | −4.8 | 5 | 7.8 | 31 | 14 | 0 | 16.1 | *28* | 10 | 7.0 | 9 | 30.7 | *49* | | |
| Feb | 5.9 | −0.8 | 2.6 | *−2.3* | 12.1 | 17 | −6.8 | 11 | 0.1 | 10 | 6.1 | 19 | 17 | 0 | 47.7 | *111* | 15 | 9.7 | 10 | 56.2 | *71* | | |
| Mar | 9.1 | 1.2 | 5.2 | *−2.1* | 15.9 | 29 | −5.2 | 4 | 1.7 | 4 | 7.7 | 25 | 12 | 0 | 50.4 | *106* | 11 | 14.9 | 8 | 89.2 | *80* | | |
| Apr | 10.9 | 3.7 | 7.3 | *−2.0* | 17.9 | 18 | −0.1 | 26 | 5.6 | 3 | 8.1 | 18 | 1 | 0 | 79.0 | *161* | 20 | 15.8 | 9 | 84.2 | *52* | | |
| May | 17.5 | 8.8 | 13.1 | *+0.7* | 25.1 | 23 | 4.9 | 2 | 10.8 | 12 | 12.6 | 23 | 0 | 1 | 87.4 | *153* | 18 | 31.2 | 27 | 119.8 | *62* | | |
| June | 20.7 | 10.8 | 15.7 | *+0.2* | 25.9 | 27 | 8.6 | 11 | 10.8 | 10 | 14.3 | 7 | 0 | 3 | 50.5 | *103* | 8 | 27.3 | 7 | 208.0 | *109* | | |
| July | 20.2 | 11.1 | 15.7 | *−2.2* | 26.4 | 30 | 6.7 | 19 | 15.2 | 25 | 14.3 | 7 | 0 | 3 | 72.4 | *148* | 14 | 20.9 | 13 | 175.3 | *85* | | |
| Aug | 19.7 | 10.7 | 15.2 | *−2.4* | 27.4 | 30 | 6.0 | 25 | 15.7 | 12 | 14.4 | 3 | 0 | 2 | 60.7 | *111* | 15 | 13.0 | 9 | 172.3 | *88* | | |
| Sep | 17.6 | 9.0 | 13.3 | *−1.6* | 26.3 | 11 | 1.6 | 18 | 11.9 | 29 | 15.6 | 12 | 0 | 2 | 45.2 | *83* | 9 | 32.6 | 24 | 116.0 | *83* | | |
| Oct | 12.9 | 5.2 | 9.1 | *−2.2* | 14.8 | 10 | −0.1 | 14 | 8.9 | 11 | 9.7 | 16 | 1 | 0 | 62.5 | *89* | 16 | 23.0 | 27 | 87.6 | *79* | | |
| Nov | 9.6 | 3.8 | 6.7 | *−0.8* | 14.9 | 8 | −2.4 | 29 | 1.8 | 27 | 9.3 | 8 | 4 | 0 | 26.5 | *40* | 7 | 13.8 | 3 | 43.7 | *62* | | |
| Dec | 5.8 | −0.3 | 2.7 | *−2.3* | 12.2 | 17 | −7.2 | 29 | −0.4 | 3 | 9.1 | 17 | 16 | 0 | 30.1 | *48* | 11 | 8.7 | 7 | 34.7 | *65* | | |
| **1889** | 13.0 | 5.3 | 9.1 | *−1.6* | 27.4 | 8 | −7.2 | 12 | −4.8 | 1 | 15.6 | 9 | 65 | 11 | 628.5 | *95* | 154 | 32.6 | 9 | 1217.7 | *77* | | |

## OXFORD 1890

| Month | Mean max | Mean min | Mean temp | Anom | Highest max | Date | Lowest min | Date | Lowest max | Date | Highest min | Date | Air frost | Days ≥ 25 °C | Total pptn | Anom | Rain days | Wettest day | Date | Total sunshine | Anom | Sunniest day | Date |
|---|---|---|---|---|---|---|---|---|---|---|---|---|---|---|---|---|---|---|---|---|---|---|---|
| Jan | 8.7 | 3.0 | 5.8 | *+1.0* | 13.1 | 25 | −6.7 | 2 | −1.8 | 2 | 8.9 | 7 | 3 | 0 | 51.0 | *90* | 18 | 11.1 | 28 | 59.0 | *95* | | |
| Feb | 6.5 | −0.1 | 3.2 | *−1.7* | 9.6 | 1 | −3.5 | 13 | 2.9 | 13 | 7.4 | 1 | 12 | 0 | 18.7 | *44* | 4 | 17.1 | 15 | 71.6 | *91* | | |
| Mar | 10.2 | 2.4 | 6.3 | *−1.0* | 16.8 | 28 | −8.2 | 4 | 1.3 | 2 | 9.9 | 28 | 5 | 0 | 21.9 | *46* | 12 | 4.3 | 10 | 103.4 | *93* | | |
| Apr | 12.0 | 2.9 | 7.4 | *−1.9* | 17.5 | 30 | −2.4 | 2 | 6.9 | 18 | 9.2 | 22 | 6 | 0 | 25.8 | *53* | 12 | 5.3 | 22 | 142.7 | *89* | | |
| May | 17.1 | 6.6 | 11.9 | *−0.6* | 24.6 | 24 | 2.3 | 31 | 10.5 | 9 | 9.8 | 25 | 0 | 0 | 45.2 | *79* | 12 | 14.9 | 9 | 231.0 | *120* | | |
| June | 19.1 | 9.6 | 14.3 | *−1.2* | 23.6 | 24 | 2.5 | 1 | 15.0 | 3 | 13.8 | 26 | 0 | 0 | 44.6 | *91* | 17 | 9.7 | 26 | 141.7 | *74* | | |
| July | 19.5 | 11.0 | 15.3 | *−2.6* | 23.6 | 23 | 6.9 | 4 | 14.7 | 6 | 16.7 | 31 | 0 | 0 | 77.2 | *158* | 18 | 20.9 | 17 | 153.7 | *74* | | |
| Aug | 19.5 | 10.6 | 15.1 | *−2.5* | 24.7 | 5 | 4.3 | 30 | 15.8 | 26 | 15.8 | 1 | 0 | 0 | 62.7 | *114* | 20 | 12.4 | 19 | 165.4 | *84* | | |
| Sep | 20.3 | 9.8 | 15.0 | *+0.2* | 24.8 | 9 | 2.3 | 1 | 16.8 | 21 | 14.2 | 27 | 0 | 0 | 26.3 | *49* | 6 | 16.1 | 21 | 154.4 | *110* | | |
| Oct | 14.1 | 5.4 | 9.8 | *−1.5* | 20.5 | 13 | −5.0 | 28 | 5.8 | 27 | 13.3 | 7 | 1 | 0 | 33.6 | *48* | 8 | 11.8 | 7 | 132.4 | *119* | | |
| Nov | 9.3 | 2.5 | 5.9 | *−1.6* | 13.4 | 23 | −6.1 | 29 | −2.7 | 28 | 10.9 | 19 | 7 | 0 | 47.1 | *71* | 16 | 9.7 | 6 | 56.0 | *79* | | |
| Dec | 0.4 | −4.4 | −2.0 | *−7.0* | 6.3 | 4 | −13.3 | 22 | −6.7 | 22 | 1.8 | 4 | 25 | 0 | 13.3 | *21* | 8 | 9.0 | 19 | 5.0 | *9* | | |
| **1890** | 13.1 | 5.0 | 9.0 | *−1.7* | 24.8 | 9 | −13.3 | 12 | −6.7 | 12 | 16.7 | 7 | 59 | 0 | 467.4 | *71* | 151 | 20.9 | 7 | 1416.3 | *90* | | |

## OXFORD 1891

| Month | Mean max | Mean min | Mean temp | Anom | Highest max | Date | Lowest min | Date | Lowest max | Date | Highest min | Date | Air frost | Days ≥ 25 °C | Total pptn | Anom | Rain days | Wettest day | Date | Total sunshine | Anom | Sunniest day | Date |
|---|---|---|---|---|---|---|---|---|---|---|---|---|---|---|---|---|---|---|---|---|---|---|---|
| Jan | 4.3 | −2.3 | 1.0 | *−3.8* | 11.1 | 28 | −11.1 | 11 | −4.1 | 11 | 7.6 | 29 | 20 | 0 | 39.4 | *69* | 14 | 10.4 | 24 | 74.6 | *120* | | |
| Feb | 9.0 | −0.6 | 4.2 | *−0.7* | 17.4 | 28 | −5.3 | 19 | 1.1 | 21 | 6.6 | 3 | 13 | 0 | 2.5 | *6* | 1 | 2.5 | 1 | 93.6 | *119* | | |
| Mar | 8.0 | 0.7 | 4.4 | *−2.9* | 14.7 | 1 | −9.7 | 12 | 0.8 | 10 | 7.5 | 2 | 13 | 0 | 35.4 | *74* | 15 | 6.0 | 8 | 90.3 | *81* | | |
| Apr | 11.3 | 2.1 | 6.7 | *−2.6* | 16.7 | 30 | −3.4 | 1 | 5.6 | 9 | 8.7 | 30 | 5 | 0 | 24.0 | *49* | 9 | 12.5 | 4 | 107.8 | *67* | | |
| May | 14.2 | 5.5 | 9.9 | *−2.6* | 24.6 | 13 | −1.2 | 17 | 6.2 | 18 | 11.6 | 13 | 2 | 0 | 65.8 | *115* | 19 | 10.6 | 2 | 168.0 | *87* | | |
| June | 20.1 | 10.3 | 15.2 | *−0.4* | 24.3 | 19 | 5.3 | 12 | 13.3 | 7 | 15.4 | 25 | 0 | 0 | 36.3 | *74* | 9 | 10.0 | 25 | 179.1 | *94* | | |
| July | 19.5 | 10.5 | 15.0 | *−2.9* | 25.1 | 17 | 6.4 | 28 | 14.9 | 29 | 14.3 | 21 | 0 | 1 | 62.3 | *128* | 16 | 12.0 | 5 | 163.4 | *79* | | |
| Aug | 18.1 | 10.7 | 14.4 | *−3.2* | 22.4 | 14 | 5.8 | 30 | 15.8 | 25 | 15.9 | 14 | 0 | 0 | 119.7 | *218* | 20 | 26.6 | 27 | 128.3 | *65* | | |
| Sep | 18.7 | 9.7 | 14.2 | *−0.7* | 27.1 | 11 | 6.5 | 3 | 12.4 | 21 | 14.4 | 18 | 0 | 4 | 36.8 | *68* | 17 | 10.8 | 21 | 141.7 | *101* | | |
| Oct | 13.3 | 6.3 | 9.8 | *−1.5* | 17.0 | 9 | −2.3 | 31 | 8.9 | 31 | 11.8 | 9 | 1 | 0 | 150.1 | *214* | 21 | 28.9 | 22 | 117.8 | *106* | | |
| Nov | 8.5 | 2.8 | 5.6 | *−1.9* | 13.1 | 19 | −1.6 | 30 | 2.6 | 30 | 9.1 | 19 | 3 | 0 | 62.6 | *94* | 16 | 16.7 | 11 | 45.8 | *65* | | |
| Dec | 7.5 | 1.2 | 4.3 | *−0.7* | 13.7 | 5 | −9.1 | 23 | −4.2 | 22 | 7.8 | 10 | 9 | 0 | 86.1 | *136* | 19 | 16.2 | 13 | 53.4 | *100* | | |
| **1891** | 12.7 | 4.8 | 8.7 | *−2.0* | 27.1 | 9 | −11.1 | 1 | −4.2 | 12 | 15.9 | 8 | 66 | 5 | 721.0 | *109* | 176 | 28.9 | 10 | 1363.8 | *87* | | |

## OXFORD 1892

| Month | Mean max | Mean min | Mean temp | Anom | Highest max | Date | Lowest min | Date | Lowest max | Date | Highest min | Date | Air frost | Days ≥ 25 °C | Total pptn | Anom | Rain days | Wettest day | Date | Total sunshine | Anom | Sunniest day | Date |
|---|---|---|---|---|---|---|---|---|---|---|---|---|---|---|---|---|---|---|---|---|---|---|---|
| Jan | 4.8 | −0.6 | 2.1 | *−2.7* | 10.6 | 31 | −7.2 | 16 | −0.8 | 9 | 7.6 | 31 | 18 | 0 | 12.0 | *21* | 9 | 3.4 | 11 | 53.9 | *87* | | |
| Feb | 6.6 | 0.4 | 3.5 | *−1.4* | 11.1 | 7 | −8.4 | 17 | −0.7 | 19 | 7.2 | 7 | 9 | 0 | 18.6 | *43* | 12 | 4.0 | 15 | 54.3 | *69* | | |
| Mar | 7.0 | −1.1 | 2.9 | *−4.3* | 15.3 | 18 | −5.3 | 13 | 0.2 | 3 | 7.7 | 17 | 23 | 0 | 9.5 | *20* | 7 | 3.6 | 15 | 106.9 | *96* | | |
| Apr | 14.4 | 1.6 | 8.0 | *−1.3* | 21.1 | 6 | −4.9 | 15 | 5.3 | 13 | 8.8 | 22 | 10 | 0 | 19.5 | *40* | 6 | 5.0 | 13 | 208.9 | *130* | | |
| May | 17.3 | 6.9 | 12.1 | *−0.4* | 25.0 | 28 | −1.1 | 7 | 6.7 | 3 | 14.4 | 27 | 3 | 1 | 32.8 | *57* | 12 | 15.5 | 26 | 191.4 | *99* | | |
| June | 19.3 | 8.5 | 13.9 | *−1.7* | 26.3 | 10 | 1.4 | 15 | 14.7 | 17 | 15.8 | 27 | 0 | 3 | 52.7 | *107* | 11 | 18.8 | 23 | 219.5 | *115* | | |
| July | 19.1 | 10.1 | 14.6 | *−3.2* | 27.6 | 3 | 5.8 | 21 | 13.1 | 13 | 14.2 | 7 | 0 | 1 | 84.4 | *173* | 12 | 29.0 | 5 | 155.6 | *75* | | |
| Aug | 20.5 | 11.3 | 15.9 | *−1.7* | 24.8 | 17 | 3.6 | 11 | 15.6 | 2 | 15.3 | 19 | 0 | 0 | 78.7 | *144* | 15 | 13.9 | 28 | 187.6 | *95* | | |
| Sep | 17.0 | 8.8 | 12.9 | *−2.0* | 20.5 | 13 | 1.7 | 18 | 13.3 | 22 | 14.4 | 12 | 0 | 0 | 59.1 | *109* | 12 | 19.3 | 21 | 119.5 | *85* | | |
| Oct | 11.2 | 3.8 | 7.5 | *−3.8* | 14.7 | 29 | −4.0 | 26 | 6.2 | 25 | 11.8 | 28 | 4 | 0 | 85.6 | *122* | 16 | 29.5 | 27 | 105.6 | *95* | | |
| Nov | 9.4 | 3.3 | 6.4 | *−1.2* | 15.2 | 14 | −2.9 | 2 | 4.2 | 18 | 10.8 | 14 | 6 | 0 | 49.2 | *74* | 12 | 10.0 | 19 | 42.6 | *60* | | |
| Dec | 4.9 | −1.0 | 2.0 | *−3.1* | 11.6 | 15 | −10.2 | 27 | −2.6 | 27 | 8.9 | 15 | 17 | 0 | 25.3 | *40* | 11 | 9.1 | 1 | 49.5 | *93* | | |
| **1892** | 12.6 | 4.3 | 8.5 | *−2.3* | 27.6 | 7 | −10.2 | 12 | −2.6 | 12 | 15.8 | 6 | 90 | 5 | 527.4 | *80* | 135 | 29.5 | 10 | 1495.3 | *95* | | |

## OXFORD 1893

| Month | Mean max | Mean min | Mean temp | Anom | Highest max | Date | Lowest min | Date | Lowest max | Date | Highest min | Date | Air frost | Days ≥ 25 °C | Total pptn | Anom | Rain days | Wettest day | Date | Total sunshine | Anom | Sunniest day | Date |
|---|---|---|---|---|---|---|---|---|---|---|---|---|---|---|---|---|---|---|---|---|---|---|---|
| Jan | 4.5 | −0.6 | 1.9 | *−2.9* | 11.6 | 23 | −10.6 | 5 | −2.3 | 5 | 7.3 | 31 | 15 | 0 | 43.1 | *76* | 17 | 11.6 | 26 | 43.3 | *70* | | |
| Feb | 8.1 | 2.0 | 5.1 | *+0.2* | 13.8 | 19 | −2.8 | 6 | 1.6 | 24 | 8.3 | 19 | 5 | 0 | 69.4 | *162* | 22 | 13.5 | 26 | 63.5 | *80* | | |
| Mar | 13.8 | 1.5 | 7.7 | *+0.4* | 19.1 | 30 | −4.9 | 19 | 7.4 | 17 | 7.4 | 14 | 12 | 0 | 9.4 | *20* | 4 | 7.2 | 1 | 198.4 | *178* | | |
| Apr | 18.1 | 3.8 | 10.9 | *+1.6* | 26.8 | 20 | −2.9 | 14 | 10.1 | 12 | 10.9 | 20 | 2 | 4 | 1.5 | *3* | 2 | 1.3 | 17 | 252.8 | *157* | | |
| May | 19.6 | 7.6 | 13.6 | *+1.1* | 25.1 | 15 | 2.4 | 7 | 13.9 | 30 | 11.6 | 20 | 0 | 1 | 23.1 | *40* | 6 | 13.2 | 17 | 204.7 | *106* | | |
| June | 21.9 | 10.1 | 16.0 | *+0.4* | 29.9 | 19 | 1.5 | 1 | 15.2 | 24 | 14.6 | 28 | 0 | 6 | 19.6 | *40* | 11 | 7.3 | 4 | 208.4 | *109* | | |
| July | 21.8 | 12.4 | 17.1 | *−0.8* | 28.5 | 7 | 9.2 | 18 | 17.4 | 14 | 16.9 | 8 | 0 | 5 | 98.9 | *203* | 20 | 16.4 | 12 | 182.4 | *88* | | |
| Aug | 23.3 | 12.9 | 18.1 | *+0.5* | 31.3 | 18 | 5.4 | 29 | 17.6 | 27 | 20.0 | 18 | 0 | 10 | 27.7 | *51* | 12 | 9.0 | 3 | 216.7 | *110* | | |
| Sep | 18.7 | 8.7 | 13.7 | *−1.2* | 25.3 | 6 | 2.8 | 24 | 12.4 | 23 | 15.4 | 7 | 0 | 1 | 16.0 | *30* | 10 | 5.8 | 29 | 156.5 | *112* | | |
| Oct | 14.5 | 6.3 | 10.4 | *−0.9* | 19.6 | 21 | −2.0 | 31 | 6.7 | 31 | 14.6 | 15 | 1 | 0 | 73.5 | *105* | 12 | 23.3 | 7 | 130.2 | *117* | | |
| Nov | 8.2 | 1.2 | 4.7 | *−2.8* | 15.2 | 3 | −4.4 | 14 | 3.2 | 19 | 8.6 | 29 | 12 | 0 | 42.5 | *64* | 13 | 9.1 | 18 | 57.7 | *81* | | |
| Dec | 8.0 | 0.9 | 4.5 | *−0.6* | 13.3 | 13 | −6.3 | 3 | −0.2 | 31 | 7.7 | 16 | 5 | 0 | 47.7 | *76* | 15 | 11.3 | 11 | 50.5 | *95* | | |
| **1893** | 15.1 | 5.6 | 10.3 | *−0.4* | 31.3 | 8 | −10.6 | 1 | −2.3 | 1 | 20.0 | 8 | 52 | 27 | 472.4 | *72* | 144 | 23.3 | 10 | 1765.1 | *112* | | |

## OXFORD 1894

| Month | Mean max | Mean min | Mean temp | Anom | Highest max | Date | Lowest min | Date | Lowest max | Date | Highest min | Date | Air frost | Days ≥ 25 °C | Total pptn | Anom | Rain days | Wettest day | Date | Total sunshine | Anom | Sunniest day | Date |
|---|---|---|---|---|---|---|---|---|---|---|---|---|---|---|---|---|---|---|---|---|---|---|---|
| Jan | 6.1 | 0.9 | 3.5 | *−1.4* | 12.1 | 11 | −11.6 | 5 | −5.9 | 5 | 6.8 | 17 | 9 | 0 | 49.6 | *87* | 19 | 9.8 | 22 | 60.0 | *96* | | |
| Feb | 8.8 | 1.7 | 5.3 | *+0.4* | 12.3 | 7 | −5.1 | 20 | 2.9 | 22 | 10.4 | 7 | 7 | 0 | 41.3 | *96* | 12 | 21.2 | 17 | 88.3 | *112* | | |
| Mar | 11.9 | 2.0 | 7.0 | *−0.3* | 17.9 | 27 | −1.6 | 3 | 7.7 | 7 | 7.2 | 20 | 5 | 0 | 45.0 | *94* | 13 | 8.6 | 1 | 167.1 | *150* | | |
| Apr | 15.0 | 4.9 | 10.0 | *+0.7* | 20.7 | 11 | 0.4 | 22 | 10.1 | 20 | 8.8 | 12 | 0 | 0 | 46.9 | *96* | 18 | 9.1 | 2 | 141.3 | *88* | | |
| May | 14.3 | 5.2 | 9.8 | *−2.7* | 19.6 | 25 | −0.7 | 21 | 9.8 | 19 | 11.5 | 16 | 1 | 0 | 39.3 | *69* | 15 | 10.2 | 11 | 174.7 | *91* | | |
| June | 18.3 | 9.8 | 14.1 | *−1.5* | 27.2 | 30 | 4.8 | 1 | 13.1 | 6 | 14.9 | 26 | 0 | 1 | 74.0 | *150* | 16 | 38.2 | 4 | 144.2 | *76* | | |
| July | 20.8 | 11.8 | 16.3 | *−1.6* | 28.1 | 1 | 8.8 | 4 | 16.1 | 23 | 16.8 | 2 | 0 | 2 | 84.8 | *174* | 16 | 23.9 | 29 | 160.3 | *77* | | |
| Aug | 18.7 | 10.8 | 14.7 | *−2.8* | 23.5 | 31 | 6.3 | 21 | 14.1 | 24 | 14.0 | 14 | 0 | 0 | 69.7 | *127* | 18 | 19.6 | 25 | 118.3 | *60* | | |
| Sep | 16.1 | 8.1 | 12.1 | *−2.7* | 20.7 | 1 | 2.1 | 28 | 13.1 | 22 | 12.2 | 24 | 0 | 0 | 47.1 | *87* | 10 | 14.9 | 25 | 101.4 | *72* | | |
| Oct | 12.6 | 6.6 | 9.6 | *−1.7* | 16.9 | 12 | −1.1 | 22 | 8.1 | 21 | 13.7 | 11 | 1 | 0 | 89.4 | *128* | 19 | 20.2 | 30 | 49.6 | *45* | | |
| Nov | 10.7 | 4.9 | 7.8 | *+0.3* | 15.6 | 3 | 0.1 | 30 | 4.2 | 26 | 12.7 | 2 | 0 | 0 | 126.1 | *189* | 10 | 37.9 | 14 | 69.0 | *97* | | |
| Dec | 7.7 | 2.4 | 5.0 | *+0.0* | 11.4 | 14 | −2.8 | 31 | 1.3 | 31 | 8.1 | 14 | 9 | 0 | 52.6 | *83* | 17 | 18.6 | 14 | 50.6 | *95* | | |
| **1894** | 13.4 | 5.8 | 9.6 | *−1.1* | 28.1 | 7 | −11.6 | 1 | −5.9 | 1 | 16.8 | 7 | 32 | 3 | 765.8 | *116* | 183 | 38.2 | 6 | 1324.8 | *84* | | |

## OXFORD 1895

| Month | Mean max | Mean min | Mean temp | Anom | Highest max | Date | Lowest min | Date | Lowest max | Date | Highest min | Date | Air frost | Days ≥ 25 °C | Total pptn | Anom | Rain days | Wettest day | Date | Total sunshine | Anom | Sunniest day | Date |
|---|---|---|---|---|---|---|---|---|---|---|---|---|---|---|---|---|---|---|---|---|---|---|---|
| Jan | 3.1 | −2.2 | 0.5 | *−4.4* | 10.4 | 20 | −10.3 | 11 | −4.9 | 11 | 4.5 | 16 | 20 | 0 | 66.3 | *117* | 19 | 12.5 | 20 | 51.6 | *83* | | |
| Feb | 1.9 | −5.7 | −1.9 | *−6.8* | 8.1 | 28 | −13.6 | 8 | −4.6 | 7 | 1.8 | 22 | 24 | 0 | 4.4 | *10* | 4 | 2.3 | 24 | 67.9 | *86* | | |
| Mar | 10.1 | 1.8 | 5.9 | *−1.3* | 17.4 | 22 | −4.8 | 4 | 3.3 | 3 | 7.2 | 20 | 9 | 0 | 39.2 | *82* | 16 | 17.9 | 27 | 93.9 | *84* | | |
| Apr | 13.2 | 4.9 | 9.1 | *−0.2* | 16.6 | 9 | −1.6 | 1 | 5.6 | 2 | 10.6 | 23 | 1 | 0 | 46.4 | *95* | 13 | 16.4 | 25 | 126.3 | *79* | | |
| May | 19.1 | 7.2 | 13.2 | *+0.7* | 28.4 | 30 | 2.9 | 2 | 8.0 | 17 | 15.3 | 31 | 0 | 1 | 4.6 | *8* | 4 | 1.9 | 30 | 251.2 | *130* | | |
| June | 21.5 | 9.7 | 15.6 | *−0.0* | 26.5 | 9 | 3.3 | 15 | 16.9 | 1 | 15.7 | 24 | 0 | 6 | 28.6 | *58* | 5 | 19.1 | 26 | 216.5 | *113* | | |
| July | 20.5 | 12.0 | 16.3 | *−1.6* | 25.5 | 8 | 8.2 | 13 | 16.9 | 28 | 16.9 | 25 | 0 | 2 | 91.5 | *187* | 14 | 20.3 | 20 | 169.9 | *82* | | |
| Aug | 20.7 | 11.9 | 16.3 | *−1.3* | 26.3 | 22 | 6.7 | 25 | 15.4 | 3 | 17.7 | 22 | 0 | 2 | 62.5 | *114* | 13 | 17.5 | 13 | 189.0 | *96* | | |
| Sep | 22.5 | 9.8 | 16.2 | *+1.3* | 28.4 | 27 | 2.3 | 22 | 15.3 | 19 | 16.3 | 7 | 0 | 7 | 16.4 | *30* | 9 | 10.6 | 7 | 214.0 | *152* | | |
| Oct | 11.8 | 3.6 | 7.7 | *−3.6* | 21.4 | 1 | −3.8 | 26 | 4.3 | 26 | 10.6 | 1 | 8 | 0 | 76.6 | *109* | 16 | 17.8 | 8 | 79.0 | *71* | | |
| Nov | 11.2 | 4.9 | 8.0 | *+0.5* | 16.4 | 16 | −1.8 | 18 | 5.7 | 26 | 10.6 | 6 | 1 | 0 | 113.5 | *170* | 20 | 15.1 | 5 | 46.7 | *66* | | |
| Dec | 7.0 | 1.4 | 4.2 | *−0.8* | 12.4 | 30 | −4.0 | 11 | 0.6 | 21 | 9.6 | 5 | 6 | 0 | 56.5 | *90* | 21 | 12.7 | 12 | 39.7 | *75* | | |
| **1895** | 13.6 | 5.0 | 9.3 | *−1.4* | 28.4 | 5 | −13.6 | 2 | −4.9 | 1 | 17.7 | 8 | 69 | 18 | 606.5 | *92* | 154 | 20.3 | 7 | 1545.7 | *98* | | |

## OXFORD 1896

| Month | Mean max | Mean min | Mean temp | Anom | Highest max | Date | Lowest min | Date | Lowest max | Date | Highest min | Date | Air frost | Days ≥ 25 °C | Total pptn | Anom | Rain days | Wettest day | Date | Total sunshine | Anom | Sunniest day | Date |
|---|---|---|---|---|---|---|---|---|---|---|---|---|---|---|---|---|---|---|---|---|---|---|---|
| Jan | 7.2 | 2.1 | 4.7 | *−0.2* | 11.4 | 15 | −3.0 | 20 | 1.0 | 7 | 7.3 | 2 | 5 | 0 | 16.8 | *30* | 9 | 7.8 | 25 | 32.3 | *52* | | |
| Feb | 8.2 | 1.0 | 4.6 | *−0.3* | 13.3 | 28 | −5.9 | 26 | 1.5 | 25 | 8.0 | 20 | 9 | 0 | 10.1 | *24* | 4 | 3.3 | 20 | 81.0 | *103* | | |
| Mar | 11.5 | 4.1 | 7.8 | *+0.6* | 18.1 | 24 | −0.3 | 15 | 5.9 | 13 | 9.5 | 8 | 3 | 0 | 69.0 | *145* | 22 | 14.9 | 18 | 86.9 | *78* | | |
| Apr | 14.0 | 5.3 | 9.7 | *+0.4* | 18.1 | 19 | 0.5 | 2 | 10.3 | 16 | 11.6 | 27 | 0 | 0 | 17.5 | *36* | 10 | 6.6 | 16 | 136.7 | *85* | | |
| May | 18.4 | 6.5 | 12.4 | *−0.0* | 24.8 | 12 | −0.9 | 2 | 12.0 | 1 | 14.4 | 18 | 1 | 0 | 6.1 | *11* | 3 | 4.9 | 22 | 225.9 | *117* | | |
| June | 22.1 | 11.8 | 16.9 | *+1.4* | 27.6 | 15 | 6.9 | 1 | 15.4 | 7 | 16.3 | 28 | 0 | 6 | 65.0 | *132* | 9 | 12.7 | 7 | 212.7 | *111* | | |
| July | 23.0 | 12.0 | 17.5 | *−0.4* | 28.4 | 13 | 6.6 | 29 | 15.9 | 16 | 16.1 | 9 | 0 | 7 | 37.4 | *77* | 11 | 18.7 | 7 | 203.8 | *98* | | |
| Aug | 19.8 | 10.5 | 15.1 | *−2.5* | 23.9 | 23 | 6.8 | 5 | 15.2 | 26 | 15.6 | 24 | 0 | 0 | 79.8 | *146* | 12 | 19.0 | 31 | 138.7 | *71* | | |
| Sep | 17.1 | 10.6 | 13.8 | *−1.0* | 20.4 | 8 | 4.3 | 21 | 14.3 | 2 | 14.6 | 9 | 0 | 0 | 152.6 | *282* | 22 | 22.4 | 12 | 88.3 | *63* | | |
| Oct | 11.6 | 4.3 | 7.9 | *−3.4* | 17.1 | 3 | −1.1 | 28 | 6.8 | 19 | 12.7 | 3 | 3 | 0 | 81.1 | *116* | 14 | 12.7 | 25 | 101.4 | *91* | | |
| Nov | 7.9 | 1.1 | 4.5 | *−3.0* | 10.8 | 21 | −5.1 | 30 | 4.1 | 30 | 6.9 | 22 | 8 | 0 | 21.8 | *33* | 7 | 7.1 | 8 | 90.6 | *128* | | |
| Dec | 6.9 | 1.2 | 4.1 | *−1.0* | 11.4 | 26 | −4.1 | 24 | 0.4 | 18 | 6.3 | 9 | 11 | 0 | 77.3 | *123* | 22 | 8.8 | 28 | 31.6 | *59* | | |
| **1896** | 14.0 | 5.9 | 9.9 | *−0.8* | 28.4 | 7 | −5.9 | 2 | 0.4 | 12 | 16.3 | 6 | 40 | 13 | 634.5 | *96* | 145 | 22.4 | 9 | 1429.9 | *91* | | |

## OXFORD 1897

| Month | Mean max | Mean min | Mean temp | Anom | Highest max | Date | Lowest min | Date | Lowest max | Date | Highest min | Date | Air frost | Days ≥ 25 °C | Total pptn | Anom | Rain days | Wettest day | Date | Total sunshine | Anom | Sunniest day | Date |
|---|---|---|---|---|---|---|---|---|---|---|---|---|---|---|---|---|---|---|---|---|---|---|---|
| Jan | 3.9 | −0.6 | 1.7 | *−3.2* | 9.7 | 7 | −5.2 | 18 | −0.2 | 23 | 5.3 | 7 | 15 | 0 | 44.7 | *79* | 15 | 13.3 | 8 | 42.8 | *69* | | |
| Feb | 8.9 | 3.7 | 6.3 | *+1.4* | 13.7 | 22 | −0.5 | 18 | 2.5 | 2 | 9.4 | 26 | 2 | 0 | 60.0 | *140* | 15 | 17.8 | 5 | 29.9 | *38* | | |
| Mar | 10.8 | 3.9 | 7.3 | *+0.1* | 16.3 | 21 | −2.5 | 30 | 6.3 | 14 | 10.2 | 22 | 3 | 0 | 73.6 | *154* | 18 | 13.5 | 14 | 115.5 | *104* | | |
| Apr | 11.7 | 4.0 | 7.8 | *−1.5* | 20.4 | 28 | −2.2 | 5 | 6.0 | 1 | 7.6 | 27 | 2 | 0 | 53.5 | *109* | 15 | 14.5 | 21 | 133.2 | *83* | | |
| May | 15.9 | 5.6 | 10.7 | *−1.7* | 20.4 | 18 | −0.1 | 13 | 9.4 | 12 | 10.8 | 30 | 1 | 0 | 17.9 | *31* | 6 | 12.8 | 30 | 234.6 | *122* | | |
| June | 20.7 | 11.1 | 15.9 | *+0.3* | 27.6 | 13 | 6.1 | 10 | 11.7 | 9 | 16.1 | 28 | 0 | 5 | 74.1 | *150* | 11 | 19.1 | 9 | 179.3 | *94* | | |
| July | 22.6 | 12.1 | 17.4 | *−0.5* | 26.7 | 18 | 6.3 | 8 | 18.3 | 7 | 16.4 | 29 | 0 | 7 | 67.1 | *137* | 9 | 41.1 | 20 | 247.6 | *120* | | |
| Aug | 21.0 | 12.0 | 16.5 | *−1.1* | 29.8 | 4 | 9.2 | 26 | 15.9 | 31 | 16.1 | 11 | 0 | 4 | 108.9 | *199* | 17 | 23.1 | 8 | 197.1 | *100* | | |
| Sep | 16.8 | 8.8 | 12.8 | *−2.1* | 19.9 | 13 | 3.4 | 10 | 12.7 | 19 | 14.2 | 24 | 0 | 0 | 59.7 | *110* | 8 | 29.5 | 29 | 132.7 | *95* | | |
| Oct | 14.4 | 6.0 | 10.2 | *−1.1* | 18.8 | 17 | 0.0 | 7 | 10.4 | 13 | 13.1 | 17 | 0 | 0 | 36.0 | *51* | 9 | 10.5 | 19 | 119.4 | *107* | | |
| Nov | 10.3 | 3.9 | 7.1 | *−0.5* | 14.9 | 13 | −0.5 | 19 | 5.5 | 23 | 12.7 | 13 | 2 | 0 | 29.8 | *45* | 14 | 7.7 | 27 | 44.9 | *63* | | |
| Dec | 7.8 | 1.5 | 4.6 | *−0.4* | 14.3 | 16 | −6.0 | 24 | 2.1 | 22 | 8.7 | 17 | 10 | 0 | 65.5 | *104* | 14 | 8.7 | 14 | 52.7 | *99* | | |
| 1897 | 13.8 | 6.0 | 9.9 | *−0.9* | 29.8 | 8 | −6.0 | 12 | −0.2 | 1 | 16.4 | 7 | 35 | 16 | 690.8 | *105* | 151 | 41.1 | 7 | 1529.7 | *97* | | |

## OXFORD 1898

| Month | Mean max | Mean min | Mean temp | Anom | Highest max | Date | Lowest min | Date | Lowest max | Date | Highest min | Date | Air frost | Days ≥ 25 °C | Total pptn | Anom | Rain days | Wettest day | Date | Total sunshine | Anom | Sunniest day | Date |
|---|---|---|---|---|---|---|---|---|---|---|---|---|---|---|---|---|---|---|---|---|---|---|---|
| Jan | 8.6 | 3.8 | 6.2 | *+1.4* | 12.8 | 21 | −1.3 | 16 | 2.7 | 16 | 9.6 | 20 | 5 | 0 | 12.8 | *23* | 2 | 12.6 | 5 | 26.8 | *43* | | |
| Feb | 8.7 | 1.5 | 5.1 | *+0.2* | 12.4 | 1 | −5.1 | 21 | 4.2 | 21 | 7.4 | 1 | 7 | 0 | 32.9 | *77* | 15 | 6.9 | 18 | 79.2 | *100* | | |
| Mar | 8.7 | 0.4 | 4.5 | *−2.7* | 12.8 | 18 | −3.1 | 10 | 2.4 | 25 | 9.1 | 18 | 15 | 0 | 17.6 | *37* | 13 | 5.4 | 27 | 99.5 | *89* | | |
| Apr | 13.7 | 3.6 | 8.6 | *−0.7* | 19.2 | 8 | −2.5 | 5 | 10.0 | 14 | 9.3 | 7 | 4 | 0 | 36.0 | *74* | 12 | 7.0 | 14 | 152.1 | *95* | | |
| May | 14.9 | 6.7 | 10.8 | *−1.7* | 22.1 | 23 | 1.1 | 13 | 9.7 | 19 | 10.6 | 10 | 0 | 0 | 61.7 | *108* | 22 | 7.7 | 19 | 140.0 | *73* | | |
| June | 18.9 | 9.4 | 14.1 | *−1.4* | 23.3 | 17 | 3.4 | 15 | 11.3 | 13 | 15.1 | 21 | 0 | 0 | 34.8 | *71* | 13 | 11.8 | 24 | 178.7 | *94* | | |
| July | 21.5 | 11.0 | 16.2 | *−1.6* | 26.5 | 15 | 5.9 | 30 | 14.5 | 29 | 15.7 | 26 | 0 | 2 | 15.2 | *31* | 9 | 4.2 | 2 | 218.7 | *106* | | |
| Aug | 22.5 | 12.3 | 17.4 | *−0.2* | 28.4 | 22 | 8.0 | 8 | 13.8 | 7 | 16.9 | 13 | 0 | 8 | 43.8 | *80* | 9 | 26.3 | 6 | 183.1 | *93* | | |
| Sep | 22.3 | 9.9 | 16.1 | *+1.2* | 32.2 | 8 | 1.7 | 26 | 13.6 | 30 | 18.4 | 8 | 0 | 10 | 10.5 | *19* | 6 | 7.2 | 29 | 205.9 | *147* | | |
| Oct | 14.7 | 8.1 | 11.4 | *+0.1* | 20.4 | 3 | 1.1 | 13 | 9.8 | 13 | 13.3 | 5 | 0 | 0 | 125.6 | *179* | 15 | 32.8 | 29 | 64.0 | *57* | | |
| Nov | 10.4 | 4.3 | 7.4 | *−0.2* | 15.9 | 3 | −2.6 | 23 | 3.8 | 28 | 10.6 | 17 | 3 | 0 | 50.9 | *76* | 13 | 16.0 | 23 | 59.2 | *83* | | |
| Dec | 10.3 | 4.7 | 7.5 | *+2.5* | 13.5 | 5 | −1.1 | 22 | 4.9 | 20 | 12.3 | 5 | 4 | 0 | 65.2 | *103* | 11 | 16.3 | 7 | 53.1 | *100* | | |
| 1898 | 14.6 | 6.3 | 10.5 | *−0.3* | 32.2 | 9 | −5.1 | 2 | 2.4 | 3 | 18.4 | 9 | 38 | 20 | 507.0 | *77* | 140 | 32.8 | 10 | 1460.3 | *93* | | |

## OXFORD 1899

| Month | Mean max | Mean min | Mean temp | Anom | Highest max | Date | Lowest min | Date | Lowest max | Date | Highest min | Date | Air frost | Days ≥ 25 °C | Total pptn | Anom | Rain days | Wettest day | Date | Total sunshine | Anom | Sunniest day | Date |
|---|---|---|---|---|---|---|---|---|---|---|---|---|---|---|---|---|---|---|---|---|---|---|---|
| Jan | 8.5 | 2.6 | 5.5 | *+0.7* | 12.3 | 21 | −3.8 | 25 | 3.9 | 31 | 10.1 | 21 | 5 | 0 | 75.3 | *132* | 18 | 9.6 | 20 | 69.4 | *111* | | |
| Feb | 9.7 | 1.6 | 5.6 | *+0.7* | 16.6 | 10 | −5.4 | 4 | 2.1 | 5 | 9.7 | 10 | 10 | 0 | 51.2 | *120* | 12 | 11.8 | 5 | 105.6 | *134* | | |
| Mar | 10.7 | 0.2 | 5.5 | *−1.8* | 16.3 | 15 | −6.1 | 23 | 3.1 | 21 | 9.6 | 29 | 16 | 0 | 7.4 | *16* | 8 | 1.9 | 31 | 159.8 | *144* | | |
| Apr | 12.3 | 4.9 | 8.6 | *−0.7* | 18.6 | 1 | −1.2 | 17 | 5.9 | 16 | 9.7 | 29 | 2 | 0 | 48.8 | *100* | 19 | 7.3 | 24 | 124.3 | *77* | | |
| May | 15.6 | 6.1 | 10.9 | *−1.6* | 21.6 | 31 | 0.2 | 5 | 11.4 | 25 | 11.1 | 20 | 0 | 0 | 38.0 | *66* | 13 | 5.7 | 15 | 202.0 | *105* | | |
| June | 22.1 | 10.1 | 16.1 | *+0.5* | 27.7 | 5 | 4.6 | 15 | 16.1 | 14 | 15.9 | 27 | 0 | 7 | 20.3 | *41* | 6 | 6.4 | 30 | 256.8 | *134* | | |
| July | 24.0 | 13.2 | 18.6 | *+0.7* | 30.0 | 20 | 8.9 | 5 | 16.7 | 2 | 16.8 | 12 | 0 | 10 | 42.9 | *88* | 7 | 19.9 | 23 | 270.3 | *131* | | |
| Aug | 25.1 | 12.2 | 18.6 | *+1.0* | 29.5 | 25 | 8.6 | 10 | 19.9 | 29 | 15.1 | 6 | 0 | 13 | 47.6 | *87* | 7 | 22.8 | 15 | 250.3 | *127* | | |
| Sep | 18.7 | 10.1 | 14.4 | *−0.5* | 29.3 | 5 | 3.7 | 28 | 10.3 | 30 | 17.6 | 7 | 0 | 2 | 58.2 | *107* | 16 | 13.5 | 29 | 166.3 | *118* | | |
| Oct | 14.2 | 4.4 | 9.3 | *−2.0* | 18.2 | 17 | −1.2 | 19 | 8.2 | 21 | 12.4 | 27 | 2 | 0 | 70.1 | *100* | 10 | 21.1 | 27 | 116.7 | *105* | | |
| Nov | 11.4 | 5.0 | 8.2 | *+0.6* | 17.1 | 4 | −4.1 | 19 | 5.2 | 19 | 10.9 | 5 | 3 | 0 | 66.6 | *100* | 8 | 16.0 | 3 | 49.8 | *70* | | |
| Dec | 5.1 | −0.6 | 2.3 | *−2.8* | 13.2 | 6 | −9.5 | 16 | −4.9 | 15 | 8.4 | 6 | 14 | 0 | 33.2 | *53* | 14 | 7.7 | 1 | 38.3 | *72* | | |
| **1899** | 14.8 | 5.8 | 10.3 | *−0.4* | 30.0 | 7 | −9.5 | 12 | −4.9 | 12 | 17.6 | 9 | 52 | 32 | 559.6 | *85* | 138 | 22.8 | 8 | 1809.6 | *115* | | |

## OXFORD 1900

| Month | Mean max | Mean min | Mean temp | Anom | Highest max | Date | Lowest min | Date | Lowest max | Date | Highest min | Date | Air frost | Days ≥ 25 °C | Total pptn | Anom | Rain days | Wettest day | Date | Total sunshine | Anom | Sunniest day | Date |
|---|---|---|---|---|---|---|---|---|---|---|---|---|---|---|---|---|---|---|---|---|---|---|---|
| Jan | 7.2 | 2.1 | 4.7 | *−0.2* | 11.6 | 23 | −4.1 | 21 | 2.7 | 30 | 7.2 | 23 | 5 | 0 | 58.5 | *103* | 20 | 17.3 | 6 | 50.6 | *81* | | |
| Feb | 6.0 | 0.7 | 3.3 | *−1.6* | 14.0 | 23 | −9.3 | 9 | 1.0 | 12 | 9.7 | 24 | 14 | 0 | 107.0 | *250* | 19 | 16.4 | 15 | 68.5 | *87* | | |
| Mar | 7.2 | 0.9 | 4.0 | *−3.2* | 13.6 | 10 | −3.9 | 18 | 3.8 | 6 | 4.8 | 15 | 8 | 0 | 12.9 | *27* | 7 | 5.8 | 21 | 80.3 | *72* | | |
| Apr | 13.8 | 3.8 | 8.8 | *−0.5* | 23.2 | 21 | −3.1 | 2 | 8.6 | 7 | 9.4 | 30 | 3 | 0 | 22.8 | *47* | 14 | 12.9 | 3 | 160.2 | *100* | | |
| May | 15.2 | 6.3 | 10.7 | *−1.7* | 20.3 | 27 | 1.2 | 11 | 9.0 | 13 | 12.2 | 28 | 0 | 0 | 34.0 | *59* | 9 | 9.6 | 9 | 160.9 | *83* | | |
| June | 19.5 | 10.6 | 15.0 | *−0.5* | 28.7 | 11 | 5.1 | 6 | 14.5 | 2 | 15.5 | 12 | 0 | 3 | 69.4 | *141* | 17 | 8.9 | 15 | 173.2 | *91* | | |
| July | 24.3 | 13.4 | 18.8 | *+0.9* | 32.0 | 19 | 7.5 | 8 | 17.6 | 2 | 18.6 | 23 | 0 | 15 | 22.4 | *46* | 10 | 5.8 | 31 | 268.3 | *130* | | |
| Aug | 20.4 | 11.5 | 16.0 | *−1.6* | 27.2 | 13 | 5.7 | 10 | 15.2 | 9 | 15.6 | 18 | 0 | 4 | 79.8 | *146* | 14 | 13.0 | 3 | 164.4 | *84* | | |
| Sep | 19.6 | 9.2 | 14.4 | *−0.5* | 23.2 | 16 | 4.7 | 20 | 14.8 | 30 | 15.6 | 22 | 0 | 0 | 10.4 | *19* | 5 | 5.5 | 26 | 165.8 | *118* | | |
| Oct | 14.2 | 6.7 | 10.4 | *−0.8* | 21.9 | 9 | 1.6 | 22 | 9.5 | 26 | 13.2 | 7 | 0 | 0 | 59.1 | *84* | 15 | 13.4 | 29 | 125.3 | *113* | | |
| Nov | 10.3 | 5.4 | 7.8 | *+0.3* | 15.4 | 1 | −1.7 | 11 | 6.0 | 20 | 12.3 | 1 | 2 | 0 | 42.2 | *63* | 20 | 8.5 | 24 | 54.3 | *77* | | |
| Dec | 10.2 | 4.9 | 7.6 | *+2.5* | 13.4 | 5 | −3.0 | 23 | 2.2 | 23 | 9.9 | 12 | 2 | 0 | 81.9 | *130* | 20 | 31.0 | 30 | 31.7 | *60* | | |
| **1900** | 14.0 | 6.3 | 10.1 | *−0.6* | 32.0 | 7 | −9.3 | 2 | 1.0 | 2 | 18.6 | 7 | 34 | 22 | 600.4 | *91* | 170 | 31.0 | 12 | 1503.5 | *95* | | |

## OXFORD 1901

| Month | Mean max | Mean min | Mean temp | Anom | Highest max | Date | Lowest min | Date | Lowest max | Date | Highest min | Date | Air frost | Days ≥ 25 °C | Total pptn | Anom | Rain days | Wettest day | Date | Total sunshine | Anom | Sunniest day | Date |
|---|---|---|---|---|---|---|---|---|---|---|---|---|---|---|---|---|---|---|---|---|---|---|---|
| Jan | 6.0 | 1.0 | 3.5 | *−1.3* | 11.0 | 27 | −8.1 | 9 | −1.3 | 7 | 8.4 | 21 | 15 | 0 | 26.9 | *47* | 17 | 6.7 | 27 | 45.7 | *73* | | |
| Feb | 4.9 | 0.1 | 2.5 | *−2.4* | 9.7 | 28 | −7.3 | 14 | 1.4 | 14 | 5.2 | 27 | 13 | 0 | 32.8 | *77* | 12 | 11.3 | 4 | 43.3 | *55* | | |
| Mar | 7.6 | 1.1 | 4.4 | *−2.9* | 12.6 | 12 | −4.6 | 28 | 3.2 | 26 | 5.5 | 31 | 8 | 0 | 36.1 | *76* | 13 | 6.9 | 19 | 75.5 | *68* | | |
| Apr | 14.0 | 4.3 | 9.1 | *−0.1* | 23.0 | 23 | −0.1 | 29 | 8.9 | 5 | 9.6 | 7 | 1 | 0 | 53.2 | *109* | 14 | 15.2 | 3 | 199.5 | *124* | | |
| May | 17.4 | 6.2 | 11.8 | *−0.7* | 26.1 | 29 | −0.1 | 5 | 8.5 | 8 | 13.5 | 30 | 1 | 1 | 31.8 | *56* | 7 | 12.5 | 7 | 249.3 | *129* | | |
| June | 19.8 | 9.4 | 14.6 | *−1.0* | 25.2 | 29 | 4.2 | 19 | 12.6 | 13 | 14.5 | 21 | 0 | 1 | 38.6 | *78* | 12 | 19.0 | 30 | 225.4 | *118* | | |
| July | 23.8 | 13.2 | 18.5 | *+0.6* | 31.7 | 19 | 9.2 | 8 | 16.9 | 25 | 15.9 | 18 | 0 | 13 | 118.1 | *242* | 8 | 39.7 | 24 | 212.9 | *103* | | |
| Aug | 21.4 | 11.5 | 16.5 | *−1.1* | 27.6 | 25 | 6.6 | 28 | 15.4 | 27 | 16.3 | 8 | 0 | 5 | 54.3 | *99* | 10 | 16.0 | 14 | 214.0 | *109* | | |
| Sep | 18.5 | 10.0 | 14.3 | *−0.6* | 22.0 | 29 | 3.8 | 16 | 16.1 | 19 | 14.8 | 9 | 0 | 0 | 47.0 | *87* | 7 | 13.6 | 16 | 123.6 | *88* | | |
| Oct | 14.1 | 6.2 | 10.1 | *−1.1* | 21.5 | 1 | −1.7 | 26 | 11.0 | 7 | 13.2 | 1 | 3 | 0 | 30.5 | *44* | 11 | 8.2 | 16 | 98.1 | *88* | | |
| Nov | 8.1 | 1.7 | 4.9 | *−2.6* | 12.6 | 11 | −6.1 | 16 | 2.8 | 5 | 9.9 | 20 | 11 | 0 | 13.4 | *20* | 8 | 4.7 | 13 | 58.9 | *83* | | |
| Dec | 6.4 | 0.9 | 3.6 | *−1.4* | 13.3 | 30 | −6.9 | 20 | −0.1 | 22 | 9.9 | 31 | 12 | 0 | 83.5 | *132* | 19 | 29.9 | 12 | 55.5 | *104* | | |
| **1901** | 13.5 | 5.5 | 9.5 | *−1.3* | 31.7 | 7 | −8.1 | 1 | −1.3 | 1 | 16.3 | 8 | 64 | 20 | 566.2 | *86* | 138 | 39.7 | 7 | 1601.7 | *102* | | |

## OXFORD 1902

| Month | Mean max | Mean min | Mean temp | Anom | Highest max | Date | Lowest min | Date | Lowest max | Date | Highest min | Date | Air frost | Days ≥ 25 °C | Total pptn | Anom | Rain days | Wettest day | Date | Total sunshine | Anom | Sunniest day | Date |
|---|---|---|---|---|---|---|---|---|---|---|---|---|---|---|---|---|---|---|---|---|---|---|---|
| Jan | 7.8 | 2.6 | 5.2 | *+0.3* | 11.6 | 4 | −4.9 | 15 | 1.9 | 14 | 8.8 | 10 | 10 | 0 | 17.1 | *30* | 9 | 6.3 | 1 | 66.0 | *106* | | |
| Feb | 4.7 | −0.5 | 2.1 | *−2.8* | 12.0 | 28 | −7.4 | 17 | 0.2 | 14 | 7.1 | 24 | 18 | 0 | 26.9 | *63* | 10 | 6.9 | 22 | 63.9 | *81* | | |
| Mar | 11.0 | 3.8 | 7.4 | *+0.1* | 15.3 | 17 | −2.8 | 6 | 6.1 | 5 | 8.1 | 10 | 4 | 0 | 30.8 | *65* | 12 | 8.8 | 14 | 91.8 | *83* | | |
| Apr | 12.7 | 3.5 | 8.1 | *−1.2* | 18.1 | 19 | −0.8 | 9 | 6.7 | 8 | 10.0 | 22 | 4 | 0 | 30.6 | *63* | 12 | 14.8 | 15 | 173.3 | *108* | | |
| May | 13.8 | 5.4 | 9.6 | *−2.9* | 21.7 | 24 | −0.5 | 14 | 9.3 | 13 | 11.9 | 31 | 1 | 0 | 38.6 | *68* | 20 | 8.1 | 22 | 170.1 | *88* | | |
| June | 18.5 | 9.7 | 14.1 | *−1.5* | 27.7 | 28 | 4.3 | 10 | 11.5 | 13 | 13.9 | 29 | 0 | 5 | 50.1 | *102* | 15 | 12.2 | 7 | 175.2 | *92* | | |
| July | 20.5 | 11.0 | 15.7 | *−2.1* | 28.7 | 14 | 5.5 | 12 | 14.2 | 21 | 15.9 | 6 | 0 | 6 | 15.7 | *32* | 9 | 4.2 | 9 | 194.1 | *94* | | |
| Aug | 19.6 | 11.0 | 15.3 | *−2.3* | 25.2 | 16 | 6.9 | 26 | 15.9 | 11 | 15.1 | 17 | 0 | 1 | 56.0 | *102* | 19 | 9.4 | 31 | 118.4 | *60* | | |
| Sep | 18.0 | 8.5 | 13.3 | *−1.6* | 22.7 | 1 | 1.4 | 19 | 13.8 | 29 | 15.3 | 3 | 0 | 0 | 27.9 | *52* | 7 | 12.1 | 10 | 154.9 | *110* | | |
| Oct | 13.0 | 6.8 | 9.9 | *−1.4* | 17.3 | 10 | 0.3 | 19 | 9.2 | 4 | 12.8 | 13 | 0 | 0 | 39.0 | *56* | 14 | 9.7 | 9 | 70.2 | *63* | | |
| Nov | 9.6 | 4.5 | 7.1 | *−0.5* | 14.0 | 6 | −2.1 | 21 | 1.1 | 19 | 9.6 | 6 | 4 | 0 | 56.8 | *85* | 11 | 18.4 | 28 | 52.1 | *73* | | |
| Dec | 7.0 | 2.4 | 4.7 | *−0.3* | 12.8 | 17 | −6.8 | 7 | −0.7 | 6 | 8.1 | 22 | 7 | 0 | 33.9 | *54* | 11 | 8.8 | 17 | 41.2 | *77* | | |
| **1902** | 13.0 | 5.7 | 9.4 | *−1.4* | 28.7 | 7 | −7.4 | 2 | −0.7 | 12 | 15.9 | 7 | 48 | 12 | 423.4 | *64* | 149 | 18.4 | 11 | 1371.2 | *87* | | |

## OXFORD 1903

| Month | Mean max | Mean min | Mean temp | Anom | Highest max | Date | Lowest min | Date | Lowest max | Date | Highest min | Date | Air frost | Days ≥ 25 °C | Total pptn | Anom | Rain days | Wettest day | Date | Total sunshine | Anom | Sunniest day | Date |
|---|---|---|---|---|---|---|---|---|---|---|---|---|---|---|---|---|---|---|---|---|---|---|---|
| Jan | 6.9 | 2.4 | 4.7 | *−0.2* | 12.4 | 5 | −6.0 | 14 | −0.1 | 14 | 9.4 | 6 | 8 | 0 | 65.4 | *115* | 16 | 18.3 | 4 | 56.6 | *91* | | |
| Feb | 10.5 | 4.5 | 7.5 | *+2.6* | 14.8 | 19 | −3.2 | 18 | 5.8 | 2 | 10.0 | 8 | 1 | 0 | 19.8 | *46* | 7 | 7.0 | 25 | 66.6 | *84* | | |
| Mar | 11.1 | 4.3 | 7.7 | *+0.4* | 18.4 | 25 | 0.1 | 11 | 7.3 | 7 | 9.3 | 20 | 0 | 0 | 77.1 | *162* | 18 | 16.0 | 23 | 112.4 | *101* | | |
| Apr | 11.0 | 2.6 | 6.8 | *−2.5* | 14.8 | 28 | −2.3 | 17 | 7.3 | 17 | 8.4 | 29 | 9 | 0 | 58.5 | *120* | 12 | 14.3 | 28 | 145.6 | *91* | | |
| May | 16.2 | 7.2 | 11.7 | *−0.8* | 24.8 | 30 | 2.4 | 18 | 9.2 | 11 | 11.8 | 31 | 0 | 0 | 111.0 | *194* | 16 | 18.5 | 30 | 159.3 | *83* | | |
| June | 17.5 | 8.9 | 13.2 | *−2.4* | 28.1 | 27 | 3.4 | 21 | 10.4 | 14 | 17.7 | 28 | 0 | 1 | 141.8 | *288* | 10 | 51.1 | 14 | 179.6 | *94* | | |
| July | 20.7 | 11.8 | 16.3 | *−1.6* | 28.7 | 10 | 6.6 | 8 | 16.1 | 20 | 16.3 | 9 | 0 | 4 | 88.2 | *181* | 15 | 17.9 | 18 | 170.4 | *82* | | |
| Aug | 18.8 | 10.9 | 14.8 | *−2.8* | 23.4 | 8 | 5.8 | 23 | 15.9 | 25 | 13.6 | 31 | 0 | 0 | 84.7 | *155* | 16 | 15.0 | 24 | 165.3 | *84* | | |
| Sep | 17.8 | 9.7 | 13.7 | *−1.1* | 22.7 | 1 | 3.1 | 13 | 12.1 | 15 | 14.3 | 24 | 0 | 0 | 39.3 | *73* | 13 | 10.4 | 10 | 141.2 | *101* | | |
| Oct | 14.0 | 8.4 | 11.2 | *−0.1* | 18.2 | 1 | 2.7 | 24 | 10.5 | 23 | 12.3 | 3 | 0 | 0 | 163.3 | *233* | 25 | 18.9 | 27 | 83.1 | *75* | | |
| Nov | 9.6 | 3.9 | 6.7 | *−0.8* | 12.5 | 21 | −1.7 | 19 | 1.9 | 30 | 9.5 | 11 | 4 | 0 | 38.2 | *57* | 11 | 17.8 | 27 | 58.7 | *83* | | |
| Dec | 5.6 | 1.6 | 3.6 | *−1.4* | 10.2 | 9 | −4.3 | 31 | −0.2 | 31 | 6.5 | 22 | 10 | 0 | 26.5 | *42* | 11 | 7.0 | 12 | 29.3 | *55* | | |
| **1903** | 13.3 | 6.4 | 9.8 | *−0.9* | 28.7 | 7 | −6.0 | 1 | −0.2 | 12 | 17.7 | 6 | 32 | 5 | 913.8 | *138* | 170 | 51.1 | 6 | 1368.1 | *87* | | |

## OXFORD 1904

| Month | Mean max | Mean min | Mean temp | Anom | Highest max | Date | Lowest min | Date | Lowest max | Date | Highest min | Date | Air frost | Days ≥ 25 °C | Total pptn | Anom | Rain days | Wettest day | Date | Total sunshine | Anom | Sunniest day | Date |
|---|---|---|---|---|---|---|---|---|---|---|---|---|---|---|---|---|---|---|---|---|---|---|---|
| Jan | 6.8 | 1.4 | 4.1 | *−0.7* | 12.3 | 13 | −4.1 | 23 | 0.2 | 24 | 8.4 | 27 | 10 | 0 | 79.1 | *139* | 18 | 18.5 | 31 | 36.4 | *58* | | |
| Feb | 6.4 | 1.5 | 4.0 | *−0.9* | 11.9 | 21 | −4.1 | 29 | 0.4 | 29 | 8.7 | 21 | 8 | 0 | 76.0 | *178* | 19 | 19.1 | 9 | 53.9 | *68* | | |
| Mar | 8.3 | 1.6 | 5.0 | *−2.3* | 15.3 | 9 | −2.9 | 12 | 1.9 | 4 | 9.1 | 20 | 7 | 0 | 26.4 | *55* | 13 | 6.7 | 7 | 88.6 | *80* | | |
| Apr | 13.8 | 5.4 | 9.6 | *+0.3* | 17.5 | 14 | 1.8 | 11 | 9.8 | 22 | 9.9 | 30 | 0 | 0 | 20.4 | *42* | 13 | 7.9 | 15 | 162.6 | *101* | | |
| May | 15.6 | 7.4 | 11.5 | *−1.0* | 21.2 | 29 | 2.3 | 8 | 8.7 | 7 | 13.6 | 27 | 0 | 0 | 81.7 | *143* | 16 | 35.3 | 27 | 137.6 | *71* | | |
| June | 18.8 | 9.1 | 14.0 | *−1.6* | 24.4 | 30 | 5.9 | 4 | 12.9 | 2 | 12.6 | 14 | 0 | 0 | 26.5 | *54* | 8 | 8.1 | 14 | 200.2 | *105* | | |
| July | 23.2 | 13.0 | 18.1 | *+0.3* | 29.2 | 17 | 9.7 | 8 | 17.4 | 27 | 16.5 | 30 | 0 | 9 | 85.1 | *174* | 11 | 28.9 | 26 | 235.4 | *114* | | |
| Aug | 20.6 | 11.0 | 15.8 | *−1.8* | 28.8 | 4 | 5.6 | 25 | 14.9 | 22 | 16.3 | 4 | 0 | 4 | 38.7 | *71* | 9 | 13.1 | 31 | 214.4 | *109* | | |
| Sep | 17.3 | 7.9 | 12.6 | *−2.3* | 21.1 | 5 | 1.8 | 30 | 13.3 | 14 | 14.3 | 6 | 0 | 0 | 31.6 | *58* | 10 | 8.6 | 14 | 165.3 | *118* | | |
| Oct | 13.5 | 6.3 | 9.9 | *−1.4* | 19.3 | 18 | −1.0 | 15 | 8.6 | 27 | 13.4 | 18 | 1 | 0 | 21.8 | *31* | 15 | 5.5 | 16 | 84.7 | *76* | | |
| Nov | 8.5 | 1.9 | 5.2 | *−2.3* | 14.7 | 9 | −8.4 | 24 | −1.3 | 26 | 8.8 | 9 | 11 | 0 | 39.5 | *59* | 17 | 9.6 | 10 | 64.4 | *91* | | |
| Dec | 6.4 | 1.6 | 4.0 | *−1.0* | 13.1 | 16 | −5.0 | 22 | −2.3 | 23 | 11.0 | 17 | 11 | 0 | 46.9 | *74* | 18 | 11.9 | 6 | 33.8 | *64* | | |
| **1904** | 13.3 | 5.7 | 9.5 | *−1.3* | 29.2 | 7 | −8.4 | 11 | −2.3 | 12 | 16.5 | 7 | 48 | 13 | 573.7 | *87* | 167 | 35.3 | 5 | 1477.3 | *94* | | |

## OXFORD 1905

| Month | Mean max | Mean min | Mean temp | Anom | Highest max | Date | Lowest min | Date | Lowest max | Date | Highest min | Date | Air frost | Days ≥ 25 °C | Total pptn | Anom | Rain days | Wettest day | Date | Total sunshine | Anom | Sunniest day | Date |
|---|---|---|---|---|---|---|---|---|---|---|---|---|---|---|---|---|---|---|---|---|---|---|---|
| Jan | 6.3 | 0.4 | 3.3 | *−1.5* | 12.5 | 6 | −5.8 | 20 | −0.8 | 16 | 8.4 | 7 | 15 | 0 | 19.9 | *35* | 12 | 7.2 | 4 | 82.0 | *132* | | |
| Feb | 8.4 | 3.2 | 5.8 | *+0.9* | 11.9 | 14 | −1.0 | 20 | 2.9 | 12 | 7.7 | 5 | 2 | 0 | 10.2 | *24* | 11 | 3.7 | 26 | 70.3 | *89* | | |
| Mar | 11.1 | 3.9 | 7.5 | *+0.2* | 15.9 | 22 | −1.0 | 3 | 6.0 | 3 | 7.4 | 18 | 2 | 0 | 75.7 | *159* | 20 | 9.7 | 12 | 120.4 | *108* | | |
| Apr | 11.5 | 4.5 | 8.0 | *−1.3* | 15.9 | 16 | −0.6 | 8 | 8.1 | 6 | 9.8 | 14 | 1 | 0 | 49.0 | *100* | 18 | 10.4 | 30 | 91.1 | *57* | | |
| May | 16.7 | 6.1 | 11.4 | *−1.1* | 23.8 | 29 | −0.1 | 23 | 11.2 | 21 | 12.2 | 30 | 1 | 0 | 19.5 | *34* | 8 | 8.9 | 30 | 222.3 | *115* | | |
| June | 19.5 | 10.9 | 15.2 | *−0.4* | 26.4 | 22 | 7.4 | 10 | 10.3 | 6 | 14.9 | 30 | 0 | 3 | 105.1 | *213* | 13 | 24.4 | 30 | 142.7 | *75* | | |
| July | 23.3 | 13.4 | 18.4 | *+0.5* | 27.3 | 21 | 7.7 | 7 | 18.4 | 6 | 17.4 | 26 | 0 | 7 | 4.0 | *8* | 6 | 1.9 | 1 | 231.1 | *112* | | |
| Aug | 19.7 | 11.4 | 15.5 | *−2.0* | 23.8 | 14 | 7.3 | 24 | 15.3 | 30 | 15.6 | 4 | 0 | 0 | 80.4 | *147* | 16 | 30.4 | 28 | 155.0 | *79* | | |
| Sep | 16.8 | 9.4 | 13.1 | *−1.8* | 22.0 | 3 | 3.3 | 15 | 12.4 | 27 | 15.7 | 6 | 0 | 0 | 32.5 | *60* | 13 | 7.1 | 28 | 116.6 | *83* | | |
| Oct | 11.4 | 3.8 | 7.6 | *−3.7* | 14.8 | 1 | −3.2 | 25 | 7.7 | 26 | 10.5 | 4 | 6 | 0 | 35.4 | *51* | 12 | 9.7 | 29 | 114.2 | *103* | | |
| Nov | 8.3 | 2.2 | 5.2 | *−2.3* | 12.9 | 26 | −3.8 | 21 | 1.3 | 18 | 8.3 | 11 | 7 | 0 | 75.5 | *113* | 20 | 15.3 | 10 | 58.9 | *83* | | |
| Dec | 7.0 | 2.2 | 4.6 | *−0.4* | 12.0 | 7 | −3.2 | 11 | 2.0 | 31 | 8.3 | 7 | 7 | 0 | 25.1 | *40* | 14 | 6.0 | 7 | 34.8 | *65* | | |
| **1905** | 13.3 | 5.9 | 9.6 | *−1.1* | 27.3 | 7 | −5.8 | 1 | −0.8 | 1 | 17.4 | 7 | 41 | 10 | 532.3 | *81* | 163 | 30.4 | 8 | 1439.4 | *91* | | |

## OXFORD 1906

| Month | Mean max | Mean min | Mean temp | Anom | Highest max | Date | Lowest min | Date | Lowest max | Date | Highest min | Date | Air frost | Days ≥ 25 °C | Total pptn | Anom | Rain days | Wettest day | Date | Total sunshine | Anom | Sunniest day | Date |
|---|---|---|---|---|---|---|---|---|---|---|---|---|---|---|---|---|---|---|---|---|---|---|---|
| Jan | 8.4 | 3.1 | 5.8 | *+0.9* | 13.2 | 26 | −2.3 | 23 | 1.1 | 1 | 7.7 | 28 | 3 | 0 | 89.3 | *157* | 16 | 14.7 | 6 | 69.2 | *111* | | |
| Feb | 6.6 | 0.7 | 3.6 | *−1.3* | 9.8 | 10 | −4.7 | 6 | 2.5 | 6 | 4.7 | 2 | 10 | 0 | 40.4 | *94* | 14 | 6.6 | 10 | 84.5 | *107* | | |
| Mar | 9.3 | 2.1 | 5.7 | *−1.6* | 18.7 | 7 | −2.9 | 23 | 5.1 | 26 | 8.3 | 16 | 9 | 0 | 32.2 | *68* | 17 | 9.0 | 10 | 105.6 | *95* | | |
| Apr | 13.8 | 2.1 | 7.9 | *−1.3* | 21.1 | 12 | −1.6 | 27 | 8.7 | 18 | 6.2 | 21 | 9 | 0 | 9.0 | *18* | 9 | 3.6 | 28 | 201.7 | *125* | | |
| May | 15.5 | 7.2 | 11.4 | *−1.1* | 23.7 | 8 | 0.6 | 18 | 9.9 | 17 | 12.2 | 28 | 0 | 0 | 45.0 | *79* | 17 | 11.2 | 20 | 126.0 | *65* | | |
| June | 20.2 | 9.3 | 14.8 | *−0.8* | 27.2 | 23 | 3.4 | 5 | 14.0 | 1 | 16.2 | 21 | 0 | 1 | 86.9 | *176* | 9 | 34.5 | 29 | 217.1 | *114* | | |
| July | 22.2 | 11.6 | 16.9 | *−1.0* | 27.0 | 30 | 6.3 | 1 | 17.6 | 13 | 16.7 | 22 | 0 | 4 | 26.7 | *55* | 9 | 9.9 | 29 | 241.6 | *117* | | |
| Aug | 23.4 | 12.6 | 18.0 | *+0.4* | 32.4 | 31 | 7.2 | 29 | 18.4 | 18 | 17.3 | 21 | 0 | 8 | 25.4 | *46* | 9 | 6.6 | 2 | 238.7 | *121* | | |
| Sep | 20.4 | 9.4 | 14.9 | *+0.0* | 33.1 | 1 | 1.8 | 28 | 14.8 | 23 | 16.2 | 7 | 0 | 4 | 25.0 | *46* | 7 | 10.7 | 13 | 196.1 | *140* | | |
| Oct | 14.9 | 8.3 | 11.6 | *+0.3* | 20.7 | 1 | 0.1 | 26 | 8.1 | 30 | 14.4 | 5 | 0 | 0 | 104.8 | *150* | 20 | 28.9 | 2 | 82.3 | *74* | | |
| Nov | 10.5 | 4.7 | 7.6 | *+0.1* | 16.2 | 22 | −2.2 | 19 | 6.4 | 19 | 11.2 | 22 | 3 | 0 | 82.4 | *123* | 13 | 32.9 | 8 | 49.4 | *70* | | |
| Dec | 6.2 | 0.6 | 3.4 | *−1.7* | 12.9 | 4 | −5.8 | 26 | 1.3 | 26 | 8.6 | 3 | 13 | 0 | 48.3 | *77* | 18 | 9.1 | 26 | 55.9 | *105* | | |
| **1906** | 14.3 | 6.0 | 10.1 | *−0.6* | 33.1 | 9 | −5.8 | 12 | 1.1 | 1 | 17.3 | 8 | 47 | 17 | 615.4 | *93* | 158 | 34.5 | 6 | 1668.1 | *106* | | |

## OXFORD 1907

| Month | Mean max | Mean min | Mean temp | Anom | Highest max | Date | Lowest min | Date | Lowest max | Date | Highest min | Date | Air frost | Days ≥ 25 °C | Total pptn | Anom | Rain days | Wettest day | Date | Total sunshine | Anom | Sunniest day | Date |
|---|---|---|---|---|---|---|---|---|---|---|---|---|---|---|---|---|---|---|---|---|---|---|---|
| Jan | 6.3 | 1.1 | 3.7 | *−1.2* | 10.9 | 2 | −6.6 | 25 | −2.9 | 24 | 6.2 | 14 | 10 | 0 | 15.5 | *27* | 8 | 5.9 | 1 | 73.6 | *118* | | |
| Feb | 6.3 | 0.3 | 3.3 | *−1.6* | 11.5 | 17 | −6.5 | 2 | 0.2 | 3 | 6.9 | 18 | 13 | 0 | 28.6 | *67* | 13 | 9.7 | 12 | 90.2 | *114* | | |
| Mar | 12.1 | 1.8 | 6.9 | *−0.3* | 20.3 | 28 | −3.8 | 5 | 6.8 | 5 | 7.9 | 16 | 8 | 0 | 21.3 | *45* | 12 | 3.8 | 5 | 185.3 | *167* | | |
| Apr | 12.6 | 4.1 | 8.4 | *−0.9* | 20.9 | 24 | −1.2 | 18 | 9.7 | 26 | 10.1 | 24 | 1 | 0 | 57.4 | *117* | 13 | 13.7 | 8 | 132.6 | *82* | | |
| May | 15.9 | 7.3 | 11.6 | *−0.9* | 25.8 | 12 | 1.6 | 19 | 10.6 | 19 | 12.2 | 27 | 0 | 1 | 59.6 | *104* | 17 | 9.1 | 13 | 132.3 | *69* | | |
| June | 16.8 | 9.5 | 13.2 | *−2.4* | 22.1 | 9 | 5.7 | 17 | 13.1 | 3 | 12.6 | 27 | 0 | 0 | 75.1 | *153* | 19 | 33.4 | 1 | 130.0 | *68* | | |
| July | 19.2 | 10.4 | 14.8 | *−3.1* | 25.7 | 19 | 5.3 | 11 | 14.2 | 10 | 14.1 | 27 | 0 | 1 | 86.0 | *176* | 16 | 47.0 | 22 | 162.1 | *78* | | |
| Aug | 19.3 | 11.0 | 15.2 | *−2.4* | 22.4 | 29 | 6.6 | 28 | 15.5 | 20 | 16.1 | 9 | 0 | 0 | 34.9 | *64* | 17 | 9.1 | 18 | 168.8 | *86* | | |
| Sep | 19.7 | 9.0 | 14.3 | *−0.6* | 24.8 | 25 | 1.4 | 24 | 13.9 | 1 | 15.2 | 6 | 0 | 0 | 18.1 | *33* | 6 | 6.5 | 2 | 162.0 | *115* | | |
| Oct | 13.7 | 7.0 | 10.4 | *−0.9* | 18.3 | 1 | 0.6 | 24 | 9.5 | 27 | 12.8 | 1 | 0 | 0 | 135.9 | *194* | 22 | 21.9 | 16 | 97.8 | *88* | | |
| Nov | 10.2 | 4.2 | 7.2 | *−0.4* | 15.1 | 9 | −2.5 | 30 | 5.2 | 30 | 9.9 | 2 | 2 | 0 | 57.5 | *86* | 15 | 13.3 | 26 | 56.9 | *80* | | |
| Dec | 7.8 | 3.0 | 5.4 | *+0.4* | 13.7 | 8 | −1.1 | 30 | 0.6 | 30 | 10.6 | 19 | 3 | 0 | 97.7 | *155* | 18 | 18.0 | 12 | 40.1 | *75* | | |
| **1907** | 13.3 | 5.7 | 9.5 | *−1.2* | 25.8 | 5 | −6.6 | 1 | −2.9 | 1 | 16.1 | 8 | 37 | 2 | 687.6 | *104* | 176 | 47.0 | 7 | 1431.7 | *91* | | |

## OXFORD 1908

| Month | Mean max | Mean min | Mean temp | Anom | Highest max | Date | Lowest min | Date | Lowest max | Date | Highest min | Date | Air frost | Days ≥ 25 °C | Total pptn | Anom | Rain days | Wettest day | Date | Total sunshine | Anom | Sunniest day | Date |
|---|---|---|---|---|---|---|---|---|---|---|---|---|---|---|---|---|---|---|---|---|---|---|---|
| Jan | 5.5 | −0.4 | 2.5 | *−2.3* | 12.6 | 17 | −7.7 | 12 | −2.7 | 5 | 9.4 | 17 | 18 | 0 | 42.4 | *75* | 10 | 25.3 | 7 | 58.5 | *94* | | |
| Feb | 8.6 | 2.7 | 5.6 | *+0.7* | 11.4 | 19 | −4.7 | 13 | 4.7 | 5 | 6.6 | 11 | 4 | 0 | 22.2 | *52* | 12 | 9.4 | 17 | 67.7 | *86* | | |
| Mar | 8.2 | 1.5 | 4.9 | *−2.4* | 12.1 | 23 | −3.0 | 20 | 3.9 | 3 | 4.4 | 25 | 6 | 0 | 67.7 | *142* | 19 | 17.7 | 25 | 79.4 | *71* | | |
| Apr | 10.4 | 2.4 | 6.4 | *−2.9* | 16.3 | 9 | −2.7 | 24 | 1.3 | 25 | 9.3 | 30 | 10 | 0 | 109.4 | *224* | 11 | 46.7 | 25 | 131.3 | *82* | | |
| May | 17.6 | 9.0 | 13.3 | *+0.8* | 24.7 | 2 | 3.4 | 11 | 11.3 | 14 | 13.1 | 31 | 0 | 0 | 37.2 | *65* | 13 | 10.9 | 3 | 193.7 | *100* | | |
| June | 20.5 | 9.8 | 15.2 | *−0.4* | 27.4 | 4 | 4.4 | 7 | 15.1 | 6 | 15.8 | 1 | 0 | 3 | 43.1 | *88* | 8 | 24.6 | 17 | 256.7 | *134* | | |
| July | 21.5 | 12.2 | 16.9 | *−1.0* | 28.7 | 3 | 9.3 | 20 | 17.1 | 16 | 15.7 | 25 | 0 | 4 | 54.2 | *111* | 11 | 10.4 | 9 | 204.0 | *99* | | |
| Aug | 20.0 | 10.7 | 15.3 | *−2.3* | 27.3 | 3 | 6.1 | 12 | 15.0 | 23 | 14.4 | 22 | 0 | 2 | 77.0 | *141* | 15 | 22.3 | 23 | 208.4 | *106* | | |
| Sep | 17.0 | 9.4 | 13.2 | *−1.7* | 25.3 | 30 | 3.9 | 13 | 12.1 | 4 | 15.8 | 29 | 0 | 1 | 41.9 | *77* | 15 | 7.4 | 3 | 123.6 | *88* | | |
| Oct | 16.1 | 8.1 | 12.1 | *+0.9* | 24.4 | 3 | −0.7 | 25 | 8.3 | 24 | 13.4 | 8 | 1 | 0 | 29.7 | *42* | 16 | 10.8 | 21 | 113.3 | *102* | | |
| Nov | 11.1 | 5.0 | 8.0 | *+0.5* | 15.3 | 1 | −5.9 | 10 | 6.7 | 10 | 10.0 | 12 | 3 | 0 | 28.6 | *43* | 14 | 5.9 | 29 | 70.5 | *99* | | |
| Dec | 6.5 | 1.6 | 4.0 | *−1.0* | 11.4 | 13 | −12.6 | 30 | −5.7 | 30 | 8.8 | 21 | 5 | 0 | 49.9 | *79* | 17 | 12.4 | 29 | 29.1 | *55* | | |
| **1908** | 13.6 | 6.0 | 9.8 | *−0.9* | 28.7 | 7 | −12.6 | 12 | −5.7 | 12 | 15.8 | 6 | 47 | 10 | 603.3 | *91* | 161 | 46.7 | 4 | 1536.2 | *97* | | |

## OXFORD 1909

| Month | Mean max | Mean min | Mean temp | Anom | Highest max | Date | Lowest min | Date | Lowest max | Date | Highest min | Date | Air frost | Days ≥ 25 °C | Total pptn | Anom | Rain days | Wettest day | Date | Total sunshine | Anom | Sunniest day | Date |
|---|---|---|---|---|---|---|---|---|---|---|---|---|---|---|---|---|---|---|---|---|---|---|---|
| Jan | 6.1 | 1.1 | 3.6 | *−1.2* | 10.2 | 10 | −5.9 | 28 | −2.6 | 28 | 7.6 | 18 | 12 | 0 | 20.9 | *37* | 12 | 4.8 | 10 | 59.9 | *96* | | |
| Feb | 6.5 | −0.6 | 2.9 | *−2.0* | 12.6 | 4 | −7.2 | 23 | 1.4 | 28 | 9.4 | 4 | 17 | 0 | 10.6 | *25* | 8 | 4.6 | 10 | 93.4 | *118* | | |
| Mar | 7.1 | 1.2 | 4.1 | *−3.1* | 14.9 | 29 | −7.1 | 5 | 0.7 | 3 | 8.7 | 29 | 9 | 0 | 63.7 | *134* | 23 | 18.3 | 6 | 73.2 | *66* | | |
| Apr | 15.1 | 3.5 | 9.3 | *+0.0* | 22.2 | 10 | −2.9 | 2 | 8.2 | 1 | 8.9 | 24 | 5 | 0 | 50.2 | *103* | 14 | 18.5 | 29 | 226.9 | *141* | | |
| May | 17.3 | 5.4 | 11.4 | *−1.1* | 26.7 | 21 | 0.5 | 2 | 9.9 | 1 | 12.9 | 23 | 0 | 2 | 48.1 | *84* | 7 | 23.4 | 25 | 293.6 | *152* | | |
| June | 16.1 | 8.8 | 12.5 | *−3.1* | 21.0 | 14 | 5.3 | 12 | 9.8 | 4 | 13.8 | 20 | 0 | 0 | 101.4 | *206* | 16 | 17.7 | 1 | 109.4 | *57* | | |
| July | 19.6 | 11.7 | 15.6 | *−2.2* | 23.0 | 17 | 7.3 | 1 | 15.4 | 11 | 15.6 | 17 | 0 | 0 | 61.5 | *126* | 15 | 14.1 | 27 | 188.4 | *91* | | |
| Aug | 21.2 | 11.7 | 16.5 | *−1.1* | 29.3 | 12 | 7.1 | 28 | 15.2 | 31 | 16.6 | 16 | 0 | 9 | 83.7 | *153* | 12 | 30.0 | 17 | 225.5 | *115* | | |
| Sep | 16.1 | 9.1 | 12.6 | *−2.3* | 20.6 | 22 | 3.7 | 2 | 11.9 | 28 | 13.0 | 25 | 0 | 0 | 79.1 | *146* | 17 | 25.0 | 17 | 92.7 | *66* | | |
| Oct | 14.1 | 8.1 | 11.1 | *−0.2* | 19.2 | 3 | −0.9 | 30 | 6.2 | 30 | 14.6 | 4 | 2 | 0 | 90.8 | *130* | 23 | 19.4 | 26 | 94.4 | *85* | | |
| Nov | 8.3 | 2.3 | 5.3 | *−2.2* | 12.4 | 5 | −2.6 | 9 | 2.9 | 14 | 9.7 | 4 | 10 | 0 | 16.1 | *24* | 9 | 5.4 | 14 | 86.8 | *122* | | |
| Dec | 7.3 | 1.9 | 4.6 | *−0.4* | 11.7 | 3 | −4.8 | 21 | 1.9 | 21 | 9.3 | 23 | 9 | 0 | 78.4 | *124* | 21 | 13.3 | 10 | 58.8 | *111* | | |
| **1909** | 12.9 | 5.4 | 9.1 | *−1.6* | 29.3 | 8 | −7.2 | 2 | −2.6 | 1 | 16.6 | 8 | 64 | 11 | 704.5 | *107* | 177 | 30.0 | 8 | 1603.0 | *102* | | |

## OXFORD 1910

| Month | Mean max | Mean min | Mean temp | Anom | Highest max | Date | Lowest min | Date | Lowest max | Date | Highest min | Date | Air frost | Days ≥ 25 °C | Total pptn | Anom | Rain days | Wettest day | Date | Total sunshine | Anom | Sunniest day | Date |
|---|---|---|---|---|---|---|---|---|---|---|---|---|---|---|---|---|---|---|---|---|---|---|---|
| Jan | 6.6 | 1.6 | 4.1 | *−0.7* | 11.7 | 9 | −7.6 | 27 | 1.4 | 22 | 7.6 | 3 | 9 | 0 | 40.4 | *71* | 13 | 11.7 | 28 | 76.6 | *123* | | |
| Feb | 8.6 | 2.8 | 5.7 | *+0.8* | 12.6 | 17 | −1.4 | 4 | 4.9 | 2 | 9.0 | 6 | 4 | 0 | 74.8 | *175* | 22 | 12.2 | 22 | 80.2 | *102* | | |
| Mar | 10.6 | 1.9 | 6.2 | *−1.1* | 15.4 | 30 | −2.5 | 27 | 5.8 | 12 | 7.3 | 6 | 10 | 0 | 15.4 | *32* | 6 | 6.7 | 9 | 154.6 | *139* | | |
| Apr | 11.8 | 4.2 | 8.0 | *−1.3* | 17.3 | 21 | −1.7 | 3 | 6.5 | 5 | 10.0 | 19 | 3 | 0 | 73.0 | *149* | 19 | 23.3 | 16 | 114.8 | *71* | | |
| May | 16.5 | 7.3 | 11.9 | *−0.6* | 23.7 | 23 | −0.1 | 11 | 9.7 | 8 | 13.4 | 20 | 1 | 0 | 52.6 | *92* | 17 | 18.0 | 31 | 197.4 | *102* | | |
| June | 19.9 | 11.3 | 15.6 | *+0.0* | 26.1 | 20 | 7.0 | 15 | 15.6 | 27 | 16.2 | 21 | 0 | 2 | 75.7 | *154* | 15 | 16.6 | 10 | 183.4 | *96* | | |
| July | 18.3 | 11.1 | 14.7 | *−3.2* | 23.2 | 28 | 6.9 | 19 | 13.2 | 7 | 15.1 | 21 | 0 | 0 | 51.1 | *105* | 15 | 14.9 | 5 | 130.0 | *63* | | |
| Aug | 19.4 | 12.0 | 15.7 | *−1.8* | 23.1 | 14 | 9.0 | 11 | 16.4 | 2 | 15.5 | 15 | 0 | 0 | 48.0 | *88* | 16 | 13.2 | 28 | 136.9 | *70* | | |
| Sep | 17.1 | 9.1 | 13.1 | *−1.7* | 23.1 | 28 | 2.7 | 21 | 13.1 | 14 | 12.9 | 29 | 0 | 0 | 11.2 | *21* | 5 | 8.8 | 14 | 121.9 | *87* | | |
| Oct | 14.2 | 8.4 | 11.3 | *+0.0* | 20.7 | 2 | 5.3 | 15 | 10.1 | 29 | 13.0 | 2 | 0 | 0 | 94.6 | *135* | 12 | 40.7 | 11 | 64.6 | *58* | | |
| Nov | 6.9 | 0.4 | 3.7 | *−3.9* | 12.8 | 1 | −6.2 | 23 | 1.3 | 22 | 7.4 | 13 | 17 | 0 | 78.6 | *118* | 18 | 22.7 | 27 | 83.2 | *117* | | |
| Dec | 8.8 | 4.7 | 6.7 | *+1.7* | 12.4 | 16 | −2.7 | 28 | 3.7 | 28 | 8.1 | 9 | 1 | 0 | 125.9 | *200* | 19 | 20.4 | 21 | 29.7 | *56* | | |
| **1910** | 13.2 | 6.2 | 9.7 | *−1.0* | 26.1 | 6 | −7.6 | 1 | 1.3 | 11 | 16.2 | 6 | 45 | 2 | 741.3 | *112* | 177 | 40.7 | 10 | 1373.3 | *87* | | |

## OXFORD 1911

| Month | Mean max | Mean min | Mean temp | Anom | Highest max | Date | Lowest min | Date | Lowest max | Date | Highest min | Date | Air frost | Days ≥ 25 °C | Total pptn | Anom | Rain days | Wettest day | Date | Total sunshine | Anom | Sunniest day | Date |
|---|---|---|---|---|---|---|---|---|---|---|---|---|---|---|---|---|---|---|---|---|---|---|---|
| Jan | 6.3 | 1.3 | 3.8 | *−1.0* | 12.6 | 28 | −4.1 | 31 | 2.7 | 31 | 7.9 | 26 | 9 | 0 | 20.4 | *36* | 10 | 6.5 | 9 | 63.5 | *102* | | |
| Feb | 8.0 | 2.4 | 5.2 | *+0.3* | 13.1 | 17 | −6.6 | 1 | 0.2 | 2 | 10.0 | 18 | 6 | 0 | 35.3 | *82* | 15 | 9.4 | 28 | 72.8 | *92* | | |
| Mar | 8.5 | 2.7 | 5.6 | *−1.7* | 14.9 | 22 | −0.9 | 17 | 3.8 | 27 | 8.1 | 3 | 2 | 0 | 41.7 | *87* | 17 | 10.9 | 12 | 80.6 | *72* | | |
| Apr | 12.2 | 4.1 | 8.2 | *−1.1* | 17.8 | 22 | −3.2 | 6 | 3.0 | 5 | 10.0 | 23 | 6 | 0 | 29.6 | *61* | 13 | 9.2 | 2 | 151.1 | *94* | | |
| May | 18.6 | 8.4 | 13.5 | *+1.0* | 24.7 | 29 | 4.2 | 6 | 10.3 | 20 | 12.8 | 27 | 0 | 0 | 56.0 | *98* | 7 | 20.7 | 26 | 185.6 | *96* | | |
| June | 20.1 | 10.5 | 15.3 | *−0.3* | 27.5 | 5 | 4.2 | 14 | 12.9 | 25 | 14.8 | 5 | 0 | 4 | 30.2 | *61* | 9 | 9.0 | 25 | 208.5 | *109* | | |
| July | 26.1 | 13.1 | 19.6 | *+1.7* | 32.5 | 29 | 7.7 | 10 | 17.7 | 2 | 17.7 | 29 | 0 | 19 | 10.8 | *22* | 3 | 7.9 | 29 | 310.5 | *150* | | |
| Aug | 24.9 | 14.0 | 19.4 | *+1.8* | 34.8 | 9 | 8.3 | 31 | 18.4 | 22 | 17.4 | 13 | 0 | 14 | 26.1 | *48* | 11 | 8.8 | 5 | 224.5 | *114* | | |
| Sep | 21.4 | 9.4 | 15.4 | *+0.5* | 33.4 | 8 | 2.9 | 22 | 14.9 | 29 | 14.9 | 12 | 0 | 8 | 36.5 | *67* | 10 | 9.2 | 13 | 222.6 | *159* | | |
| Oct | 13.4 | 6.6 | 10.0 | *−1.2* | 17.2 | 13 | −1.7 | 29 | 8.3 | 29 | 12.2 | 14 | 1 | 0 | 62.6 | *89* | 15 | 10.5 | 4 | 104.1 | *94* | | |
| Nov | 9.3 | 3.6 | 6.4 | *−1.1* | 14.4 | 5 | −2.2 | 22 | 3.1 | 27 | 9.1 | 16 | 4 | 0 | 61.9 | *93* | 19 | 11.9 | 17 | 60.3 | *85* | | |
| Dec | 9.1 | 4.6 | 6.9 | *+1.8* | 11.9 | 19 | −1.1 | 6 | 5.7 | 8 | 9.7 | 17 | 1 | 0 | 126.9 | *201* | 27 | 15.9 | 22 | 58.2 | *109* | | |
| **1911** | 14.8 | 6.7 | 10.8 | *+0.0* | 34.8 | 8 | −6.6 | 2 | 0.2 | 2 | 17.7 | 7 | 29 | 45 | 538.0 | *81* | 156 | 20.7 | 5 | 1742.3 | *111* | | |

## OXFORD 1912

| Month | Mean max | Mean min | Mean temp | Anom | Highest max | Date | Lowest min | Date | Lowest max | Date | Highest min | Date | Air frost | Days ≥ 25 °C | Total pptn | Anom | Rain days | Wettest day | Date | Total sunshine | Anom | Sunniest day | Date |
|---|---|---|---|---|---|---|---|---|---|---|---|---|---|---|---|---|---|---|---|---|---|---|---|
| Jan | 6.7 | 1.7 | 4.2 | *−0.6* | 10.5 | 9 | −7.4 | 29 | 1.0 | 18 | 8.3 | 4 | 9 | 0 | 111.5 | *196* | 22 | 25.2 | 17 | 51.8 | *83* | | |
| Feb | 8.8 | 3.7 | 6.2 | *+1.3* | 14.4 | 28 | −7.7 | 5 | −2.1 | 4 | 9.8 | 23 | 6 | 0 | 54.2 | *127* | 18 | 13.0 | 23 | 45.9 | *58* | | |
| Mar | 10.8 | 5.0 | 7.9 | *+0.6* | 16.1 | 25 | 1.1 | 16 | 6.1 | 11 | 10.1 | 25 | 0 | 0 | 83.3 | *175* | 24 | 11.4 | 4 | 86.5 | *78* | | |
| Apr | 14.7 | 3.8 | 9.3 | *−0.0* | 21.1 | 21 | −2.7 | 12 | 8.8 | 9 | 9.6 | 6 | 2 | 0 | 0.6 | *1* | 2 | 0.3 | 9 | 239.1 | *149* | | |
| May | 17.5 | 8.8 | 13.2 | *+0.7* | 25.0 | 11 | 1.8 | 1 | 11.5 | 23 | 13.2 | 8 | 0 | 1 | 55.2 | *97* | 12 | 15.8 | 22 | 150.0 | *78* | | |
| June | 18.6 | 10.2 | 14.4 | *−1.2* | 25.5 | 22 | 5.5 | 3 | 14.2 | 3 | 14.3 | 23 | 0 | 1 | 83.0 | *169* | 20 | 21.2 | 10 | 178.3 | *93* | | |
| July | 21.1 | 12.5 | 16.8 | *−1.1* | 30.3 | 15 | 7.3 | 9 | 14.8 | 19 | 15.9 | 14 | 0 | 4 | 78.7 | *161* | 17 | 20.0 | 23 | 137.8 | *67* | | |
| Aug | 17.1 | 10.0 | 13.5 | *−4.0* | 20.6 | 4 | 5.1 | 3 | 14.4 | 6 | 14.5 | 24 | 0 | 0 | 123.6 | *226* | 23 | 20.3 | 26 | 97.7 | *50* | | |
| Sep | 15.2 | 7.7 | 11.5 | *−3.4* | 19.1 | 8 | 3.2 | 22 | 11.2 | 11 | 12.8 | 4 | 0 | 0 | 29.8 | *55* | 9 | 19.4 | 30 | 112.3 | *80* | | |
| Oct | 13.3 | 3.9 | 8.6 | *−2.7* | 17.4 | 14 | −2.2 | 5 | 8.6 | 23 | 12.9 | 27 | 7 | 0 | 80.8 | *115* | 16 | 15.3 | 30 | 122.1 | *110* | | |
| Nov | 9.2 | 4.1 | 6.6 | *−0.9* | 13.2 | 9 | −2.3 | 3 | 0.8 | 30 | 10.8 | 8 | 3 | 0 | 47.2 | *71* | 16 | 11.2 | 29 | 31.6 | *45* | | |
| Dec | 10.1 | 5.0 | 7.5 | *+2.5* | 13.2 | 28 | −3.4 | 1 | 6.2 | 17 | 11.3 | 14 | 1 | 0 | 85.9 | *136* | 22 | 13.0 | 29 | 43.8 | *82* | | |
| **1912** | 13.6 | 6.4 | 10.0 | *−0.8* | 30.3 | 7 | −7.7 | 2 | −2.1 | 2 | 15.9 | 7 | 28 | 6 | 833.8 | *126* | 201 | 25.2 | 1 | 1296.9 | *82* | | |

## OXFORD 1913

| Month | Mean max | Mean min | Mean temp | Anom | Highest max | Date | Lowest min | Date | Lowest max | Date | Highest min | Date | Air frost | Days ≥ 25 °C | Total pptn | Anom | Rain days | Wettest day | Date | Total sunshine | Anom | Sunniest day | Date |
|---|---|---|---|---|---|---|---|---|---|---|---|---|---|---|---|---|---|---|---|---|---|---|---|
| Jan | 7.6 | 3.0 | 5.3 | *+0.4* | 10.8 | 5 | −1.7 | 13 | 1.6 | 13 | 7.4 | 4 | 5 | 0 | 77.0 | *135* | 20 | 14.7 | 19 | 51.3 | *82* | | |
| Feb | 7.8 | 1.9 | 4.8 | *−0.1* | 12.6 | 9 | −2.3 | 19 | 2.1 | 20 | 8.3 | 4 | 9 | 0 | 24.9 | *58* | 14 | 6.1 | 9 | 57.6 | *73* | | |
| Mar | 10.4 | 3.3 | 6.9 | *−0.4* | 13.1 | 11 | −2.2 | 18 | 6.1 | 1 | 8.2 | 5 | 4 | 0 | 68.1 | *143* | 18 | 10.6 | 17 | 97.3 | *87* | | |
| Apr | 12.2 | 4.7 | 8.4 | *−0.9* | 18.9 | 23 | −3.1 | 13 | 8.0 | 8 | 10.0 | 22 | 1 | 0 | 96.8 | *198* | 19 | 35.7 | 29 | 106.5 | *66* | | |
| May | 17.0 | 7.8 | 12.4 | *−0.1* | 26.7 | 26 | 2.2 | 7 | 10.6 | 8 | 13.8 | 27 | 0 | 3 | 68.4 | *120* | 15 | 10.7 | 1 | 183.8 | *95* | | |
| June | 19.9 | 10.0 | 15.0 | *−0.6* | 28.1 | 16 | 5.5 | 1 | 15.3 | 24 | 14.1 | 17 | 0 | 2 | 14.5 | *29* | 10 | 4.0 | 8 | 204.2 | *107* | | |
| July | 19.6 | 11.4 | 15.5 | *−2.4* | 25.6 | 28 | 7.9 | 8 | 16.0 | 23 | 14.7 | 18 | 0 | 2 | 20.1 | *41* | 13 | 4.6 | 15 | 98.6 | *48* | | |
| Aug | 21.0 | 11.1 | 16.1 | *−1.5* | 27.2 | 28 | 6.1 | 5 | 15.7 | 19 | 16.1 | 22 | 0 | 3 | 20.4 | *37* | 12 | 5.6 | 24 | 151.6 | *77* | | |
| Sep | 18.6 | 10.8 | 14.7 | *−0.2* | 25.1 | 27 | 4.6 | 16 | 13.3 | 2 | 15.7 | 27 | 0 | 1 | 47.6 | *88* | 13 | 11.5 | 1 | 117.5 | *84* | | |
| Oct | 15.0 | 8.5 | 11.8 | *+0.5* | 19.2 | 1 | −0.2 | 25 | 8.6 | 24 | 12.3 | 1 | 1 | 0 | 120.9 | *173* | 21 | 27.2 | 2 | 91.9 | *83* | | |
| Nov | 11.9 | 5.6 | 8.8 | *+1.2* | 14.9 | 2 | −3.8 | 23 | 8.4 | 23 | 10.6 | 17 | 2 | 0 | 63.6 | *95* | 15 | 12.4 | 10 | 88.0 | *124* | | |
| Dec | 7.5 | 2.8 | 5.2 | *+0.1* | 12.0 | 3 | −4.6 | 31 | 1.3 | 30 | 9.4 | 3 | 10 | 0 | 24.2 | *38* | 14 | 5.6 | 6 | 50.9 | *96* | | |
| **1913** | 14.0 | 6.7 | 10.4 | *−0.4* | 28.1 | 6 | −4.6 | 12 | 1.3 | 12 | 16.1 | 8 | 32 | 11 | 646.5 | *98* | 184 | 35.7 | 4 | 1299.2 | *82* | | |

## OXFORD 1914

| Month | Mean max | Mean min | Mean temp | Anom | Highest max | Date | Lowest min | Date | Lowest max | Date | Highest min | Date | Air frost | Days ≥ 25 °C | Total pptn | Anom | Rain days | Wettest day | Date | Total sunshine | Anom | Sunniest day | Date |
|---|---|---|---|---|---|---|---|---|---|---|---|---|---|---|---|---|---|---|---|---|---|---|---|
| Jan | 6.0 | 1.2 | 3.6 | *−1.2* | 12.4 | 9 | −5.9 | 24 | 0.7 | 20 | 10.1 | 9 | 13 | 0 | 13.9 | *24* | 9 | 6.3 | 5 | 44.9 | *72* | | |
| Feb | 10.4 | 3.8 | 7.1 | *+2.2* | 13.9 | 14 | −2.8 | 27 | 8.0 | 16 | 9.3 | 1 | 2 | 0 | 51.4 | *120* | 19 | 7.4 | 13 | 72.4 | *92* | | |
| Mar | 10.1 | 3.3 | 6.7 | *−0.6* | 18.1 | 31 | −1.7 | 11 | 4.3 | 20 | 10.7 | 31 | 6 | 0 | 92.0 | *193* | 26 | 15.9 | 9 | 89.4 | *80* | | |
| Apr | 15.9 | 4.7 | 10.3 | *+1.0* | 23.0 | 21 | 0.6 | 16 | 9.9 | 7 | 9.1 | 2 | 0 | 0 | 29.8 | *61* | 9 | 12.5 | 5 | 243.8 | *152* | | |
| May | 16.5 | 6.3 | 11.4 | *−1.0* | 24.3 | 22 | −1.2 | 2 | 8.8 | 9 | 11.7 | 4 | 2 | 0 | 37.1 | *65* | 13 | 7.3 | 7 | 192.5 | *100* | | |
| June | 20.8 | 10.0 | 15.4 | *−0.2* | 29.2 | 30 | 5.0 | 8 | 13.0 | 8 | 14.2 | 30 | 0 | 5 | 77.4 | *157* | 10 | 22.2 | 18 | 248.0 | *130* | | |
| July | 20.8 | 12.2 | 16.5 | *−1.4* | 30.3 | 1 | 7.1 | 4 | 16.6 | 23 | 16.3 | 2 | 0 | 3 | 77.7 | *159* | 19 | 24.6 | 3 | 164.5 | *79* | | |
| Aug | 21.8 | 12.1 | 17.0 | *−0.6* | 27.4 | 13 | 7.6 | 18 | 17.4 | 6 | 16.7 | 9 | 0 | 4 | 47.3 | *86* | 12 | 12.0 | 5 | 188.2 | *96* | | |
| Sep | 19.5 | 8.3 | 13.9 | *−1.0* | 27.6 | 3 | 0.4 | 30 | 14.1 | 20 | 14.7 | 10 | 0 | 3 | 22.2 | *41* | 8 | 7.0 | 11 | 201.8 | *144* | | |
| Oct | 14.6 | 7.2 | 10.9 | *−0.4* | 18.9 | 3 | 1.1 | 28 | 11.4 | 31 | 11.6 | 25 | 0 | 0 | 74.8 | *107* | 10 | 23.6 | 14 | 91.1 | *82* | | |
| Nov | 10.2 | 4.0 | 7.1 | *−0.4* | 14.9 | 9 | −3.2 | 19 | 3.4 | 21 | 10.6 | 3 | 6 | 0 | 74.1 | *111* | 19 | 14.5 | 5 | 67.5 | *95* | | |
| Dec | 8.0 | 2.6 | 5.3 | *+0.3* | 13.2 | 7 | −3.4 | 25 | 1.8 | 25 | 8.3 | 7 | 6 | 0 | 147.9 | *234* | 26 | 22.3 | 28 | 55.8 | *105* | | |
| **1914** | 14.6 | 6.3 | 10.4 | *−0.3* | 30.3 | 7 | −5.9 | 1 | 0.7 | 1 | 16.7 | 8 | 35 | 15 | 745.6 | *113* | 180 | 24.6 | 7 | 1659.9 | *105* | | |

## OXFORD 1915

| Month | Mean max | Mean min | Mean temp | Anom | Highest max | Date | Lowest min | Date | Lowest max | Date | Highest min | Date | Air frost | Days ≥ 25 °C | Total pptn | Anom | Rain days | Wettest day | Date | Total sunshine | Anom | Sunniest day | Date |
|---|---|---|---|---|---|---|---|---|---|---|---|---|---|---|---|---|---|---|---|---|---|---|---|
| Jan | 6.5 | 2.2 | 4.3 | *−0.5* | 10.9 | 13 | −0.7 | 27 | 2.1 | 28 | 7.6 | 14 | 6 | 0 | 81.1 | *143* | 20 | 20.2 | 3 | 44.8 | *72* | | |
| Feb | 7.9 | 1.9 | 4.9 | *−0.0* | 10.8 | 3 | −3.0 | 25 | 4.1 | 24 | 7.0 | 4 | 7 | 0 | 91.7 | *214* | 20 | 14.9 | 17 | 91.9 | *116* | | |
| Mar | 9.3 | 1.8 | 5.5 | *−1.7* | 13.1 | 6 | −4.0 | 30 | 5.1 | 19 | 8.3 | 5 | 11 | 0 | 33.1 | *69* | 15 | 11.3 | 22 | 102.8 | *92* | | |
| Apr | 12.9 | 3.5 | 8.2 | *−1.1* | 21.9 | 30 | −0.1 | 24 | 9.2 | 6 | 6.8 | 4 | 1 | 0 | 21.0 | *43* | 13 | 8.6 | 12 | 155.5 | *97* | | |
| May | 17.1 | 6.4 | 11.7 | *−0.7* | 23.6 | 25 | 0.4 | 15 | 7.1 | 13 | 11.2 | 22 | 0 | 0 | 90.3 | *158* | 11 | 36.5 | 13 | 229.2 | *119* | | |
| June | 20.6 | 9.5 | 15.1 | *−0.5* | 27.8 | 8 | 3.9 | 19 | 12.6 | 24 | 15.7 | 8 | 0 | 1 | 35.1 | *71* | 8 | 9.6 | 27 | 247.4 | *130* | | |
| July | 19.7 | 11.3 | 15.5 | *−2.4* | 24.7 | 3 | 8.4 | 15 | 15.2 | 16 | 14.8 | 4 | 0 | 0 | 114.1 | *234* | 18 | 24.1 | 6 | 172.4 | *83* | | |
| Aug | 20.8 | 12.2 | 16.5 | *−1.1* | 24.4 | 26 | 6.2 | 30 | 16.0 | 29 | 16.4 | 8 | 0 | 0 | 35.2 | *64* | 12 | 10.0 | 3 | 176.0 | *90* | | |
| Sep | 18.8 | 8.6 | 13.7 | *−1.2* | 24.6 | 18 | 2.2 | 30 | 10.9 | 29 | 17.7 | 16 | 0 | 0 | 74.3 | *137* | 7 | 43.7 | 24 | 186.6 | *133* | | |
| Oct | 12.7 | 6.3 | 9.5 | *−1.8* | 17.2 | 12 | −1.6 | 30 | 7.6 | 27 | 12.1 | 15 | 2 | 0 | 53.3 | *76* | 10 | 13.9 | 31 | 57.6 | *52* | | |
| Nov | 6.9 | 0.1 | 3.5 | *−4.0* | 13.6 | 12 | −7.5 | 28 | −1.9 | 27 | 6.3 | 1 | 15 | 0 | 64.6 | *97* | 12 | 19.4 | 12 | 84.7 | *119* | | |
| Dec | 9.0 | 3.5 | 6.2 | *+1.2* | 13.4 | 10 | −2.1 | 13 | 3.2 | 12 | 8.9 | 10 | 6 | 0 | 127.1 | *201* | 25 | 18.1 | 9 | 37.7 | *71* | | |
| **1915** | 13.5 | 5.6 | 9.5 | *−1.2* | 27.8 | 6 | −7.5 | 11 | −1.9 | 11 | 17.7 | 9 | 48 | 1 | 820.9 | *124* | 171 | 43.7 | 9 | 1586.6 | *101* | | |

## OXFORD 1916

| Month | Mean max | Mean min | Mean temp | Anom | Highest max | Date | Lowest min | Date | Lowest max | Date | Highest min | Date | Air frost | Days ≥ 25 °C | Total pptn | Anom | Rain days | Wettest day | Date | Total sunshine | Anom | Sunniest day | Date |
|---|---|---|---|---|---|---|---|---|---|---|---|---|---|---|---|---|---|---|---|---|---|---|---|
| Jan | 10.1 | 5.1 | 7.6 | *+2.8* | 13.9 | 1 | 0.1 | 14 | 5.8 | 14 | 8.4 | 1 | 0 | 0 | 38.1 | *67* | 17 | 11.3 | 2 | 54.2 | *87* | | |
| Feb | 6.7 | 0.9 | 3.8 | *−1.1* | 11.8 | 16 | −3.3 | 25 | 0.2 | 24 | 5.8 | 13 | 10 | 0 | 97.3 | *227* | 19 | 12.1 | 4 | 73.0 | *92* | | |
| Mar | 6.6 | 0.9 | 3.8 | *−3.5* | 12.6 | 19 | −3.3 | 9 | 1.9 | 8 | 6.2 | 31 | 9 | 0 | 131.8 | *277* | 19 | 32.6 | 28 | 61.9 | *56* | | |
| Apr | 14.1 | 4.2 | 9.1 | *−0.2* | 22.9 | 27 | −0.4 | 8 | 9.2 | 16 | 8.4 | 28 | 1 | 0 | 25.2 | *52* | 13 | 7.9 | 19 | 204.6 | *127* | | |
| May | 17.4 | 7.7 | 12.5 | *+0.1* | 26.5 | 21 | 2.2 | 10 | 8.7 | 2 | 11.3 | 21 | 0 | 2 | 51.6 | *90* | 16 | 11.3 | 2 | 188.0 | *97* | | |
| June | 16.1 | 7.9 | 12.0 | *−3.6* | 19.0 | 17 | 2.9 | 17 | 11.2 | 12 | 13.4 | 23 | 0 | 0 | 46.9 | *95* | 17 | 8.2 | 24 | 132.8 | *70* | | |
| July | 21.1 | 10.8 | 15.9 | *−2.0* | 27.2 | 30 | 7.9 | 15 | 16.3 | 1 | 14.3 | 31 | 0 | 7 | 45.3 | *93* | 12 | 19.4 | 7 | 191.8 | *93* | | |
| Aug | 22.0 | 12.4 | 17.2 | *−0.4* | 28.1 | 1 | 6.2 | 31 | 14.2 | 29 | 16.3 | 25 | 0 | 7 | 102.9 | *188* | 15 | 23.6 | 30 | 189.8 | *97* | | |
| Sep | 17.6 | 9.0 | 13.3 | *−1.6* | 21.2 | 28 | 2.3 | 21 | 12.9 | 19 | 14.3 | 1 | 0 | 0 | 23.3 | *43* | 13 | 7.9 | 17 | 114.2 | *81* | | |
| Oct | 14.8 | 8.0 | 11.4 | *+0.1* | 19.3 | 4 | −1.9 | 21 | 7.8 | 22 | 15.1 | 13 | 2 | 0 | 93.7 | *134* | 21 | 11.3 | 2 | 98.0 | *88* | | |
| Nov | 9.5 | 3.5 | 6.5 | *−1.0* | 14.2 | 24 | −2.8 | 28 | 1.4 | 18 | 9.6 | 11 | 4 | 0 | 85.6 | *128* | 14 | 21.8 | 19 | 67.4 | *95* | | |
| Dec | 4.6 | −1.1 | 1.7 | *−3.3* | 13.2 | 29 | −7.2 | 17 | −1.4 | 14 | 8.7 | 29 | 18 | 0 | 72.3 | *115* | 16 | 26.3 | 21 | 36.3 | *68* | | |
| **1916** | 13.4 | 5.8 | 9.6 | *−1.1* | 28.1 | 8 | −7.2 | 12 | −1.4 | 12 | 16.3 | 8 | 44 | 16 | 814.0 | *123* | 192 | 32.6 | 3 | 1412.0 | *90* | | |

## OXFORD 1917

| Month | Mean max | Mean min | Mean temp | Anom | Highest max | Date | Lowest min | Date | Lowest max | Date | Highest min | Date | Air frost | Days ≥ 25 °C | Total pptn | Anom | Rain days | Wettest day | Date | Total sunshine | Anom | Sunniest day | Date |
|---|---|---|---|---|---|---|---|---|---|---|---|---|---|---|---|---|---|---|---|---|---|---|---|
| Jan | 2.9 | 0.1 | 1.5 | *−3.4* | 10.9 | 1 | −3.9 | 29 | −1.1 | 29 | 8.4 | 3 | 18 | 0 | 36.5 | *64* | 13 | 8.8 | 8 | 24.2 | *39* | | |
| Feb | 4.5 | −2.0 | 1.2 | *−3.7* | 9.8 | 27 | −10.4 | 7 | −0.3 | 5 | 4.9 | 21 | 16 | 0 | 32.0 | *75* | 12 | 9.5 | 20 | 56.7 | *72* | | |
| Mar | 7.0 | −0.4 | 3.3 | *−4.0* | 13.5 | 17 | −6.2 | 9 | 0.8 | 8 | 5.2 | 11 | 15 | 0 | 45.5 | *95* | 20 | 8.6 | 6 | 79.8 | *72* | | |
| Apr | 10.6 | 1.7 | 6.1 | *−3.1* | 16.6 | 30 | −2.5 | 7 | 3.6 | 10 | 8.6 | 29 | 11 | 0 | 41.5 | *85* | 15 | 14.1 | 11 | 152.0 | *94* | | |
| May | 19.6 | 8.0 | 13.8 | *+1.4* | 25.5 | 26 | 0.2 | 7 | 10.2 | 17 | 13.9 | 27 | 0 | 1 | 42.0 | *73* | 10 | 18.7 | 17 | 199.0 | *103* | | |
| June | 21.5 | 10.7 | 16.1 | *+0.5* | 31.4 | 17 | 6.9 | 23 | 15.1 | 29 | 15.7 | 18 | 0 | 4 | 56.4 | *115* | 8 | 41.8 | 29 | 210.8 | *110* | | |
| July | 21.2 | 11.7 | 16.5 | *−1.4* | 24.6 | 3 | 7.2 | 2 | 14.0 | 31 | 15.8 | 26 | 0 | 0 | 82.9 | *170* | 9 | 30.3 | 8 | 207.2 | *100* | | |
| Aug | 19.1 | 12.6 | 15.9 | *−1.7* | 23.3 | 22 | 9.2 | 27 | 13.1 | 1 | 15.3 | 6 | 0 | 0 | 101.3 | *185* | 23 | 23.7 | 1 | 136.4 | *69* | | |
| Sep | 18.7 | 10.6 | 14.6 | *−0.3* | 22.4 | 5 | 6.4 | 30 | 14.7 | 13 | 14.2 | 6 | 0 | 0 | 43.1 | *80* | 10 | 13.3 | 19 | 145.5 | *104* | | |
| Oct | 12.2 | 3.8 | 8.0 | *−3.3* | 20.2 | 1 | −3.4 | 28 | 7.3 | 28 | 10.4 | 3 | 1 | 0 | 106.0 | *151* | 19 | 21.1 | 4 | 143.3 | *129* | | |
| Nov | 10.8 | 4.6 | 7.7 | *+0.2* | 14.6 | 21 | −1.6 | 26 | 5.9 | 26 | 11.3 | 21 | 2 | 0 | 17.3 | *26* | 14 | 3.8 | 6 | 52.4 | *74* | | |
| Dec | 4.9 | −1.2 | 1.9 | *−3.1* | 11.1 | 7 | −8.3 | 19 | −2.1 | 19 | 6.1 | 7 | 18 | 0 | 24.5 | *39* | 14 | 8.7 | 16 | 75.0 | *141* | | |
| 1917 | 12.8 | 5.1 | 8.9 | *−1.8* | 31.4 | 6 | −10.4 | 2 | −2.1 | 12 | 15.8 | 7 | 81 | 5 | 629.0 | *95* | 167 | 41.8 | 6 | 1482.3 | *94* | | |

## OXFORD 1918

| Month | Mean max | Mean min | Mean temp | Anom | Highest max | Date | Lowest min | Date | Lowest max | Date | Highest min | Date | Air frost | Days ≥ 25 °C | Total pptn | Anom | Rain days | Wettest day | Date | Total sunshine | Anom | Sunniest day | Date |
|---|---|---|---|---|---|---|---|---|---|---|---|---|---|---|---|---|---|---|---|---|---|---|---|
| Jan | 6.7 | 0.8 | 3.8 | *−1.1* | 12.5 | 21 | −8.4 | 9 | −2.1 | 8 | 8.9 | 24 | 13 | 0 | 69.6 | *122* | 15 | 26.9 | 15 | 65.0 | *104* | | |
| Feb | 9.3 | 3.4 | 6.3 | *+1.4* | 13.8 | 23 | −5.3 | 18 | 3.7 | 18 | 9.3 | 23 | 5 | 0 | 27.9 | *65* | 13 | 4.3 | 9 | 71.2 | *90* | | |
| Mar | 11.1 | 1.8 | 6.5 | *−0.8* | 21.1 | 23 | −3.7 | 9 | 1.9 | 3 | 6.1 | 28 | 7 | 0 | 19.8 | *42* | 13 | 6.0 | 30 | 129.5 | *116* | | |
| Apr | 10.8 | 3.2 | 7.0 | *−2.3* | 18.4 | 25 | −0.6 | 3 | 3.6 | 21 | 7.5 | 26 | 2 | 0 | 81.5 | *167* | 17 | 30.3 | 16 | 98.5 | *61* | | |
| May | 19.1 | 8.0 | 13.6 | *+1.1* | 27.6 | 22 | 3.2 | 3 | 10.2 | 1 | 11.1 | 20 | 0 | 4 | 47.0 | *82* | 10 | 12.9 | 23 | 228.6 | *119* | | |
| June | 19.0 | 8.3 | 13.7 | *−1.9* | 26.3 | 2 | 4.1 | 16 | 15.0 | 18 | 12.8 | 7 | 0 | 3 | 33.3 | *68* | 11 | 9.4 | 18 | 225.9 | *118* | | |
| July | 21.0 | 11.5 | 16.2 | *−1.6* | 26.3 | 31 | 7.1 | 4 | 14.8 | 26 | 16.3 | 16 | 0 | 4 | 100.6 | *206* | 16 | 19.9 | 14 | 202.3 | *98* | | |
| Aug | 21.1 | 11.8 | 16.5 | *−1.1* | 29.0 | 22 | 6.1 | 24 | 15.4 | 27 | 16.7 | 21 | 0 | 4 | 30.5 | *56* | 12 | 11.2 | 25 | 167.1 | *85* | | |
| Sep | 16.2 | 9.1 | 12.6 | *−2.2* | 20.7 | 7 | 2.8 | 30 | 7.9 | 29 | 14.8 | 17 | 0 | 0 | 126.1 | *233* | 25 | 28.5 | 29 | 145.0 | *103* | | |
| Oct | 12.6 | 6.2 | 9.4 | *−1.9* | 16.0 | 10 | 0.4 | 1 | 9.6 | 24 | 13.7 | 10 | 0 | 0 | 44.3 | *63* | 16 | 9.0 | 17 | 69.8 | *63* | | |
| Nov | 9.3 | 1.9 | 5.6 | *−2.0* | 14.5 | 2 | −3.7 | 14 | 3.1 | 20 | 10.6 | 1 | 11 | 0 | 49.5 | *74* | 17 | 10.7 | 4 | 88.0 | *124* | | |
| Dec | 10.0 | 4.8 | 7.4 | *+2.4* | 13.8 | 3 | −2.3 | 26 | 4.5 | 21 | 10.3 | 4 | 2 | 0 | 59.5 | *94* | 23 | 9.0 | 18 | 36.5 | *69* | | |
| 1918 | 13.9 | 5.9 | 9.9 | *−0.8* | 29.0 | 8 | −8.4 | 1 | −2.1 | 1 | 16.7 | 8 | 40 | 15 | 689.6 | *104* | 188 | 30.3 | 4 | 1527.4 | *97* | | |

## OXFORD 1919

| Month | Mean max | Mean min | Mean temp | Anom | Highest max | Date | Lowest min | Date | Lowest max | Date | Highest min | Date | Air frost | Days ≥ 25 °C | Total pptn | Anom | Rain days | Wettest day | Date | Total sunshine | Anom | Sunniest day | Date |
|---|---|---|---|---|---|---|---|---|---|---|---|---|---|---|---|---|---|---|---|---|---|---|---|
| Jan | 5.5 | 0.2 | 2.9 | *−2.0* | 11.3 | 15 | −5.1 | 25 | −1.9 | 31 | 6.9 | 15 | 12 | 0 | 85.5 | *150* | 21 | 22.3 | 3 | 51.0 | *82* | | |
| Feb | 4.6 | −1.3 | 1.6 | *−3.3* | 10.7 | 20 | −9.1 | 9 | −0.2 | 8 | 7.2 | 21 | 15 | 0 | 55.7 | *130* | 14 | 13.9 | 16 | 64.8 | *82* | | |
| Mar | 7.6 | 1.0 | 4.3 | *−2.9* | 13.9 | 2 | −3.3 | 23 | 3.3 | 21 | 6.3 | 11 | 11 | 0 | 97.9 | *205* | 17 | 18.3 | 19 | 106.8 | *96* | | |
| Apr | 11.8 | 3.2 | 7.5 | *−1.8* | 19.8 | 19 | −1.8 | 1 | 4.6 | 27 | 9.3 | 11 | 7 | 0 | 57.1 | *117* | 13 | 19.5 | 27 | 122.6 | *76* | | |
| May | 19.6 | 7.6 | 13.6 | *+1.2* | 25.7 | 31 | 4.6 | 20 | 10.3 | 7 | 11.1 | 14 | 0 | 2 | 15.6 | *27* | 6 | 8.2 | 6 | 250.0 | *130* | | |
| June | 20.4 | 9.7 | 15.1 | *−0.5* | 26.8 | 11 | 4.9 | 3 | 14.6 | 24 | 14.8 | 5 | 0 | 4 | 17.9 | *36* | 8 | 7.0 | 20 | 220.2 | *115* | | |
| July | 18.7 | 10.0 | 14.4 | *−3.5* | 25.7 | 11 | 7.0 | 9 | 13.6 | 1 | 14.6 | 17 | 0 | 1 | 59.7 | *122* | 11 | 15.5 | 19 | 130.0 | *63* | | |
| Aug | 22.4 | 11.9 | 17.1 | *−0.4* | 28.9 | 9 | 6.4 | 30 | 15.6 | 30 | 17.0 | 16 | 0 | 10 | 72.5 | *132* | 12 | 27.7 | 28 | 221.8 | *113* | | |
| Sep | 18.0 | 8.6 | 13.3 | *−1.6* | 29.9 | 11 | −0.4 | 29 | 12.0 | 20 | 16.4 | 5 | 1 | 1 | 31.5 | *58* | 11 | 12.6 | 3 | 154.0 | *110* | | |
| Oct | 12.2 | 2.0 | 7.1 | *−4.2* | 16.9 | 5 | −1.2 | 23 | 6.4 | 28 | 8.7 | 6 | 5 | 0 | 38.2 | *55* | 11 | 23.9 | 24 | 147.1 | *132* | | |
| Nov | 6.3 | 0.7 | 3.5 | *−4.1* | 13.1 | 24 | −4.6 | 12 | 0.8 | 28 | 9.7 | 23 | 12 | 0 | 21.8 | *33* | 16 | 3.6 | 29 | 53.5 | *75* | | |
| Dec | 8.9 | 2.8 | 5.8 | *+0.8* | 12.1 | 23 | −1.6 | 9 | 3.5 | 9 | 8.2 | 30 | 3 | 0 | 112.3 | *178* | 22 | 25.5 | 2 | 45.9 | *86* | | |
| **1919** | 13.1 | 4.8 | 8.9 | *−1.8* | 29.9 | 9 | −9.1 | 2 | −1.9 | 1 | 17.0 | 8 | 66 | 18 | 665.7 | *101* | 162 | 27.7 | 8 | 1567.7 | *99* | | |

## OXFORD 1920

| Month | Mean max | Mean min | Mean temp | Anom | Highest max | Date | Lowest min | Date | Lowest max | Date | Highest min | Date | Air frost | Days ≥ 25 °C | Total pptn | Anom | Rain days | Wettest day | Date | Total sunshine | Anom | Sunniest day | Date |
|---|---|---|---|---|---|---|---|---|---|---|---|---|---|---|---|---|---|---|---|---|---|---|---|
| Jan | 8.7 | 2.0 | 5.4 | *+0.5* | 12.9 | 13 | −5.9 | 7 | 1.2 | 6 | 9.3 | 18 | 6 | 0 | 54.3 | *95* | 18 | 12.9 | 11 | 57.7 | *93* | | |
| Feb | 9.8 | 1.8 | 5.8 | *+0.9* | 15.9 | 19 | −2.9 | 5 | 5.4 | 5 | 8.2 | 10 | 6 | 0 | 16.1 | *38* | 11 | 9.2 | 20 | 75.0 | *95* | | |
| Mar | 12.0 | 3.4 | 7.7 | *+0.4* | 18.7 | 22 | −2.9 | 9 | 5.7 | 8 | 10.9 | 28 | 3 | 0 | 41.8 | *88* | 17 | 9.6 | 15 | 135.4 | *122* | | |
| Apr | 12.1 | 5.8 | 8.9 | *−0.3* | 16.1 | 23 | 2.2 | 29 | 8.0 | 3 | 8.7 | 16 | 0 | 0 | 78.5 | *160* | 23 | 11.0 | 9 | 87.1 | *54* | | |
| May | 17.3 | 7.7 | 12.5 | *−0.0* | 26.3 | 25 | 2.2 | 5 | 11.3 | 6 | 13.3 | 27 | 0 | 1 | 42.9 | *75* | 14 | 9.4 | 1 | 192.0 | *100* | | |
| June | 19.6 | 10.0 | 14.8 | *−0.8* | 24.2 | 17 | 3.9 | 9 | 14.8 | 10 | 14.3 | 29 | 0 | 0 | 81.2 | *165* | 12 | 33.2 | 12 | 195.7 | *102* | | |
| July | 18.5 | 10.9 | 14.7 | *−3.2* | 21.8 | 19 | 7.3 | 25 | 11.8 | 5 | 15.6 | 21 | 0 | 0 | 119.0 | *244* | 19 | 16.6 | 5 | 132.0 | *64* | | |
| Aug | 18.1 | 9.6 | 13.8 | *−3.8* | 22.8 | 8 | 5.4 | 20 | 13.2 | 23 | 13.3 | 12 | 0 | 0 | 28.5 | *52* | 8 | 7.4 | 18 | 139.9 | *71* | | |
| Sep | 18.3 | 8.8 | 13.5 | *−1.3* | 23.8 | 12 | 3.2 | 20 | 11.5 | 21 | 13.8 | 6 | 0 | 0 | 45.0 | *83* | 13 | 10.8 | 18 | 111.8 | *80* | | |
| Oct | 14.7 | 6.4 | 10.5 | *−0.7* | 21.0 | 7 | −0.8 | 30 | 6.8 | 27 | 12.9 | 4 | 1 | 0 | 48.9 | *70* | 15 | 18.2 | 31 | 96.6 | *87* | | |
| Nov | 9.9 | 2.5 | 6.2 | *−1.3* | 14.7 | 15 | −5.1 | 22 | 3.1 | 22 | 9.5 | 10 | 7 | 0 | 35.1 | *53* | 12 | 9.0 | 1 | 59.2 | *83* | | |
| Dec | 7.2 | 1.8 | 4.5 | *−0.5* | 13.3 | 3 | −8.1 | 13 | −2.8 | 13 | 8.9 | 25 | 9 | 0 | 58.9 | *93* | 19 | 11.2 | 31 | 40.7 | *77* | | |
| **1920** | 13.9 | 5.9 | 9.9 | *−0.9* | 26.3 | 5 | −8.1 | 12 | −2.8 | 12 | 15.6 | 7 | 32 | 1 | 650.2 | *98* | 181 | 33.2 | 6 | 1323.1 | *84* | | |

## OXFORD 1921

| Month | Mean max | Mean min | Mean temp | Anom | Highest max | Date | Lowest min | Date | Lowest max | Date | Highest min | Date | Air frost | Days ≥ 25 °C | Total pptn | Anom | Rain days | Wettest day | Date | Total sunshine | Anom | Sunniest day | Date |
|---|---|---|---|---|---|---|---|---|---|---|---|---|---|---|---|---|---|---|---|---|---|---|---|
| Jan | 10.0 | 4.9 | 7.4 | *+2.6* | 12.7 | 10 | −1.6 | 15 | 4.4 | 14 | 9.8 | 4 | 2 | 0 | 62.8 | *110* | 18 | 11.9 | 7 | 33.9 | *54* | 5.1 | 14 |
| Feb | 8.7 | 0.9 | 4.8 | *−0.1* | 15.6 | 24 | −2.9 | 21 | 1.8 | 7 | 6.5 | 17 | 7 | 0 | 14.3 | *33* | 5 | 8.1 | 25 | 78.0 | *99* | 9.4 | 26 |
| Mar | 11.9 | 3.8 | 7.9 | *+0.6* | 17.8 | 24 | −0.6 | 3 | 8.1 | 7 | 9.6 | 16 | 1 | 0 | 23.6 | *50* | 17 | 3.8 | 28 | 127.1 | *114* | 10.7 | 18 |
| Apr | 14.3 | 3.8 | 9.0 | *−0.3* | 21.8 | 28 | −1.7 | 16 | 6.3 | 15 | 7.6 | 6 | 3 | 0 | 23.2 | *47* | 9 | 7.0 | 13 | 195.7 | *122* | 13.5 | 30 |
| May | 17.7 | 6.8 | 12.3 | *−0.2* | 24.9 | 25 | −0.2 | 5 | 9.5 | 4 | 11.4 | 7 | 1 | 0 | 38.1 | *67* | 17 | 9.2 | 14 | 228.8 | *119* | 14.5 | 22 |
| June | 21.5 | 9.3 | 15.4 | *−0.2* | 29.6 | 25 | 4.5 | 19 | 14.8 | 19 | 14.4 | 26 | 0 | 5 | 8.7 | *18* | 6 | 4.1 | 26 | 244.9 | *128* | 15.9 | 28 |
| July | 26.3 | 13.2 | 19.7 | *+1.9* | 31.7 | 10 | 7.1 | 5 | 19.0 | 4 | 18.7 | 23 | 0 | 16 | 7.1 | *15* | 6 | 2.8 | 23 | 258.9 | *125* | 15.0 | 11 |
| Aug | 21.0 | 11.9 | 16.4 | *−1.2* | 27.0 | 18 | 6.0 | 31 | 15.2 | 12 | 17.9 | 1 | 0 | 2 | 42.5 | *78* | 14 | 17.1 | 16 | 146.1 | *74* | 12.7 | 18 |
| Sep | 20.7 | 9.4 | 15.0 | *+0.1* | 28.5 | 8 | 2.2 | 29 | 15.1 | 19 | 15.5 | 14 | 0 | 4 | 43.3 | *80* | 8 | 27.9 | 11 | 171.6 | *122* | 11.4 | 7 |
| Oct | 18.9 | 7.9 | 13.4 | *+2.1* | 27.2 | 5 | 0.6 | 25 | 9.8 | 23 | 15.4 | 3 | 0 | 3 | 32.4 | *46* | 9 | 10.7 | 18 | 159.7 | *143* | 9.6 | 9 |
| Nov | 7.9 | 1.2 | 4.5 | *−3.0* | 14.2 | 1 | −6.2 | 13 | 2.9 | 27 | 9.4 | 4 | 12 | 0 | 52.0 | *78* | 13 | 9.1 | 2 | 61.5 | *87* | 7.6 | 8 |
| Dec | 9.9 | 3.0 | 6.4 | *+1.4* | 13.9 | 28 | −3.1 | 26 | 1.8 | 4 | 10.1 | 19 | 5 | 0 | 32.7 | *52* | 14 | 5.6 | 26 | 48.0 | *90* | 6.6 | 20 |
| **1921** | 15.8 | 6.4 | 11.1 | *+0.3* | 31.7 | 7 | −6.2 | 11 | 1.8 | 2 | 18.7 | 7 | 31 | 30 | 380.7 | *58* | 136 | 27.9 | 9 | 1753.9 | *111* | 15.9 | 6 |

## OXFORD 1922

| Month | Mean max | Mean min | Mean temp | Anom | Highest max | Date | Lowest min | Date | Lowest max | Date | Highest min | Date | Air frost | Days ≥ 25 °C | Total pptn | Anom | Rain days | Wettest day | Date | Total sunshine | Anom | Sunniest day | Date |
|---|---|---|---|---|---|---|---|---|---|---|---|---|---|---|---|---|---|---|---|---|---|---|---|
| Jan | 7.2 | 1.0 | 4.1 | *−0.8* | 13.8 | 2 | −4.7 | 24 | 1.2 | 15 | 7.3 | 9 | 13 | 0 | 61.1 | *107* | 22 | 9.7 | 13 | 55.2 | *89* | 6.9 | 16 |
| Feb | 8.3 | 1.0 | 4.6 | *−0.3* | 15.7 | 25 | −8.8 | 6 | −0.8 | 5 | 10.7 | 24 | 11 | 0 | 61.1 | *143* | 17 | 9.1 | 27 | 92.0 | *117* | 8.5 | 18 |
| Mar | 8.6 | 1.9 | 5.3 | *−2.0* | 14.6 | 3 | −4.5 | 26 | 3.8 | 31 | 7.6 | 6 | 8 | 0 | 47.3 | *99* | 15 | 7.7 | 6 | 109.0 | *98* | 10.2 | 23 |
| Apr | 10.4 | 1.8 | 6.1 | *−3.2* | 17.9 | 14 | −4.6 | 2 | 2.7 | 3 | 7.7 | 15 | 7 | 0 | 71.2 | *146* | 18 | 12.4 | 13 | 149.9 | *93* | 11.3 | 2 |
| May | 20.0 | 8.0 | 14.0 | *+1.5* | 29.8 | 23 | 0.4 | 13 | 10.4 | 1 | 15.6 | 22 | 0 | 8 | 36.8 | *64* | 7 | 23.5 | 24 | 268.8 | *139* | 14.7 | 27 |
| June | 19.9 | 9.2 | 14.5 | *−1.1* | 28.2 | 1 | 4.7 | 16 | 14.4 | 14 | 14.4 | 19 | 0 | 3 | 30.2 | *61* | 9 | 9.7 | 13 | 227.6 | *119* | 15.5 | 7 |
| July | 18.6 | 10.3 | 14.5 | *−3.4* | 22.9 | 12 | 6.7 | 16 | 14.8 | 23 | 14.4 | 21 | 0 | 0 | 89.7 | *184* | 15 | 31.5 | 5 | 153.7 | *74* | 13.0 | 19 |
| Aug | 18.2 | 10.0 | 14.1 | *−3.5* | 20.9 | 29 | 5.0 | 26 | 16.1 | 12 | 13.3 | 28 | 0 | 0 | 130.3 | *238* | 17 | 70.8 | 6 | 111.1 | *57* | 11.7 | 26 |
| Sep | 17.1 | 8.6 | 12.9 | *−2.0* | 21.7 | 21 | 3.4 | 12 | 13.4 | 1 | 13.3 | 5 | 0 | 0 | 35.7 | *66* | 13 | 7.1 | 14 | 110.8 | *79* | 10.9 | 18 |
| Oct | 13.1 | 3.8 | 8.5 | *−2.8* | 18.5 | 14 | −2.4 | 29 | 5.5 | 29 | 12.5 | 4 | 6 | 0 | 13.3 | *19* | 11 | 3.8 | 29 | 138.9 | *125* | 9.8 | 15 |
| Nov | 9.5 | 1.5 | 5.5 | *−2.0* | 12.9 | 7 | −3.9 | 25 | 2.7 | 14 | 7.9 | 18 | 8 | 0 | 29.0 | *43* | 11 | 11.4 | 5 | 61.6 | *87* | 7.2 | 4 |
| Dec | 8.7 | 3.2 | 6.0 | *+0.9* | 11.8 | 13 | −1.3 | 9 | 4.4 | 9 | 8.3 | 13 | 1 | 0 | 65.3 | *103* | 16 | 13.1 | 19 | 49.6 | *93* | 5.5 | 26 |
| **1922** | 13.3 | 5.0 | 9.2 | *−1.6* | 29.8 | 5 | −8.8 | 2 | −0.8 | 2 | 15.6 | 5 | 54 | 11 | 671.0 | *102* | 171 | 70.8 | 8 | 1528.0 | *97* | 15.5 | 6 |

## OXFORD 1923

| Month | Mean max | Mean min | Mean temp | Anom | Highest max | Date | Lowest min | Date | Lowest max | Date | Highest min | Date | Air frost | Days ≥ 25 °C | Total pptn | Anom | Rain days | Wettest day | Date | Total sunshine | Anom | Sunniest day | Date |
|---|---|---|---|---|---|---|---|---|---|---|---|---|---|---|---|---|---|---|---|---|---|---|---|
| Jan | 8.7 | 1.8 | 5.2 | *+0.4* | 12.1 | 25 | −3.3 | 17 | 4.7 | 13 | 7.3 | 30 | 4 | 0 | 40.7 | *72* | 15 | 14.1 | 6 | 66.5 | *107* | 7.3 | 23 |
| Feb | 8.9 | 3.7 | 6.3 | *+1.4* | 13.4 | 26 | −1.9 | 5 | 2.3 | 20 | 10.8 | 2 | 2 | 0 | 90.0 | *210* | 22 | 11.3 | 7 | 54.6 | *69* | 7.3 | 5 |
| Mar | 9.8 | 3.4 | 6.6 | *−0.6* | 19.3 | 27 | −1.7 | 12 | 4.3 | 15 | 8.9 | 27 | 1 | 0 | 67.3 | *141* | 19 | 12.4 | 20 | 69.2 | *62* | 7.8 | 3 |
| Apr | 12.1 | 4.2 | 8.1 | *−1.2* | 16.7 | 4 | −1.0 | 23 | 6.8 | 9 | 9.3 | 12 | 2 | 0 | 58.3 | *119* | 14 | 14.6 | 10 | 122.2 | *76* | 10.8 | 8 |
| May | 14.7 | 6.2 | 10.5 | *−2.0* | 23.9 | 4 | −0.1 | 12 | 10.2 | 16 | 11.1 | 4 | 1 | 0 | 36.1 | *63* | 19 | 6.1 | 23 | 165.2 | *86* | 13.5 | 4 |
| June | 17.5 | 9.0 | 13.2 | *−2.4* | 24.3 | 29 | 2.2 | 3 | 11.8 | 2 | 12.7 | 22 | 0 | 0 | 10.9 | *22* | 7 | 4.6 | 15 | 126.1 | *66* | 13.7 | 29 |
| July | 23.5 | 13.5 | 18.5 | *+0.6* | 33.9 | 12 | 8.4 | 27 | 16.9 | 1 | 19.2 | 13 | 0 | 10 | 37.6 | *77* | 10 | 13.3 | 31 | 201.7 | *97* | 13.9 | 6 |
| Aug | 21.4 | 11.1 | 16.3 | *−1.3* | 28.2 | 9 | 6.6 | 31 | 15.6 | 29 | 15.3 | 14 | 0 | 6 | 85.3 | *156* | 13 | 22.0 | 29 | 241.7 | *123* | 12.8 | 4 |
| Sep | 18.0 | 8.5 | 13.3 | *−1.6* | 23.4 | 7 | 4.4 | 17 | 13.5 | 19 | 15.8 | 30 | 0 | 0 | 47.5 | *88* | 13 | 19.6 | 14 | 171.0 | *122* | 11.2 | 3 |
| Oct | 13.7 | 7.2 | 10.4 | *−0.8* | 18.1 | 9 | 0.6 | 5 | 9.4 | 5 | 11.6 | 1 | 0 | 0 | 99.4 | *142* | 23 | 20.1 | 12 | 92.9 | *83* | 8.7 | 4 |
| Nov | 6.8 | −0.4 | 3.2 | *−4.3* | 13.6 | 3 | −8.8 | 26 | −1.4 | 25 | 6.6 | 3 | 13 | 0 | 34.6 | *52* | 13 | 11.7 | 13 | 107.9 | *152* | 8.3 | 4 |
| Dec | 7.5 | 0.8 | 4.1 | *−0.9* | 10.8 | 17 | −3.5 | 25 | 1.5 | 21 | 7.1 | 11 | 14 | 0 | 61.2 | *97* | 17 | 10.4 | 25 | 51.9 | *98* | 6.5 | 3 |
| **1923** | 13.6 | 5.8 | 9.7 | *−1.1* | 33.9 | 7 | −8.8 | 11 | −1.4 | 11 | 19.2 | 7 | 37 | 16 | 668.9 | *101* | 185 | 22.0 | 8 | 1470.7 | *93* | 13.9 | 7 |

## OXFORD 1924

| Month | Mean max | Mean min | Mean temp | Anom | Highest max | Date | Lowest min | Date | Lowest max | Date | Highest min | Date | Air frost | Days ≥ 25 °C | Total pptn | Anom | Rain days | Wettest day | Date | Total sunshine | Anom | Sunniest day | Date |
|---|---|---|---|---|---|---|---|---|---|---|---|---|---|---|---|---|---|---|---|---|---|---|---|
| Jan | 7.8 | 2.6 | 5.2 | *+0.4* | 10.9 | 19 | −4.4 | 9 | 0.6 | 9 | 6.9 | 13 | 5 | 0 | 81.1 | *143* | 17 | 11.2 | 21 | 59.1 | *95* | 7.5 | 25 |
| Feb | 6.0 | 0.4 | 3.2 | *−1.7* | 10.7 | 6 | −6.2 | 17 | 1.3 | 14 | 7.4 | 6 | 12 | 0 | 6.0 | *14* | 6 | 3.1 | 24 | 54.5 | *69* | 8.5 | 17 |
| Mar | 9.7 | −0.3 | 4.7 | *−2.6* | 15.3 | 12 | −5.1 | 10 | 5.1 | 3 | 9.3 | 24 | 21 | 0 | 24.0 | *50* | 8 | 6.7 | 2 | 175.5 | *158* | 10.7 | 19 |
| Apr | 12.5 | 3.6 | 8.1 | *−1.2* | 21.4 | 21 | −1.7 | 6 | 3.2 | 10 | 9.9 | 25 | 5 | 0 | 67.3 | *138* | 11 | 19.1 | 13 | 155.9 | *97* | 11.9 | 17 |
| May | 16.9 | 8.0 | 12.5 | *−0.0* | 22.4 | 29 | 2.6 | 5 | 11.8 | 5 | 13.9 | 31 | 0 | 0 | 105.3 | *184* | 22 | 17.6 | 31 | 171.9 | *89* | 12.0 | 16 |
| June | 19.1 | 10.4 | 14.7 | *−0.9* | 24.2 | 26 | 3.3 | 14 | 12.9 | 4 | 14.0 | 25 | 0 | 0 | 62.0 | *126* | 13 | 12.8 | 4 | 182.6 | *96* | 14.7 | 14 |
| July | 20.7 | 10.9 | 15.8 | *−2.1* | 29.8 | 12 | 6.1 | 1 | 16.1 | 17 | 14.3 | 13 | 0 | 2 | 93.8 | *192* | 15 | 23.1 | 17 | 219.8 | *106* | 15.0 | 11 |
| Aug | 18.9 | 10.5 | 14.7 | *−2.9* | 24.6 | 11 | 6.7 | 9 | 15.6 | 24 | 14.7 | 5 | 0 | 0 | 49.3 | *90* | 20 | 6.8 | 30 | 149.1 | *76* | 12.0 | 11 |
| Sep | 17.4 | 10.5 | 14.0 | *−0.9* | 21.3 | 6 | 2.9 | 28 | 12.2 | 25 | 14.7 | 7 | 0 | 0 | 112.1 | *207* | 20 | 20.6 | 25 | 107.0 | *76* | 10.1 | 14 |
| Oct | 14.1 | 7.1 | 10.6 | *−0.7* | 21.7 | 13 | −1.2 | 24 | 7.8 | 25 | 12.8 | 10 | 1 | 0 | 97.8 | *140* | 20 | 12.5 | 6 | 85.7 | *77* | 8.8 | 24 |
| Nov | 10.1 | 4.6 | 7.3 | *−0.2* | 15.3 | 1 | −4.8 | 18 | 5.1 | 17 | 10.1 | 2 | 4 | 0 | 47.8 | *72* | 12 | 8.9 | 1 | 43.0 | *61* | 7.8 | 4 |
| Dec | 9.4 | 4.2 | 6.8 | *+1.8* | 13.3 | 5 | −0.6 | 11 | 1.4 | 11 | 9.0 | 19 | 1 | 0 | 111.4 | *177* | 17 | 20.3 | 15 | 49.8 | *94* | 5.8 | 14 |
| **1924** | 13.6 | 6.1 | 9.8 | *−0.9* | 29.8 | 7 | −6.2 | 2 | 0.6 | 1 | 14.7 | 8 | 49 | 2 | 857.9 | *130* | 181 | 23.1 | 7 | 1453.5 | *92* | 15.0 | 7 |

## OXFORD 1925

| Month | Mean max | Mean min | Mean temp | Anom | Highest max | Date | Lowest min | Date | Lowest max | Date | Highest min | Date | Air frost | Days ≥ 25 °C | Total pptn | Anom | Rain days | Wettest day | Date | Total sunshine | Anom | Sunniest day | Date |
|---|---|---|---|---|---|---|---|---|---|---|---|---|---|---|---|---|---|---|---|---|---|---|---|
| Jan | 8.4 | 2.3 | 5.4 | *+0.5* | 12.5 | 2 | −2.3 | 25 | 4.2 | 21 | 9.2 | 31 | 6 | 0 | 40.9 | *72* | 13 | 15.1 | 1 | 55.4 | *89* | 6.5 | 5 |
| Feb | 9.0 | 2.5 | 5.8 | *+0.9* | 12.2 | 10 | −0.4 | 24 | 4.9 | 23 | 8.4 | 11 | 2 | 0 | 86.4 | *202* | 20 | 16.9 | 25 | 74.5 | *94* | 7.3 | 15 |
| Mar | 8.9 | 1.5 | 5.2 | *−2.1* | 13.1 | 15 | −4.2 | 13 | 5.0 | 12 | 6.6 | 7 | 13 | 0 | 11.9 | *25* | 10 | 2.2 | 14 | 85.0 | *76* | 9.0 | 9 |
| Apr | 12.2 | 3.6 | 7.9 | *−1.4* | 17.2 | 8 | −3.4 | 4 | 8.0 | 4 | 7.3 | 23 | 1 | 0 | 56.6 | *116* | 18 | 11.0 | 26 | 142.3 | *88* | 10.8 | 12 |
| May | 16.8 | 7.8 | 12.3 | *−0.1* | 24.0 | 16 | 1.2 | 1 | 10.1 | 1 | 13.6 | 19 | 0 | 0 | 68.0 | *119* | 21 | 10.0 | 3 | 187.6 | *97* | 11.5 | 14 |
| June | 21.5 | 9.9 | 15.7 | *+0.1* | 29.1 | 11 | 5.5 | 2 | 14.3 | 24 | 15.3 | 12 | 0 | 7 | 1.7 | *3* | 2 | 1.5 | 26 | 267.4 | *140* | 15.5 | 10 |
| July | 22.5 | 12.3 | 17.4 | *−0.4* | 29.2 | 22 | 7.9 | 9 | 15.2 | 27 | 16.7 | 23 | 0 | 8 | 98.9 | *203* | 14 | 26.2 | 22 | 178.3 | *86* | 12.7 | 14 |
| Aug | 20.1 | 12.0 | 16.0 | *−1.6* | 26.2 | 17 | 8.2 | 26 | 16.2 | 12 | 15.7 | 29 | 0 | 2 | 59.5 | *109* | 16 | 10.1 | 23 | 138.5 | *70* | 13.1 | 16 |
| Sep | 16.3 | 7.8 | 12.1 | *−2.8* | 21.5 | 30 | 2.0 | 13 | 13.5 | 5 | 14.6 | 1 | 0 | 0 | 82.0 | *151* | 19 | 20.5 | 19 | 133.0 | *95* | 10.3 | 2 |
| Oct | 15.1 | 6.9 | 11.0 | *−0.3* | 20.4 | 3 | −0.2 | 10 | 10.1 | 14 | 13.8 | 1 | 2 | 0 | 71.4 | *102* | 16 | 19.1 | 19 | 102.1 | *92* | 8.8 | 14 |
| Nov | 7.7 | 1.3 | 4.5 | *−3.0* | 16.9 | 3 | −4.5 | 27 | 2.6 | 28 | 9.9 | 1 | 16 | 0 | 47.6 | *71* | 15 | 15.4 | 2 | 104.5 | *147* | 7.7 | 12 |
| Dec | 6.0 | 0.4 | 3.2 | *−1.8* | 13.7 | 29 | −6.9 | 6 | −0.3 | 4 | 9.9 | 30 | 15 | 0 | 51.5 | *82* | 14 | 12.8 | 22 | 57.5 | *108* | 6.5 | 5 |
| **1925** | 13.7 | 5.7 | 9.7 | *−1.0* | 29.2 | 7 | −6.9 | 12 | −0.3 | 12 | 16.7 | 7 | 55 | 17 | 676.4 | *102* | 178 | 26.2 | 7 | 1525.8 | *97* | 15.5 | 6 |

## OXFORD 1926

| Month | Mean max | Mean min | Mean temp | Anom | Highest max | Date | Lowest min | Date | Lowest max | Date | Highest min | Date | Air frost | Days ≥ 25 °C | Total pptn | Anom | Rain days | Wettest day | Date | Total sunshine | Anom | Sunniest day | Date |
|---|---|---|---|---|---|---|---|---|---|---|---|---|---|---|---|---|---|---|---|---|---|---|---|
| Jan | 7.5 | 1.5 | 4.5 | *−0.4* | 11.9 | 25 | −14.3 | 17 | −2.6 | 15 | 7.6 | 6 | 9 | 0 | 101.0 | *178* | 21 | 33.5 | 1 | 46.5 | *75* | 6.6 | 20 |
| Feb | 10.2 | 4.7 | 7.4 | *+2.5* | 14.2 | 26 | −1.9 | 14 | 2.2 | 10 | 9.6 | 20 | 1 | 0 | 54.1 | *126* | 15 | 11.9 | 6 | 40.9 | *52* | 8.0 | 28 |
| Mar | 10.6 | 3.3 | 6.9 | *−0.3* | 15.6 | 6 | −2.3 | 22 | 5.6 | 22 | 9.8 | 7 | 2 | 0 | 16.8 | *35* | 9 | 7.6 | 27 | 111.7 | *100* | 9.2 | 10 |
| Apr | 14.1 | 5.2 | 9.7 | *+0.4* | 22.0 | 2 | 0.0 | 13 | 10.2 | 16 | 11.6 | 4 | 0 | 0 | 68.8 | *141* | 19 | 11.2 | 25 | 121.3 | *75* | 12.0 | 13 |
| May | 15.1 | 6.6 | 10.8 | *−1.6* | 23.8 | 26 | 1.1 | 9 | 9.4 | 14 | 12.4 | 25 | 0 | 0 | 79.5 | *139* | 20 | 22.2 | 30 | 141.8 | *74* | 13.1 | 3 |
| June | 18.8 | 9.4 | 14.1 | *−1.5* | 23.1 | 7 | 5.0 | 4 | 12.4 | 2 | 15.9 | 21 | 0 | 0 | 61.9 | *126* | 11 | 17.3 | 17 | 200.0 | *105* | 14.1 | 7 |
| July | 21.6 | 12.9 | 17.2 | *−0.7* | 28.9 | 14 | 7.8 | 27 | 14.1 | 26 | 18.3 | 12 | 0 | 5 | 67.6 | *138* | 14 | 16.9 | 26 | 182.9 | *88* | 13.7 | 14 |
| Aug | 21.6 | 12.1 | 16.9 | *−0.7* | 27.8 | 30 | 7.5 | 27 | 18.1 | 6 | 17.8 | 25 | 0 | 1 | 30.2 | *55* | 14 | 8.4 | 13 | 199.7 | *102* | 12.7 | 1 |
| Sep | 19.6 | 11.0 | 15.3 | *+0.4* | 29.6 | 19 | 2.8 | 27 | 10.9 | 26 | 16.3 | 6 | 0 | 3 | 35.7 | *66* | 12 | 10.6 | 2 | 111.8 | *80* | 10.6 | 13 |
| Oct | 12.6 | 4.5 | 8.6 | *−2.7* | 21.4 | 3 | −4.0 | 19 | 5.3 | 21 | 13.2 | 14 | 7 | 0 | 55.0 | *79* | 15 | 15.2 | 28 | 108.0 | *97* | 9.3 | 18 |
| Nov | 10.0 | 3.6 | 6.8 | *−0.8* | 13.3 | 15 | −3.1 | 1 | 5.3 | 29 | 8.8 | 16 | 5 | 0 | 113.5 | *170* | 26 | 13.4 | 6 | 49.0 | *69* | 7.5 | 12 |
| Dec | 6.9 | 1.3 | 4.1 | *−0.9* | 11.3 | 31 | −1.9 | 16 | 2.7 | 23 | 6.6 | 31 | 10 | 0 | 11.3 | *18* | 9 | 6.3 | 5 | 63.8 | *120* | 6.2 | 15 |
| **1926** | 14.0 | 6.3 | 10.2 | *−0.6* | 29.6 | 9 | −14.3 | 1 | −2.6 | 1 | 18.3 | 7 | 34 | 9 | 695.4 | *105* | 185 | 33.5 | 1 | 1377.0 | *87* | 14.1 | 6 |

## OXFORD 1927

| Month | Mean max | Mean min | Mean temp | Anom | Highest max | Date | Lowest min | Date | Lowest max | Date | Highest min | Date | Air frost | Days ≥ 25 °C | Total pptn | Anom | Rain days | Wettest day | Date | Total sunshine | Anom | Sunniest day | Date |
|---|---|---|---|---|---|---|---|---|---|---|---|---|---|---|---|---|---|---|---|---|---|---|---|
| Jan | 7.7 | 1.8 | 4.8 | *−0.1* | 12.5 | 9 | −3.3 | 20 | 2.6 | 17 | 6.7 | 3 | 7 | 0 | 60.6 | *107* | 21 | 27.4 | 28 | 59.3 | *95* | 5.8 | 31 |
| Feb | 7.0 | 1.3 | 4.1 | *−0.8* | 11.8 | 21 | −3.9 | 11 | −0.9 | 11 | 6.7 | 27 | 12 | 0 | 97.9 | *229* | 15 | 24.1 | 1 | 50.2 | *64* | 7.8 | 2 |
| Mar | 11.5 | 4.4 | 8.0 | *+0.7* | 19.1 | 21 | −1.3 | 15 | 4.9 | 14 | 9.9 | 22 | 1 | 0 | 60.4 | *127* | 19 | 12.7 | 24 | 133.6 | *120* | 9.3 | 28 |
| Apr | 13.1 | 4.5 | 8.8 | *−0.5* | 19.6 | 21 | −2.2 | 30 | 8.7 | 7 | 9.0 | 22 | 1 | 0 | 44.1 | *90* | 12 | 11.8 | 5 | 179.6 | *112* | 13.4 | 30 |
| May | 17.6 | 6.8 | 12.2 | *−0.2* | 23.6 | 7 | −0.9 | 1 | 11.3 | 27 | 11.6 | 5 | 1 | 0 | 21.4 | *37* | 9 | 5.8 | 26 | 215.5 | *112* | 12.5 | 24 |
| June | 17.9 | 8.8 | 13.4 | *−2.2* | 26.5 | 16 | 4.1 | 4 | 14.9 | 11 | 12.6 | 17 | 0 | 1 | 81.1 | *165* | 19 | 12.7 | 18 | 170.4 | *89* | 13.9 | 22 |
| July | 19.6 | 12.6 | 16.1 | *−1.8* | 25.8 | 10 | 8.4 | 18 | 15.2 | 14 | 15.6 | 26 | 0 | 1 | 78.4 | *161* | 17 | 23.4 | 11 | 105.5 | *51* | 11.5 | 10 |
| Aug | 20.0 | 12.2 | 16.1 | *−1.5* | 24.7 | 31 | 8.1 | 25 | 15.9 | 15 | 15.5 | 8 | 0 | 0 | 105.4 | *192* | 22 | 12.2 | 15 | 152.7 | *78* | 13.4 | 3 |
| Sep | 16.4 | 9.6 | 13.0 | *−1.9* | 22.2 | 2 | 1.1 | 27 | 12.3 | 23 | 15.0 | 1 | 0 | 0 | 133.8 | *247* | 19 | 39.4 | 14 | 96.5 | *69* | 9.7 | 26 |
| Oct | 15.0 | 6.8 | 10.9 | *−0.4* | 19.2 | 7 | 0.5 | 24 | 10.3 | 12 | 14.8 | 27 | 0 | 0 | 48.1 | *69* | 16 | 15.5 | 30 | 96.5 | *87* | 10.3 | 3 |
| Nov | 9.5 | 3.5 | 6.5 | *−1.0* | 17.2 | 3 | −2.3 | 8 | 3.4 | 8 | 12.8 | 3 | 8 | 0 | 67.1 | *101* | 16 | 12.4 | 28 | 54.9 | *77* | 7.8 | 11 |
| Dec | 4.0 | 0.3 | 2.1 | *−2.9* | 12.0 | 6 | −7.8 | 19 | −2.6 | 19 | 8.0 | 23 | 12 | 0 | 87.0 | *138* | 11 | 34.1 | 25 | 25.7 | *48* | 6.7 | 28 |
| **1927** | 13.3 | 6.1 | 9.7 | *−1.1* | 26.5 | 6 | −7.8 | 12 | −2.6 | 12 | 15.6 | 7 | 42 | 2 | 885.3 | *134* | 196 | 39.4 | 9 | 1340.1 | *85* | 13.9 | 6 |

## OXFORD 1928

| Month | Mean max | Mean min | Mean temp | Anom | Highest max | Date | Lowest min | Date | Lowest max | Date | Highest min | Date | Air frost | Days ≥ 25 °C | Total pptn | Anom | Rain days | Wettest day | Date | Total sunshine | Anom | Sunniest day | Date |
|---|---|---|---|---|---|---|---|---|---|---|---|---|---|---|---|---|---|---|---|---|---|---|---|
| Jan | 9.0 | 2.2 | 5.6 | *+0.7* | 13.2 | 6 | −4.4 | 1 | 5.2 | 1 | 7.0 | 8 | 3 | 0 | 97.3 | *171* | 23 | 18.2 | 1 | 70.1 | *113* | 7.0 | 19 |
| Feb | 10.3 | 2.2 | 6.2 | *+1.3* | 13.0 | 13 | −1.2 | 20 | 5.6 | 24 | 10.7 | 16 | 7 | 0 | 44.9 | *105* | 12 | 9.9 | 12 | 91.7 | *116* | 9.7 | 27 |
| Mar | 10.6 | 3.4 | 7.0 | *−0.3* | 18.4 | 4 | −3.9 | 12 | 1.6 | 12 | 10.1 | 20 | 6 | 0 | 39.0 | *82* | 18 | 8.4 | 5 | 89.7 | *81* | 9.5 | 26 |
| Apr | 13.0 | 4.5 | 8.8 | *−0.5* | 22.0 | 26 | −1.8 | 17 | 5.0 | 15 | 12.2 | 10 | 3 | 0 | 22.5 | *46* | 12 | 4.4 | 3 | 129.2 | *80* | 12.9 | 24 |
| May | 16.5 | 6.4 | 11.4 | *−1.0* | 24.9 | 28 | −0.2 | 10 | 9.7 | 22 | 13.8 | 29 | 1 | 0 | 14.8 | *26* | 12 | 3.9 | 3 | 167.8 | *87* | 13.7 | 6 |
| June | 18.6 | 9.1 | 13.8 | *−1.8* | 24.2 | 13 | 4.3 | 3 | 15.0 | 16 | 12.9 | 22 | 0 | 0 | 45.1 | *92* | 15 | 14.0 | 13 | 220.8 | *116* | 15.3 | 2 |
| July | 23.2 | 12.4 | 17.8 | *−0.1* | 30.7 | 15 | 8.3 | 29 | 16.7 | 3 | 17.8 | 25 | 0 | 12 | 48.0 | *98* | 8 | 18.6 | 27 | 247.1 | *119* | 15.4 | 12 |
| Aug | 20.2 | 11.7 | 16.0 | *−1.6* | 24.3 | 11 | 7.3 | 19 | 15.3 | 3 | 16.1 | 12 | 0 | 0 | 33.3 | *61* | 12 | 6.1 | 14 | 165.4 | *84* | 11.9 | 30 |
| Sep | 19.2 | 7.9 | 13.5 | *−1.4* | 27.4 | 8 | 2.7 | 27 | 12.5 | 28 | 15.0 | 9 | 0 | 3 | 16.0 | *30* | 7 | 7.6 | 9 | 204.8 | *146* | 12.2 | 4 |
| Oct | 14.8 | 6.8 | 10.8 | *−0.5* | 19.3 | 8 | −0.7 | 1 | 11.3 | 26 | 12.8 | 8 | 1 | 0 | 83.9 | *120* | 19 | 14.7 | 22 | 122.2 | *110* | 10.5 | 1 |
| Nov | 11.5 | 4.6 | 8.0 | *+0.5* | 15.6 | 13 | −3.4 | 10 | 5.6 | 28 | 12.2 | 13 | 2 | 0 | 70.8 | *106* | 19 | 12.2 | 21 | 67.6 | *95* | 7.3 | 17 |
| Dec | 6.6 | 0.4 | 3.5 | *−1.5* | 12.6 | 5 | −6.4 | 15 | −1.8 | 14 | 9.9 | 1 | 9 | 0 | 55.8 | *88* | 15 | 14.1 | 27 | 41.2 | *77* | 6.7 | 7 |
| **1928** | 14.4 | 5.9 | 10.2 | *−0.6* | 30.7 | 7 | −6.4 | 12 | −1.8 | 12 | 17.8 | 7 | 32 | 15 | 571.4 | *87* | 172 | 18.6 | 7 | 1617.4 | *103* | 15.4 | 7 |

## OXFORD 1929

| Month | Mean max | Mean min | Mean temp | Anom | Highest max | Date | Lowest min | Date | Lowest max | Date | Highest min | Date | Air frost | Days ≥ 25 °C | Total pptn | Anom | Rain days | Wettest day | Date | Total sunshine | Anom | Sunniest day | Date |
|---|---|---|---|---|---|---|---|---|---|---|---|---|---|---|---|---|---|---|---|---|---|---|---|
| Jan | 3.7 | −1.1 | 1.3 | *−3.6* | 11.9 | 30 | −5.0 | 13 | −0.7 | 6 | 7.2 | 31 | 21 | 0 | 29.8 | *52* | 14 | 8.7 | 31 | 43.8 | *70* | 6.4 | 16 |
| Feb | 3.3 | −3.0 | 0.2 | *−4.7* | 11.1 | 1 | −11.9 | 15 | −3.8 | 13 | 7.2 | 1 | 19 | 0 | 14.1 | *33* | 8 | 3.4 | 9 | 60.6 | *77* | 8.9 | 28 |
| Mar | 12.9 | 0.0 | 6.4 | *−0.8* | 22.0 | 30 | −6.9 | 2 | 4.3 | 1 | 8.5 | 22 | 17 | 0 | 1.4 | *3* | 3 | 0.6 | 25 | 190.2 | *171* | 10.9 | 30 |
| Apr | 11.8 | 2.5 | 7.1 | *−2.1* | 22.4 | 19 | −3.6 | 6 | 4.4 | 12 | 6.7 | 28 | 6 | 0 | 34.3 | *70* | 11 | 12.9 | 27 | 144.7 | *90* | 12.5 | 19 |
| May | 17.1 | 6.4 | 11.7 | *−0.7* | 26.7 | 23 | −1.4 | 2 | 10.8 | 1 | 12.9 | 24 | 2 | 1 | 28.4 | *50* | 9 | 9.7 | 4 | 240.9 | *125* | 15.0 | 20 |
| June | 18.9 | 8.8 | 13.9 | *−1.7* | 25.1 | 19 | 4.1 | 5 | 13.6 | 5 | 14.8 | 23 | 0 | 1 | 24.3 | *49* | 11 | 8.3 | 5 | 210.3 | *110* | 15.1 | 18 |
| July | 22.7 | 11.2 | 17.0 | *−0.9* | 30.3 | 16 | 5.9 | 8 | 14.8 | 1 | 16.3 | 21 | 0 | 10 | 34.8 | *71* | 10 | 9.9 | 28 | 219.7 | *106* | 15.2 | 15 |
| Aug | 21.2 | 11.6 | 16.4 | *−1.2* | 27.4 | 27 | 7.7 | 26 | 16.2 | 3 | 16.1 | 24 | 0 | 2 | 25.4 | *46* | 11 | 6.4 | 3 | 176.4 | *90* | 12.7 | 26 |
| Sep | 23.2 | 10.7 | 17.0 | *+2.1* | 30.0 | 4 | 2.6 | 26 | 17.1 | 30 | 13.9 | 5 | 0 | 8 | 2.6 | *5* | 2 | 1.4 | 30 | 199.1 | *142* | 11.7 | 8 |
| Oct | 14.1 | 6.3 | 10.2 | *−1.1* | 17.7 | 12 | −1.7 | 27 | 8.9 | 27 | 11.4 | 14 | 2 | 0 | 77.7 | *111* | 14 | 15.9 | 24 | 109.2 | *98* | 9.1 | 21 |
| Nov | 11.0 | 3.3 | 7.1 | *−0.4* | 15.7 | 8 | −3.9 | 15 | 4.6 | 15 | 9.3 | 30 | 9 | 0 | 165.9 | *249* | 19 | 22.6 | 23 | 78.7 | *111* | 7.4 | 3 |
| Dec | 9.2 | 3.4 | 6.3 | *+1.3* | 13.1 | 13 | −1.8 | 18 | 3.4 | 18 | 8.7 | 14 | 5 | 0 | 131.7 | *209* | 23 | 14.0 | 23 | 67.0 | *126* | 6.5 | 19 |
| **1929** | 14.1 | 5.0 | 9.5 | *−1.2* | 30.3 | 7 | −11.9 | 2 | −3.8 | 2 | 16.3 | 7 | 81 | 22 | 570.4 | *86* | 135 | 22.6 | 11 | 1740.3 | *110* | 15.2 | 7 |

## OXFORD 1930

| Month | Mean max | Mean min | Mean temp | Anom | Highest max | Date | Lowest min | Date | Lowest max | Date | Highest min | Date | Air frost | Days ≥ 25 °C | Total pptn | Anom | Rain days | Wettest day | Date | Total sunshine | Anom | Sunniest day | Date |
|---|---|---|---|---|---|---|---|---|---|---|---|---|---|---|---|---|---|---|---|---|---|---|---|
| Jan | 9.0 | 3.4 | 6.2 | *+1.3* | 14.7 | 19 | −0.3 | 26 | 5.4 | 28 | 8.3 | 19 | 1 | 0 | 73.0 | *128* | 20 | 10.1 | 10 | 49.9 | *80* | 6.3 | 6 |
| Feb | 5.4 | 0.3 | 2.9 | *−2.0* | 9.7 | 27 | −3.4 | 25 | 1.6 | 7 | 4.6 | 2 | 13 | 0 | 13.7 | *32* | 10 | 3.6 | 1 | 54.4 | *69* | 8.7 | 9 |
| Mar | 9.8 | 1.8 | 5.8 | *−1.5* | 14.4 | 26 | −5.6 | 20 | 5.7 | 19 | 7.3 | 28 | 10 | 0 | 31.0 | *65* | 10 | 9.5 | 9 | 109.7 | *99* | 11.2 | 29 |
| Apr | 12.5 | 5.0 | 8.7 | *−0.6* | 19.7 | 25 | −0.7 | 22 | 7.4 | 19 | 9.4 | 2 | 1 | 0 | 55.1 | *113* | 17 | 9.6 | 3 | 102.0 | *63* | 13.8 | 30 |
| May | 15.6 | 7.0 | 11.3 | *−1.2* | 20.3 | 27 | 1.9 | 1 | 11.0 | 6 | 10.9 | 30 | 0 | 0 | 54.9 | *96* | 20 | 13.5 | 31 | 134.5 | *70* | 13.0 | 1 |
| June | 21.1 | 10.8 | 16.0 | *+0.4* | 27.6 | 30 | 6.2 | 8 | 15.8 | 1 | 15.6 | 21 | 0 | 3 | 20.1 | *41* | 8 | 6.2 | 30 | 211.2 | *111* | 14.5 | 6 |
| July | 20.0 | 11.7 | 15.9 | *−2.0* | 25.4 | 5 | 9.0 | 12 | 12.9 | 23 | 16.8 | 1 | 0 | 1 | 79.5 | *163* | 15 | 15.3 | 14 | 174.2 | *84* | 15.1 | 5 |
| Aug | 21.2 | 12.0 | 16.6 | *−1.0* | 31.8 | 28 | 7.9 | 16 | 16.9 | 5 | 17.2 | 30 | 0 | 5 | 51.9 | *95* | 13 | 8.9 | 20 | 207.5 | *106* | 13.3 | 16 |
| Sep | 17.9 | 10.4 | 14.2 | *−0.7* | 22.8 | 2 | 6.9 | 26 | 11.6 | 29 | 14.4 | 24 | 0 | 0 | 90.8 | *168* | 16 | 18.5 | 29 | 108.2 | *77* | 10.5 | 2 |
| Oct | 14.6 | 7.5 | 11.0 | *−0.2* | 18.6 | 16 | 1.1 | 27 | 10.0 | 27 | 13.9 | 15 | 0 | 0 | 32.0 | *46* | 19 | 5.6 | 22 | 113.5 | *102* | 8.4 | 6 |
| Nov | 10.2 | 2.8 | 6.5 | *−1.0* | 14.4 | 8 | −6.2 | 17 | 1.7 | 17 | 9.1 | 22 | 8 | 0 | 106.1 | *159* | 18 | 23.1 | 28 | 68.7 | *97* | 8.4 | 4 |
| Dec | 7.0 | 1.9 | 4.4 | *−0.6* | 11.1 | 12 | −2.8 | 10 | 1.1 | 5 | 6.7 | 1 | 9 | 0 | 76.8 | *122* | 21 | 11.7 | 10 | 23.3 | *44* | 4.3 | 14 |
| **1930** | 13.7 | 6.2 | 10.0 | *−0.8* | 31.8 | 8 | −6.2 | 11 | 1.1 | 12 | 17.2 | 8 | 42 | 9 | 684.9 | *104* | 187 | 23.1 | 11 | 1356.8 | *86* | 15.1 | 7 |

# OXFORD 1931

| Month | Mean max | Mean min | Mean temp | Anom | Highest max | Date | Lowest min | Date | Lowest max | Date | Highest min | Date | Air frost | Days ≥ 25 °C | Total pptn | Anom | Rain days | Wettest day | Date | Total sunshine | Anom | Sunniest day | Date |
|---|---|---|---|---|---|---|---|---|---|---|---|---|---|---|---|---|---|---|---|---|---|---|---|
| Jan | 6.5 | 0.8 | 3.7 | *−1.2* | 11.1 | 16 | −5.0 | 7 | 1.1 | 7 | 7.8 | 20 | 14 | 0 | 36.2 | *64* | 21 | 4.6 | 11 | 63.7 | *102* | 7.2 | 25 |
| Feb | 7.4 | 1.3 | 4.3 | *−0.6* | 11.7 | 9 | −1.7 | 5 | 2.2 | 17 | 7.8 | 10 | 6 | 0 | 43.7 | *102* | 20 | 9.4 | 27 | 72.0 | *91* | 7.7 | 14 |
| Mar | 9.2 | 0.3 | 4.7 | *−2.5* | 18.3 | 20 | −6.1 | 10 | 0.6 | 7 | 9.4 | 20 | 16 | 0 | 3.7 | *8* | 6 | 1.8 | 18 | 142.3 | *128* | 9.7 | 27 |
| Apr | 11.8 | 4.7 | 8.2 | *−1.1* | 18.9 | 11 | 1.1 | 18 | 6.1 | 19 | 8.9 | 9 | 0 | 0 | 75.8 | *155* | 21 | 19.0 | 2 | 93.5 | *58* | 12.3 | 13 |
| May | 16.1 | 7.8 | 11.9 | *−0.5* | 23.9 | 27 | 1.7 | 21 | 11.1 | 19 | 12.8 | 28 | 0 | 0 | 72.8 | *127* | 17 | 16.8 | 27 | 150.1 | *78* | 14.2 | 25 |
| June | 19.5 | 11.0 | 15.3 | *−0.3* | 23.3 | 22 | 5.0 | 26 | 14.4 | 24 | 15.0 | 12 | 0 | 0 | 89.7 | *182* | 14 | 22.9 | 5 | 158.7 | *83* | 14.2 | 29 |
| July | 19.6 | 12.1 | 15.9 | *−2.0* | 22.8 | 1 | 6.7 | 21 | 16.7 | 8 | 14.4 | 12 | 0 | 0 | 80.6 | *165* | 19 | 17.0 | 25 | 127.9 | *62* | 12.5 | 3 |
| Aug | 19.0 | 11.2 | 15.1 | *−2.5* | 23.9 | 4 | 6.1 | 24 | 13.9 | 24 | 15.6 | 3 | 0 | 0 | 90.4 | *165* | 15 | 28.4 | 4 | 135.5 | *69* | 11.5 | 15 |
| Sep | 15.7 | 8.3 | 12.0 | *−2.8* | 20.6 | 16 | 1.7 | 8 | 12.2 | 5 | 14.4 | 2 | 0 | 0 | 62.9 | *116* | 13 | 24.1 | 2 | 92.3 | *66* | 10.3 | 8 |
| Oct | 13.4 | 5.2 | 9.3 | *−2.0* | 18.3 | 2 | −3.3 | 27 | 8.3 | 24 | 14.4 | 5 | 8 | 0 | 18.1 | *26* | 9 | 7.1 | 6 | 102.5 | *92* | 8.7 | 26 |
| Nov | 11.0 | 5.1 | 8.0 | *+0.5* | 15.0 | 3 | −0.6 | 30 | 3.9 | 29 | 13.3 | 4 | 1 | 0 | 90.6 | *136* | 19 | 15.5 | 3 | 62.4 | *88* | 7.0 | 12 |
| Dec | 7.8 | 2.8 | 5.3 | *+0.3* | 14.4 | 4 | −5.0 | 19 | 1.1 | 18 | 8.9 | 26 | 6 | 0 | 28.8 | *46* | 10 | 11.2 | 5 | 38.5 | *72* | 6.8 | 7 |
| **1931** | 13.1 | 5.9 | 9.5 | *−1.3* | 23.9 | 5 | −6.1 | 3 | 0.6 | 3 | 15.6 | 8 | 51 | 0 | 693.3 | *105* | 184 | 28.4 | 8 | 1239.4 | *79* | 14.2 | 5 |

# OXFORD 1932

| Month | Mean max | Mean min | Mean temp | Anom | Highest max | Date | Lowest min | Date | Lowest max | Date | Highest min | Date | Air frost | Days ≥ 25 °C | Total pptn | Anom | Rain days | Wettest day | Date | Total sunshine | Anom | Sunniest day | Date |
|---|---|---|---|---|---|---|---|---|---|---|---|---|---|---|---|---|---|---|---|---|---|---|---|
| Jan | 9.4 | 3.5 | 6.4 | *+1.6* | 13.3 | 6 | −2.2 | 1 | 3.3 | 28 | 10.6 | 3 | 5 | 0 | 38.3 | *67* | 14 | 7.9 | 6 | 51.0 | *82* | 6.1 | 8 |
| Feb | 5.8 | −0.3 | 2.7 | *−2.2* | 10.0 | 22 | −3.9 | 10 | −1.1 | 10 | 5.0 | 23 | 17 | 0 | 3.5 | *8* | 7 | 0.8 | 1 | 57.9 | *73* | 8.9 | 28 |
| Mar | 9.6 | 0.4 | 5.0 | *−2.3* | 13.3 | 20 | −5.6 | 13 | 5.0 | 12 | 7.2 | 30 | 15 | 0 | 54.4 | *114* | 11 | 17.8 | 28 | 133.4 | *120* | 10.3 | 12 |
| Apr | 11.3 | 3.8 | 7.6 | *−1.7* | 17.2 | 30 | −0.6 | 13 | 6.1 | 16 | 7.8 | 10 | 1 | 0 | 62.1 | *127* | 23 | 9.4 | 3 | 110.9 | *69* | 9.7 | 12 |
| May | 14.8 | 7.5 | 11.1 | *−1.3* | 21.7 | 20 | 0.6 | 6 | 8.3 | 4 | 13.9 | 21 | 0 | 0 | 139.1 | *243* | 24 | 32.8 | 21 | 112.4 | *58* | 12.1 | 17 |
| June | 19.2 | 9.5 | 14.3 | *−1.2* | 25.6 | 27 | 3.3 | 6 | 11.7 | 4 | 14.4 | 28 | 0 | 1 | 14.6 | *30* | 5 | 8.4 | 30 | 199.3 | *104* | 15.5 | 17 |
| July | 20.5 | 12.7 | 16.6 | *−1.2* | 27.9 | 10 | 7.1 | 19 | 16.0 | 18 | 16.1 | 13 | 0 | 3 | 61.6 | *126* | 19 | 27.7 | 24 | 124.0 | *60* | 11.7 | 9 |
| Aug | 22.9 | 13.6 | 18.2 | *+0.7* | 35.1 | 19 | 9.9 | 23 | 18.2 | 22 | 17.8 | 20 | 0 | 7 | 64.7 | *118* | 8 | 35.6 | 20 | 178.3 | *91* | 12.7 | 11 |
| Sep | 17.7 | 9.7 | 13.7 | *−1.2* | 24.4 | 14 | 2.2 | 24 | 10.0 | 23 | 17.2 | 3 | 0 | 0 | 66.0 | *122* | 20 | 14.7 | 18 | 102.1 | *73* | 10.7 | 21 |
| Oct | 13.1 | 5.9 | 9.5 | *−1.8* | 15.6 | 7 | −0.6 | 29 | 7.2 | 28 | 12.8 | 1 | 1 | 0 | 120.8 | *173* | 24 | 18.8 | 20 | 100.7 | *90* | 8.9 | 4 |
| Nov | 9.6 | 4.0 | 6.8 | *−0.7* | 15.0 | 3 | −1.1 | 10 | 3.9 | 18 | 11.1 | 3 | 1 | 0 | 31.5 | *47* | 15 | 6.6 | 22 | 55.7 | *78* | 8.1 | 7 |
| Dec | 8.2 | 3.5 | 5.9 | *+0.8* | 12.8 | 18 | −1.7 | 6 | 2.8 | 11 | 10.0 | 18 | 4 | 0 | 17.7 | *28* | 13 | 9.1 | 30 | 58.0 | *109* | 5.9 | 19 |
| **1932** | 13.5 | 6.1 | 9.8 | *−0.9* | 35.1 | 8 | −5.6 | 3 | −1.1 | 2 | 17.8 | 8 | 44 | 11 | 674.3 | *102* | 183 | 35.6 | 8 | 1283.7 | *81* | 15.5 | 6 |

## OXFORD 1933

| Month | Mean max | Mean min | Mean temp | Anom | Highest max | Date | Lowest min | Date | Lowest max | Date | Highest min | Date | Air frost | Days ≥ 25 °C | Total pptn | Anom | Rain days | Wettest day | Date | Total sunshine | Anom | Sunniest day | Date |
|---|---|---|---|---|---|---|---|---|---|---|---|---|---|---|---|---|---|---|---|---|---|---|---|
| Jan | 5.3 | −0.7 | 2.3 | *−2.6* | 11.7 | 2 | −6.7 | 23 | −1.1 | 24 | 8.9 | 3 | 19 | 0 | 44.5 | *78* | 18 | 7.1 | 15 | 69.5 | *112* | 7.7 | 26 |
| Feb | 7.9 | 1.4 | 4.6 | *−0.2* | 13.3 | 5 | −3.9 | 20 | 2.8 | 19 | 10.0 | 9 | 14 | 0 | 88.6 | *207* | 15 | 30.7 | 25 | 94.3 | *119* | 8.7 | 23 |
| Mar | 12.9 | 3.3 | 8.1 | *+0.9* | 19.4 | 28 | −1.7 | 26 | 8.9 | 1 | 7.8 | 4 | 4 | 0 | 60.6 | *127* | 16 | 17.3 | 16 | 187.6 | *169* | 11.7 | 27 |
| Apr | 14.4 | 4.5 | 9.5 | *+0.2* | 20.6 | 7 | −2.2 | 19 | 8.9 | 19 | 10.0 | 27 | 2 | 0 | 25.3 | *52* | 6 | 7.1 | 25 | 152.5 | *95* | 11.9 | 14 |
| May | 17.7 | 8.1 | 12.9 | *+0.5* | 25.0 | 22 | 3.9 | 15 | 11.7 | 2 | 12.2 | 20 | 0 | 1 | 44.9 | *79* | 16 | 8.4 | 2 | 170.2 | *88* | 13.3 | 23 |
| June | 21.4 | 10.6 | 16.0 | *+0.4* | 29.4 | 5 | 7.2 | 12 | 15.0 | 18 | 14.4 | 6 | 0 | 7 | 34.1 | *69* | 12 | 10.2 | 20 | 226.9 | *119* | 15.0 | 5 |
| July | 23.9 | 13.6 | 18.7 | *+0.8* | 31.1 | 27 | 9.4 | 1 | 18.3 | 13 | 16.1 | 4 | 0 | 15 | 47.5 | *97* | 11 | 13.7 | 10 | 237.6 | *115* | 14.8 | 5 |
| Aug | 24.5 | 12.7 | 18.6 | *+1.0* | 31.7 | 6 | 7.8 | 31 | 18.3 | 21 | 17.8 | 3 | 0 | 13 | 20.5 | *37* | 6 | 7.1 | 11 | 242.7 | *124* | 12.7 | 26 |
| Sep | 20.6 | 10.7 | 15.7 | *+0.8* | 26.7 | 4 | 3.3 | 16 | 12.8 | 24 | 16.7 | 18 | 0 | 4 | 52.8 | *97* | 14 | 15.2 | 12 | 177.3 | *126* | 11.2 | 5 |
| Oct | 14.0 | 6.7 | 10.3 | *−1.0* | 19.4 | 7 | −1.1 | 28 | 6.7 | 27 | 13.3 | 1 | 1 | 0 | 36.3 | *52* | 16 | 6.6 | 10 | 101.3 | *91* | 9.4 | 4 |
| Nov | 8.7 | 2.8 | 5.8 | *−1.8* | 15.6 | 20 | −2.8 | 13 | 4.4 | 26 | 7.8 | 20 | 6 | 0 | 28.5 | *43* | 16 | 10.7 | 14 | 53.9 | *76* | 6.9 | 14 |
| Dec | 3.5 | −0.7 | 1.4 | *−3.6* | 7.8 | 22 | −5.0 | 6 | −0.6 | 9 | 3.3 | 2 | 16 | 0 | 7.8 | *12* | 8 | 3.3 | 30 | 59.0 | *111* | 6.8 | 4 |
| **1933** | 14.6 | 6.1 | 10.3 | *−0.4* | 31.7 | 8 | −6.7 | 1 | −1.1 | 1 | 17.8 | 8 | 62 | 40 | 491.4 | *74* | 154 | 30.7 | 2 | 1772.8 | *112* | 15.0 | 6 |

## OXFORD 1934

| Month | Mean max | Mean min | Mean temp | Anom | Highest max | Date | Lowest min | Date | Lowest max | Date | Highest min | Date | Air frost | Days ≥ 25 °C | Total pptn | Anom | Rain days | Wettest day | Date | Total sunshine | Anom | Sunniest day | Date |
|---|---|---|---|---|---|---|---|---|---|---|---|---|---|---|---|---|---|---|---|---|---|---|---|
| Jan | 7.5 | 0.6 | 4.0 | *−0.8* | 12.2 | 17 | −6.1 | 24 | 1.1 | 24 | 9.4 | 18 | 12 | 0 | 47.8 | *84* | 20 | 7.6 | 11 | 65.9 | *106* | 7.0 | 20 |
| Feb | 7.5 | −0.5 | 3.5 | *−1.4* | 12.2 | 16 | −4.4 | 3 | 2.8 | 2 | 3.9 | 5 | 14 | 0 | 7.7 | *18* | 5 | 5.6 | 24 | 96.6 | *122* | 8.7 | 22 |
| Mar | 9.6 | 1.3 | 5.4 | *−1.8* | 15.6 | 25 | −3.9 | 1 | 6.1 | 6 | 6.1 | 25 | 8 | 0 | 45.2 | *95* | 16 | 9.7 | 11 | 122.5 | *110* | 11.3 | 31 |
| Apr | 12.7 | 4.5 | 8.6 | *−0.7* | 22.8 | 15 | −1.1 | 8 | 4.4 | 6 | 10.0 | 15 | 3 | 0 | 53.0 | *108* | 17 | 12.2 | 11 | 124.9 | *78* | 11.0 | 20 |
| May | 17.6 | 6.9 | 12.2 | *−0.2* | 26.1 | 12 | 0.0 | 17 | 12.8 | 15 | 12.2 | 22 | 0 | 1 | 18.0 | *31* | 7 | 9.7 | 6 | 216.3 | *112* | 14.1 | 30 |
| June | 21.1 | 10.6 | 15.8 | *+0.2* | 28.9 | 17 | 6.7 | 12 | 16.1 | 5 | 15.0 | 16 | 0 | 6 | 36.6 | *74* | 11 | 9.9 | 23 | 192.7 | *101* | 14.4 | 17 |
| July | 25.3 | 12.9 | 19.1 | *+1.3* | 30.0 | 8 | 10.0 | 25 | 20.0 | 24 | 16.7 | 12 | 0 | 17 | 24.0 | *49* | 8 | 15.7 | 24 | 269.3 | *130* | 15.5 | 8 |
| Aug | 21.1 | 11.3 | 16.2 | *−1.4* | 26.1 | 18 | 5.0 | 31 | 16.7 | 29 | 15.6 | 17 | 0 | 1 | 46.2 | *84* | 13 | 14.2 | 28 | 184.9 | *94* | 12.7 | 26 |
| Sep | 20.2 | 10.7 | 15.4 | *+0.5* | 27.2 | 14 | 6.1 | 22 | 15.6 | 24 | 16.1 | 8 | 0 | 3 | 61.4 | *113* | 15 | 17.3 | 2 | 149.1 | *106* | 10.2 | 5 |
| Oct | 14.5 | 7.6 | 11.1 | *−0.2* | 18.9 | 1 | −1.7 | 31 | 3.3 | 31 | 13.3 | 1 | 1 | 0 | 34.9 | *50* | 17 | 6.9 | 3 | 81.5 | *73* | 8.3 | 22 |
| Nov | 9.3 | 3.9 | 6.6 | *−0.9* | 14.4 | 27 | −2.8 | 1 | 6.1 | 5 | 9.4 | 28 | 3 | 0 | 41.4 | *62* | 14 | 17.3 | 9 | 48.7 | *69* | 6.5 | 3 |
| Dec | 10.5 | 5.9 | 8.2 | *+3.2* | 13.3 | 3 | 1.1 | 24 | 5.6 | 24 | 11.7 | 4 | 0 | 0 | 142.2 | *225* | 28 | 17.0 | 13 | 31.3 | *59* | 5.1 | 17 |
| **1934** | 14.7 | 6.3 | 10.5 | *−0.2* | 30.0 | 7 | −6.1 | 1 | 1.1 | 1 | 16.7 | 7 | 41 | 28 | 558.4 | *85* | 171 | 17.3 | 9 | 1583.7 | *100* | 15.5 | 7 |

## OXFORD 1935

| Month | Mean max | Mean min | Mean temp | Anom | Highest max | Date | Lowest min | Date | Lowest max | Date | Highest min | Date | Air frost | Days ≥ 25 °C | Total pptn | Anom | Rain days | Wettest day | Date | Total sunshine | Anom | Sunniest day | Date |
|---|---|---|---|---|---|---|---|---|---|---|---|---|---|---|---|---|---|---|---|---|---|---|---|
| Jan | 7.2 | 2.5 | 4.9 | *+0.0* | 12.2 | 1 | −3.9 | 10 | 2.8 | 8 | 9.4 | 3 | 5 | 0 | 13.9 | *24* | 7 | 6.6 | 11 | 48.1 | *77* | 5.3 | 27 |
| Feb | 9.3 | 2.9 | 6.1 | *+1.2* | 13.9 | 2 | −3.9 | 9 | 2.2 | 9 | 8.3 | 20 | 4 | 0 | 52.0 | *121* | 16 | 14.5 | 5 | 62.0 | *79* | 9.3 | 26 |
| Mar | 11.1 | 2.8 | 6.9 | *−0.3* | 18.3 | 20 | −3.3 | 9 | 2.8 | 9 | 8.3 | 22 | 6 | 0 | 11.4 | *24* | 9 | 4.1 | 16 | 121.5 | *109* | 10.4 | 12 |
| Apr | 12.6 | 4.4 | 8.5 | *−0.8* | 17.2 | 23 | −1.1 | 5 | 7.8 | 3 | 11.1 | 10 | 3 | 0 | 91.8 | *188* | 18 | 11.7 | 7 | 127.3 | *79* | 9.5 | 6 |
| May | 15.1 | 5.4 | 10.3 | *−2.2* | 23.3 | 6 | −1.7 | 18 | 8.9 | 16 | 10.0 | 28 | 2 | 0 | 35.9 | *63* | 10 | 8.9 | 28 | 188.5 | *98* | 13.9 | 10 |
| June | 20.0 | 11.2 | 15.6 | *+0.0* | 29.4 | 22 | 6.7 | 9 | 15.0 | 5 | 16.1 | 25 | 0 | 6 | 114.6 | *233* | 21 | 28.7 | 25 | 196.3 | *103* | 15.1 | 24 |
| July | 23.9 | 12.7 | 18.3 | *+0.4* | 30.0 | 13 | 7.2 | 31 | 18.9 | 18 | 16.7 | 28 | 0 | 14 | 12.9 | *26* | 6 | 5.8 | 18 | 274.3 | *133* | 14.1 | 31 |
| Aug | 23.3 | 12.1 | 17.7 | *+0.1* | 28.9 | 7 | 6.1 | 28 | 16.1 | 23 | 15.6 | 8 | 0 | 12 | 45.3 | *83* | 7 | 24.9 | 30 | 199.9 | *102* | 13.6 | 10 |
| Sep | 18.7 | 10.2 | 14.5 | *−0.4* | 21.1 | 2 | 4.4 | 26 | 14.4 | 30 | 15.6 | 28 | 0 | 0 | 116.3 | *215* | 21 | 17.8 | 3 | 144.7 | *103* | 10.9 | 7 |
| Oct | 14.0 | 6.0 | 10.0 | *−1.3* | 17.2 | 15 | −2.2 | 21 | 8.3 | 22 | 13.3 | 28 | 3 | 0 | 94.9 | *136* | 19 | 21.3 | 9 | 102.6 | *92* | 8.1 | 7 |
| Nov | 10.4 | 4.3 | 7.3 | *−0.2* | 16.7 | 3 | −1.1 | 25 | 6.1 | 24 | 11.7 | 3 | 2 | 0 | 103.1 | *155* | 23 | 16.0 | 16 | 65.4 | *92* | 6.9 | 23 |
| Dec | 5.6 | 0.9 | 3.3 | *−1.8* | 11.1 | 31 | −9.4 | 24 | −0.6 | 20 | 7.8 | 27 | 12 | 0 | 78.8 | *125* | 21 | 19.3 | 27 | 48.9 | *92* | 6.3 | 2 |
| **1935** | 14.3 | 6.3 | 10.3 | *−0.5* | 30.0 | 7 | −9.4 | 12 | −0.6 | 12 | 16.7 | 7 | 37 | 32 | 770.9 | *117* | 178 | 28.7 | 6 | 1579.5 | *100* | 15.1 | 6 |

## OXFORD 1936

| Month | Mean max | Mean min | Mean temp | Anom | Highest max | Date | Lowest min | Date | Lowest max | Date | Highest min | Date | Air frost | Days ≥ 25 °C | Total pptn | Anom | Rain days | Wettest day | Date | Total sunshine | Anom | Sunniest day | Date |
|---|---|---|---|---|---|---|---|---|---|---|---|---|---|---|---|---|---|---|---|---|---|---|---|
| Jan | 7.1 | 1.7 | 4.4 | *−0.4* | 13.9 | 9 | −6.1 | 19 | 1.1 | 14 | 9.4 | 10 | 12 | 0 | 80.3 | *141* | 22 | 15.5 | 28 | 43.4 | *70* | 5.7 | 21 |
| Feb | 6.2 | −0.3 | 3.0 | *−1.9* | 12.2 | 18 | −6.7 | 13 | 0.0 | 11 | 6.1 | 1 | 15 | 0 | 39.6 | *93* | 12 | 12.4 | 22 | 89.2 | *113* | 8.0 | 4 |
| Mar | 10.9 | 4.1 | 7.5 | *+0.2* | 16.1 | 21 | −2.2 | 4 | 4.4 | 1 | 10.6 | 22 | 4 | 0 | 46.5 | *98* | 17 | 7.1 | 8 | 71.4 | *64* | 8.1 | 5 |
| Apr | 10.7 | 2.9 | 6.8 | *−2.5* | 17.2 | 28 | −1.7 | 23 | 6.1 | 5 | 10.6 | 25 | 4 | 0 | 41.8 | *85* | 15 | 10.2 | 21 | 133.2 | *83* | 12.4 | 18 |
| May | 17.1 | 6.8 | 12.0 | *−0.5* | 25.6 | 18 | 1.1 | 2 | 11.7 | 9 | 13.3 | 16 | 0 | 2 | 10.7 | *19* | 4 | 6.1 | 22 | 179.5 | *93* | 14.5 | 19 |
| June | 20.0 | 10.6 | 15.3 | *−0.3* | 28.3 | 20 | 2.8 | 1 | 13.3 | 3 | 16.7 | 21 | 0 | 4 | 91.6 | *186* | 17 | 15.7 | 29 | 166.2 | *87* | 15.3 | 17 |
| July | 19.6 | 12.0 | 15.8 | *−2.1* | 23.3 | 5 | 7.8 | 22 | 15.6 | 9 | 15.6 | 7 | 0 | 0 | 87.1 | *178* | 22 | 16.0 | 14 | 119.2 | *58* | 9.2 | 8 |
| Aug | 22.2 | 11.7 | 17.0 | *−0.6* | 28.3 | 29 | 7.2 | 23 | 17.2 | 7 | 15.6 | 18 | 0 | 5 | 21.0 | *38* | 6 | 15.0 | 10 | 184.7 | *94* | 12.8 | 23 |
| Sep | 18.7 | 11.3 | 15.0 | *+0.1* | 23.3 | 13 | 2.8 | 29 | 13.3 | 30 | 15.6 | 25 | 0 | 0 | 83.7 | *155* | 16 | 38.1 | 20 | 79.3 | *57* | 8.5 | 28 |
| Oct | 13.8 | 5.2 | 9.5 | *−1.8* | 19.4 | 15 | 0.0 | 4 | 8.3 | 31 | 10.6 | 17 | 0 | 0 | 33.6 | *48* | 11 | 11.2 | 30 | 103.1 | *93* | 10.0 | 3 |
| Nov | 9.0 | 2.2 | 5.6 | *−1.9* | 14.4 | 17 | −2.2 | 24 | −0.6 | 24 | 9.4 | 17 | 10 | 0 | 67.7 | *101* | 17 | 15.2 | 8 | 48.9 | *69* | 5.4 | 9 |
| Dec | 8.5 | 2.2 | 5.4 | *+0.3* | 13.9 | 17 | −3.3 | 7 | 2.2 | 9 | 8.3 | 3 | 8 | 0 | 50.7 | *80* | 20 | 8.4 | 14 | 62.3 | *117* | 6.6 | 4 |
| **1936** | 13.7 | 5.9 | 9.8 | *−1.0* | 28.3 | 6 | −6.7 | 2 | −0.6 | 11 | 16.7 | 6 | 53 | 11 | 654.3 | *99* | 179 | 38.1 | 9 | 1280.4 | *81* | 15.3 | 6 |

## OXFORD 1937

| Month | Mean max | Mean min | Mean temp | Anom | Highest max | Date | Lowest min | Date | Lowest max | Date | Highest min | Date | Air frost | Days ≥ 25 °C | Total pptn | Anom | Rain days | Wettest day | Date | Total sunshine | Anom | Sunniest day | Date |
|---|---|---|---|---|---|---|---|---|---|---|---|---|---|---|---|---|---|---|---|---|---|---|---|
| Jan | 8.4 | 2.5 | 5.4 | *+0.6* | 11.7 | 6 | −2.8 | 29 | −1.7 | 29 | 8.3 | 4 | 6 | 0 | 84.8 | *149* | 22 | 9.1 | 18 | 50.9 | *82* | 6.5 | 10 |
| Feb | 9.5 | 3.5 | 6.5 | *+1.6* | 13.3 | 14 | −1.1 | 28 | 2.2 | 28 | 8.9 | 16 | 2 | 0 | 120.3 | *281* | 24 | 18.3 | 7 | 66.6 | *84* | 8.5 | 23 |
| Mar | 7.7 | 0.8 | 4.2 | *−3.0* | 13.3 | 20 | −2.8 | 10 | 2.8 | 7 | 7.2 | 18 | 12 | 0 | 77.5 | *163* | 16 | 11.7 | 10 | 99.7 | *90* | 10.3 | 15 |
| Apr | 13.5 | 5.8 | 9.7 | *+0.4* | 17.2 | 23 | −1.7 | 1 | 9.4 | 14 | 11.1 | 10 | 1 | 0 | 66.9 | *137* | 17 | 11.9 | 16 | 103.1 | *64* | 13.1 | 25 |
| May | 17.5 | 8.2 | 12.8 | *+0.4* | 26.1 | 29 | 2.2 | 1 | 10.6 | 13 | 14.4 | 26 | 0 | 1 | 66.7 | *117* | 13 | 23.9 | 25 | 185.0 | *96* | 13.3 | 18 |
| June | 20.0 | 10.0 | 15.0 | *−0.6* | 26.7 | 11 | 5.6 | 2 | 16.1 | 19 | 15.0 | 11 | 0 | 1 | 39.8 | *81* | 11 | 10.2 | 12 | 181.3 | *95* | 13.7 | 6 |
| July | 21.2 | 12.6 | 16.9 | *−1.0* | 28.3 | 3 | 7.8 | 8 | 16.7 | 4 | 17.2 | 3 | 0 | 3 | 35.2 | *72* | 11 | 17.3 | 15 | 140.1 | *68* | 13.9 | 16 |
| Aug | 23.6 | 12.5 | 18.1 | *+0.5* | 30.6 | 6 | 8.3 | 15 | 18.3 | 14 | 16.1 | 10 | 0 | 13 | 11.2 | *20* | 3 | 5.6 | 14 | 202.3 | *103* | 12.3 | 6 |
| Sep | 19.3 | 9.1 | 14.2 | *−0.7* | 26.7 | 7 | 4.4 | 23 | 13.9 | 9 | 13.9 | 1 | 0 | 2 | 50.4 | *93* | 13 | 13.0 | 17 | 143.0 | *102* | 10.3 | 5 |
| Oct | 14.8 | 7.2 | 11.0 | *−0.3* | 20.6 | 2 | 1.7 | 18 | 11.1 | 18 | 11.1 | 30 | 0 | 0 | 68.6 | *98* | 12 | 23.9 | 27 | 79.1 | *71* | 6.9 | 16 |
| Nov | 8.6 | 1.6 | 5.1 | *−2.5* | 12.2 | 4 | −6.1 | 21 | 4.4 | 21 | 8.3 | 9 | 11 | 0 | 31.0 | *46* | 10 | 7.1 | 22 | 65.9 | *93* | 6.9 | 11 |
| Dec | 6.0 | 0.7 | 3.3 | *−1.7* | 12.2 | 22 | −3.3 | 18 | 1.7 | 7 | 8.9 | 23 | 14 | 0 | 48.5 | *77* | 21 | 11.4 | 4 | 41.2 | *77* | 5.7 | 23 |
| **1937** | 14.2 | 6.2 | 10.2 | *−0.6* | 30.6 | 8 | −6.1 | 11 | −1.7 | 1 | 17.2 | 7 | 46 | 20 | 700.9 | *106* | 173 | 23.9 | 5 | 1358.2 | *86* | 13.9 | 7 |

## OXFORD 1938

| Month | Mean max | Mean min | Mean temp | Anom | Highest max | Date | Lowest min | Date | Lowest max | Date | Highest min | Date | Air frost | Days ≥ 25 °C | Total pptn | Anom | Rain days | Wettest day | Date | Total sunshine | Anom | Sunniest day | Date |
|---|---|---|---|---|---|---|---|---|---|---|---|---|---|---|---|---|---|---|---|---|---|---|---|
| Jan | 8.8 | 3.4 | 6.1 | *+1.2* | 12.2 | 16 | 0.0 | 11 | 2.8 | 3 | 7.8 | 21 | 0 | 0 | 67.3 | *118* | 20 | 8.6 | 16 | 41.2 | *66* | 6.2 | 24 |
| Feb | 8.2 | 2.0 | 5.1 | *+0.2* | 15.0 | 25 | −2.2 | 25 | 2.8 | 16 | 6.7 | 4 | 6 | 0 | 21.3 | *50* | 12 | 8.1 | 26 | 66.9 | *85* | 7.8 | 11 |
| Mar | 15.5 | 4.3 | 9.9 | *+2.6* | 19.4 | 20 | −1.7 | 7 | 10.6 | 26 | 10.6 | 25 | 5 | 0 | 8.9 | *19* | 3 | 4.6 | 24 | 172.3 | *155* | 10.3 | 14 |
| Apr | 13.0 | 2.5 | 7.8 | *−1.5* | 18.9 | 6 | −2.2 | 11 | 8.3 | 9 | 7.2 | 1 | 5 | 0 | 2.7 | *6* | 5 | 1.0 | 2 | 180.9 | *112* | 12.0 | 12 |
| May | 15.9 | 6.5 | 11.2 | *−1.3* | 21.7 | 22 | −1.7 | 8 | 11.7 | 1 | 11.7 | 15 | 1 | 0 | 50.1 | *88* | 15 | 9.9 | 27 | 144.9 | *75* | 13.9 | 5 |
| June | 20.3 | 10.8 | 15.5 | *−0.1* | 26.1 | 17 | 5.0 | 11 | 14.4 | 2 | 15.6 | 24 | 0 | 2 | 24.3 | *49* | 10 | 7.4 | 10 | 190.6 | *100* | 13.9 | 14 |
| July | 21.0 | 11.7 | 16.4 | *−1.5* | 25.6 | 20 | 6.1 | 2 | 16.1 | 15 | 17.2 | 31 | 0 | 3 | 30.4 | *62* | 14 | 5.6 | 7 | 133.5 | *64* | 11.1 | 2 |
| Aug | 21.9 | 12.5 | 17.2 | *−0.4* | 28.9 | 1 | 6.7 | 30 | 16.1 | 28 | 16.7 | 11 | 0 | 8 | 91.8 | *168* | 14 | 35.6 | 7 | 151.3 | *77* | 9.9 | 20 |
| Sep | 19.0 | 10.1 | 14.5 | *−0.4* | 25.0 | 12 | 3.9 | 16 | 15.6 | 9 | 14.4 | 18 | 0 | 1 | 54.5 | *101* | 15 | 16.3 | 27 | 110.9 | *79* | 11.9 | 2 |
| Oct | 14.7 | 7.0 | 10.9 | *−0.4* | 18.3 | 13 | 0.0 | 25 | 9.4 | 27 | 12.2 | 13 | 0 | 0 | 72.0 | *103* | 19 | 15.5 | 3 | 118.1 | *106* | 8.8 | 21 |
| Nov | 12.9 | 6.9 | 9.9 | *+2.3* | 18.3 | 5 | 0.6 | 27 | 7.8 | 21 | 13.3 | 13 | 0 | 0 | 80.6 | *121* | 18 | 26.7 | 20 | 70.0 | *99* | 6.9 | 22 |
| Dec | 7.2 | 1.5 | 4.4 | *−0.7* | 13.3 | 12 | −6.7 | 20 | −3.3 | 20 | 9.4 | 12 | 11 | 0 | 73.5 | *116* | 22 | 13.5 | 9 | 52.4 | *99* | 5.0 | 18 |
| **1938** | 14.9 | 6.6 | 10.7 | *−0.0* | 28.9 | 8 | −6.7 | 12 | −3.3 | 12 | 17.2 | 7 | 28 | 14 | 577.4 | *87* | 167 | 35.6 | 8 | 1433.0 | *91* | 13.9 | 5 |

## OXFORD 1939

| Month | Mean max | Mean min | Mean temp | Anom | Highest max | Date | Lowest min | Date | Lowest max | Date | Highest min | Date | Air frost | Days ≥ 25 °C | Total pptn | Anom | Rain days | Wettest day | Date | Total sunshine | Anom | Sunniest day | Date |
|---|---|---|---|---|---|---|---|---|---|---|---|---|---|---|---|---|---|---|---|---|---|---|---|
| Jan | 7.3 | 2.3 | 4.8 | *−0.0* | 12.8 | 14 | −4.4 | 6 | 1.7 | 25 | 8.9 | 9 | 6 | 0 | 113.0 | *199* | 22 | 19.3 | 25 | 46.4 | *75* | 6.9 | 13 |
| Feb | 9.5 | 2.8 | 6.2 | *+1.3* | 14.4 | 11 | −2.2 | 3 | 2.2 | 1 | 9.4 | 11 | 4 | 0 | 26.6 | *62* | 12 | 5.3 | 22 | 104.9 | *133* | 8.1 | 6 |
| Mar | 9.6 | 2.8 | 6.2 | *−1.1* | 16.1 | 3 | −1.7 | 10 | 4.4 | 27 | 8.9 | 4 | 4 | 0 | 40.0 | *84* | 15 | 8.4 | 21 | 86.2 | *78* | 9.6 | 7 |
| Apr | 14.1 | 4.9 | 9.5 | *+0.2* | 23.9 | 11 | −0.6 | 26 | 6.7 | 30 | 11.7 | 12 | 1 | 0 | 76.8 | *157* | 16 | 15.5 | 30 | 171.7 | *107* | 12.7 | 20 |
| May | 16.3 | 7.3 | 11.8 | *−0.7* | 23.9 | 24 | 4.4 | 2 | 7.2 | 1 | 12.8 | 10 | 0 | 0 | 30.1 | *53* | 9 | 8.9 | 5 | 184.9 | *96* | 13.9 | 13 |
| June | 19.5 | 9.7 | 14.6 | *−1.0* | 28.9 | 6 | 5.0 | 13 | 12.8 | 24 | 13.9 | 7 | 0 | 5 | 49.9 | *101* | 12 | 21.8 | 15 | 227.9 | *119* | 15.5 | 5 |
| July | 19.8 | 12.1 | 15.9 | *−2.0* | 24.4 | 4 | 7.2 | 25 | 16.7 | 20 | 17.2 | 30 | 0 | 0 | 106.2 | *217* | 21 | 18.8 | 19 | 139.7 | *67* | 12.3 | 9 |
| Aug | 21.5 | 12.4 | 17.0 | *−0.6* | 25.6 | 21 | 8.3 | 14 | 16.1 | 3 | 16.1 | 27 | 0 | 2 | 42.6 | *78* | 12 | 9.1 | 3 | 153.7 | *78* | 11.7 | 17 |
| Sep | 19.3 | 10.8 | 15.0 | *+0.2* | 26.7 | 8 | 2.2 | 28 | 14.4 | 25 | 17.8 | 2 | 0 | 2 | 13.8 | *25* | 6 | 10.9 | 2 | 147.0 | *105* | 10.9 | 8 |
| Oct | 12.5 | 5.2 | 8.9 | *−2.4* | 18.3 | 5 | 0.0 | 26 | 6.1 | 29 | 11.1 | 6 | 0 | 0 | 134.0 | *191* | 17 | 40.6 | 17 | 94.3 | *85* | 8.2 | 2 |
| Nov | 12.2 | 6.5 | 9.4 | *+1.8* | 15.0 | 8 | 0.6 | 25 | 7.8 | 24 | 12.8 | 30 | 0 | 0 | 104.3 | *156* | 22 | 14.0 | 23 | 41.1 | *58* | 5.9 | 20 |
| Dec | 5.6 | 0.6 | 3.1 | *−1.9* | 12.8 | 1 | −6.7 | 30 | 0.0 | 28 | 11.7 | 1 | 11 | 0 | 36.3 | *58* | 16 | 10.2 | 8 | 46.0 | *86* | 5.7 | 2 |
| **1939** | 13.9 | 6.5 | 10.2 | *−0.5* | 28.9 | 6 | −6.7 | 12 | 0.0 | 12 | 17.8 | 9 | 26 | 9 | 773.6 | *117* | 180 | 40.6 | 10 | 1443.8 | *92* | 15.5 | 6 |

## OXFORD 1940

| Month | Mean max | Mean min | Mean temp | Anom | Highest max | Date | Lowest min | Date | Lowest max | Date | Highest min | Date | Air frost | Days ≥ 25 °C | Total pptn | Anom | Rain days | Wettest day | Date | Total sunshine | Anom | Sunniest day | Date |
|---|---|---|---|---|---|---|---|---|---|---|---|---|---|---|---|---|---|---|---|---|---|---|---|
| Jan | 1.9 | −4.6 | −1.3 | *−6.2* | 10.6 | 7 | −10.6 | 21 | −3.9 | 20 | 2.2 | 8 | 26 | 0 | 74.2 | *130* | 12 | 29.2 | 26 | 83.1 | *133* | 7.2 | 12 |
| Feb | 5.6 | 0.2 | 2.9 | *−2.0* | 13.3 | 23 | −8.3 | 18 | −0.6 | 9 | 8.9 | 28 | 13 | 0 | 55.8 | *130* | 17 | 10.4 | 18 | 22.8 | *29* | 6.9 | 24 |
| Mar | 10.8 | 2.5 | 6.7 | *−0.6* | 17.2 | 11 | −4.4 | 7 | 5.6 | 14 | 8.3 | 13 | 7 | 0 | 64.3 | *135* | 14 | 18.0 | 13 | 128.3 | *115* | 11.1 | 27 |
| Apr | 13.9 | 5.0 | 9.4 | *+0.1* | 22.2 | 22 | −2.2 | 11 | 9.4 | 10 | 11.1 | 23 | 5 | 0 | 45.2 | *92* | 18 | 9.7 | 18 | 118.1 | *73* | 11.5 | 6 |
| May | 18.6 | 7.8 | 13.2 | *+0.8* | 23.9 | 25 | 3.9 | 21 | 13.9 | 6 | 11.7 | 26 | 0 | 0 | 38.0 | *66* | 14 | 13.2 | 22 | 215.4 | *112* | 14.7 | 19 |
| June | 22.6 | 11.1 | 16.9 | *+1.3* | 28.3 | 8 | 7.8 | 26 | 13.3 | 22 | 13.9 | 9 | 0 | 6 | 15.3 | *31* | 4 | 8.1 | 22 | 273.4 | *143* | 15.5 | 5 |
| July | 20.5 | 11.1 | 15.8 | *−2.1* | 25.6 | 2 | 7.2 | 14 | 16.1 | 12 | 16.1 | 10 | 0 | 2 | 61.0 | *125* | 18 | 21.3 | 10 | 165.1 | *80* | 12.1 | 1 |
| Aug | 22.1 | 11.1 | 16.6 | *−1.0* | 27.2 | 17 | 6.1 | 24 | 18.3 | 21 | 15.0 | 10 | 0 | 5 | 1.8 | *3* | 2 | 1.0 | 10 | 199.5 | *102* | 12.5 | 12 |
| Sep | 19.2 | 8.6 | 13.9 | *−1.0* | 29.4 | 4 | 3.3 | 26 | 13.3 | 29 | 13.9 | 13 | 0 | 4 | 24.1 | *44* | 9 | 5.6 | 12 | 179.5 | *128* | 11.7 | 6 |
| Oct | 13.9 | 6.0 | 10.0 | *−1.3* | 18.3 | 20 | −2.2 | 12 | 8.3 | 29 | 12.2 | 6 | 1 | 0 | 67.1 | *96* | 17 | 13.7 | 16 | 95.5 | *86* | 8.7 | 11 |
| Nov | 10.3 | 4.3 | 7.3 | *−0.3* | 14.4 | 1 | −3.9 | 30 | 1.7 | 30 | 9.4 | 3 | 2 | 0 | 175.5 | *263* | 20 | 30.7 | 2 | 76.8 | *108* | 7.3 | 14 |
| Dec | 6.6 | 0.8 | 3.7 | *−1.3* | 11.7 | 16 | −5.0 | 13 | 1.1 | 22 | 6.1 | 4 | 11 | 0 | 32.8 | *52* | 17 | 7.9 | 29 | 46.5 | *87* | 6.1 | 8 |
| **1940** | 13.8 | 5.3 | 9.6 | *−1.2* | 29.4 | 9 | −10.6 | 1 | −3.9 | 1 | 16.1 | 7 | 65 | 17 | 655.1 | *99* | 162 | 30.7 | 11 | 1604.0 | *102* | 15.5 | 6 |

## OXFORD 1941

| Month | Mean max | Mean min | Mean temp | Anom | Highest max | Date | Lowest min | Date | Lowest max | Date | Highest min | Date | Air frost | Days ≥ 25 °C | Total pptn | Anom | Rain days | Wettest day | Date | Total sunshine | Anom | Sunniest day | Date |
|---|---|---|---|---|---|---|---|---|---|---|---|---|---|---|---|---|---|---|---|---|---|---|---|
| Jan | 3.0 | −1.2 | 0.9 | *−4.0* | 8.9 | 22 | −8.3 | 17 | −1.7 | 2 | 2.2 | 12 | 17 | 0 | 64.4 | *113* | 22 | 10.4 | 24 | 41.7 | *67* | 6.8 | 9 |
| Feb | 7.3 | 1.3 | 4.3 | *−0.6* | 12.8 | 9 | −5.0 | 26 | 1.1 | 3 | 10.0 | 9 | 11 | 0 | 63.9 | *149* | 17 | 8.6 | 18 | 67.9 | *86* | 7.2 | 23 |
| Mar | 9.4 | 1.4 | 5.4 | *−1.8* | 14.4 | 14 | −3.3 | 19 | 4.4 | 12 | 7.8 | 26 | 10 | 0 | 64.5 | *135* | 13 | 20.8 | 7 | 118.7 | *107* | 10.5 | 13 |
| Apr | 11.2 | 3.1 | 7.2 | *−2.1* | 16.7 | 12 | −1.7 | 28 | 5.6 | 1 | 8.3 | 14 | 3 | 0 | 26.3 | *54* | 12 | 9.1 | 18 | 106.9 | *66* | 11.5 | 26 |
| May | 14.1 | 5.0 | 9.6 | *−2.9* | 19.4 | 18 | −1.7 | 4 | 9.4 | 8 | 10.0 | 13 | 5 | 0 | 43.7 | *76* | 14 | 12.4 | 29 | 141.9 | *74* | 13.1 | 3 |
| June | 21.0 | 10.9 | 15.9 | *+0.4* | 31.7 | 22 | 6.1 | 2 | 10.6 | 2 | 18.9 | 22 | 0 | 9 | 52.9 | *107* | 7 | 29.7 | 9 | 210.3 | *110* | 15.0 | 25 |
| July | 23.6 | 13.1 | 18.3 | *+0.4* | 30.6 | 7 | 8.9 | 20 | 17.2 | 19 | 17.8 | 13 | 0 | 13 | 112.7 | *231* | 15 | 20.8 | 13 | 227.5 | *110* | 15.4 | 6 |
| Aug | 19.3 | 11.3 | 15.3 | *−2.3* | 23.3 | 2 | 7.2 | 7 | 16.1 | 5 | 15.0 | 10 | 0 | 0 | 85.1 | *155* | 23 | 14.7 | 23 | 163.7 | *83* | 12.4 | 31 |
| Sep | 19.0 | 10.4 | 14.7 | *−0.2* | 25.0 | 4 | 2.8 | 16 | 14.4 | 15 | 16.1 | 3 | 0 | 1 | 10.6 | *20* | 5 | 3.6 | 28 | 123.7 | *88* | 10.1 | 7 |
| Oct | 14.6 | 6.4 | 10.5 | *−0.8* | 21.7 | 7 | 0.0 | 13 | 6.7 | 29 | 13.9 | 10 | 0 | 0 | 28.8 | *41* | 13 | 8.1 | 9 | 120.2 | *108* | 9.1 | 12 |
| Nov | 10.0 | 4.2 | 7.1 | *−0.4* | 13.9 | 10 | −2.8 | 5 | 4.4 | 15 | 9.4 | 21 | 2 | 0 | 66.6 | *100* | 20 | 18.5 | 13 | 47.1 | *66* | 8.0 | 7 |
| Dec | 8.3 | 2.7 | 5.5 | *+0.5* | 14.4 | 14 | −6.1 | 29 | 2.8 | 29 | 8.9 | 10 | 6 | 0 | 29.8 | *47* | 12 | 13.0 | 6 | 46.9 | *88* | 5.5 | 16 |
| **1941** | 13.4 | 5.7 | 9.6 | *−1.2* | 31.7 | 6 | −8.3 | 1 | −1.7 | 1 | 18.9 | 6 | 54 | 23 | 649.3 | *98* | 173 | 29.7 | 6 | 1416.5 | *90* | 15.4 | 7 |

## OXFORD 1942

| Month | Mean max | Mean min | Mean temp | Anom | Highest max | Date | Lowest min | Date | Lowest max | Date | Highest min | Date | Air frost | Days ≥ 25 °C | Total pptn | Anom | Rain days | Wettest day | Date | Total sunshine | Anom | Sunniest day | Date |
|---|---|---|---|---|---|---|---|---|---|---|---|---|---|---|---|---|---|---|---|---|---|---|---|
| Jan | 3.5 | −2.0 | 0.7 | *−4.1* | 10.0 | 3 | −9.4 | 22 | −2.8 | 20 | 7.2 | 4 | 20 | 0 | 87.5 | *154* | 16 | 18.8 | 19 | 45.1 | *72* | 6.1 | 25 |
| Feb | 2.8 | −2.6 | 0.1 | *−4.8* | 8.3 | 28 | −7.2 | 22 | −1.1 | 7 | 2.2 | 13 | 25 | 0 | 16.8 | *39* | 8 | 7.9 | 1 | 53.9 | *68* | 8.3 | 14 |
| Mar | 9.4 | 1.9 | 5.6 | *−1.6* | 17.2 | 25 | −4.4 | 7 | −1.1 | 6 | 10.0 | 17 | 9 | 0 | 60.3 | *127* | 15 | 9.9 | 4 | 69.4 | *62* | 9.7 | 25 |
| Apr | 14.3 | 5.1 | 9.7 | *+0.4* | 20.0 | 12 | 1.7 | 30 | 10.6 | 24 | 9.4 | 13 | 0 | 0 | 26.0 | *53* | 9 | 6.3 | 7 | 202.5 | *126* | 13.3 | 30 |
| May | 16.8 | 6.8 | 11.8 | *−0.7* | 22.8 | 7 | 1.1 | 4 | 9.4 | 12 | 12.2 | 18 | 0 | 0 | 78.8 | *138* | 16 | 15.5 | 10 | 209.4 | *109* | 13.4 | 4 |
| June | 21.5 | 9.5 | 15.5 | *−0.1* | 30.0 | 6 | 3.3 | 11 | 13.3 | 12 | 15.0 | 30 | 0 | 9 | 4.1 | *8* | 3 | 2.3 | 12 | 234.0 | *123* | 14.9 | 3 |
| July | 20.8 | 11.8 | 16.3 | *−1.6* | 25.6 | 2 | 7.2 | 12 | 13.9 | 17 | 15.6 | 4 | 0 | 3 | 46.8 | *96* | 17 | 19.3 | 26 | 150.1 | *73* | 12.7 | 30 |
| Aug | 21.8 | 12.8 | 17.3 | *−0.3* | 31.1 | 28 | 7.2 | 18 | 17.8 | 4 | 17.2 | 26 | 0 | 5 | 94.2 | *172* | 17 | 38.1 | 29 | 132.7 | *68* | 12.1 | 17 |
| Sep | 18.6 | 10.0 | 14.3 | *−0.6* | 23.3 | 11 | 3.3 | 27 | 13.3 | 25 | 15.6 | 1 | 0 | 0 | 28.4 | *52* | 13 | 4.6 | 19 | 124.5 | *89* | 8.9 | 6 |
| Oct | 14.5 | 7.3 | 10.9 | *−0.4* | 21.1 | 4 | 0.0 | 28 | 7.2 | 26 | 13.9 | 23 | 0 | 0 | 89.4 | *128* | 13 | 20.3 | 26 | 84.6 | *76* | 8.7 | 2 |
| Nov | 8.4 | 1.9 | 5.2 | *−2.4* | 13.3 | 3 | −3.9 | 22 | 3.9 | 1 | 6.1 | 7 | 6 | 0 | 50.0 | *75* | 14 | 21.3 | 5 | 57.8 | *81* | 6.6 | 22 |
| Dec | 9.4 | 4.7 | 7.0 | *+2.0* | 12.8 | 13 | −1.7 | 3 | 3.9 | 26 | 10.0 | 11 | 2 | 0 | 85.9 | *136* | 21 | 14.5 | 17 | 51.0 | *96* | 5.4 | 23 |
| **1942** | 13.5 | 5.6 | 9.5 | *−1.2* | 31.1 | 8 | −9.4 | 1 | −2.8 | 1 | 17.2 | 8 | 62 | 17 | 668.2 | *101* | 162 | 38.1 | 8 | 1415.0 | *90* | 14.9 | 6 |

## OXFORD 1943

| Month | Mean max | Mean min | Mean temp | Anom | Highest max | Date | Lowest min | Date | Lowest max | Date | Highest min | Date | Air frost | Days ≥ 25 °C | Total pptn | Anom | Rain days | Wettest day | Date | Total sunshine | Anom | Sunniest day | Date |
|---|---|---|---|---|---|---|---|---|---|---|---|---|---|---|---|---|---|---|---|---|---|---|---|
| Jan | 8.3 | 2.9 | 5.6 | *+0.7* | 12.8 | 29 | −2.8 | 8 | 1.7 | 4 | 7.8 | 29 | 7 | 0 | 96.5 | *170* | 24 | 20.6 | 31 | 54.7 | *88* | 6.3 | 15 |
| Feb | 9.5 | 2.7 | 6.1 | *+1.2* | 13.3 | 14 | −2.2 | 8 | 4.4 | 22 | 8.3 | 12 | 3 | 0 | 23.7 | *55* | 10 | 8.6 | 8 | 79.3 | *100* | 7.7 | 26 |
| Mar | 11.7 | 2.2 | 7.0 | *−0.3* | 15.6 | 27 | −2.8 | 8 | 6.7 | 19 | 8.3 | 30 | 8 | 0 | 19.6 | *41* | 6 | 13.2 | 24 | 132.9 | *119* | 10.5 | 28 |
| Apr | 16.2 | 6.9 | 11.6 | *+2.3* | 24.4 | 16 | 1.7 | 21 | 11.1 | 8 | 10.6 | 1 | 0 | 0 | 25.7 | *53* | 12 | 3.8 | 23 | 166.6 | *104* | 13.0 | 20 |
| May | 18.2 | 7.7 | 12.9 | *+0.5* | 25.0 | 14 | 1.1 | 7 | 11.1 | 8 | 13.9 | 27 | 0 | 1 | 57.4 | *100* | 12 | 15.2 | 10 | 244.1 | *127* | 14.9 | 18 |
| June | 20.1 | 10.3 | 15.2 | *−0.4* | 25.0 | 9 | 6.7 | 17 | 16.1 | 1 | 14.4 | 11 | 0 | 2 | 43.0 | *87* | 11 | 10.2 | 1 | 192.9 | *101* | 12.4 | 24 |
| July | 23.0 | 11.9 | 17.4 | *−0.5* | 33.3 | 31 | 5.6 | 24 | 16.7 | 20 | 16.7 | 28 | 0 | 11 | 25.0 | *51* | 10 | 13.0 | 20 | 196.7 | *95* | 14.4 | 16 |
| Aug | 21.8 | 12.4 | 17.1 | *−0.5* | 28.3 | 17 | 6.7 | 12 | 18.3 | 7 | 17.8 | 18 | 0 | 6 | 56.9 | *104* | 12 | 12.4 | 26 | 178.3 | *91* | 13.1 | 16 |
| Sep | 18.8 | 9.2 | 14.0 | *−0.9* | 24.4 | 13 | 1.7 | 27 | 13.3 | 26 | 15.0 | 11 | 0 | 0 | 44.0 | *81* | 12 | 7.4 | 12 | 142.5 | *102* | 10.1 | 22 |
| Oct | 15.0 | 6.8 | 10.9 | *−0.4* | 17.8 | 9 | 1.1 | 26 | 7.8 | 26 | 13.3 | 1 | 0 | 0 | 62.0 | *89* | 18 | 17.3 | 22 | 84.9 | *76* | 8.5 | 2 |
| Nov | 9.6 | 3.1 | 6.4 | *−1.2* | 15.6 | 3 | −2.8 | 20 | 3.9 | 16 | 11.7 | 2 | 5 | 0 | 35.1 | *53* | 18 | 7.1 | 12 | 72.3 | *102* | 6.9 | 10 |
| Dec | 6.3 | 0.6 | 3.5 | *−1.6* | 10.6 | 18 | −5.0 | 14 | −0.6 | 14 | 5.6 | 27 | 12 | 0 | 26.0 | *41* | 14 | 13.0 | 18 | 49.5 | *93* | 6.5 | 4 |
| 1943 | 14.9 | 6.4 | 10.6 | *−0.1* | 33.3 | 7 | −5.0 | 12 | −0.6 | 12 | 17.8 | 8 | 35 | 20 | 514.9 | *78* | 159 | 20.6 | 1 | 1594.7 | *101* | 14.9 | 5 |

## OXFORD 1944

| Month | Mean max | Mean min | Mean temp | Anom | Highest max | Date | Lowest min | Date | Lowest max | Date | Highest min | Date | Air frost | Days ≥ 25 °C | Total pptn | Anom | Rain days | Wettest day | Date | Total sunshine | Anom | Sunniest day | Date |
|---|---|---|---|---|---|---|---|---|---|---|---|---|---|---|---|---|---|---|---|---|---|---|---|
| Jan | 9.2 | 3.0 | 6.1 | *+1.2* | 13.3 | 27 | −4.4 | 16 | 2.2 | 15 | 9.4 | 28 | 6 | 0 | 36.3 | *64* | 16 | 7.6 | 11 | 41.2 | *66* | 6.5 | 14 |
| Feb | 6.4 | 0.9 | 3.6 | *−1.3* | 12.8 | 7 | −4.4 | 29 | 1.7 | 19 | 8.9 | 2 | 10 | 0 | 23.2 | *54* | 16 | 7.9 | 15 | 62.4 | *79* | 8.9 | 24 |
| Mar | 10.3 | 1.2 | 5.7 | *−1.6* | 21.7 | 26 | −2.8 | 3 | 4.4 | 4 | 6.7 | 13 | 10 | 0 | 6.5 | *14* | 7 | 1.8 | 14 | 142.7 | *128* | 11.4 | 26 |
| Apr | 15.8 | 6.1 | 11.0 | *+1.7* | 22.8 | 30 | 0.0 | 1 | 7.2 | 1 | 9.4 | 4 | 0 | 0 | 36.2 | *74* | 10 | 9.7 | 3 | 147.0 | *91* | 13.0 | 27 |
| May | 18.3 | 6.5 | 12.4 | *−0.1* | 30.6 | 29 | −1.1 | 7 | 10.6 | 21 | 15.0 | 30 | 1 | 5 | 18.1 | *32* | 8 | 7.6 | 13 | 235.6 | *122* | 13.6 | 7 |
| June | 19.0 | 9.6 | 14.3 | *−1.3* | 24.4 | 24 | 5.0 | 18 | 16.1 | 5 | 14.4 | 29 | 0 | 0 | 52.2 | *106* | 15 | 11.7 | 26 | 170.3 | *89* | 14.5 | 24 |
| July | 21.5 | 13.1 | 17.3 | *−0.6* | 26.1 | 17 | 10.0 | 25 | 17.2 | 21 | 16.1 | 7 | 0 | 2 | 56.2 | *115* | 18 | 23.4 | 2 | 97.5 | *47* | 10.1 | 6 |
| Aug | 23.2 | 13.0 | 18.1 | *+0.5* | 29.4 | 16 | 8.9 | 15 | 14.4 | 21 | 17.2 | 12 | 0 | 12 | 49.0 | *89* | 14 | 19.0 | 19 | 185.2 | *94* | 13.1 | 16 |
| Sep | 17.9 | 8.6 | 13.2 | *−1.7* | 21.1 | 17 | 2.8 | 10 | 13.9 | 24 | 14.4 | 15 | 0 | 0 | 64.8 | *120* | 14 | 15.2 | 1 | 143.8 | *102* | 11.1 | 12 |
| Oct | 13.2 | 6.6 | 9.9 | *−1.4* | 15.6 | 6 | 0.6 | 30 | 7.8 | 29 | 9.4 | 9 | 0 | 0 | 85.6 | *122* | 21 | 12.4 | 17 | 102.0 | *92* | 9.5 | 2 |
| Nov | 9.8 | 3.6 | 6.7 | *−0.8* | 14.4 | 5 | −1.7 | 10 | 4.4 | 14 | 12.2 | 23 | 4 | 0 | 91.0 | *136* | 23 | 12.2 | 16 | 49.3 | *69* | 6.8 | 21 |
| Dec | 6.0 | 0.8 | 3.4 | *−1.6* | 12.8 | 17 | −7.8 | 27 | −2.8 | 25 | 7.8 | 18 | 11 | 0 | 40.5 | *64* | 20 | 12.2 | 16 | 64.1 | *121* | 6.7 | 2 |
| 1944 | 14.2 | 6.1 | 10.1 | *−0.6* | 30.6 | 5 | −7.8 | 12 | −2.8 | 12 | 17.2 | 8 | 42 | 19 | 559.6 | *85* | 182 | 23.4 | 7 | 1441.1 | *91* | 14.5 | 6 |

## OXFORD 1945

| Month | Mean max | Mean min | Mean temp | Anom | Highest max | Date | Lowest min | Date | Lowest max | Date | Highest min | Date | Air frost | Days ≥ 25 °C | Total pptn | Anom | Rain days | Wettest day | Date | Total sunshine | Anom | Sunniest day | Date |
|---|---|---|---|---|---|---|---|---|---|---|---|---|---|---|---|---|---|---|---|---|---|---|---|
| Jan | 3.2 | −2.0 | 0.6 | *−4.3* | 10.0 | 31 | −9.4 | 29 | −4.4 | 26 | 4.4 | 18 | 22 | 0 | 49.9 | *88* | 24 | 17.0 | 29 | 51.5 | *83* | 6.9 | 21 |
| Feb | 10.9 | 4.7 | 7.8 | *+2.9* | 16.7 | 18 | 0.0 | 21 | 6.1 | 10 | 10.0 | 13 | 0 | 0 | 46.8 | *109* | 14 | 13.7 | 11 | 65.7 | *83* | 7.1 | 14 |
| Mar | 13.1 | 3.9 | 8.5 | *+1.2* | 21.1 | 23 | −3.3 | 3 | 6.7 | 2 | 10.0 | 29 | 3 | 0 | 20.4 | *43* | 8 | 10.2 | 19 | 152.6 | *137* | 10.8 | 23 |
| Apr | 16.1 | 5.6 | 10.9 | *+1.6* | 24.4 | 17 | −1.7 | 9 | 7.8 | 30 | 12.8 | 15 | 2 | 0 | 32.6 | *67* | 11 | 15.0 | 28 | 209.1 | *130* | 12.5 | 22 |
| May | 18.0 | 8.2 | 13.1 | *+0.6* | 27.2 | 12 | −1.1 | 1 | 8.9 | 3 | 16.7 | 9 | 2 | 2 | 54.0 | *94* | 17 | 14.2 | 20 | 156.1 | *81* | 12.4 | 15 |
| June | 19.7 | 10.7 | 15.2 | *−0.4* | 26.1 | 19 | 6.7 | 17 | 14.4 | 28 | 15.6 | 20 | 0 | 2 | 66.2 | *134* | 18 | 11.2 | 5 | 182.4 | *96* | 14.1 | 18 |
| July | 22.1 | 12.8 | 17.4 | *−0.4* | 28.9 | 15 | 8.9 | 8 | 16.1 | 26 | 17.2 | 15 | 0 | 5 | 52.0 | *106* | 13 | 18.0 | 10 | 159.0 | *77* | 14.1 | 23 |
| Aug | 21.3 | 11.7 | 16.5 | *−1.1* | 28.9 | 4 | 7.2 | 2 | 15.0 | 7 | 15.6 | 25 | 0 | 5 | 37.2 | *68* | 14 | 9.4 | 22 | 144.8 | *74* | 13.5 | 4 |
| Sep | 18.6 | 11.5 | 15.0 | *+0.2* | 24.4 | 12 | 5.6 | 29 | 14.4 | 4 | 16.7 | 12 | 0 | 0 | 47.5 | *88* | 16 | 11.7 | 3 | 64.9 | *46* | 7.7 | 22 |
| Oct | 16.5 | 8.0 | 12.3 | *+1.0* | 22.2 | 11 | 2.8 | 19 | 13.3 | 25 | 13.3 | 21 | 0 | 0 | 69.4 | *99* | 12 | 19.0 | 23 | 111.7 | *100* | 8.6 | 19 |
| Nov | 10.0 | 4.4 | 7.2 | *−0.3* | 15.0 | 4 | −2.2 | 15 | 6.7 | 15 | 10.0 | 3 | 4 | 0 | 4.8 | *7* | 4 | 2.8 | 4 | 31.0 | *44* | 6.1 | 11 |
| Dec | 8.1 | 2.5 | 5.3 | *+0.3* | 12.8 | 16 | −4.4 | 31 | 0.6 | 9 | 10.0 | 17 | 4 | 0 | 88.9 | *141* | 19 | 22.1 | 27 | 46.6 | *88* | 6.1 | 3 |
| **1945** | 14.8 | 6.8 | 10.8 | *+0.1* | 28.9 | 7 | −9.4 | 1 | −4.4 | 1 | 17.2 | 7 | 37 | 14 | 569.7 | *86* | 170 | 22.1 | 12 | 1375.4 | *87* | 14.1 | 6 |

## OXFORD 1946

| Month | Mean max | Mean min | Mean temp | Anom | Highest max | Date | Lowest min | Date | Lowest max | Date | Highest min | Date | Air frost | Days ≥ 25 °C | Total pptn | Anom | Rain days | Wettest day | Date | Total sunshine | Anom | Sunniest day | Date |
|---|---|---|---|---|---|---|---|---|---|---|---|---|---|---|---|---|---|---|---|---|---|---|---|
| Jan | 5.9 | 0.1 | 3.0 | *−1.9* | 13.9 | 11 | −8.3 | 21 | −2.8 | 20 | 8.3 | 11 | 16 | 0 | 35.2 | *62* | 15 | 6.6 | 11 | 49.9 | *80* | 7.3 | 17 |
| Feb | 9.4 | 4.0 | 6.7 | *+1.8* | 13.3 | 6 | −5.0 | 28 | 0.6 | 26 | 10.6 | 8 | 3 | 0 | 56.1 | *131* | 20 | 11.9 | 3 | 78.7 | *100* | 9.0 | 27 |
| Mar | 9.6 | 1.3 | 5.5 | *−1.8* | 21.7 | 29 | −5.6 | 10 | 1.7 | 4 | 9.4 | 19 | 9 | 0 | 24.4 | *51* | 7 | 9.9 | 21 | 107.2 | *96* | 11.1 | 26 |
| Apr | 16.1 | 5.0 | 10.5 | *+1.2* | 25.0 | 3 | 0.0 | 10 | 8.9 | 5 | 8.9 | 5 | 0 | 1 | 34.4 | *70* | 12 | 9.1 | 4 | 203.3 | *126* | 12.9 | 19 |
| May | 15.7 | 6.2 | 11.0 | *−1.5* | 20.0 | 10 | 0.6 | 15 | 11.1 | 14 | 11.1 | 27 | 0 | 0 | 78.6 | *137* | 15 | 20.3 | 26 | 162.9 | *84* | 12.7 | 10 |
| June | 18.1 | 9.5 | 13.8 | *−1.8* | 24.4 | 23 | 6.1 | 11 | 14.4 | 4 | 13.3 | 30 | 0 | 0 | 58.3 | *118* | 20 | 7.9 | 12 | 149.8 | *78* | 12.9 | 13 |
| July | 22.3 | 12.7 | 17.5 | *−0.4* | 28.9 | 12 | 7.2 | 17 | 16.1 | 16 | 16.7 | 4 | 0 | 10 | 32.2 | *66* | 9 | 10.4 | 3 | 197.6 | *95* | 14.5 | 7 |
| Aug | 20.1 | 11.3 | 15.7 | *−1.9* | 26.7 | 5 | 5.6 | 18 | 17.2 | 16 | 15.0 | 7 | 0 | 2 | 92.2 | *168* | 20 | 24.9 | 9 | 134.8 | *69* | 11.7 | 18 |
| Sep | 18.4 | 11.0 | 14.7 | *−0.2* | 24.4 | 28 | 7.2 | 16 | 15.6 | 3 | 15.6 | 26 | 0 | 0 | 85.4 | *158* | 19 | 15.5 | 3 | 98.5 | *70* | 10.9 | 15 |
| Oct | 13.7 | 7.1 | 10.4 | *−0.9* | 20.6 | 1 | −2.8 | 29 | 7.2 | 25 | 13.3 | 1 | 3 | 0 | 20.2 | *29* | 9 | 7.9 | 26 | 79.3 | *71* | 9.5 | 11 |
| Nov | 11.2 | 5.7 | 8.4 | *+0.9* | 18.9 | 4 | −0.6 | 2 | 6.1 | 1 | 10.0 | 28 | 1 | 0 | 114.1 | *171* | 20 | 19.8 | 27 | 42.9 | *60* | 7.7 | 4 |
| Dec | 6.1 | −0.1 | 3.0 | *−2.0* | 11.1 | 1 | −7.8 | 21 | 0.6 | 15 | 5.0 | 2 | 13 | 0 | 53.8 | *85* | 21 | 11.2 | 30 | 74.7 | *140* | 6.7 | 20 |
| **1946** | 13.9 | 6.1 | 10.0 | *−0.7* | 28.9 | 7 | −8.3 | 1 | −2.8 | 1 | 16.7 | 7 | 45 | 13 | 684.9 | *104* | 187 | 24.9 | 8 | 1379.6 | *88* | 14.5 | 7 |

## OXFORD 1947

| Month | Mean max | Mean min | Mean temp | Anom | Highest max | Date | Lowest min | Date | Lowest max | Date | Highest min | Date | Air frost | Days ≥ 25 °C | Total pptn | Anom | Rain days | Wettest day | Date | Total sunshine | Anom | Sunniest day | Date |
|---|---|---|---|---|---|---|---|---|---|---|---|---|---|---|---|---|---|---|---|---|---|---|---|
| Jan | 4.9 | −0.6 | 2.2 | *−2.7* | 13.3 | 16 | −12.4 | 29 | −4.1 | 29 | 8.3 | 15 | 15 | 0 | 35.2 | *62* | 22 | 5.3 | 10 | 59.5 | *96* | 6.3 | 2 |
| Feb | −0.1 | −4.3 | −2.2 | *−7.1* | 6.1 | 26 | −16.1 | 25 | −2.9 | 17 | 1.0 | 3 | 26 | 0 | 42.5 | *99* | 14 | 11.7 | 8 | 26.3 | *33* | 7.5 | 24 |
| Mar | 7.7 | 1.6 | 4.7 | *−2.6* | 14.9 | 29 | −10.8 | 7 | 0.1 | 5 | 8.5 | 30 | 14 | 0 | 132.9 | *279* | 25 | 21.1 | 5 | 70.9 | *64* | 9.8 | 1 |
| Apr | 13.8 | 5.2 | 9.5 | *+0.2* | 21.1 | 16 | 1.7 | 3 | 7.2 | 2 | 9.4 | 15 | 0 | 0 | 46.9 | *96* | 16 | 7.5 | 3 | 150.5 | *94* | 12.1 | 26 |
| May | 19.2 | 9.0 | 14.1 | *+1.6* | 30.6 | 30 | 4.0 | 2 | 10.6 | 1 | 14.8 | 30 | 0 | 4 | 47.8 | *84* | 18 | 15.8 | 23 | 147.3 | *76* | 14.7 | 28 |
| June | 21.5 | 11.5 | 16.5 | *+0.9* | 32.1 | 2 | 5.6 | 13 | 14.9 | 15 | 17.6 | 27 | 0 | 5 | 27.7 | *56* | 11 | 7.5 | 14 | 185.9 | *97* | 14.5 | 12 |
| July | 22.6 | 13.7 | 18.1 | *+0.3* | 30.7 | 27 | 8.4 | 12 | 16.1 | 8 | 18.3 | 26 | 0 | 9 | 43.0 | *88* | 14 | 8.9 | 28 | 144.5 | *70* | 11.3 | 7 |
| Aug | 25.8 | 13.0 | 19.4 | *+1.8* | 31.1 | 16 | 8.4 | 8 | 18.9 | 5 | 16.7 | 2 | 0 | 19 | 15.2 | *28* | 2 | 14.2 | 4 | 274.9 | *140* | 13.5 | 14 |
| Sep | 20.7 | 11.1 | 15.9 | *+1.0* | 26.1 | 4 | 3.3 | 25 | 15.6 | 23 | 16.7 | 8 | 0 | 4 | 29.2 | *54* | 9 | 9.2 | 19 | 159.9 | *114* | 10.7 | 5 |
| Oct | 15.8 | 5.6 | 10.7 | *−0.6* | 21.7 | 6 | −2.2 | 21 | 8.3 | 29 | 12.2 | 8 | 2 | 0 | 10.5 | *15* | 7 | 3.3 | 22 | 107.3 | *96* | 10.3 | 3 |
| Nov | 10.9 | 4.5 | 7.7 | *+0.2* | 16.7 | 1 | −3.3 | 27 | 3.3 | 27 | 15.0 | 22 | 8 | 0 | 32.2 | *48* | 19 | 9.1 | 18 | 75.4 | *106* | 7.3 | 5 |
| Dec | 8.1 | 2.8 | 5.5 | *+0.4* | 12.2 | 26 | −7.8 | 1 | 2.2 | 30 | 7.8 | 13 | 6 | 0 | 44.9 | *71* | 15 | 9.1 | 4 | 37.9 | *71* | 6.8 | 29 |
| 1947 | 14.2 | 6.1 | 10.2 | *−0.6* | 32.1 | 6 | −16.1 | 2 | −4.1 | 1 | 18.3 | 7 | 71 | 41 | 508.0 | *77* | 172 | 21.1 | 3 | 1440.3 | *91* | 14.7 | 5 |

## OXFORD 1948

| Month | Mean max | Mean min | Mean temp | Anom | Highest max | Date | Lowest min | Date | Lowest max | Date | Highest min | Date | Air frost | Days ≥ 25 °C | Total pptn | Anom | Rain days | Wettest day | Date | Total sunshine | Anom | Sunniest day | Date |
|---|---|---|---|---|---|---|---|---|---|---|---|---|---|---|---|---|---|---|---|---|---|---|---|
| Jan | 9.1 | 3.2 | 6.1 | *+1.3* | 13.9 | 30 | −2.2 | 20 | 3.9 | 16 | 10.6 | 4 | 4 | 0 | 127.3 | *224* | 23 | 16.8 | 29 | 45.7 | *73* | 6.1 | 18 |
| Feb | 7.9 | 2.2 | 5.0 | *+0.1* | 16.7 | 29 | −6.1 | 22 | −1.1 | 20 | 10.0 | 8 | 11 | 0 | 22.6 | *53* | 13 | 4.3 | 5 | 92.9 | *118* | 9.2 | 26 |
| Mar | 14.0 | 3.5 | 8.8 | *+1.5* | 21.1 | 9 | −1.1 | 5 | 3.3 | 5 | 8.9 | 8 | 4 | 0 | 18.5 | *39* | 8 | 9.9 | 31 | 160.2 | *144* | 10.5 | 28 |
| Apr | 14.8 | 4.4 | 9.6 | *+0.3* | 21.1 | 21 | −0.6 | 5 | 9.4 | 4 | 12.2 | 22 | 2 | 0 | 42.6 | *87* | 14 | 9.4 | 17 | 205.2 | *128* | 12.4 | 15 |
| May | 17.5 | 6.5 | 12.0 | *−0.5* | 25.0 | 18 | 0.6 | 3 | 8.3 | 23 | 10.6 | 11 | 0 | 1 | 126.0 | *220* | 14 | 25.4 | 11 | 237.1 | *123* | 14.7 | 18 |
| June | 18.5 | 9.9 | 14.2 | *−1.4* | 25.0 | 14 | 6.1 | 3 | 13.9 | 3 | 13.9 | 26 | 0 | 1 | 53.7 | *109* | 18 | 9.1 | 3 | 124.2 | *65* | 12.7 | 26 |
| July | 21.1 | 12.1 | 16.6 | *−1.3* | 32.2 | 29 | 7.2 | 6 | 15.0 | 4 | 18.3 | 29 | 0 | 6 | 21.7 | *44* | 10 | 7.4 | 4 | 174.7 | *84* | 13.5 | 27 |
| Aug | 20.1 | 11.9 | 16.0 | *−1.6* | 23.9 | 1 | 7.8 | 20 | 15.6 | 3 | 16.7 | 2 | 0 | 0 | 87.3 | *159* | 17 | 21.1 | 2 | 133.7 | *68* | 12.1 | 29 |
| Sep | 19.1 | 10.6 | 14.9 | *−0.0* | 24.4 | 9 | 2.2 | 22 | 13.3 | 12 | 16.1 | 28 | 0 | 0 | 63.1 | *116* | 12 | 31.2 | 12 | 146.4 | *104* | 9.7 | 21 |
| Oct | 14.3 | 6.3 | 10.3 | *−1.0* | 19.4 | 2 | −3.9 | 27 | 7.8 | 27 | 13.3 | 2 | 3 | 0 | 59.3 | *85* | 11 | 16.5 | 17 | 93.7 | *84* | 8.9 | 26 |
| Nov | 10.5 | 4.2 | 7.4 | *−0.2* | 16.1 | 13 | −2.8 | 9 | 2.8 | 23 | 12.2 | 19 | 7 | 0 | 38.3 | *57* | 15 | 9.1 | 6 | 68.6 | *97* | 8.1 | 5 |
| Dec | 9.1 | 3.3 | 6.2 | *+1.2* | 15.0 | 3 | −5.6 | 26 | 0.0 | 26 | 10.0 | 3 | 5 | 0 | 68.7 | *109* | 20 | 17.0 | 30 | 65.4 | *123* | 5.9 | 4 |
| 1948 | 14.7 | 6.5 | 10.6 | *−0.2* | 32.2 | 7 | −6.1 | 2 | −1.1 | 2 | 18.3 | 7 | 36 | 8 | 729.1 | *110* | 175 | 31.2 | 9 | 1547.8 | *98* | 14.7 | 5 |

## OXFORD 1949

| Month | Mean max | Mean min | Mean temp | Anom | Highest max | Date | Lowest min | Date | Lowest max | Date | Highest min | Date | Air frost | Days ≥ 25 °C | Total pptn | Anom | Rain days | Wettest day | Date | Total sunshine | Anom | Sunniest day | Date |
|---|---|---|---|---|---|---|---|---|---|---|---|---|---|---|---|---|---|---|---|---|---|---|---|
| Jan | 8.6 | 3.1 | 5.8 | *+1.0* | 11.7 | 27 | −1.7 | 29 | 4.4 | 3 | 8.3 | 7 | 5 | 0 | 23.4 | *41* | 17 | 8.1 | 3 | 51.0 | *82* | 6.0 | 9 |
| Feb | 10.3 | 1.6 | 5.9 | *+1.0* | 13.3 | 17 | −6.1 | 4 | 4.4 | 3 | 8.9 | 15 | 8 | 0 | 25.8 | *60* | 9 | 12.2 | 8 | 115.1 | *146* | 8.5 | 19 |
| Mar | 9.0 | 1.7 | 5.3 | *−1.9* | 18.3 | 26 | −2.8 | 3 | 2.2 | 6 | 8.9 | 23 | 9 | 0 | 43.2 | *91* | 8 | 18.8 | 4 | 112.5 | *101* | 10.9 | 24 |
| Apr | 15.5 | 6.1 | 10.8 | *+1.5* | 25.6 | 16 | −0.6 | 9 | 9.4 | 7 | 11.1 | 6 | 1 | 1 | 33.4 | *68* | 9 | 10.4 | 5 | 182.5 | *113* | 12.9 | 20 |
| May | 16.7 | 6.6 | 11.7 | *−0.8* | 21.1 | 21 | 0.0 | 10 | 11.7 | 9 | 11.1 | 23 | 0 | 0 | 61.4 | *107* | 13 | 23.6 | 23 | 233.3 | *121* | 14.3 | 10 |
| June | 21.5 | 10.3 | 15.9 | *+0.3* | 29.4 | 27 | 5.0 | 2 | 13.9 | 3 | 17.2 | 29 | 0 | 5 | 11.7 | *24* | 3 | 5.6 | 1 | 242.1 | *127* | 14.6 | 26 |
| July | 24.7 | 12.8 | 18.7 | *+0.9* | 31.7 | 12 | 7.2 | 8 | 15.6 | 7 | 18.9 | 25 | 0 | 17 | 23.0 | *47* | 7 | 7.4 | 17 | 262.5 | *127* | 15.0 | 1 |
| Aug | 22.9 | 12.6 | 17.8 | *+0.2* | 28.9 | 14 | 6.7 | 12 | 18.3 | 7 | 17.2 | 20 | 0 | 9 | 37.7 | *69* | 9 | 16.8 | 1 | 210.3 | *107* | 13.9 | 13 |
| Sep | 22.0 | 12.5 | 17.3 | *+2.4* | 29.4 | 4 | 7.2 | 18 | 17.8 | 21 | 20.6 | 5 | 0 | 3 | 58.3 | *108* | 9 | 38.9 | 22 | 151.1 | *108* | 10.1 | 6 |
| Oct | 16.3 | 8.4 | 12.4 | *+1.1* | 22.2 | 3 | −1.7 | 28 | 8.3 | 26 | 14.4 | 11 | 2 | 0 | 162.3 | *232* | 17 | 36.1 | 23 | 125.0 | *112* | 9.1 | 27 |
| Nov | 9.9 | 3.4 | 6.7 | *−0.9* | 15.0 | 4 | −3.3 | 2 | 5.0 | 19 | 10.0 | 5 | 5 | 0 | 64.2 | *96* | 17 | 20.3 | 5 | 76.2 | *107* | 7.4 | 8 |
| Dec | 9.1 | 3.3 | 6.2 | *+1.2* | 12.8 | 3 | −2.2 | 12 | 3.3 | 9 | 10.0 | 27 | 4 | 0 | 34.6 | *55* | 20 | 9.4 | 18 | 65.4 | *123* | 6.5 | 4 |
| **1949** | 15.5 | 6.9 | 11.2 | *+0.5* | 31.7 | 7 | −6.1 | 2 | 2.2 | 3 | 20.6 | 9 | 34 | 35 | 579.0 | *88* | 138 | 38.9 | 9 | 1827.0 | *116* | 15.0 | 7 |

## OXFORD 1950

| Month | Mean max | Mean min | Mean temp | Anom | Highest max | Date | Lowest min | Date | Lowest max | Date | Highest min | Date | Air frost | Days ≥ 25 °C | Total pptn | Anom | Rain days | Wettest day | Date | Total sunshine | Anom | Sunniest day | Date |
|---|---|---|---|---|---|---|---|---|---|---|---|---|---|---|---|---|---|---|---|---|---|---|---|
| Jan | 6.9 | 1.5 | 4.2 | *−0.6* | 12.2 | 7 | −6.1 | 26 | −0.6 | 25 | 8.9 | 4 | 12 | 0 | 14.2 | *25* | 9 | 4.1 | 30 | 40.6 | *65* | 6.7 | 21 |
| Feb | 9.5 | 2.7 | 6.1 | *+1.2* | 15.6 | 17 | −3.9 | 27 | 4.4 | 8 | 10.0 | 16 | 5 | 0 | 103.9 | *243* | 19 | 16.3 | 2 | 67.5 | *86* | 8.3 | 28 |
| Mar | 12.1 | 3.6 | 7.9 | *+0.6* | 15.6 | 7 | −4.4 | 1 | 6.7 | 10 | 10.0 | 23 | 3 | 0 | 22.3 | *47* | 13 | 3.8 | 19 | 138.8 | *125* | 10.3 | 25 |
| Apr | 12.5 | 4.2 | 8.4 | *−0.9* | 18.3 | 21 | −0.6 | 16 | 7.2 | 25 | 9.4 | 30 | 2 | 0 | 54.4 | *111* | 17 | 15.2 | 17 | 163.7 | *102* | 12.1 | 20 |
| May | 16.0 | 7.1 | 11.6 | *−0.9* | 22.8 | 30 | 1.7 | 17 | 10.0 | 25 | 10.6 | 22 | 0 | 0 | 55.5 | *97* | 10 | 11.9 | 21 | 174.1 | *90* | 14.5 | 11 |
| June | 22.1 | 11.8 | 16.9 | *+1.4* | 30.0 | 6 | 7.2 | 15 | 13.9 | 14 | 16.7 | 27 | 0 | 6 | 39.1 | *79* | 9 | 11.2 | 22 | 244.5 | *128* | 14.2 | 9 |
| July | 20.8 | 12.5 | 16.6 | *−1.2* | 27.2 | 9 | 9.4 | 1 | 13.9 | 3 | 17.2 | 19 | 0 | 1 | 135.5 | *277* | 19 | 32.8 | 3 | 174.0 | *84* | 13.1 | 9 |
| Aug | 20.8 | 12.2 | 16.5 | *−1.1* | 25.0 | 6 | 8.3 | 14 | 16.1 | 17 | 15.6 | 9 | 0 | 2 | 77.6 | *142* | 17 | 21.8 | 30 | 171.1 | *87* | 12.7 | 4 |
| Sep | 17.3 | 10.0 | 13.6 | *−1.2* | 20.6 | 1 | 3.3 | 27 | 13.3 | 26 | 14.4 | 10 | 0 | 0 | 73.0 | *135* | 23 | 14.0 | 30 | 112.7 | *80* | 10.5 | 2 |
| Oct | 13.4 | 6.6 | 10.0 | *−1.3* | 22.2 | 5 | −3.9 | 28 | 6.1 | 27 | 12.2 | 5 | 3 | 0 | 12.9 | *18* | 9 | 4.6 | 30 | 96.6 | *87* | 9.9 | 5 |
| Nov | 8.8 | 3.4 | 6.1 | *−1.5* | 13.3 | 27 | −2.8 | 25 | 0.0 | 26 | 9.4 | 9 | 4 | 0 | 112.3 | *168* | 22 | 17.3 | 20 | 56.9 | *80* | 7.6 | 7 |
| Dec | 3.7 | −0.9 | 1.4 | *−3.6* | 10.6 | 1 | −5.6 | 6 | 0.0 | 15 | 7.8 | 1 | 21 | 0 | 33.3 | *53* | 17 | 7.6 | 10 | 55.3 | *104* | 6.1 | 5 |
| **1950** | 13.7 | 6.2 | 9.9 | *−0.8* | 30.0 | 6 | −6.1 | 1 | −0.6 | 1 | 17.2 | 7 | 50 | 9 | 734.0 | *111* | 184 | 32.8 | 7 | 1495.8 | *95* | 14.5 | 5 |

## OXFORD 1951

| Month | Mean max | Mean min | Mean temp | Anom | Highest max | Date | Lowest min | Date | Lowest max | Date | Highest min | Date | Air frost | Days ≥ 25 °C | Total pptn | Anom | Rain days | Wettest day | Date | Total sunshine | Anom | Sunniest day | Date |
|---|---|---|---|---|---|---|---|---|---|---|---|---|---|---|---|---|---|---|---|---|---|---|---|
| Jan | 7.1 | 2.0 | 4.5 | *−0.3* | 11.7 | 16 | −4.4 | 31 | 0.6 | 29 | 7.2 | 6 | 8 | 0 | 78.0 | *137* | 19 | 19.8 | 5 | 48.9 | *79* | 7.3 | 15 |
| Feb | 6.9 | 1.3 | 4.1 | *−0.8* | 10.6 | 12 | −2.8 | 1 | 4.4 | 11 | 5.0 | 3 | 4 | 0 | 87.5 | *204* | 20 | 13.0 | 4 | 58.6 | *74* | 6.3 | 19 |
| Mar | 8.0 | 1.6 | 4.8 | *−2.5* | 12.8 | 17 | −1.7 | 3 | 2.2 | 9 | 8.9 | 23 | 9 | 0 | 90.2 | *189* | 22 | 16.3 | 21 | 92.6 | *83* | 10.8 | 24 |
| Apr | 12.0 | 2.9 | 7.5 | *−1.8* | 22.8 | 24 | −0.6 | 22 | 6.7 | 9 | 7.8 | 26 | 1 | 0 | 64.8 | *132* | 15 | 18.5 | 8 | 200.4 | *125* | 13.3 | 22 |
| May | 14.5 | 6.5 | 10.5 | *−2.0* | 21.7 | 24 | 2.2 | 1 | 8.3 | 8 | 11.7 | 24 | 0 | 0 | 53.4 | *93* | 17 | 10.4 | 26 | 152.1 | *79* | 14.9 | 31 |
| June | 19.7 | 9.4 | 14.6 | *−1.0* | 23.3 | 21 | 5.0 | 8 | 14.4 | 26 | 14.4 | 12 | 0 | 0 | 35.9 | *73* | 11 | 10.9 | 21 | 251.9 | *132* | 15.2 | 8 |
| July | 22.3 | 12.1 | 17.2 | *−0.6* | 27.8 | 19 | 7.8 | 14 | 15.0 | 23 | 16.7 | 31 | 0 | 6 | 34.2 | *70* | 10 | 21.8 | 22 | 216.7 | *105* | 14.0 | 2 |
| Aug | 19.9 | 11.7 | 15.8 | *−1.8* | 24.4 | 3 | 6.1 | 16 | 16.1 | 30 | 15.6 | 4 | 0 | 0 | 93.7 | *171* | 22 | 20.1 | 11 | 174.7 | *89* | 12.5 | 2 |
| Sep | 18.8 | 10.7 | 14.8 | *−0.1* | 23.9 | 4 | 4.4 | 30 | 16.1 | 18 | 17.2 | 5 | 0 | 0 | 125.1 | *231* | 13 | 84.8 | 6 | 110.8 | *79* | 8.7 | 14 |
| Oct | 13.8 | 5.0 | 9.4 | *−1.9* | 17.8 | 3 | −2.8 | 23 | 8.3 | 22 | 12.8 | 2 | 4 | 0 | 22.9 | *33* | 12 | 9.7 | 29 | 105.6 | *95* | 8.7 | 23 |
| Nov | 11.7 | 6.2 | 8.9 | *+1.4* | 15.0 | 7 | −0.6 | 26 | 7.2 | 26 | 10.0 | 7 | 1 | 0 | 132.1 | *198* | 25 | 30.7 | 5 | 75.1 | *106* | 7.5 | 1 |
| Dec | 9.0 | 3.0 | 6.0 | *+1.0* | 12.2 | 24 | −6.1 | 13 | 2.8 | 11 | 9.4 | 5 | 5 | 0 | 39.8 | *63* | 16 | 10.9 | 28 | 69.8 | *131* | 6.6 | 1 |
| **1951** | 13.7 | 6.0 | 9.8 | *−0.9* | 27.8 | 7 | −6.1 | 12 | 0.6 | 1 | 17.2 | 9 | 32 | 6 | 857.6 | *130* | 202 | 84.8 | 9 | 1557.2 | *99* | 15.2 | 6 |

## OXFORD 1952

| Month | Mean max | Mean min | Mean temp | Anom | Highest max | Date | Lowest min | Date | Lowest max | Date | Highest min | Date | Air frost | Days ≥ 25 °C | Total pptn | Anom | Rain days | Wettest day | Date | Total sunshine | Anom | Sunniest day | Date |
|---|---|---|---|---|---|---|---|---|---|---|---|---|---|---|---|---|---|---|---|---|---|---|---|
| Jan | 6.0 | 0.3 | 3.2 | *−1.7* | 11.1 | 15 | −8.3 | 27 | 0.6 | 27 | 7.8 | 7 | 13 | 0 | 46.4 | *82* | 17 | 8.4 | 2 | 94.1 | *151* | 7.4 | 27 |
| Feb | 6.9 | 0.3 | 3.6 | *−1.3* | 12.2 | 29 | −3.3 | 5 | 3.3 | 11 | 3.9 | 19 | 13 | 0 | 13.3 | *31* | 11 | 5.1 | 10 | 96.7 | *123* | 7.7 | 12 |
| Mar | 10.5 | 3.9 | 7.2 | *−0.1* | 15.6 | 8 | −2.2 | 27 | 1.1 | 29 | 9.4 | 22 | 4 | 0 | 61.1 | *128* | 17 | 12.4 | 29 | 105.0 | *94* | 9.2 | 14 |
| Apr | 15.2 | 5.3 | 10.3 | *+1.0* | 23.3 | 19 | −1.1 | 3 | 7.2 | 1 | 10.0 | 10 | 2 | 0 | 41.2 | *84* | 18 | 8.9 | 7 | 178.2 | *111* | 12.6 | 29 |
| May | 19.1 | 9.5 | 14.3 | *+1.8* | 28.3 | 18 | 5.0 | 7 | 13.3 | 3 | 14.4 | 19 | 0 | 4 | 56.6 | *99* | 16 | 19.0 | 4 | 207.0 | *107* | 14.3 | 17 |
| June | 20.5 | 10.5 | 15.5 | *−0.1* | 27.8 | 28 | 5.6 | 8 | 13.3 | 8 | 17.8 | 30 | 0 | 5 | 34.0 | *69* | 12 | 9.4 | 12 | 217.6 | *114* | 15.2 | 23 |
| July | 23.0 | 12.9 | 17.9 | *+0.0* | 31.7 | 1 | 6.7 | 16 | 16.1 | 3 | 17.2 | 2 | 0 | 10 | 7.5 | *15* | 6 | 3.8 | 11 | 194.0 | *94* | 13.5 | 5 |
| Aug | 21.1 | 12.6 | 16.9 | *−0.7* | 24.4 | 26 | 8.3 | 21 | 15.0 | 18 | 16.7 | 12 | 0 | 0 | 108.5 | *198* | 16 | 21.3 | 6 | 180.5 | *92* | 12.5 | 22 |
| Sep | 15.7 | 7.2 | 11.4 | *−3.4* | 21.7 | 23 | 1.7 | 20 | 10.6 | 30 | 12.2 | 2 | 0 | 0 | 43.0 | *79* | 19 | 12.2 | 9 | 140.3 | *100* | 10.4 | 13 |
| Oct | 12.8 | 5.5 | 9.2 | *−2.1* | 15.6 | 4 | −3.3 | 12 | 7.8 | 20 | 11.1 | 29 | 4 | 0 | 95.5 | *136* | 19 | 20.3 | 12 | 105.0 | *94* | 9.3 | 5 |
| Nov | 7.0 | 1.3 | 4.2 | *−3.4* | 13.3 | 2 | −6.7 | 25 | 1.7 | 29 | 8.3 | 2 | 7 | 0 | 96.8 | *145* | 18 | 18.8 | 21 | 72.1 | *102* | 7.8 | 3 |
| Dec | 6.0 | −0.6 | 2.7 | *−2.3* | 11.7 | 23 | −8.9 | 5 | −2.2 | 5 | 6.1 | 11 | 16 | 0 | 55.2 | *87* | 20 | 11.2 | 18 | 77.7 | *146* | 7.0 | 1 |
| **1952** | 13.6 | 5.7 | 9.7 | *−1.1* | 31.7 | 7 | −8.9 | 12 | −2.2 | 12 | 17.8 | 6 | 59 | 19 | 659.1 | *100* | 189 | 21.3 | 8 | 1668.2 | *106* | 15.2 | 6 |

## OXFORD 1953

| Month | Mean max | Mean min | Mean temp | Anom | Highest max | Date | Lowest min | Date | Lowest max | Date | Highest min | Date | Air frost | Days ≥ 25 °C | Total pptn | Anom | Rain days | Wettest day | Date | Total sunshine | Anom | Sunniest day | Date |
|---|---|---|---|---|---|---|---|---|---|---|---|---|---|---|---|---|---|---|---|---|---|---|---|
| Jan | 5.4 | 0.8 | 3.1 | *−1.8* | 12.8 | 29 | −4.4 | 20 | 1.1 | 19 | 8.3 | 29 | 13 | 0 | 21.8 | *38* | 15 | 9.1 | 5 | 42.5 | *68* | 7.0 | 20 |
| Feb | 7.3 | 1.1 | 4.2 | *−0.7* | 15.0 | 27 | −4.4 | 15 | 1.7 | 11 | 8.3 | 22 | 11 | 0 | 40.4 | *94* | 13 | 15.7 | 9 | 75.9 | *96* | 7.7 | 27 |
| Mar | 10.8 | 0.3 | 5.6 | *−1.7* | 20.6 | 25 | −5.0 | 3 | 2.8 | 2 | 9.4 | 29 | 17 | 0 | 17.2 | *36* | 9 | 7.6 | 29 | 140.7 | *127* | 9.9 | 15 |
| Apr | 12.5 | 3.5 | 8.0 | *−1.3* | 18.3 | 22 | −0.6 | 8 | 10.0 | 5 | 6.7 | 17 | 1 | 0 | 53.0 | *108* | 16 | 9.9 | 29 | 183.3 | *114* | 13.1 | 22 |
| May | 17.9 | 8.2 | 13.1 | *+0.6* | 29.4 | 25 | 0.6 | 11 | 11.1 | 1 | 14.4 | 25 | 0 | 2 | 42.8 | *75* | 11 | 15.7 | 18 | 223.0 | *116* | 14.4 | 3 |
| June | 18.7 | 10.5 | 14.6 | *−1.0* | 26.7 | 26 | 5.0 | 2 | 10.0 | 3 | 15.0 | 25 | 0 | 2 | 40.5 | *82* | 15 | 7.9 | 21 | 139.7 | *73* | 11.5 | 9 |
| July | 20.4 | 11.7 | 16.0 | *−1.8* | 24.4 | 5 | 7.2 | 11 | 17.2 | 31 | 15.6 | 6 | 0 | 0 | 82.1 | *168* | 21 | 13.5 | 12 | 203.4 | *98* | 13.3 | 5 |
| Aug | 21.7 | 12.2 | 17.0 | *−0.6* | 32.2 | 12 | 7.8 | 19 | 17.2 | 22 | 17.2 | 7 | 0 | 4 | 74.1 | *135* | 9 | 19.6 | 23 | 237.4 | *121* | 13.7 | 3 |
| Sep | 18.6 | 10.0 | 14.3 | *−0.6* | 24.4 | 8 | 5.6 | 25 | 15.6 | 28 | 15.0 | 2 | 0 | 0 | 47.3 | *87* | 16 | 13.0 | 21 | 151.7 | *108* | 11.1 | 1 |
| Oct | 14.0 | 6.5 | 10.2 | *−1.0* | 20.0 | 1 | 1.7 | 6 | 9.4 | 14 | 15.0 | 1 | 0 | 0 | 56.0 | *80* | 13 | 18.3 | 12 | 84.1 | *76* | 8.7 | 4 |
| Nov | 11.3 | 5.9 | 8.6 | *+1.0* | 14.4 | 28 | 1.7 | 3 | 7.2 | 22 | 10.6 | 13 | 0 | 0 | 39.7 | *59* | 16 | 18.5 | 1 | 54.4 | *77* | 8.3 | 4 |
| Dec | 9.5 | 5.0 | 7.2 | *+2.2* | 13.9 | 3 | 0.6 | 21 | 5.6 | 9 | 11.1 | 4 | 0 | 0 | 18.0 | *29* | 14 | 4.6 | 30 | 28.4 | *53* | 6.0 | 24 |
| **1953** | 14.0 | 6.3 | 10.1 | *−0.6* | 32.2 | 8 | −5.0 | 3 | 1.1 | 1 | 17.2 | 8 | 42 | 8 | 532.9 | *81* | 168 | 19.6 | 8 | 1564.5 | *99* | 14.4 | 5 |

## OXFORD 1954

| Month | Mean max | Mean min | Mean temp | Anom | Highest max | Date | Lowest min | Date | Lowest max | Date | Highest min | Date | Air frost | Days ≥ 25 °C | Total pptn | Anom | Rain days | Wettest day | Date | Total sunshine | Anom | Sunniest day | Date |
|---|---|---|---|---|---|---|---|---|---|---|---|---|---|---|---|---|---|---|---|---|---|---|---|
| Jan | 5.6 | 0.3 | 2.9 | *−1.9* | 13.3 | 15 | −8.3 | 28 | −2.2 | 28 | 10.6 | 21 | 16 | 0 | 33.0 | *58* | 11 | 10.2 | 13 | 75.7 | *122* | 6.7 | 16 |
| Feb | 6.5 | 0.2 | 3.3 | *−1.6* | 12.8 | 22 | −8.9 | 2 | −2.2 | 1 | 8.3 | 23 | 10 | 0 | 59.7 | *139* | 18 | 10.7 | 9 | 79.7 | *101* | 8.7 | 20 |
| Mar | 10.1 | 2.9 | 6.5 | *−0.8* | 16.7 | 11 | −3.9 | 2 | 2.8 | 15 | 8.9 | 23 | 4 | 0 | 64.4 | *135* | 17 | 15.2 | 30 | 107.9 | *97* | 9.7 | 30 |
| Apr | 12.8 | 2.8 | 7.8 | *−1.5* | 16.1 | 30 | −1.1 | 7 | 9.4 | 6 | 10.0 | 15 | 5 | 0 | 10.1 | *21* | 9 | 5.3 | 1 | 202.4 | *126* | 13.7 | 27 |
| May | 15.9 | 7.2 | 11.5 | *−0.9* | 25.6 | 12 | 1.1 | 8 | 9.4 | 2 | 13.9 | 13 | 0 | 1 | 53.3 | *93* | 17 | 11.7 | 2 | 145.9 | *76* | 13.4 | 8 |
| June | 17.9 | 10.3 | 14.1 | *−1.5* | 21.7 | 4 | 7.2 | 14 | 12.8 | 1 | 15.0 | 18 | 0 | 0 | 92.3 | *187* | 17 | 31.0 | 12 | 157.1 | *82* | 13.1 | 23 |
| July | 18.6 | 11.3 | 15.0 | *−2.9* | 22.2 | 20 | 6.1 | 7 | 15.0 | 5 | 15.6 | 14 | 0 | 0 | 55.8 | *114* | 16 | 15.5 | 25 | 157.1 | *76* | 10.5 | 20 |
| Aug | 19.2 | 11.3 | 15.3 | *−2.3* | 25.0 | 31 | 7.2 | 17 | 14.4 | 19 | 15.6 | 5 | 0 | 1 | 80.1 | *146* | 20 | 11.2 | 4 | 123.5 | *63* | 12.3 | 31 |
| Sep | 17.6 | 9.4 | 13.5 | *−1.4* | 28.3 | 1 | 2.2 | 23 | 12.8 | 29 | 15.6 | 3 | 0 | 1 | 66.3 | *122* | 22 | 19.8 | 23 | 168.8 | *120* | 10.4 | 20 |
| Oct | 15.8 | 9.3 | 12.5 | *+1.2* | 19.4 | 1 | −1.7 | 26 | 11.7 | 25 | 15.6 | 18 | 1 | 0 | 54.9 | *78* | 20 | 11.9 | 31 | 94.5 | *85* | 9.1 | 6 |
| Nov | 10.9 | 3.8 | 7.4 | *−0.2* | 15.0 | 11 | −3.3 | 18 | 6.7 | 15 | 10.0 | 12 | 5 | 0 | 112.2 | *168* | 25 | 16.5 | 26 | 60.1 | *85* | 8.1 | 9 |
| Dec | 9.6 | 4.4 | 7.0 | *+2.0* | 14.4 | 2 | −2.2 | 11 | 3.9 | 7 | 10.6 | 3 | 3 | 0 | 51.4 | *81* | 14 | 28.7 | 8 | 54.5 | *102* | 6.7 | 1 |
| **1954** | 13.4 | 6.1 | 9.7 | *−1.0* | 28.3 | 9 | −8.9 | 2 | −2.2 | 1 | 15.6 | 7 | 44 | 3 | 733.5 | *111* | 206 | 31.0 | 6 | 1427.2 | *91* | 13.7 | 4 |

# OXFORD 1955

| Month | Mean max | Mean min | Mean temp | Anom | Highest max | Date | Lowest min | Date | Lowest max | Date | Highest min | Date | Air frost | Days ≥ 25 °C | Total pptn | Anom | Rain days | Wettest day | Date | Total sunshine | Anom | Sunniest day | Date |
|---|---|---|---|---|---|---|---|---|---|---|---|---|---|---|---|---|---|---|---|---|---|---|---|
| Jan | 5.4 | 0.3 | 2.9 | *−2.0* | 11.7 | 10 | −8.3 | 20 | 0.6 | 4 | 8.3 | 30 | 13 | 0 | 62.7 | *110* | 18 | 11.4 | 16 | 40.1 | *64* | 6.7 | 18 |
| Feb | 4.5 | −1.1 | 1.7 | *−3.2* | 12.2 | 7 | −8.3 | 28 | −0.6 | 26 | 5.6 | 1 | 18 | 0 | 36.3 | *85* | 18 | 6.9 | 4 | 86.3 | *109* | 7.3 | 28 |
| Mar | 7.5 | −0.4 | 3.5 | *−3.8* | 14.4 | 25 | −5.6 | 31 | 1.7 | 7 | 8.3 | 24 | 19 | 0 | 36.5 | *77* | 12 | 11.9 | 25 | 161.4 | *145* | 11.2 | 30 |
| Apr | 14.6 | 5.6 | 10.1 | *+0.8* | 18.3 | 11 | 0.6 | 16 | 10.0 | 18 | 11.1 | 29 | 0 | 0 | 16.6 | *34* | 11 | 5.6 | 7 | 173.7 | *108* | 12.9 | 24 |
| May | 14.6 | 6.0 | 10.3 | *−2.2* | 19.4 | 30 | 1.1 | 18 | 7.8 | 17 | 11.1 | 24 | 0 | 0 | 113.5 | *198* | 20 | 21.6 | 16 | 217.5 | *113* | 15.2 | 30 |
| June | 19.3 | 10.1 | 14.7 | *−0.9* | 23.9 | 6 | 3.3 | 10 | 12.2 | 9 | 15.0 | 23 | 0 | 0 | 107.4 | *218* | 13 | 39.4 | 8 | 155.3 | *81* | 12.3 | 25 |
| July | 23.7 | 12.6 | 18.1 | *+0.3* | 30.0 | 17 | 7.2 | 4 | 16.7 | 3 | 17.8 | 17 | 0 | 12 | 7.4 | *15* | 2 | 3.8 | 3 | 268.3 | *130* | 14.0 | 12 |
| Aug | 24.3 | 13.6 | 18.9 | *+1.3* | 30.6 | 22 | 6.1 | 8 | 17.8 | 9 | 17.8 | 19 | 0 | 15 | 11.5 | *21* | 6 | 5.1 | 17 | 188.0 | *96* | 13.1 | 8 |
| Sep | 19.4 | 10.0 | 14.7 | *−0.2* | 23.9 | 2 | 3.3 | 19 | 15.0 | 14 | 16.1 | 2 | 0 | 0 | 38.3 | *71* | 11 | 11.9 | 22 | 153.1 | *109* | 10.0 | 7 |
| Oct | 13.9 | 5.4 | 9.6 | *−1.7* | 20.0 | 9 | −3.3 | 31 | 7.8 | 31 | 11.7 | 15 | 4 | 0 | 40.1 | *57* | 15 | 11.2 | 19 | 119.7 | *108* | 8.9 | 16 |
| Nov | 10.4 | 3.7 | 7.0 | *−0.5* | 17.2 | 4 | −4.4 | 1 | 6.1 | 16 | 10.0 | 7 | 6 | 0 | 35.4 | *53* | 12 | 8.4 | 6 | 71.5 | *101* | 7.9 | 1 |
| Dec | 9.4 | 2.2 | 5.8 | *+0.8* | 13.9 | 28 | −6.1 | 19 | 3.3 | 18 | 10.0 | 7 | 6 | 0 | 79.1 | *125* | 26 | 12.7 | 23 | 47.3 | *89* | 5.0 | 12 |
| **1955** | 13.9 | 5.6 | 9.8 | *−1.0* | 30.6 | 8 | −8.3 | 1 | −0.6 | 2 | 17.8 | 7 | 66 | 27 | 584.8 | *89* | 164 | 39.4 | 6 | 1682.2 | *107* | 15.2 | 5 |

# OXFORD 1956

| Month | Mean max | Mean min | Mean temp | Anom | Highest max | Date | Lowest min | Date | Lowest max | Date | Highest min | Date | Air frost | Days ≥ 25 °C | Total pptn | Anom | Rain days | Wettest day | Date | Total sunshine | Anom | Sunniest day | Date |
|---|---|---|---|---|---|---|---|---|---|---|---|---|---|---|---|---|---|---|---|---|---|---|---|
| Jan | 6.7 | 1.0 | 3.8 | *−1.0* | 12.2 | 29 | −4.4 | 5 | −1.7 | 5 | 6.1 | 27 | 13 | 0 | 99.5 | *175* | 21 | 23.9 | 30 | 50.3 | *81* | 6.8 | 18 |
| Feb | 2.5 | −3.5 | −0.5 | *−5.4* | 12.2 | 28 | −8.9 | 1 | −4.4 | 1 | 5.6 | 29 | 21 | 0 | 11.4 | *27* | 17 | 1.3 | 6 | 78.9 | *100* | 8.5 | 25 |
| Mar | 10.8 | 2.2 | 6.5 | *−0.7* | 16.7 | 26 | −3.9 | 11 | 3.9 | 13 | 8.9 | 3 | 8 | 0 | 17.3 | *36* | 9 | 5.6 | 22 | 149.7 | *135* | 9.9 | 9 |
| Apr | 12.2 | 2.6 | 7.4 | *−1.9* | 17.8 | 10 | −0.6 | 20 | 5.6 | 14 | 6.1 | 11 | 2 | 0 | 48.1 | *98* | 13 | 13.7 | 13 | 158.7 | *99* | 12.4 | 21 |
| May | 18.2 | 7.1 | 12.6 | *+0.2* | 23.3 | 23 | 0.6 | 19 | 12.8 | 18 | 12.2 | 9 | 0 | 0 | 8.5 | *15* | 8 | 2.8 | 9 | 246.6 | *128* | 14.0 | 27 |
| June | 17.9 | 9.6 | 13.7 | *−1.9* | 23.9 | 25 | 5.0 | 15 | 12.8 | 8 | 13.9 | 22 | 0 | 0 | 62.7 | *127* | 13 | 11.9 | 8 | 145.0 | *76* | 13.3 | 27 |
| July | 20.8 | 12.5 | 16.7 | *−1.2* | 27.8 | 26 | 8.3 | 31 | 14.4 | 13 | 16.7 | 8 | 0 | 5 | 62.0 | *127* | 14 | 18.0 | 8 | 168.0 | *81* | 12.7 | 2 |
| Aug | 18.3 | 10.6 | 14.5 | *−3.1* | 20.6 | 9 | 6.7 | 5 | 13.3 | 30 | 15.0 | 11 | 0 | 0 | 114.9 | *210* | 23 | 16.5 | 26 | 161.4 | *82* | 12.9 | 14 |
| Sep | 18.4 | 11.1 | 14.8 | *−0.1* | 25.0 | 24 | 6.7 | 1 | 15.0 | 1 | 16.1 | 23 | 0 | 1 | 60.0 | *111* | 12 | 16.5 | 4 | 94.3 | *67* | 9.1 | 24 |
| Oct | 13.6 | 5.8 | 9.7 | *−1.6* | 17.2 | 15 | 0.0 | 28 | 8.9 | 30 | 12.8 | 17 | 0 | 0 | 45.9 | *66* | 14 | 12.7 | 2 | 116.5 | *105* | 8.5 | 6 |
| Nov | 9.1 | 2.9 | 6.0 | *−1.5* | 15.0 | 8 | −5.6 | 22 | 2.2 | 22 | 8.9 | 26 | 4 | 0 | 19.7 | *30* | 15 | 7.9 | 8 | 77.9 | *110* | 6.5 | 6 |
| Dec | 8.2 | 3.5 | 5.8 | *+0.8* | 13.9 | 12 | −2.8 | 22 | 0.0 | 21 | 8.9 | 16 | 10 | 0 | 80.7 | *128* | 20 | 13.0 | 23 | 18.2 | *34* | 4.3 | 13 |
| **1956** | 13.1 | 5.5 | 9.3 | *−1.5* | 27.8 | 7 | −8.9 | 2 | −4.4 | 2 | 16.7 | 7 | 58 | 6 | 630.7 | *96* | 179 | 23.9 | 1 | 1465.5 | *93* | 14.0 | 5 |

## OXFORD 1957

| Month | Mean max | Mean min | Mean temp | Anom | Highest max | Date | Lowest min | Date | Lowest max | Date | Highest min | Date | Air frost | Days ≥ 25 °C | Total pptn | Anom | Rain days | Wettest day | Date | Total sunshine | Anom | Sunniest day | Date |
|---|---|---|---|---|---|---|---|---|---|---|---|---|---|---|---|---|---|---|---|---|---|---|---|
| Jan | 8.6 | 3.0 | 5.8 | *+0.9* | 13.9 | 5 | −1.1 | 2 | 2.2 | 14 | 11.7 | 5 | 4 | 0 | 41.0 | *72* | 18 | 7.9 | 31 | 50.8 | *82* | 6.1 | 10 |
| Feb | 9.1 | 3.2 | 6.1 | *+1.2* | 12.8 | 1 | −2.2 | 20 | 3.3 | 19 | 10.0 | 1 | 6 | 0 | 71.7 | *167* | 17 | 14.2 | 7 | 83.1 | *105* | 8.8 | 16 |
| Mar | 13.6 | 5.9 | 9.8 | *+2.5* | 18.3 | 11 | −2.2 | 4 | 9.4 | 5 | 10.6 | 10 | 1 | 0 | 43.1 | *90* | 13 | 10.4 | 9 | 110.2 | *99* | 9.4 | 22 |
| Apr | 13.5 | 5.1 | 9.3 | *+0.0* | 18.3 | 4 | −0.6 | 12 | 8.3 | 8 | 8.9 | 18 | 1 | 0 | 9.9 | *20* | 6 | 6.3 | 22 | 168.0 | *104* | 13.7 | 29 |
| May | 15.4 | 6.2 | 10.8 | *−1.7* | 22.2 | 31 | 0.6 | 6 | 8.3 | 7 | 10.6 | 11 | 0 | 0 | 42.4 | *74* | 13 | 10.4 | 11 | 205.6 | *107* | 15.0 | 26 |
| June | 22.2 | 10.1 | 16.1 | *+0.6* | 30.0 | 28 | 4.4 | 12 | 16.7 | 6 | 16.1 | 30 | 0 | 10 | 55.7 | *113* | 11 | 33.5 | 30 | 297.1 | *156* | 15.5 | 15 |
| July | 21.4 | 13.4 | 17.4 | *−0.5* | 29.4 | 6 | 9.4 | 11 | 17.2 | 15 | 17.2 | 6 | 0 | 6 | 85.4 | *175* | 21 | 18.0 | 11 | 157.9 | *76* | 12.9 | 1 |
| Aug | 20.4 | 12.1 | 16.2 | *−1.4* | 26.7 | 5 | 7.8 | 27 | 15.0 | 15 | 16.7 | 6 | 0 | 3 | 95.9 | *175* | 16 | 56.1 | 12 | 170.5 | *87* | 12.2 | 24 |
| Sep | 16.9 | 9.7 | 13.3 | *−1.6* | 20.6 | 22 | 2.8 | 30 | 11.7 | 30 | 13.9 | 1 | 0 | 0 | 101.5 | *187* | 19 | 24.4 | 25 | 102.6 | *73* | 8.9 | 12 |
| Oct | 15.4 | 7.6 | 11.5 | *+0.2* | 18.9 | 14 | 2.2 | 20 | 12.2 | 2 | 13.3 | 30 | 0 | 0 | 52.6 | *75* | 14 | 13.0 | 16 | 92.9 | *83* | 8.1 | 4 |
| Nov | 8.8 | 3.7 | 6.3 | *−1.3* | 12.8 | 3 | −1.7 | 8 | 5.0 | 14 | 7.8 | 12 | 3 | 0 | 50.5 | *76* | 9 | 19.0 | 3 | 76.4 | *108* | 6.5 | 4 |
| Dec | 7.4 | 1.6 | 4.5 | *−0.5* | 13.3 | 20 | −5.0 | 2 | 0.6 | 4 | 7.2 | 8 | 10 | 0 | 64.4 | *102* | 14 | 17.3 | 12 | 62.3 | *117* | 6.0 | 15 |
| **1957** | 14.4 | 6.8 | 10.6 | *−0.2* | 30.0 | 6 | −5.0 | 12 | 0.6 | 12 | 17.2 | 7 | 25 | 19 | 714.1 | *108* | 171 | 56.1 | 8 | 1577.4 | *100* | 15.5 | 6 |

## OXFORD 1958

| Month | Mean max | Mean min | Mean temp | Anom | Highest max | Date | Lowest min | Date | Lowest max | Date | Highest min | Date | Air frost | Days ≥ 25 °C | Total pptn | Anom | Rain days | Wettest day | Date | Total sunshine | Anom | Sunniest day | Date |
|---|---|---|---|---|---|---|---|---|---|---|---|---|---|---|---|---|---|---|---|---|---|---|---|
| Jan | 6.6 | 1.0 | 3.8 | *−1.1* | 12.8 | 6 | −6.1 | 24 | 0.6 | 21 | 8.3 | 28 | 11 | 0 | 90.2 | *159* | 17 | 36.1 | 28 | 54.6 | *88* | 6.7 | 24 |
| Feb | 9.0 | 2.3 | 5.6 | *+0.7* | 15.0 | 14 | −2.8 | 8 | 0.6 | 25 | 11.1 | 15 | 8 | 0 | 65.4 | *153* | 17 | 16.0 | 24 | 55.4 | *70* | 8.8 | 26 |
| Mar | 7.8 | 0.9 | 4.3 | *−2.9* | 13.9 | 4 | −5.6 | 10 | 1.1 | 10 | 8.9 | 28 | 14 | 0 | 34.6 | *73* | 12 | 8.6 | 25 | 110.7 | *100* | 8.2 | 9 |
| Apr | 12.1 | 3.6 | 7.9 | *−1.4* | 21.7 | 30 | −2.8 | 2 | 5.0 | 3 | 9.4 | 26 | 6 | 0 | 27.5 | *56* | 8 | 9.1 | 4 | 145.2 | *90* | 12.6 | 12 |
| May | 16.4 | 7.9 | 12.2 | *−0.3* | 23.9 | 2 | 3.9 | 27 | 11.7 | 24 | 12.8 | 6 | 0 | 0 | 54.4 | *95* | 19 | 11.7 | 15 | 196.0 | *102* | 12.7 | 21 |
| June | 19.1 | 10.7 | 14.9 | *−0.7* | 23.9 | 16 | 7.2 | 25 | 15.0 | 2 | 13.3 | 7 | 0 | 0 | 84.2 | *171* | 18 | 23.1 | 2 | 145.7 | *76* | 12.8 | 18 |
| July | 20.8 | 12.3 | 16.6 | *−1.3* | 26.7 | 8 | 7.2 | 23 | 16.7 | 5 | 15.0 | 20 | 0 | 4 | 63.6 | *130* | 20 | 10.9 | 25 | 188.1 | *91* | 13.6 | 8 |
| Aug | 20.2 | 12.7 | 16.5 | *−1.1* | 26.7 | 10 | 8.3 | 6 | 17.2 | 21 | 16.1 | 10 | 0 | 1 | 67.3 | *123* | 21 | 13.0 | 18 | 119.6 | *61* | 12.0 | 2 |
| Sep | 19.3 | 11.8 | 15.5 | *+0.6* | 23.9 | 1 | 4.4 | 27 | 16.1 | 27 | 17.2 | 6 | 0 | 0 | 80.3 | *148* | 14 | 22.1 | 23 | 118.0 | *84* | 9.5 | 1 |
| Oct | 14.4 | 8.1 | 11.2 | *−0.0* | 17.8 | 8 | 2.2 | 29 | 11.1 | 28 | 11.7 | 4 | 0 | 0 | 72.5 | *104* | 15 | 14.5 | 2 | 93.7 | *84* | 8.8 | 16 |
| Nov | 9.2 | 4.2 | 6.7 | *−0.8* | 13.9 | 2 | −0.6 | 14 | 5.0 | 25 | 11.1 | 3 | 1 | 0 | 77.0 | *115* | 14 | 26.9 | 1 | 51.8 | *73* | 7.5 | 13 |
| Dec | 7.4 | 2.7 | 5.0 | *+0.0* | 12.8 | 27 | −1.1 | 2 | 2.8 | 23 | 9.4 | 20 | 4 | 0 | 84.4 | *134* | 21 | 14.5 | 30 | 29.5 | *55* | 5.5 | 14 |
| **1958** | 13.5 | 6.5 | 10.0 | *−0.7* | 26.7 | 7 | −6.1 | 1 | 0.6 | 1 | 17.2 | 9 | 44 | 5 | 801.4 | *121* | 196 | 36.1 | 1 | 1308.3 | *83* | 13.6 | 7 |

## OXFORD 1959

| Month | Mean max | Mean min | Mean temp | Anom | Highest max | Date | Lowest min | Date | Lowest max | Date | Highest min | Date | Air frost | Days ≥ 25 °C | Total pptn | Anom | Rain days | Wettest day | Date | Total sunshine | Anom | Sunniest day | Date |
|---|---|---|---|---|---|---|---|---|---|---|---|---|---|---|---|---|---|---|---|---|---|---|---|
| Jan | 4.9 | −0.6 | 2.1 | *−2.7* | 11.1 | 22 | −5.6 | 16 | −1.7 | 14 | 7.2 | 20 | 20 | 0 | 94.1 | *165* | 18 | 18.3 | 6 | 85.8 | *138* | 7.5 | 27 |
| Feb | 7.1 | 1.3 | 4.2 | *−0.7* | 17.8 | 28 | −3.3 | 6 | 0.6 | 6 | 6.7 | 22 | 12 | 0 | 2.2 | *5* | 6 | 0.5 | 9 | 65.5 | *83* | 9.7 | 28 |
| Mar | 11.4 | 4.1 | 7.7 | *+0.5* | 16.1 | 1 | 0.0 | 19 | 3.9 | 18 | 8.9 | 3 | 0 | 0 | 86.6 | *182* | 21 | 24.4 | 3 | 102.0 | *92* | 10.2 | 28 |
| Apr | 14.2 | 6.3 | 10.2 | *+0.9* | 19.4 | 4 | 2.2 | 19 | 9.4 | 19 | 10.0 | 3 | 0 | 0 | 52.2 | *107* | 16 | 13.7 | 16 | 153.8 | *96* | 11.7 | 29 |
| May | 18.3 | 7.7 | 13.0 | *+0.5* | 26.1 | 12 | 1.7 | 5 | 12.2 | 19 | 13.9 | 12 | 0 | 2 | 25.8 | *45* | 9 | 10.4 | 21 | 228.0 | *118* | 14.4 | 15 |
| June | 21.0 | 10.9 | 16.0 | *+0.4* | 26.1 | 24 | 7.2 | 2 | 16.7 | 6 | 16.1 | 23 | 0 | 2 | 25.2 | *51* | 10 | 8.1 | 28 | 226.3 | *118* | 15.7 | 17 |
| July | 23.9 | 12.8 | 18.4 | *+0.5* | 32.8 | 5 | 7.8 | 14 | 19.4 | 13 | 16.7 | 5 | 0 | 12 | 64.9 | *133* | 10 | 19.6 | 27 | 272.7 | *132* | 15.3 | 4 |
| Aug | 23.3 | 13.0 | 18.1 | *+0.6* | 28.9 | 20 | 7.2 | 29 | 18.3 | 1 | 17.2 | 10 | 0 | 10 | 44.0 | *80* | 7 | 11.7 | 10 | 245.6 | *125* | 13.2 | 18 |
| Sep | 21.8 | 9.4 | 15.6 | *+0.7* | 27.2 | 10 | 2.2 | 28 | 18.3 | 18 | 14.4 | 22 | 0 | 6 | 5.4 | *10* | 3 | 4.1 | 21 | 213.9 | *152* | 11.1 | 9 |
| Oct | 17.3 | 7.8 | 12.6 | *+1.3* | 25.6 | 3 | 1.7 | 29 | 9.4 | 29 | 13.3 | 2 | 0 | 1 | 49.1 | *70* | 15 | 11.7 | 26 | 144.2 | *130* | 10.0 | 6 |
| Nov | 10.6 | 3.7 | 7.1 | *−0.4* | 15.0 | 1 | −5.6 | 12 | 6.1 | 15 | 10.0 | 1 | 7 | 0 | 44.9 | *67* | 16 | 9.9 | 17 | 70.6 | *99* | 7.7 | 6 |
| Dec | 9.2 | 3.4 | 6.3 | *+1.3* | 12.2 | 16 | −1.1 | 2 | 5.0 | 1 | 6.7 | 17 | 2 | 0 | 124.0 | *197* | 26 | 30.0 | 14 | 45.0 | *85* | 5.1 | 21 |
| **1959** | 15.2 | 6.7 | 10.9 | *+0.2* | 32.8 | 7 | −5.6 | 1 | −1.7 | 1 | 17.2 | 8 | 41 | 33 | 618.4 | *94* | 157 | 30.0 | 12 | 1853.4 | *118* | 15.7 | 6 |

## OXFORD 1960

| Month | Mean max | Mean min | Mean temp | Anom | Highest max | Date | Lowest min | Date | Lowest max | Date | Highest min | Date | Air frost | Days ≥ 25 °C | Total pptn | Anom | Rain days | Wettest day | Date | Total sunshine | Anom | Sunniest day | Date |
|---|---|---|---|---|---|---|---|---|---|---|---|---|---|---|---|---|---|---|---|---|---|---|---|
| Jan | 6.5 | 1.6 | 4.1 | *−0.8* | 12.2 | 4 | −6.7 | 14 | −1.1 | 13 | 10.6 | 23 | 12 | 0 | 82.3 | *145* | 25 | 21.6 | 27 | 37.8 | *61* | 6.0 | 6 |
| Feb | 7.5 | 1.9 | 4.7 | *−0.2* | 16.7 | 28 | −3.9 | 24 | 1.7 | 13 | 10.6 | 29 | 6 | 0 | 50.6 | *118* | 13 | 11.2 | 24 | 82.3 | *104* | 8.6 | 19 |
| Mar | 9.8 | 3.9 | 6.8 | *−0.4* | 15.6 | 24 | −0.6 | 22 | 2.8 | 8 | 8.9 | 2 | 1 | 0 | 24.9 | *52* | 14 | 10.2 | 29 | 78.0 | *70* | 9.2 | 22 |
| Apr | 14.0 | 4.8 | 9.4 | *+0.1* | 18.9 | 20 | −1.1 | 1 | 9.4 | 15 | 10.0 | 6 | 3 | 0 | 16.9 | *35* | 12 | 5.6 | 2 | 164.9 | *102* | 11.5 | 17 |
| May | 17.8 | 8.8 | 13.3 | *+0.8* | 23.9 | 8 | 4.4 | 2 | 10.6 | 1 | 14.4 | 9 | 0 | 0 | 61.4 | *107* | 9 | 20.3 | 12 | 182.3 | *95* | 14.9 | 24 |
| June | 21.8 | 11.9 | 16.8 | *+1.3* | 27.8 | 5 | 8.3 | 15 | 15.6 | 24 | 16.1 | 18 | 0 | 7 | 111.0 | *225* | 15 | 81.3 | 22 | 274.9 | *144* | 15.3 | 20 |
| July | 19.6 | 11.9 | 15.7 | *−2.1* | 21.7 | 4 | 8.3 | 1 | 16.7 | 23 | 15.0 | 4 | 0 | 0 | 110.1 | *225* | 21 | 17.0 | 10 | 129.7 | *63* | 9.7 | 31 |
| Aug | 20.0 | 11.4 | 15.7 | *−1.9* | 23.3 | 22 | 7.2 | 13 | 14.4 | 11 | 15.6 | 22 | 0 | 0 | 50.5 | *92* | 15 | 11.7 | 24 | 147.1 | *75* | 9.4 | 15 |
| Sep | 17.8 | 9.8 | 13.8 | *−1.1* | 23.3 | 11 | 4.4 | 24 | 11.7 | 30 | 15.6 | 3 | 0 | 0 | 104.8 | *193* | 12 | 25.4 | 1 | 122.8 | *87* | 10.3 | 11 |
| Oct | 13.8 | 7.9 | 10.9 | *−0.4* | 19.4 | 2 | 0.6 | 13 | 10.0 | 29 | 12.2 | 4 | 0 | 0 | 138.6 | *198* | 25 | 19.6 | 31 | 60.9 | *55* | 9.7 | 12 |
| Nov | 10.8 | 4.7 | 7.8 | *+0.2* | 15.6 | 1 | −1.7 | 8 | 4.4 | 8 | 8.3 | 21 | 1 | 0 | 106.1 | *159* | 26 | 18.0 | 17 | 73.4 | *103* | 7.4 | 16 |
| Dec | 6.6 | 2.2 | 4.4 | *−0.6* | 12.2 | 1 | −3.3 | 13 | 1.7 | 13 | 8.9 | 1 | 2 | 0 | 107.5 | *170* | 22 | 28.7 | 3 | 54.5 | *102* | 6.8 | 2 |
| **1960** | 13.8 | 6.7 | 10.3 | *−0.5* | 27.8 | 6 | −6.7 | 1 | −1.1 | 1 | 16.1 | 6 | 25 | 7 | 964.7 | *146* | 209 | 81.3 | 6 | 1408.6 | *89* | 15.3 | 6 |

## OXFORD 1961

| Month | Mean max | Mean min | Mean temp | Anom | Highest max | Date | Lowest min | Date | Lowest max | Date | Highest min | Date | Air frost | Days ≥ 25 °C | Total pptn | Anom | Rain days | Wettest day | Date | Total sunshine | Anom | Sunniest day | Date |
|---|---|---|---|---|---|---|---|---|---|---|---|---|---|---|---|---|---|---|---|---|---|---|---|
| Jan | 6.6 | 2.0 | 4.3 | *−0.6* | 11.7 | 29 | −1.7 | 12 | 1.0 | 25 | 5.8 | 30 | 5 | 0 | 86.1 | *151* | 17 | 17.5 | 28 | 43.6 | *70* | 5.6 | 11 |
| Feb | 10.3 | 5.0 | 7.7 | *+2.8* | 16.1 | 14 | 2.2 | 23 | 5.8 | 21 | 8.5 | 9 | 0 | 0 | 53.4 | *125* | 15 | 10.4 | 27 | 65.1 | *82* | 7.8 | 9 |
| Mar | 13.6 | 3.8 | 8.7 | *+1.4* | 20.7 | 15 | −1.7 | 20 | 7.5 | 19 | 9.3 | 8 | 3 | 0 | 4.5 | *9* | 6 | 1.8 | 18 | 173.2 | *156* | 10.1 | 16 |
| Apr | 14.4 | 7.1 | 10.8 | *+1.5* | 17.9 | 18 | 3.3 | 4 | 6.8 | 3 | 10.3 | 21 | 0 | 0 | 75.9 | *155* | 21 | 11.7 | 10 | 86.9 | *54* | 10.5 | 24 |
| May | 16.3 | 7.2 | 11.8 | *−0.7* | 23.9 | 13 | 1.6 | 27 | 12.3 | 15 | 10.6 | 5 | 0 | 0 | 28.0 | *49* | 10 | 10.4 | 3 | 225.2 | *117* | 14.4 | 31 |
| June | 20.9 | 10.3 | 15.6 | *−0.0* | 29.6 | 30 | 6.3 | 14 | 14.9 | 12 | 14.6 | 18 | 0 | 5 | 33.9 | *69* | 9 | 25.4 | 12 | 231.2 | *121* | 15.2 | 29 |
| July | 21.0 | 12.0 | 16.5 | *−1.4* | 28.7 | 1 | 7.6 | 6 | 15.6 | 16 | 17.8 | 2 | 0 | 1 | 54.3 | *111* | 9 | 17.5 | 12 | 196.5 | *95* | 14.0 | 30 |
| Aug | 21.1 | 12.1 | 16.6 | *−1.0* | 30.7 | 29 | 8.2 | 3 | 17.3 | 11 | 15.6 | 4 | 0 | 3 | 59.1 | *108* | 18 | 16.8 | 10 | 185.7 | *94* | 12.7 | 30 |
| Sep | 20.4 | 11.6 | 16.0 | *+1.1* | 28.4 | 2 | 6.1 | 26 | 15.8 | 6 | 17.1 | 16 | 0 | 2 | 56.9 | *105* | 16 | 18.8 | 13 | 135.8 | *97* | 10.9 | 9 |
| Oct | 15.0 | 7.6 | 11.3 | *+0.0* | 20.3 | 10 | −0.8 | 30 | 9.4 | 19 | 14.7 | 10 | 1 | 0 | 67.7 | *97* | 21 | 12.2 | 23 | 118.2 | *106* | 8.6 | 5 |
| Nov | 9.2 | 3.1 | 6.2 | *−1.4* | 15.6 | 1 | −2.8 | 27 | 5.5 | 26 | 9.5 | 2 | 7 | 0 | 36.9 | *55* | 15 | 7.9 | 10 | 63.4 | *89* | 8.4 | 4 |
| Dec | 5.2 | −0.2 | 2.5 | *−2.5* | 14.6 | 10 | −9.3 | 29 | −1.6 | 28 | 11.3 | 11 | 18 | 0 | 97.8 | *155* | 14 | 26.7 | 29 | 76.5 | *144* | 7.2 | 2 |
| **1961** | 14.5 | 6.8 | 10.7 | *−0.1* | 30.7 | 8 | −9.3 | 12 | −1.6 | 12 | 17.8 | 7 | 34 | 11 | 654.5 | *99* | 171 | 26.7 | 12 | 1601.3 | *102* | 15.2 | 6 |

## OXFORD 1962

| Month | Mean max | Mean min | Mean temp | Anom | Highest max | Date | Lowest min | Date | Lowest max | Date | Highest min | Date | Air frost | Days ≥ 25 °C | Total pptn | Anom | Rain days | Wettest day | Date | Total sunshine | Anom | Sunniest day | Date |
|---|---|---|---|---|---|---|---|---|---|---|---|---|---|---|---|---|---|---|---|---|---|---|---|
| Jan | 7.5 | 1.7 | 4.6 | *−0.2* | 12.7 | 24 | −10.0 | 2 | −2.2 | 1 | 9.0 | 26 | 7 | 0 | 94.7 | *167* | 18 | 17.0 | 21 | 70.0 | *112* | 7.2 | 13 |
| Feb | 7.5 | 1.6 | 4.6 | *−0.3* | 11.8 | 10 | −2.9 | 26 | −0.1 | 26 | 6.7 | 7 | 8 | 0 | 10.3 | *24* | 9 | 2.5 | 12 | 70.6 | *89* | 8.6 | 13 |
| Mar | 7.3 | −0.7 | 3.3 | *−4.0* | 14.7 | 29 | −6.1 | 17 | 2.3 | 1 | 7.1 | 10 | 18 | 0 | 32.4 | *68* | 12 | 16.5 | 28 | 126.3 | *114* | 10.4 | 13 |
| Apr | 12.0 | 4.5 | 8.3 | *−1.0* | 20.7 | 26 | −0.2 | 14 | 7.2 | 15 | 8.3 | 24 | 1 | 0 | 50.7 | *104* | 15 | 8.9 | 4 | 143.3 | *89* | 12.6 | 25 |
| May | 14.8 | 7.0 | 10.9 | *−1.6* | 18.8 | 6 | 0.6 | 1 | 8.9 | 26 | 12.2 | 6 | 0 | 0 | 38.6 | *68* | 18 | 11.7 | 20 | 177.5 | *92* | 14.9 | 31 |
| June | 19.8 | 9.3 | 14.6 | *−1.0* | 25.0 | 9 | 1.6 | 1 | 13.8 | 1 | 15.9 | 18 | 0 | 1 | 5.5 | *11* | 6 | 2.8 | 28 | 280.3 | *147* | 15.5 | 8 |
| July | 20.2 | 11.8 | 16.0 | *−1.9* | 24.4 | 25 | 7.1 | 7 | 14.2 | 26 | 15.0 | 19 | 0 | 0 | 54.6 | *112* | 11 | 20.6 | 14 | 137.3 | *66* | 11.4 | 21 |
| Aug | 19.3 | 11.7 | 15.5 | *−2.1* | 22.2 | 1 | 7.8 | 9 | 16.1 | 6 | 15.7 | 20 | 0 | 0 | 96.1 | *175* | 15 | 53.6 | 6 | 154.5 | *79* | 11.8 | 4 |
| Sep | 17.0 | 9.5 | 13.2 | *−1.6* | 23.4 | 2 | 4.1 | 18 | 12.3 | 21 | 14.7 | 12 | 0 | 0 | 115.0 | *212* | 17 | 25.7 | 28 | 142.3 | *101* | 10.8 | 9 |
| Oct | 14.9 | 6.8 | 10.8 | *−0.5* | 19.7 | 5 | −0.7 | 27 | 9.3 | 26 | 11.9 | 1 | 1 | 0 | 29.6 | *42* | 9 | 11.9 | 25 | 108.8 | *98* | 7.4 | 8 |
| Nov | 8.3 | 2.9 | 5.6 | *−1.9* | 13.0 | 4 | −2.9 | 20 | 2.5 | 21 | 8.8 | 10 | 7 | 0 | 55.1 | *83* | 14 | 12.4 | 1 | 32.7 | *46* | 7.5 | 15 |
| Dec | 4.2 | −1.5 | 1.3 | *−3.7* | 13.1 | 15 | −8.1 | 6 | −2.2 | 3 | 6.2 | 9 | 18 | 0 | 50.4 | *80* | 13 | 9.7 | 8 | 72.6 | *137* | 6.6 | 24 |
| **1962** | 12.7 | 5.4 | 9.1 | *−1.7* | 25.0 | 6 | −10.0 | 1 | −2.2 | 1 | 15.9 | 6 | 60 | 1 | 633.0 | *96* | 157 | 53.6 | 8 | 1516.2 | *96* | 15.5 | 6 |

## OXFORD 1963

| Month | Mean max | Mean min | Mean temp | Anom | Highest max | Date | Lowest min | Date | Lowest max | Date | Highest min | Date | Air frost | Days ≥ 25 °C | Total pptn | Anom | Rain days | Wettest day | Date | Total sunshine | Anom | Sunniest day | Date |
|---|---|---|---|---|---|---|---|---|---|---|---|---|---|---|---|---|---|---|---|---|---|---|---|
| Jan | −0.3 | −5.8 | −3.0 | *−7.9* | 4.9 | 26 | −14.2 | 23 | −6.7 | 23 | 1.7 | 29 | 28 | 0 | 27.9 | *49* | 12 | 10.2 | 3 | 58.9 | *95* | 7.3 | 21 |
| Feb | 1.7 | −3.1 | −0.7 | *−5.6* | 6.8 | 28 | −9.3 | 25 | −2.1 | 2 | 1.7 | 8 | 25 | 0 | 10.4 | *24* | 11 | 3.3 | 6 | 66.4 | *84* | 9.0 | 27 |
| Mar | 9.8 | 3.0 | 6.4 | *−0.8* | 15.6 | 14 | −5.7 | 2 | 4.7 | 22 | 8.9 | 7 | 5 | 0 | 75.9 | *159* | 20 | 9.4 | 13 | 85.8 | *77* | 9.3 | 1 |
| Apr | 13.2 | 5.7 | 9.4 | *+0.1* | 18.3 | 28 | 1.3 | 12 | 7.6 | 5 | 11.1 | 29 | 0 | 0 | 50.4 | *103* | 20 | 5.8 | 11 | 119.1 | *74* | 11.0 | 22 |
| May | 15.6 | 6.9 | 11.3 | *−1.2* | 26.1 | 31 | 2.7 | 2 | 11.4 | 1 | 10.5 | 26 | 0 | 1 | 43.6 | *76* | 14 | 9.1 | 20 | 193.4 | *100* | 13.8 | 31 |
| June | 20.0 | 11.0 | 15.5 | *−0.1* | 25.8 | 9 | 7.6 | 15 | 14.6 | 29 | 14.1 | 13 | 0 | 1 | 65.9 | *134* | 17 | 13.2 | 13 | 199.4 | *104* | 15.0 | 2 |
| July | 20.4 | 11.4 | 15.9 | *−2.0* | 27.3 | 30 | 7.2 | 28 | 14.7 | 6 | 16.0 | 24 | 0 | 4 | 48.0 | *98* | 14 | 12.2 | 6 | 188.2 | *91* | 14.5 | 27 |
| Aug | 18.7 | 11.1 | 14.9 | *−2.7* | 23.5 | 1 | 7.2 | 29 | 15.7 | 30 | 14.9 | 5 | 0 | 0 | 81.4 | *149* | 23 | 20.6 | 4 | 142.6 | *73* | 13.6 | 1 |
| Sep | 17.5 | 9.7 | 13.6 | *−1.3* | 23.6 | 15 | 4.7 | 13 | 14.2 | 7 | 14.1 | 18 | 0 | 0 | 58.0 | *107* | 12 | 16.5 | 2 | 129.7 | *92* | 10.5 | 15 |
| Oct | 14.8 | 8.5 | 11.6 | *+0.4* | 21.6 | 12 | 3.8 | 2 | 10.8 | 29 | 12.6 | 20 | 0 | 0 | 44.4 | *63* | 14 | 10.9 | 31 | 78.8 | *71* | 8.8 | 9 |
| Nov | 11.3 | 6.3 | 8.8 | *+1.3* | 15.8 | 10 | 0.8 | 21 | 6.4 | 30 | 10.9 | 11 | 0 | 0 | 115.4 | *173* | 25 | 31.5 | 17 | 60.4 | *85* | 6.5 | 8 |
| Dec | 4.9 | 0.0 | 2.5 | *−2.6* | 10.6 | 29 | −6.8 | 24 | 0.9 | 23 | 5.6 | 2 | 15 | 0 | 18.6 | *29* | 14 | 6.6 | 24 | 49.8 | *94* | 6.5 | 21 |
| **1963** | 12.3 | 5.4 | 8.8 | *−1.9* | 27.3 | 7 | −14.2 | 1 | −6.7 | 1 | 16.0 | 7 | 73 | 6 | 639.9 | *97* | 196 | 31.5 | 11 | 1372.5 | *87* | 15.0 | 6 |

## OXFORD 1964

| Month | Mean max | Mean min | Mean temp | Anom | Highest max | Date | Lowest min | Date | Lowest max | Date | Highest min | Date | Air frost | Days ≥ 25 °C | Total pptn | Anom | Rain days | Wettest day | Date | Total sunshine | Anom | Sunniest day | Date |
|---|---|---|---|---|---|---|---|---|---|---|---|---|---|---|---|---|---|---|---|---|---|---|---|
| Jan | 5.6 | 1.1 | 3.3 | *−1.5* | 12.2 | 31 | −7.9 | 14 | 0.3 | 12 | 5.4 | 2 | 9 | 0 | 13.0 | *23* | 8 | 4.6 | 12 | 37.2 | *60* | 6.5 | 30 |
| Feb | 7.4 | 2.5 | 4.9 | *+0.0* | 13.3 | 27 | −4.1 | 6 | 1.2 | 19 | 8.6 | 3 | 7 | 0 | 22.2 | *52* | 11 | 6.1 | 15 | 57.3 | *73* | 7.8 | 5 |
| Mar | 7.2 | 2.1 | 4.6 | *−2.6* | 13.9 | 13 | −2.8 | 11 | 1.9 | 7 | 7.2 | 14 | 7 | 0 | 92.6 | *194* | 14 | 31.2 | 14 | 65.9 | *59* | 10.5 | 26 |
| Apr | 12.6 | 5.6 | 9.1 | *−0.2* | 20.4 | 27 | −0.4 | 6 | 3.9 | 2 | 11.1 | 28 | 1 | 0 | 59.5 | *122* | 16 | 7.6 | 15 | 128.5 | *80* | 11.1 | 8 |
| May | 18.6 | 9.3 | 14.0 | *+1.5* | 24.8 | 12 | 4.7 | 15 | 13.4 | 1 | 13.0 | 18 | 0 | 0 | 44.4 | *78* | 15 | 8.6 | 21 | 190.8 | *99* | 14.3 | 14 |
| June | 18.6 | 10.7 | 14.6 | *−1.0* | 24.4 | 26 | 4.4 | 20 | 10.8 | 2 | 15.4 | 28 | 0 | 0 | 49.3 | *100* | 12 | 11.7 | 1 | 156.1 | *82* | 12.6 | 29 |
| July | 22.0 | 12.6 | 17.3 | *−0.6* | 26.8 | 17 | 7.8 | 4 | 17.8 | 10 | 15.9 | 15 | 0 | 5 | 21.2 | *43* | 7 | 5.1 | 10 | 214.7 | *104* | 14.2 | 30 |
| Aug | 21.7 | 11.8 | 16.7 | *−0.9* | 30.3 | 26 | 5.1 | 21 | 16.1 | 19 | 16.3 | 2 | 0 | 6 | 14.6 | *27* | 9 | 4.8 | 18 | 217.7 | *111* | 13.2 | 25 |
| Sep | 20.5 | 9.7 | 15.1 | *+0.2* | 25.7 | 5 | 4.5 | 29 | 16.8 | 19 | 15.8 | 11 | 0 | 1 | 15.5 | *29* | 6 | 4.1 | 16 | 210.4 | *150* | 11.0 | 12 |
| Oct | 13.3 | 4.9 | 9.1 | *−2.2* | 19.8 | 4 | −1.9 | 13 | 8.3 | 23 | 12.3 | 7 | 2 | 0 | 17.5 | *25* | 14 | 4.3 | 5 | 134.5 | *121* | 9.8 | 1 |
| Nov | 10.5 | 5.1 | 7.8 | *+0.3* | 14.4 | 18 | −3.9 | 30 | 3.9 | 10 | 11.7 | 18 | 1 | 0 | 25.0 | *37* | 11 | 5.6 | 13 | 55.0 | *77* | 7.3 | 15 |
| Dec | 6.8 | 0.7 | 3.7 | *−1.3* | 14.4 | 8 | −6.6 | 29 | 0.7 | 16 | 11.0 | 13 | 13 | 0 | 43.7 | *69* | 15 | 6.4 | 9 | 51.1 | *96* | 6.8 | 4 |
| **1964** | 13.7 | 6.3 | 10.0 | *−0.7* | 30.3 | 8 | −7.9 | 1 | 0.3 | 1 | 16.3 | 8 | 40 | 12 | 418.5 | *63* | 138 | 31.2 | 3 | 1519.2 | *96* | 14.3 | 5 |

## OXFORD 1965

| Month | Mean max | Mean min | Mean temp | Anom | Highest max | Date | Lowest min | Date | Lowest max | Date | Highest min | Date | Air frost | Days ≥ 25 °C | Total pptn | Anom | Rain days | Wettest day | Date | Total sunshine | Anom | Sunniest day | Date |
|---|---|---|---|---|---|---|---|---|---|---|---|---|---|---|---|---|---|---|---|---|---|---|---|
| Jan | 6.3 | 1.4 | 3.9 | *−1.0* | 11.7 | 16 | −3.6 | 5 | 1.6 | 29 | 8.8 | 11 | 14 | 0 | 56.0 | *98* | 16 | 8.9 | 17 | 83.7 | *134* | 6.0 | 14 |
| Feb | 5.6 | 0.6 | 3.1 | *−1.8* | 9.7 | 12 | −5.6 | 3 | 1.6 | 20 | 4.9 | 8 | 12 | 0 | 5.8 | *14* | 9 | 2.0 | 19 | 31.1 | *39* | 8.0 | 14 |
| Mar | 10.1 | 1.8 | 5.9 | *−1.3* | 22.1 | 29 | −6.8 | 3 | −0.9 | 4 | 9.4 | 27 | 10 | 0 | 44.2 | *93* | 18 | 13.2 | 20 | 137.7 | *124* | 11.3 | 30 |
| Apr | 13.0 | 4.5 | 8.8 | *−0.5* | 18.5 | 3 | −1.2 | 1 | 8.4 | 20 | 10.1 | 17 | 2 | 0 | 49.4 | *101* | 16 | 6.9 | 11 | 141.6 | *88* | 10.9 | 13 |
| May | 16.3 | 8.4 | 12.3 | *−0.1* | 27.0 | 14 | 2.8 | 19 | 10.1 | 28 | 12.3 | 17 | 0 | 1 | 63.8 | *112* | 19 | 17.3 | 17 | 179.6 | *93* | 13.7 | 12 |
| June | 19.2 | 10.8 | 15.0 | *−0.6* | 23.9 | 11 | 6.8 | 2 | 14.2 | 8 | 15.2 | 13 | 0 | 0 | 53.4 | *108* | 13 | 10.4 | 17 | 188.7 | *99* | 15.2 | 28 |
| July | 18.5 | 11.1 | 14.8 | *−3.1* | 23.3 | 14 | 4.4 | 4 | 15.1 | 15 | 15.1 | 13 | 0 | 0 | 90.0 | *184* | 17 | 22.4 | 21 | 122.8 | *59* | 12.2 | 1 |
| Aug | 20.2 | 11.0 | 15.6 | *−2.0* | 25.1 | 12 | 7.6 | 8 | 16.1 | 2 | 15.1 | 19 | 0 | 1 | 61.7 | *113* | 14 | 32.3 | 2 | 181.2 | *92* | 13.3 | 7 |
| Sep | 16.9 | 9.1 | 13.0 | *−1.9* | 21.7 | 22 | 3.9 | 5 | 12.3 | 30 | 14.8 | 16 | 0 | 0 | 75.4 | *139* | 15 | 26.2 | 25 | 122.9 | *88* | 10.1 | 21 |
| Oct | 15.2 | 7.0 | 11.1 | *−0.2* | 22.8 | 5 | 1.0 | 17 | 7.4 | 23 | 13.8 | 8 | 0 | 0 | 12.0 | *17* | 7 | 3.8 | 31 | 109.9 | *99* | 9.3 | 10 |
| Nov | 8.0 | 1.9 | 5.0 | *−2.6* | 14.1 | 8 | −5.6 | 15 | 0.1 | 15 | 9.4 | 9 | 11 | 0 | 51.2 | *77* | 17 | 9.9 | 28 | 77.4 | *109* | 7.8 | 6 |
| Dec | 8.1 | 2.4 | 5.3 | *+0.3* | 13.7 | 18 | −6.1 | 28 | 0.0 | 27 | 11.6 | 18 | 5 | 0 | 95.5 | *151* | 19 | 14.5 | 22 | 71.9 | *135* | 6.7 | 26 |
| **1965** | 13.1 | 5.8 | 9.5 | *−1.3* | 27.0 | 5 | −6.8 | 3 | −0.9 | 3 | 15.2 | 6 | 54 | 2 | 658.4 | *100* | 180 | 32.3 | 8 | 1448.5 | *92* | 15.2 | 6 |

## OXFORD 1966

| Month | Mean max | Mean min | Mean temp | Anom | Highest max | Date | Lowest min | Date | Lowest max | Date | Highest min | Date | Air frost | Days ≥ 25 °C | Total pptn | Anom | Rain days | Wettest day | Date | Total sunshine | Anom | Sunniest day | Date |
|---|---|---|---|---|---|---|---|---|---|---|---|---|---|---|---|---|---|---|---|---|---|---|---|
| Jan | 5.0 | 0.9 | 2.9 | *−1.9* | 12.2 | 29 | −9.6 | 19 | −2.0 | 19 | 8.8 | 29 | 15 | 0 | 30.8 | *54* | 13 | 8.9 | 1 | 42.5 | *68* | 5.8 | 14 |
| Feb | 8.9 | 4.6 | 6.8 | *+1.9* | 13.9 | 5 | −0.4 | 14 | 1.3 | 13 | 9.2 | 2 | 1 | 0 | 74.9 | *175* | 15 | 11.4 | 9 | 41.5 | *53* | 6.7 | 27 |
| Mar | 10.6 | 3.2 | 6.9 | *−0.4* | 13.9 | 20 | −3.8 | 20 | 7.2 | 13 | 8.0 | 2 | 3 | 0 | 9.9 | *21* | 7 | 2.5 | 28 | 121.5 | *109* | 10.2 | 18 |
| Apr | 11.2 | 5.0 | 8.1 | *−1.2* | 21.1 | 30 | −0.9 | 14 | 1.3 | 14 | 10.0 | 27 | 3 | 0 | 81.7 | *167* | 21 | 14.0 | 18 | 81.5 | *51* | 13.6 | 30 |
| May | 16.3 | 7.2 | 11.7 | *−0.7* | 25.0 | 1 | 2.6 | 10 | 11.1 | 8 | 11.2 | 22 | 0 | 1 | 54.0 | *94* | 15 | 19.6 | 11 | 234.9 | *122* | 15.1 | 30 |
| June | 20.9 | 11.3 | 16.1 | *+0.5* | 26.7 | 9 | 4.4 | 1 | 16.5 | 21 | 14.3 | 27 | 0 | 2 | 56.6 | *115* | 15 | 19.3 | 22 | 199.0 | *104* | 13.7 | 3 |
| July | 19.8 | 11.6 | 15.7 | *−2.2* | 24.4 | 3 | 5.8 | 18 | 15.7 | 16 | 14.9 | 9 | 0 | 0 | 80.9 | *166* | 17 | 14.0 | 29 | 149.7 | *72* | 14.5 | 23 |
| Aug | 19.9 | 11.1 | 15.5 | *−2.1* | 26.6 | 18 | 6.1 | 24 | 15.1 | 1 | 17.2 | 12 | 0 | 3 | 78.8 | *144* | 14 | 34.5 | 29 | 188.6 | *96* | 13.7 | 17 |
| Sep | 18.5 | 10.3 | 14.4 | *−0.5* | 21.9 | 9 | 5.1 | 27 | 12.2 | 25 | 16.4 | 4 | 0 | 0 | 52.5 | *97* | 9 | 20.1 | 29 | 168.0 | *120* | 11.2 | 6 |
| Oct | 14.2 | 7.9 | 11.1 | *−0.2* | 19.6 | 3 | 1.6 | 26 | 9.5 | 27 | 14.5 | 4 | 0 | 0 | 122.7 | *175* | 19 | 16.3 | 1 | 90.4 | *81* | 7.9 | 11 |
| Nov | 8.2 | 2.9 | 5.6 | *−2.0* | 14.3 | 7 | −2.8 | 24 | 4.1 | 24 | 9.4 | 8 | 4 | 0 | 28.0 | *42* | 15 | 10.9 | 5 | 61.0 | *86* | 7.0 | 2 |
| Dec | 9.3 | 3.2 | 6.2 | *+1.2* | 12.5 | 29 | −1.4 | 26 | 3.9 | 25 | 8.3 | 18 | 2 | 0 | 80.4 | *127* | 21 | 12.4 | 9 | 41.7 | *78* | 5.5 | 25 |
| **1966** | 13.6 | 6.6 | 10.1 | *−0.7* | 26.7 | 6 | −9.6 | 1 | −2.0 | 1 | 17.2 | 8 | 28 | 6 | 751.2 | *114* | 181 | 34.5 | 8 | 1420.3 | *90* | 15.1 | 5 |

## OXFORD 1967

| Month | Mean max | Mean min | Mean temp | Anom | Highest max | Date | Lowest min | Date | Lowest max | Date | Highest min | Date | Air frost | Days ≥ 25 °C | Total pptn | Anom | Rain days | Wettest day | Date | Total sunshine | Anom | Sunniest day | Date |
|---|---|---|---|---|---|---|---|---|---|---|---|---|---|---|---|---|---|---|---|---|---|---|---|
| Jan | 7.1 | 2.6 | 4.9 | *+0.0* | 12.8 | 27 | −4.9 | 9 | −0.6 | 8 | 9.1 | 30 | 8 | 0 | 35.5 | *62* | 14 | 8.9 | 6 | 64.2 | *103* | 6.6 | 3 |
| Feb | 8.9 | 2.9 | 5.9 | *+1.0* | 12.4 | 3 | −2.3 | 13 | 2.6 | 13 | 10.0 | 2 | 5 | 0 | 64.3 | *150* | 11 | 16.8 | 27 | 88.2 | *112* | 8.5 | 21 |
| Mar | 11.2 | 4.2 | 7.7 | *+0.4* | 16.7 | 21 | −1.3 | 30 | 7.8 | 11 | 7.5 | 7 | 2 | 0 | 30.7 | *64* | 11 | 14.0 | 8 | 174.6 | *157* | 10.7 | 29 |
| Apr | 12.0 | 4.8 | 8.4 | *−0.9* | 20.6 | 17 | −2.8 | 1 | 6.0 | 10 | 9.9 | 29 | 1 | 0 | 40.4 | *83* | 14 | 15.2 | 10 | 130.8 | *81* | 11.4 | 16 |
| May | 15.6 | 7.4 | 11.5 | *−0.9* | 25.6 | 11 | 0.1 | 3 | 10.6 | 1 | 12.1 | 13 | 0 | 1 | 85.6 | *150* | 25 | 21.3 | 14 | 184.2 | *95* | 13.3 | 31 |
| June | 19.6 | 9.9 | 14.7 | *−0.9* | 23.6 | 14 | 5.3 | 8 | 16.1 | 10 | 14.5 | 29 | 0 | 0 | 49.3 | *100* | 6 | 25.4 | 25 | 232.0 | *121* | 14.7 | 11 |
| July | 22.6 | 13.2 | 17.9 | *+0.0* | 28.2 | 17 | 8.1 | 9 | 18.6 | 4 | 17.6 | 31 | 0 | 6 | 76.6 | *157* | 12 | 44.5 | 22 | 234.4 | *113* | 15.2 | 9 |
| Aug | 20.6 | 12.4 | 16.5 | *−1.1* | 24.9 | 22 | 8.3 | 20 | 16.3 | 12 | 16.1 | 1 | 0 | 0 | 44.0 | *80* | 15 | 10.9 | 2 | 167.1 | *85* | 12.7 | 4 |
| Sep | 17.7 | 11.0 | 14.4 | *−0.5* | 20.2 | 25 | 3.5 | 21 | 15.0 | 16 | 15.0 | 29 | 0 | 0 | 59.4 | *110* | 18 | 11.7 | 25 | 116.5 | *83* | 9.4 | 4 |
| Oct | 14.2 | 8.8 | 11.5 | *+0.3* | 18.8 | 7 | 2.4 | 18 | 8.5 | 28 | 16.2 | 9 | 0 | 0 | 121.7 | *174* | 22 | 23.9 | 16 | 95.1 | *85* | 9.7 | 2 |
| Nov | 8.8 | 3.0 | 5.9 | *−1.6* | 15.0 | 11 | −3.3 | 17 | 5.1 | 8 | 8.4 | 14 | 5 | 0 | 29.2 | *44* | 15 | 6.4 | 2 | 69.2 | *97* | 7.1 | 26 |
| Dec | 7.1 | 2.0 | 4.5 | *−0.5* | 13.1 | 22 | −7.1 | 9 | −1.4 | 8 | 10.7 | 23 | 12 | 0 | 63.3 | *100* | 17 | 21.6 | 18 | 70.6 | *133* | 6.2 | 7 |
| **1967** | 13.8 | 6.9 | 10.3 | *−0.4* | 28.2 | 7 | −7.1 | 12 | −1.4 | 12 | 17.6 | 7 | 33 | 7 | 700.0 | *106* | 180 | 44.5 | 7 | 1626.9 | *103* | 15.2 | 7 |

## OXFORD 1968

| Month | Mean max | Mean min | Mean temp | Anom | Highest max | Date | Lowest min | Date | Lowest max | Date | Highest min | Date | Air frost | Days ≥ 25 °C | Total pptn | Anom | Rain days | Wettest day | Date | Total sunshine | Anom | Sunniest day | Date |
|---|---|---|---|---|---|---|---|---|---|---|---|---|---|---|---|---|---|---|---|---|---|---|---|
| Jan | 7.3 | 2.1 | 4.7 | *−0.2* | 12.9 | 14 | −7.2 | 10 | 0.7 | 9 | 8.9 | 15 | 9 | 0 | 57.9 | *102* | 17 | 13.5 | 8 | 50.0 | *80* | 7.1 | 28 |
| Feb | 4.8 | −0.1 | 2.3 | *−2.6* | 8.3 | 13 | −3.3 | 20 | −0.2 | 20 | 4.4 | 14 | 16 | 0 | 26.8 | *63* | 9 | 9.9 | 5 | 54.0 | *68* | 9.0 | 25 |
| Mar | 10.6 | 3.6 | 7.1 | *−0.1* | 21.2 | 29 | −1.5 | 1 | 2.3 | 1 | 10.6 | 29 | 5 | 0 | 19.5 | *41* | 14 | 4.3 | 24 | 148.5 | *134* | 11.1 | 26 |
| Apr | 13.0 | 4.4 | 8.7 | *−0.6* | 22.7 | 21 | −3.1 | 8 | 3.7 | 2 | 11.4 | 20 | 6 | 0 | 46.1 | *94* | 15 | 7.4 | 27 | 184.5 | *115* | 12.8 | 26 |
| May | 14.4 | 6.6 | 10.5 | *−1.9* | 22.6 | 31 | 1.2 | 19 | 8.7 | 5 | 11.0 | 15 | 0 | 0 | 66.2 | *116* | 19 | 15.5 | 17 | 155.0 | *80* | 13.0 | 31 |
| June | 19.9 | 11.4 | 15.6 | *+0.1* | 26.4 | 30 | 7.2 | 1 | 16.2 | 22 | 15.2 | 18 | 0 | 1 | 69.2 | *141* | 15 | 18.0 | 27 | 206.6 | *108* | 15.3 | 13 |
| July | 19.7 | 12.0 | 15.8 | *−2.0* | 31.6 | 1 | 7.4 | 28 | 15.1 | 10 | 19.9 | 1 | 0 | 1 | 141.6 | *290* | 11 | 87.9 | 10 | 136.2 | *66* | 15.4 | 3 |
| Aug | 19.4 | 12.3 | 15.9 | *−1.7* | 27.0 | 23 | 8.7 | 19 | 15.0 | 29 | 15.2 | 13 | 0 | 2 | 91.7 | *167* | 16 | 17.8 | 8 | 124.0 | *63* | 11.8 | 18 |
| Sep | 18.0 | 11.1 | 14.5 | *−0.3* | 22.8 | 9 | 6.7 | 18 | 14.1 | 14 | 14.5 | 27 | 0 | 0 | 108.3 | *200* | 20 | 21.3 | 14 | 114.9 | *82* | 10.5 | 9 |
| Oct | 16.0 | 10.7 | 13.4 | *+2.1* | 19.3 | 16 | 5.9 | 15 | 11.2 | 23 | 15.0 | 3 | 0 | 0 | 67.7 | *97* | 15 | 18.8 | 8 | 66.5 | *60* | 8.0 | 10 |
| Nov | 9.0 | 4.6 | 6.8 | *−0.7* | 13.9 | 1 | −2.4 | 9 | 3.8 | 9 | 11.0 | 1 | 3 | 0 | 54.5 | *82* | 15 | 18.0 | 1 | 47.9 | *67* | 7.3 | 4 |
| Dec | 5.1 | 1.1 | 3.1 | *−1.9* | 11.2 | 21 | −3.8 | 15 | −0.1 | 13 | 7.1 | 2 | 11 | 0 | 63.6 | *101* | 10 | 19.1 | 15 | 35.9 | *68* | 6.2 | 27 |
| **1968** | 13.1 | 6.6 | 9.9 | *−0.9* | 31.6 | 7 | −7.2 | 1 | −0.2 | 2 | 19.9 | 7 | 50 | 4 | 813.1 | *123* | 176 | 87.9 | 7 | 1324.0 | *84* | 15.4 | 7 |

## OXFORD 1969

| Month | Mean max | Mean min | Mean temp | Anom | Highest max | Date | Lowest min | Date | Lowest max | Date | Highest min | Date | Air frost | Days ≥ 25 °C | Total pptn | Anom | Rain days | Wettest day | Date | Total sunshine | Anom | Sunniest day | Date |
|---|---|---|---|---|---|---|---|---|---|---|---|---|---|---|---|---|---|---|---|---|---|---|---|
| Jan | 8.6 | 3.7 | 6.1 | *+1.3* | 13.1 | 22 | −2.9 | 1 | 5.1 | 16 | 9.3 | 23 | 5 | 0 | 74.9 | *132* | 22 | 13.0 | 17 | 40.0 | *64* | 7.2 | 31 |
| Feb | 4.2 | −0.9 | 1.6 | *−3.3* | 11.4 | 23 | −6.2 | 8 | −0.4 | 15 | 4.2 | 24 | 16 | 0 | 39.9 | *93* | 17 | 9.4 | 19 | 78.1 | *99* | 8.7 | 8 |
| Mar | 7.2 | 1.2 | 4.2 | *−3.1* | 12.8 | 8 | −5.2 | 6 | 2.5 | 2 | 6.3 | 15 | 8 | 0 | 48.3 | *101* | 12 | 17.3 | 12 | 70.5 | *63* | 9.9 | 5 |
| Apr | 13.1 | 3.7 | 8.4 | *−0.9* | 20.1 | 8 | −1.7 | 3 | 7.2 | 2 | 10.8 | 10 | 5 | 0 | 37.0 | *76* | 13 | 15.2 | 21 | 221.3 | *138* | 11.7 | 4 |
| May | 16.4 | 8.2 | 12.3 | *−0.2* | 23.9 | 13 | 3.3 | 19 | 11.0 | 18 | 14.7 | 13 | 0 | 0 | 88.8 | *155* | 23 | 15.7 | 25 | 161.9 | *84* | 12.5 | 22 |
| June | 19.9 | 9.5 | 14.7 | *−0.9* | 26.7 | 14 | 3.7 | 5 | 12.9 | 4 | 14.4 | 16 | 0 | 2 | 20.7 | *42* | 11 | 3.8 | 17 | 276.9 | *145* | 14.7 | 7 |
| July | 23.2 | 13.1 | 18.1 | *+0.2* | 30.9 | 16 | 7.7 | 30 | 14.4 | 6 | 17.3 | 24 | 0 | 11 | 49.0 | *100* | 6 | 29.0 | 28 | 243.8 | *118* | 14.9 | 4 |
| Aug | 21.1 | 13.2 | 17.2 | *−0.4* | 28.6 | 11 | 7.9 | 24 | 17.1 | 29 | 17.8 | 19 | 0 | 3 | 87.9 | *160* | 15 | 43.4 | 2 | 157.5 | *80* | 13.4 | 7 |
| Sep | 17.7 | 10.7 | 14.2 | *−0.7* | 20.7 | 11 | 0.2 | 30 | 13.3 | 12 | 15.5 | 11 | 0 | 0 | 32.8 | *61* | 10 | 10.9 | 11 | 97.4 | *69* | 10.5 | 22 |
| Oct | 16.7 | 9.9 | 13.3 | *+2.0* | 23.7 | 9 | 2.1 | 30 | 11.7 | 30 | 15.6 | 8 | 0 | 0 | 5.0 | *7* | 7 | 2.0 | 19 | 112.0 | *101* | 9.5 | 9 |
| Nov | 9.0 | 3.2 | 6.1 | *−1.5* | 16.8 | 2 | −3.6 | 29 | 1.2 | 28 | 13.9 | 3 | 5 | 0 | 52.0 | *78* | 16 | 12.7 | 14 | 84.6 | *119* | 8.0 | 8 |
| Dec | 6.1 | 1.2 | 3.7 | *−1.4* | 12.2 | 21 | −4.3 | 19 | 0.7 | 31 | 7.5 | 22 | 9 | 0 | 72.9 | *116* | 22 | 16.0 | 11 | 35.6 | *67* | 6.4 | 10 |
| **1969** | 13.6 | 6.4 | 10.0 | *−0.8* | 30.9 | 7 | −6.2 | 2 | −0.4 | 2 | 17.8 | 8 | 48 | 16 | 609.2 | *92* | 174 | 43.4 | 8 | 1579.6 | *100* | 14.9 | 7 |

## OXFORD 1970

| Month | Mean max | Mean min | Mean temp | Anom | Highest max | Date | Lowest min | Date | Lowest max | Date | Highest min | Date | Air frost | Days ≥ 25 °C | Total pptn | Anom | Rain days | Wettest day | Date | Total sunshine | Anom | Sunniest day | Date |
|---|---|---|---|---|---|---|---|---|---|---|---|---|---|---|---|---|---|---|---|---|---|---|---|
| Jan | 6.6 | 1.9 | 4.2 | *−0.6* | 10.3 | 22 | −6.7 | 5 | 0.9 | 7 | 7.8 | 12 | 10 | 0 | 66.6 | *117* | 23 | 10.2 | 29 | 34.5 | *55* | 7.2 | 6 |
| Feb | 6.8 | 0.6 | 3.7 | *−1.2* | 11.3 | 2 | −5.6 | 15 | 1.1 | 13 | 8.7 | 22 | 11 | 0 | 49.2 | *115* | 17 | 6.9 | 22 | 115.9 | *147* | 9.2 | 20 |
| Mar | 7.6 | 0.7 | 4.1 | *−3.1* | 14.1 | 20 | −5.9 | 9 | 2.7 | 5 | 7.8 | 21 | 12 | 0 | 45.8 | *96* | 14 | 12.7 | 4 | 120.8 | *109* | 9.7 | 9 |
| Apr | 11.3 | 3.8 | 7.5 | *−1.7* | 17.7 | 15 | −3.0 | 2 | 5.6 | 9 | 11.1 | 16 | 3 | 0 | 75.9 | *155* | 21 | 20.1 | 25 | 132.7 | *82* | 12.6 | 29 |
| May | 18.8 | 8.6 | 13.7 | *+1.3* | 26.0 | 5 | 4.8 | 18 | 12.2 | 10 | 13.5 | 31 | 0 | 1 | 22.9 | *40* | 8 | 12.7 | 9 | 224.8 | *117* | 13.7 | 24 |
| June | 22.4 | 11.6 | 17.0 | *+1.4* | 28.2 | 10 | 7.4 | 30 | 16.0 | 14 | 15.6 | 11 | 0 | 7 | 43.0 | *87* | 7 | 22.1 | 11 | 270.9 | *142* | 14.8 | 4 |
| July | 20.4 | 12.1 | 16.3 | *−1.6* | 31.2 | 7 | 7.8 | 22 | 16.3 | 1 | 16.7 | 8 | 0 | 2 | 42.3 | *87* | 15 | 9.9 | 13 | 192.4 | *93* | 14.1 | 30 |
| Aug | 21.0 | 12.2 | 16.6 | *−1.0* | 27.4 | 3 | 7.3 | 18 | 13.9 | 19 | 16.1 | 6 | 0 | 4 | 51.9 | *95* | 11 | 19.6 | 19 | 185.6 | *94* | 13.7 | 11 |
| Sep | 19.5 | 11.2 | 15.3 | *+0.4* | 25.8 | 28 | 6.1 | 16 | 13.5 | 12 | 15.1 | 8 | 0 | 1 | 40.3 | *74* | 11 | 8.1 | 12 | 157.3 | *112* | 11.1 | 1 |
| Oct | 15.0 | 7.5 | 11.2 | *−0.1* | 19.9 | 4 | 1.2 | 8 | 9.0 | 21 | 13.4 | 5 | 0 | 0 | 19.5 | *28* | 7 | 5.6 | 25 | 106.2 | *95* | 8.9 | 8 |
| Nov | 11.3 | 5.6 | 8.5 | *+0.9* | 16.1 | 2 | −0.6 | 16 | 6.4 | 14 | 10.7 | 2 | 1 | 0 | 150.3 | *225* | 23 | 20.6 | 6 | 64.7 | *91* | 7.1 | 15 |
| Dec | 6.6 | 1.7 | 4.2 | *−0.8* | 12.3 | 18 | −3.9 | 28 | −0.6 | 27 | 9.5 | 19 | 13 | 0 | 33.9 | *54* | 12 | 12.2 | 3 | 43.8 | *82* | 6.7 | 24 |
| **1970** | 13.9 | 6.5 | 10.2 | *−0.5* | 31.2 | 7 | −6.7 | 1 | −0.6 | 12 | 16.7 | 7 | 50 | 15 | 641.6 | *97* | 169 | 22.1 | 6 | 1649.6 | *105* | 14.8 | 6 |

## OXFORD 1971

| Month | Mean max | Mean min | Mean temp | Anom | Highest max | Date | Lowest min | Date | Lowest max | Date | Highest min | Date | Air frost | Days ≥ 25 °C | Total pptn | Anom | Rain days | Wettest day | Date | Total sunshine | Anom | Sunniest day | Date |
|---|---|---|---|---|---|---|---|---|---|---|---|---|---|---|---|---|---|---|---|---|---|---|---|
| Jan | 7.2 | 2.5 | 4.8 | *−0.0* | 12.3 | 10 | −5.8 | 4 | −2.8 | 4 | 9.2 | 8 | 7 | 0 | 105.3 | *185* | 17 | 19.3 | 31 | 42.1 | *68* | 6.9 | 10 |
| Feb | 7.9 | 1.5 | 4.7 | *−0.2* | 12.2 | 20 | −2.2 | 16 | 2.1 | 16 | 5.4 | 21 | 8 | 0 | 22.1 | *52* | 8 | 7.4 | 14 | 69.5 | *88* | 7.5 | 19 |
| Mar | 8.7 | 2.2 | 5.5 | *−1.8* | 12.3 | 30 | −6.0 | 7 | 1.6 | 6 | 6.7 | 29 | 8 | 0 | 42.9 | *90* | 15 | 13.0 | 17 | 106.0 | *95* | 8.7 | 10 |
| Apr | 12.0 | 4.3 | 8.1 | *−1.2* | 20.6 | 22 | −0.5 | 27 | 7.2 | 26 | 12.2 | 23 | 1 | 0 | 44.8 | *92* | 10 | 22.2 | 23 | 122.9 | *76* | 12.3 | 12 |
| May | 17.3 | 7.4 | 12.3 | *−0.1* | 21.3 | 6 | 0.8 | 1 | 10.3 | 25 | 13.3 | 12 | 0 | 0 | 46.3 | *81* | 12 | 8.7 | 24 | 228.9 | *119* | 14.4 | 19 |
| June | 16.7 | 9.4 | 13.0 | *−2.6* | 21.8 | 24 | 4.9 | 15 | 11.7 | 14 | 12.2 | 20 | 0 | 0 | 135.8 | *276* | 13 | 38.0 | 10 | 143.5 | *75* | 15.0 | 22 |
| July | 23.1 | 13.1 | 18.1 | *+0.2* | 28.8 | 11 | 7.2 | 18 | 19.4 | 16 | 17.3 | 12 | 0 | 9 | 37.3 | *76* | 7 | 22.5 | 27 | 259.3 | *125* | 14.9 | 13 |
| Aug | 20.3 | 12.7 | 16.5 | *−1.1* | 23.1 | 9 | 7.8 | 17 | 15.6 | 15 | 16.6 | 29 | 0 | 0 | 133.2 | *243* | 20 | 33.3 | 3 | 134.2 | *68* | 13.1 | 17 |
| Sep | 19.5 | 9.8 | 14.7 | *−0.2* | 23.6 | 8 | 5.9 | 19 | 14.9 | 15 | 13.7 | 30 | 0 | 0 | 13.4 | *25* | 4 | 4.7 | 25 | 166.5 | *119* | 11.4 | 9 |
| Oct | 15.8 | 7.4 | 11.6 | *+0.4* | 22.3 | 2 | 1.1 | 15 | 8.7 | 29 | 14.4 | 12 | 0 | 0 | 74.2 | *106* | 10 | 15.7 | 13 | 135.2 | *121* | 9.6 | 6 |
| Nov | 9.8 | 3.0 | 6.4 | *−1.2* | 16.6 | 2 | −2.8 | 11 | 3.1 | 23 | 8.7 | 27 | 7 | 0 | 61.3 | *92* | 14 | 20.0 | 20 | 102.0 | *144* | 7.6 | 10 |
| Dec | 8.7 | 4.6 | 6.6 | *+1.6* | 13.6 | 21 | −3.5 | 1 | 2.9 | 29 | 12.4 | 21 | 3 | 0 | 24.0 | *38* | 9 | 8.5 | 19 | 29.8 | *56* | 6.0 | 28 |
| **1971** | 13.9 | 6.5 | 10.2 | *−0.5* | 28.8 | 7 | −6.0 | 3 | −2.8 | 1 | 17.3 | 7 | 34 | 9 | 740.6 | *112* | 139 | 38.0 | 6 | 1539.9 | *98* | 15.0 | 6 |

## OXFORD 1972

| Month | Mean max | Mean min | Mean temp | Anom | Highest max | Date | Lowest min | Date | Lowest max | Date | Highest min | Date | Air frost | Days ≥ 25 °C | Total pptn | Anom | Rain days | Wettest day | Date | Total sunshine | Anom | Sunniest day | Date |
|---|---|---|---|---|---|---|---|---|---|---|---|---|---|---|---|---|---|---|---|---|---|---|---|
| Jan | 6.4 | 2.0 | 4.2 | *−0.7* | 11.0 | 23 | −8.0 | 31 | −1.6 | 30 | 8.0 | 11 | 6 | 0 | 60.5 | *106* | 20 | 8.7 | 10 | 41.8 | *67* | 6.4 | 21 |
| Feb | 7.0 | 2.5 | 4.8 | *−0.1* | 9.8 | 29 | −4.4 | 1 | 3.3 | 20 | 4.9 | 3 | 1 | 0 | 50.8 | *119* | 19 | 13.4 | 3 | 27.3 | *35* | 6.6 | 14 |
| Mar | 11.8 | 2.8 | 7.3 | *+0.0* | 18.8 | 17 | −3.0 | 2 | 2.7 | 11 | 9.3 | 31 | 4 | 0 | 63.2 | *133* | 16 | 22.7 | 4 | 144.0 | *129* | 11.2 | 24 |
| Apr | 12.1 | 5.4 | 8.8 | *−0.5* | 15.2 | 2 | 0.3 | 25 | 8.6 | 23 | 9.7 | 1 | 0 | 0 | 45.8 | *94* | 16 | 7.6 | 4 | 117.6 | *73* | 13.2 | 25 |
| May | 15.1 | 7.4 | 11.2 | *−1.2* | 18.4 | 3 | 2.3 | 17 | 11.9 | 5 | 11.3 | 22 | 0 | 0 | 49.8 | *87* | 20 | 12.2 | 9 | 141.9 | *74* | 12.2 | 1 |
| June | 16.5 | 8.8 | 12.7 | *−2.9* | 19.4 | 17 | 4.9 | 2 | 13.0 | 10 | 12.6 | 25 | 0 | 0 | 55.8 | *113* | 15 | 16.2 | 9 | 150.4 | *79* | 13.7 | 30 |
| July | 21.0 | 11.8 | 16.4 | *−1.5* | 25.1 | 23 | 7.1 | 12 | 15.2 | 8 | 15.1 | 22 | 0 | 1 | 23.3 | *48* | 10 | 5.7 | 31 | 168.7 | *82* | 13.6 | 13 |
| Aug | 20.7 | 11.6 | 16.2 | *−1.4* | 24.7 | 14 | 6.8 | 11 | 16.4 | 18 | 15.0 | 8 | 0 | 0 | 29.8 | *54* | 5 | 19.3 | 1 | 160.9 | *82* | 12.6 | 11 |
| Sep | 16.5 | 8.3 | 12.4 | *−2.5* | 21.6 | 1 | 3.4 | 26 | 12.7 | 9 | 12.7 | 4 | 0 | 0 | 31.0 | *57* | 6 | 22.3 | 8 | 129.1 | *92* | 8.6 | 21 |
| Oct | 14.8 | 7.4 | 11.1 | *−0.2* | 21.4 | 7 | 0.9 | 21 | 10.3 | 19 | 12.7 | 9 | 0 | 0 | 25.9 | *37* | 6 | 14.3 | 9 | 91.6 | *82* | 8.9 | 28 |
| Nov | 9.5 | 3.8 | 6.7 | *−0.9* | 15.5 | 6 | −1.7 | 18 | 3.6 | 17 | 13.2 | 7 | 7 | 0 | 65.8 | *99* | 11 | 23.7 | 19 | 80.6 | *114* | 7.2 | 15 |
| Dec | 8.7 | 3.5 | 6.1 | *+1.1* | 13.8 | 13 | −3.0 | 31 | 2.1 | 22 | 11.1 | 14 | 2 | 0 | 74.0 | *117* | 17 | 22.2 | 1 | 47.6 | *90* | 6.2 | 8 |
| **1972** | 13.3 | 6.3 | 9.8 | *−0.9* | 25.1 | 7 | −8.0 | 1 | −1.6 | 1 | 15.1 | 7 | 20 | 1 | 575.7 | *87* | 161 | 23.7 | 11 | 1301.5 | *83* | 13.7 | 6 |

## OXFORD 1973

| Month | Mean max | Mean min | Mean temp | Anom | Highest max | Date | Lowest min | Date | Lowest max | Date | Highest min | Date | Air frost | Days ≥ 25 °C | Total pptn | Anom | Rain days | Wettest day | Date | Total sunshine | Anom | Sunniest day | Date |
|---|---|---|---|---|---|---|---|---|---|---|---|---|---|---|---|---|---|---|---|---|---|---|---|
| Jan | 6.8 | 2.1 | 4.5 | *−0.4* | 11.0 | 29 | −4.3 | 19 | 2.0 | 17 | 8.4 | 4 | 10 | 0 | 24.4 | *43* | 12 | 8.4 | 14 | 38.8 | *62* | 6.6 | 22 |
| Feb | 7.7 | 1.3 | 4.5 | *−0.4* | 12.4 | 20 | −4.7 | 15 | 3.7 | 25 | 7.6 | 21 | 11 | 0 | 15.0 | *35* | 10 | 4.2 | 12 | 74.1 | *94* | 8.3 | 27 |
| Mar | 11.0 | 2.1 | 6.6 | *−0.7* | 17.5 | 23 | −2.2 | 11 | 6.2 | 14 | 7.8 | 5 | 4 | 0 | 10.8 | *23* | 4 | 6.0 | 5 | 150.7 | *135* | 11.2 | 29 |
| Apr | 12.0 | 4.2 | 8.1 | *−1.2* | 17.6 | 26 | −1.3 | 9 | 7.2 | 9 | 7.4 | 28 | 2 | 0 | 42.4 | *87* | 14 | 9.0 | 1 | 167.7 | *104* | 13.0 | 26 |
| May | 16.2 | 8.1 | 12.2 | *−0.3* | 23.4 | 26 | 2.3 | 2 | 12.1 | 7 | 14.0 | 28 | 0 | 0 | 71.2 | *125* | 19 | 14.1 | 21 | 168.0 | *87* | 13.9 | 16 |
| June | 20.9 | 10.9 | 15.9 | *+0.3* | 26.6 | 16 | 5.3 | 1 | 14.2 | 20 | 14.8 | 26 | 0 | 3 | 109.3 | *222* | 5 | 67.3 | 27 | 249.5 | *131* | 15.6 | 23 |
| July | 20.8 | 12.2 | 16.5 | *−1.4* | 25.8 | 5 | 7.0 | 24 | 15.9 | 23 | 16.4 | 6 | 0 | 1 | 60.0 | *123* | 12 | 11.6 | 20 | 171.1 | *83* | 14.4 | 4 |
| Aug | 22.5 | 12.8 | 17.7 | *+0.1* | 29.7 | 16 | 9.4 | 23 | 16.7 | 21 | 18.5 | 17 | 0 | 6 | 28.8 | *53* | 6 | 16.1 | 5 | 192.0 | *98* | 12.3 | 14 |
| Sep | 19.6 | 11.0 | 15.3 | *+0.5* | 28.6 | 5 | 4.8 | 27 | 12.8 | 30 | 17.2 | 5 | 0 | 4 | 32.7 | *60* | 10 | 11.1 | 27 | 153.9 | *110* | 10.9 | 12 |
| Oct | 12.8 | 5.5 | 9.2 | *−2.1* | 19.4 | 5 | −0.4 | 30 | 6.8 | 30 | 12.5 | 7 | 1 | 0 | 40.8 | *58* | 11 | 12.5 | 19 | 95.1 | *85* | 10.1 | 1 |
| Nov | 9.8 | 3.2 | 6.5 | *−1.1* | 15.2 | 3 | −3.7 | 21 | 1.4 | 30 | 11.1 | 4 | 10 | 0 | 28.3 | *42* | 12 | 11.1 | 28 | 89.1 | *126* | 8.2 | 6 |
| Dec | 8.3 | 2.3 | 5.3 | *+0.3* | 12.4 | 4 | −4.8 | 2 | 2.0 | 1 | 8.5 | 29 | 8 | 0 | 30.9 | *49* | 17 | 6.8 | 7 | 62.9 | *118* | 6.6 | 9 |
| **1973** | 14.0 | 6.3 | 10.2 | *−0.6* | 29.7 | 8 | −4.8 | 12 | 1.4 | 11 | 18.5 | 8 | 46 | 14 | 494.6 | *75* | 132 | 67.3 | 6 | 1612.9 | *102* | 15.6 | 6 |

## OXFORD 1974

| Month | Mean max | Mean min | Mean temp | Anom | Highest max | Date | Lowest min | Date | Lowest max | Date | Highest min | Date | Air frost | Days ≥ 25 °C | Total pptn | Anom | Rain days | Wettest day | Date | Total sunshine | Anom | Sunniest day | Date |
|---|---|---|---|---|---|---|---|---|---|---|---|---|---|---|---|---|---|---|---|---|---|---|---|
| Jan | 9.3 | 3.5 | 6.4 | *+1.6* | 13.0 | 15 | −5.1 | 1 | 1.1 | 1 | 7.2 | 5 | 3 | 0 | 65.0 | *114* | 21 | 10.9 | 6 | 56.3 | *90* | 7.2 | 29 |
| Feb | 8.7 | 3.1 | 5.9 | *+1.0* | 12.1 | 11 | −0.9 | 25 | 2.8 | 27 | 8.6 | 11 | 4 | 0 | 77.7 | *182* | 17 | 15.8 | 10 | 58.3 | *74* | 8.5 | 24 |
| Mar | 9.3 | 2.3 | 5.8 | *−1.4* | 16.7 | 25 | −2.5 | 13 | 3.1 | 9 | 6.2 | 16 | 6 | 0 | 36.7 | *77* | 11 | 14.4 | 9 | 101.7 | *91* | 10.7 | 31 |
| Apr | 13.0 | 3.8 | 8.4 | *−0.9* | 18.6 | 4 | 0.2 | 15 | 8.2 | 17 | 6.0 | 3 | 0 | 0 | 4.2 | *9* | 4 | 2.0 | 27 | 142.1 | *88* | 12.4 | 14 |
| May | 16.1 | 6.8 | 11.5 | *−1.0* | 21.2 | 17 | 1.3 | 8 | 10.6 | 4 | 10.7 | 17 | 0 | 0 | 31.6 | *55* | 10 | 9.2 | 11 | 191.4 | *99* | 14.2 | 15 |
| June | 19.1 | 10.1 | 14.6 | *−1.0* | 25.0 | 16 | 5.2 | 9 | 11.5 | 27 | 14.3 | 17 | 0 | 1 | 78.7 | *160* | 10 | 39.0 | 16 | 206.2 | *108* | 14.7 | 21 |
| July | 20.1 | 12.2 | 16.1 | *−1.7* | 24.4 | 21 | 8.3 | 25 | 16.2 | 2 | 16.1 | 20 | 0 | 0 | 41.7 | *85* | 12 | 12.8 | 15 | 187.8 | *91* | 14.0 | 6 |
| Aug | 20.6 | 11.3 | 16.0 | *−1.6* | 24.6 | 15 | 6.6 | 28 | 13.3 | 4 | 15.8 | 15 | 0 | 0 | 97.2 | *177* | 17 | 24.6 | 25 | 198.7 | *101* | 11.7 | 2 |
| Sep | 16.5 | 9.1 | 12.8 | *−2.1* | 22.1 | 12 | 2.7 | 30 | 10.1 | 27 | 15.9 | 13 | 0 | 0 | 156.1 | *288* | 23 | 20.0 | 12 | 146.5 | *104* | 9.8 | 18 |
| Oct | 11.0 | 4.9 | 8.0 | *−3.3* | 13.6 | 9 | −0.4 | 31 | 6.0 | 30 | 9.2 | 27 | 1 | 0 | 66.4 | *95* | 19 | 16.2 | 6 | 105.4 | *95* | 7.9 | 21 |
| Nov | 10.1 | 5.0 | 7.5 | *−0.0* | 14.1 | 9 | 1.6 | 19 | 5.8 | 18 | 8.6 | 8 | 0 | 0 | 95.1 | *143* | 20 | 21.8 | 13 | 52.9 | *75* | 6.9 | 15 |
| Dec | 10.6 | 5.9 | 8.2 | *+3.2* | 13.4 | 2 | 0.1 | 13 | 5.2 | 11 | 11.9 | 3 | 0 | 0 | 33.4 | *53* | 12 | 10.2 | 26 | 58.8 | *111* | 5.3 | 11 |
| **1974** | 13.7 | 6.5 | 10.1 | *−0.7* | 25.0 | 6 | −5.1 | 1 | 1.1 | 1 | 16.1 | 7 | 14 | 1 | 783.8 | *119* | 176 | 39.0 | 6 | 1506.1 | *96* | 14.7 | 6 |

## OXFORD 1975

| Month | Mean max | Mean min | Mean temp | Anom | Highest max | Date | Lowest min | Date | Lowest max | Date | Highest min | Date | Air frost | Days ≥ 25 °C | Total pptn | Anom | Rain days | Wettest day | Date | Total sunshine | Anom | Sunniest day | Date |
|---|---|---|---|---|---|---|---|---|---|---|---|---|---|---|---|---|---|---|---|---|---|---|---|
| Jan | 9.8 | 4.6 | 7.2 | *+2.4* | 14.0 | 15 | −0.2 | 19 | 4.2 | 18 | 9.6 | 31 | 1 | 0 | 69.4 | *122* | 21 | 12.6 | 19 | 44.7 | *72* | 6.2 | 19 |
| Feb | 7.9 | 1.8 | 4.8 | *−0.1* | 12.5 | 17 | −2.2 | 27 | 4.5 | 4 | 7.1 | 12 | 4 | 0 | 29.7 | *69* | 8 | 14.2 | 14 | 57.8 | *73* | 7.2 | 28 |
| Mar | 7.7 | 2.5 | 5.1 | *−2.1* | 12.8 | 1 | −1.8 | 19 | 4.5 | 13 | 7.1 | 3 | 6 | 0 | 95.2 | *200* | 18 | 19.2 | 8 | 72.7 | *65* | 8.5 | 18 |
| Apr | 12.8 | 5.3 | 9.1 | *−0.2* | 20.8 | 24 | −2.0 | 5 | 5.2 | 3 | 12.9 | 26 | 4 | 0 | 46.9 | *96* | 18 | 12.1 | 18 | 131.4 | *82* | 11.8 | 24 |
| May | 14.3 | 6.4 | 10.3 | *−2.1* | 21.2 | 20 | 0.8 | 31 | 8.7 | 17 | 10.3 | 2 | 0 | 0 | 49.0 | *86* | 10 | 13.7 | 17 | 159.1 | *82* | 13.0 | 20 |
| June | 21.5 | 9.4 | 15.5 | *−0.1* | 27.4 | 12 | 3.5 | 4 | 13.1 | 2 | 14.0 | 14 | 0 | 7 | 18.0 | *37* | 4 | 5.8 | 2 | 301.0 | *158* | 15.4 | 25 |
| July | 23.8 | 13.4 | 18.6 | *+0.7* | 30.1 | 30 | 9.7 | 25 | 19.1 | 24 | 18.2 | 14 | 0 | 7 | 34.1 | *70* | 11 | 14.6 | 7 | 242.4 | *117* | 13.7 | 28 |
| Aug | 25.1 | 14.1 | 19.6 | *+2.0* | 32.9 | 4 | 7.8 | 23 | 16.8 | 31 | 20.1 | 5 | 0 | 17 | 30.2 | *55* | 10 | 5.6 | 14 | 218.6 | *111* | 13.9 | 3 |
| Sep | 18.7 | 9.9 | 14.3 | *−0.6* | 23.1 | 1 | 2.6 | 16 | 11.0 | 14 | 15.1 | 9 | 0 | 0 | 95.3 | *176* | 15 | 39.5 | 13 | 147.0 | *105* | 10.9 | 4 |
| Oct | 14.2 | 6.5 | 10.3 | *−0.9* | 17.8 | 6 | −0.2 | 14 | 11.1 | 17 | 11.4 | 5 | 1 | 0 | 11.0 | *16* | 6 | 4.2 | 9 | 120.7 | *108* | 9.7 | 7 |
| Nov | 9.8 | 2.5 | 6.2 | *−1.4* | 14.3 | 15 | −3.8 | 14 | 2.4 | 30 | 8.4 | 20 | 7 | 0 | 37.9 | *57* | 15 | 10.2 | 15 | 85.1 | *120* | 7.4 | 3 |
| Dec | 7.5 | 1.7 | 4.6 | *−0.4* | 12.5 | 1 | −4.1 | 14 | 1.8 | 14 | 6.9 | 31 | 10 | 0 | 21.7 | *34* | 10 | 8.7 | 1 | 34.4 | *65* | 6.7 | 24 |
| **1975** | 14.4 | 6.5 | 10.5 | *−0.3* | 32.9 | 8 | −4.1 | 12 | 1.8 | 12 | 20.1 | 8 | 33 | 31 | 538.4 | *82* | 146 | 39.5 | 9 | 1614.9 | *102* | 15.4 | 6 |

## OXFORD 1976

| Month | Mean max | Mean min | Mean temp | Anom | Highest max | Date | Lowest min | Date | Lowest max | Date | Highest min | Date | Air frost | Days ≥ 25 °C | Total pptn | Anom | Rain days | Wettest day | Date | Total sunshine | Anom | Sunniest day | Date |
|---|---|---|---|---|---|---|---|---|---|---|---|---|---|---|---|---|---|---|---|---|---|---|---|
| Jan | 8.5 | 3.4 | 5.9 | *+1.1* | 13.2 | 2 | −3.7 | 30 | 0.0 | 31 | 9.0 | 11 | 8 | 0 | 18.9 | *33* | 8 | 7.0 | 4 | 70.3 | *113* | 6.6 | 30 |
| Feb | 7.5 | 2.3 | 4.9 | *−0.0* | 14.4 | 29 | −2.5 | 1 | 0.1 | 3 | 9.2 | 24 | 8 | 0 | 18.8 | *44* | 13 | 11.0 | 12 | 42.6 | *54* | 8.1 | 10 |
| Mar | 9.0 | 1.7 | 5.3 | *−1.9* | 13.4 | 25 | −3.2 | 7 | 3.4 | 7 | 8.0 | 26 | 11 | 0 | 16.0 | *34* | 9 | 3.7 | 15 | 107.8 | *97* | 9.8 | 1 |
| Apr | 12.9 | 4.1 | 8.5 | *−0.8* | 18.8 | 20 | −1.8 | 29 | 7.9 | 27 | 7.9 | 1 | 4 | 0 | 8.1 | *17* | 5 | 4.0 | 14 | 148.7 | *92* | 13.0 | 24 |
| May | 17.8 | 8.3 | 13.1 | *+0.6* | 25.3 | 7 | 4.3 | 1 | 12.8 | 19 | 12.3 | 8 | 0 | 1 | 44.2 | *77* | 12 | 9.4 | 28 | 172.0 | *89* | 12.6 | 7 |
| June | 24.4 | 12.6 | 18.5 | *+2.9* | 34.3 | 27 | 7.8 | 3 | 15.2 | 16 | 18.7 | 28 | 0 | 15 | 17.1 | *35* | 5 | 11.7 | 19 | 261.4 | *137* | 15.2 | 30 |
| July | 25.9 | 13.2 | 19.6 | *+1.7* | 33.4 | 3 | 9.0 | 29 | 19.1 | 31 | 20.3 | 4 | 0 | 15 | 14.3 | *29* | 4 | 7.8 | 15 | 254.2 | *123* | 14.8 | 1 |
| Aug | 24.4 | 11.4 | 17.9 | *+0.3* | 29.6 | 25 | 6.6 | 1 | 18.1 | 30 | 14.1 | 30 | 0 | 15 | 23.9 | *44* | 3 | 20.8 | 29 | 257.4 | *131* | 13.5 | 4 |
| Sep | 18.1 | 10.5 | 14.3 | *−0.6* | 23.6 | 7 | 5.7 | 17 | 12.7 | 15 | 14.9 | 25 | 0 | 0 | 90.7 | *167* | 16 | 17.0 | 10 | 118.2 | *84* | 12.1 | 3 |
| Oct | 14.1 | 8.3 | 11.2 | *−0.1* | 18.7 | 11 | 2.3 | 31 | 9.9 | 29 | 13.5 | 1 | 0 | 0 | 99.5 | *142* | 21 | 18.5 | 3 | 43.6 | *39* | 7.7 | 7 |
| Nov | 9.2 | 3.4 | 6.3 | *−1.3* | 12.7 | 5 | −3.0 | 14 | 5.3 | 12 | 7.4 | 6 | 5 | 0 | 55.9 | *84* | 13 | 15.4 | 30 | 54.1 | *76* | 7.0 | 7 |
| Dec | 4.6 | −0.4 | 2.1 | *−2.9* | 9.0 | 22 | −8.9 | 29 | 1.1 | 28 | 4.8 | 23 | 13 | 0 | 101.8 | *161* | 15 | 17.6 | 31 | 58.6 | *110* | 6.8 | 9 |
| **1976** | 14.7 | 6.6 | 10.6 | *−0.1* | 34.3 | 6 | −8.9 | 12 | 0.0 | 1 | 20.3 | 7 | 49 | 46 | 509.2 | *77* | 124 | 20.8 | 8 | 1588.9 | *101* | 15.2 | 6 |

## OXFORD 1977

| Month | Mean max | Mean min | Mean temp | Anom | Highest max | Date | Lowest min | Date | Lowest max | Date | Highest min | Date | Air frost | Days ≥ 25 °C | Total pptn | Anom | Rain days | Wettest day | Date | Total sunshine | Anom | Sunniest day | Date |
|---|---|---|---|---|---|---|---|---|---|---|---|---|---|---|---|---|---|---|---|---|---|---|---|
| Jan | 5.4 | 0.7 | 3.0 | *−1.8* | 12.6 | 25 | −4.4 | 30 | −0.1 | 12 | 5.6 | 21 | 11 | 0 | 75.9 | *133* | 19 | 18.4 | 13 | 35.4 | *57* | 5.9 | 3 |
| Feb | 8.9 | 3.3 | 6.1 | *+1.2* | 11.7 | 5 | −2.6 | 28 | 4.1 | 1 | 6.8 | 7 | 4 | 0 | 114.4 | *267* | 23 | 15.2 | 19 | 74.4 | *94* | 8.6 | 28 |
| Mar | 10.4 | 4.3 | 7.4 | *+0.1* | 17.2 | 2 | −2.6 | 30 | 5.9 | 21 | 8.4 | 3 | 4 | 0 | 59.8 | *125* | 19 | 8.4 | 16 | 87.2 | *78* | 9.9 | 28 |
| Apr | 11.6 | 3.8 | 7.7 | *−1.6* | 16.8 | 25 | −2.4 | 8 | 5.1 | 7 | 11.1 | 26 | 4 | 0 | 40.7 | *83* | 17 | 10.5 | 26 | 157.9 | *98* | 13.8 | 27 |
| May | 15.7 | 6.1 | 10.9 | *−1.6* | 23.7 | 28 | 0.8 | 1 | 10.7 | 13 | 12.4 | 11 | 0 | 0 | 43.1 | *75* | 13 | 14.1 | 2 | 217.5 | *113* | 14.9 | 27 |
| June | 16.7 | 8.5 | 12.6 | *−3.0* | 24.8 | 3 | 4.0 | 1 | 12.3 | 19 | 12.4 | 25 | 0 | 0 | 86.9 | *176* | 15 | 29.9 | 13 | 123.2 | *65* | 15.1 | 1 |
| July | 21.0 | 12.0 | 16.5 | *−1.4* | 26.7 | 3 | 6.6 | 15 | 15.6 | 13 | 16.1 | 23 | 0 | 7 | 5.8 | *12* | 8 | 1.9 | 25 | 193.6 | *94* | 15.2 | 3 |
| Aug | 19.7 | 12.2 | 15.9 | *−1.7* | 25.9 | 1 | 7.8 | 28 | 13.5 | 19 | 15.4 | 3 | 0 | 2 | 129.0 | *235* | 20 | 38.0 | 16 | 123.4 | *63* | 12.2 | 1 |
| Sep | 17.5 | 10.4 | 14.0 | *−0.9* | 21.8 | 6 | 5.9 | 18 | 12.6 | 21 | 14.7 | 2 | 0 | 0 | 7.8 | *14* | 6 | 2.1 | 27 | 101.8 | *73* | 11.2 | 9 |
| Oct | 15.4 | 8.3 | 11.9 | *+0.6* | 17.9 | 20 | 4.5 | 8 | 13.1 | 8 | 14.3 | 21 | 0 | 0 | 32.8 | *47* | 17 | 8.2 | 8 | 107.3 | *96* | 9.4 | 1 |
| Nov | 9.6 | 3.8 | 6.7 | *−0.9* | 16.9 | 10 | −2.5 | 30 | 4.2 | 26 | 12.3 | 10 | 7 | 0 | 46.6 | *70* | 16 | 9.1 | 1 | 102.4 | *144* | 8.0 | 15 |
| Dec | 8.9 | 3.9 | 6.4 | *+1.3* | 14.9 | 23 | −0.4 | 3 | 3.4 | 3 | 9.7 | 24 | 2 | 0 | 67.8 | *107* | 15 | 16.4 | 7 | 45.2 | *85* | 5.1 | 4 |
| **1977** | 13.4 | 6.4 | 9.9 | *−0.8* | 26.7 | 7 | −4.4 | 1 | −0.1 | 1 | 16.1 | 7 | 32 | 9 | 710.6 | *108* | 188 | 38.0 | 8 | 1369.3 | *87* | 15.2 | 7 |

## OXFORD 1978

| Month | Mean max | Mean min | Mean temp | Anom | Highest max | Date | Lowest min | Date | Lowest max | Date | Highest min | Date | Air frost | Days ≥ 25 °C | Total pptn | Anom | Rain days | Wettest day | Date | Total sunshine | Anom | Sunniest day | Date |
|---|---|---|---|---|---|---|---|---|---|---|---|---|---|---|---|---|---|---|---|---|---|---|---|
| Jan | 6.1 | 0.7 | 3.4 | *−1.4* | 9.8 | 23 | −6.5 | 18 | 2.6 | 12 | 5.6 | 1 | 11 | 0 | 68.2 | *120* | 18 | 15.9 | 27 | 53.6 | *86* | 7.4 | 25 |
| Feb | 5.5 | 0.4 | 2.9 | *−2.0* | 12.6 | 24 | −6.1 | 11 | −0.2 | 9 | 8.1 | 25 | 15 | 0 | 45.6 | *107* | 16 | 7.4 | 25 | 58.9 | *75* | 7.7 | 11 |
| Mar | 10.8 | 3.3 | 7.0 | *−0.2* | 16.2 | 10 | −2.0 | 6 | 5.7 | 17 | 7.6 | 12 | 3 | 0 | 52.3 | *110* | 18 | 6.0 | 22 | 128.1 | *115* | 10.6 | 30 |
| Apr | 10.2 | 3.3 | 6.7 | *−2.5* | 16.5 | 24 | −3.2 | 11 | 4.9 | 10 | 7.7 | 22 | 5 | 0 | 45.2 | *92* | 16 | 8.1 | 19 | 113.6 | *71* | 10.7 | 23 |
| May | 16.9 | 7.1 | 12.0 | *−0.4* | 26.5 | 31 | 4.1 | 10 | 9.4 | 1 | 11.9 | 31 | 0 | 1 | 48.5 | *85* | 10 | 17.1 | 5 | 202.8 | *105* | 14.9 | 26 |
| June | 19.1 | 9.6 | 14.4 | *−1.2* | 26.4 | 1 | 6.2 | 27 | 12.7 | 13 | 13.3 | 29 | 0 | 2 | 39.9 | *81* | 11 | 7.2 | 29 | 171.7 | *90* | 13.1 | 11 |
| July | 20.0 | 11.5 | 15.8 | *−2.1* | 25.6 | 29 | 7.6 | 11 | 14.3 | 5 | 17.0 | 30 | 0 | 1 | 82.4 | *169* | 11 | 29.9 | 31 | 146.4 | *71* | 13.5 | 25 |
| Aug | 20.2 | 11.3 | 15.8 | *−1.8* | 24.9 | 19 | 6.4 | 31 | 14.7 | 31 | 14.5 | 20 | 0 | 0 | 32.6 | *60* | 15 | 11.8 | 1 | 157.4 | *80* | 12.7 | 19 |
| Sep | 19.1 | 10.6 | 14.8 | *−0.0* | 23.0 | 23 | 5.0 | 20 | 13.2 | 30 | 15.8 | 10 | 0 | 0 | 24.4 | *45* | 7 | 9.8 | 27 | 159.3 | *113* | 10.4 | 12 |
| Oct | 16.2 | 8.1 | 12.1 | *+0.8* | 23.9 | 12 | 1.8 | 18 | 11.9 | 17 | 14.2 | 11 | 0 | 0 | 5.4 | *8* | 6 | 3.1 | 17 | 87.5 | *79* | 8.7 | 23 |
| Nov | 11.6 | 5.5 | 8.6 | *+1.0* | 16.9 | 8 | −6.9 | 30 | −1.1 | 29 | 13.4 | 19 | 5 | 0 | 22.8 | *34* | 9 | 5.9 | 19 | 77.3 | *109* | 7.4 | 20 |
| Dec | 7.1 | 2.2 | 4.6 | *−0.4* | 14.4 | 10 | −7.1 | 1 | −2.6 | 31 | 10.2 | 11 | 12 | 0 | 109.5 | *174* | 21 | 14.9 | 10 | 34.3 | *64* | 5.6 | 17 |
| **1978** | 13.6 | 6.1 | 9.9 | *−0.9* | 26.5 | 5 | −7.1 | 12 | −2.6 | 12 | 17.0 | 7 | 51 | 4 | 576.8 | *87* | 158 | 29.9 | 7 | 1390.9 | *88* | 14.9 | 5 |

## OXFORD 1979

| Month | Mean max | Mean min | Mean temp | Anom | Highest max | Date | Lowest min | Date | Lowest max | Date | Highest min | Date | Air frost | Days ≥ 25 °C | Total pptn | Anom | Rain days | Wettest day | Date | Total sunshine | Anom | Sunniest day | Date |
|---|---|---|---|---|---|---|---|---|---|---|---|---|---|---|---|---|---|---|---|---|---|---|---|
| Jan | 3.2 | −3.1 | 0.1 | *−4.8* | 8.7 | 7 | −9.8 | 28 | −2.7 | 1 | 5.6 | 8 | 23 | 0 | 56.8 | *100* | 18 | 8.7 | 9 | 72.0 | *116* | 7.2 | 13 |
| Feb | 3.8 | −0.9 | 1.5 | *−3.4* | 9.1 | 1 | −5.5 | 15 | −0.8 | 15 | 2.0 | 1 | 18 | 0 | 56.2 | *131* | 16 | 15.7 | 1 | 67.3 | *85* | 8.7 | 25 |
| Mar | 8.4 | 2.1 | 5.2 | *−2.0* | 13.2 | 3 | −1.6 | 23 | 1.2 | 16 | 8.8 | 3 | 2 | 0 | 110.3 | *231* | 23 | 18.9 | 16 | 92.9 | *84* | 10.5 | 23 |
| Apr | 12.3 | 4.8 | 8.6 | *−0.7* | 21.9 | 15 | −2.0 | 4 | 7.1 | 5 | 10.0 | 11 | 3 | 0 | 38.9 | *80* | 15 | 6.2 | 23 | 126.4 | *79* | 11.0 | 30 |
| May | 15.2 | 6.8 | 11.0 | *−1.5* | 25.8 | 15 | −0.1 | 3 | 8.5 | 1 | 13.3 | 16 | 2 | 2 | 124.8 | *218* | 16 | 21.2 | 10 | 190.9 | *99* | 14.1 | 19 |
| June | 19.0 | 10.7 | 14.8 | *−0.7* | 25.7 | 20 | 6.9 | 1 | 14.5 | 7 | 14.5 | 21 | 0 | 2 | 43.6 | *89* | 11 | 17.8 | 24 | 166.6 | *87* | 14.3 | 22 |
| July | 22.5 | 13.0 | 17.7 | *−0.1* | 29.5 | 27 | 9.1 | 1 | 17.1 | 22 | 17.5 | 29 | 0 | 6 | 16.6 | *34* | 4 | 9.7 | 30 | 192.2 | *93* | 13.9 | 5 |
| Aug | 20.3 | 11.7 | 16.0 | *−1.6* | 25.5 | 12 | 5.6 | 28 | 15.5 | 23 | 17.4 | 13 | 0 | 1 | 60.8 | *111* | 16 | 14.3 | 8 | 167.4 | *85* | 12.7 | 28 |
| Sep | 18.6 | 9.6 | 14.1 | *−0.8* | 23.8 | 6 | 2.7 | 22 | 12.5 | 23 | 16.4 | 19 | 0 | 0 | 17.0 | *31* | 7 | 5.1 | 18 | 171.0 | *122* | 11.2 | 16 |
| Oct | 15.5 | 8.0 | 11.8 | *+0.5* | 20.8 | 8 | 0.3 | 28 | 8.9 | 28 | 16.0 | 9 | 0 | 0 | 59.8 | *85* | 13 | 12.5 | 25 | 117.0 | *105* | 9.2 | 18 |
| Nov | 10.5 | 3.9 | 7.2 | *−0.4* | 15.5 | 3 | −3.5 | 13 | 4.0 | 20 | 11.8 | 4 | 4 | 0 | 39.4 | *59* | 17 | 8.4 | 26 | 70.0 | *99* | 8.3 | 2 |
| Dec | 8.9 | 3.8 | 6.3 | *+1.3* | 15.0 | 4 | −4.0 | 25 | 2.0 | 22 | 9.7 | 7 | 3 | 0 | 127.7 | *202* | 18 | 33.2 | 27 | 59.7 | *112* | 6.2 | 19 |
| 1979 | 13.2 | 5.9 | 9.5 | *−1.2* | 29.5 | 7 | −9.8 | 1 | −2.7 | 1 | 17.5 | 7 | 55 | 11 | 751.9 | *114* | 174 | 33.2 | 12 | 1493.4 | *95* | 14.3 | 6 |

## OXFORD 1980

| Month | Mean max | Mean min | Mean temp | Anom | Highest max | Date | Lowest min | Date | Lowest max | Date | Highest min | Date | Air frost | Days ≥ 25 °C | Total pptn | Anom | Rain days | Wettest day | Date | Total sunshine | Anom | Sunniest day | Date |
|---|---|---|---|---|---|---|---|---|---|---|---|---|---|---|---|---|---|---|---|---|---|---|---|
| Jan | 5.3 | −0.3 | 2.5 | *−2.4* | 10.9 | 3 | −5.2 | 14 | 0.7 | 14 | 6.2 | 31 | 15 | 0 | 36.7 | *65* | 12 | 11.7 | 3 | 84.0 | *135* | 7.3 | 20 |
| Feb | 9.5 | 3.2 | 6.3 | *+1.4* | 13.5 | 17 | −1.8 | 26 | 6.1 | 27 | 8.2 | 9 | 5 | 0 | 39.8 | *93* | 13 | 6.3 | 4 | 58.8 | *75* | 8.5 | 10 |
| Mar | 8.5 | 2.3 | 5.4 | *−1.9* | 12.7 | 27 | −2.8 | 23 | 2.6 | 19 | 6.1 | 29 | 7 | 0 | 69.8 | *146* | 18 | 9.8 | 12 | 86.7 | *78* | 9.3 | 3 |
| Apr | 13.7 | 4.9 | 9.3 | *+0.0* | 20.9 | 14 | 0.1 | 4 | 9.2 | 20 | 9.7 | 17 | 0 | 0 | 14.1 | *29* | 6 | 8.7 | 1 | 165.4 | *103* | 12.1 | 21 |
| May | 16.6 | 6.3 | 11.5 | *−1.0* | 25.6 | 12 | −0.5 | 9 | 10.3 | 8 | 10.9 | 20 | 1 | 1 | 35.5 | *62* | 6 | 11.8 | 20 | 228.7 | *119* | 14.8 | 16 |
| June | 19.0 | 10.7 | 14.8 | *−0.7* | 28.4 | 4 | 7.2 | 23 | 14.7 | 22 | 17.3 | 5 | 0 | 2 | 89.4 | *182* | 19 | 12.2 | 10 | 186.7 | *98* | 12.7 | 6 |
| July | 19.7 | 11.2 | 15.4 | *−2.4* | 27.7 | 25 | 6.7 | 15 | 15.1 | 9 | 16.7 | 26 | 0 | 3 | 59.5 | *122* | 16 | 15.3 | 13 | 168.9 | *82* | 14.5 | 22 |
| Aug | 21.2 | 12.8 | 17.0 | *−0.6* | 25.1 | 13 | 7.2 | 24 | 17.5 | 23 | 17.1 | 14 | 0 | 1 | 63.6 | *116* | 12 | 32.6 | 14 | 169.0 | *86* | 11.3 | 21 |
| Sep | 19.2 | 12.2 | 15.7 | *+0.9* | 24.4 | 3 | 7.1 | 26 | 15.3 | 28 | 15.2 | 21 | 0 | 0 | 65.6 | *121* | 14 | 39.8 | 20 | 136.4 | *97* | 11.9 | 3 |
| Oct | 13.0 | 6.0 | 9.5 | *−1.8* | 17.6 | 27 | 0.0 | 31 | 7.2 | 16 | 14.3 | 28 | 0 | 0 | 84.9 | *121* | 20 | 23.8 | 15 | 111.8 | *100* | 9.5 | 18 |
| Nov | 9.1 | 4.4 | 6.8 | *−0.8* | 15.1 | 15 | −1.5 | 3 | 2.9 | 28 | 13.3 | 23 | 3 | 0 | 42.5 | *64* | 16 | 12.5 | 17 | 61.4 | *87* | 7.9 | 12 |
| Dec | 8.5 | 2.9 | 5.7 | *+0.7* | 13.0 | 12 | −4.1 | 1 | 3.5 | 7 | 9.8 | 24 | 7 | 0 | 32.9 | *52* | 16 | 7.2 | 19 | 64.4 | *121* | 6.9 | 7 |
| 1980 | 13.6 | 6.4 | 10.0 | *−0.7* | 28.4 | 6 | −5.2 | 1 | 0.7 | 1 | 17.3 | 6 | 38 | 7 | 634.3 | *96* | 168 | 39.8 | 9 | 1522.2 | *97* | 14.8 | 5 |

## OXFORD 1981

| Month | Mean max | Mean min | Mean temp | Anom | Highest max | Date | Lowest min | Date | Lowest max | Date | Highest min | Date | Air frost | Days ≥ 25 °C | Total pptn | Anom | Rain days | Wettest day | Date | Total sunshine | Anom | Sunniest day | Date |
|---|---|---|---|---|---|---|---|---|---|---|---|---|---|---|---|---|---|---|---|---|---|---|---|
| Jan | 7.3 | 2.4 | 4.9 | *+0.0* | 12.1 | 21 | −2.6 | 31 | 2.3 | 31 | 8.2 | 3 | 8 | 0 | 39.2 | *69* | 16 | 6.8 | 9 | 50.9 | *82* | 5.7 | 1 |
| Feb | 6.2 | 0.2 | 3.2 | *−1.7* | 12.5 | 6 | −5.3 | 14 | 2.5 | 26 | 9.4 | 8 | 15 | 0 | 20.9 | *49* | 8 | 6.0 | 27 | 75.3 | *95* | 8.7 | 13 |
| Mar | 11.1 | 6.4 | 8.7 | *+1.5* | 15.9 | 28 | 1.1 | 17 | 3.5 | 4 | 11.3 | 11 | 0 | 0 | 129.7 | *272* | 25 | 19.8 | 1 | 57.8 | *52* | 7.8 | 26 |
| Apr | 12.0 | 4.4 | 8.2 | *−1.1* | 21.4 | 10 | −1.0 | 23 | 5.6 | 26 | 13.2 | 11 | 2 | 0 | 43.9 | *90* | 13 | 22.1 | 25 | 110.4 | *69* | 12.7 | 18 |
| May | 15.7 | 8.2 | 12.0 | *−0.5* | 21.6 | 31 | 0.9 | 3 | 8.9 | 3 | 12.6 | 31 | 0 | 0 | 87.7 | *153* | 24 | 14.1 | 25 | 98.3 | *51* | 8.2 | 24 |
| June | 18.1 | 10.4 | 14.2 | *−1.4* | 24.2 | 14 | 5.3 | 29 | 13.1 | 26 | 14.8 | 15 | 0 | 0 | 26.1 | *53* | 8 | 11.4 | 1 | 122.8 | *64* | 12.0 | 22 |
| July | 21.3 | 12.3 | 16.8 | *−1.1* | 25.8 | 28 | 7.9 | 25 | 15.3 | 24 | 15.5 | 9 | 0 | 3 | 34.0 | *70* | 9 | 15.3 | 31 | 129.0 | *62* | 10.7 | 7 |
| Aug | 22.2 | 12.7 | 17.4 | *−0.2* | 27.8 | 5 | 7.7 | 2 | 15.4 | 7 | 17.1 | 15 | 0 | 6 | 38.8 | *71* | 5 | 28.5 | 6 | 199.8 | *102* | 13.7 | 2 |
| Sep | 19.8 | 11.0 | 15.4 | *+0.5* | 24.8 | 5 | 7.5 | 29 | 15.3 | 26 | 15.6 | 8 | 0 | 0 | 92.5 | *171* | 14 | 22.0 | 25 | 157.3 | *112* | 11.2 | 5 |
| Oct | 12.3 | 5.7 | 9.0 | *−2.3* | 17.3 | 1 | −0.1 | 16 | 7.6 | 25 | 15.9 | 1 | 1 | 0 | 64.7 | *92* | 19 | 21.9 | 19 | 101.9 | *92* | 8.5 | 7 |
| Nov | 10.8 | 4.9 | 7.9 | *+0.3* | 15.9 | 1 | −0.9 | 9 | 6.6 | 7 | 13.2 | 2 | 2 | 0 | 31.4 | *47* | 13 | 10.6 | 19 | 43.7 | *62* | 6.8 | 9 |
| Dec | 3.6 | −2.0 | 0.8 | *−4.2* | 10.3 | 3 | −16.1 | 13 | −3.4 | 12 | 4.6 | 5 | 18 | 0 | 74.0 | *117* | 18 | 15.3 | 28 | 57.5 | *108* | 7.1 | 12 |
| 1981 | 13.4 | 6.4 | 9.9 | *−0.9* | 27.8 | 8 | −16.1 | 12 | −3.4 | 12 | 17.1 | 8 | 46 | 9 | 682.9 | *103* | 172 | 28.5 | 8 | 1204.7 | *76* | 13.7 | 8 |

## OXFORD 1982

| Month | Mean max | Mean min | Mean temp | Anom | Highest max | Date | Lowest min | Date | Lowest max | Date | Highest min | Date | Air frost | Days ≥ 25 °C | Total pptn | Anom | Rain days | Wettest day | Date | Total sunshine | Anom | Sunniest day | Date |
|---|---|---|---|---|---|---|---|---|---|---|---|---|---|---|---|---|---|---|---|---|---|---|---|
| Jan | 5.5 | −0.1 | 2.7 | *−2.1* | 12.3 | 2 | −16.6 | 14 | −7.1 | 13 | 9.0 | 4 | 13 | 0 | 57.2 | *101* | 17 | 12.0 | 7 | 55.0 | *88* | 7.4 | 27 |
| Feb | 7.9 | 2.7 | 5.3 | *+0.4* | 13.9 | 9 | −3.0 | 24 | 1.9 | 20 | 7.7 | 6 | 5 | 0 | 32.0 | *75* | 12 | 6.7 | 5 | 33.8 | *43* | 5.4 | 2 |
| Mar | 10.3 | 2.9 | 6.6 | *−0.7* | 16.9 | 27 | −1.6 | 5 | 7.1 | 20 | 7.8 | 1 | 2 | 0 | 96.2 | *202* | 19 | 29.8 | 6 | 144.7 | *130* | 10.7 | 24 |
| Apr | 13.8 | 4.9 | 9.3 | *+0.0* | 17.6 | 25 | 0.3 | 13 | 8.4 | 12 | 9.0 | 7 | 0 | 0 | 27.4 | *56* | 6 | 13.9 | 5 | 164.4 | *102* | 12.1 | 28 |
| May | 17.4 | 7.1 | 12.3 | *−0.2* | 25.9 | 31 | −0.2 | 5 | 10.9 | 5 | 12.5 | 21 | 1 | 2 | 35.9 | *63* | 10 | 10.0 | 23 | 187.2 | *97* | 15.0 | 29 |
| June | 21.3 | 12.7 | 17.0 | *+1.4* | 28.7 | 5 | 7.9 | 14 | 15.9 | 13 | 18.2 | 8 | 0 | 6 | 69.4 | *141* | 16 | 12.9 | 22 | 148.4 | *78* | 11.7 | 5 |
| July | 22.0 | 12.8 | 17.4 | *−0.5* | 27.7 | 8 | 8.5 | 28 | 17.6 | 23 | 15.9 | 10 | 0 | 6 | 23.5 | *48* | 6 | 9.4 | 13 | 151.4 | *73* | 12.4 | 8 |
| Aug | 21.2 | 12.8 | 17.0 | *−0.6* | 27.9 | 3 | 7.1 | 28 | 16.5 | 24 | 17.4 | 6 | 0 | 2 | 37.9 | *69* | 15 | 11.4 | 5 | 159.8 | *81* | 12.7 | 10 |
| Sep | 20.1 | 10.6 | 15.3 | *+0.5* | 27.5 | 18 | 4.5 | 15 | 13.0 | 29 | 15.8 | 20 | 0 | 3 | 51.9 | *96* | 13 | 8.8 | 5 | 141.0 | *100* | 12.5 | 2 |
| Oct | 13.6 | 7.8 | 10.7 | *−0.6* | 16.3 | 1 | 2.0 | 24 | 10.4 | 22 | 13.6 | 2 | 0 | 0 | 83.4 | *119* | 18 | 14.7 | 2 | 66.3 | *60* | 8.0 | 27 |
| Nov | 10.8 | 6.1 | 8.4 | *+0.9* | 16.5 | 1 | −4.7 | 30 | 2.1 | 27 | 12.7 | 1 | 4 | 0 | 77.3 | *116* | 21 | 14.3 | 21 | 68.5 | *97* | 7.3 | 15 |
| Dec | 7.7 | 1.9 | 4.8 | *−0.2* | 12.1 | 7 | −3.5 | 1 | 3.0 | 12 | 8.5 | 26 | 9 | 0 | 62.4 | *99* | 14 | 20.7 | 9 | 60.4 | *114* | 6.1 | 11 |
| 1982 | 14.3 | 6.9 | 10.6 | *−0.2* | 28.7 | 6 | −16.6 | 1 | −7.1 | 1 | 18.2 | 6 | 34 | 19 | 654.5 | *99* | 167 | 29.8 | 3 | 1380.9 | *88* | 15.0 | 5 |

## OXFORD 1983

| Month | Mean max | Mean min | Mean temp | Anom | Highest max | Date | Lowest min | Date | Lowest max | Date | Highest min | Date | Air frost | Days ≥ 25 °C | Total pptn | Anom | Rain days | Wettest day | Date | Total sunshine | Anom | Sunniest day | Date |
|---|---|---|---|---|---|---|---|---|---|---|---|---|---|---|---|---|---|---|---|---|---|---|---|
| Jan | 9.8 | 4.7 | 7.2 | *+2.4* | 13.9 | 5 | −2.1 | 20 | 4.5 | 30 | 11.8 | 6 | 2 | 0 | 45.7 | *80* | 12 | 10.4 | 31 | 54.9 | *88* | 6.3 | 8 |
| Feb | 4.8 | −0.7 | 2.0 | *−2.9* | 12.1 | 26 | −5.4 | 4 | 1.4 | 9 | 5.0 | 26 | 17 | 0 | 23.8 | *56* | 16 | 4.0 | 5 | 74.0 | *94* | 8.8 | 22 |
| Mar | 10.3 | 3.3 | 6.8 | *−0.5* | 14.4 | 10 | −1.5 | 28 | 6.6 | 24 | 9.5 | 18 | 4 | 0 | 42.5 | *89* | 18 | 9.2 | 23 | 89.5 | *80* | 9.1 | 15 |
| Apr | 11.8 | 3.7 | 7.8 | *−1.5* | 17.3 | 16 | −1.4 | 3 | 4.5 | 18 | 8.0 | 25 | 2 | 0 | 89.5 | *183* | 23 | 12.5 | 24 | 150.5 | *94* | 12.0 | 15 |
| May | 14.8 | 7.8 | 11.3 | *−1.2* | 20.4 | 31 | 3.9 | 10 | 10.1 | 2 | 11.9 | 6 | 0 | 0 | 106.3 | *186* | 22 | 19.9 | 21 | 126.8 | *66* | 10.5 | 14 |
| June | 19.6 | 11.0 | 15.3 | *−0.3* | 25.5 | 7 | 7.0 | 15 | 14.1 | 24 | 17.1 | 8 | 0 | 1 | 19.4 | *39* | 11 | 5.6 | 23 | 180.3 | *94* | 15.1 | 19 |
| July | 26.8 | 15.4 | 21.1 | *+3.3* | 31.9 | 14 | 9.8 | 3 | 20.5 | 20 | 19.1 | 15 | 0 | 22 | 21.7 | *44* | 7 | 13.5 | 31 | 257.5 | *124* | 14.6 | 3 |
| Aug | 23.5 | 13.3 | 18.4 | *+0.8* | 29.7 | 19 | 6.9 | 30 | 18.5 | 2 | 18.0 | 20 | 0 | 9 | 24.0 | *44* | 5 | 20.3 | 21 | 214.5 | *109* | 13.2 | 3 |
| Sep | 18.0 | 11.1 | 14.6 | *−0.3* | 21.7 | 1 | 5.9 | 22 | 11.8 | 21 | 13.9 | 27 | 0 | 0 | 59.4 | *110* | 13 | 9.0 | 2 | 92.4 | *66* | 9.7 | 5 |
| Oct | 14.3 | 7.7 | 11.0 | *−0.3* | 20.1 | 4 | −1.2 | 30 | 8.8 | 29 | 15.6 | 4 | 1 | 0 | 53.0 | *76* | 15 | 15.1 | 15 | 119.7 | *108* | 8.7 | 22 |
| Nov | 10.4 | 5.2 | 7.8 | *+0.2* | 16.2 | 9 | −7.0 | 23 | 2.0 | 23 | 12.8 | 2 | 6 | 0 | 42.6 | *64* | 8 | 25.2 | 26 | 44.3 | *62* | 7.2 | 14 |
| Dec | 8.7 | 3.2 | 5.9 | *+0.9* | 13.1 | 24 | −5.0 | 7 | 1.2 | 11 | 10.4 | 25 | 8 | 0 | 49.8 | *79* | 16 | 9.9 | 20 | 62.9 | *118* | 6.6 | 4 |
| 1983 | 14.4 | 7.1 | 10.8 | *+0.0* | 31.9 | 7 | −7.0 | 11 | 1.2 | 12 | 19.1 | 7 | 40 | 32 | 577.7 | *87* | 166 | 25.2 | 11 | 1467.3 | *93* | 15.1 | 6 |

## OXFORD 1984

| Month | Mean max | Mean min | Mean temp | Anom | Highest max | Date | Lowest min | Date | Lowest max | Date | Highest min | Date | Air frost | Days ≥ 25 °C | Total pptn | Anom | Rain days | Wettest day | Date | Total sunshine | Anom | Sunniest day | Date |
|---|---|---|---|---|---|---|---|---|---|---|---|---|---|---|---|---|---|---|---|---|---|---|---|
| Jan | 7.5 | 1.5 | 4.5 | *−0.3* | 12.8 | 12 | −3.0 | 20 | 2.6 | 19 | 8.2 | 11 | 7 | 0 | 81.5 | *143* | 23 | 9.9 | 2 | 86.1 | *138* | 6.5 | 18 |
| Feb | 6.4 | 1.0 | 3.7 | *−1.2* | 11.5 | 4 | −4.5 | 16 | 0.3 | 15 | 5.1 | 3 | 7 | 0 | 35.4 | *83* | 14 | 8.2 | 6 | 66.0 | *84* | 8.2 | 9 |
| Mar | 8.2 | 2.4 | 5.3 | *−2.0* | 13.0 | 1 | −0.3 | 19 | 4.1 | 16 | 6.6 | 6 | 1 | 0 | 50.3 | *106* | 17 | 19.6 | 23 | 48.8 | *44* | 8.4 | 3 |
| Apr | 13.8 | 3.4 | 8.6 | *−0.7* | 22.6 | 21 | −2.9 | 3 | 6.0 | 1 | 11.1 | 22 | 5 | 0 | 1.6 | *3* | 6 | 0.3 | 11 | 236.8 | *147* | 13.2 | 28 |
| May | 14.5 | 6.0 | 10.3 | *−2.2* | 22.5 | 24 | 0.5 | 9 | 9.4 | 26 | 9.9 | 24 | 0 | 0 | 72.0 | *126* | 12 | 21.1 | 21 | 143.4 | *74* | 14.2 | 13 |
| June | 20.8 | 10.8 | 15.8 | *+0.2* | 26.8 | 20 | 5.0 | 4 | 15.2 | 1 | 17.7 | 20 | 0 | 4 | 30.0 | *61* | 6 | 9.8 | 6 | 244.3 | *128* | 14.7 | 9 |
| July | 23.9 | 12.2 | 18.1 | *+0.2* | 31.5 | 28 | 7.5 | 3 | 17.8 | 2 | 16.5 | 18 | 0 | 11 | 13.6 | *28* | 8 | 3.9 | 30 | 249.4 | *120* | 14.6 | 5 |
| Aug | 23.6 | 13.6 | 18.6 | *+1.0* | 29.2 | 21 | 7.8 | 11 | 17.6 | 7 | 17.7 | 31 | 0 | 9 | 32.7 | *60* | 10 | 10.0 | 3 | 199.8 | *102* | 13.1 | 19 |
| Sep | 17.9 | 11.1 | 14.5 | *−0.4* | 25.8 | 2 | 6.6 | 26 | 12.9 | 23 | 16.4 | 1 | 0 | 1 | 91.1 | *168* | 16 | 26.3 | 19 | 99.6 | *71* | 9.9 | 26 |
| Oct | 15.0 | 8.6 | 11.8 | *+0.5* | 19.8 | 8 | −1.0 | 27 | 11.0 | 26 | 14.8 | 9 | 1 | 0 | 54.4 | *78* | 19 | 9.4 | 22 | 88.6 | *80* | 8.6 | 14 |
| Nov | 11.3 | 6.0 | 8.7 | *+1.1* | 16.7 | 1 | 0.4 | 19 | 5.4 | 19 | 12.0 | 2 | 0 | 0 | 101.6 | *152* | 19 | 12.4 | 12 | 53.8 | *76* | 7.7 | 10 |
| Dec | 8.3 | 2.7 | 5.5 | *+0.5* | 12.9 | 5 | −2.6 | 28 | 2.3 | 28 | 8.8 | 1 | 8 | 0 | 44.5 | *71* | 18 | 8.4 | 15 | 57.6 | *108* | 6.7 | 9 |
| 1984 | 14.3 | 6.6 | 10.4 | *−0.3* | 31.5 | 7 | −4.5 | 2 | 0.3 | 2 | 17.7 | 6 | 29 | 25 | 608.7 | *92* | 168 | 26.3 | 9 | 1574.2 | *100* | 14.7 | 6 |

## OXFORD 1985

| Month | Mean max | Mean min | Mean temp | Anom | Highest max | Date | Lowest min | Date | Lowest max | Date | Highest min | Date | Air frost | Days ≥ 25 °C | Total pptn | Anom | Rain days | Wettest day | Date | Total sunshine | Anom | Sunniest day | Date |
|---|---|---|---|---|---|---|---|---|---|---|---|---|---|---|---|---|---|---|---|---|---|---|---|
| Jan | 3.9 | −1.8 | 1.0 | *−3.9* | 11.9 | 31 | −10.0 | 17 | −2.0 | 7 | 6.5 | 31 | 22 | 0 | 44.4 | *78* | 17 | 11.1 | 25 | 51.1 | *82* | 6.9 | 22 |
| Feb | 5.5 | −0.8 | 2.3 | *−2.6* | 12.9 | 24 | −9.7 | 16 | −3.0 | 13 | 7.9 | 1 | 15 | 0 | 34.7 | *81* | 6 | 12.9 | 8 | 78.0 | *99* | 9.2 | 16 |
| Mar | 8.9 | 1.4 | 5.1 | *−2.1* | 14.5 | 31 | −4.4 | 20 | 4.1 | 16 | 7.9 | 31 | 10 | 0 | 36.7 | *77* | 17 | 5.0 | 26 | 112.4 | *101* | 9.7 | 18 |
| Apr | 13.1 | 5.1 | 9.1 | *−0.2* | 19.4 | 19 | −1.5 | 24 | 8.5 | 25 | 11.1 | 3 | 3 | 0 | 32.7 | *67* | 16 | 7.6 | 28 | 138.3 | *86* | 13.2 | 24 |
| May | 15.7 | 7.3 | 11.5 | *−1.0* | 21.0 | 18 | 3.5 | 16 | 10.4 | 11 | 14.1 | 26 | 0 | 0 | 80.5 | *141* | 11 | 39.5 | 14 | 178.3 | *92* | 15.5 | 30 |
| June | 17.5 | 9.0 | 13.3 | *−2.3* | 23.7 | 3 | 4.5 | 1 | 11.7 | 6 | 13.5 | 29 | 0 | 0 | 122.5 | *249* | 20 | 24.5 | 6 | 165.0 | *86* | 14.4 | 3 |
| July | 21.6 | 12.7 | 17.1 | *−0.7* | 27.7 | 25 | 8.8 | 7 | 16.8 | 29 | 16.9 | 26 | 0 | 5 | 45.8 | *94* | 13 | 11.3 | 28 | 216.0 | *104* | 14.8 | 23 |
| Aug | 19.0 | 12.1 | 15.5 | *−2.0* | 24.4 | 29 | 9.3 | 25 | 16.4 | 2 | 16.7 | 29 | 0 | 0 | 67.0 | *122* | 21 | 16.5 | 4 | 164.9 | *84* | 11.1 | 21 |
| Sep | 19.4 | 11.2 | 15.3 | *+0.4* | 25.5 | 12 | 5.6 | 7 | 14.8 | 2 | 15.7 | 19 | 0 | 1 | 19.4 | *36* | 7 | 11.0 | 2 | 143.3 | *102* | 11.8 | 6 |
| Oct | 14.5 | 7.9 | 11.2 | *−0.1* | 26.4 | 1 | 2.4 | 21 | 8.6 | 26 | 16.8 | 3 | 0 | 1 | 29.1 | *42* | 6 | 17.5 | 6 | 100.8 | *91* | 8.7 | 13 |
| Nov | 7.4 | 1.7 | 4.6 | *−3.0* | 16.0 | 8 | −3.5 | 13 | 1.8 | 19 | 9.7 | 9 | 10 | 0 | 41.1 | *62* | 14 | 11.9 | 29 | 92.7 | *131* | 8.3 | 3 |
| Dec | 9.4 | 5.0 | 7.2 | *+2.2* | 15.2 | 2 | −6.0 | 28 | 0.0 | 28 | 12.5 | 3 | 5 | 0 | 108.1 | *171* | 20 | 38.3 | 23 | 44.2 | *83* | 6.3 | 23 |
| **1985** | 13.0 | 5.9 | 9.4 | *−1.3* | 27.7 | 7 | −10.0 | 1 | −3.0 | 2 | 16.9 | 7 | 65 | 7 | 662.0 | *100* | 168 | 39.5 | 5 | 1485.0 | *94* | 15.5 | 5 |

## OXFORD 1986

| Month | Mean max | Mean min | Mean temp | Anom | Highest max | Date | Lowest min | Date | Lowest max | Date | Highest min | Date | Air frost | Days ≥ 25 °C | Total pptn | Anom | Rain days | Wettest day | Date | Total sunshine | Anom | Sunniest day | Date |
|---|---|---|---|---|---|---|---|---|---|---|---|---|---|---|---|---|---|---|---|---|---|---|---|
| Jan | 6.7 | 1.1 | 3.9 | *−0.9* | 11.8 | 10 | −5.6 | 26 | 2.1 | 7 | 8.6 | 19 | 9 | 0 | 67.3 | *118* | 21 | 12.8 | 7 | 76.4 | *123* | 8.5 | 25 |
| Feb | 1.0 | −4.0 | −1.5 | *−6.4* | 3.5 | 13 | −10.7 | 21 | −1.7 | 25 | 2.0 | 2 | 24 | 0 | 7.3 | *17* | 5 | 3.2 | 1 | 71.2 | *90* | 9.4 | 27 |
| Mar | 9.2 | 2.1 | 5.7 | *−1.6* | 13.4 | 27 | −5.6 | 3 | 0.8 | 1 | 6.0 | 22 | 4 | 0 | 58.0 | *122* | 17 | 8.9 | 18 | 127.3 | *114* | 9.5 | 2 |
| Apr | 10.2 | 3.1 | 6.7 | *−2.6* | 15.6 | 26 | −1.3 | 11 | 4.9 | 8 | 7.4 | 27 | 2 | 0 | 70.4 | *144* | 23 | 10.8 | 19 | 139.6 | *87* | 11.9 | 26 |
| May | 15.7 | 8.0 | 11.8 | *−0.6* | 20.1 | 2 | 3.5 | 16 | 12.4 | 7 | 13.0 | 20 | 0 | 0 | 70.7 | *124* | 19 | 20.3 | 14 | 205.6 | *107* | 13.4 | 29 |
| June | 20.6 | 10.5 | 15.5 | *−0.0* | 28.8 | 28 | 6.0 | 5 | 13.6 | 5 | 15.5 | 17 | 0 | 7 | 22.3 | *45* | 8 | 8.6 | 23 | 222.7 | *117* | 14.8 | 14 |
| July | 21.5 | 12.8 | 17.2 | *−0.7* | 28.4 | 15 | 7.5 | 24 | 17.1 | 30 | 17.2 | 16 | 0 | 4 | 37.1 | *76* | 13 | 9.4 | 4 | 201.2 | *97* | 14.5 | 1 |
| Aug | 18.4 | 11.0 | 14.7 | *−2.9* | 21.4 | 6 | 5.4 | 29 | 14.3 | 3 | 15.7 | 14 | 0 | 0 | 114.3 | *209* | 18 | 29.3 | 25 | 147.0 | *75* | 10.4 | 24 |
| Sep | 16.6 | 7.5 | 12.0 | *−2.9* | 21.0 | 2 | 2.3 | 19 | 9.9 | 15 | 12.8 | 29 | 0 | 0 | 39.0 | *72* | 3 | 34.7 | 13 | 171.6 | *122* | 11.6 | 4 |
| Oct | 15.3 | 7.8 | 11.5 | *+0.3* | 20.2 | 7 | 0.8 | 17 | 10.6 | 23 | 12.5 | 1 | 0 | 0 | 75.1 | *107* | 15 | 24.4 | 19 | 119.7 | *108* | 9.5 | 16 |
| Nov | 11.4 | 5.4 | 8.4 | *+0.9* | 14.8 | 7 | 1.8 | 2 | 5.9 | 29 | 11.5 | 25 | 0 | 0 | 78.4 | *117* | 18 | 14.3 | 20 | 77.9 | *110* | 8.2 | 6 |
| Dec | 9.2 | 3.7 | 6.5 | *+1.5* | 14.2 | 4 | −0.4 | 23 | 3.5 | 22 | 12.4 | 5 | 1 | 0 | 64.2 | *102* | 18 | 10.7 | 15 | 64.8 | *122* | 7.1 | 6 |
| **1986** | 13.0 | 5.7 | 9.4 | *−1.4* | 28.8 | 6 | −10.7 | 2 | −1.7 | 2 | 17.2 | 7 | 40 | 11 | 704.1 | *107* | 178 | 34.7 | 9 | 1625.0 | *103* | 14.8 | 6 |

## OXFORD 1987

| Month | Mean max | Mean min | Mean temp | Anom | Highest max | Date | Lowest min | Date | Lowest max | Date | Highest min | Date | Air frost | Days ≥ 25 °C | Total pptn | Anom | Rain days | Wettest day | Date | Total sunshine | Anom | Sunniest day | Date |
|---|---|---|---|---|---|---|---|---|---|---|---|---|---|---|---|---|---|---|---|---|---|---|---|
| Jan | 3.0 | −1.4 | 0.8 | *−4.1* | 10.3 | 1 | −10.5 | 13 | −6.2 | 12 | 6.9 | 1 | 20 | 0 | 9.9 | *17* | 7 | 4.2 | 4 | 68.8 | *110* | 8.4 | 31 |
| Feb | 7.1 | 1.0 | 4.1 | *−0.8* | 13.1 | 28 | −6.5 | 1 | 2.3 | 16 | 9.2 | 28 | 13 | 0 | 34.0 | *79* | 14 | 8.1 | 26 | 67.8 | *86* | 8.8 | 20 |
| Mar | 8.2 | 1.5 | 4.9 | *−2.4* | 12.7 | 1 | −3.1 | 12 | 3.0 | 7 | 8.9 | 25 | 13 | 0 | 53.9 | *113* | 18 | 13.6 | 26 | 111.6 | *100* | 10.1 | 29 |
| Apr | 15.2 | 6.6 | 10.9 | *+1.6* | 22.5 | 28 | 0.8 | 2 | 8.6 | 4 | 12.5 | 30 | 0 | 0 | 51.4 | *105* | 14 | 12.5 | 4 | 158.7 | *99* | 12.6 | 23 |
| May | 15.2 | 6.4 | 10.8 | *−1.7* | 22.0 | 9 | 1.1 | 4 | 10.4 | 23 | 12.3 | 1 | 0 | 0 | 48.7 | *85* | 13 | 10.4 | 22 | 168.7 | *87* | 13.8 | 8 |
| June | 17.8 | 10.4 | 14.1 | *−1.5* | 27.2 | 29 | 4.6 | 16 | 13.0 | 9 | 18.1 | 29 | 0 | 1 | 86.0 | *175* | 20 | 15.8 | 25 | 129.3 | *68* | 9.8 | 30 |
| July | 21.2 | 12.7 | 16.9 | *−0.9* | 27.8 | 6 | 7.1 | 26 | 15.3 | 21 | 16.8 | 14 | 0 | 5 | 29.3 | *60* | 14 | 6.8 | 18 | 179.4 | *87* | 15.1 | 5 |
| Aug | 21.0 | 12.5 | 16.8 | *−0.8* | 28.1 | 20 | 7.7 | 10 | 14.8 | 24 | 17.5 | 20 | 0 | 5 | 34.0 | *62* | 12 | 8.0 | 22 | 158.2 | *81* | 12.5 | 16 |
| Sep | 18.6 | 10.9 | 14.7 | *−0.1* | 24.5 | 1 | 3.3 | 29 | 13.6 | 28 | 15.9 | 21 | 0 | 0 | 25.4 | *47* | 15 | 5.6 | 19 | 132.3 | *94* | 8.6 | 22 |
| Oct | 14.0 | 6.7 | 10.3 | *−1.0* | 19.2 | 5 | −2.2 | 25 | 8.7 | 10 | 13.0 | 5 | 1 | 0 | 138.8 | *198* | 22 | 32.7 | 9 | 109.5 | *98* | 9.3 | 1 |
| Nov | 9.0 | 4.3 | 6.7 | *−0.9* | 12.6 | 15 | −3.8 | 29 | 4.2 | 29 | 9.7 | 1 | 2 | 0 | 58.2 | *87* | 10 | 15.9 | 19 | 43.9 | *62* | 7.8 | 14 |
| Dec | 8.0 | 3.7 | 5.9 | *+0.8* | 14.2 | 17 | −4.3 | 11 | 2.0 | 14 | 11.7 | 29 | 7 | 0 | 26.0 | *41* | 10 | 11.0 | 15 | 42.9 | *81* | 6.9 | 10 |
| 1987 | 13.2 | 6.3 | 9.7 | *−1.0* | 28.1 | 8 | −10.5 | 1 | −6.2 | 1 | 18.1 | 6 | 56 | 11 | 595.6 | *90* | 169 | 32.7 | 10 | 1371.1 | *87* | 15.1 | 7 |

## OXFORD 1988

| Month | Mean max | Mean min | Mean temp | Anom | Highest max | Date | Lowest min | Date | Lowest max | Date | Highest min | Date | Air frost | Days ≥ 25 °C | Total pptn | Anom | Rain days | Wettest day | Date | Total sunshine | Anom | Sunniest day | Date |
|---|---|---|---|---|---|---|---|---|---|---|---|---|---|---|---|---|---|---|---|---|---|---|---|
| Jan | 8.1 | 3.3 | 5.7 | *+0.8* | 13.0 | 1 | −1.8 | 17 | 2.5 | 16 | 8.4 | 2 | 4 | 0 | 100.0 | *176* | 23 | 17.9 | 21 | 47.7 | *77* | 6.6 | 7 |
| Feb | 8.1 | 2.3 | 5.2 | *+0.3* | 13.8 | 15 | −1.8 | 26 | 4.3 | 29 | 6.7 | 14 | 5 | 0 | 33.8 | *79* | 11 | 9.3 | 3 | 103.7 | *131* | 8.2 | 8 |
| Mar | 10.2 | 3.8 | 7.0 | *−0.3* | 13.7 | 21 | −3.3 | 2 | 5.2 | 4 | 7.7 | 20 | 4 | 0 | 59.1 | *124* | 17 | 11.6 | 24 | 85.2 | *77* | 9.6 | 4 |
| Apr | 12.6 | 4.6 | 8.6 | *−0.7* | 17.7 | 18 | −0.8 | 10 | 5.7 | 9 | 11.7 | 19 | 3 | 0 | 30.2 | *62* | 9 | 7.3 | 30 | 132.0 | *82* | 12.9 | 24 |
| May | 17.1 | 7.8 | 12.4 | *−0.0* | 22.8 | 14 | 3.0 | 20 | 11.9 | 18 | 10.9 | 9 | 0 | 0 | 32.0 | *56* | 12 | 8.4 | 29 | 178.1 | *92* | 12.9 | 16 |
| June | 18.7 | 10.3 | 14.5 | *−1.1* | 24.1 | 19 | 6.9 | 5 | 11.6 | 9 | 14.6 | 26 | 0 | 0 | 59.6 | *121* | 13 | 18.6 | 25 | 142.3 | *75* | 13.4 | 12 |
| July | 18.9 | 11.6 | 15.2 | *−2.6* | 23.5 | 20 | 7.8 | 16 | 15.5 | 3 | 17.2 | 22 | 0 | 0 | 96.5 | *198* | 23 | 20.8 | 3 | 138.8 | *67* | 12.2 | 11 |
| Aug | 20.4 | 11.9 | 16.1 | *−1.5* | 28.2 | 7 | 7.6 | 3 | 16.8 | 25 | 15.2 | 9 | 0 | 4 | 38.8 | *71* | 14 | 15.3 | 31 | 178.8 | *91* | 14.1 | 7 |
| Sep | 17.9 | 10.4 | 14.2 | *−0.7* | 25.4 | 7 | 3.3 | 30 | 13.4 | 20 | 15.8 | 8 | 0 | 2 | 39.6 | *73* | 10 | 16.7 | 24 | 136.9 | *98* | 10.8 | 7 |
| Oct | 14.5 | 8.0 | 11.2 | *−0.1* | 17.4 | 19 | −0.7 | 30 | 8.8 | 30 | 14.1 | 19 | 2 | 0 | 38.6 | *55* | 13 | 9.6 | 8 | 120.0 | *108* | 10.0 | 2 |
| Nov | 9.1 | 1.6 | 5.3 | *−2.2* | 15.3 | 9 | −3.8 | 22 | 3.3 | 21 | 12.7 | 10 | 15 | 0 | 29.5 | *44* | 10 | 13.1 | 29 | 91.4 | *129* | 8.6 | 3 |
| Dec | 10.1 | 5.4 | 7.7 | *+2.7* | 14.1 | 26 | 0.0 | 3 | 3.2 | 2 | 10.5 | 26 | 0 | 0 | 13.4 | *21* | 7 | 8.4 | 3 | 46.7 | *88* | 5.9 | 6 |
| 1988 | 13.8 | 6.7 | 10.3 | *−0.5* | 28.2 | 8 | −3.8 | 11 | 2.5 | 1 | 17.2 | 7 | 33 | 6 | 571.1 | *86* | 162 | 20.8 | 7 | 1401.6 | *89* | 14.1 | 8 |

## OXFORD 1989

| Month | Mean max | Mean min | Mean temp | Anom | Highest max | Date | Lowest min | Date | Lowest max | Date | Highest min | Date | Air frost | Days ≥ 25 °C | Total pptn | Anom | Rain days | Wettest day | Date | Total sunshine | Anom | Sunniest day | Date |
|---|---|---|---|---|---|---|---|---|---|---|---|---|---|---|---|---|---|---|---|---|---|---|---|
| Jan | 9.1 | 3.6 | 6.3 | *+1.5* | 12.5 | 13 | −2.7 | 29 | 5.0 | 19 | 9.4 | 28 | 5 | 0 | 33.4 | *59* | 10 | 8.2 | 11 | 68.2 | *110* | 7.5 | 22 |
| Feb | 9.9 | 3.0 | 6.5 | *+1.6* | 14.6 | 6 | −3.2 | 2 | 5.1 | 1 | 9.4 | 19 | 4 | 0 | 55.1 | *129* | 16 | 14.5 | 26 | 107.0 | *136* | 9.6 | 28 |
| Mar | 12.0 | 4.6 | 8.3 | *+1.0* | 17.7 | 27 | −0.6 | 18 | 4.0 | 16 | 9.9 | 28 | 1 | 0 | 47.0 | *99* | 18 | 7.9 | 2 | 100.9 | *91* | 11.0 | 30 |
| Apr | 11.0 | 3.6 | 7.3 | *−2.0* | 15.6 | 1 | −1.3 | 26 | 4.6 | 5 | 7.5 | 16 | 1 | 0 | 63.8 | *130* | 17 | 14.9 | 5 | 133.9 | *83* | 10.7 | 15 |
| May | 19.7 | 8.6 | 14.1 | *+1.7* | 27.2 | 23 | 3.4 | 31 | 13.5 | 12 | 14.7 | 24 | 0 | 4 | 29.2 | *51* | 3 | 27.9 | 24 | 300.8 | *156* | 14.1 | 28 |
| June | 20.8 | 10.5 | 15.6 | *+0.0* | 29.5 | 20 | 2.8 | 2 | 11.2 | 1 | 16.7 | 13 | 0 | 9 | 45.9 | *93* | 10 | 11.0 | 5 | 244.4 | *128* | 15.0 | 18 |
| July | 24.9 | 14.0 | 19.5 | *+1.6* | 32.4 | 22 | 9.5 | 3 | 17.8 | 31 | 19.6 | 22 | 0 | 15 | 29.4 | *60* | 6 | 15.5 | 7 | 280.4 | *135* | 15.3 | 4 |
| Aug | 23.4 | 12.4 | 17.9 | *+0.3* | 28.2 | 6 | 7.1 | 28 | 18.5 | 27 | 15.9 | 25 | 0 | 9 | 43.3 | *79* | 9 | 16.7 | 9 | 269.5 | *137* | 14.1 | 1 |
| Sep | 19.9 | 11.9 | 15.9 | *+1.0* | 26.2 | 7 | 6.0 | 29 | 15.2 | 9 | 17.3 | 22 | 0 | 3 | 23.8 | *44* | 8 | 9.6 | 16 | 141.6 | *101* | 11.0 | 5 |
| Oct | 15.9 | 8.9 | 12.4 | *+1.1* | 18.9 | 17 | 3.5 | 3 | 13.3 | 28 | 12.0 | 28 | 0 | 0 | 53.9 | *77* | 17 | 10.5 | 20 | 90.8 | *82* | 9.2 | 17 |
| Nov | 9.8 | 3.2 | 6.5 | *−1.0* | 15.1 | 11 | −4.5 | 26 | 4.0 | 30 | 10.4 | 11 | 7 | 0 | 32.5 | *49* | 9 | 7.5 | 2 | 104.1 | *147* | 7.5 | 12 |
| Dec | 8.2 | 3.3 | 5.8 | *+0.7* | 13.8 | 16 | −3.5 | 1 | 1.7 | 1 | 10.4 | 21 | 4 | 0 | 141.5 | *224* | 14 | 24.8 | 13 | 26.5 | *50* | 4.6 | 22 |
| **1989** | 15.4 | 7.3 | 11.3 | *+0.6* | 32.4 | 7 | −4.5 | 11 | 1.7 | 12 | 19.6 | 7 | 22 | 40 | 598.8 | *91* | 137 | 27.9 | 5 | 1868.1 | *118* | 15.3 | 7 |

## OXFORD 1990

| Month | Mean max | Mean min | Mean temp | Anom | Highest max | Date | Lowest min | Date | Lowest max | Date | Highest min | Date | Air frost | Days ≥ 25 °C | Total pptn | Anom | Rain days | Wettest day | Date | Total sunshine | Anom | Sunniest day | Date |
|---|---|---|---|---|---|---|---|---|---|---|---|---|---|---|---|---|---|---|---|---|---|---|---|
| Jan | 10.0 | 4.5 | 7.3 | *+2.4* | 12.6 | 21 | −0.2 | 2 | 5.4 | 3 | 10.4 | 16 | 1 | 0 | 75.0 | *132* | 20 | 11.9 | 6 | 58.3 | *94* | 7.3 | 24 |
| Feb | 11.3 | 5.2 | 8.3 | *+3.4* | 15.6 | 23 | 0.4 | 16 | 5.9 | 14 | 10.7 | 20 | 0 | 0 | 94.2 | *220* | 18 | 18.0 | 7 | 95.7 | *121* | 8.6 | 8 |
| Mar | 12.7 | 5.0 | 8.9 | *+1.6* | 20.9 | 17 | −1.6 | 27 | 7.7 | 2 | 11.1 | 19 | 2 | 0 | 18.4 | *39* | 7 | 12.0 | 19 | 144.1 | *130* | 10.2 | 30 |
| Apr | 13.7 | 3.2 | 8.5 | *−0.8* | 22.2 | 30 | −2.4 | 5 | 8.0 | 4 | 8.8 | 30 | 5 | 0 | 20.4 | *42* | 11 | 5.5 | 2 | 234.4 | *146* | 13.9 | 30 |
| May | 19.2 | 7.8 | 13.5 | *+1.0* | 25.9 | 3 | 3.6 | 26 | 14.1 | 9 | 11.3 | 29 | 0 | 2 | 8.7 | *15* | 4 | 7.3 | 14 | 285.0 | *148* | 14.7 | 26 |
| June | 18.2 | 10.7 | 14.4 | *−1.1* | 23.7 | 26 | 6.2 | 12 | 13.4 | 12 | 16.1 | 26 | 0 | 0 | 48.8 | *99* | 12 | 13.7 | 30 | 121.1 | *63* | 11.7 | 19 |
| July | 23.8 | 12.2 | 18.0 | *+0.2* | 31.1 | 21 | 6.3 | 3 | 15.7 | 5 | 15.6 | 8 | 0 | 11 | 17.7 | *36* | 6 | 8.1 | 29 | 268.5 | *130* | 15.1 | 14 |
| Aug | 24.7 | 14.0 | 19.4 | *+1.8* | 35.1 | 3 | 10.2 | 31 | 18.6 | 18 | 18.9 | 4 | 0 | 14 | 26.5 | *48* | 5 | 10.4 | 18 | 236.6 | *120* | 13.6 | 1 |
| Sep | 18.6 | 9.4 | 14.0 | *−0.8* | 24.2 | 2 | 3.8 | 27 | 14.0 | 23 | 14.5 | 2 | 0 | 0 | 41.1 | *76* | 10 | 24.4 | 29 | 164.7 | *117* | 11.1 | 9 |
| Oct | 15.8 | 9.3 | 12.6 | *+1.3* | 22.3 | 12 | 1.5 | 8 | 9.3 | 29 | 14.4 | 15 | 0 | 0 | 45.9 | *66* | 13 | 13.6 | 25 | 124.4 | *112* | 9.5 | 8 |
| Nov | 9.8 | 4.2 | 7.0 | *−0.5* | 15.4 | 17 | −0.6 | 6 | 3.3 | 22 | 13.5 | 17 | 1 | 0 | 20.0 | *30* | 12 | 4.5 | 23 | 83.7 | *118* | 8.9 | 1 |
| Dec | 6.9 | 1.7 | 4.3 | *−0.7* | 12.4 | 26 | −4.9 | 16 | 2.1 | 18 | 9.2 | 22 | 10 | 0 | 56.3 | *89* | 17 | 8.0 | 7 | 58.8 | *111* | 6.5 | 27 |
| **1990** | 15.4 | 7.3 | 11.3 | *+0.6* | 35.1 | 8 | −4.9 | 12 | 2.1 | 12 | 18.9 | 8 | 19 | 27 | 473.0 | *72* | 135 | 24.4 | 9 | 1875.3 | *119* | 15.1 | 7 |

## OXFORD 1991

| Month | Mean max | Mean min | Mean temp | Anom | Highest max | Date | Lowest min | Date | Lowest max | Date | Highest min | Date | Air frost | Days ≥ 25 °C | Total pptn | Anom | Rain days | Wettest day | Date | Total sunshine | Anom | Sunniest day | Date |
|---|---|---|---|---|---|---|---|---|---|---|---|---|---|---|---|---|---|---|---|---|---|---|---|
| Jan | 6.4 | 1.2 | 3.8 | *−1.1* | 12.8 | 1 | −3.5 | 16 | 1.1 | 30 | 5.6 | 10 | 8 | 0 | 60.9 | *107* | 15 | 11.0 | 6 | 66.9 | *107* | 7.2 | 13 |
| Feb | 4.7 | −1.6 | 1.5 | *−3.4* | 12.5 | 23 | −7.9 | 7 | −3.6 | 9 | 9.7 | 24 | 18 | 0 | 20.5 | *48* | 11 | 4.4 | 27 | 60.0 | *76* | 8.1 | 16 |
| Mar | 11.5 | 5.1 | 8.3 | *+1.0* | 16.0 | 13 | −1.4 | 2 | 6.7 | 1 | 10.0 | 20 | 2 | 0 | 46.1 | *97* | 15 | 12.2 | 6 | 87.2 | *78* | 10.2 | 27 |
| Apr | 12.2 | 4.6 | 8.4 | *−0.9* | 18.6 | 12 | −0.7 | 20 | 7.7 | 17 | 9.7 | 12 | 1 | 0 | 56.3 | *115* | 14 | 26.5 | 29 | 156.6 | *97* | 11.3 | 28 |
| May | 15.3 | 7.4 | 11.3 | *−1.1* | 23.6 | 21 | 2.1 | 2 | 9.3 | 1 | 12.2 | 26 | 0 | 0 | 8.6 | *15* | 7 | 3.5 | 15 | 142.9 | *74* | 13.0 | 9 |
| June | 16.8 | 9.0 | 12.9 | *−2.7* | 21.7 | 30 | 1.8 | 2 | 13.1 | 1 | 13.3 | 30 | 0 | 0 | 74.8 | *152* | 23 | 12.2 | 23 | 150.0 | *79* | 12.0 | 4 |
| July | 22.2 | 13.7 | 18.0 | *+0.1* | 27.4 | 29 | 10.6 | 2 | 18.2 | 19 | 17.7 | 6 | 0 | 5 | 92.8 | *190* | 11 | 32.5 | 30 | 217.3 | *105* | 13.3 | 10 |
| Aug | 23.2 | 13.3 | 18.2 | *+0.7* | 26.7 | 21 | 9.4 | 19 | 16.5 | 23 | 17.3 | 22 | 0 | 5 | 11.6 | *21* | 4 | 6.9 | 22 | 241.4 | *123* | 12.8 | 18 |
| Sep | 20.2 | 10.6 | 15.4 | *+0.5* | 26.2 | 1 | 4.2 | 21 | 12.2 | 27 | 15.2 | 3 | 0 | 2 | 42.9 | *79* | 10 | 19.1 | 28 | 165.4 | *118* | 11.5 | 7 |
| Oct | 13.4 | 7.3 | 10.3 | *−1.0* | 19.6 | 11 | 0.3 | 22 | 9.3 | 18 | 12.6 | 12 | 0 | 0 | 34.2 | *49* | 15 | 12.8 | 29 | 79.4 | *71* | 9.2 | 1 |
| Nov | 9.9 | 4.5 | 7.2 | *−0.4* | 15.1 | 1 | −1.7 | 21 | 5.0 | 16 | 11.7 | 1 | 2 | 0 | 62.2 | *93* | 13 | 24.2 | 18 | 57.4 | *81* | 6.3 | 20 |
| Dec | 7.0 | 1.6 | 4.3 | *−0.8* | 13.0 | 22 | −9.7 | 12 | −1.5 | 12 | 11.2 | 23 | 11 | 0 | 15.9 | *25* | 8 | 6.2 | 17 | 58.4 | *110* | 6.0 | 11 |
| **1991** | 13.5 | 6.4 | 10.0 | *−0.8* | 27.4 | 7 | −9.7 | 12 | −3.6 | 2 | 17.7 | 7 | 42 | 12 | 526.8 | *80* | 146 | 32.5 | 7 | 1482.9 | *94* | 13.3 | 7 |

## OXFORD 1992

| Month | Mean max | Mean min | Mean temp | Anom | Highest max | Date | Lowest min | Date | Lowest max | Date | Highest min | Date | Air frost | Days ≥ 25 °C | Total pptn | Anom | Rain days | Wettest day | Date | Total sunshine | Anom | Sunniest day | Date |
|---|---|---|---|---|---|---|---|---|---|---|---|---|---|---|---|---|---|---|---|---|---|---|---|
| Jan | 5.9 | 1.5 | 3.7 | *−1.1* | 12.2 | 5 | −5.6 | 23 | 0.2 | 23 | 9.8 | 6 | 12 | 0 | 30.4 | *53* | 9 | 11.8 | 8 | 46.8 | *75* | 6.9 | 22 |
| Feb | 8.9 | 2.1 | 5.5 | *+0.6* | 14.1 | 27 | −3.5 | 1 | 0.3 | 1 | 7.7 | 27 | 8 | 0 | 22.0 | *51* | 10 | 5.7 | 17 | 67.3 | *85* | 7.0 | 17 |
| Mar | 11.0 | 4.8 | 7.9 | *+0.6* | 14.7 | 22 | −1.4 | 9 | 7.0 | 30 | 8.5 | 18 | 1 | 0 | 47.7 | *100* | 24 | 6.6 | 29 | 73.3 | *66* | 6.9 | 21 |
| Apr | 13.2 | 5.6 | 9.4 | *+0.1* | 18.0 | 18 | −0.4 | 5 | 7.8 | 2 | 10.9 | 21 | 2 | 0 | 57.7 | *118* | 15 | 20.6 | 14 | 139.2 | *87* | 11.7 | 20 |
| May | 19.7 | 9.1 | 14.4 | *+1.9* | 26.1 | 14 | 3.4 | 2 | 12.6 | 2 | 15.1 | 14 | 0 | 4 | 62.2 | *109* | 11 | 25.2 | 29 | 262.7 | *136* | 13.7 | 16 |
| June | 21.4 | 11.7 | 16.6 | *+1.0* | 28.7 | 29 | 6.7 | 19 | 14.2 | 5 | 17.2 | 30 | 0 | 5 | 43.4 | *88* | 8 | 14.8 | 30 | 212.2 | *111* | 14.8 | 13 |
| July | 21.1 | 13.5 | 17.3 | *−0.6* | 26.9 | 31 | 9.3 | 28 | 16.7 | 1 | 17.1 | 17 | 0 | 2 | 70.1 | *144* | 16 | 18.1 | 20 | 164.0 | *79* | 13.7 | 22 |
| Aug | 20.3 | 12.8 | 16.5 | *−1.1* | 25.2 | 7 | 7.2 | 31 | 15.6 | 13 | 16.0 | 9 | 0 | 1 | 108.9 | *199* | 18 | 23.0 | 8 | 173.6 | *88* | 11.8 | 4 |
| Sep | 17.6 | 10.8 | 14.2 | *−0.7* | 22.0 | 27 | 6.6 | 14 | 13.7 | 25 | 15.6 | 18 | 0 | 0 | 106.9 | *197* | 18 | 41.2 | 22 | 120.2 | *86* | 10.1 | 7 |
| Oct | 11.2 | 5.2 | 8.2 | *−3.1* | 16.1 | 2 | −0.5 | 17 | 7.7 | 20 | 10.8 | 4 | 2 | 0 | 75.7 | *108* | 17 | 20.5 | 19 | 101.3 | *91* | 9.1 | 13 |
| Nov | 10.9 | 4.7 | 7.8 | *+0.3* | 15.8 | 6 | 0.2 | 13 | 6.8 | 28 | 10.6 | 23 | 0 | 0 | 119.9 | *180* | 22 | 19.1 | 27 | 61.5 | *87* | 7.0 | 12 |
| Dec | 6.4 | 1.3 | 3.8 | *−1.2* | 13.2 | 2 | −6.5 | 30 | −2.1 | 29 | 9.6 | 2 | 11 | 0 | 50.7 | *80* | 11 | 19.9 | 6 | 47.2 | *89* | 6.0 | 22 |
| **1992** | 14.0 | 6.9 | 10.4 | *−0.3* | 28.7 | 6 | −6.5 | 12 | −2.1 | 12 | 17.2 | 6 | 36 | 12 | 795.6 | *120* | 179 | 41.2 | 9 | 1469.3 | *93* | 14.8 | 6 |

## OXFORD 1993

| Month | Mean max | Mean min | Mean temp | Anom | Highest max | Date | Lowest min | Date | Lowest max | Date | Highest min | Date | Air frost | Days ≥ 25 °C | Total pptn | Anom | Rain days | Wettest day | Date | Total sunshine | Anom | Sunniest day | Date |
|---|---|---|---|---|---|---|---|---|---|---|---|---|---|---|---|---|---|---|---|---|---|---|---|
| Jan | 9.6 | 3.5 | 6.5 | *+1.7* | 13.2 | 16 | −6.2 | 3 | −1.2 | 2 | 9.1 | 17 | 5 | 0 | 82.4 | *145* | 20 | 13.9 | 10 | 38.0 | *61* | 6.0 | 14 |
| Feb | 7.1 | 2.4 | 4.7 | *−0.2* | 10.8 | 17 | −3.0 | 28 | 2.9 | 11 | 7.5 | 8 | 4 | 0 | 7.1 | *17* | 5 | 5.4 | 25 | 54.7 | *69* | 8.1 | 27 |
| Mar | 11.0 | 3.9 | 7.5 | *+0.2* | 18.2 | 15 | −1.8 | 26 | 3.1 | 2 | 10.2 | 18 | 3 | 0 | 32.7 | *69* | 6 | 15.4 | 31 | 135.0 | *121* | 10.3 | 19 |
| Apr | 13.2 | 6.6 | 9.9 | *+0.6* | 19.0 | 29 | 0.9 | 4 | 7.7 | 1 | 10.7 | 23 | 0 | 0 | 69.2 | *141* | 15 | 14.4 | 9 | 111.9 | *70* | 10.7 | 4 |
| May | 16.6 | 8.0 | 12.3 | *−0.2* | 24.2 | 24 | 2.5 | 6 | 12.5 | 20 | 13.6 | 24 | 0 | 0 | 56.6 | *99* | 15 | 15.5 | 9 | 196.7 | *102* | 13.9 | 6 |
| June | 20.6 | 11.5 | 16.0 | *+0.5* | 26.4 | 8 | 7.0 | 24 | 16.3 | 3 | 17.3 | 10 | 0 | 3 | 40.1 | *81* | 11 | 7.4 | 12 | 231.4 | *121* | 15.1 | 27 |
| July | 20.8 | 12.3 | 16.6 | *−1.3* | 26.8 | 3 | 6.6 | 11 | 15.7 | 9 | 17.4 | 29 | 0 | 2 | 51.3 | *105* | 16 | 9.8 | 13 | 190.0 | *92* | 11.1 | 3 |
| Aug | 20.5 | 11.1 | 15.8 | *−1.8* | 25.1 | 19 | 6.7 | 27 | 16.4 | 23 | 15.5 | 8 | 0 | 1 | 26.7 | *49* | 7 | 10.3 | 11 | 238.3 | *121* | 12.8 | 16 |
| Sep | 16.7 | 9.3 | 13.0 | *−1.9* | 24.3 | 1 | 4.3 | 26 | 9.8 | 27 | 14.6 | 9 | 0 | 0 | 96.1 | *177* | 16 | 19.8 | 7 | 107.7 | *77* | 9.7 | 1 |
| Oct | 11.9 | 6.0 | 8.9 | *−2.3* | 17.1 | 11 | −2.8 | 17 | 8.2 | 16 | 10.7 | 12 | 3 | 0 | 112.4 | *161* | 16 | 28.2 | 12 | 120.0 | *108* | 9.5 | 16 |
| Nov | 8.0 | 2.6 | 5.3 | *−2.2* | 14.1 | 3 | −4.3 | 23 | 1.2 | 23 | 10.4 | 5 | 8 | 0 | 51.0 | *76* | 9 | 17.8 | 13 | 76.3 | *107* | 7.6 | 19 |
| Dec | 8.6 | 3.4 | 6.0 | *+1.0* | 13.5 | 18 | −3.4 | 27 | 2.1 | 26 | 10.8 | 19 | 4 | 0 | 98.0 | *155* | 23 | 15.9 | 12 | 55.0 | *103* | 5.8 | 16 |
| **1993** | 13.7 | 6.7 | 10.2 | *−0.5* | 26.8 | 7 | −6.2 | 1 | −1.2 | 1 | 17.4 | 7 | 27 | 6 | 723.6 | *110* | 159 | 28.2 | 10 | 1555.0 | *99* | 15.1 | 6 |

## OXFORD 1994

| Month | Mean max | Mean min | Mean temp | Anom | Highest max | Date | Lowest min | Date | Lowest max | Date | Highest min | Date | Air frost | Days ≥ 25 °C | Total pptn | Anom | Rain days | Wettest day | Date | Total sunshine | Anom | Sunniest day | Date |
|---|---|---|---|---|---|---|---|---|---|---|---|---|---|---|---|---|---|---|---|---|---|---|---|
| Jan | 8.5 | 3.3 | 5.9 | *+1.0* | 12.6 | 12 | −0.7 | 18 | 2.4 | 6 | 7.2 | 13 | 2 | 0 | 82.2 | *145* | 23 | 17.6 | 4 | 87.0 | *140* | 7.9 | 28 |
| Feb | 6.7 | 1.3 | 4.0 | *−0.9* | 12.8 | 27 | −3.6 | 15 | 2.1 | 21 | 9.1 | 27 | 8 | 0 | 62.4 | *146* | 17 | 12.3 | 22 | 77.6 | *98* | 8.8 | 21 |
| Mar | 11.6 | 5.3 | 8.5 | *+1.2* | 15.5 | 30 | 0.2 | 20 | 8.6 | 1 | 9.8 | 9 | 0 | 0 | 41.9 | *88* | 23 | 8.7 | 31 | 127.6 | *115* | 10.3 | 26 |
| Apr | 12.5 | 5.1 | 8.8 | *−0.5* | 22.1 | 29 | 0.6 | 3 | 8.2 | 14 | 12.6 | 28 | 0 | 0 | 52.0 | *106* | 12 | 12.8 | 3 | 177.7 | *110* | 12.8 | 30 |
| May | 15.3 | 7.6 | 11.4 | *−1.0* | 20.7 | 14 | 2.7 | 29 | 8.5 | 17 | 10.6 | 12 | 0 | 0 | 84.5 | *148* | 17 | 12.3 | 25 | 162.1 | *84* | 14.9 | 31 |
| June | 20.6 | 11.0 | 15.8 | *+0.2* | 28.3 | 24 | 5.2 | 5 | 13.1 | 4 | 15.5 | 25 | 0 | 2 | 13.5 | *27* | 5 | 5.7 | 4 | 254.9 | *133* | 15.4 | 16 |
| July | 24.8 | 13.8 | 19.3 | *+1.4* | 31.7 | 12 | 10.0 | 1 | 19.1 | 6 | 18.5 | 27 | 0 | 14 | 54.2 | *111* | 9 | 35.5 | 24 | 248.3 | *120* | 15.4 | 11 |
| Aug | 21.1 | 13.0 | 17.1 | *−0.5* | 26.6 | 3 | 8.3 | 14 | 15.3 | 31 | 19.6 | 4 | 0 | 2 | 52.5 | *96* | 14 | 11.1 | 31 | 191.8 | *98* | 13.0 | 14 |
| Sep | 16.6 | 10.4 | 13.5 | *−1.4* | 21.2 | 2 | 6.2 | 18 | 10.5 | 15 | 13.9 | 4 | 0 | 0 | 62.6 | *116* | 16 | 12.0 | 14 | 113.1 | *81* | 11.8 | 2 |
| Oct | 14.1 | 6.9 | 10.5 | *−0.8* | 18.6 | 10 | 0.4 | 4 | 10.9 | 4 | 13.2 | 22 | 0 | 0 | 67.4 | *96* | 15 | 13.6 | 30 | 137.5 | *124* | 10.4 | 4 |
| Nov | 12.7 | 8.4 | 10.6 | *+3.0* | 16.2 | 3 | 1.9 | 2 | 8.4 | 30 | 14.0 | 20 | 0 | 0 | 55.7 | *83* | 15 | 21.8 | 4 | 46.7 | *66* | 8.0 | 1 |
| Dec | 9.6 | 3.9 | 6.8 | *+1.8* | 14.3 | 10 | −7.8 | 24 | −1.6 | 23 | 11.9 | 12 | 5 | 0 | 74.6 | *118* | 19 | 10.9 | 3 | 68.2 | *128* | 6.1 | 22 |
| **1994** | 14.5 | 7.5 | 11.0 | *+0.3* | 31.7 | 7 | −7.8 | 12 | −1.6 | 12 | 19.6 | 8 | 15 | 18 | 703.5 | *107* | 185 | 35.5 | 7 | 1692.5 | *107* | 15.4 | 6 |

## OXFORD 1995

| Month | Mean max | Mean min | Mean temp | Anom | Highest max | Date | Lowest min | Date | Lowest max | Date | Highest min | Date | Air frost | Days ≥ 25 °C | Total pptn | Anom | Rain days | Wettest day | Date | Total sunshine | Anom | Sunniest day | Date |
|---|---|---|---|---|---|---|---|---|---|---|---|---|---|---|---|---|---|---|---|---|---|---|---|
| Jan | 8.3 | 2.2 | 5.3 | *+0.4* | 13.4 | 10 | −4.9 | 3 | 2.5 | 31 | 7.9 | 16 | 7 | 0 | 116.3 | *204* | 20 | 15.5 | 19 | 61.8 | *99* | 7.4 | 23 |
| Feb | 10.4 | 4.4 | 7.4 | *+2.5* | 12.6 | 11 | −0.2 | 27 | 7.2 | 9 | 8.6 | 7 | 1 | 0 | 66.0 | *154* | 19 | 8.2 | 16 | 74.9 | *95* | 9.1 | 26 |
| Mar | 10.5 | 2.0 | 6.3 | *−1.0* | 17.1 | 31 | −3.9 | 4 | 5.2 | 3 | 7.9 | 26 | 6 | 0 | 49.3 | *103* | 14 | 14.4 | 2 | 198.7 | *179* | 11.4 | 13 |
| Apr | 14.3 | 5.4 | 9.9 | *+0.6* | 19.7 | 6 | −1.4 | 20 | 10.2 | 26 | 10.1 | 6 | 3 | 0 | 18.9 | *39* | 8 | 4.5 | 22 | 190.1 | *118* | 12.4 | 13 |
| May | 17.8 | 7.8 | 12.8 | *+0.3* | 26.3 | 6 | 1.9 | 14 | 9.1 | 17 | 13.7 | 28 | 0 | 4 | 54.1 | *95* | 10 | 16.1 | 17 | 233.5 | *121* | 13.5 | 4 |
| June | 20.0 | 10.4 | 15.2 | *−0.4* | 32.2 | 30 | 7.1 | 8 | 13.3 | 13 | 13.4 | 20 | 0 | 4 | 6.6 | *13* | 8 | 2.3 | 3 | 194.4 | *102* | 15.5 | 23 |
| July | 25.1 | 13.9 | 19.5 | *+1.6* | 31.9 | 31 | 9.5 | 4 | 17.7 | 2 | 18.5 | 19 | 0 | 14 | 37.8 | *77* | 9 | 16.1 | 2 | 247.6 | *120* | 15.0 | 22 |
| Aug | 26.4 | 13.9 | 20.1 | *+2.5* | 33.3 | 2 | 7.8 | 9 | 18.1 | 29 | 18.6 | 2 | 0 | 18 | 4.4 | *8* | 4 | 2.2 | 29 | 285.1 | *145* | 13.7 | 3 |
| Sep | 18.3 | 10.3 | 14.3 | *−0.6* | 21.2 | 9 | 4.8 | 30 | 14.9 | 27 | 14.4 | 2 | 0 | 0 | 98.8 | *182* | 18 | 20.0 | 10 | 135.3 | *96* | 10.9 | 9 |
| Oct | 17.3 | 10.3 | 13.8 | *+2.5* | 23.4 | 8 | 2.1 | 31 | 12.7 | 20 | 14.8 | 4 | 0 | 0 | 33.2 | *47* | 8 | 8.1 | 26 | 139.9 | *126* | 9.9 | 8 |
| Nov | 11.2 | 5.2 | 8.2 | *+0.7* | 15.1 | 11 | −2.0 | 18 | 5.1 | 17 | 11.5 | 12 | 5 | 0 | 98.2 | *147* | 14 | 22.6 | 26 | 78.1 | *110* | 8.6 | 4 |
| Dec | 4.6 | 0.2 | 2.4 | *−2.6* | 12.5 | 3 | −7.5 | 29 | −2.5 | 28 | 10.6 | 3 | 15 | 0 | 99.8 | *158* | 17 | 32.0 | 19 | 40.9 | *77* | 6.9 | 9 |
| **1995** | 15.4 | 7.2 | 11.3 | *+0.5* | 33.3 | 8 | −7.5 | 12 | −2.5 | 12 | 18.6 | 8 | 37 | 40 | 683.4 | *103* | 149 | 32.0 | 12 | 1880.3 | *119* | 15.5 | 6 |

## OXFORD 1996

| Month | Mean max | Mean min | Mean temp | Anom | Highest max | Date | Lowest min | Date | Lowest max | Date | Highest min | Date | Air frost | Days ≥ 25 °C | Total pptn | Anom | Rain days | Wettest day | Date | Total sunshine | Anom | Sunniest day | Date |
|---|---|---|---|---|---|---|---|---|---|---|---|---|---|---|---|---|---|---|---|---|---|---|---|
| Jan | 6.6 | 3.1 | 4.8 | *−0.0* | 12.2 | 8 | −4.6 | 27 | −1.1 | 25 | 9.2 | 13 | 7 | 0 | 34.4 | *60* | 13 | 14.5 | 8 | 29.3 | *47* | 6.5 | 31 |
| Feb | 6.2 | −0.1 | 3.1 | *−1.8* | 12.8 | 16 | −2.9 | 21 | 0.1 | 6 | 5.6 | 11 | 17 | 0 | 70.8 | *165* | 17 | 14.5 | 11 | 103.4 | *131* | 9.7 | 28 |
| Mar | 8.4 | 1.8 | 5.1 | *−2.2* | 15.6 | 23 | −3.5 | 11 | 2.7 | 12 | 7.0 | 24 | 9 | 0 | 29.3 | *61* | 14 | 5.8 | 11 | 76.4 | *69* | 12.4 | 27 |
| Apr | 13.8 | 4.8 | 9.3 | *+0.0* | 21.1 | 21 | −2.9 | 2 | 7.2 | 12 | 10.4 | 16 | 5 | 0 | 49.3 | *101* | 12 | 18.0 | 22 | 148.3 | *92* | 11.6 | 4 |
| May | 14.1 | 5.6 | 9.8 | *−2.7* | 24.1 | 30 | −0.1 | 4 | 8.3 | 17 | 13.3 | 30 | 1 | 0 | 35.7 | *62* | 15 | 10.0 | 18 | 185.7 | *96* | 13.6 | 8 |
| June | 21.2 | 10.2 | 15.7 | *+0.1* | 28.8 | 6 | 5.8 | 23 | 15.6 | 20 | 16.5 | 7 | 0 | 5 | 31.3 | *64* | 8 | 24.4 | 7 | 290.7 | *152* | 15.4 | 15 |
| July | 23.3 | 12.6 | 18.0 | *+0.1* | 30.8 | 22 | 7.6 | 18 | 17.8 | 1 | 16.6 | 14 | 0 | 9 | 22.9 | *47* | 10 | 6.1 | 28 | 256.6 | *124* | 14.5 | 15 |
| Aug | 22.4 | 12.5 | 17.4 | *−0.1* | 29.9 | 18 | 8.5 | 30 | 17.3 | 30 | 15.6 | 20 | 0 | 6 | 43.6 | *80* | 13 | 12.7 | 22 | 211.6 | *108* | 13.1 | 4 |
| Sep | 18.4 | 10.3 | 14.3 | *−0.5* | 22.1 | 5 | 3.7 | 14 | 14.4 | 30 | 15.3 | 3 | 0 | 0 | 19.9 | *37* | 9 | 10.9 | 24 | 125.3 | *89* | 11.1 | 16 |
| Oct | 15.8 | 8.6 | 12.2 | *+0.9* | 19.1 | 23 | 1.0 | 17 | 11.8 | 30 | 13.0 | 28 | 0 | 0 | 38.5 | *55* | 19 | 5.7 | 8 | 130.8 | *118* | 9.6 | 2 |
| Nov | 9.6 | 2.9 | 6.2 | *−1.3* | 17.1 | 2 | −2.2 | 28 | 5.6 | 20 | 14.7 | 3 | 10 | 0 | 73.7 | *110* | 18 | 16.1 | 3 | 101.0 | *142* | 8.0 | 21 |
| Dec | 5.5 | 0.9 | 3.2 | *−1.9* | 12.2 | 3 | −4.4 | 31 | −1.1 | 31 | 7.8 | 19 | 13 | 0 | 21.4 | *34* | 11 | 9.7 | 2 | 55.5 | *104* | 6.7 | 23 |
| **1996** | 13.8 | 6.1 | 9.9 | *−0.8* | 30.8 | 7 | −4.6 | 1 | −1.1 | 1 | 16.6 | 7 | 62 | 20 | 470.8 | *71* | 159 | 24.4 | 6 | 1714.6 | *109* | 15.4 | 6 |

## OXFORD 1997

| Month | Mean max | Mean min | Mean temp | Anom | Highest max | Date | Lowest min | Date | Lowest max | Date | Highest min | Date | Air frost | Days ≥ 25 °C | Total pptn | Anom | Rain days | Wettest day | Date | Total sunshine | Anom | Sunniest day | Date |
|---|---|---|---|---|---|---|---|---|---|---|---|---|---|---|---|---|---|---|---|---|---|---|---|
| Jan | 5.1 | −0.2 | 2.4 | *−2.4* | 10.9 | 25 | −7.4 | 2 | −1.3 | 1 | 5.4 | 18 | 16 | 0 | 11.9 | *21* | 5 | 7.4 | 21 | 51.4 | *83* | 8.0 | 26 |
| Feb | 10.4 | 4.2 | 7.3 | *+2.4* | 12.9 | 21 | −2.5 | 3 | 5.5 | 1 | 9.2 | 7 | 3 | 0 | 76.5 | *179* | 18 | 17.4 | 24 | 64.2 | *81* | 8.8 | 15 |
| Mar | 13.3 | 5.5 | 9.4 | *+2.1* | 17.1 | 15 | 1.0 | 11 | 9.2 | 8 | 8.7 | 15 | 0 | 0 | 9.2 | *19* | 8 | 2.3 | 3 | 150.2 | *135* | 11.7 | 31 |
| Apr | 14.5 | 4.9 | 9.7 | *+0.4* | 19.8 | 30 | −2.0 | 21 | 8.1 | 20 | 10.2 | 29 | 2 | 0 | 15.8 | *32* | 3 | 12.7 | 25 | 189.9 | *118* | 13.5 | 30 |
| May | 17.4 | 7.4 | 12.4 | *−0.1* | 24.6 | 2 | −0.2 | 7 | 9.8 | 6 | 14.0 | 18 | 1 | 0 | 67.0 | *117* | 16 | 9.8 | 16 | 261.4 | *136* | 15.4 | 30 |
| June | 18.9 | 11.3 | 15.1 | *−0.5* | 25.2 | 5 | 7.8 | 1 | 11.9 | 26 | 15.4 | 7 | 0 | 1 | 81.8 | *166* | 20 | 13.7 | 25 | 137.1 | *72* | 15.3 | 1 |
| July | 22.8 | 12.8 | 17.8 | *−0.0* | 27.5 | 8 | 9.2 | 5 | 16.3 | 3 | 17.4 | 9 | 0 | 7 | 25.2 | *52* | 11 | 5.1 | 3 | 231.7 | *112* | 15.3 | 20 |
| Aug | 24.9 | 15.7 | 20.3 | *+2.7* | 30.2 | 10 | 11.7 | 3 | 18.8 | 30 | 20.5 | 11 | 0 | 17 | 74.0 | *135* | 15 | 17.8 | 24 | 172.0 | *88* | 12.4 | 16 |
| Sep | 19.7 | 10.7 | 15.2 | *+0.3* | 23.6 | 18 | 6.3 | 22 | 16.5 | 19 | 15.1 | 3 | 0 | 0 | 18.5 | *34* | 8 | 6.6 | 3 | 169.7 | *121* | 12.0 | 4 |
| Oct | 14.9 | 6.6 | 10.7 | *−0.5* | 23.5 | 1 | −3.9 | 29 | 9.3 | 28 | 16.3 | 1 | 6 | 0 | 57.2 | *82* | 9 | 22.2 | 11 | 153.8 | *138* | 10.2 | 2 |
| Nov | 11.7 | 6.1 | 8.9 | *+1.3* | 15.6 | 16 | −0.5 | 1 | 8.0 | 30 | 12.7 | 16 | 1 | 0 | 94.6 | *142* | 21 | 24.2 | 5 | 47.1 | *66* | 6.5 | 10 |
| Dec | 9.0 | 3.6 | 6.3 | *+1.2* | 13.7 | 9 | −1.2 | 5 | 2.5 | 16 | 11.0 | 11 | 6 | 0 | 58.7 | *93* | 14 | 11.6 | 25 | 55.7 | *105* | 6.9 | 12 |
| **1997** | 15.2 | 7.4 | 11.3 | *+0.6* | 30.2 | 8 | −7.4 | 1 | −1.3 | 1 | 20.5 | 8 | 35 | 25 | 590.4 | *89* | 148 | 24.2 | 11 | 1684.2 | *107* | 15.4 | 5 |

## OXFORD 1998

| Month | Mean max | Mean min | Mean temp | Anom | Highest max | Date | Lowest min | Date | Lowest max | Date | Highest min | Date | Air frost | Days ≥ 25 °C | Total pptn | Anom | Rain days | Wettest day | Date | Total sunshine | Anom | Sunniest day | Date |
|---|---|---|---|---|---|---|---|---|---|---|---|---|---|---|---|---|---|---|---|---|---|---|---|
| Jan | 8.3 | 2.9 | 5.6 | *+0.7* | 14.6 | 9 | −2.4 | 23 | 3.5 | 29 | 10.5 | 9 | 9 | 0 | 60.0 | *105* | 16 | 8.7 | 2 | 64.0 | *103* | 6.9 | 26 |
| Feb | 11.5 | 4.1 | 7.8 | *+2.9* | 18.5 | 13 | −3.1 | 1 | 2.8 | 1 | 8.8 | 21 | 4 | 0 | 8.7 | *20* | 7 | 3.4 | 6 | 113.0 | *143* | 8.9 | 22 |
| Mar | 11.8 | 5.4 | 8.6 | *+1.3* | 16.5 | 28 | −1.5 | 9 | 8.5 | 24 | 12.8 | 30 | 1 | 0 | 57.4 | *120* | 13 | 10.4 | 5 | 77.5 | *70* | 10.1 | 9 |
| Apr | 11.9 | 5.1 | 8.5 | *−0.7* | 19.3 | 22 | −0.7 | 12 | 6.0 | 15 | 10.8 | 23 | 1 | 0 | 107.7 | *220* | 22 | 14.7 | 10 | 105.8 | *66* | 8.6 | 25 |
| May | 18.7 | 9.1 | 13.9 | *+1.4* | 24.7 | 20 | 2.0 | 4 | 11.9 | 1 | 13.9 | 9 | 0 | 0 | 20.9 | *37* | 7 | 6.4 | 26 | 199.8 | *104* | 14.6 | 18 |
| June | 19.1 | 11.5 | 15.3 | *−0.2* | 27.1 | 20 | 5.8 | 12 | 13.2 | 11 | 15.7 | 20 | 0 | 1 | 75.9 | *154* | 22 | 15.0 | 1 | 114.0 | *60* | 9.7 | 21 |
| July | 21.0 | 12.4 | 16.7 | *−1.2* | 25.0 | 20 | 8.6 | 25 | 16.1 | 2 | 15.8 | 6 | 0 | 1 | 29.5 | *60* | 13 | 9.1 | 11 | 157.0 | *76* | 12.1 | 13 |
| Aug | 22.4 | 11.9 | 17.1 | *−0.5* | 29.3 | 10 | 7.5 | 28 | 15.6 | 26 | 15.4 | 21 | 0 | 6 | 26.8 | *49* | 9 | 12.4 | 1 | 225.0 | *114* | 13.5 | 4 |
| Sep | 19.6 | 11.8 | 15.7 | *+0.8* | 23.5 | 1 | 6.2 | 13 | 15.2 | 12 | 15.7 | 19 | 0 | 0 | 119.4 | *220* | 18 | 46.4 | 26 | 122.2 | *87* | 8.6 | 13 |
| Oct | 14.4 | 8.2 | 11.3 | *−0.0* | 18.2 | 13 | 2.2 | 18 | 10.5 | 31 | 14.6 | 23 | 0 | 0 | 116.4 | *166* | 21 | 24.0 | 31 | 94.5 | *85* | 8.4 | 18 |
| Nov | 9.3 | 2.2 | 5.7 | *−1.8* | 16.1 | 8 | −2.6 | 18 | 4.0 | 17 | 8.4 | 9 | 7 | 0 | 69.9 | *105* | 16 | 19.2 | 2 | 66.3 | *93* | 7.7 | 10 |
| Dec | 9.4 | 3.2 | 6.3 | *+1.2* | 14.3 | 14 | −3.5 | 6 | 2.8 | 6 | 11.7 | 15 | 8 | 0 | 72.7 | *115* | 22 | 10.4 | 25 | 38.1 | *72* | 6.2 | 6 |
| **1998** | 14.8 | 7.3 | 11.0 | *+0.3* | 29.3 | 8 | −3.5 | 12 | 2.8 | 2 | 15.8 | 7 | 30 | 8 | 765.3 | *116* | 186 | 46.4 | 9 | 1377.2 | *87* | 14.6 | 5 |

## OXFORD 1999

| Month | Mean max | Mean min | Mean temp | Anom | Highest max | Date | Lowest min | Date | Lowest max | Date | Highest min | Date | Air frost | Days ≥ 25 °C | Total pptn | Anom | Rain days | Wettest day | Date | Total sunshine | Anom | Sunniest day | Date |
|---|---|---|---|---|---|---|---|---|---|---|---|---|---|---|---|---|---|---|---|---|---|---|---|
| Jan | 9.0 | 3.4 | 6.2 | *+1.4* | 14.5 | 6 | −2.4 | 23 | 3.5 | 10 | 10.5 | 6 | 4 | 0 | 91.9 | *162* | 22 | 13.1 | 15 | 51.1 | *82* | 6.9 | 17 |
| Feb | 8.6 | 2.9 | 5.7 | *+0.8* | 12.9 | 19 | −3.6 | 11 | 2.8 | 8 | 8.9 | 19 | 5 | 0 | 32.9 | *77* | 14 | 7.4 | 6 | 81.3 | *103* | 8.5 | 9 |
| Mar | 11.6 | 4.8 | 8.2 | *+0.9* | 18.9 | 17 | −1.5 | 11 | 4.9 | 7 | 10.5 | 2 | 1 | 0 | 28.8 | *60* | 12 | 4.2 | 3 | 97.4 | *88* | 10.8 | 27 |
| Apr | 14.3 | 6.1 | 10.2 | *+0.9* | 20.4 | 30 | −1.9 | 15 | 8.0 | 14 | 12.4 | 5 | 2 | 0 | 57.6 | *118* | 17 | 11.8 | 23 | 147.6 | *92* | 13.5 | 30 |
| May | 18.2 | 9.7 | 13.9 | *+1.5* | 24.8 | 27 | 6.3 | 1 | 12.3 | 30 | 14.9 | 28 | 0 | 0 | 50.9 | *89* | 14 | 13.8 | 19 | 153.6 | *80* | 13.5 | 25 |
| June | 19.7 | 10.1 | 14.9 | *−0.7* | 26.1 | 26 | 5.6 | 8 | 15.2 | 5 | 14.5 | 20 | 0 | 2 | 65.3 | *133* | 13 | 18.5 | 28 | 202.1 | *106* | 14.8 | 25 |
| July | 24.2 | 13.3 | 18.7 | *+0.9* | 30.4 | 31 | 8.6 | 27 | 18.6 | 14 | 17.5 | 3 | 0 | 13 | 7.2 | *15* | 6 | 2.9 | 20 | 235.4 | *114* | 13.4 | 24 |
| Aug | 21.6 | 12.9 | 17.3 | *−0.3* | 31.2 | 1 | 7.5 | 22 | 16.6 | 17 | 18.0 | 3 | 0 | 4 | 98.3 | *179* | 16 | 16.4 | 8 | 159.2 | *81* | 12.6 | 21 |
| Sep | 20.8 | 12.4 | 16.6 | *+1.8* | 27.9 | 11 | 7.8 | 10 | 14.9 | 14 | 18.0 | 11 | 0 | 6 | 102.1 | *188* | 18 | 30.0 | 19 | 155.2 | *111* | 11.4 | 2 |
| Oct | 15.1 | 7.1 | 11.1 | *−0.2* | 18.3 | 10 | 0.6 | 6 | 9.4 | 20 | 13.2 | 10 | 0 | 0 | 70.8 | *101* | 13 | 15.5 | 23 | 144.8 | *130* | 9.7 | 4 |
| Nov | 10.8 | 5.3 | 8.1 | *+0.5* | 16.6 | 1 | 0.6 | 21 | 5.9 | 18 | 11.4 | 1 | 0 | 0 | 41.1 | *62* | 16 | 10.3 | 5 | 81.1 | *114* | 7.7 | 17 |
| Dec | 8.3 | 1.7 | 5.0 | *−0.1* | 13.1 | 24 | −7.5 | 20 | −0.1 | 19 | 7.5 | 7 | 9 | 0 | 89.8 | *142* | 22 | 11.6 | 24 | 60.9 | *115* | 7.0 | 19 |
| 1999 | 15.2 | 7.5 | 11.3 | *+0.6* | 31.2 | 8 | −7.5 | 12 | −0.1 | 12 | 18.0 | 8 | 21 | 25 | 736.7 | *112* | 183 | 30.0 | 9 | 1569.7 | *100* | 14.8 | 6 |

## OXFORD 2000

| Month | Mean max | Mean min | Mean temp | Anom | Highest max | Date | Lowest min | Date | Lowest max | Date | Highest min | Date | Air frost | Days ≥ 25 °C | Total pptn | Anom | Rain days | Wettest day | Date | Total sunshine | Anom | Sunniest day | Date |
|---|---|---|---|---|---|---|---|---|---|---|---|---|---|---|---|---|---|---|---|---|---|---|---|
| Jan | 8.1 | 2.0 | 5.0 | *+0.2* | 13.1 | 29 | −3.5 | 26 | 1.1 | 26 | 9.7 | 31 | 9 | 0 | 19.8 | *35* | 10 | 7.8 | 3 | 82.1 | *132* | 6.9 | 27 |
| Feb | 10.2 | 3.8 | 7.0 | *+2.1* | 13.4 | 8 | −2.2 | 20 | 7.7 | 16 | 8.9 | 1 | 3 | 0 | 78.3 | *183* | 18 | 14.1 | 28 | 104.9 | *133* | 8.7 | 25 |
| Mar | 11.7 | 4.2 | 7.9 | *+0.7* | 15.4 | 13 | −1.4 | 5 | 6.6 | 29 | 10.2 | 8 | 2 | 0 | 11.9 | *25* | 8 | 4.5 | 2 | 112.6 | *101* | 9.6 | 4 |
| Apr | 12.4 | 4.8 | 8.6 | *−0.7* | 17.7 | 29 | −0.1 | 6 | 5.7 | 12 | 9.6 | 25 | 2 | 0 | 137.2 | *280* | 23 | 38.0 | 11 | 143.4 | *89* | 12.5 | 29 |
| May | 17.2 | 8.6 | 12.9 | *+0.4* | 25.3 | 14 | 4.9 | 29 | 11.3 | 3 | 14.1 | 15 | 0 | 2 | 67.7 | *118* | 15 | 20.5 | 27 | 189.8 | *98* | 13.0 | 25 |
| June | 20.5 | 12.2 | 16.4 | *+0.8* | 30.3 | 19 | 7.6 | 10 | 15.9 | 5 | 18.8 | 19 | 0 | 3 | 19.4 | *39* | 8 | 9.3 | 29 | 164.6 | *86* | 14.2 | 18 |
| July | 20.8 | 12.3 | 16.5 | *−1.3* | 25.7 | 21 | 7.2 | 17 | 15.9 | 14 | 15.0 | 27 | 0 | 3 | 28.5 | *58* | 8 | 11.6 | 29 | 161.2 | *78* | 13.4 | 21 |
| Aug | 22.7 | 13.3 | 18.0 | *+0.4* | 27.2 | 25 | 9.3 | 21 | 18.6 | 18 | 17.1 | 14 | 0 | 5 | 58.3 | *106* | 16 | 20.4 | 18 | 209.4 | *107* | 12.4 | 12 |
| Sep | 19.2 | 12.1 | 15.7 | *+0.8* | 27.1 | 11 | 5.2 | 4 | 14.2 | 19 | 15.9 | 10 | 0 | 2 | 87.2 | *161* | 19 | 17.8 | 19 | 126.1 | *90* | 11.1 | 3 |
| Oct | 14.3 | 8.2 | 11.3 | *−0.0* | 18.5 | 1 | 4.8 | 17 | 10.2 | 30 | 11.5 | 4 | 0 | 0 | 116.1 | *166* | 20 | 33.8 | 29 | 83.9 | *75* | 8.6 | 19 |
| Nov | 10.6 | 4.8 | 7.7 | *+0.1* | 15.0 | 28 | 0.5 | 14 | 7.3 | 20 | 11.0 | 29 | 0 | 0 | 99.1 | *149* | 22 | 11.7 | 5 | 69.2 | *97* | 7.5 | 4 |
| Dec | 8.6 | 4.2 | 6.4 | *+1.4* | 13.9 | 11 | −6.7 | 29 | 1.4 | 29 | 12.1 | 12 | 7 | 0 | 99.8 | *158* | 19 | 16.8 | 12 | 51.9 | *98* | 7.1 | 15 |
| 2000 | 14.7 | 7.5 | 11.1 | *+0.4* | 30.3 | 6 | −6.7 | 12 | 1.1 | 1 | 18.8 | 6 | 23 | 15 | 823.3 | *125* | 186 | 38.0 | 4 | 1499.1 | *95* | 14.2 | 6 |

## OXFORD 2001

| Month | Mean max | Mean min | Mean temp | Anom | Highest max | Date | Lowest min | Date | Lowest max | Date | Highest min | Date | Air frost | Days ≥ 25 °C | Total pptn | Anom | Rain days | Wettest day | Date | Total sunshine | Anom | Sunniest day | Date |
|---|---|---|---|---|---|---|---|---|---|---|---|---|---|---|---|---|---|---|---|---|---|---|---|
| Jan | 6.4 | 1.1 | 3.8 | *−1.1* | 12.1 | 23 | −3.6 | 16 | 1.6 | 17 | 7.3 | 2 | 13 | 0 | 57.4 | *101* | 14 | 15.2 | 26 | 83.3 | *134* | 7.0 | 13 |
| Feb | 8.8 | 2.0 | 5.4 | *+0.5* | 13.4 | 11 | −2.2 | 25 | 4.0 | 28 | 10.7 | 12 | 10 | 0 | 68.6 | *160* | 13 | 24.1 | 12 | 85.0 | *108* | 9.3 | 25 |
| Mar | 9.0 | 3.1 | 6.1 | *−1.2* | 15.7 | 31 | −5.1 | 5 | 3.0 | 17 | 9.2 | 11 | 5 | 0 | 75.1 | *158* | 20 | 13.1 | 16 | 74.1 | *67* | 10.1 | 30 |
| Apr | 12.9 | 4.8 | 8.9 | *−0.4* | 17.5 | 2 | 0.1 | 19 | 9.3 | 18 | 9.2 | 2 | 0 | 0 | 72.7 | *149* | 19 | 16.4 | 3 | 148.2 | *92* | 11.4 | 30 |
| May | 18.5 | 7.9 | 13.2 | *+0.7* | 26.2 | 11 | 2.5 | 5 | 11.8 | 1 | 15.6 | 28 | 0 | 2 | 35.5 | *62* | 7 | 15.1 | 15 | 198.3 | *103* | 14.1 | 21 |
| June | 20.7 | 10.6 | 15.7 | *+0.1* | 31.6 | 26 | 4.5 | 9 | 15.6 | 9 | 16.6 | 27 | 0 | 3 | 23.1 | *47* | 8 | 10.7 | 15 | 226.8 | *119* | 14.8 | 21 |
| July | 23.3 | 13.4 | 18.4 | *+0.5* | 30.4 | 29 | 6.8 | 16 | 17.4 | 19 | 19.7 | 29 | 0 | 12 | 57.2 | *117* | 10 | 17.6 | 9 | 206.1 | *100* | 13.0 | 26 |
| Aug | 22.7 | 12.9 | 17.8 | *+0.2* | 29.7 | 15 | 7.9 | 29 | 16.9 | 9 | 17.4 | 15 | 0 | 5 | 92.0 | *168* | 14 | 33.0 | 9 | 184.1 | *94* | 13.1 | 27 |
| Sep | 17.9 | 10.5 | 14.2 | *−0.7* | 22.7 | 28 | 5.8 | 23 | 14.3 | 19 | 15.4 | 2 | 0 | 0 | 35.0 | *65* | 16 | 9.8 | 28 | 123.0 | *88* | 10.2 | 14 |
| Oct | 17.0 | 11.1 | 14.1 | *+2.8* | 21.8 | 13 | 7.2 | 22 | 11.4 | 21 | 14.2 | 15 | 0 | 0 | 92.4 | *132* | 17 | 14.1 | 7 | 100.9 | *91* | 8.1 | 3 |
| Nov | 11.2 | 3.9 | 7.6 | *+0.0* | 15.4 | 30 | −2.0 | 15 | 6.6 | 9 | 12.1 | 30 | 5 | 0 | 38.5 | *58* | 11 | 8.1 | 7 | 86.9 | *122* | 8.2 | 1 |
| Dec | 6.8 | 0.2 | 3.5 | *−1.5* | 13.9 | 5 | −4.4 | 23 | 2.9 | 29 | 9.2 | 1 | 17 | 0 | 23.5 | *37* | 11 | 9.8 | 4 | 79.2 | *149* | 7.0 | 31 |
| **2001** | 14.6 | 6.8 | 10.7 | *−0.0* | 31.6 | 6 | −5.1 | 3 | 1.6 | 1 | 19.7 | 7 | 50 | 22 | 671.0 | *102* | 160 | 33.0 | 8 | 1595.9 | *101* | 14.8 | 6 |

## OXFORD 2002

| Month | Mean max | Mean min | Mean temp | Anom | Highest max | Date | Lowest min | Date | Lowest max | Date | Highest min | Date | Air frost | Days ≥ 25 °C | Total pptn | Anom | Rain days | Wettest day | Date | Total sunshine | Anom | Sunniest day | Date |
|---|---|---|---|---|---|---|---|---|---|---|---|---|---|---|---|---|---|---|---|---|---|---|---|
| Jan | 9.2 | 3.4 | 6.3 | *+1.4* | 14.0 | 29 | −6.4 | 2 | 3.3 | 1 | 10.6 | 21 | 5 | 0 | 57.4 | *101* | 21 | 10.9 | 26 | 44.4 | *71* | 7.2 | 1 |
| Feb | 11.0 | 4.4 | 7.7 | *+2.8* | 13.9 | 8 | −1.5 | 15 | 6.7 | 27 | 11.4 | 2 | 3 | 0 | 70.9 | *166* | 20 | 15.3 | 11 | 80.5 | *102* | 8.8 | 18 |
| Mar | 12.4 | 4.6 | 8.5 | *+1.2* | 16.9 | 29 | −1.7 | 2 | 7.3 | 13 | 9.9 | 21 | 3 | 0 | 33.0 | *69* | 13 | 6.4 | 15 | 101.1 | *91* | 9.9 | 29 |
| Apr | 15.2 | 5.2 | 10.2 | *+0.9* | 21.7 | 24 | 0.9 | 12 | 10.1 | 9 | 10.6 | 3 | 0 | 0 | 43.0 | *88* | 9 | 13.0 | 30 | 210.0 | *131* | 12.6 | 23 |
| May | 16.8 | 8.4 | 12.6 | *+0.1* | 25.0 | 16 | 2.6 | 4 | 12.5 | 9 | 14.3 | 17 | 0 | 1 | 58.3 | *102* | 17 | 13.6 | 13 | 177.5 | *92* | 13.0 | 1 |
| June | 20.0 | 10.3 | 15.2 | *−0.4* | 25.3 | 2 | 6.8 | 1 | 15.5 | 9 | 14.5 | 17 | 0 | 1 | 42.4 | *86* | 13 | 23.9 | 5 | 163.3 | *86* | 14.4 | 1 |
| July | 22.1 | 12.0 | 17.1 | *−0.8* | 30.1 | 29 | 8.1 | 4 | 16.6 | 1 | 16.8 | 30 | 0 | 7 | 89.6 | *183* | 12 | 23.7 | 8 | 178.8 | *86* | 14.3 | 13 |
| Aug | 22.8 | 12.9 | 17.8 | *+0.3* | 28.5 | 17 | 9.0 | 26 | 17.2 | 26 | 16.0 | 19 | 0 | 7 | 53.2 | *97* | 11 | 25.4 | 9 | 166.2 | *85* | 13.1 | 2 |
| Sep | 19.7 | 10.0 | 14.8 | *−0.1* | 24.1 | 13 | 4.9 | 24 | 15.9 | 9 | 15.1 | 12 | 0 | 0 | 24.8 | *46* | 4 | 20.5 | 9 | 167.1 | *119* | 12.6 | 1 |
| Oct | 14.8 | 7.1 | 10.9 | *−0.4* | 21.4 | 1 | −0.1 | 19 | 10.2 | 30 | 12.8 | 2 | 1 | 0 | 109.4 | *156* | 18 | 27.1 | 15 | 104.8 | *94* | 9.1 | 4 |
| Nov | 11.8 | 6.4 | 9.1 | *+1.6* | 15.7 | 2 | 1.0 | 18 | 6.3 | 25 | 10.9 | 6 | 0 | 0 | 133.4 | *200* | 26 | 14.3 | 5 | 52.3 | *74* | 7.4 | 7 |
| Dec | 8.7 | 4.1 | 6.4 | *+1.4* | 13.6 | 24 | −1.6 | 19 | 2.4 | 10 | 9.7 | 24 | 4 | 0 | 98.1 | *155* | 24 | 22.1 | 21 | 41.3 | *78* | 6.2 | 18 |
| **2002** | 15.4 | 7.4 | 11.4 | *+0.6* | 30.1 | 7 | −6.4 | 1 | 2.4 | 12 | 16.8 | 7 | 16 | 16 | 813.5 | *123* | 188 | 27.1 | 10 | 1487.3 | *94* | 14.4 | 6 |

## OXFORD 2003

| Month | Mean max | Mean min | Mean temp | Anom | Highest max | Date | Lowest min | Date | Lowest max | Date | Highest min | Date | Air frost | Days ≥ 25 °C | Total pptn | Anom | Rain days | Wettest day | Date | Total sunshine | Anom | Sunniest day | Date |
|---|---|---|---|---|---|---|---|---|---|---|---|---|---|---|---|---|---|---|---|---|---|---|---|
| Jan | 7.6 | 1.8 | 4.7 | *−0.2* | 12.6 | 26 | −5.2 | 8 | 1.2 | 8 | 7.8 | 15 | 11 | 0 | 81.4 | *143* | 17 | 21.5 | 1 | 87.7 | *141* | 7.1 | 11 |
| Feb | 8.4 | 0.6 | 4.5 | *−0.4* | 14.4 | 27 | −5.2 | 18 | 1.9 | 13 | 7.1 | 8 | 15 | 0 | 17.3 | *40* | 8 | 6.3 | 10 | 94.4 | *120* | 8.7 | 18 |
| Mar | 13.3 | 2.7 | 8.0 | *+0.7* | 18.3 | 23 | −2.0 | 16 | 7.5 | 13 | 9.3 | 5 | 8 | 0 | 25.2 | *53* | 7 | 13.6 | 7 | 154.8 | *139* | 10.5 | 31 |
| Apr | 15.3 | 4.6 | 9.9 | *+0.7* | 26.3 | 16 | −2.6 | 8 | 7.7 | 10 | 12.2 | 28 | 5 | 1 | 36.8 | *75* | 8 | 10.6 | 27 | 169.8 | *106* | 12.0 | 17 |
| May | 17.5 | 8.1 | 12.8 | *+0.3* | 27.6 | 31 | 2.0 | 15 | 13.2 | 2 | 13.9 | 28 | 0 | 3 | 48.1 | *84* | 17 | 8.8 | 16 | 191.2 | *99* | 12.9 | 31 |
| June | 22.4 | 11.8 | 17.1 | *+1.5* | 27.7 | 22 | 8.3 | 21 | 18.6 | 30 | 15.8 | 19 | 0 | 5 | 55.5 | *113* | 10 | 19.9 | 22 | 213.6 | *112* | 14.0 | 20 |
| July | 23.7 | 13.4 | 18.6 | *+0.7* | 32.5 | 15 | 10.2 | 4 | 18.8 | 17 | 15.8 | 16 | 0 | 8 | 53.8 | *110* | 12 | 10.7 | 26 | 192.0 | *93* | 14.7 | 12 |
| Aug | 25.5 | 13.5 | 19.5 | *+1.9* | 34.6 | 9 | 4.5 | 30 | 19.0 | 28 | 18.9 | 10 | 0 | 15 | 3.0 | *5* | 3 | 2.0 | 28 | 228.4 | *116* | 13.6 | 2 |
| Sep | 21.3 | 8.1 | 14.7 | *−0.2* | 27.1 | 17 | 1.2 | 25 | 15.1 | 23 | 14.4 | 22 | 0 | 5 | 17.4 | *32* | 5 | 7.4 | 22 | 175.3 | *125* | 10.8 | 13 |
| Oct | 14.0 | 3.9 | 9.0 | *−2.3* | 19.2 | 9 | −1.9 | 25 | 9.3 | 31 | 11.6 | 10 | 5 | 0 | 31.7 | *45* | 13 | 11.3 | 30 | 134.8 | *121* | 9.3 | 11 |
| Nov | 12.0 | 5.7 | 8.8 | *+1.3* | 15.4 | 6 | −0.4 | 27 | 6.6 | 23 | 11.3 | 19 | 1 | 0 | 89.6 | *134* | 16 | 21.6 | 25 | 78.9 | *111* | 8.0 | 1 |
| Dec | 8.6 | 1.8 | 5.2 | *+0.2* | 13.8 | 12 | −5.0 | 8 | 3.4 | 29 | 8.9 | 26 | 10 | 0 | 70.7 | *112* | 18 | 10.8 | 1 | 51.6 | *97* | 7.0 | 7 |
| **2003** | 15.8 | 6.3 | 11.1 | *+0.3* | 34.6 | 8 | −5.2 | 1 | 1.2 | 1 | 18.9 | 8 | 55 | 37 | 530.5 | *80* | 134 | 21.6 | 11 | 1772.5 | *112* | 14.7 | 7 |

## OXFORD 2004

| Month | Mean max | Mean min | Mean temp | Anom | Highest max | Date | Lowest min | Date | Lowest max | Date | Highest min | Date | Air frost | Days ≥ 25 °C | Total pptn | Anom | Rain days | Wettest day | Date | Total sunshine | Anom | Sunniest day | Date |
|---|---|---|---|---|---|---|---|---|---|---|---|---|---|---|---|---|---|---|---|---|---|---|---|
| Jan | 8.8 | 2.7 | 5.7 | *+0.9* | 13.2 | 31 | −3.7 | 29 | 4.4 | 26 | 7.8 | 21 | 9 | 0 | 66.1 | *116* | 21 | 10.6 | 30 | 62.7 | *101* | 7.9 | 29 |
| Feb | 8.7 | 2.9 | 5.8 | *+0.9* | 16.7 | 4 | −4.3 | 26 | 4.1 | 25 | 12.0 | 4 | 11 | 0 | 25.5 | *60* | 10 | 10.0 | 5 | 79.1 | *100* | 8.3 | 19 |
| Mar | 10.9 | 3.4 | 7.1 | *−0.1* | 18.6 | 31 | −4.4 | 1 | 5.3 | 10 | 10.6 | 16 | 2 | 0 | 51.8 | *109* | 19 | 11.1 | 12 | 101.5 | *91* | 9.4 | 1 |
| Apr | 14.7 | 6.2 | 10.4 | *+1.1* | 21.4 | 24 | 1.8 | 9 | 8.8 | 29 | 9.8 | 1 | 0 | 0 | 68.6 | *140* | 17 | 13.0 | 18 | 150.3 | *93* | 13.7 | 23 |
| May | 17.9 | 8.6 | 13.3 | *+0.8* | 24.0 | 17 | 3.2 | 22 | 11.9 | 8 | 13.6 | 29 | 0 | 0 | 41.1 | *72* | 14 | 11.1 | 31 | 195.5 | *101* | 14.3 | 16 |
| June | 21.8 | 12.0 | 16.9 | *+1.3* | 29.3 | 8 | 6.4 | 19 | 17.4 | 1 | 15.9 | 5 | 0 | 6 | 23.8 | *48* | 9 | 13.0 | 22 | 223.6 | *117* | 14.0 | 13 |
| July | 22.0 | 12.2 | 17.1 | *−0.7* | 28.2 | 29 | 7.8 | 6 | 18.2 | 11 | 16.8 | 16 | 0 | 4 | 83.1 | *170* | 10 | 50.0 | 8 | 169.5 | *82* | 14.0 | 23 |
| Aug | 23.4 | 14.2 | 18.8 | *+1.2* | 29.4 | 8 | 9.5 | 28 | 18.6 | 29 | 19.6 | 9 | 0 | 10 | 135.0 | *246* | 18 | 35.2 | 9 | 194.1 | *99* | 12.4 | 7 |
| Sep | 20.1 | 11.7 | 15.9 | *+1.0* | 29.9 | 5 | 5.8 | 25 | 14.9 | 29 | 15.5 | 18 | 0 | 1 | 28.9 | *53* | 14 | 8.9 | 12 | 174.4 | *124* | 11.9 | 1 |
| Oct | 14.7 | 8.6 | 11.7 | *+0.4* | 17.7 | 24 | 5.0 | 27 | 11.7 | 15 | 13.8 | 23 | 0 | 0 | 131.4 | *188* | 20 | 26.2 | 13 | 102.4 | *92* | 8.5 | 21 |
| Nov | 10.9 | 5.8 | 8.4 | *+0.8* | 14.2 | 12 | −0.8 | 14 | 4.9 | 20 | 10.2 | 8 | 1 | 0 | 34.2 | *51* | 12 | 16.7 | 18 | 52.4 | *74* | 8.2 | 13 |
| Dec | 8.6 | 2.8 | 5.7 | *+0.7* | 12.7 | 30 | −3.2 | 20 | 2.8 | 25 | 9.6 | 31 | 6 | 0 | 37.7 | *60* | 9 | 9.6 | 27 | 58.3 | *110* | 6.8 | 25 |
| **2004** | 15.2 | 7.6 | 11.4 | *+0.7* | 29.9 | 9 | −4.4 | 3 | 2.8 | 12 | 19.6 | 8 | 29 | 21 | 727.2 | *110* | 173 | 50.0 | 7 | 1563.8 | *99* | 14.3 | 5 |

## OXFORD 2005

| Month | Mean max | Mean min | Mean temp | Anom | Highest max | Date | Lowest min | Date | Lowest max | Date | Highest min | Date | Air frost | Days ≥ 25 °C | Total pptn | Anom | Rain days | Wettest day | Date | Total sunshine | Anom | Sunniest day | Date |
|---|---|---|---|---|---|---|---|---|---|---|---|---|---|---|---|---|---|---|---|---|---|---|---|
| Jan | 9.4 | 3.8 | 6.6 | *+1.8* | 14.2 | 7 | −0.8 | 26 | 4.6 | 25 | 7.2 | 4 | 2 | 0 | 18.2 | *32* | 14 | 2.8 | 7 | 65.4 | *105* | 7.7 | 23 |
| Feb | 7.3 | 2.1 | 4.7 | *−0.2* | 12.7 | 12 | −4.0 | 28 | 2.2 | 24 | 8.3 | 1 | 10 | 0 | 20.0 | *47* | 12 | 3.6 | 24 | 66.3 | *84* | 8.3 | 19 |
| Mar | 11.2 | 4.7 | 7.9 | *+0.7* | 17.4 | 18 | −2.7 | 4 | 4.2 | 3 | 12.3 | 22 | 6 | 0 | 44.5 | *93* | 12 | 22.5 | 29 | 79.3 | *71* | 8.8 | 3 |
| Apr | 14.3 | 5.7 | 10.0 | *+0.7* | 20.7 | 30 | −1.7 | 9 | 8.8 | 8 | 12.3 | 30 | 1 | 0 | 50.2 | *103* | 15 | 11.0 | 26 | 136.3 | *85* | 11.3 | 27 |
| May | 17.2 | 7.5 | 12.4 | *−0.1* | 28.5 | 27 | 0.2 | 11 | 12.9 | 10 | 14.1 | 28 | 0 | 1 | 31.5 | *55* | 12 | 13.5 | 21 | 226.4 | *117* | 14.0 | 15 |
| June | 21.6 | 12.0 | 16.8 | *+1.2* | 31.2 | 19 | 4.0 | 7 | 16.0 | 6 | 18.3 | 20 | 0 | 7 | 93.4 | *190* | 9 | 26.9 | 28 | 177.6 | *93* | 15.5 | 8 |
| July | 22.3 | 13.4 | 17.8 | *−0.0* | 29.6 | 14 | 9.9 | 5 | 15.8 | 25 | 16.4 | 2 | 0 | 7 | 34.9 | *71* | 10 | 13.4 | 24 | 192.0 | *93* | 14.9 | 17 |
| Aug | 23.1 | 12.2 | 17.6 | *+0.0* | 29.9 | 31 | 8.6 | 8 | 17.6 | 24 | 15.6 | 22 | 0 | 8 | 31.3 | *57* | 10 | 6.5 | 22 | 235.7 | *120* | 13.9 | 7 |
| Sep | 20.9 | 11.9 | 16.4 | *+1.6* | 28.3 | 4 | 3.5 | 17 | 15.0 | 16 | 17.8 | 5 | 0 | 4 | 38.5 | *71* | 13 | 9.1 | 15 | 155.4 | *111* | 12.4 | 2 |
| Oct | 17.1 | 10.9 | 14.0 | *+2.7* | 22.3 | 11 | 5.7 | 5 | 13.6 | 18 | 16.7 | 12 | 0 | 0 | 53.4 | *76* | 13 | 14.3 | 23 | 92.0 | *83* | 8.7 | 1 |
| Nov | 9.9 | 3.1 | 6.5 | *−1.0* | 18.0 | 2 | −4.2 | 18 | 0.9 | 20 | 13.7 | 3 | 11 | 0 | 38.6 | *58* | 10 | 10.4 | 6 | 88.0 | *124* | 7.4 | 1 |
| Dec | 7.7 | 1.8 | 4.8 | *−0.3* | 11.2 | 15 | −3.7 | 18 | 1.0 | 28 | 7.1 | 2 | 11 | 0 | 50.2 | *80* | 8 | 14.9 | 2 | 55.7 | *105* | 6.8 | 17 |
| **2005** | 15.2 | 7.4 | 11.3 | *+0.6* | 31.2 | 6 | −4.2 | 11 | 0.9 | 11 | 18.3 | 6 | 41 | 27 | 504.7 | *76* | 138 | 26.9 | 6 | 1570.1 | *100* | 15.5 | 6 |

## OXFORD 2006

| Month | Mean max | Mean min | Mean temp | Anom | Highest max | Date | Lowest min | Date | Lowest max | Date | Highest min | Date | Air frost | Days ≥ 25 °C | Total pptn | Anom | Rain days | Wettest day | Date | Total sunshine | Anom | Sunniest day | Date |
|---|---|---|---|---|---|---|---|---|---|---|---|---|---|---|---|---|---|---|---|---|---|---|---|
| Jan | 7.2 | 2.2 | 4.7 | *−0.1* | 14.3 | 18 | −4.8 | 23 | 2.8 | 6 | 8.8 | 19 | 8 | 0 | 16.6 | *29* | 10 | 5.6 | 8 | 54.9 | *88* | 7.9 | 29 |
| Feb | 6.6 | 0.8 | 3.7 | *−1.2* | 12.0 | 15 | −3.7 | 11 | −0.8 | 2 | 7.1 | 15 | 12 | 0 | 33.0 | *77* | 11 | 8.8 | 14 | 73.1 | *93* | 8.4 | 8 |
| Mar | 9.2 | 2.0 | 5.6 | *−1.7* | 15.9 | 30 | −5.2 | 4 | 3.8 | 16 | 10.8 | 27 | 10 | 0 | 59.3 | *124* | 16 | 16.2 | 30 | 95.5 | *86* | 9.2 | 4 |
| Apr | 13.9 | 5.4 | 9.7 | *+0.4* | 17.2 | 21 | −2.3 | 5 | 9.0 | 4 | 9.6 | 20 | 2 | 0 | 36.3 | *74* | 15 | 12.9 | 30 | 152.5 | *95* | 11.6 | 5 |
| May | 17.6 | 9.0 | 13.3 | *+0.8* | 26.7 | 4 | 4.5 | 31 | 13.4 | 24 | 12.7 | 27 | 0 | 1 | 104.7 | *183* | 21 | 19.8 | 22 | 165.7 | *86* | 13.0 | 11 |
| June | 23.1 | 11.6 | 17.4 | *+1.8* | 30.4 | 12 | 5.0 | 1 | 16.2 | 26 | 18.1 | 11 | 0 | 10 | 5.3 | *11* | 5 | 2.2 | 13 | 246.2 | *129* | 14.6 | 3 |
| July | 27.1 | 14.9 | 21.0 | *+3.2* | 34.8 | 19 | 9.2 | 14 | 21.5 | 10 | 20.0 | 20 | 0 | 18 | 87.3 | *179* | 12 | 44.2 | 22 | 303.7 | *147* | 14.6 | 17 |
| Aug | 21.7 | 13.1 | 17.4 | *−0.2* | 27.9 | 6 | 8.5 | 30 | 18.4 | 13 | 18.0 | 6 | 0 | 3 | 58.7 | *107* | 16 | 12.3 | 23 | 167.2 | *85* | 10.3 | 27 |
| Sep | 22.1 | 13.5 | 17.8 | *+2.9* | 28.9 | 11 | 7.2 | 8 | 17.6 | 22 | 17.5 | 5 | 0 | 4 | 90.8 | *168* | 17 | 28.4 | 13 | 156.5 | *112* | 11.2 | 8 |
| Oct | 17.2 | 10.6 | 13.9 | *+2.6* | 19.9 | 10 | 6.4 | 27 | 13.0 | 23 | 15.4 | 11 | 0 | 0 | 118.0 | *169* | 15 | 35.0 | 19 | 109.0 | *98* | 9.1 | 7 |
| Nov | 12.1 | 5.1 | 8.6 | *+1.1* | 16.0 | 13 | −1.3 | 3 | 8.0 | 6 | 11.4 | 15 | 4 | 0 | 86.4 | *129* | 17 | 14.6 | 24 | 101.2 | *143* | 8.3 | 1 |
| Dec | 9.4 | 4.3 | 6.9 | *+1.8* | 14.0 | 4 | −3.0 | 20 | 0.1 | 20 | 10.9 | 15 | 6 | 0 | 68.7 | *109* | 20 | 8.8 | 30 | 42.7 | *80* | 6.1 | 9 |
| **2006** | 15.6 | 7.7 | 11.7 | *+0.9* | 34.8 | 7 | −5.2 | 3 | −0.8 | 2 | 20.0 | 7 | 42 | 36 | 765.1 | *116* | 175 | 44.2 | 7 | 1668.2 | *106* | 14.6 | 6 |

## OXFORD 2007

| Month | Mean max | Mean min | Mean temp | Anom | Highest max | Date | Lowest min | Date | Lowest max | Date | Highest min | Date | Air frost | Days ≥ 25 °C | Total pptn | Anom | Rain days | Wettest day | Date | Total sunshine | Anom | Sunniest day | Date |
|---|---|---|---|---|---|---|---|---|---|---|---|---|---|---|---|---|---|---|---|---|---|---|---|
| Jan | 10.4 | 4.8 | 7.6 | *+2.7* | 13.7 | 9 | −2.0 | 26 | 3.9 | 23 | 10.1 | 10 | 4 | 0 | 73.3 | *129* | 18 | 14.1 | 9 | 79.8 | *128* | 7.6 | 25 |
| Feb | 9.7 | 3.7 | 6.7 | *+1.8* | 12.4 | 27 | −5.4 | 7 | 2.4 | 8 | 8.4 | 20 | 7 | 0 | 81.0 | *189* | 19 | 9.6 | 24 | 67.6 | *86* | 8.1 | 3 |
| Mar | 12.0 | 3.7 | 7.9 | *+0.6* | 16.4 | 11 | −1.2 | 21 | 6.5 | 20 | 8.2 | 17 | 2 | 0 | 36.9 | *77* | 13 | 8.9 | 4 | 165.4 | *149* | 9.5 | 13 |
| Apr | 17.8 | 5.6 | 11.7 | *+2.4* | 23.0 | 15 | 1.1 | 4 | 10.2 | 3 | 11.2 | 25 | 0 | 0 | 1.8 | *4* | 3 | 0.9 | 23 | 210.7 | *131* | 13.1 | 30 |
| May | 17.5 | 7.1 | 12.3 | *−0.2* | 24.2 | 24 | 2.0 | 29 | 10.7 | 27 | 11.9 | 18 | 0 | 0 | 135.2 | *236* | 19 | 40.7 | 27 | 165.5 | *86* | 13.1 | 1 |
| June | 20.5 | 11.3 | 15.9 | *+0.3* | 24.8 | 9 | 6.2 | 1 | 16.0 | 6 | 14.8 | 12 | 0 | 0 | 78.5 | *159* | 16 | 11.6 | 30 | 149.0 | *78* | 13.1 | 2 |
| July | 20.6 | 12.2 | 16.4 | *−1.5* | 23.1 | 14 | 8.0 | 31 | 17.7 | 20 | 15.8 | 16 | 0 | 0 | 110.2 | *226* | 18 | 34.2 | 20 | 195.1 | *94* | 13.2 | 8 |
| Aug | 21.7 | 12.0 | 16.8 | *−0.8* | 28.8 | 5 | 9.0 | 10 | 16.3 | 20 | 15.1 | 15 | 0 | 4 | 38.2 | *70* | 9 | 10.8 | 18 | 209.2 | *106* | 13.3 | 25 |
| Sep | 19.2 | 10.3 | 14.7 | *−0.2* | 23.5 | 6 | 4.9 | 30 | 13.0 | 26 | 14.5 | 7 | 0 | 0 | 17.7 | *33* | 9 | 7.9 | 23 | 142.6 | *102* | 11.1 | 15 |
| Oct | 15.1 | 7.8 | 11.5 | *+0.2* | 18.5 | 13 | 0.2 | 21 | 10.3 | 25 | 12.8 | 13 | 0 | 0 | 69.5 | *99* | 6 | 40.6 | 16 | 102.5 | *92* | 9.7 | 5 |
| Nov | 11.3 | 4.3 | 7.8 | *+0.3* | 17.0 | 1 | −2.6 | 16 | 7.7 | 18 | 9.8 | 8 | 4 | 0 | 61.1 | *92* | 16 | 14.9 | 19 | 86.9 | *122* | 7.6 | 15 |
| Dec | 8.3 | 2.4 | 5.3 | *+0.3* | 14.4 | 6 | −5.6 | 20 | 2.6 | 20 | 10.0 | 28 | 11 | 0 | 60.6 | *96* | 12 | 11.7 | 24 | 57.2 | *108* | 6.4 | 13 |
| **2007** | 15.3 | 7.1 | 11.2 | *+0.5* | 28.8 | 8 | −5.6 | 12 | 2.4 | 2 | 15.8 | 7 | 28 | 4 | 764.0 | *116* | 158 | 40.7 | 5 | 1631.5 | *103* | 13.3 | 8 |

## OXFORD 2008

| Month | Mean max | Mean min | Mean temp | Anom | Highest max | Date | Lowest min | Date | Lowest max | Date | Highest min | Date | Air frost | Days ≥ 25 °C | Total pptn | Anom | Rain days | Wettest day | Date | Total sunshine | Anom | Sunniest day | Date |
|---|---|---|---|---|---|---|---|---|---|---|---|---|---|---|---|---|---|---|---|---|---|---|---|
| Jan | 10.2 | 4.7 | 7.5 | *+2.6* | 13.5 | 20 | 0.5 | 3 | 4.5 | 3 | 12.1 | 20 | 0 | 0 | 79.1 | *139* | 19 | 12.6 | 15 | 62.4 | *100* | 6.7 | 30 |
| Feb | 10.5 | 1.7 | 6.1 | *+1.2* | 13.4 | 9 | −5.0 | 17 | 4.6 | 14 | 8.6 | 24 | 11 | 0 | 16.9 | *39* | 6 | 6.7 | 3 | 124.1 | *157* | 9.0 | 16 |
| Mar | 10.6 | 3.5 | 7.1 | *−0.2* | 14.1 | 11 | −2.5 | 5 | 6.5 | 22 | 9.0 | 2 | 2 | 0 | 90.3 | *189* | 17 | 42.1 | 15 | 115.2 | *104* | 8.7 | 27 |
| Apr | 13.5 | 4.7 | 9.1 | *−0.2* | 19.5 | 26 | −0.7 | 6 | 6.2 | 6 | 11.0 | 27 | 3 | 0 | 47.6 | *97* | 14 | 11.0 | 29 | 161.2 | *100* | 10.5 | 12 |
| May | 17.4 | 9.5 | 13.4 | *+1.0* | 22.3 | 4 | 2.2 | 19 | 11.5 | 26 | 14.5 | 4 | 0 | 0 | 106.3 | *186* | 16 | 20.0 | 26 | 173.2 | *90* | 12.8 | 6 |
| June | 20.1 | 11.1 | 15.6 | *−0.0* | 26.0 | 9 | 7.2 | 16 | 15.8 | 12 | 15.7 | 28 | 0 | 1 | 67.3 | *137* | 9 | 27.9 | 3 | 223.8 | *117* | 14.0 | 9 |
| July | 22.0 | 12.8 | 17.4 | *−0.5* | 28.3 | 27 | 8.4 | 4 | 17.5 | 9 | 16.0 | 15 | 0 | 6 | 90.0 | *184* | 12 | 24.5 | 28 | 198.5 | *96* | 12.5 | 21 |
| Aug | 20.8 | 13.9 | 17.4 | *−0.2* | 25.4 | 30 | 9.0 | 15 | 18.0 | 9 | 18.2 | 30 | 0 | 1 | 76.9 | *140* | 14 | 18.4 | 11 | 141.8 | *72* | 12.6 | 14 |
| Sep | 17.8 | 10.2 | 14.0 | *−0.9* | 20.4 | 10 | 5.0 | 28 | 15.2 | 29 | 13.6 | 11 | 0 | 0 | 72.9 | *135* | 16 | 16.2 | 6 | 108.6 | *77* | 9.3 | 26 |
| Oct | 14.6 | 5.7 | 10.1 | *−1.2* | 21.0 | 12 | −2.8 | 29 | 7.8 | 29 | 12.0 | 15 | 1 | 0 | 39.4 | *56* | 13 | 13.9 | 20 | 132.2 | *119* | 9.0 | 21 |
| Nov | 10.1 | 4.8 | 7.4 | *−0.1* | 14.5 | 15 | −1.5 | 23 | 3.9 | 29 | 10.0 | 16 | 3 | 0 | 94.2 | *141* | 21 | 34.9 | 1 | 67.0 | *94* | 6.6 | 10 |
| Dec | 6.9 | 1.3 | 4.1 | *−0.9* | 12.0 | 20 | −3.4 | 31 | 0.6 | 31 | 9.6 | 22 | 11 | 0 | 37.3 | *59* | 13 | 13.0 | 12 | 73.5 | *138* | 7.0 | 6 |
| **2008** | 14.5 | 7.0 | 10.8 | *+0.0* | 28.3 | 7 | −5.0 | 2 | 0.6 | 12 | 18.2 | 8 | 31 | 8 | 818.2 | *124* | 170 | 42.1 | 3 | 1581.5 | *100* | 14.0 | 6 |

## OXFORD 2009

| Month | Mean max | Mean min | Mean temp | Anom | Highest max | Date | Lowest min | Date | Lowest max | Date | Highest min | Date | Air frost | Days ≥ 25 °C | Total pptn | Anom | Rain days | Wettest day | Date | Total sunshine | Anom | Sunniest day | Date |
|---|---|---|---|---|---|---|---|---|---|---|---|---|---|---|---|---|---|---|---|---|---|---|---|
| Jan | 6.3 | 0.2 | 3.2 | *−1.6* | 9.9 | 22 | −6.3 | 7 | −1.9 | 9 | 6.4 | 17 | 16 | 0 | 46.0 | *81* | 16 | 14.3 | 22 | 69.3 | *111* | 7.3 | 6 |
| Feb | 7.9 | 2.0 | 4.9 | *+0.0* | 13.7 | 28 | −3.6 | 7 | 0.6 | 2 | 7.7 | 23 | 10 | 0 | 54.5 | *127* | 12 | 25.9 | 9 | 64.5 | *82* | 8.5 | 27 |
| Mar | 12.4 | 3.2 | 7.8 | *+0.5* | 16.1 | 16 | −2.0 | 30 | 8.0 | 5 | 6.9 | 14 | 3 | 0 | 20.2 | *42* | 12 | 10.5 | 3 | 161.4 | *145* | 9.6 | 6 |
| Apr | 15.6 | 6.1 | 10.9 | *+1.6* | 21.0 | 15 | 2.3 | 5 | 10.5 | 17 | 9.8 | 10 | 0 | 0 | 32.0 | *65* | 15 | 7.6 | 28 | 168.4 | *105* | 12.0 | 29 |
| May | 18.2 | 8.3 | 13.3 | *+0.8* | 24.9 | 31 | 3.5 | 4 | 14.2 | 4 | 11.6 | 29 | 0 | 0 | 46.3 | *81* | 14 | 19.3 | 15 | 226.1 | *117* | 14.7 | 30 |
| June | 22.0 | 10.7 | 16.3 | *+0.7* | 30.3 | 29 | 5.2 | 12 | 13.2 | 6 | 17.2 | 29 | 0 | 7 | 62.3 | *127* | 12 | 27.7 | 7 | 203.3 | *106* | 14.8 | 24 |
| July | 22.4 | 13.1 | 17.7 | *−0.1* | 31.0 | 1 | 9.1 | 10 | 18.1 | 29 | 19.4 | 1 | 0 | 4 | 72.8 | *149* | 25 | 12.3 | 29 | 212.3 | *103* | 12.7 | 2 |
| Aug | 22.7 | 13.3 | 18.0 | *+0.4* | 26.7 | 11 | 9.1 | 29 | 18.4 | 30 | 17.8 | 5 | 0 | 5 | 46.6 | *85* | 13 | 22.2 | 6 | 190.6 | *97* | 11.4 | 2 |
| Sep | 20.2 | 10.6 | 15.4 | *+0.5* | 25.8 | 8 | 6.3 | 27 | 16.6 | 17 | 17.1 | 8 | 0 | 1 | 9.0 | *17* | 4 | 4.3 | 3 | 163.7 | *117* | 11.2 | 10 |
| Oct | 16.2 | 8.4 | 12.3 | *+1.0* | 20.3 | 6 | 2.5 | 18 | 12.0 | 17 | 12.0 | 31 | 0 | 0 | 51.6 | *74* | 18 | 9.0 | 31 | 109.7 | *99* | 9.7 | 12 |
| Nov | 12.5 | 7.0 | 9.7 | *+2.2* | 17.0 | 1 | 2.4 | 8 | 6.5 | 30 | 12.9 | 1 | 0 | 0 | 109.2 | *164* | 26 | 15.1 | 25 | 73.5 | *104* | 6.3 | 7 |
| Dec | 6.8 | 0.8 | 3.8 | *−1.2* | 13.1 | 5 | −5.4 | 23 | 2.6 | 21 | 7.4 | 6 | 12 | 0 | 92.8 | *147* | 22 | 19.7 | 29 | 61.5 | *116* | 6.8 | 20 |
| **2009** | 15.3 | 7.0 | 11.1 | *+0.4* | 31.0 | 7 | −6.3 | 1 | −1.9 | 1 | 19.4 | 7 | 41 | 17 | 643.3 | *97* | 189 | 27.7 | 6 | 1704.3 | *108* | 14.8 | 6 |

## OXFORD 2010

| Month | Mean max | Mean min | Mean temp | Anom | Highest max | Date | Lowest min | Date | Lowest max | Date | Highest min | Date | Air frost | Days ≥ 25 °C | Total pptn | Anom | Rain days | Wettest day | Date | Total sunshine | Anom | Sunniest day | Date |
|---|---|---|---|---|---|---|---|---|---|---|---|---|---|---|---|---|---|---|---|---|---|---|---|
| Jan | 4.5 | −1.0 | 1.8 | *−3.1* | 10.7 | 18 | −8.2 | 7 | 0.2 | 4 | 5.0 | 19 | 17 | 0 | 66.9 | *118* | 16 | 13.6 | 24 | 62.7 | *101* | 7.0 | 1 |
| Feb | 7.1 | 1.3 | 4.2 | *−0.7* | 11.4 | 5 | −1.5 | 1 | 2.7 | 8 | 6.3 | 25 | 7 | 0 | 80.1 | *187* | 23 | 20.6 | 19 | 59.3 | *75* | 6.6 | 11 |
| Mar | 11.3 | 3.2 | 7.2 | *−0.0* | 17.8 | 18 | −4.4 | 7 | 5.6 | 7 | 9.8 | 20 | 8 | 0 | 47.6 | *100* | 14 | 10.4 | 24 | 130.2 | *117* | 10.1 | 7 |
| Apr | 15.8 | 4.9 | 10.3 | *+1.1* | 21.8 | 28 | 0.7 | 18 | 9.6 | 1 | 11.1 | 29 | 0 | 0 | 25.5 | *52* | 7 | 8.8 | 3 | 209.5 | *130* | 12.9 | 17 |
| May | 17.6 | 7.3 | 12.4 | *−0.0* | 28.9 | 24 | 0.8 | 4 | 10.5 | 8 | 13.8 | 20 | 0 | 3 | 28.6 | *50* | 10 | 11.7 | 1 | 207.4 | *108* | 14.8 | 22 |
| June | 23.0 | 11.1 | 17.1 | *+1.5* | 29.6 | 27 | 6.2 | 17 | 14.6 | 1 | 16.1 | 29 | 0 | 12 | 43.6 | *89* | 10 | 11.7 | 8 | 230.5 | *121* | 15.0 | 17 |
| July | 23.9 | 14.2 | 19.0 | *+1.1* | 28.9 | 20 | 10.6 | 6 | 19.2 | 13 | 17.7 | 26 | 0 | 8 | 17.9 | *37* | 7 | 8.2 | 22 | 181.0 | *87* | 13.8 | 21 |
| Aug | 21.4 | 12.1 | 16.8 | *−0.8* | 26.0 | 16 | 6.9 | 31 | 15.6 | 26 | 18.5 | 21 | 0 | 3 | 146.2 | *267* | 20 | 51.1 | 25 | 141.7 | *72* | 11.4 | 24 |
| Sep | 18.8 | 10.3 | 14.5 | *−0.4* | 22.4 | 4 | 4.2 | 25 | 13.0 | 24 | 16.0 | 11 | 0 | 0 | 52.5 | *97* | 14 | 12.6 | 6 | 123.1 | *88* | 11.3 | 1 |
| Oct | 14.5 | 7.4 | 11.0 | *−0.3* | 21.2 | 8 | −1.1 | 25 | 9.2 | 20 | 14.3 | 9 | 2 | 0 | 44.2 | *63* | 16 | 8.6 | 1 | 123.5 | *111* | 9.2 | 20 |
| Nov | 8.2 | 2.9 | 5.5 | *−2.0* | 15.5 | 5 | −4.8 | 29 | −1.2 | 28 | 13.6 | 5 | 10 | 0 | 38.6 | *58* | 13 | 14.4 | 5 | 53.6 | *76* | 8.3 | 10 |
| Dec | 2.7 | −2.2 | 0.2 | *−4.8* | 8.0 | 29 | −10.9 | 20 | −4.4 | 19 | 4.9 | 30 | 25 | 0 | 31.8 | *50* | 11 | 12.6 | 17 | 20.4 | *38* | 4.3 | 5 |
| **2010** | 14.1 | 6.0 | 10.0 | *−0.7* | 29.6 | 6 | −10.9 | 12 | −4.4 | 12 | 18.5 | 8 | 69 | 26 | 623.5 | *94* | 161 | 51.1 | 8 | 1542.9 | *98* | 15.0 | 6 |

## OXFORD 2011

| Month | Mean max | Mean min | Mean temp | Anom | Highest max | Date | Lowest min | Date | Lowest max | Date | Highest min | Date | Air frost | Days ≥ 25 °C | Total pptn | Anom | Rain days | Wettest day | Date | Total sunshine | Anom | Sunniest day | Date |
|---|---|---|---|---|---|---|---|---|---|---|---|---|---|---|---|---|---|---|---|---|---|---|---|
| Jan | 7.0 | 1.9 | 4.4 | *−0.4* | 12.8 | 15 | −3.3 | 31 | 1.8 | 29 | 10.5 | 13 | 12 | 0 | 58.8 | *103* | 17 | 11.1 | 7 | 49.3 | *79* | 7.2 | 9 |
| Feb | 10.4 | 4.4 | 7.4 | *+2.5* | 14.8 | 25 | −1.4 | 1 | 5.8 | 28 | 10.7 | 5 | 1 | 0 | 50.2 | *117* | 16 | 13.5 | 27 | 34.2 | *43* | 6.2 | 8 |
| Mar | 12.5 | 4.8 | 8.7 | *+1.4* | 17.7 | 24 | −3.4 | 8 | 4.2 | 3 | 8.9 | 23 | 3 | 0 | 9.4 | *20* | 4 | 5.0 | 30 | 142.5 | *128* | 10.4 | 23 |
| Apr | 19.4 | 7.3 | 13.3 | *+4.0* | 27.6 | 23 | 3.2 | 28 | 12.5 | 13 | 11.1 | 6 | 0 | 2 | 0.5 | *1* | 2 | 0.2 | 28 | 211.4 | *131* | 12.7 | 22 |
| May | 18.9 | 8.4 | 13.7 | *+1.2* | 24.2 | 6 | 1.0 | 4 | 16.1 | 3 | 14.1 | 7 | 0 | 0 | 42.5 | *74* | 10 | 15.7 | 7 | 208.7 | *108* | 13.3 | 24 |
| June | 19.3 | 10.2 | 14.8 | *−0.8* | 27.8 | 26 | 6.5 | 1 | 14.5 | 17 | 19.6 | 27 | 0 | 2 | 54.1 | *110* | 14 | 15.2 | 12 | 188.5 | *99* | 12.8 | 2 |
| July | 21.4 | 10.2 | 15.8 | *−2.1* | 25.8 | 28 | 7.6 | 14 | 17.5 | 18 | 13.4 | 27 | 0 | 1 | 43.1 | *88* | 12 | 9.5 | 15 | 173.9 | *84* | 13.0 | 24 |
| Aug | 20.7 | 11.3 | 16.0 | *−1.6* | 27.2 | 2 | 7.2 | 19 | 15.1 | 18 | 15.5 | 2 | 0 | 3 | 65.6 | *120* | 13 | 16.6 | 18 | 157.8 | *80* | 12.3 | 2 |
| Sep | 20.7 | 10.7 | 15.7 | *+0.8* | 27.9 | 30 | 5.2 | 15 | 16.5 | 18 | 14.5 | 12 | 0 | 2 | 31.8 | *59* | 11 | 10.3 | 20 | 151.4 | *108* | 10.2 | 28 |
| Oct | 17.7 | 7.9 | 12.8 | *+1.5* | 29.1 | 1 | 0.8 | 20 | 12.7 | 20 | 13.8 | 5 | 0 | 3 | 26.4 | *38* | 10 | 8.6 | 26 | 136.5 | *123* | 10.3 | 2 |
| Nov | 13.5 | 6.3 | 9.9 | *+2.4* | 17.3 | 13 | −0.8 | 28 | 10.4 | 14 | 12.0 | 3 | 1 | 0 | 28.4 | *43* | 11 | 9.9 | 3 | 59.7 | *84* | 6.3 | 13 |
| Dec | 9.3 | 3.1 | 6.2 | *+1.2* | 13.0 | 8 | −1.4 | 3 | 4.0 | 16 | 10.2 | 26 | 4 | 0 | 71.2 | *113* | 17 | 11.5 | 12 | 61.0 | *115* | 5.5 | 18 |
| 2011 | 15.9 | 7.2 | 11.6 | *+0.8* | 29.1 | 10 | −3.4 | 3 | 1.8 | 1 | 19.6 | 6 | 21 | 13 | 482.0 | *73* | 137 | 16.6 | 8 | 1574.9 | *100* | 13.3 | 5 |

## OXFORD 2012

| Month | Mean max | Mean min | Mean temp | Anom | Highest max | Date | Lowest min | Date | Lowest max | Date | Highest min | Date | Air frost | Days ≥ 25 °C | Total pptn | Anom | Rain days | Wettest day | Date | Total sunshine | Anom | Sunniest day | Date |
|---|---|---|---|---|---|---|---|---|---|---|---|---|---|---|---|---|---|---|---|---|---|---|---|
| Jan | 9.8 | 3.5 | 6.6 | *+1.8* | 13.2 | 12 | −3.4 | 15 | 3.6 | 29 | 10.6 | 1 | 6 | 0 | 34.6 | *61* | 9 | 7.0 | 23 | 85.7 | *138* | 7.3 | 15 |
| Feb | 7.3 | 1.2 | 4.3 | *−0.6* | 14.0 | 24 | −7.2 | 11 | −0.5 | 8 | 9.0 | 24 | 12 | 0 | 21.3 | *50* | 13 | 9.0 | 4 | 71.7 | *91* | 8.3 | 2 |
| Mar | 14.0 | 3.5 | 8.8 | *+1.5* | 21.3 | 28 | −1.1 | 15 | 7.8 | 4 | 8.4 | 10 | 2 | 0 | 24.3 | *51* | 8 | 11.6 | 4 | 164.3 | *148* | 11.5 | 27 |
| Apr | 12.6 | 4.5 | 8.6 | *−0.7* | 18.3 | 30 | −1.2 | 5 | 8.5 | 28 | 9.7 | 27 | 2 | 0 | 143.0 | *292* | 21 | 20.0 | 27 | 141.0 | *88* | 9.8 | 30 |
| May | 17.7 | 9.7 | 13.7 | *+1.2* | 26.5 | 27 | 4.1 | 16 | 8.9 | 3 | 14.5 | 25 | 0 | 5 | 55.5 | *97* | 12 | 12.1 | 7 | 174.9 | *91* | 14.5 | 26 |
| June | 18.9 | 11.6 | 15.3 | *−0.3* | 28.3 | 27 | 7.0 | 4 | 12.1 | 11 | 18.3 | 28 | 0 | 2 | 151.7 | *308* | 20 | 21.2 | 11 | 134.5 | *70* | 12.9 | 13 |
| July | 21.4 | 13.3 | 17.3 | *−0.5* | 29.0 | 25 | 9.8 | 21 | 16.9 | 6 | 17.0 | 4 | 0 | 5 | 101.3 | *207* | 18 | 27.8 | 6 | 175.4 | *85* | 14.2 | 24 |
| Aug | 23.1 | 14.3 | 18.7 | *+1.1* | 29.1 | 18 | 7.2 | 31 | 17.0 | 29 | 16.8 | 4 | 0 | 11 | 79.1 | *144* | 13 | 26.0 | 5 | 172.7 | *88* | 12.2 | 10 |
| Sep | 18.2 | 9.3 | 13.7 | *−1.1* | 22.8 | 9 | 3.2 | 22 | 14.5 | 23 | 15.2 | 4 | 0 | 0 | 57.7 | *107* | 12 | 32.8 | 23 | 173.2 | *123* | 11.0 | 13 |
| Oct | 13.4 | 7.1 | 10.3 | *−1.0* | 17.9 | 1 | 0.8 | 14 | 8.6 | 27 | 12.6 | 1 | 0 | 0 | 124.2 | *177* | 21 | 37.0 | 17 | 85.5 | *77* | 9.0 | 14 |
| Nov | 10.1 | 4.2 | 7.2 | *−0.4* | 14.8 | 13 | −1.3 | 30 | 3.8 | 30 | 11.0 | 14 | 1 | 0 | 83.6 | *125* | 16 | 19.8 | 24 | 74.0 | *104* | 8.3 | 11 |
| Dec | 8.6 | 2.6 | 5.6 | *+0.6* | 13.0 | 22 | −4.8 | 12 | −0.2 | 11 | 8.9 | 29 | 9 | 0 | 103.2 | *164* | 21 | 17.9 | 19 | 61.7 | *116* | 6.5 | 5 |
| 2012 | 14.6 | 7.1 | 10.8 | *+0.1* | 29.1 | 8 | −7.2 | 2 | −0.5 | 2 | 18.3 | 6 | 32 | 23 | 979.5 | *148* | 184 | 37.0 | 10 | 1514.6 | *96* | 14.5 | 5 |

## OXFORD 2013

| Month | Mean max | Mean min | Mean temp | Anom | Highest max | Date | Lowest min | Date | Lowest max | Date | Highest min | Date | Air frost | Days ≥ 25 °C | Total pptn | Anom | Rain days | Wettest day | Date | Total sunshine | Anom | Sunniest day | Date |
|---|---|---|---|---|---|---|---|---|---|---|---|---|---|---|---|---|---|---|---|---|---|---|---|
| Jan | 6.4 | 1.9 | 4.1 | *−0.7* | 14.3 | 29 | −3.8 | 16 | −0.2 | 16 | 8.7 | 30 | 14 | 0 | 64.0 | *113* | 17 | 16.8 | 26 | 47.8 | *77* | 6.5 | 9 |
| Feb | 6.2 | 0.9 | 3.6 | *−1.3* | 10.8 | 14 | −2.4 | 19 | 1.2 | 11 | 4.6 | 1 | 5 | 0 | 47.4 | *111* | 9 | 23.5 | 10 | 72.3 | *92* | 8.4 | 2 |
| Mar | 6.3 | 0.2 | 3.3 | *−4.0* | 15.1 | 5 | −4.1 | 31 | 1.0 | 24 | 7.6 | 8 | 16 | 0 | 77.4 | *162* | 13 | 15.5 | 16 | 60.7 | *55* | 9.2 | 5 |
| Apr | 12.9 | 3.9 | 8.4 | *−0.9* | 19.8 | 23 | −2.2 | 2 | 4.5 | 4 | 10.8 | 25 | 6 | 0 | 24.9 | *51* | 10 | 10.3 | 11 | 177.0 | *110* | 13.4 | 20 |
| May | 15.9 | 6.7 | 11.3 | *−1.1* | 22.4 | 7 | 0.8 | 1 | 10.9 | 14 | 11.6 | 21 | 0 | 0 | 66.2 | *116* | 15 | 16.3 | 28 | 193.6 | *100* | 14.5 | 26 |
| June | 19.3 | 10.2 | 14.8 | *−0.8* | 24.8 | 19 | 5.8 | 10 | 15.5 | 8 | 15.7 | 20 | 0 | 0 | 17.3 | *35* | 10 | 4.8 | 27 | 181.3 | *95* | 14.5 | 3 |
| July | 25.5 | 13.9 | 19.7 | *+1.8* | 31.6 | 22 | 9.8 | 12 | 16.3 | 2 | 18.3 | 23 | 0 | 19 | 45.6 | *93* | 9 | 11.7 | 27 | 297.3 | *144* | 15.0 | 19 |
| Aug | 23.1 | 13.1 | 18.1 | *+0.5* | 31.3 | 1 | 8.9 | 8 | 19.0 | 13 | 17.1 | 1 | 0 | 4 | 19.3 | *35* | 12 | 4.4 | 5 | 205.9 | *105* | 12.3 | 1 |
| Sep | 19.0 | 9.9 | 14.4 | *−0.4* | 27.6 | 5 | 5.4 | 15 | 13.5 | 17 | 14.4 | 13 | 0 | 2 | 40.6 | *75* | 11 | 16.1 | 13 | 123.1 | *88* | 11.4 | 5 |
| Oct | 16.3 | 9.9 | 13.1 | *+1.8* | 20.2 | 2 | 2.5 | 30 | 11.0 | 13 | 14.7 | 22 | 0 | 0 | 86.7 | *124* | 23 | 14.6 | 27 | 101.8 | *91* | 8.3 | 30 |
| Nov | 9.9 | 3.8 | 6.8 | *−0.7* | 14.6 | 6 | −2.0 | 26 | 6.0 | 23 | 9.9 | 1 | 2 | 0 | 114.1 | *171* | 19 | 19.0 | 21 | 85.7 | *121* | 8.4 | 4 |
| Dec | 10.0 | 3.6 | 6.8 | *+1.8* | 13.6 | 15 | 0.6 | 5 | 6.7 | 11 | 8.7 | 16 | 0 | 0 | 97.7 | *155* | 18 | 26.3 | 23 | 63.6 | *120* | 5.5 | 24 |
| 2013 | 14.2 | 6.5 | 10.4 | *−0.4* | 31.6 | 7 | −4.1 | 3 | −0.2 | 1 | 18.3 | 7 | 43 | 25 | 701.2 | *106* | 166 | 26.3 | 12 | 1610.1 | *102* | 15.0 | 7 |

## OXFORD 2014

| Month | Mean max | Mean min | Mean temp | Anom | Highest max | Date | Lowest min | Date | Lowest max | Date | Highest min | Date | Air frost | Days ≥ 25 °C | Total pptn | Anom | Rain days | Wettest day | Date | Total sunshine | Anom | Sunniest day | Date |
|---|---|---|---|---|---|---|---|---|---|---|---|---|---|---|---|---|---|---|---|---|---|---|---|
| Jan | 9.3 | 3.5 | 6.4 | *+1.6* | 12.8 | 5 | −1.5 | 12 | 4.7 | 30 | 7.0 | 3 | 3 | 0 | 146.9 | *258* | 30 | 14.5 | 3 | 77.7 | *125* | 7.8 | 11 |
| Feb | 9.7 | 4.2 | 7.0 | *+2.1* | 12.1 | 23 | 1.7 | 13 | 5.5 | 28 | 7.5 | 24 | 0 | 0 | 90.1 | *211* | 23 | 25.1 | 6 | 97.4 | *123* | 8.8 | 26 |
| Mar | 12.8 | 3.4 | 8.1 | *+0.9* | 19.0 | 9 | −0.8 | 1 | 8.2 | 25 | 8.9 | 31 | 2 | 0 | 39.2 | *82* | 11 | 11.2 | 2 | 135.6 | *122* | 10.0 | 9 |
| Apr | 15.3 | 6.3 | 10.8 | *+1.5* | 19.4 | 6 | 0.9 | 19 | 11.3 | 19 | 12.3 | 6 | 0 | 0 | 50.2 | *103* | 13 | 11.9 | 24 | 149.5 | *93* | 12.8 | 14 |
| May | 16.9 | 8.9 | 12.9 | *+0.5* | 20.6 | 20 | 0.5 | 3 | 11.0 | 3 | 13.9 | 20 | 0 | 0 | 90.3 | *158* | 18 | 19.5 | 1 | 185.7 | *96* | 14.6 | 25 |
| June | 21.5 | 11.8 | 16.7 | *+1.1* | 25.6 | 9 | 8.3 | 1 | 15.7 | 4 | 15.5 | 23 | 0 | 3 | 37.0 | *75* | 9 | 15.0 | 13 | 236.3 | *124* | 13.8 | 10 |
| July | 24.9 | 13.5 | 19.2 | *+1.4* | 28.5 | 24 | 9.3 | 2 | 20.2 | 8 | 17.6 | 19 | 0 | 16 | 45.5 | *93* | 13 | 10.8 | 19 | 264.6 | *128* | 14.5 | 22 |
| Aug | 20.8 | 11.7 | 16.2 | *−1.3* | 25.1 | 7 | 6.7 | 21 | 16.7 | 25 | 16.0 | 6 | 0 | 1 | 85.6 | *156* | 19 | 16.5 | 9 | 190.5 | *97* | 12.3 | 3 |
| Sep | 21.1 | 11.4 | 16.3 | *+1.4* | 24.6 | 18 | 5.0 | 22 | 18.1 | 11 | 15.7 | 5 | 0 | 0 | 4.1 | *8* | 3 | 2.2 | 23 | 143.7 | *102* | 10.4 | 8 |
| Oct | 16.8 | 9.9 | 13.4 | *+2.1* | 21.6 | 31 | 3.2 | 5 | 12.2 | 13 | 15.4 | 18 | 0 | 0 | 65.6 | *94* | 20 | 11.8 | 13 | 116.9 | *105* | 8.9 | 7 |
| Nov | 11.5 | 5.7 | 8.6 | *+1.0* | 16.8 | 1 | −2.1 | 25 | 5.9 | 24 | 12.1 | 1 | 2 | 0 | 98.6 | *148* | 22 | 18.0 | 13 | 60.9 | *86* | 6.9 | 1 |
| Dec | 8.7 | 2.4 | 5.5 | *+0.5* | 13.4 | 18 | −3.4 | 31 | 3.9 | 27 | 10.0 | 18 | 9 | 0 | 44.0 | *70* | 17 | 12.3 | 16 | 96.9 | *182* | 7.3 | 6 |
| 2014 | 15.8 | 7.7 | 11.8 | *+1.0* | 28.5 | 7 | −3.4 | 12 | 3.9 | 12 | 17.6 | 7 | 16 | 20 | 797.1 | *121* | 198 | 25.1 | 2 | 1755.7 | *111* | 14.6 | 5 |

## OXFORD 2015

| Month | Mean max | Mean min | Mean temp | Anom | Highest max | Date | Lowest min | Date | Lowest max | Date | Highest min | Date | Air frost | Days ≥ 25 °C | Total pptn | Anom | Rain days | Wettest day | Date | Total sunshine | Anom | Sunniest day | Date |
|---|---|---|---|---|---|---|---|---|---|---|---|---|---|---|---|---|---|---|---|---|---|---|---|
| Jan | 8.4 | 1.6 | 5.0 | *+0.2* | 13.8 | 9 | −4.5 | 23 | 3.7 | 18 | 9.5 | 10 | 10 | 0 | 55.5 | *98* | 22 | 10.7 | 2 | 81.0 | *130* | 6.7 | 19 |
| Feb | 7.5 | 1.0 | 4.3 | *−0.6* | 13.2 | 25 | −3.5 | 2 | 2.6 | 2 | 6.8 | 26 | 10 | 0 | 46.4 | *108* | 13 | 11.8 | 16 | 80.8 | *102* | 9.0 | 18 |
| Mar | 11.3 | 3.3 | 7.3 | *+0.1* | 15.4 | 7 | −1.2 | 25 | 6.9 | 15 | 8.2 | 29 | 1 | 0 | 25.8 | *54* | 12 | 5.8 | 25 | 122.9 | *110* | 8.8 | 10 |
| Apr | 15.9 | 4.5 | 10.2 | *+0.9* | 22.5 | 15 | 0.2 | 27 | 10.2 | 4 | 10.9 | 25 | 0 | 0 | 22.6 | *46* | 8 | 10.1 | 25 | 214.1 | *133* | 13.1 | 21 |
| May | 16.8 | 7.7 | 12.2 | *−0.2* | 20.8 | 11 | 3.0 | 1 | 11.8 | 14 | 11.0 | 5 | 0 | 0 | 55.6 | *97* | 17 | 10.5 | 14 | 183.3 | *95* | 13.7 | 21 |
| June | 21.1 | 10.2 | 15.6 | *+0.1* | 29.7 | 30 | 4.9 | 9 | 15.0 | 9 | 14.3 | 28 | 0 | 2 | 28.9 | *59* | 11 | 7.9 | 20 | 242.9 | *127* | 15.4 | 7 |
| July | 22.6 | 12.5 | 17.5 | *−0.3* | 33.5 | 1 | 5.6 | 31 | 15.9 | 24 | 16.5 | 1 | 0 | 5 | 47.8 | *98* | 10 | 31.9 | 24 | 201.0 | *97* | 15.0 | 9 |
| Aug | 21.3 | 13.0 | 17.1 | *−0.5* | 29.2 | 22 | 8.7 | 7 | 14.8 | 31 | 17.5 | 21 | 0 | 4 | 57.2 | *104* | 12 | 17.8 | 13 | 135.3 | *69* | 11.1 | 7 |
| Sep | 18.4 | 8.9 | 13.6 | *−1.3* | 21.5 | 10 | 4.6 | 26 | 14.8 | 22 | 14.8 | 12 | 0 | 0 | 36.8 | *68* | 13 | 12.3 | 16 | 164.3 | *117* | 11.9 | 6 |
| Oct | 15.1 | 7.9 | 11.5 | *+0.2* | 18.5 | 1 | 2.7 | 26 | 11.5 | 15 | 13.7 | 6 | 0 | 0 | 45.1 | *64* | 11 | 13.0 | 5 | 84.1 | *76* | 8.5 | 4 |
| Nov | 13.1 | 7.3 | 10.2 | *+2.7* | 17.3 | 6 | −1.6 | 23 | 5.1 | 21 | 13.3 | 11 | 2 | 0 | 65.3 | *98* | 22 | 6.9 | 23 | 33.8 | *48* | 5.7 | 1 |
| Dec | 13.4 | 8.2 | 10.8 | *+5.8* | 15.9 | 18 | 3.5 | 12 | 9.4 | 11 | 12.2 | 7 | 0 | 0 | 62.0 | *98* | 20 | 13.5 | 30 | 38.3 | *72* | 6.1 | 9 |
| **2015** | 15.4 | 7.2 | 11.3 | *+0.5* | 33.5 | 7 | −4.5 | 1 | 2.6 | 2 | 17.5 | 8 | 23 | 11 | 548.9 | *83* | 171 | 31.9 | 7 | 1581.8 | *100* | 15.4 | 6 |

## OXFORD 2016

| Month | Mean max | Mean min | Mean temp | Anom | Highest max | Date | Lowest min | Date | Lowest max | Date | Highest min | Date | Air frost | Days ≥ 25 °C | Total pptn | Anom | Rain days | Wettest day | Date | Total sunshine | Anom | Sunniest day | Date |
|---|---|---|---|---|---|---|---|---|---|---|---|---|---|---|---|---|---|---|---|---|---|---|---|
| Jan | 9.4 | 3.0 | 6.2 | *+1.3* | 15.9 | 24 | −4.8 | 20 | 3.9 | 19 | 9.5 | 27 | 5 | 0 | 83.7 | *147* | 22 | 14.3 | 8 | 59.1 | *95* | 7.6 | 15 |
| Feb | 9.1 | 2.7 | 5.9 | *+1.0* | 13.9 | 4 | −2.5 | 16 | 4.5 | 13 | 9.7 | 21 | 6 | 0 | 47.6 | *111* | 9 | 12.6 | 7 | 112.9 | *143* | 8.5 | 18 |
| Mar | 10.2 | 2.1 | 6.2 | *−1.1* | 15.4 | 25 | −2.1 | 8 | 7.0 | 9 | 6.0 | 26 | 4 | 0 | 74.2 | *156* | 13 | 28.5 | 8 | 123.8 | *111* | 11.0 | 7 |
| Apr | 12.9 | 4.0 | 8.5 | *−0.8* | 17.5 | 13 | −0.3 | 28 | 8.6 | 16 | 8.7 | 4 | 1 | 0 | 53.2 | *109* | 21 | 9.1 | 10 | 163.7 | *102* | 12.8 | 20 |
| May | 18.4 | 8.6 | 13.5 | *+1.0* | 26.2 | 8 | 1.5 | 1 | 13.0 | 25 | 14.3 | 10 | 0 | 1 | 81.4 | *142* | 11 | 26.7 | 10 | 202.9 | *105* | 13.1 | 4 |
| June | 20.0 | 11.9 | 16.0 | *+0.4* | 25.6 | 6 | 8.8 | 3 | 13.0 | 1 | 15.1 | 11 | 0 | 2 | 95.7 | *194* | 18 | 16.5 | 19 | 100.4 | *53* | 12.8 | 6 |
| July | 23.3 | 13.7 | 18.5 | *+0.6* | 32.3 | 19 | 9.1 | 6 | 18.5 | 12 | 21.2 | 20 | 0 | 7 | 3.6 | *7* | 7 | 1.1 | 10 | 227.6 | *110* | 15.0 | 19 |
| Aug | 23.5 | 13.6 | 18.5 | *+0.9* | 29.5 | 24 | 9.1 | 15 | 19.0 | 1 | 18.3 | 7 | 0 | 7 | 41.2 | *75* | 13 | 11.5 | 1 | 204.5 | *104* | 13.3 | 15 |
| Sep | 20.9 | 13.0 | 17.0 | *+2.1* | 29.2 | 13 | 7.0 | 23 | 16.2 | 17 | 18.1 | 7 | 0 | 3 | 41.7 | *77* | 13 | 12.8 | 19 | 113.2 | *81* | 11.0 | 11 |
| Oct | 15.1 | 7.3 | 11.2 | *−0.1* | 18.5 | 4 | 2.3 | 11 | 12.7 | 21 | 12.1 | 29 | 0 | 0 | 26.5 | *38* | 10 | 15.2 | 15 | 112.2 | *101* | 10.0 | 3 |
| Nov | 9.4 | 3.3 | 6.4 | *−1.2* | 14.4 | 15 | −4.7 | 30 | 4.7 | 30 | 10.4 | 15 | 3 | 0 | 78.2 | *117* | 17 | 20.0 | 8 | 88.3 | *124* | 8.5 | 7 |
| Dec | 9.6 | 2.9 | 6.3 | *+1.2* | 14.6 | 8 | −4.0 | 28 | 5.4 | 27 | 11.0 | 9 | 10 | 0 | 32.1 | *51* | 15 | 16.4 | 10 | 62.3 | *117* | 7.5 | 1 |
| **2016** | 15.1 | 7.2 | 11.2 | *+0.4* | 32.3 | 7 | −4.8 | 1 | 3.9 | 1 | 21.2 | 7 | 29 | 20 | 659.1 | *100* | 169 | 28.5 | 3 | 1570.9 | *100* | 15.0 | 7 |

## OXFORD 2017

| Month | Mean max | Mean min | Mean temp | Anom | Highest max | Date | Lowest min | Date | Lowest max | Date | Highest min | Date | Air frost | Days ≥ 25 °C | Total pptn | Anom | Rain days | Wettest day | Date | Total sunshine | Anom | Sunniest day | Date |
|---|---|---|---|---|---|---|---|---|---|---|---|---|---|---|---|---|---|---|---|---|---|---|---|
| Jan | 7.3 | 0.8 | 4.0 | *−0.8* | 11.2 | 7 | −3.9 | 22 | 1.2 | 26 | 8.1 | 8 | 16 | 0 | 64.1 | *113* | 17 | 11.4 | 29 | 65.0 | *104* | 7.8 | 20 |
| Feb | 9.4 | 4.0 | 6.7 | *+1.8* | 15.6 | 20 | −1.7 | 6 | 2.3 | 9 | 10.5 | 22 | 1 | 0 | 42.8 | *100* | 17 | 8.6 | 7 | 60.9 | *77* | 8.1 | 13 |
| Mar | 13.7 | 5.7 | 9.7 | *+2.5* | 21.2 | 30 | 1.6 | 25 | 9.7 | 1 | 12.2 | 30 | 0 | 0 | 44.1 | *93* | 19 | 10.8 | 3 | 115.6 | *104* | 10.7 | 25 |
| Apr | 15.1 | 4.7 | 9.9 | *+0.6* | 24.1 | 9 | 0.0 | 27 | 10.3 | 26 | 9.4 | 4 | 0 | 0 | 5.4 | *11* | 5 | 2.6 | 30 | 183.6 | *114* | 11.5 | 8 |
| May | 19.2 | 9.4 | 14.3 | *+1.8* | 27.0 | 26 | 0.6 | 10 | 11.7 | 9 | 14.4 | 17 | 0 | 3 | 63.6 | *111* | 14 | 22.6 | 17 | 169.4 | *88* | 14.0 | 21 |
| June | 22.7 | 12.9 | 17.8 | *+2.2* | 32.5 | 21 | 8.3 | 4 | 16.4 | 28 | 17.8 | 19 | 0 | 6 | 30.5 | *62* | 8 | 17.6 | 5 | 203.3 | *106* | 14.0 | 17 |
| July | 23.3 | 13.9 | 18.6 | *+0.7* | 32.2 | 6 | 10.6 | 17 | 18.8 | 22 | 17.8 | 9 | 0 | 9 | 62.3 | *128* | 14 | 15.2 | 11 | 181.5 | *88* | 15.0 | 5 |
| Aug | 21.4 | 12.6 | 17.0 | *−0.6* | 27.6 | 28 | 8.9 | 13 | 15.3 | 8 | 16.2 | 22 | 0 | 2 | 62.8 | *115* | 17 | 11.7 | 7 | 176.1 | *90* | 12.6 | 10 |
| Sep | 18.6 | 10.5 | 14.6 | *−0.3* | 23.3 | 4 | 4.8 | 22 | 15.3 | 15 | 15.3 | 5 | 0 | 0 | 60.7 | *112* | 21 | 14.2 | 8 | 119.9 | *85* | 9.8 | 1 |
| Oct | 16.3 | 9.9 | 13.1 | *+1.8* | 21.2 | 16 | 1.1 | 30 | 11.6 | 30 | 14.6 | 14 | 0 | 0 | 17.1 | *24* | 14 | 3.2 | 19 | 97.7 | *88* | 8.7 | 5 |
| Nov | 10.7 | 3.5 | 7.1 | *−0.5* | 15.4 | 1 | −1.2 | 30 | 4.1 | 30 | 11.6 | 22 | 4 | 0 | 48.4 | *73* | 14 | 12.1 | 10 | 99.6 | *140* | 8.3 | 6 |
| Dec | 8.2 | 2.6 | 5.4 | *+0.4* | 13.2 | 7 | −4.1 | 12 | 1.9 | 10 | 9.3 | 22 | 9 | 0 | 76.1 | *121* | 18 | 24.2 | 26 | 54.5 | *102* | 6.9 | 28 |
| 2017 | 15.5 | 7.5 | 11.5 | *+0.8* | 32.5 | 6 | −4.1 | 12 | 1.2 | 1 | 17.8 | 6 | 30 | 20 | 577.9 | *88* | 178 | 24.2 | 12 | 1527.1 | *97* | 15.0 | 7 |

## OXFORD 2018

| Month | Mean max | Mean min | Mean temp | Anom | Highest max | Date | Lowest min | Date | Lowest max | Date | Highest min | Date | Air frost | Days [3] 25 °C | Total pptn | Anom | Rain days | Wettest day | Date | Total sunshine | Anom | Sunniest day | Date |
|---|---|---|---|---|---|---|---|---|---|---|---|---|---|---|---|---|---|---|---|---|---|---|---|
| Jan | 9.0 | 3.0 | 6.0 | *+1.2* | 14.6 | 23 | -1.3 | 6 | 3.7 | 8 | 9.8 | 24 | 4 | 0 | 58.6 | *103* | 20 | 15.5 | 21 | 73.0 | *117* | 7.0 | 7 |
| Feb | 6.4 | 0.3 | 3.3 | *-1.6* | 11.4 | 19 | -5.2 | 28 | -1.1 | 28 | 7.4 | 20 | 14 | 0 | 24.1 | *56* | 11 | 6.5 | 14 | 117.6 | *149* | 9.0 | 25 |
| Mar | 9.1 | 2.5 | 5.8 | *-1.4* | 14.0 | 10 | -6.1 | 1 | -1.9 | 1 | 7.3 | 15 | 6 | 0 | 83.4 | *175* | 23 | 12.8 | 30 | 79.1 | *71* | 9.6 | 19 |
| Apr | 14.2 | 7.5 | 10.8 | *+1.6* | 26.9 | 19 | 1.3 | 2 | 7.9 | 12 | 13.8 | 22 | 0 | 1 | 52.8 | *108* | 16 | 14.3 | 1 | 120.5 | *75* | 12.7 | 20 |
| May | 19.9 | 8.7 | 14.3 | *+1.8* | 27.2 | 7 | 1.9 | 1 | 13.4 | 2 | 13.8 | 28 | 0 | 3 | 86.2 | *151* | 12 | 36.4 | 31 | 244.8 | *127* | 14.4 | 14 |
| June | 22.8 | 11.6 | 17.2 | *+1.6* | 28.8 | 25 | 7.2 | 22 | 16.9 | 5 | 16.3 | 20 | 0 | 7 | 2.5 | *5* | 4 | 1.1 | 1 | 250.9 | *131* | 15.2 | 22 |
| July | 27.4 | 14.7 | 21.0 | *+3.2* | 32.4 | 26 | 11.3 | 3 | 20.6 | 29 | 18.5 | 27 | 0 | 26 | 23.2 | *48* | 5 | 13.0 | 28 | 281.9 | *136* | 15.3 | 3 |
| Aug | 23.4 | 13.4 | 18.4 | *+0.8* | 30.2 | 5 | 6.9 | 25 | 18.6 | 12 | 16.9 | 19 | 0 | 8 | 43.2 | *79* | 10 | 14.2 | 26 | 200.3 | *102* | 13.4 | 5 |
| Sep | 20.4 | 10.1 | 15.2 | *+0.4* | 25.5 | 1 | 1.6 | 29 | 12.8 | 22 | 16.5 | 17 | 0 | 3 | 30.3 | *56* | 5 | 14.2 | 22 | 169.8 | *121* | 10.7 | 7 |
| Oct | 15.6 | 7.3 | 11.5 | *+0.2* | 23.9 | 10 | -1.4 | 29 | 6.6 | 27 | 17.0 | 13 | 2 | 0 | 50.1 | *72* | 13 | 18.1 | 14 | 117.3 | *105* | 9.8 | 9 |
| Nov | 11.7 | 5.5 | 8.6 | *+1.1* | 15.7 | 5 | -2.7 | 22 | 5.5 | 22 | 11.4 | 7 | 2 | 0 | 61.9 | *93* | 13 | 13.1 | 10 | 82.7 | *117* | 8.4 | 2 |
| Dec | 10.2 | 4.7 | 7.4 | *+2.4* | 14.0 | 2 | -2.9 | 14 | 3.1 | 14 | 10.5 | 3 | 3 | 0 | 67.7 | *107* | 15 | 12.2 | 15 | 53.7 | *101* | 5.3 | 13 |
| **2018** | 15.8 | 7.4 | 11.6 | *+0.9* | 32.4 | 7 | -6.1 | 3 | -1.9 | 3 | 18.5 | 7 | 31 | 48 | 584.0 | *88* | 147 | 36.4 | 5 | 1791.7 | *114* | 15.3 | 7 |

# References and further reading


1. Bridgland, D. R., 1994. *Quaternary of the Thames*. Dordrecht: Springer Science
2. Newell, A. J., 2008. *Morphology and Quaternary Geology of the Thames Floodplain around Oxford*. Groundwater Resources Programme Open Report OR/08/030. Keyworth: British Geological Survey. Accessed 20 January 2019; available from http://nora.nerc.ac.uk/id/eprint/511162/1/OR_08_030%20Morphology%20%26%20Quaternary%20geology%20of%20the%20Thames%20floodplain%20around%20Oxford.pdf
3. Lawrence, E. N., 1972. The earliest known journal of the weather. *Weather*, **27**: 494–501
4. Lawrence, E. N., 1973. Merle's weather observations (Letter to the Editor). *Weather*, **28**: 127–130
5. Meaden, G. T., 1973. Merle's weather diary and its motivation. *Weather*, **28**: 210–211
6. Symons, G. J., 1891. *Merle's MS: Consideraciones Temperiei Pro 7 Annis, 1337–1344*. London: E. Stanford
7. Mortimer, R., 1981. William Merle's weather diary and the reliability of historical evidence for medieval climate. *Climate Monitor*, **10**: 42–45
8. Pribyl, K., 2014. The study of the climate of medieval England: A review of historical climatology's past achievements and future potential. *Weather*, **69**: 116–120
9. Middleton, W. E. K., 1966. *A History of the Thermometer and Its Use in Meteorology*. Baltimore: Johns Hopkins Press. 249 pp
10. Patterson, L. D., 1953. The Royal Society's standard thermometer, 1663–1709. *Isis (Harvard Univ. History Sci., Cambridge, Mass.)*, **44**: 55–64
11. Camuffo, D., et al., 2010. The earliest daily barometric pressure readings in Italy: Pisa AD 1657–1658 and Modena AD 1694, and the weather over Europe. *The Holocene*, **20**: 337–349
12. Wallis, J., and J. Beale, 1669. Some observations concerning the baroscope and thermoscope, made and communicated by Doctor J. Wallis at Oxford, and Dr J Beale at Yeovil in Somerset, deliver'd here according to the several dates, when they were imparted. *Philosophical Transactions*, **4**(55): 1113–1120
13. Wallis, J., 1665. A Relation of an Accident by Thunder and Lightning, at Oxford [10 May 1666, OS]. *Philosophical Transactions*, **1**(1–22): 222–226
14. Plot, R. and J. Bobart, 1684. A discourse concerning the effects of the great frost, on trees and other plants anno 1683. drawn from the answers to fame Queries sent into divers Countries by Dr Rob. Plot S. R. S. and from several observations made at Oxford, by the skilful Botanist Mr Jacob Bobart. *Philosophical Transactions*, **14**(165): 766–779
15. Plot, R., 1685. Observations of the Wind, Weather, and Height of the Mercury in the Barometer, through out the year 1684; Taken in the Musaeum Ashmoleanum at Oxford, by Robert Plot, LLD. to which is Prefixt a Letter from Him, to Dr Martin Lister, F. of the R. S. concerning the Use of This and Such Like Historys of the Weather. *Philosophical Transactions*, **15**(167–178): 930–943
16. Tyack, G., 2005. The making of the Radcliffe Observatory, in *A History of the Radcliffe Observatory Oxford: The Biography of a Building*, J. Burley and K. Plenderleith, eds. Oxford: Green College at the Radcliffe Observatory, pp. 1–10

17. Guest, I., 1991. *Dr John Radcliffe and His Trust*. London: The Radcliffe Trust. 595 pp
18. Smith, C. G., 1968. The Radcliffe Meteorological Station. *Weather*, **23**: 362–367
19. Smith, C. G., 1994. The Radcliffe Observatory, Oxford, in *Observatories and Climatological Research*, B. D. Giles and J. M. Kenworthy, eds. Durham: University of Durham, Dept of Geography.
20. Tyack, G., 2005. The making of the Radcliffe Observatory, in *A History of the Radcliffe Observatory, Oxford: The Biography of a Building*, J. Burley and K. Plenderleith, eds. Oxford: Green College.
21. Wallace, J. G., 1997. *Meteorological Observations at the Radcliffe Observatory, Oxford: 1815–1995*. Oxford: University of Oxford School of Geography. 77 pp
22. Wallace, G., 2005. Meteorological observation at the Radcliffe Observatory, in *A History of the Radcliffe Observatory Oxford: The Biography of a Building*, J. Burley and K. Plenderleith, eds. Oxford: Green College, pp. 102–128
23. Smith, C. G., 1985. Meteorological observations at the Radcliffe Observatory, Oxford. *Bulletin of the Scientific Instrument Society*, **5**: 11–13
24. Craddock, J. M. and E. Craddock, 1977. Rainfall at Oxford from 1767 to 1814, estimated from the records of Dr Thomas Hornsby and others. *Meteorological Magazine*, **106**: 361–372
25. Smith, C. G., 1979. The cold winters of 1767–68, 1776 and 1814, as observed at Oxford. *Weather*, **34**: 346–358
26. Craddock, J. M. and C. G. Smith, 1978. An investigation into rainfall recording at Oxford. *Meteorological Magazine*, **107**: 257–271
27. Bilham, E. G., 1932. The swan song of the Radcliffe Observatory. *Meteorological Magazine*, **67**: 191–193
28. Oliver, H. R. and A. S. Goudie, 2000. Meteorologist's profile: C. G. Smith. *Weather*, **55**: 92–93
29. MacDonald, L. T., 2018. *Kew Observatory and the Evolution of Victorian Science, 1840–1910*. Pittsburgh: University of Pittsburgh Press. 336 pp
30. Knox-Shaw, H. and J. G. Balk, 1932. Appendix A: Explanation of the tables, in *Results of Meteorological Observations Made at the Radcliffe Observatory, Oxford, in the Five Years 1926–30.*. Oxford: University of Oxford
31. Smith, C. G. and G. Manley, 1975. Central England Temperatures; monthly means of the Radcliffe Meteorological Station, Oxford. *Quarterly Journal of the Royal Meteorological Society*, **101**: 385–389
32. Pirie, M., 2005. A history of the gardens and grounds of the Radcliffe Observatory, in *A History of the Radcliffe Observatory Oxford: The Biography of a Building*, J. Burley and K. Plenderleith, eds. Oxford: Green College, pp. 31–61.
33. Chance, E., et al., 1979. Modern Oxford, in *A History of the County of Oxford: Volume 4, The City of Oxford*, A. Crossley and C. R. Elrington, eds. London: Victoria County History, pp. 181–259. Accessed 24 April 2018; available from http://www.british-history.ac.uk/vch/oxon/vol4/pp181-259.
34. Manley, G., 1953. The mean temperature of central England, 1698–1952. *Quarterly Journal of the Royal Meteorological Society*, **79**: 558–567
35. Manley, G., 1974. Central England temperatures: Monthly means 1659 to 1973. *Quarterly Journal of the Royal Meteorological Society*, **100**: 389–405
36. Parker, D. E., T. P. Legg and C. K. Folland, 1992. A new daily central England temperature series, 1772–1991. *International Journal of Climatology*, **12**: 317–342
37. Parker, D. and B. Horton, 2005. Uncertainties in central England temperature 1878–2003 and some improvements to the maximum and minimum series. *International Journal of Climatology*, **25**: 1173–1188

38. Perryman, S. A. M., et al., 2018. The electronic Rothamsted Archive (e-RA), an online resource for data from the Rothamsted long-term experiments. *Nature Scientific Data*, **5**: 180072
39. Barry, R. G. and R. J. Chorley, 2010. *Atmosphere, Weather and Climate: Ninth Edition*. London: Routledge. 516 pp.
40. Burt, T. P., 1998. The climate of Oxfordshire, in *The Flora of Oxfordshire*, J. Killick, R. Perry and S. Woodell, eds. Newbury: Pisces Publications, pp. 24–28
41. Jones, P. D., T. Jónsson and D. Wheeler, 1997. Extension to the North Atlantic oscillation using early instrumental pressure observations from Gibraltar and south-west Iceland. *International Journal of Climatology*, **17**: 1433–1450
42. Ropelewski, C.F.and P. D. Jones, 1987. An extension of the Tahiti–Darwin Southern Oscillation Index. *Monthly Weather Review*, **115**: 2161–2165
43. Burt, T. P., N. J. K. Howden and F. Worrall, 2014. On the importance of very long-term water quality records. *Wiley Interdisciplinary Reviews: WIREs Water*, **1**: 41–48
44. Kalnay, E., et al., 1996. The NCEP/NCAR 40-year reanalysis project. *Bulletin of the American Meteorological Society*, **77**: 437–472
45. Compo, G. P., et al., 2011. The Twentieth Century Reanalysis project. *Quarterly Journal of the Royal Meteorological Society*, **137**: 1–28
46. Jones, P. D., C. Harpham and K. R. Briffa, 2013. Lamb weather types derived from reanalysis products. *International Journal of Climatology*, **33**: 1129–1139
47. Lamb, H. H., 1972. *British Isles Weather Types and a Register of Daily Sequence of Circulation Patterns, 1861–1971*. Meteorological Office Geophysical Memoirs No. 116. London: Her Majesty's Stationary Office
48. Hurrell, J. W., 1995. Decadal trends in the North Atlantic Oscillation and relationships to regional temperature and precipitation. *Science* **269**: 676–679
49. Alexander, L. V. and P. D. Jones, 2000. Updated precipitation series for the U.K. and discussion of recent extremes. *Atmospheric Science Letters*, **1**: 142–150
50. Shaw, M.S., 1977. The exceptional heat-wave of 23 June to 8 July 1976. *Meteorological Magazine*, **106**: 329–346
51. Smith, C. G., 1977. The weather at Oxford during 1976: An unprecedented drought and the finest summer on record. *Journal of Meteorology, UK*, **2**: 107–109
52. Doornkamp, J. C., K. J. Gregory and A. S. Burn, eds, 1980. *Atlas of Drought in Britain, 1975–1976*. London: Institute of British Geographers. 82 pp.
53. Eden, P., 2011. December 2010: Coldest December since 1890. *Weather*, **66**: i–iv
54. Prior, J. and M. Kendon, 2011. The UK winter of 2009/2010 compared with severe winters of the last 100 years. *Weather*, **66**: 4–10
55. Kendon, M. and M. McCarthy, 2015. The UK's wet and stormy winter of 2013/2014. *Weather*, **70**: 40–47
56. Muchan, K., et al., 2015. The winter storms of 2013/2014 in the UK: Hydrological responses and impacts. *Weather*, **70**: 55–61
57. Priestley, M. D. K., et al., 2017. The role of cyclone clustering during the stormy winter of 2013/2014. *Weather*, **72**: 187–192
58. Burt, S. and M. Kendon, 2016. December 2015: An exceptionally mild month in the United Kingdom. *Weather*, **71**: 314–320
59. Burt, S., 2016. New extreme monthly rainfall totals for the United Kingdom and Ireland: December 2015. *Weather*, **71**: 333–338
60. Camuffo, D. and C. Bertolin, 2013. The world's earliest instrumental temperature records, from 1632 to 1648, claimed by G. Libri, are reality or myth? *Climatic Change*, **119**: 647–657

61. Cornes, R., 2008. The barometer measurements of the Royal Society of London: 1774–1842. *Weather*, **63**: 230–235
62. Cornes, R. C., et al., 2012. A daily series of mean sea-level pressure for London, 1692–2007. *International Journal of Climatology*, **32**: 641–656
63. Cornes, R., 2010. *Early Meteorological Data from London and Paris*. PhD thesis, University of East Anglia
64. Cornes, R. C., et al., 2012. A daily series of mean sea-level pressure for Paris, 1670–2007. *International Journal of Climatology*, **32**: 1135–1150
65. Bergström, H. and A. Moberg, 2002. Daily air temperature and pressure series for Uppsala (1722–1998). *Climatic Change*, **53**: 213–252
66. Cocheo, C. and D. Camuffo, 2002. Corrections of systematic errors and data homogenisation in the daily temperature Padova series (1725–1998). *Climatic Change*, **53**: 77–100
67. Moberg, A., et al., 2002. Daily air temperature and pressure series for Stockholm (1756–1998). *Climatic Change*, **53**: 171–212
68. Maugeri, M., L. Buffoni and F. Chlistovsky, 2002. Daily Milan temperature and pressure series (1763–1998): History of the observations and data and metadata recovery. *Climatic Change*, **53**: 101–117
69. Maugeri, M., et al., 2002. Daily Milan temperature and pressure series (1763–1998): Completing and homogenising the data. *Climatic Change*, **53**: 119–149
70. Whipple, F. J. W., 1937. Some aspects of the early history of Kew Observatory. *Quarterly Journal of the Royal Meteorological Society*, **63**: 127–135
71. Jacobs, L., 1969. The 200-years story of Kew Observatory. *Meteorological Magazine*, **98**: 162–171
72. MacDonald, L. T., 2015. Making Kew Observatory: The Royal Society, the British Association and the politics of early Victorian science. *British Society for the History of Science*, **48**: 409–433
73. Mayes, J., 1994. Kew Observatory 1769–1980: Climatological implications of an observatory closure, in *Observatories and Climatological Research*, B. D. Giles and J. M. Kenworthy, eds. Durham: University of Durham, Dept of Geography, pp. 152–159
74. Galvin, J. F. P., 2003. Kew Observatory. *Weather*, **58**: 478–484
75. Czech Hydrometeorological Institute, 2018. *Meteorological Observations at the Prague Clementinum*. Prague: Czech Hydrometeorological Institute. Accessed 20 July 2018; available from http://portal.chmi.cz/historicka-data/pocasi/praha-klementinum?l=en
76. Deutscher Wetterdienst, 2018. *Hohenpeissenberg Meteorological Observatory (MOHp)*. Offenbach: Deutscher Wetterdienst. Accessed 20 July 2018; available from https://www.dwd.de/EN/aboutus/locations/observatories/mohp/mohp.html
77. Butler, C. J., et al., 2005. Air temperatures at Armagh Observatory, Northern Ireland, from 1796 to 2002. *International Journal of Climatology*, **25**: 1055–1079
78. Manley, G., 1941. The Durham meteorological record, 1847–1940. *Quarterly Journal of the Royal Meteorological Society*, **67**: 363–380
79. Burt, T. P., P. D. Jones and N. J. K. Howden, 2015. An analysis of rainfall across the British Isles in the 1870s. *International Journal of Climatology*, **35**: 2934–2947
80. Kenworthy, J. M., T. P. Burt and N. J. Cox, 2007. Durham University Observatory and its meteorological record. *Weather*, **62**: 265–269
81. O'Sullivan, J., 1992. *Valentia Observatory: A History of the Early Years*. Dublin: Irish Meteorological Service
82. Keane, T., 2012. *Establishment of the Meteorological Service in Ireland. The Foynes Years, 1936–1945*. Dublin: Met Éireann.

83. Cookson, G., 2003. *The Cable: The Wire That Changed the World.* Stroud: Tempus Publishin 160 pp
84. Burt, T. P., 1994. Long-term study of the natural environment: perceptive science or mindless monitoring? *Progress in Physical Geography*, **18**: 475–496
85. Burt, S., 2012. *The Weather Observer's Handbook*. New York: Cambridge University Press. 444 pp.
86. Cram, T. A., et al., 2015. The International Surface Pressure Databank version 2. *Geoscience Data Journal*, **2**(1): 31–46
87. Brohan, P., et al., 2012. Constraining the temperature history of the past millennium using early instrumental observations. *Climate of the Past*, **8**: 1551–1563
88. Hawkins, E., et al., 2019. Hourly weather observations from the Scottish Highlands (1883–1904) rescued by volunteer citizen scientists. Submitted to *Geoscience Data Journal*
89. Lamb, H. H., 1950. Types and spells of weather around the year in the British Isles: Annual trends, seasonal structure of the year, singularities. *Quarterly Journal of the Royal Meteorological Society*, **76**: 393–429
90. Perry, A. H., 1976. Synoptic climatology, in *The Climate of the British Isles*, T. J. Chandler and S. Gregory, eds. London: Logman, pp. 8–38
91. Smith, C. G., 1979. *The Gale of 2nd January 1976*. School of Geography Research Paper No. 21. Oxford: University of Oxford
92. Webb, J. D. C., 2011. Violent thunderstorms in the Thames Valley and south Midlands in early June 1910. *Weather*, **66**: 153–155
93. Burt, S., 1979. Heavy rainfall and thunderstorms at Rugby, 9 to 14 June 1977 (Letter to the Editor). *Weather*, **34**: 123–124
94. Smith, C. G., 1952. Heavy thunderstorm at Oxford [7 September 1951]. *Weather*, **7**: 163–169
95. McFarlane, D. and C.G. Smith, 1968. Remarkable rainfall in Oxford. *Meteorological Magazine*, **97**: 235–245
96. Webb, J. D. C. and D. M. Elsom, 1994. The great hailstorm of August 1843: The severest recorded in Britain? *Weather*, **49**: 266–273
97. Marsh, T. J., B. J. Greenfield and J. A. Hannaford, 2005. The 1894 Thames flood: A reappraisal. *Proceedings of the Institution of Civil Engineers*, **158**(WM3): 103–110
98. Burt, T. P. and N. J. K. Howden, 2011. A homogenous daily rainfall record for the Radcliffe Observatory, Oxford, from the 1820s. *Water Resources Research*, **47**: 1–6
99. Burt, S., 2010. British rainfall 1860–1993. *Weather*, **65**: 121–128
100. Burt, S., 2007. The lowest of the lows...: Extremes of barometric pressure in the British Isles, Part 1: The deepest depressions. *Weather*, **62**: 4–14
101. Burt, S., 2007. The highest of the highs...: Extremes of barometric pressure in the British Isles, Part 2: The most intense anticyclones. *Weather*, **62**: 31–41
102. Brugge, R. and S. Burt, 2015. *One Hundred Years of Reading Weather*. Reading: Climatological Observers Link/University of Reading. 193 pp.
103. Tout, D. G., 1976. Temperature, in *The Climate of the British Isles*, T.J. Chandler and S. Gregory, eds. London: Longman, pp. 96–128
104. Smith, C. G., 1969. Winters at Oxford since 1815. *Weather*, **24**: 23–28
105. Burt, T. P. and M. Shahgedanova, 1992. Long-period rainfall deficits at Oxford. *Journal of Meteorology, UK*, **17**: 223–226
106. Davis, N. E., 1968. An optimum summer weather index. *Weather*, **23**: 305–317
107. Stokes, G. G., 1880. Description of the card supporter for sunshine recorders adopted at the Meteorological Office. *Quarterly Journal of the Royal Meteorological Society*, **6**: 83–94

108. Ellis, W., 1880. On the Greenwich sunshine records, 1876–1880. *Quarterly Journal of the Royal Meteorological Society*, **6**: 126–135
109. Brodie, F. J., 1891. Some remarkable features in the winter of 1890–91. *Quarterly Journal of the Royal Meteorological Society*, **17**: 155–167
110. Pachauri, R. K., et al., eds, 2014. *Climate Change 2014: Synthesis Report. Contribution of Working Groups I, II and III to the Fifth Assessment Report of the Intergovernmental Panel on Climate Change*. Geneva: IPCC. 151 pp.
111. Wuebbles, D. J., et al., 2017. *Climate Science Special Report: Fourth National Climate Assessment, Volume I*. Washington, DC: U.S. Global Change Research Program. 470 pp.
112. Johnson, H. and J. Robinson, 2013. *The World Atlas of Wine: 7th Edition*. London: Mitchell Beazley
113. Fitter, A. H. and R. S. R. Fitter, 2002. Rapid changes in flowering time in British plants. *Science*, **296**: 1689–1691
114. Burt, T. P., N. J. K. Howden and F. Worrall, 2016. The changing water cycle: Hydroclimatic extremes in the British Isles. *Wiley Interdisciplinary Reviews: WIREs Water*, **3**: 854–870
115. Burt, T. P., et al., 2016. More rain, less soil: Long-term changes in rainfall intensity with climate change. *Earth Surface Processes and Landforms*, **41**: 563–566
116. Phillips, I. D. and G. R. McGregor, 1998. The utility of a drought index for assessing the drought hazard in Devon and Cornwall, South West England. *Meteorological Applications*, **5**: 359–372
117. Burt, T. P. and M. Shahgedanova, 1998. An historical record of evaporation losses since 1815 calculated using long-term observations from the Radcliffe Meteorological Station, Oxford, England. *Journal of Hydrology*, **205**(1–2): 101–111
118. Bill, E. G. W., 1965. *Christ Church Meadow*. Oxford: Oxford University Press. 39 pp.
119. British Hydrological Society, 2018. *British Hydrological Society Chronology of British Hydrological Events*. London: British Hydrological. Accessed 23 July 2018; available from http://www.cbhe.hydrology.org.uk/index.php
120. Meaden, G. T., 1979. Point deluge and tornado at Oxford, May 1682. *Weather*, **34**: 358–361
121. Elsom, D. M., 1980. Point deluge and tornado at Oxford, May 1682 (Letter to the Editor). *Weather*, **35**: 310–311
122. Griffiths, P. P., 1983. *A Chronology of Thames Floods (Second Edition)*. Water Resources Report No. 73. Reading: Thames Water Authority *(now Environment Agency)*
123. Black, A. R. and F. M. Law, 2004. Development and utilization of a national web-based chronology of hydrological events. *Hydrological Sciences Journal*, **49**: 237–246
124. White, G., 1789. *The Natural History of Selborne*. London: Benjamin White
125. Wales-Smith, B. G., 1971. Monthly and annual totals of rainfall representative of Kew, Surrey, from 1697 to 1970. *Meteorological Magazine*, **100**: 345–362
126. Barker, T., 1775. XVIII. Extract of a register of the barometer, thermometer, and rain, at Lyndon, in Rutland, 1774. *Philosophical Transactions*, **65**: 199–201
127. Craddock, J. M. and B. G. Wales-Smith, 1977. Monthly rainfall totals representing the East Midlands for the years 1726 to 1975. *Meteorological Magazine*, **106**: 97–111
128. Rowntree, P. R., 2018. The exceptional cold spell of January 1776. *Royal Meteorological Society History Group Newsletter*, **2018**(1): 3–7
129. Pearson, M., 1977. The Great Frost of January 1776. *Weather*, **32**: 310
130. White, G., 1977. The great frost of January 1776. *Weather*, **32**: 106–108
131. Raible, C. C., et al., 2016. Tambora 1815 as a test case for high impact volcanic eruptions: Earth system effects. *Wiley Interdisciplinary Reviews: Climate Change*, **7**: 569–589

132. Harington, C. R., ed., 1992. *The Year without a Summer? World Climate in 1816*. Ottawa: Canadian Museum of Nature
133. Wood, G. D., 2014. *Tambora: The Eruption That Changed the World*. Princeton: Princeton University Press. 293 pp.
134. Klingaman, W. K. and N. P. Klingaman, 2013. *The Year without Summer: 1816 and the Volcano that Darkened the World and Changed History*. New York: St Martin's Press. 338 pp
135. Briffa, K. R., et al., 1998. Influence of volcanic eruptions on Northern Hemisphere summer temperature over the past 600 years. *Nature*, **393**: 450–455
136. Oppenheimer, C., 2003. Climatic, environmental and human consequences of the largest known historic eruption: Tambora volcano (Indonesia) 1815. *Progress in Physical Geography: Earth and Environment*, **27**(2): 230–259
137. Post, J. D., 1977. *The Last Great Subsistence Crisis in the Western World*. Baltimore: Johns Hopkins. 240 pp
138. Veale, L. and G. H. Endfield, 2016. Situating 1816, the 'year without summer', in the UK. *The Geographical Journal*, **182**: 318–330
139. Bryson, B., 2010. *At Home*. New York: Doubleday
140. Elsom, D. M., 1987. *Taming the Rivers of Oxford*. Kingston Bagpuize: Thematic Trails
141. Doherty, H. B., 1968. The weather on *Alice in Wonderland* Day, 4 July 1862. *Weather*, **23**: 75–78
142. Burt, P. J. A., 2004. The great storm and the fall of the first Tay Rail Bridge. *Weather*, **59**: 347–350
143. Anon, 1908. The Easter snowstorm of 1908. *Symons's Monthly Meteorological Magazine*, **43**: 65–74
144. Pike, W. S., 2008. A note on the snowfall of 25 April 1908, UK. *The International Journal of Meteorology*, **33**: 340–342
145. Burt, S., 2004. The August 2003 heatwave in the United Kingdom: Part 1: Maximum temperatures and historical precedents. *Weather*, **59**: 199–208
146. Meaden, G. T., 1977. Method of describing and reporting tornadoes and other whirlwinds: Illustrated with regard to the Oxford tornado of 16 October 1966. *Journal of Meteorology, UK*, **2**: 103–106
147. Hanwell, J. D. and M. D. Newson, 1970. *The Great Storms and Floods of July 1968 on Mendip*. Wessex Cave Club Occasional Publication, Vol. 2. Pangbourne: Wessex Cave Club
148. Bleasdale, A., 1974. The year 1968: An outstanding one for multiple events with exceptionally heavy and widespread rainfall, in *British Rainfall 1968*, Meteorological Office. London: Her Majesty's Stationary Office, pp. 223–231
149. Lawrence, M. B., 1987. Annual summary of the 1986 Atlantic hurricane season. *Monthly Weather Review*, **115**: 2158–2160
150. Shawyer, M. S., 1987. The rainfall of 22–26 August 1986. *Weather*, **42**: 114–117
151. Burt, S. D. and D. A. Mansfield, 1988. The Great Storm of 15–16 October 1987. *Weather*, **43**: 90–110
152. Defoe, D., 2003 [1704]. *The Storm*, ed. Richard Hamblyn. London: Allen Lane
153. Clow, D. G., 1988. Daniel Defoe's account of the storm of 1703. *Weather*, **43**: 140–141
154. Browning, K. A., 2004. The sting at the end of the tail: Damaging winds associated with extratropical cyclones. *Quarterly Journal of the Royal Meteorological Society*, **130**: 375–399
155. Burt, S., 1992. Weather and streamflow in central southern England during water year 1989/90. *Weather*, **47**: 2–10
156. Burt, S., 1992. The exceptional hot spell of early August 1990 in the United Kingdom. *International Journal of Climatology*, **12**: 547–567

157. Prior, J. and M. Beswick, 2007. The record-breaking heat and sunshine of July 2006. *Weather*, **62**: 174–182
158. Lane, S. N., 2008. Climate change and the summer 2007 floods in the UK. *Geography*, **93**: 91–97
159. Burt, T. P. and S. N. Lane, 2008. Extreme UK floods. *Geography Review*, **22**: 2–5
160. Johnson, M. J. and R. M. Main, 1861. *Astronomical and Meteorological Observations Made at the Radcliffe Observatory in the Year 1858*, Vol. 19. Oxford: Radcliffe Trustees
161. Knox-Shaw, H., 1926. *Results of Meteorological Observations Made at the Radcliffe Observatory, 1921–25*. Oxford: University of Oxford
162. Knox-Shaw, H. and J. G. Balk, 1933. The swan song of the Radcliffe Observatory. *Meteorological Magazine*, **67**: 286–288
163. Wigley, T. M. L., J. M. Lough and P. D. Jones, 1984. Spatial patterns of precipitation in England and Wales and a revised, homogeneous England and Wales precipitation series. *Journal of Climatology*, **4**: 1–25
164. Searcy, T. K. and C. H. Hardison, 1960. *Double-Mass Curves*. Geological Survey Water-Supply Paper 1541-B. Washington, D. C.: United States Government Printing Office. Accessed 22 January 2019; available from http://pubs.usgs.gov/wsp/1541b/report.pdf
165. Wood, C. R. and R. G. Harrison, 2011. Scorch marks from the sky. *Weather*, **66**: 39–41
166. Kenworthy, J., 2007. Meteorologist's profile: Wilfred George Kendrew (1884–1962). *Weather*, **62**: 49–52
167. Gomez, B. and C. G. Smith, 1984. Atmospheric pollution and fog frequency in Oxford, 1926–1980. *Weather*, **39**: 379–384

# Index